Ninth International Conference on Mathematical Modeling and Computational Physics (MMCP 2017)

EPJ Web of Conferences Volume 173 (2018)

Dubna, Russia
3-7 July 2017

Editors:

G. Adam
J. Busa
Michal Hnatic
D. Podgainy

ISBN: 978-1-5108-5831-2

Printed from e-media with permission by:

Curran Associates, Inc.
57 Morehouse Lane
Red Hook, NY 12571

Some format issues inherent in the e-media version may also appear in this print version.

Printed by Curran Associates, Inc. (2018)

For additional information, please contact EDP Sciences – Web of Conferences at the address below.

EDP Sciences – Web of Conferences
17, Avenue du Hoggar
Parc d'Activité de Courtabœuf
BP 112
F-91944 Les Ulis Cedex A
France

Phone: +33 (0) 1 69 18 75 75
Fax: +33 (0) 1 69 28 84 91

contact-edps@webofconferences.org

Additional copies of this publication are available from:

Curran Associates, Inc.
57 Morehouse Lane
Red Hook, NY 12571 USA
Phone: 845-758-0400
Fax: 845-758-2633
Email: curran@proceedings.com
Web: www.proceedings.com

TABLE OF CONTENTS

Preface

The ninth international conference in the series "Mathematical Modeling and Computational Physics", MMCP 2017 (http://mmcp2017.jinr.ru), was continuing the firmly established traditions of providing an open forum for exchange of ideas, learning and communication, promotion of future scientific cooperation through strengthened personal relations. On the one hand, the participants' grasp of new research topics was enhanced by presentations of forerunner reports on challenges in the contemporary research; the excellent perspectives of the mathematical modeling and computational physics have been illustrated with topics generated by big science projects, cross-disciplinary research, the study of new systems and phenomena, the deep knowledge of special functions chapters. On the other hand, this MMCP conference has shown that we may look confidently into the future. The young authors' original reports covered a significant fraction of the contributions to the conference. For novices who have not finished their academic studies yet, the students' school "Mathematical Modeling for NICA", organized as a satellite event of MMCP 2017, has brought a noticeable contribution to the growth of the interest of the young generation to NICA, the flagman project of the Joint Institute for Nuclear Research (JINR) in Dubna, Russia.

The MMCP 2017 and the satellite students' school took place in Dubna, on July 3–7, 2017. Together with the permanent organizer of the MMCP conferences, the Laboratory of Information Technologies (LIT) of the JINR, the co-organizers of the ninth MMCP edition have been the Horia Hulubei National Institute for R&D in Physics and Nuclear Engineering (IFIN-HH), Bucharest, Romania, and three Institutions from Košice, Slovakia (the Institute of Experimental Physics (IEP) SAS, the Technical University (FEI TU), and the Pavol Jozef Šafárik University (NSF UPJŠ)). The myriad of technical and organizational problems raised during MMCP 2017 have been overcome thanks to the abnegation of the Local Organizing Committee members and of several tens of volunteers. We are grateful to all of them for the in-time and high-level solution of all the foreseen and unforeseen tasks.

The coverage of MMCP 2017 (involving plenary lectures and contributions in parallel sections) included: distributed and parallel computing in science and technology; mathematical methods and application software for complex systems; bioinformatics methods and computational biophysics; mathematical methods and software for experimental data processing; computer algebra and quantum computing.

The unprecedented interest for the conference: 257 participants (80 expected) submitting a total of 215 contributions (invited and proposed) (60 expected), together with an attendance of 54 (30 expected) at the heterogeneous cluster HybriLIT based (http://hybrilit.jinr.ru) Students' School asked for extraordinary organizational measures. With direct consequences for the present volume was the draconian limitation of the number of pages per submitted manuscript.

As a result of a meticulously passed refereeing process (coordinated by the TEXnical editor Ján Buša, who contacted 212 potential referees), out of the 127 submitted manuscripts, 12 plenary lectures and 88 contributions have been accepted for publication under the covers of the present issue of the EPJ-WoC. Gheorghe Adam made the final critical reading of each accepted manuscript (the authors have had, however, the last word in deciding to accept or not the proposed English corrections).

Apart from the already made remarks, contributions to the success of the MMCP 2017 have been brought by a great many colleagues from various research institutions. Special mention is to be made of the role of the Conference Chairman and all the members of the International Program Committee for their contribution to the definition of the basic aims and the scope of the conference, the proposal, selection, and promotion of invited lectures of the highest interest, the coverage of an important fraction of the reviewing process. The contribution of the 146 independent referees to the quality assessment of the submitted manuscripts is gratefully acknowledged.

To facilitate the lecture, we have found it useful to organize the content of this volume into six chapters: plenary lectures, mathematical modeling and methods, numerical modeling and methods, mathematical and computational support of the experiments, computing tools and software services, optimization and simulation. Within each chapter, the paper are ordered alphabetically following the name of the first/single author.

There is a threefold message emerging from the lecture of these papers. First, the computing is pervasive in the modern physics research. Second, maximum advantage from the possibilities offered by the existing multi-core, many core, many-GPU hardware facilities can be obtained provided specialized software tools are built up. Third, the derivation of computing-based sound solutions needs the concurrence of three competences: deep understanding of the underlying physics, grasp of rigorous mathematical methods, flexibility in the use of the reality of the hardware and software environments.

The editors are convinced of the usefulness of the job done. The readers are warmly invited to explore this resource of valuable information for their current and future activity. We hope that the high standard of the MMCP conferences will be preserved in the future and that these conferences will remain a strong attractor of the worldwide scientific community.

Gheorghe Adam (LIT JINR, IFIN-HH)
Ján Buša (FEI TU)
Michal Hnatič (BLTP JINR, IEP SAS, NSF UPJŠ)
Dmitry Podgainy (LIT JINR)

MMCP 2017

July 3 – 7, 2017

Dubna, Russia

Organizers

Laboratory of Information Technologies of the JINR, Dubna, Russia
Institute of Experimental Physics SAS, Košice, Slovakia
University of Pavol Jozef Šafárik, Košice, Slovakia
Technical University, Košice, Slovakia
IFIN-HH, Bucharest, Romania

Organizing Committee

Chairman and Vice-Chairmen

V.V. Korenkov (LIT JINR), Gh. Adam (LIT JINR, IFIN-HH), M. Hnatič (BLTP JINR, IEP, UPJŠ)

D.V. Podgainy (LIT) – Scientific Secretary, S. Adam, G. Adamov, M.V. Aristarkhova, A.S. Ayriyan, M. Bashashin, D. Belyakov, A. Bondyakov, J. Buša (TU) – scientific secretary, Yu. Butenko, O.Yu. Derenovskaya, A. Dolbilov, A. Grafov, H. Grigoryan, T.I. Katraseva, M. Kirakosyan, A. Kondratyev, M. Matveyev, A. Mayorov, Ye. Mazhitova, A. Nechaevskiy, I. Pelevanyuk, M. Plyashkevich, A. Polyushkevich, E.M. Rudneva, O.Yu. Rumyantseva, O.I. Streltsova, T.A. Strizh, A.G. Torosyan, Sh.G. Torosyan, A. Vorontsov, N. Voytishin, A.G. Zaikina, T.N. Zaikina, P.V. Zrelov, M.I. Zuev (all LIT JINR)

International Program Committee

Gh. Adam (JINR, IFIN-HH), Ch.-M. Fu (Taiwan), N.Z. Akopov (Armenia), E.A. Ayryan (JINR), O.V. Belov (JINR), A.V. Bogdanov (Russia), S. Budnyam (Mongolia), J. Buša (Slovakia), B.N. Chetverushkin (Russia), S. Dimova (Bulgaria), M. Dulea (Romania), Yu.G. Evtushenko (Russia), V. Friese (Germany), V. Gerdt (JINR), U.H.E. Hansmann (USA), J. Honkonen (Finland), Ch.-K. Hu (Taiwan), S. Incerti (France), A. Isar (Romania), V.V. Ivanov (JINR), F. Jakab (Slovakia), I. Kisel (Germany), N. Kolkovska (Bulgaria), P. Kopčanský (Slovakia), N.A. Kudryashov (Russia), V.D. Lakhno (Russia), R. Lazarov (USA), I.V. Puzynin (JINR), S.V. Polyakov (Russia), G.Yu. Riznichenko (Russia), H. Safouhi (Canada), S. Scott (North Ireland), G. Semanišin (Slovakia), L.A. Sevastianov (Russia), T.A. Strizh (JINR), P.N. Vabishchevich (Russia), V.V. Voevodin (Russia), T. Zhanlav (Mongolia), P. Zinterhof (Austria), P.V. Zrelov (JINR)

Statement of Peer review

In submitting conference proceedings to *Web of Conferences*, the editors of the proceedings certify to the Publisher that

1. They adhere to its **Policy on Publishing Integrity** in order to safeguard good scientific practice in publishing.
2. All articles have been subjected to peer review administered by the proceedings editors.
3. Reviews have been conducted by expert referees, who have been requested to provide unbiased and constructive comments aimed, whenever possible, at improving the work.
4. Proceedings editors have taken all reasonable steps to ensure the quality of the materials they publish and their decision to accept or reject a paper for publication has been based only on the merits of the work and the relevance to the journal.

Title, date and place of the conference

Mathematical Modeling and Computational Physics 2017 (MMCP 2017), July 3-7, 2017, LIT-JINR, Dubna, Russia

Proceedings editor(s):

Gheorghe Adam, Ján Buša, Michal Hnatič, Dmitry Podgainy

Date and editor's signature

29. 01. 2018
29. 01. 2018
29. 01. 2018

17, avenue du Hoggar - PA de Courtabœuf – BP 112 - 91944 Les Ulis Cedex A (France)
Tél. : 33 (0)1 69 18 75 75 - Fax : 33(0)1 69 07 45 17 – www.edpsciences.org

Disentangling Complexity in Bayesian Automatic Adaptive Quadrature

Gheorghe Adam[1,2,⋆] and *Sanda* Adam[1,2,⋆⋆]

[1] *Laboratory of Information Technologies, Joint Institute for Nuclear Research, 6, Joliot Curie St., 141980 Dubna, Moscow Region, Russia*
[2] *Horia Hulubei National Institute for Physics and Nuclear Engineering (IFIN-HH), 30, Reactorului St., Măgurele – Bucharest, 077125, Romania*

Abstract. The paper describes a Bayesian automatic adaptive quadrature (BAAQ) solution for numerical integration which is simultaneously robust, reliable, and efficient. Detailed discussion is provided of three main factors which contribute to the enhancement of these features: (1) refinement of the m-panel automatic adaptive scheme through the use of integration-domain-length-scale-adapted quadrature sums; (2) fast early problem complexity assessment – enables the non-transitive choice among three execution paths: (i) immediate termination (exceptional cases); (ii) pessimistic – involves time and resource consuming Bayesian inference resulting in radical reformulation of the problem to be solved; (iii) optimistic – asks exclusively for subrange subdivision by bisection; (3) use of the weaker accuracy target from the two possible ones (the input accuracy specifications and the intrinsic integrand properties respectively) – results in maximum possible solution accuracy under minimum possible computing time.

1 Introduction

The present paper reports results along the lines of our recent investigations [1, 2] concerning the Bayesian automatic adaptive quadrature (BAAQ) solution [3] of definite Riemann integrals.

The essential requirements to the BAAQ algorithms are robustness, reliability, and efficiency under floating point computations. Their implementation asks for three main BAAQ developments.

(a) The refinement of the m-panel scheme of the standard adaptive quadrature (described, e.g., in [4–7]). This development was suggested by the fact that, under floating point computations, the algebraic degree of precision of an interpolatory quadrature sum ceases to be a characteristic invariant feature of it. The floating point degree of precision is the adequate quantity characterizing a quadrature sum [8]. A convenient way of taking into account this floating point feature was [1] to modify the m-panel scheme by the definition of three classes of integration domain length scales: *macroscopic, mesoscopic*, and *microscopic* and to use characteristic quadrature sums over each scale.

(b) The fast early problem complexity assessment is a new undertaking the need of which follows from the empirical fact that a single algorithm cannot efficiently cope with the myriad of existing

⋆e-mail: adamg@jinr.ru,adamg@theory.nipne.ro
⋆⋆e-mail: adams@jinr.ru,adams@theory.nipne.ro

Riemann integrals. However, instead of leaving with the user the expert decision on the choice among the large number of existing computing codes for numerical solution of the integrals (see, [4–7]), an attempt was made to investigate the possibility to define decision paths based on Bayesian inferences which are tailoring automatically the best way to follow to the derivation of accurate output.

(c) Finally, new results are reported concerning the code optimization in connection with the user accuracy requirements. A useful by-product of the problem complexity assessment is the possibility to define an integrand adapted upper accuracy target. Whenever this is less stringent than the target following from the user defined accuracy requirements, the latter is superseded by the former one. This decision has two desirable effects: it provides at output the best possible accuracy for the problem at hand, while sparing the unnecessary computations done beyond the maximum available accuracy.

With this general plan in mind, the discussion below is organized in four sections. Section 2 provides a summary of the BAAQ approach enabling independent lecture of this paper. Section 3 adds progress to the three class m-panel scheme proposed in [1]. Section 4 describes in detail the complexity assessment over macroscopic ranges. It also summarizes the integrand induced accuracy limitations. The paper ends with concluding remarks in section 5.

2 Summary of the Bayesian automatic adaptive quadrature

A BAAQ numerical solution of the (proper or improper) Riemann integral

$$I \equiv I_{[a,b]}[f] = \int_a^b g(x)f(x)\,\mathrm{d}x\,, \qquad -\infty < a < b < \infty, \tag{1}$$

is sought under the assumption that the real valued *integrand function* $f(x)$ is continuous almost everywhere on $[a, b]$ such that (1) exists and is finite. Here the *weight function* $g(x)$ either absorbs an analytically integrable difficult factor in the integrand (e.g., endpoint singularity or oscillatory function), or else $g(x) \equiv 1,\ \forall x \in [a, b]$.

Following [1], the subsequent BAAQ developments implement a solution of (1) within a *class dependent* m-panel rule approach. We remind that the m-panel rule idea was proposed within the standard automatic adaptive quadrature (AAQ) (see, e.g., [4–6]) as a means to adjust the discretization of the original integration domain $[a, b]$ to the rate of variation of $f(x)$ inside $[a, b]$. Within AAQ, over any integration range $[\alpha, \beta] \subseteq [a, b]$, a *fixed quadrature rule* is used irrespective of the width of $[\alpha, \beta]$.

The quadrature rule yields an $[\alpha, \beta]$ – dependent pair $\{q, e > 0\}$ where $q \equiv q_{[\alpha,\beta]}[f]$ denotes an approximation of $I_{[\alpha,\beta]}[f]$ produced by an *interpolatory quadrature sum*, while $e \equiv e_{[\alpha,\beta]}[f]$ denotes the *error estimate* associated to q, which is intended to provide an upper bound to the modulus of the remainder $r_{[\alpha,\beta]}[f] = I_{[\alpha,\beta]}[f] - q_{[\alpha,\beta]}[f]$. As pointed out in [1], the quadrature sum q is to be computed by *different* quadrature sums over three classes of integration domain widths: *macroscopic*, *mesoscopic*, and *microscopic*, respectively.

An AAQ algorithm works iteratively. The quadrature rule at hand provides a global output *initialization* $\{Q_N, E_N | N = 1\}$ over $[a, b]$. If the obtained approximation Q_1 of (1) is not accurate enough, then global output *refinement* is obtained by gradual subdivision of $[a, b]$ into subranges $[x^{i-1}, x^i] \subset [a, b]$; $i = 1, 2, \ldots, N$; $N > 1$, over which *local outputs* $\{q_i, e_i > 0\}$ are computed. The summation of the local outputs over the existing subranges results in a *global output* update $\{Q_N, E_N > 0\}$. After each derived global output $\{Q_N, E_N | N \geq 1\}$, the fulfilment of the *global termination criterion* is checked [4],

$$|I - Q_N| < E_N < \max\{\varepsilon_a, \varepsilon_r |I|\} \simeq \max\{\varepsilon_a, \varepsilon_r |Q_N|\}, \tag{2}$$

where, ε_a and ε_r denote the *absolute accuracy*, respectively *relative accuracy* requested at input.

The termination of the iterative process may end successfully, or it may not. The reason for failure is the evaluation of local error estimates by use of *probabilistic arguments* which might result in one or more persistently spurious e_i values.

The BAAQ advancement to the solution rests on Bayesian inference based on four pillars: the theory of the Riemann integral, the theory of the numerical integration (quadrature), features of the floating point computation, the accumulated empirical evidence. Essentially, the probabilistic character of the AAQ approach to the derivation of local error estimates $e > 0$ is preserved. However, each step of the gradual advancement to the solution is scrutinized based on a set of hierarchically ordered criteria which enable decision taking in terms of the established diagnostics [3].

3 Refinement of the m-panel AAQ scheme

The inadequacy of the algebraic degree of precision for the characterization of the outputs of a quadrature sum in floating point computations [8] is illustrated in the figure 1.

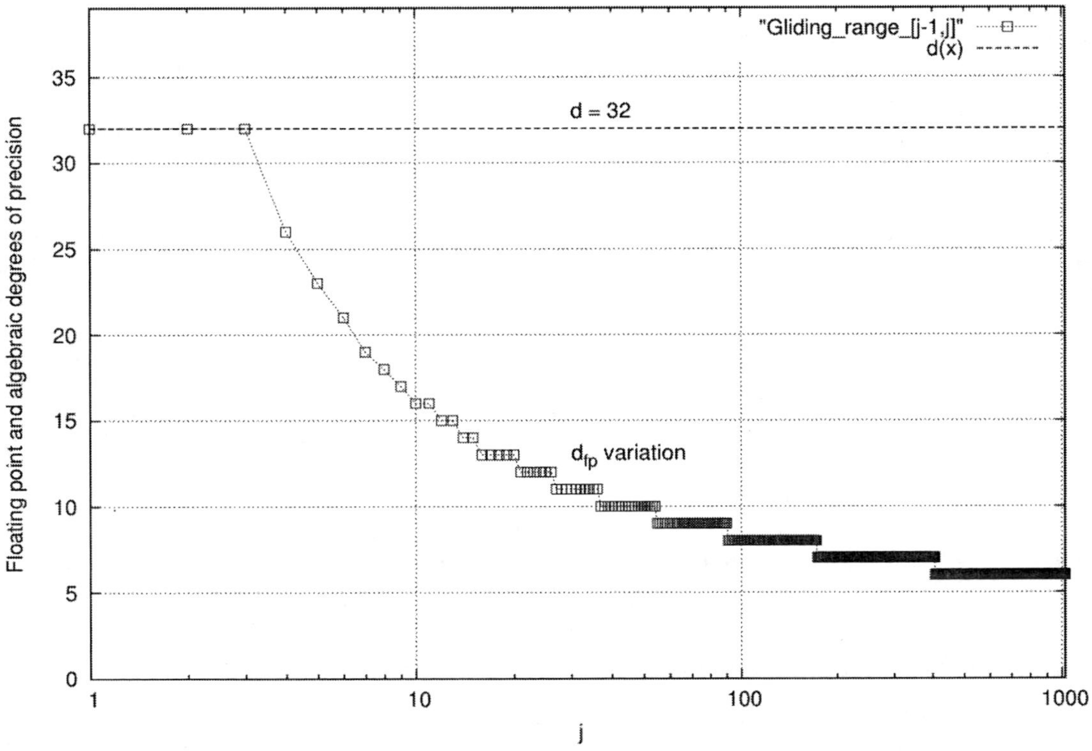

Figure 1. Variation of the floating point degree of precision of the CC-32 local quadrature rule over the gliding range $[0, 1]$ versus its distance $j - 1$ from the origin. It is found that $d_{fp} = d = 32$ at low j values ($j = 1, 2, 3$), then d_{fp} abruptly decreases at larger but small enough j, to show slower decreasing rates under the displacement of $[0, 1]$ far away from the origin, and reaching a minimal value $d_{fp} = 6$ at $407 \leq j \leq 1023$.

The quadrature sum of interest, abridged CC-32, implements Clenshaw-Curtis (CC) quadrature [9] at the 33-knot set at which the Chebyshev polynomial of the first kind and of polynomial degree 32 gets extremal values inside the standard reduced interval $[-1, 1]$. The CC-32 is of special interest since it is the best choice [2] among the CC-like quadrature sums of algebraic degree of precision 2^m, $m \in \mathbb{N}^+$.

The evidence reported in fig. 1 contrasts the constant value ($d = 32$) of the algebraic degree of precision of the CC-32 and the dramatic variation of its floating point degree of precision in the case of the unit length range $[j-1, j]$ gliding along the real axis through integer values, $j = 1, 2, \ldots, 1023$.

Extensive numerical tests have confirmed the usefulness of the m-panel AAQ scheme refinement proposed in [1], concerning the definition of three length scales of the integration domain widths: *macroscopic, mesoscopic*, and *microscopic*, respectively.

However the *separation boundaries* between adjacent classes have been shifted toward larger values: $\tau_\mu = 2^{-20}$ (in-between microscopic and mesoscopic lengths) and $\tau_m = 2^{-6}$ (in-between mesoscopic and macroscopic lengths). The motivation stems from the need to start, over microscopic and mesoscopic length ranges, with computed integrand values over large enough sets of equally spaced quadrature knots. This enables a preliminary analysis of the complexity of the integral to be solved similar to that detailed below in the case of macroscopic ranges.

Over *ranges of microscopic length*, the minimal set of equally spaced quadrature knots is to consist of $2 \times 2 + 1 = 5$ elements. Over such a set, a two-term *composite Simpson rule* provides the reference quadrature sum output characterized by the highest possible algebraic degree of precision. The alternative knot grouping which consists of one central triplet for Simpson rule and two lateral doublets for composite trapezoidal rule yields a secondary composite quadrature sum output. From the difference of the two quadrature outputs and use of the triangle inequality rule, an associated error estimate results as a superposition of integrand curvatures at the three inner knots.

Over *ranges of mesoscopic length*, the use of a set of $4 \times 2 + 1 = 9$ equally spaced quadrature knots yields a reference quadrature sum from a two-term *composite 4-interval Newton rule.* Combination of a 4-interval Newton rule over the central knot quintuple with two lateral triplets for composite Simpson rule yields a secondary quadrature rule. Similar to the microscopic case, quadrature error estimates are obtained as a superposition of integrand curvatures at the seven inner knots.

Precision loss originating in cancellation by subtraction over microscopic and mesoscopic ranges. As noticed in [1], the specification of the integration domain ends, which enters the standard input formulation provided by the equation (1), results in an *unavoidable precision loss* due to the cancellation by subtraction involved in the computation of the integration domain length as the difference of two (arbitrary) machine numbers sharing a set of common most significant digits.

If, however, the integration domain length is specified to machine accuracy either separately or within the definition of the integral to be solved, then no precision loss originating in cancellation by subtraction will happen.

4 Early Bayesian diagnostics over macroscopic ranges

4.1 Enhanced accuracy Clenshaw-Curtis Quadrature

The recent monograph [10] provides a detailed rigorous presentation of the attractive mathematical properties of the CC quadrature. In the frame of the BAAQ, the selection of the CC quadrature sums for the approximation of the Riemann integrals (1) over macroscopic integration ranges is motivated by several features of the spanning Chebyshev polynomials: closeness to the polynomials of the best approximation, symmetry properties, easiness of the analytic computation of the quadrature weights for different weight functions [4, 9, 11–13].

The interpolatory polynomial of the method is given by the truncated Chebyshev series expansion

$$L_n^\phi(y) = \sum_{k=0}^{n}{}'' b_k^n T_k(y), \tag{3}$$

(where the double prime shows that the first and the last terms of the sum are halved) of the reduced integrand

$$\phi(y) = f(c + hy), \quad c = (b + a)/2, \quad h = (b - a)/2, \quad y \in [-1, 1]. \tag{4}$$

$L_n^{\phi}(y)$ equates $\phi(y)$ at the set of $(n + 1)$ standard reduced CC quadrature knots

$$\{y_j^n = \cos(j\pi/n) \mid j = 0, 1, \ldots, n\}. \tag{5}$$

The computation of the Chebyshev expansion coefficients in (3) is the most computer intensive task of the method. The elucidation of the binary tree structures unveiling inheritance properties in the family of spanning Chebyshev polynomials of degrees $n = 2^m$ $(m = 1, 2, \ldots)$ [2] allowed the derivation of a fast algorithm for the computation of these coefficients inside which the number of multiplications is kept at the lowest possible value.

The scrutiny of the derived algorithm has shown that the CC quadrature spanned by the 32-nd degree Chebyshev polynomial provides the best CC solution (rich integrand sampling and still economical algorithm). An undesirable feature of the reduction procedure (4) to the standard integration domain $[-1, 1]$ is the sizeable precision loss, due to the cancellation by subtraction, of the floating point values of the distances in-between neighbouring quadrature knots for the overwhelming fraction of the emerging pairs. This unfavourably impinges on the floating point accuracy of the divided differences entering the *integrand profile* (IP) analysis enabling Bayesian inferences [14].

The solution which avoids the above-mentioned precision loss is to use *modified reduced abscissas* (MRA), which are defined as distances of the standard reduced abscissas y_j^n to the nearest integration domain ends. The implementation of the MRA computation was done to machine accuracy. The computation of the distances between neighbouring MRA abscissas can be straightforwardly done, without precision loss by subtraction.

There is a set of 17 MRA entering the CC-32:

$$0 = \eta_0 < \eta_1 < \cdots < \eta_k < \cdots < \eta_{16} = 1, \quad \eta_{k+\mu} - \eta_{k+\mu-1} > \eta_k - \eta_{k-1}, \quad \mu > 0. \tag{6}$$

The price to be paid for this accuracy increase is the separate computation and storage into two separate vectors of the integrand values at MRA over the left and the right halves of an integration domain $[a, b]$ of width $h = (b - a)/2$: $f_k^l = f(a + h\eta_k)$; $f_k^r = f(b - h\eta_k)$; $k = 0, 1, \ldots, K = 16$, such that $f_{16}^l = f(a + h) = f_{16}^r = f(b - h)$.

4.2 Flow chart of early Bayesian inference

There are two main results established in this subsection: (1) the early Bayesian inference enables *consistent accommodation of the standard* AAQ *approach within the* BAAQ and (2) *robust termination criteria* are defined which efficiently stop the computations while returning optimum output for the input integral (with a problem defined accuracy ceiling which supersedes an unreachable ceiling following from the user accuracy requirements).

The flow chart below summarizes the eight steps of the proposed early Bayesian inference.

Input required:

- the accuracy parameter values $\varepsilon_a^{(i)}, \varepsilon_r^{(i)}$
- the integration domain (ID) $[a, b]$; its centre $c = (b + a)/2$; half width $h = (b - a)/2$
- the integrand function $f(x)$
- the MRA set $\{\eta_k\}$, computed to machine accuracy
- the computed integrand profile (IP) $\{(\eta_k, f_k^l) \,|k = 0, 1, \ldots, K = 16\} \cup \{(\eta_k, f_k^r) \,|k = 0, 1, \ldots, K = 16\}$

Main steps:

- **Step 0**: *Set grey diagnostic* (GD) *as default* (postponed decision)

- **Step 1**: *Integrand boundedness check over the input* IP enables:
  - Definition of extremal (max,min) integrand values, together with their location inside the IP
  - End of computation (EOC) under detection of exceptional cases: (computationally) constant integrand; odd integrand with respect to the ID centre $c = (b + a)/2$
- **Step 2**: `If(.NOT.EOC)` *Computation of Riemann sums over sublattices*
  - Two CC-32 sublattices are defined respectively by:
    - ▹ the 17 inherited CC-16 reduced abscissas (CC16)
    - ▹ the 16 newly added Fejer reduced abscissas (FJ16)
  - Pairs of Riemann sums over sublattices: $q_{\mathrm{CC16}}[(*)]$ and $q_{\mathrm{FJ16}}[(*)]$, where $(*)$ stays either for f or $|f|$
  - Riemann sum quadrature rule outputs for CC-32 IP:
    - ▹ trapezoidal rule quadrature sums: $q_{\mathrm{CC32}}[(*)] = (q_{\mathrm{CC16}}[(*)] + q_{\mathrm{FJ16}}[(*)])/2$
    - ▹ rough error estimates: $e_{\mathrm{CC32}}[(*)] = |q_{\mathrm{CC16}}[(*)] - q_{\mathrm{FJ16}}[(*)]|$
- **Step 3**: EOC *under detection of catastrophic cancellation by subtraction:* $|q_{\mathrm{CC32}}[(f)]| < \tau \cdot q_{\mathrm{CC32}}[|f|]$, τ close to machine epsilon with respect to addition.
- **Step 4**: `If (.NOT. EOC)` *check for problem dependent update of accuracy parameters* $\varepsilon_a^{(o)}, \varepsilon_r^{(o)}$
- **Step 5**: `If (.NOT. EOC)` decide on an *ill-conditioning* (IC) *diagnostic* `iff`
  - either $|q_{\mathrm{CC32}}[(f)]| < 2.0 \cdot q_{\mathrm{CC32}}[|f|]$
  - or $q_M > t_i \cdot q_m$, where $t_i \sim 100$ is an empirical threshold for the comparison of q_M with q_m, where $q_M = \max\{|q_{\mathrm{CC16}}[f]|, |q_{\mathrm{FJ16}}[f]|\}$ and $q_m = \min\{|q_{\mathrm{CC16}}[f]|, |q_{\mathrm{FJ16}}[f]|\}$
- **Step 6**: GD *diagnostic refinement* based on the outputs got for the Chebyshev series expansion coefficients
  - GD is changed to EOC iff negligible highest label even-rank and odd-rank CC-32 coefficients
  - GD is changed to IC iff either
    - (1) monotonicity is infringed for suitably chosen binary tree structure dependent subsets of CC-32 coefficients, or
    - (2) non-convergence or slow convergence is established to occurr for the CC-32 coefficients
- **Step 7**: *Check for further accuracy parameter updates,* $\varepsilon_a^{(o)}, \varepsilon_r^{(o)}$
- **Step 8**: *Path to subrange subdivision:*
  - `If (IC) then` define offending IP abscissas and refine IC diagnostic at these abscissas; subdivide current subrange into diagnostic-dependent finer subranges
  - `elseif (GD) then` proceed along the standard AAQ scheme.

4.3 Two illustrative examples

The accumulated empirical evidence plays a very important role in the decision taking process based on Bayesian inference. The data reported in [2] pointed to three different patterns of behaviour of the magnitudes of the Chebyshev expansion coefficients: *fast convergence* (suggesting rapid end of computations), *moderate convergence* (suggesting output improvement based on subrange bisection along the lines of the standard AAQ scheme), and *ill-conditioning* (irregular behaviour inside both the even-rank and the odd-rank coefficient subsets which is asking for IP analysis to resolve the offending integrand features).

The above-mentioned framework is refined by the scrutiny of the dependence of the even-rank (e) and odd-rank (o) Chebyshev expansion coefficients on the position of a singularity with respect to the ends of the current integration domain (figure 2).

Figure 2–left reports data for the integrand ([4], p.110) $f_1 : [0, 1] \to \mathbb{R}$, $f_1(x) = |x^2 + 2x - 2|^{-1/2}$, which shows an inner singularity at $x_s = \sqrt{3} - 1$.

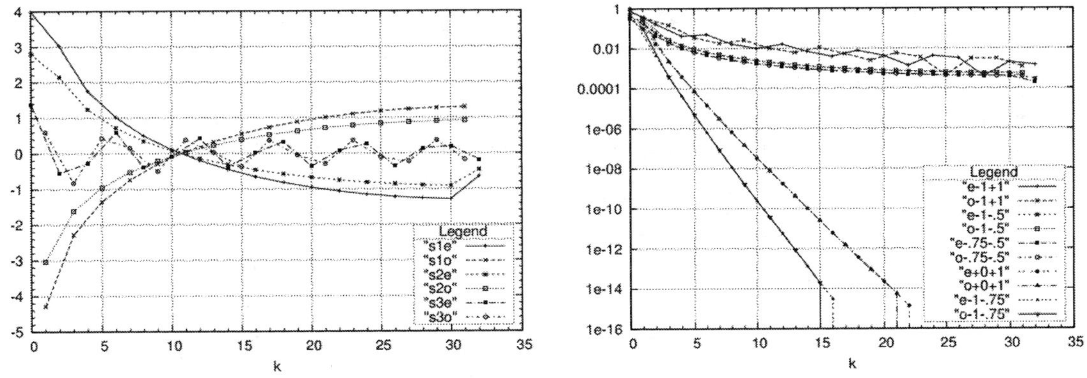

Figure 2. Patterns of variation of the Chebyshev expansion coefficients within the even and odd rank subsets versus the coefficient labels. In the file names in the left figure, "s" points to singular, "e" to even-rank coefficient subset, "3" to the subrange $[0, 1]$, "2" to the subrange $[x_s, 1]$, and "1" to the subrange $[x_s, \sqrt{3}/2]$. In the file names in the right figure "e" points to even-rank coefficient subset, "o" to odd-rank coefficient subset, the remaining figure pairs point to the left and the right ends of the subranges, respectively.

Figure 2–right reports outputs got for the integrand ([9]) $f_2 : [-1, 1] \to \mathbb{R}$, $f_2(x) = |x + 0.5|^{1/2}$, which shows an inner derivative singularity at $x_d = -0.5$.

The meanings of the notations used in the figure 2 are explained in the figure caption.

In the left figure, the behaviour of the "e" and "o" coefficient subsets over the range $[0, 1]$ is irregular due to the presence of an inner singularity at x_s, in agreement with [2]. In this case, an ill-conditioning diagnostic is issued at step 8 of Sec. 4.2, which is asking for the localization of the offending integrand feature to machine accuracy. However, for the ranges $[x_s, 1]$ and $[x_s, \sqrt{3}/2]$, both characterized by endpoint singularities, a *regular variation* of both the "e" and "o" coefficient subsets is noticed and this asks in the decision path for the activation of a convergence acceleration procedure.

The data in figure 2–right are given in semi-logarithmic scale. The behavioural patterns over the inner derivative singularity range $[-1, 1]$ as well as over the outer derivative singularity ranges $[0, 1]$ and $[-1, -0.75]$ are consistent with the data reported in Fig. 2–left and in [2]. The irregular behaviour pattern over the range $[-1, 1]$ asks for localization of the offending abscissa to machine accuracy. Over the ranges $[-1, x_d]$ and $[-0.75, x_d]$, characterized by endpoint derivative singularities, the noticed slowly converging monotonic behaviours of the coefficient subsets ask for the activation of the convergence acceleration procedure. Far from singularity (the other two cases), the good convergence properties shift the decision path to the standard `AAQ` scheme.

In both the left and right figures 2, the smaller the extension of the integration domain with an endpoint singularity, the larger are the ranges of variation of the two Chebyshev expansion coefficient subsets.

5 Conclusions

The present report discusses a Bayesian automatic adaptive quadrature (BAAQ) solution for numerical integration which is simultaneously robust, reliable, and efficient, yielding maximum possible output accuracy in numerical experiments under arbitrary behaviour of the integrand function.

An essential ingredient of the solution is the multiscale approach.

An early decision path to the integrand profile (IP) scrutiny enables the identification of trivial or manifestly unsolvable problems as well as the need to relax the user requested accuracy parameters.

Within the Clenshaw-Curtis quadrature over macroscopic ranges, the scrutiny of the Chebyshev expansion coefficients enable further identification of unresolved ill-conditioned features.

We are thus left either with a hopefully well-conditioned integral, for which the standard automatic adaptive quadrature can be used, or with a manifestly ill-conditioned problem for which an improved version of the full BAAQ machinery is activated.

Acknowledgement

Work supported within the Hulubei-Meshcheryakov Programme on 2017, JINR Orders 218/10.04.2017 (pp. 22, 23) and 219/10.04.2017 (pp. 93–95). The authors are grateful to an anonymous referee for the careful reading of the manuscript and his precise comments which helped improving details of the presentation.

References

[1] Gh. Adam and S. Adam, in Gh. Adam, J. Buša, and M. Hnatič (Eds.), MMCP 2015 *Mathematical Modeling and Computational Physics*, EPJ Web of Conferences **108**, 02002 (p1–p6) 2016, https://doi.org/10.1051/epjconf/201610802002

[2] S. Adam and Gh. Adam, in Gh. Adam, J. Buša, and M. Hnatič (Eds.), MMCP 2015 *Mathematical Modeling and Computational Physics*, EPJ Web of Conferences **108**, 02003 (p1–p6) 2016, https://doi.org/10.1051/epjconf/201610802003

[3] Gh. Adam and S. Adam, in *Mathematical Modeling and Computational Science* (MMCP2011), Gh. Adam, J. Buša, and M. Hnatič (Eds.) (Springer, Heidelberg, LNCS 7125, 2012) pp. 1–16

[4] R. Piessens, E. de Doncker-Kapenga, C. W. Überhuber, and D. K. Kahaner, QUADPACK, *a subroutine package for automatic integration* (Springer Verlag, Berlin, 1983)

[5] P.J. Davis and P. Rabinowitz, *Methods of Numerical Integration* (2-nd Ed., Academic Press, Orlando (Fla), USA, 1984)

[6] A.R. Krommer and C.W. Ueberhuber, *Computational Integration* (SIAM, Philadelphia, 1998)

[7] NAG *Library Manual*, Mark 24, section D01: http://www.nag.co.uk/numeric/fl/nagdoc_fl24/html/D01/d01conts.html (The Numerical Algorithms Group Ltd., Oxford, UK, 2012)

[8] S. Adam and Gh. Adam, in *Mathematical Modeling and Computational Science* (MMCP2011), Gh. Adam, J. Buša, and M. Hnatič (Eds.) (Springer, Heidelberg, LNCS 7125, 2012) pp. 189–194

[9] C.W. Clenshaw and A.R. Curtis, Numer. Math. **2**, 197–205 (1960)

[10] H. Brass and K. Petras, *Quadrature Theory. The Theory of Numerical Integration on a Compact Interval* (Math. Surv. and Monogr. vol. 178, American Mathematical Society, Providence, 2011)

[11] W.M. Gentleman, Comm. Assoc. Comp. Mach. **15**, 337–342, 343–346 (1972)

[12] R. Piessens and M. Branders, J. Comp. Appl. Math. **1**, 55, 153 (1975)

[13] Gh. Adam and A. Nobile, IMA J. Numer. Analysis **11**, 271–296 (1991)

[14] Gh. Adam and S. Adam, Numerical Methods and Programming: Advanced Computing **10**, 391–397 (2009) (http://num-meth.srcc.msu.ru/english/zhurnal/tom_2009/pdf/v10r145.pdf)

Online Event Reconstruction in the CBM Experiment at FAIR

Valentina Akishina[1,*] and *Ivan* Kisel[1,2,3,**] for the CBM collaboration

[1] *Goethe-Universität, Frankfurt am Main, Germany*
[2] *FIAS, Frankfurt am Main, Germany*
[3] *GSI Helmholtz Center for Heavy Ion Research GmbH, Darmstadt, Germany*

Abstract. Targeting for rare observables, the CBM experiment will operate at high interaction rates of up to 10 MHz, which is unprecedented in heavy-ion experiments so far. It requires a novel free-streaming readout system and a new concept of data processing. The huge data rates of the CBM experiment will be reduced online to the recordable rate before saving the data to the mass storage. Full collision reconstruction and selection will be performed online in a dedicated processor farm. In order to make an efficient event selection online a clean sample of particles has to be provided by the reconstruction package called First Level Event Selection (FLES).
The FLES reconstruction and selection package consists of several modules: track finding, track fitting, event building, short-lived particles finding, and event selection. Since detector measurements contain also time information, the event building is done at all stages of the reconstruction process. The input data are distributed within the FLES farm in a form of time-slices. A time-slice is reconstructed in parallel between processor cores. After all tracks of the whole time-slice are found and fitted, they are collected into clusters of tracks originated from common primary vertices, which then are fitted, thus identifying the interaction points. Secondary tracks are associated with primary vertices according to their estimated production time. After that short-lived particles are found and the full event building process is finished. The last stage of the FLES package is a selection of events according to the requested trigger signatures. The event reconstruction procedure and the results of its application to simulated collisions in the CBM detector setup are presented and discussed in detail.

1 Introduction

One of the main objectives of the physics program for the future heavy-ion experiment CBM (FAIR, Germany) [1] is to understand the properties of strongly interacting matter at very high baryonic densities. The physics program of CBM includes the measurement of yields, spectral distributions, correlations and fluctuations of the particles produced in the heavy-ion reactions. The CBM detector setup is designed to measure not only bulk hadrons, but very rare probes like multi-strange hyperons or charmed hadrons as well. In order to measure rare probes with sufficient statistics the experiment will operate at high interaction rates of up to 10 MHz, which puts strong constraints on the data

*e-mail: V.Akishina@gsi.de
**e-mail: I.Kisel@compeng.uni-frankfurt.de

processing. It requires very fast and radiation-hard detectors, and a novel data read-out and analysis concept based on free streaming front-end electronics and a high-performance computing cluster for online event selection. In order to make an efficient event selection online a clean sample of particles has to be provided by the reconstruction package called First Level Event Selection (FLES) [2].

2 Reconstruction of free-streaming data in the CBM experiment

The huge data rates force the CBM experiment to reduce data online before saving it to the mass storage. In order to reduce data in real time, the experiment has to implement triggers. Simple hardware triggers typical for traditional latency-limited trigger architectures are inapplicable for the case of CBM, since the experiment is targeted to events with complicated patterns, which require full event reconstruction. Therefore, CBM plans to implement sophisticated software trigger, which will process raw detector data, reconstruct collision patterns online and take a trigger decision based on the reconstructed information.

To implement the described scheme CBM will benefit from a new data acquisition concept with autonomous, self-triggered front-end electronics. Every detector channel in this case will deliver time-stamped data signals upon its activation. These messages will be transported to the readout buffer, the size of which will be adjusted to the power of a large computer farm, where it will be reconstructed.

The detector data obtained during a certain period of time is called a time-slice. As a consequence of the read-out concept, the detector signals inside a time-slice are not associated with physical collisions. However, in order to obtain correct physics interpretations, one needs to extract information on the individual collisions of ions from the continuous time-slice data. Thus, the reconstruction routines of CBM should deliver event-associated data in the end. The process of event-association is called event building and should be a part of the standard reconstruction chain.

The event building is not a trivial task for CBM, especially for the cases of the highest iteration rates, when information on different collisions overlaps in time. This case requires sophisticated algorithms, which will allow to reliably resolve different collisions from each other, using not only time, but also spacial information. This is a task for a FLES reconstruction package.

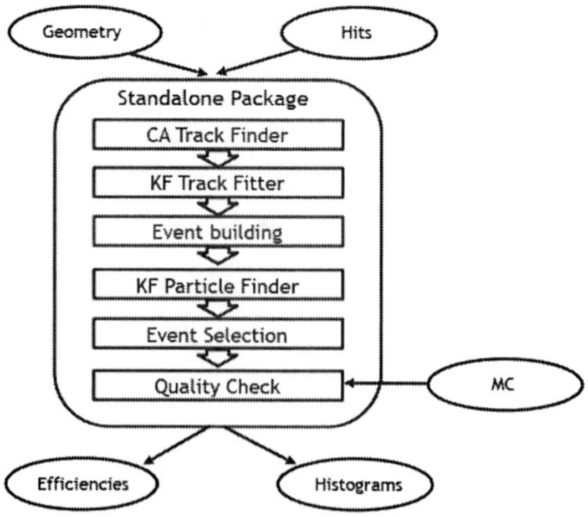

Figure 1. Block diagram of the FLES package.

The FLES package consists of several modules (see figure 1): track finder, track fitter, event builder, particle finder and physics selection. As an input the FLES package receives a simplified

Table 1. Track reconstruction efficiency performance for one hundred minimum bias AuAu UrQMD collisions at 25 *A*GeV for the event-by-event reconstruction and the time-slice-based reconstruction at 0.1 MHz, 1 MHz and 10 MHz interaction rates, presented for different track categories.

Track category, %	E-by-E	10^5Hz	10^6Hz	10^7Hz
All tracks	92.1	92.6	92.6	92.2
Primary high-p	97.9	98.2	98.2	97.9
Primary low-p	93.6	94.1	94.1	93.5
Secondary high-p	92.0	92.7	92.7	92.0
Secondary low-p	65.7	66.7	66.6	65.9
Clone level	2.8	0.3	0.3	3.1
Ghost level	4.9	3.5	3.5	4.2
MC tracks found	145	146	146	145

geometry of the tracking detectors and the hits, which are created by the charged particles crossing the detectors. Tracks of the charged particles are reconstructed by the Cellular Automaton (CA) track finder [3]. The Kalman filter (KF) based track fit [4] is used for precise estimation of the track parameters. The event builder algorithm takes as input fitted tracks and based on this information builds events via splitting tracks into event-corresponding groups. The next step of the data processing chain is particle reconstruction. The short-lived particles, which decay before the tracking detectors, can be reconstructed via their decay products only. The KF Particle Finder [5], which is based on the KF Particle package [5, 6] is used in order to find and reconstruct the parameters of short-lived particles by combining already found tracks of the long-lived charged particles. The algorithm takes as input events delivered by the event-builder algorithm. The KF Particle Finder also selects particle-candidates from a large number of random combinations. In addition, a quality check module is implemented, that allows to monitor and control the reconstruction process at all stages.

The FLES package in the CBM experiment will be executed for the online selection and the off-line analysis on a dedicated many-core CPU/GPU farm. The package is platform and operating system independent. The farm is currently estimated to have a compute power equivalent to 60 000 modern CPU cores. The FLES algorithms have to be therefore local and parallel with respect to data and use the full potential of modern and future many-core CPU/GPU architectures.

3 4D Cellular Automaton track finder

The track finder algorithm used to reconstruct tracks in the tracking detectors of CBM is based on the Cellular Automaton method [7]. The CA track finder takes as an input hit measurements from the Silicon Tracking System (STS detector) in the form of a time-slice, which includes time and spacial measurements. The track finding procedure starts with combining the hits, delivered by the reconstruction algorithms of STS detector, into triplets – combination of three hits on adjacent stations. The triplet structure was chosen, since it allows to estimate the momentum of particle, which could produce it. The triplets with two common hits are combined into track candidates. In order to take into account detector inefficiency, a missing hit in one station is tolerated for a reconstructed track. The track candidates should survive a dedicated selection based on the track length and calculated χ^2-value to be accepted to the reconstructed tracks.

In order to optimize and speed up the process of track reconstruction the track finding procedure is performed in several iterations [8]. At first the track finder algorithm is searching for the tracks, which are easier to reconstruct, for instance, high momentum and primary tracks. Having found these tracks, the algorithm removes hits, which were used to construct these tracks, from the input information. Thus, the combinatorics, which is higher for the iterations when searching for low-momenta and secondary tracks gets significantly reduced. Therefore, in the first iteration only high-momentum primary tracks are reconstructed, in the second one – low-momentum primary tracks, and then – all other tracks.

Input time information is used in the algorithm to the same extent and in similar manner as it is done with the spacial coordinates. The same logic is used while constructing triplets: the hits in the triplet should belong to the same particle, therefore they should correlate not only in space, but also in time. Since the estimated time of flight between two consecutive stations is negligible in comparison to the detector time precision, the hits, belonging to the same track, should coincide within three σ-s of the detector time precision.

In order to estimate the performance of the algorithm, the efficiency calculation tests based on the generated Monte-Carlo information for one hundred AuAu minimum bias UrQMD [9, 10] events at 25*A* GeV have been performed on a server with Intel Xeon E7-4860 CPUs. For evaluation purposes the particle is called reconstructable if it has crossed at least 4 stations of the STS detector. A reconstructed track is assigned to a simulated particle, if at least 70% of its hits have been caused by the same particle. If a certain particle is assigned to more than one track, such additionally reconstructed tracks are regarded as clones. If a track was not assigned to any particle it is called a ghost. The resulting track reconstruction efficiencies for the cases of event-by-event analysis (so-called 3D analysis) as well as for the 4D case (with included time measurement, as well as 3-dimensional spacial information) while reconstructing time-slices, produced out of one hundred AuAu minimum bias UrQMD events at 25*A* GeV, are presented in table 1 [8].

The inclusion of the time measurement have made it possible to achieve an efficiency performance comparable to the case of the event-by-event analysis. Moreover, the track reconstruction efficiency has been improved after taking into account the STS time measurement in comparison to the event-based performance. The effect of efficiency improvement is present even for the extreme case of 10 MHz interaction rate. It can be explained by the presence of slow particles, which create random combinations of hits in the 3D case. These random combinations can be rejected in the case of time-slices thanks to the hit time measurement cut, thus improving the performance [8].

The algorithm has been included into the CBMROOT framework [11]. The simulation of detector response in the framework provides a time measurement, taking into account the anticipated behavior of the detector. The performance for the algorithm included into CBMROOT framework is presented in table 1. It is comparable to the 3D case.

The CA track finder is both vectorized (using SIMD instructions) and parallelized (between CPU cores) with the help of OpenMP interfaces [12]. The algorithm shows strong scalability on many-core systems. A speed-up factor of 10.6 was achieved on an Intel Xeon E7-4860 CPU with 10 hyper-threaded physical cores [13].

4 Kalman filter track fit

The Kalman filter is intensively used in the CA track finder for obtaining parameters for partially reconstructed track segments, therefore its fast implementation on modern CPU/GPU computer systems and stability in single precision are crucial not only for the fitting procedure, but for the track finder as well. A massive optimization was performed with the initial purpose of vectorisation of the algorithm [2]. To optimize the memory usage, a magnetic field approximation was implemented for particle propagation instead of using the full magnetic field map, which takes about 70 MB and therefore does not fit into the CPU cache memory. The magnetic field is approximated with a polynomial function of fifth order at each detector station. During the fit of a track the field behavior between the stations is approximated with a parabola taking field values at the three closest measurements along the track. The first measurement is processed with a dedicated procedure, which increases the

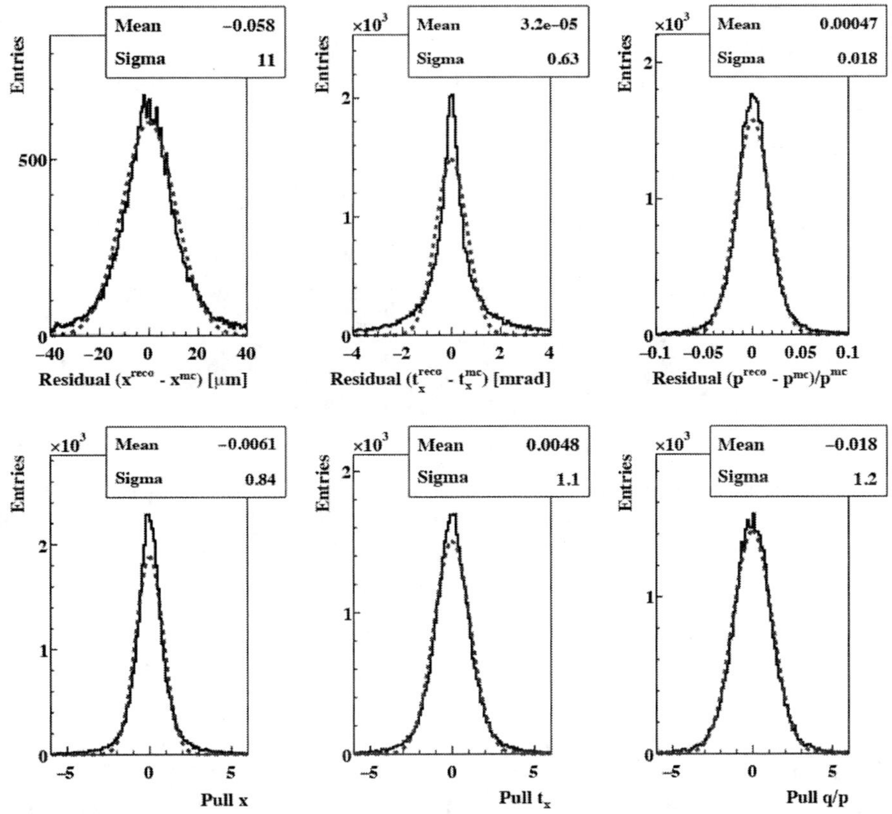

Figure 2. Residuals and pulls distributions for the x, t_x and q/p track parameters, reconstructed with 4D CA track finder and Kalman Filter track fit. The dashed lines correspond to Gaussian fits of the distributions.

numerical stability of the method in single precision. The track propagation in the non-homogeneous magnetic field is done by an analytic formula based on the Taylor expansion [14]. The analytic formula allows to obtain the same track fit quality as the standard fourth order Runge-Kutta method, while being 40% faster. Operator overloading has been used to keep flexibility of the algorithm with respect to different CPU/GPU architectures. All these changes have increased the processing speed of the SIMD KF track fit algorithm down to 1 μs per track. This is an improvement by a factor 10000 with respect to the original scalar version of the algorithm [4].

Since CBM is a forward geometry experiment, the following track parameters, which are measured at a certain z position, are used: x and y coordinates, t_x and t_y are the track slopes in the XZ- and YZ-planes, q/p is the inverse particle momentum with the sign of the particle charge and t is the time. The time parameter is fully included in the fitting procedure and treated the same way as the spacial parameters. The obtained fitted time parameter is particularly important and intensively used during the next element of reconstruction chain – event builder. In order to demonstrate the fit quality the residuals and pulls of the parameters in the track state vector have been calculated (see figure 2). The residuals of the track parameters are defined as differences between the reconstructed parameters and the true Monte-Carlo values. The pulls are defined as the residuals normalized by the estimated errors of the track parameters. In the ideal case these should be unbiased and Gaussian distributed with width of 1.0.

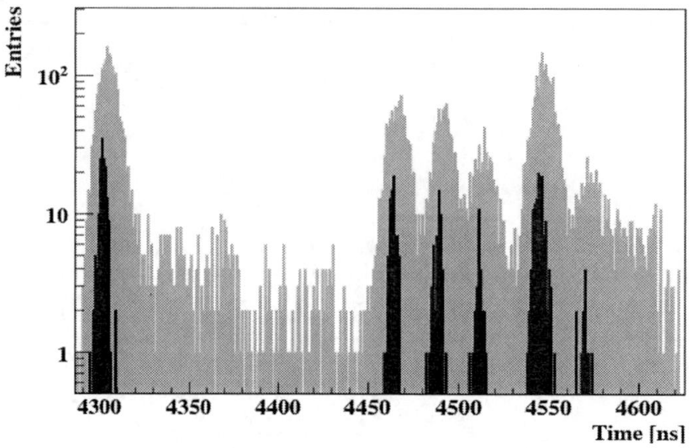

Figure 3. The distribution of the STS hit time measurement in the time-slice with one hundred minimum bias AuAu UrQMD events at 25 *A*GeV at 10 MHz is shown with a light blue color, the distribution of the fitted track time is shown with black color: the reconstructed tracks are clearly grouped into the clusters of events (color online).

5 Event builder

The event building algorithm based on found tracks will deliver reconstructed events for the case of the highest CBM interaction rate of 10 MHz. The first version of the algorithm is implemented based on the information from the STS detector alone and will be extended in future for other sub-detector systems. For instance, the Time-of-Flight (ToF) detector, the time resolution of which is several orders of magnitude better than the STS precision, is expected to drastically improve the ability to resolve the events, which contain ToF information [8].

The track-based event builder splits the reconstructed tracks into event-corresponding groups based on the fitted time parameter, extrapolated to the region of primary vertex, and the estimated errors, obtained after the Kalman Filter track fit [15]. The algorithm starts grouping tracks around the best fitted track with the smallest estimated time errors. All tracks which coincide in the reconstructed time within 3 σ of estimated time errors are assigned to the same event. The procedure is repeated for all tracks until none of them are left.

The initial distribution of hits measurements illustrating the complexity of defining event borders in a time-slice at the interaction rate of 10 MHz is shown in figure 3 in a light blue (light gray in print). This is a typical example of the fact that at such an extreme interaction rate there are no isolated events. After the 4D time-slice-based reconstruction we assign to each track a time measurement, which is calculated as the average of its hits measurements. As a next step we need to group the reconstructed tracks in time into clusters of tracks belonging to same collision, identifying thus physical events. The resulting time distribution of track clusters is shown in a black color in figure 3. The reconstructed track clusters represent groups corresponding to collisions the tracks of which originate from. Even in the area of the severe initial event overlap on a hit measurement level the time-based CA track finder allows to resolve tracks from different collisions in time.

The tests performed with one hundred AuAu minimum bias UrQMD events at 25*A* GeV have shown that 83% of events were reconstructed without any event merging, 17% of events were reconstructed in merged double event clusters. For evaluation purpose an event is called merged if at least one track from the reconstructed event was assigned to a wrong event. Absence of event splitting was achieved. Due to properties of Poisson distribution, which determines the time distance between

Table 2. The reconstruction efficiency for minimum bias AuAu UrQMD collisions at 10 *A*GeV in the cases of event-by-event reconstruction and time-slice-based reconstruction at 10 MHz interaction rate for K_s^o, Λ and Ξ^-.

Particle/ case	3D, %	10 MHz, %
K_s^o	22.9	21.2
Λ	21.9	19.6
Ξ^-	7.8	6.3

consecutive collisions, the method based on time information alone can not guarantee total resolution of collisions since the probability of events coinciding in time within time precision is never zero. Therefore, for the extreme scenario of interaction rate of 10 MHz the remaining 17% of events are to be resolved later with the use of the multi-primary-vertex analysis and the full event topology.

6 KF Particle Finder

The next step in the FLES data processing chain is the reconstruction of particles. It is done in two stages. First, particles registered in the CBM detector system are reconstructed. Then short-lived particles decayed before or inside the setup are searched based on both their charged and neutral daughter particles.

The KF Particle Finder package [2, 5] is used in CBM to reconstruct short-lived particles. The event builder algorithm provides a set of tracks grouped into the event to the KF Particle Finder as an input, thus, allowing the physics analysis for the time-stamped data. Since the FLES package is developed to run on many-core computer architectures, the reconstruction of particles is done in parallel that provides a possibility for a global competition between particle candidates. Such a global event reconstruction significantly improves suppression of a combinatorial background and provides for further physics analysis a very clean sample of particles produced at different stages of heavy-ion collision. Recently, the package has been extended to hyper-nuclei and weak decays involving neutral particles in the final state.

The resulting reconstruction efficiencies normalised to the total number of produced particles are given in table 2. The resulting efficiency of short-lived particle reconstruction for the 10 MHz interaction rate is similar to the event-based analysis. The loss of particle reconstruction efficiency despite stable track reconstruction efficiency, which is normalised to the number of reconstructable tracks, for the case of time-based reconstruction can be explained by the smaller number of produced hits due to the strip dead time.

The search for the short-lived particles is done in one go, thus minimizing access to the memory. Together with the optimization and vectorization of the code this allows to achieve a high speed even in the presence of a huge combinatorics. The speed of the KF Particle Finder per core on a server with Intel Xeon E7-4860 CPUs is 1.4 ms per minimum bias AuAu UrQMD collision and 10.5 ms per central AuAu UrQMD collision at 25 *A*GeV.

7 Conclusions

The reconstruction chain of CBM being developed enables the free-streaming data processing by dedicated time-based algorithms. It allows to reconstruct individual events out of continuous time-based data. The reconstructed events, each of which in the ideal case corresponds to one real interaction in the target, serve as input for high-level physics analysis. The problem of overlapping events at highest interaction rates will be addressed by studying the possibility to reconstruct multiple primary vertices inside one event.

The FLES package for the CBM experiment contains all reconstruction stages: track finding, track fitting, short-lived particles finding, event building and event selection. Time-based 4D Cellular Automaton and the Kalman filter algorithms are used for finding and fitting tracks both in space and time, that allows to achieve a high track reconstruction efficiency up to 92% and the track parameters quality with about 1.5% of the momentum resolution.

The 4D event building has been implemented for the CBM experiment. It allows to resolve the major part of overlapping on a hit level events and group tracks into event-corresponding clusters without event splitting. About 17% of events are merged and can not be separated using only the time information at the interaction rate of 10 MHz. The first version of the algorithm is implemented and will be enhanced with multi-vertex reconstruction and information from the TOF detector.

The KF Particle Finder package shows a high reconstruction efficiency with an optimal signal to background ratio. The physics simulations and performance studies were extended to hyper-nuclei and weak decays involving neutral particles in the final state.

Thus, the first version of the full reconstruction chain for the CBM experiment is established. All the algorithms are developed in a simple and general form to be experiment specifics independent with efficient use of many-core architectures. The package is vectorized using SIMD instructions and parallelized between CPU cores. All algorithms are optimized with respect to the memory usage and speed.

Acknowledgements

This work was supported by HICforFAIR, FIAS and BMBF.

References

[1] B. Friman, C. Hohne, J. Knoll, S. Leupold, J. Randrup, R. Rapp, and P. Senger, *The CBM physics book: Compressed baryonic matter in laboratory experiments* (Lect. Notes Phys. 814, 2011) 980
[2] I. Kisel, I. Kulakov, and M. Zyzak, IEEE Transactions on Nuclear Science **60**, 5, 3703 (2013)
[3] V. Akishina, *4D Event Reconstruction in the CBM Experiment* (Dissertation thesis, Goethe university, Frankfurt am Main, 2017) 181 pp.
[4] S. Gorbunov, U. Kebschull, I. Kisel, V. Lindenstruth, and W.F.J. Müller, Computer Physics Communications **178**, 5, 374–383 (2008)
[5] M. Zyzak, *Online Selection of Short-Lived Particles on Many-Core Computer Architectures in the CBM Experiment at FAIR* (Dissertation thesis, Goethe university, Frankfurt am Main, 2016) 159 pp.
[6] S. Gorbunov, *On-line reconstruction algorithms for the CBM and ALICE experiments* (Dissertation thesis, Goethe University of Frankfurt, 2012) 104 pp.
[7] I. Kisel, Nucl. Instr. and Meth. **A566**, 85–88 (2005)
[8] V. Akishina, I. Kisel, I. Vassiliev, and M. Zyzak "Time-based Reconstruction of Free-streaming Data in CBM," this journal
[9] S. A. Bass, *et al.*, Prog. Part. Nucl. Phys. **41** (1998)
[10] M. Bleicher, *et al.*, J. Phys. G: Nucl. Part. Phys. **25** (1999)
[11] CBMROOT framework: https://redmine.cbm.gsi.de/projects/cbmroot
[12] The OpenMP API specification for parallel programming http://openmp.org/wp
[13] V. Akishina and I. Kisel, J. Phys. Conf. Ser. **599**, 1, 012024 (2015)
[14] S. Gorbunov and I. Kisel, Nucl. Instr. and Meth. **A559**, 148–152 (2006)
[15] V. Akishina and I. Kisel, IEEE Trans. Nucl. Sci. **62**, 6, 3172 (2015)

EPJ Web of Conferences **173**, 01003 (2018) https://doi.org/10.1051/epjconf/201817301003
Mathematical Modeling and Computational Physics 2017

Shape Approximation Based on Higher-Degree Polynomials

Nikolay Dikusar[1] *

[1] *Laboratory of Information Technology, Joint Institute for Nuclear Research, Dubna, Russia*

Abstract. The planar shape (contour) of an object is a fundamental source of information in a pattern recognition problem. Obtaining the relevant information set rests on difficult procedures and is a key problem in pattern recognition. A method is proposed for the segmentation of contours with a complex geometrical form. It is based on a parametric piecewise approximation of 12th order spanned by a polynomial model defined by basic elements. Higher-order polynomial approximation allows to optimize the number of segments on the contour and to obtain analytically the dependence of the curvature for more exact calculation of informative signs that are invariant to geometrical transformations. The algorithm based on this method as well as specific examples are described in detail.

1 Introduction

Not the goal is the subject of decision,
but the means to the goal ...
Aristoteles

The planar shape (contour) of the object is a fundamental source of information not only in pattern recognition but also in the information technologies as well. For example, to recognize objects at complex image in real time, the processor's performance should be approximately 10^8–10^{14} elementary operations per second. Therefore, instead of processing each point of the object, only its contour is processed.

Analytical curves such as pieces of straight lines, circles and spirals arches, cubic splines, etc. are used more often in the approximation of contour lines (CL). In this case the efficiency of the CL approximation is decreased because a big number of short segments [1–3].

The goal of this work is to increase the efficiency of the pattern recognition and image processing algorithms using *high-degree* polynomials for the approximation and smoothing of planar shapes.

An approximation by a higher-degree polynomial has several advantages over the approximation by a low-degree polynomial. For example, if the length of intervals *increases*, the errors of the approximation *are decreased*. The number of segments on the contour *is decreased*, informative signs *are calculate more precisely*.

The higher degree of the polynomial, the smaller the number of the grid nodes, and the accuracy and quality of the approximation is better. *Increasing the degree* of the polynomial *reduces* the number of the nodes and *improves* the smoothness and quality of the solution inside of the interval of approximation.

*e-mail: dnd@jinr.ru

At the same time, an increase in the degree of the polynomial usually leads to: *loss of stability*, considerable error increase of the predicted value at *extrapolation*, *increase* of the order of the derivatives used, *increase* of the computational complexity. Therefore formulas of smoothing by higher-degree polynomials are avoided almost altogether.

To reduce these difficulties, a new proposal for the choice of the polynomial spanning basis, called the Basic Element Method (BEM) has been developed recently in our Laboratory [4–7].

A higher-order polynomial approximation allows to optimize the number of segments on the contour line and to obtain analytically the dependence of the curvature for it more precise calculation of informative signs that are invariant with respect to geometrical transformations.

The efficiency of the polynomial approximation and smoothing depends of many factors including *the form* of the polynomial.

The new polynomial form has been constructed in the frame of BEM (BEM-polynomial). Higher-degree BEM-polynomials simplify the function approximation and data smoothing.

The following conception underlies the construction of the BEM-polynomial: *"analytical connection between nodes of a three-point grid and corresponding triplets of pivot points of the BEM-polynomial and of its derivatives"*.

2 What is a BEM-polynomial?

But very often it is important for an error to be reduced to zero within the limits of the interval [8]
P. L. Chebyshev

The BEM-polynomial construction is based on Chebyshev's idea about the approximation of a smooth function by a polynomial on an interval of limited length (1853) [8].

The nth degree BEM-polynomial is expressed via four basic elements (BE): one cubic parabola (Q) and three quadratic parabolas (w_1, w_2, w_3).

The four functions Q, w_1, w_2, w_3 are defined by three nodes of *a three-point grid* $\Delta_3^{\alpha\beta}$: $x_\alpha < x_0 < x_\beta$ and depend on τ, α, β, where $\tau = x - x_0$, $\alpha = x_\alpha - x_0$, $\beta = x_\beta - x_0$; $x, x_\alpha, x_\beta, x_0 \in R$.

The basic elements w_1, w_2, w_3 and Q depend continuously on the parameters x_0, α and β. These parameters are related to the independent variable τ by the special cross-ratio rule $[\xi_1\xi_2\xi_3\xi_4] = [13]/[23] : [34]/[14]$, $[ij] = \xi_j - \xi_i$, $\xi_j \neq \xi_i$. For the quadruple $[\tau\alpha\beta 0]$, where the first point is fixed, this rule generates three fractional rational functions w_1, w_2, w_3 with respect to the parameters α, β and τ (quadratic functions relative to τ) [4]:

$$w_1 = \frac{-\tau(\tau-\beta)}{\alpha\gamma}, \quad w_2 = \frac{\tau(\tau-\alpha)}{\beta\gamma}, \quad w_3 = \frac{(\tau-\alpha)(\tau-\beta)}{\alpha\beta}, \quad \gamma = \beta - \alpha, \quad \alpha\beta\gamma \neq 0, \quad \alpha\beta < 0.$$

The fourth element Q is *a zeroing* cubic parabola [4]

$$Q = \alpha\beta\tau w_3 = \tau(\tau-\alpha)(\tau-\beta), \qquad \tau,\ \alpha,\ \beta \in R.$$

The polynomial $P_n(x) = \sum_{i=1}^{n} a_i x^i$ can be written in the BEM basis form as follows:

$$P_{n\downarrow m}(x) = \sum_{j=0}^{m} Q^j \Pi_j, \qquad m = \lfloor n/3 \rfloor, \qquad m, n \in Z, \tag{1}$$

where the quadratic parabolas Π_j in Eq. (1) are convolutions of the vector $\mathbf{w} = [w_1, w_2, w_3]^T$ with the vectors of the coefficients $\mathbf{r}_j = [r_{j1}, r_{j2}, r_{j3}]^T$, i.e. $\Pi_j = \mathbf{w}^T \mathbf{r}_j$, $j = \overline{0, m}$.

The basis functions (BF) in Eq. (1) $\mathbf{b}_j(\tau,\alpha,\beta) = Q^j \mathbf{w}_j$ are polynomials of the degree $3j+2$ having zeros at the nodes of the three-point grid $\Delta_3^{\alpha\beta}$. Graphs of the BF are shown in Fig. 1, left.

The basic elements define *a structure* and fulfill properties such as

- *natural normalization* $w_1 + w_2 + w_3 = 1$;
- *partial symmetry* with respect to the permutation of the parameters $\alpha \leftrightarrow \beta$:
  $w_1 \leftrightarrow w_2,\ w_3 \leftrightarrow w_3, Q \leftrightarrow Q$;
- *scale invariance* of w_j: $w_j(\mu\tau,\mu\alpha,\mu\beta) = w_j(\tau,\alpha,\beta)$ and
- *uniformity* of Q : $Q(\mu\tau,\mu\alpha,\mu\beta) = \mu^3 Q(\tau,\alpha,\beta)$.

These properties allow to optimize the parameters of a piecewise polynomial approximation (PPA) [7].

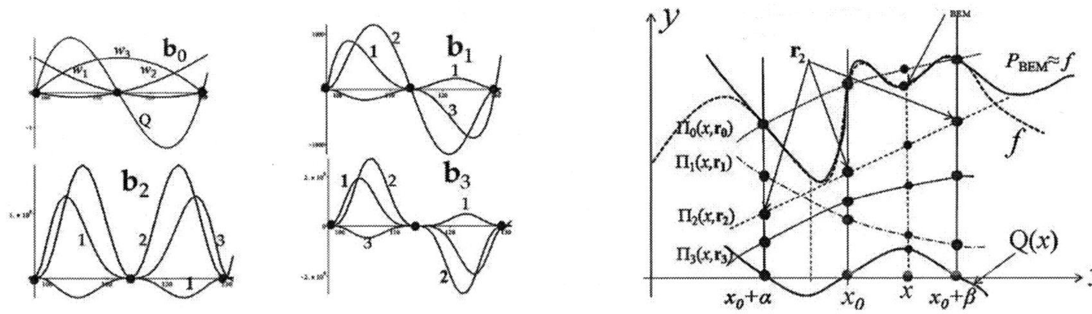

Figure 1. Graphs of the basis functions and geometric sense of the P_{BEM}-approximation

In terms of the BF the BEM-polynomial (1) can be written as

$$P_{\mathrm{BEM}} = \sum_{j=0}^{m} \mathbf{b}_j^T \mathbf{r}_j = \sum_{j=0}^{m} \sum_{i=1}^{3} b_{ji} r_{ji}, \qquad b_{ji} = Q^j w_i. \tag{2}$$

The geometrical sense of Eq. (1) is shown in Fig. 1, right. The construction of the BEM-polynomial presents a synthesis of the properties of the Taylor polynomial and of the second degree Lagrange polynomial defined over the same three-point grid.

2.1 Properties of the BEM-Polynomials

The BEM-polynomials have some advantages with respect to other polynomials: they are *defined on three-point grids*; all their continuous parameters are *measurable values*; the errors *associated to them vanish at the boundaries of the interval*; it is possible to *change the basis functions and the regression matrices* via the parameters α,β [4]; the polynomial approximation of continuous functions asks for the $\lfloor n/3 \rfloor$ times smaller *order of derivatives*, where n is the degree of the polynomial; the coefficients of the BEM-polynomial depend on the parameters α,β by *recurrent formulas* [6]; the BEM-polynomials *decrease the computing complexity* and provide *stability of calculations with respect to the input errors* [4]; the control parameters allow to optimize the solution of *local and global* piecewise polynomial approximations [7], etc.

3 Mean-Square Piecewise Polynomial Approximation Algorithm

In this section an algorithm for smoothing a planar CL is described using mean-squares piecewise BEM-polynomials of 11th degree (BEM-MSPPA).

Data smoothing using the piecewise BEM-polynomial allows to optimize the solution via the control parameters α, β, x_0. The order of the derivatives is decreased substantially in comparison with the degree of the approximating polynomial in the computation of $r_{j\nu}$, $\nu = \alpha, \beta, 0$, $j = \overline{0, m}$ [6], [7].

The algorithm BEM-MSPPA is executed in four stages:

1. An interval $[a, b]$ and input data set $\{\tilde{S}\} = \left\{\tilde{f}_i\right\}_{i=1}^{N}$ are splitted on equal (except Kth) $K \ll N$ subintervals provided that $x_{\beta_{k-1}} \equiv x_{\alpha_k}$

$$[a, b] = \bigcup_{k=1}^{K} [x_{\alpha_k}, x_{\beta_k}], \qquad \{\tilde{S}\} = \bigcup_{k=1}^{K} \{\tilde{f}_{i_k}\}_{i_k=1}^{N_s}, \qquad N_s = \lfloor N/K \rfloor, \qquad N_s \geq 13. \tag{3}$$

Setting parameters $x_{0_k}, \alpha_k, \beta_k$ and $\gamma_k = \beta_k - \alpha_k$ and preparation of nodes of the three-point grids:

$$x_{0_k} = x_{0_k}(\alpha_k, \beta_k, N_s), \qquad x_{\alpha_k} = x_{0_k} + \alpha_k, \qquad x_{\beta_k} = x_{0_k} + \beta_k, \qquad k = \overline{1, K},$$

where N_s is the number of points of the kth segment on the grid $\Delta_3^{\alpha_k\beta_k}$.

The data of each segment on the grid $\Delta_3^{\alpha_k\beta_k}$ is smoothed by the 11th degree BEM-polynomial $P_{\mathrm{BEM}}(x) = \sum_{j=0}^{3} \mathbf{b}_j^T \mathbf{r}_j$ where the components of $\mathbf{r}_0$ are pivotal points and $\mathbf{r}_j$ is free, $j = \overline{1, 3}$ on a curve to be approximated [4] (the index k of the segment is dropped here and below for easy reading).

2. Calculation of $\hat{\mathbf{r}}_0 = [\bar{f}_\alpha, \bar{f}_\beta, \bar{f}_0]^T$ using $2M + 1$ coordinates nearest the nodes of the grid $\Delta_3^{\alpha\beta}$

$$\bar{f}_\nu = \frac{1}{2M + 1} \sum_{\mu=-M}^{M} \tilde{f}_\mu, \qquad \nu = \alpha, \beta, 0; \qquad f_\nu = f(\nu)$$

and transformation of S-data into u-data as follows: $\tilde{u}_i = \tilde{f}_i - \mathbf{w}^T \hat{\mathbf{r}}_0$, $i = \overline{1, N_s}$ (see Fig. 2a).

Conditions $x_{\beta_{k-1}} \equiv x_{\alpha_k}$ and $\bar{f}_{\beta_{k-1}} \equiv \bar{f}_{\alpha_k}$, $k = \overline{2, K}$ ensure C^0 smoothness at the breakpoints.

3. Calculation of nine coefficients $\hat{\mathbf{r}}_j$ using the Least Squares Method (LSM) criterion

$$\frac{\partial}{\partial \mathbf{r}_j} \Phi(\mathbf{r}_j) = 0, \quad \text{where} \quad \Phi(\mathbf{r}_j) = \sum_{i=1}^{N_s} [\tilde{u}_i - \sum_{j=1}^{3} \mathbf{b}_j^T(\tau_i) \mathbf{r}_j]^2, \qquad \tau_i \in [\alpha, \beta],$$

$$\hat{\mathbf{r}}_j = [\mathbf{B}^T \mathbf{B}]^{-1} \mathbf{B}^T \tilde{\mathbf{u}},$$

where $\tilde{\mathbf{u}} = [\tilde{u}_1, \tilde{u}_2, \ldots, \tilde{u}_{N_s}]^T$ and $[\mathbf{B}^T \mathbf{B}]^{-1}$ is normal matrix 9×9.

4. Estimates $\hat{\Pi}_j = \mathbf{w}_j^T \hat{\mathbf{r}}_j$, $j = \overline{1, 3}$ are used for the calculation of $\hat{f}(x)$, $x \in [x_\alpha, x_\beta]$ by the Horner type scheme:

$$\hat{f} = \hat{\Pi}_0 + \underbrace{Q(\hat{\Pi}_1 + Q(\hat{\Pi}_2 + Q\hat{\Pi}_3))}_{\hat{u}}.$$

Testings of the BEM-approximation algorithm are presented in [4]–[7]. The example in Fig. 2 shows the four stages of the BEM-MSPPA algorithm for data smoothing at the grid $\Delta_3^{\alpha\beta}$ [6].

Data $S = \{\tilde{f}_i\}_{i=1}^{250}$ was obtained by scattering points around the curve

$$f(x) = \exp\{-(x - 1.3)^2 / [2(0.025x - 0.0275)^2]\} + 1, \qquad x \in [1.25, 1.5]$$

using the `stats[random,normald[0,0.1]](1)` from a sub-package of Maple. S-data have been smoothed by the procedure

$\hat{\varphi} =$ `LeastSquares(xdat,ydat,t,curve=a_0+a_1t+...+a_{11}t^{11})`.

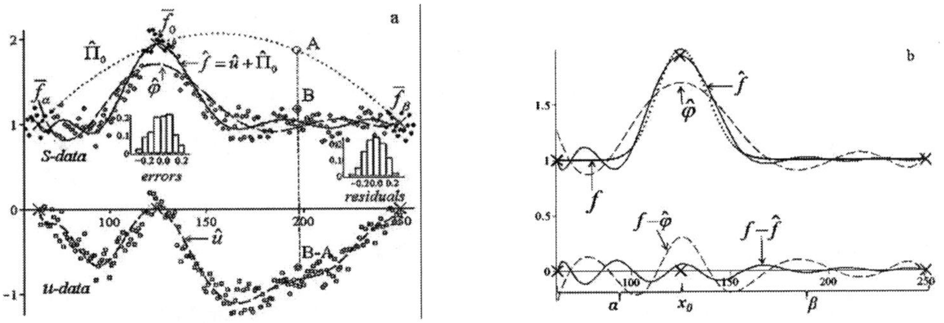

Figure 2. Four stages of the BEM-smoothing algorithm (a) and results of the BEM-smoothing (b)

The same data have been processed by BEM-MSPPA algorithm for $\alpha = 0.025$, $x_0 = 1.3$, $\beta = 0.05$, $M = 7$. The stages of BEM-smoothing are illustrated in Fig. 2a. $2M + 1$ samples are visible as black points. Three nodes and estimates of the three pivot points $\bar{f}_\alpha, \bar{f}_0, \bar{f}_\beta$ are marked by crosses. The graphics of Maple-estimate and BEM-estimate are denoted as $\hat{\varphi}$ and $\hat{f}$ respectively.

The deviation $f - \hat{f}$ is made smaller than the deviation $f - \hat{\varphi}$ at the expense of choosing the parameter x_0 in a zone of peak position (Figs. 2a, 2b).

4 Approximation of Planar Shapes

A planar CL is parametrically definable in the form of two functions $x(s)$ and $y(s)$ depending on a parameter s (length of the arch): $\Gamma(x(s), y(s))$, $s \in [0, L]$.

In practice, a contour line (CL) of complex topology is *measured with errors*: $\tilde{x} = x+e_x, \tilde{y} = y+e_y$, $e \sim N(0, \sigma)$. BEM-MSPPA has been used for smoothing $\{\tilde{x}_i\}$ and $\{\tilde{y}_i\}$, $i = \overline{1, N}$.

The algorithm BEM-MSPPA has five parameters for the optimization of the solution: the shift parameter (x_0), three parameters of *local smoothing* (α, β, M) and one parameter of *global smoothing* (K), where $2M+1$ is the number of points nearest to the nodes of the three-point grid for the calculation of components $\mathbf{r_0}$ while K stands for the number of segments [7].

There are hot difficulties of PPA: the optimum arrangement of the nodes on the grid and the order of the smoothness at the breakpoints.

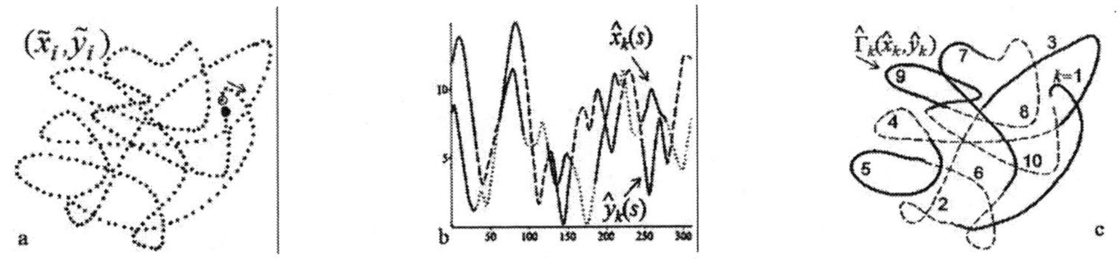

Figure 3. Parametric smoothing of a curve of complex topology using 11th degree BEM-polynomials: measurement points (a), estimates of $\hat{x}_k$ and $\hat{y}_k$ (b) and 10 smoothed segments (c)

The curve approximating contour of an object should lie *near the measurement points*, and its representation will be *the most effective* if we use *the minimum possible number of segments K*. The BEM-polynomials fulfill these requirements. The analytical dependence of the borders of complex geometric shapes is obtained through approximation (smoothing) of 2D curves.

Results of BEM-smoothing of a complex topology CL that was made in pencil by hand (Fig. 3).

The planar curve has 21 self-intersection points. Fig. 3a shows the start and direction of measurement of the points $\{\tilde{x}_i, \tilde{y}_i\}_{i=1}^N$, $N = 311$. These coordinates have been divided in 10 parts and the length of the curve has been separated by groups of three-point grids $s_{\alpha_k} \leq s_{0_k} \leq s_{\beta_k}$, $k = \overline{1,10}$: [1, 15, 31], [31, 46, 62], [62, 77, 93], [93, 108, 124], [124, 139, 155], [155, 170, 186], [186, 201, 217], [217, 232, 248], [248, 263, 279], [279, 294, 311].

The estimates $\hat{x}_k(s)$, $\hat{y}_k(s)$, $\alpha_k \leq s \leq s_{\beta_k}$, $k = \overline{1,10}$ are 11th degree polynomials (Fig. 3b). 10 segments $\Gamma_k(\hat{x}_k, \hat{y}_k)$ are shown in Fig. 3c.

4.1 Computation of Phase Portraits and Signatures

Signature curves are very useful for developments in computer vision applications, such as artificial intelligence, medical imaging devices, the study of DNA, etc [9], [10]. Informative signs of the shape are the length of the contour line L, the zeros of the curvature $\kappa(s)$, differential informative signs – the phase curve $\Phi(\kappa(s), \kappa_s(s)) = \kappa(s)/\kappa_s(s)$, where κ is the curvature $\kappa(s) = \sqrt{\ddot{x}(s) + \ddot{y}(s)}$ and κ_s is the first derivative with respect to the arc length s: $\kappa_s(s) = \dot{\kappa}(s)$, etc.

Both κ and κ_s (as well as all higher order arc length derivatives) are Euclidean differential invariants, meaning that they are unchanged under rigid motion (*isometric* transformation (TI)) [11], but they change under *similarity* (TS) and *affine* (TA) transformations.

It is possible to use various functions depending on the regular points ($\kappa_s(s) \neq 0$) on the phase curve $\Phi(\kappa, \kappa_s; K)$. To compute the signature of the shape at the nodes s_{0_k} we used two values:

$$\text{sgnt} = \frac{1}{K}\sum_{k=1}^{K}\frac{\kappa(s_{0_k})}{\kappa_s(s_{0_k})} \quad \text{and} \quad \text{SGNT} = \frac{1}{N_T}\sum_{i=1}^{N_T}\text{sgnt}_i,$$

where $s_{0_k} \in [s_{\alpha_k}, s_{\beta_k}]$. SGNT is average of N_T values of sign obtained at TS- or TA- transformations. The length of the interval $[s_\alpha, s_\beta]$ *increases* with *the degree increase* of the polynomial. The inner

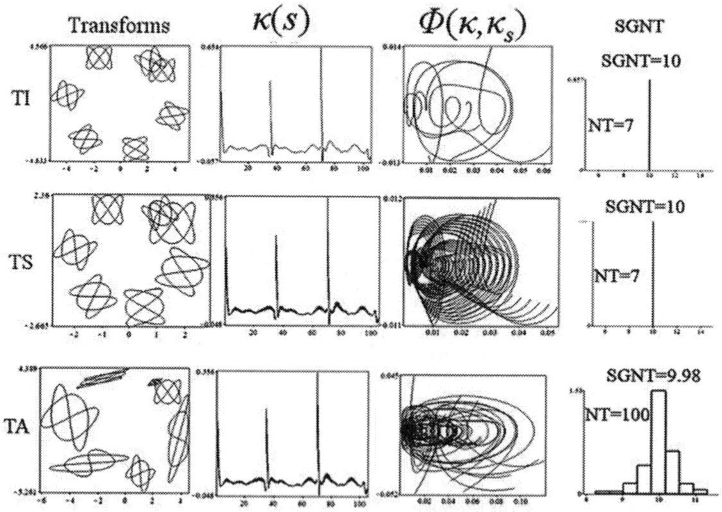

Figure 4. Graphics pointing to the change of the curvature, phase curves and histograms of SGNT under isometric (TI), similarity (TS) and affine (TA) transformations

nodes $s_{0_k} \in C^{11}$ will be equidistant and away from breakpoints. Therefore *the order of the smoothness* of segments at breakpoints is not critical under such choice of the internal nodes. The request of the

 https://doi.org/10.1051/epjconf/201817301003

second order smoothness at breakpoints is extremely important in the case of cubic splines with very short segments [2], [3].

An example of smoothing of 105 coordinates on the Lissajous figure by three BEM-polynomials of 11th degree is presented in Fig. 4. The graphics show the change of curvatures and phase curves under similarity and affine transformations of Lissajous forms. Histograms of SGNT are presented as well. Graphics of the phase curves of 5 first segments coincide under for isometric transformations ($N_T = 7$). Fig. 5 shows N point data smoothing on contours of a cat ($N = 239$) and of a dog ($N = 261$) respectively using the following parameters:

$K = 7$, $M = 0$, $\alpha_1 = -16$, $\beta_1 = 18$, $\alpha_k = -19$, $\beta_k = 18$, $k = \overline{2,7}$ with knots $1, 35, 69, 103, 137, 171, 205, 239$ and

$K = 9$, $M = 0$, $\alpha_1 = -13$, $\beta_1 = 15$, $\alpha_k = -16$, $\beta_k = 15$, $k = \overline{2,9}$ with knots $1, 29, 58, 87, 116, 145, 174, 203, 232, 261$.

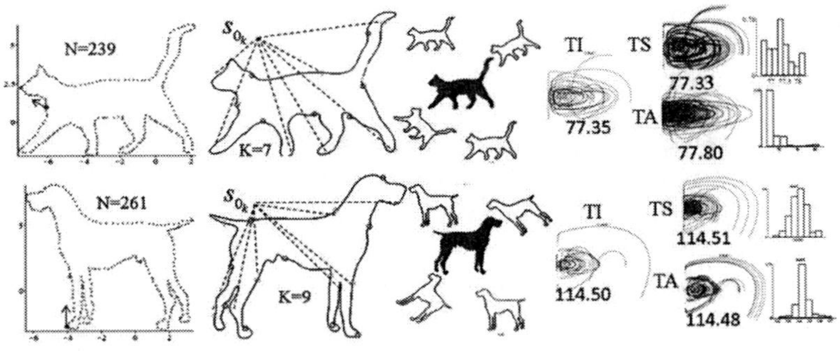

Figure 5. Measurement points, segments, phase portraits (TI,TS,TA) and SGNT computed for $N_T = 100$. Histograms of sign for TS and TA are shown on the right

Curvatures $\kappa(s)$ and phase portraits $\Phi(\kappa, \kappa_s)$ and sign are shown in Fig. 6 under similarity and affine transformations.

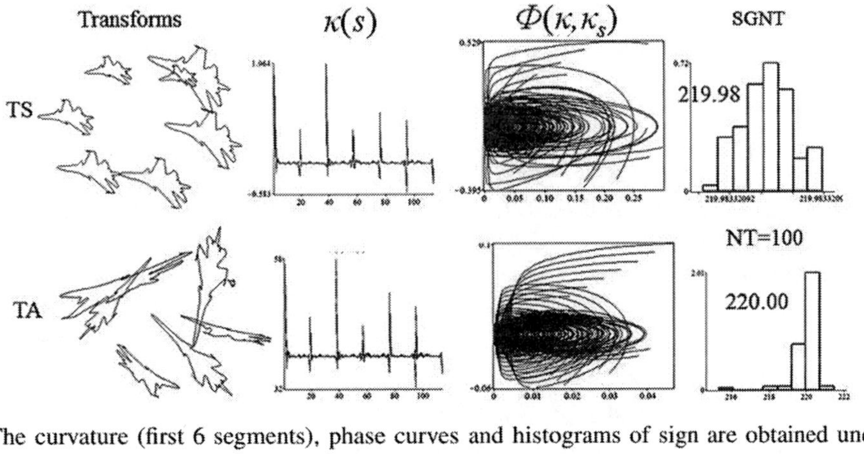

Figure 6. The curvature (first 6 segments), phase curves and histograms of sign are obtained under TS- and TA-transformations ($N_T = 100$)

The value SGNT is computed for TS and TA as average values sign obtained at N_T repeated transformations of the object. It can be seen in Figs. 4, 5 and 6 that the values of SGNT are equal or almost equal to the signature sign of the isometric transformation.

The value SGNT is an informative sign. It can be used for objects classification, as shown in Fig. 5, where $SGNT_{cat} \ll SGNT_{dog}$.

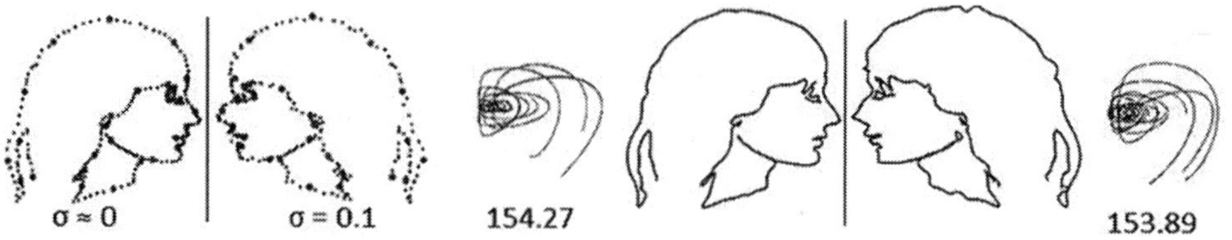

Figure 7. Mirror reflection of a face without and with errors, phase curves and SGNT

An example of BEM-segmentation of contours with errors ($\sigma = 0.1$) and without ($\sigma \approx 0$) in input data is visualized in Fig. 7. The comparison of the phase portraits and signatures SGNT is shown for mirror reflection transformations.

5 Conclusion

In this paper, we have discussed the problem of higher-degree piecewise polynomial smoothing of digitized 1D and 2D curves and the segmentation of planar curves. A mean-squares approximation algorithm is proposed. It is spanned by a new polynomial basis, called BEM-polynomial, which is defined on a three-point grid. The construction of BEM polynomials comprises the properties of Taylor polynomial expansions around the nodes of a three-point grid and Lagrange polynomials of degree 2.

The segmentation of complex curves by cubic splines or pieces of straight lines, circles and spirals arches, etc. yields a big number of short segments. In comparison with these approaches, the segmentation by high-degree BEM-polynomials gives a much smaller number of segments ensuring a significantly higher order of smoothness at larger length of the interval of the three-point grid.

Examples of the segmentation of the boundary curves of objects and computations of their differential invariant signatures confirm the efficiency of the BEM-polynomials in accuracy and quality of approximations. The algorithm based on BEM-polynomials can be used for the classification of the contours under several geometric transformations, even in the presence of input errors.

References

[1] M. Worring, *Shape Analysis of Digital Curves* (Febodruk, Enschede, 1993)

[2] D. Ben-Haim, G. Harary, and A. Tal, *Proceedings of the 14th ACM Symposium on Solid and Physical Modelling*, 201–206, (2010)

[3] M.L. Torrente, S. Anzellotti, C. Finocchiaro, and C. Fontanari, arXiv:1507.03865v1 [math.NA] 14 Jul 2015.

[4] N.D. Dikusar, Mathematical Models and Computer Simulations **3**, 4, 492–507 (2011)

[5] N.D. Dikusar, Mathematical Models and Computer Simulations **6**, 5, 509–522 (2014)

[6] N.D. Dikusar, Mathematical Models and Computer Simulations **8**, 2, 183–200 (2016)

[7] N.D. Dikusar, http://wwwinfo.jinr.ru/publish/Preprints/2016/085(P11-2016-85).pdf [in Russian]

[8] P. L. Chebyshev, *Selected Works* (Akademia Nauk SSSR, Moscow, 1955), p. 639 [in Russian]

[9] E. Calabi, P.J. Olver, C. Shakiban, A. Tannenbaum, and S. Haker, Int. J. Comput. Vis. **26**, 107–135 (1998)

[10] M. Boutin, Int. J. Comput. Vision **40** (3), 235–248 (2000)

[11] D.J. Hoff and P.J. Olver, J. Math. Imaging Vision **45**, 2, 176–185 (2013)

Application of Power Geometry and Normal Form Methods to the Study of Nonlinear ODEs

Victor Edneral[1,2,⋆]

[1] *Skobeltsyn Institute of Nuclear Physics, Lomonosov Moscow State University, Leninskie Gory 1 (2), Moscow, 119991, Russian Federation*
[2] *Peoples's Friendship University of Russia (RUDN University), Miklukho-Maklaya street 6, Moscow, 117198, Russian Federation*

Abstract. This paper describes power transformations of degenerate autonomous polynomial systems of ordinary differential equations which reduce such systems to a non-degenerative form. Example of creating exact first integrals of motion of some planar degenerate system in a closed form is given.

1 Introduction

We consider an autonomous system of ordinary differential equations of the form

$$\frac{dx_i}{dt} \stackrel{\text{def}}{=} \dot{x}_i = \varphi_i(X, P), \qquad i = 1, 2, \tag{1}$$

where $X = (x_1, x_2) \in \mathbb{C}^2$ and $\varphi_i(X, P)$ are polynomials in the variables X and involve the parameters $P = (b, a_0, a_1, b_0, b_1)$.

A method of analysis of the integrability of the system (1) based on Power Transformations [1] and computation of normal forms near stationary solutions of transformed systems (see [2] and Ch. II in [3]) was proposed in [4, 5].

In a neighborhood of the stationary point $X = 0$, the system (1) can be written in the form

$$\dot{X} = \hat{A}\, X + \tilde{\Phi}(X, P), \tag{2}$$

where $\tilde{\Phi}(X, P)$ has no terms linear in X.

Let λ_1 and λ_2 be eigenvalues of the matrix $\hat{A}$. If at least one is different from zero, then the stationary point $X = 0$ is called *elementary* stationary point. In this case, the system (2) has a normal form (see, e.g. Ch. II in [3]). If both eigenvalues vanish, then the stationary point $X = 0$ is called *nonelementary* stationary point. In this case, there is no normal form for the system (2). But using power transformations, we can split the nonelementary stationary point $X = 0$ to a set of elementary stationary points [1]. For each of these elementary stationary points, we can compute the normal form and obtain the conditions of local integrability.

⋆e-mail: edneral_vf@rudn.university

In the present paper, we demonstrate how this approach can be applied to the study of local and global integrability in the case of a planar system near the stationary point $X^0 = 0$ of high degeneracy in the case of the system

$$\begin{aligned} \dot{x} &= \alpha\, y^3 + \beta\, x^3\, y + (a_0\, x^5 + a_1\, x^2 y^2)\,, \\ \dot{y} &= \gamma\, x^2\, y^2 + \delta\, x^5 + (b_0\, x^4 y + b_1\, x\, y^3)\,. \end{aligned} \tag{3}$$

This system is a subfamily of the system introduced in [6].

Systems with a nilpotent matrix of the linear part were thoroughly studied by Lyapunov and others. In the system (3), there is no linear part, and the first approximation is not homogeneous. This is the simplest case of a planar system without linear part and with Newton's open polygon (see Ch. II in [3]) consisting of a single edge. In the general case, such problems have not been studied yet. However, a particular case of the system of this type was considered in [6], where the authors put $-\alpha = \delta = 1$ and $3\beta + 2\gamma = 0$. Further, the authors of [6] studied the Hamiltonian subcase of this system under the additional assumption that the Hamiltonian function is expandable into the product of only square-free factors.

We study the following problem: under what conditions imposed on the parameters, the system (3) is locally or globally integrable when the system is non-Hamiltonian.

We discuss the first quasi homogeneous approximation of the system (3) and the necessary conditions of local integrability. Then we calculate the set of such conditions for the complete system (3). After that, we prove that these conditions are sufficient for the global integrability by computing corresponding the first integrals of motion. First, we do it for the case when $b^2 \neq 2/3$ in the system (6), and after that for a subcase of the case $b^2 = 2/3$.

2 Problem Formulation

We start by investigating the case when the first quasi homogeneous approximation of (3) considered in [5, 7] has the form

$$\dot{\tilde{x}} = \alpha\, \tilde{y}^3 + \beta\, \tilde{x}^3\, \tilde{y}, \qquad \dot{\tilde{y}} = \gamma\, \tilde{x}^2\, \tilde{y}^2 + \delta\, \tilde{x}^5, \tag{4}$$

where $\alpha \neq 0$ and $\delta \neq 0$. Using the linear transformation $x = \sigma\tilde{x}$, $y = \tau\tilde{y}$ we can fix two nonzero parameters in (4) and obtain the system

$$\dot{x} = -y^3 - b\, x^3\, y, \qquad \dot{y} = c\, x^2\, y^2 + x^5. \tag{5}$$

Each autonomous planar quasi homogeneous system (4) has an integral, but this integral can be non analytic. We have are interested in the analytic integrability of (3), so we look for the conditions on the parameters under which the system (5) is locally or globally analytically integrable.

The following result was proved in [5, 7]:

Theorem 1. *In the case* $D \stackrel{\text{def}}{=} (3\,b + 2\,c)^2 - 24 \neq 0$, *the system (5) is locally integrable if and only if the number* $N = (3\,b - 2\,c)/\sqrt{D}$ *is rational. The system (5) is also locally integrable in the case* $b^2 = 2/3$.

In this paper, at first we study the simple particular case $c = 1/b$, then $N = 1$ and $D = (3b - 2/b)^2 \neq 0$. Due to the Theorem 1, the first quasi homogeneous approximation has an analytic integral but it is not a Hamiltonian system. Then we study the case $b^2 = 2/3$, $D = 0$.

We study the integrability problem for the entire system (3) with the first quasi homogeneous approximation (5) representing the system in the form

$$\begin{aligned} \frac{dx}{dt} &= -y^3 - b\, x^3 y + a_0\, x^5 + a_1\, x^2 y^2, \\ \frac{dy}{dt} &= \frac{1}{b}\, x^2 y^2 + x^5 + b_0\, x^4 y + b_1\, x\, y^3. \end{aligned} \tag{6}$$

Thus, we consider the system with five arbitrary parameters a_i, b_i, $(i = 0, 1)$ and $b \neq 0$.

3 Power Transformation

First, we rewrite (6) in a non-degenerate form. This can be done by using the Blow-Up process [9]. We can use here the Power Transformation (see Ch. I, §1.8 in [3])

$$x = u\,v^2, \qquad y = u\,v^3. \tag{7}$$

The *power transformation*

$$y_1 = x_1^{\alpha_{1,1}} x_2^{\alpha_{1,2}},$$
$$y_2 = x_1^{\alpha_{2,1}} x_2^{\alpha_{2,2}},$$

has the inverse one

$$x_1 = y_1^{\beta_{1,1}} y_2^{\beta_{1,2}},$$
$$x_2 = y_1^{\beta_{2,1}} y_2^{\beta_{2,2}},$$

where $\hat\beta = \hat\alpha^{-1}$. In fact, the above transformations are linear with respect to the logarithms of the variables

$$\begin{aligned} \log y_1 &= \alpha_{1,1}\, \log x_1 + \alpha_{1,2}\, \log x_2\,, & \log x_1 &= \beta_{1,1}\, \log y_1 + \beta_{1,2}\, \log y_2\,,\\ \log y_2 &= \alpha_{2,1}\, \log x_1 + \alpha_{2,2}\, \log x_2\,, & \log x_2 &= \beta_{2,1}\, \log y_1 + \beta_{2,2}\, \log y_2\,. \end{aligned}$$

The equation (6) can be rewritten as

$$\begin{aligned} \frac{d\log x}{dt} &= -x^{-1}y^3 - b\,x^2y + a_0\,x^4 + a_1\,xy^2\,,\\ \frac{d\log y}{dt} &= \frac{1}{b}\,x^2y + x^5y^{-1} + b_0\,x^4 + b_1\,x\,y^2\,. \end{aligned} \tag{8}$$

In the two-dimension space of degrees of the variables x and y the monomials of equations (8) form two parallel lines. The task now is to perform a rotation of the lines clockwise or counterclockwise till they become parallel to the abscissa or ordinate axes. The parallel to these lines vector before and after the needed transformation is

$$\begin{pmatrix}1 & 1\\ 2 & 3\end{pmatrix}\begin{pmatrix}3\\ -2\end{pmatrix} = \begin{pmatrix}1\\ 0\end{pmatrix}$$

is parallel to the abscissa line.

The transformation (7) corresponds to the transposed matrix above

$$\begin{pmatrix}\log x\\ \log y\end{pmatrix} = \begin{pmatrix}1 & 2\\ 1 & 3\end{pmatrix}\begin{pmatrix}\log u\\ \log v\end{pmatrix}.$$

With the time rescaling $u^2v^7dt = d\tau$, the system (6) is rewritten in the form

$$\begin{aligned} \frac{du}{d\tau} &= -3\,u - [3\,b + (2/b)]u^2 - 2\,u^3 + (3\,a_1 - 2\,b_1)u^2v + (3\,a_0 - 2\,b_0)u^3v\,,\\ \frac{dv}{d\tau} &= v + \left[b + \frac{1}{b}\right]u\,v + u^2v + (b_1 - a_1)u\,v^2 + (b_0 - a_0)u^2v^2\,. \end{aligned} \tag{9}$$

The transformation of variables (7) is not unique. We are searching for a substitution such that:

- the ration of the eigenvalues of the linear part λ_1/λ_2 would be a negative rational number;
- the value of the resonance of the linear part of the resulting equation is expected to be minimal;
- the inverse transformation has integer degrees of the variables. So the wanted matrix is unimodular.

Under the power transformation (7), the point $x = y = 0$ blows up into two straight invariant lines $u = 0$ and $v = 0$. Along the line $u = 0$, system (9) has the stationary point $u = v = 0$. Along the second line $v = 0$, this system has four elementary stationary points

$$u = 0, \qquad u = -\frac{1}{b}, \qquad u = -\frac{3b}{2}, \qquad u = \infty\,. \tag{10}$$

The necessary condition for local integrability of the system (6) at the point $x = y = 0$ is the local integrability near each of the stationary point above.

Lemma 1. *The system (9) is locally integrable near the points $u = v = 0$ and $u = \infty$, $v = 0$.*

This lemma was proved in [5, 7]. Thus, conditions of local integrability are to be found at the other two stationary points (10). Then we will get the conditions of local integrability of system the (6) near the original point $(0, 0)$. For these purposes we use the condition A. Since it would take a lot of space, and because it was presented earlier in several papers (see[2, 3]) we do not repeat it here. The condition A was formulated and applied to the considered problem in [4]. For two-dimensional systems, it is sufficient condition of their local integrability. It is an algebraic condition on the coefficients of the normal form. In our case, it can be presented as a system of algebraic equations. We have computed the condition A with the program described in [8].

Let us consider the stationary point $u = -1/b$, $v = 0$. There are two solutions for the corresponding subset of equations at $b \neq 0$ [7]

$$a_0 = 0, \qquad a_1 = -b_0\, b, \qquad b_1 = 0, \qquad b^2 \neq 2/3 \tag{11}$$

and

$$a_0 = a_1\, b, \qquad b_0 = b_1\, b, \qquad b^2 \neq 2/3\,. \tag{12}$$

The consideration of the stationary point $u = -3\,b/2, v = 0$ under the condition (12) gives three additional two-parameter solutions (depending on a_1 and b), see [7]

$$\begin{array}{llll} 1)\ \ b_1 = -2\,a_1, & a_0 = a_1 b, & b_0 = b_1 b, & b^2 \neq 2/3, \\ 2)\ \ b_1 = (3/2)\,a_1, & a_0 = a_1 b, & b_0 = b_1 b, & b^2 \neq 2/3, \\ 3)\ \ b_1 = (8/3)\,a_1, & a_0 = a_1 b, & b_0 = b_1 b, & b^2 \neq 2/3. \end{array} \tag{13}$$

Thus, we proved ([5, 7])

Theorem 2. *The conditions (11) and (13) form the set of necessary conditions of local integrability of the system (9) near all its stationary points and the local integrability of the system (6) at the stationary point $x = y = 0$.*

4 First Integrals of Motion

The conditions presented in Theorem 2 as necessary conditions for the local integrability of system (6) can be considered as good candidates for sufficient conditions of global integrability. However, it is necessary to prove the sufficiency of these conditions by independent methods. It is necessary to do it for each of four conditions (11) and (13) at every stationary point $u = -3b/2$, $v = 0$ and $u = -1/b$, $v = 0$, for $b^2 \neq 2/3$.

In [10], we have calculated first integrals of system (9) for all cases (11), (13) (mainly by the Darboux method, see, e.g., [11]).

We have found four families of solutions which covered all the cases mentioned above:

1. At $a_0 = 0, \quad a_1 = -b_0\, b, \quad b_1 = 0$:

$$I_{1uv} = u^2(3\,b + 2\,u)v^6, \qquad I_{1xy} = 2\,x^3 + 3\,b\,y^2.$$

2. At $b_1 = -2a_1, \quad a_0 = a_1 b, \quad b_0 = b_1 b$:

$$I_{2uv} = u^2\,v^6\,(3\,b + u\,(2 - 6\,a_1\,b\,v)), \qquad I_{2xy} = 2\,x^3 - 6\,a_1\,b\,x^2\,y + 3\,b\,y^2\,.$$

3. At $b_1 = 3a_1/2, \quad a_0 = a_1 b, \quad b_0 = b_1 b$:

$$I_{3uv} = \frac{4 - 4a_1\,u\,v + 3^{5/6}a_1 \times {}_2F_1(2/3, 1/6; 5/3; -2u/(3b)) \times u\,v\,(3 + 2u/b)^{1/6}}{u^{1/3}v\,(3b + 2u)^{1/6}},$$

$$I_{3xy} = \frac{a_1 x^2\left(-4 + 3^{5/6}\,{}_2F_1\left(2/3, 1/6; 5/3; -2\,x^3/(3\,b\,y^2)\right) \times \left(3 + 2x^3/(b\,y^2)\right)^{1/6}\right) + 4y}{y^{4/3}(3\,b + 2\,x^3/y^2)^{1/6}},$$

4. At $b_1 = 8a_1/3, \quad a_0 = a_1 b, \quad b_0 = b_1 b$:

$$I_{4u,v} = \frac{u\,(3 + 2\,a_1^2 bu) + 6\,a_1\,b\,v}{3\,u\,[u^3(6 + a_1^2 b\,u) + 6\,a_1^2 b\,u^2 v + 9\,b\,v^2]^{1/6}} - 8\,a_1\,\sqrt{-b}/3^{5/3} B_{6+a_1\,\sqrt{-6\,b\,u}+3\,v\,\sqrt{-6\,b/u^3}}(5/6, 5/6)\,,$$

where $B_t(a, b)$ is the incomplete beta function and ${}_2F_1(a, b; c; z)$ is the hypergeometric function [12].

The first integrals and the solutions do not have any singularities at the values $b^2 = 2/3$, but the approach by which these solutions were found, has the limitation $b^2 \neq 2/3$, so there could exist possible additional integrals for these values. Thus, we need to study the case $b^2 = 2/3$ separately.

5 Case $b^2 = 2/3$, subcase $3a_0 - 2b_0 = b(3a_1 - 2b_1)$

Let us consider the case $b = \sqrt{2/3}$ (the case $b = -\sqrt{2/3}$ gives the same results). For this value of b, both stationary points $u = -3b/2, v = 0$ and $u = -1/b, v = 0$ are collapsing, and after the shift $u \to w - 1/b$, we get the degenerate system instead of (9)

$$\begin{aligned}
\frac{dw}{d\tau} &= -\tfrac{3v}{2b}\,[(3a_0 - 2b_0) - b(3a_1 - 2b_1)] + wv\left(\tfrac{27}{2\,a_0 - 3\,\sqrt{6}}\,a_1 - 9\,b_0 + 2\,\sqrt{6}\,b_1\right)\\
&\quad + \sqrt{6}\,w^2 + w^2 v\left(-9\sqrt{\tfrac{3}{2}}\,a_0 + 3\,a_1 + 3\,\sqrt{6}\,b_0 - 2\,b_1\right) - 2\,w^3 + w^3 v(3\,a_0 - 2\,b_0)\,,\\
\frac{dv}{d\tau} &= -\tfrac{\sqrt{6}}{6}wv + v^2\left(-\tfrac{3}{2}\,a_0 + \sqrt{\tfrac{3}{2}}\,a_1 + \tfrac{3}{2}\,b_0 - \sqrt{\tfrac{3}{2}}\,b_1\right)\\
&\quad + w^2 v + wv^2\left(\sqrt{6}\,a_0 - a_1 - \sqrt{6}\,b_0 + b_1\right) + w^2 v^2(-a_0 + b_0)\,.
\end{aligned} \tag{14}$$

At the stationary point $w = v = 0$ this system has zero eigenvalues, so we need to apply the power transformation once again. In [7], we applied the transformation

$$v \to r^2 w, \qquad \dot{v} \to 2\dot{r}rw + r^2\dot{w}\,, \tag{15}$$

and obtained systems with resonances of 19th and 27th orders. We have calculated the corresponding normal form with 4 free parameters till 19th order, but to find new solutions we need to compute the 27th order resonance. This resonance exists only if $3a_0 - 2b_0 \neq b(3a_1 - 2b_1)$, $b^2 = 2/3$, and its calculation is very complicated. We postpone the investigation of this subcase.

We see that in the systems (14), the coefficient of v in the linear part of the first equation is equal to zero if $3a_0 - 2b_0 = b(3a_1 - 2b_1)$. So we have the special subcase

$$3a_0 - 2b_0 = b(3a_1 - 2b_1), \qquad b^2 = 2/3\,.$$

For this subcase, we use the transformation

$$u \to w - 1/b, \qquad v \to wr, \qquad \dot{v} \to \dot{w}r + w\dot{r}, \qquad \hat{\alpha} = \begin{pmatrix} 1 & 0 \\ 1 & 1 \end{pmatrix}, \tag{16}$$

with time scaling by division of the equations by $w/\sqrt{6}$, so $\tilde{\tau} = w\tau/\sqrt{6}$. Then, from (9) we get

$$\begin{aligned}
\frac{dw}{d\tilde{\tau}} &= 6w + 3(3a_1 - 2b_1)rw - 2\sqrt{6}w^2 - 2\sqrt{6}(3a_1 - 2b_1)rw^2 + 2(3a_1 - 2b_1)rw^3\,,\\
\frac{dr}{d\tilde{\tau}} &= -7r - \left(9a_1 - \sqrt{\tfrac{3}{2}}b_0 - 5b_1\right)r^2 + 3\sqrt{6}\,rw\\
&\quad + \left(7\sqrt{6}a_1 - 2b_0 - 13\sqrt{\tfrac{2}{3}}b_1\right)r^2w - \left(8a_1 - \sqrt{\tfrac{3}{2}}b_0 - \tfrac{16}{3}b_1\right)r^2w^2\,.
\end{aligned} \tag{17}$$

This is a three-parameter system with the resonance of the 13th order at the stationary point $w = 0$, $r = 0$ along the invariant line $w = 0$. Along this line, there is also another stationary point. It is possible to prove the local integrability of the system at these points and they do not supply any additional restrictions on the parameters.

We have calculated the normal form for (17) till the 26th order and obtained two equations for the condition A. They are a13 = 0 and a26 = 0, where a13 and a26 are given in [13]. Each of these equations is homogeneous in the parameters a_1, b_0, b_1 of the system (6) of the sixth and twelfth orders correspondingly. Both a13 and a26 are equal to zero for the solutions (11) and (13).

Homogeneous algebraic equations in three variables can be reduced to inhomogeneous equations in two variables. Let us assume that $a_1 \neq 0$. In this case, we substitute $b_0 = c_0\,a_1$, $b_1 = c_1\,a_1$ and obtain the system of two equations in two variables $\mathtt{a13}(c_0, c_1) = 0$, $\mathtt{a26}(c_0, c_1) = 0$. The resultant of $\mathtt{a13}(c_0, c_1)$ and $\mathtt{a26}(c_0, c_1)$ in each of two variables c_0 and c_1 is identically equal to zero. So it is sufficient to solve only the equation $\mathtt{a13}(c_0, c_1) = 0$ and check these solutions by conditions of the higher order (i.e. by a26 = 0, a39 = 0, ...).

It is interesting that the condition A of the 19th order (a19) from [7, 14] under the equality $3a_0 - 2b_0 = b(3a_1 - 2b_1)$ is *identically equal* to a13 up to the multiplication by a constant.

The a13 can be factorized as the product of four factors including a_1^6:

$$\begin{aligned}
\mathtt{a13} &= 48(c_1 - 3/2) \times \left(c_0 - 1/12\sqrt{6}c_1 + 1/2\sqrt{6}\right)^2\\
&\times \Big[409790784c_0^3 - 104\sqrt{6}c_0^2(-9152256 + 3385633c_1) -\\
&- 208c_0\big(-10917702 + c_1(-360720 + 3319927c_1)\big) +\\
&+ \sqrt{6}\big(-718439040 + c_1\big(2461047528 + c_1(-1944898681 + 441207868c_1)\big)\big)\Big] \times a_1^6\,.
\end{aligned}$$

From the first two factors, we get two two-parametric solutions $c_1 = 3/2$ and $c_1 = 6 + 2\sqrt{6}c_0$ or

$$\begin{aligned}
b_1 &= 3a_1/2, & a_0 &= (2b_0 + b(3a_1 - 2b_1))/3, & b &= \sqrt{2/3}\,,\\
b_1 &= 6a_1 + 2\sqrt{6}b_0, & a_0 &= (2b_0 + b(3a_1 - 2b_1))/3, & b &= \sqrt{2/3}\,.
\end{aligned} \tag{18}$$

For the solutions (18), we calculate the normal form of (17) till the 36th order and obtain for each solution that it is an integrable diagonal linear system. The condition $a_1 = 0$ gives only one parametric solutions of the system `a13 = 0, a26 = 0`. The same result is obtained from the study of the cubic factor in `a13` above. Therefore we do not consider that case here.

The corresponding first integrals of (17) are

$$I_{1rw} = w^{-7/6}\left(1 - \sqrt{\tfrac{2}{3}w}\right)^{-1/3}\Big[- 9a_1 + 3\sqrt{6}b_0 - \tfrac{42}{r} - 6\left(\sqrt{6}a_1 + 5b_0\right)w + 2\left(9a_1 + 4\sqrt{6}b_0\right)w^2 - 2^{1/6}\left(9\sqrt{2}a_1 + 8\sqrt{3}b_0\right)w^{5/3}\left(-\sqrt{6} + 2w\right)^{1/3} \times {}_2F_1\left(-1/2, 1/3; 1/2; \sqrt{2/3}/w\right)\Big],$$

$$I_{2rw} = r^{3-\frac{3a_1}{3a_1+\sqrt{6}b_0}} \cdot w^{\frac{7}{3}+\frac{7b_0}{3\sqrt{6}a_1+6b_0}} \cdot \left(1 - \sqrt{2/3}w\right)^{\frac{-a_1}{3a_1+\sqrt{6}b_0}} \times \left\{\frac{-6+2\sqrt{6}w}{6a_1+3\sqrt{6}b_0} + r\left[3 + 2w(-\sqrt{6} + w)\right]\right\}.$$

Back to the original variables x, y of equation (6), these integrals up to the multiplication by a constant have the form:

$$I_{1xy} = \frac{y}{x^2}\left(\sqrt{6} + \frac{2x^3}{y^2}\right)^{-7/6}\left(\frac{x^3}{y^2}\right)^{2/3} \times \Bigg\{42\sqrt{6} + \frac{1}{xy^3}\Bigg[- 36a_1x^6 - 16\sqrt{6}b_0x^6 + 84x^4y - 24\sqrt{6}a_1x^3y^2 - 36b_0x^3y^2 + 2^{1/3}\left(\frac{x^3}{y^2}\right)^{1/3} y^2 \cdot \left(\sqrt{6} + \left(\frac{x^3}{y^2}\right)^{2/3}\right) \times \left(2(9a_1 + 4\sqrt{6}b_0)x^3 + 3(3\sqrt{6}a_1 + 8b_0)y^2\right) \times {}_2F_1\left(-1/2, 1/3; 1/2; \frac{3y^2}{3y^2+\sqrt{6}x^3}\right)\Bigg]\Bigg\},$$

$$I_{2xy} = y \cdot \left(\sqrt{2/3} + \frac{x^3}{y^2}\right)^{-\frac{1}{2}+\frac{a_1}{-6a_1-2\sqrt{6}b_0}}\left(\frac{x^2}{y}\right)^{-\frac{a_1}{3a_1+\sqrt{6}b_0}} \times \left\{3 + \frac{x^2}{y^2}\left[\sqrt{6}x + 3\left(2a_1 + \sqrt{6}b_0\right)y\right]\right\}.$$

6 Conclusions

For the five-parameter non-Hamiltonian planar system (6), we have found six sets of two-parametric necessary conditions on parameters under which the system is locally integrable near the degenerate point $x = y = 0$. These sets of conditions are also sufficient for the local and global integrability of the system (6). These six expressions for the first integrals of motion have been checked by direct substitution to the original system.

It is very interesting that the local study of the different stationary points of the system provides a step in proceeding from local to global analysis.

Acknowledgements

The publication was prepared with the support of the "RUDN University Program 5-100" and by the grant NSh-7989.2016.2 of the President of Russian Federation. The author is grateful to Professor Alexander D. Bruno from Keldysh Institute of Applied Mathematics of RAS for many fruitful ideas and to Professor Valery G. Romanovski from Maribor University (Slovenia) for significant contribution at works with Darboux factors.

References

[1] A.D. Bruno ed., *Power Geometry in Algebraic and Differential Equations* (Elsevier Science, Amsterdam, 2000) 396 pp.

[2] A.D. Bruno, Trans. Moscow Math. Soc. **25**, 131–288 (1971), **26**, 199–239 (1972)
[3] A.D. Bruno, *Local Methods in Nonlinear Differential Equations* (Springer-Verlag, Berlin, 1989) X, 348 pp.
[4] A.D. Bruno and V.F. Edneral, Doklady Mathem. **79** (1), 48–52 (2009)
[5] A.D. Bruno and V.F. Edneral, *Proceedings of CASC 2009* (LNCS **5743**, Springer, Heidelberg, 2009) 45–53
[6] A. Algaba A, E. Gamero, and C. Garcia, Nonlinearity **22**, 395–420 (2009)
[7] A.D. Bruno and V.F. Edneral, *Proceedings of CASC 2013* (LNCS **8136**, Springer, Heidelberg, 2013) 75–87
[8] V.F. Edneral, *Proceedings of CASC 2007* (LNCS **4770**, Springer, Heidelberg, 2007) 134–142
[9] F. Dumortier, J. Llibre, J.C. Artés, *Qualitative Theory of Planar Differential Systems* (Springer, Berlin, Heidelberg, New York, 2006) XVI, 302 pp.
[10] V. Edneral and V.G. Romanovski, *Proceedings of CASC 2010* (LNCS **6244**, Springer, Heidelberg, 2010) 97–105
[11] V.G. Romanovski and D.S. Shafer, *The Center and Cyclicity Problems: A Computational Algebra Approach* (Birkhäuser, Boston, 2009) 330 pp.
[12] H. Bateman and A. Erdélyi, *Higher Transcendental Functions, volume I* (McGraw-Hill Book Comp., New York, Toronto, London, 1953) 316 pp.
[13] http://theory.sinp.msu.ru/~edneral/CASC2017/a13-26.txt
[14] http://theory.sinp.msu.ru/~edneral/CASC2017/a19.txt
[15] C. Christopher, P. Mardešić, and C. Rousseau, J. Dyn. Control Sys. **9**, 311–363 (2003)

EPJ Web of Conferences **173**, 01005 (2018) https://doi.org/10.1051/epjconf/201817301005
Mathematical Modeling and Computational Physics 2017

Fractional Stochastic Field Theory

Juha Honkonen[1,⋆]

[1] *Finnish National Defence University, Helsinki, Finland*

Abstract. Models describing evolution of physical, chemical, biological, social and financial processes are often formulated as differential equations with the understanding that they are large-scale equations for averages of quantities describing intrinsically random processes. Explicit account of randomness may lead to significant changes in the asymptotic behaviour (anomalous scaling) in such models especially in low spatial dimensions, which in many cases may be captured with the use of the renormalization group. Anomalous scaling and memory effects may also be introduced with the use of fractional derivatives and fractional noise. Construction of renormalized stochastic field theory with fractional derivatives and fractional noise in the underlying stochastic differential equations and master equations and the interplay between fluctuation-induced and built-in anomalous scaling behaviour is reviewed and discussed.

1 Introduction

Description of evolution at the kinetic stage is usually expressed in the form of differential equations with the understanding that they are large-scale equations for averages of quantities describing intrinsically random processes. Account of randomness may lead to significant changes in the asymptotic behaviour in such models especially in low spatial dimensions. An important example of changes in the large-scale behaviour is anomalous diffusion, which is deviation of the mean-square path-length of a random walker from the usual linear growth in time. Anomalous diffusion may occur, for instance, due to random advection, random walks with power-law falloff of the step-length distribution or long-tailed distribution of waiting times between consecutive steps. In case of interactions between fluctuations of densities anomalous diffusion may occur due to large-scale fluctuations even in the case of short-range interactions and correlations in the underlying kinetic model as in models of critical dynamics and stochastic transport models. This report is concentrated on the anomalous scaling behaviour in diffusion-limited reactions brought about by powerlike distribution of step length and waiting times.

Lévy flights are a generalization of random walks with a powerlike falloff of the step-length distribution $p(l) \propto l^{-1-\sigma}$ characterized by the step index σ assuming values $0 < \sigma < 2$. In the generic case leading to anomalous diffusion due to spatial effects, the probability density function of the position $\mathbf{r}$ of a test particle obeys the Fokker-Planck equation

$$\frac{\partial P}{\partial t} = -D'(-\nabla^2)^{\sigma/2} P + D\nabla^2 P\,. \tag{1}$$

⋆e-mail: juha.honkonen@mil.fi

Here, the fractional power of ∇^2 may be defined through the Fourier transform i.e. by writing the diffusion operator in the Fourier space as

$$-D'(-\nabla^2)^{\sigma/2}P + D\nabla^2 P \to -D'(\mathbf{k}^2)^{\sigma/2}P - D\mathbf{k}^2 P\,.$$

In the position space the fractional power of the Laplace operator ∇^2 may be defined with the use of an integral operator (Riesz derivative). In this case the fractional differential operator must be understood in terms of distribution theory [1]. The ordinary diffusion term $D\nabla^2 P$ in (1) is brought about by the small-scale part of the step distribution under renormalization and must be taken into account, when the step index $\sigma \to 2$. For fixed $\sigma < 2$ the ordinary diffusion term is irrelevant by power counting [2].

In epidemic models there are situations in which an infected individual can infect other individuals only after a certain incubation time (waiting time) [3] giving rise to long tails in waiting-time distribution of the form

$$p\,(\Delta t) \propto (\Delta t)^{-1-\alpha}\,.$$

Long tails in waiting times give rise to memory in system dynamics. Memory effects may be described by integral operators, which in case of powerlike asymptotic behaviour of waiting-time distribution gives rise to fractional differentiation and integration.

2 Fractional derivatives

There are different definitions of fractional derivatives on a time interval. Popular choices are the fractional derivative of Riemann-Liouville

$$(\mathcal{D}^{\alpha}_{+}f)\,(t) = \frac{1}{\Gamma(1-\alpha)}\,\frac{d}{dt}\int\limits_{-\infty}^{t}\frac{f(t')\,dt'}{(t-t')^{\alpha}} \tag{2}$$

and the fractional derivative of Marchaud

$$(\mathbf{D}^{\alpha}_{+}f)\,(t) = \frac{\alpha}{\Gamma(1-\alpha)}\int\limits_{-\infty}^{t}\frac{f(t)-f(t')}{(t-t')^{1+\alpha}}\,dt'\,, \tag{3}$$

whose definitions are quoted here for the whole time axis [1]. Fractional derivatives (2) and (3) are not equivalent, in general, but they coincide on functions “decent enough” [1]. From the point of view of construction of perturbation expansion, the basic property of the fractional derivative is its Fourier transformation:

$$\mathcal{F}\,(\mathcal{D}^{\alpha}_{+}f)\,(\omega) = (-i\omega)^{\alpha}\mathcal{F}(f)(\omega)\,, \qquad \mathrm{Re}\,\alpha \geq 0\,, \tag{4}$$

where

$$\mathcal{F}(f)(\omega) = \int\limits_{-\infty}^{\infty} e^{i\omega t} f(t)\,dt\,.$$

Relation (4) allows to use standard Feynman rules in the Martin-Siggia-Rose approach to stochastic problems [4].

There are different ways to introduce the temporal fractional derivative into a kinetic equation. The most straightforward way is to replace the usual time derivative by a fractional derivative, i.e. instead of (1) write ($0 < \alpha \leq 1$)

$$(\mathcal{D}^{\alpha}_{+}P)\,(t) = -D'(-\nabla^2)^{\sigma/2}P + D\nabla^2 P\,. \tag{5}$$

In this case, however, the quantity P cannot be interpreted as a (probability) density [5], as in (1).

Conservation of probability imposes conditions on the structure of the Fokker-Planck equation and the consistent way to proceed appears to be to apply the fractional time derivative to the right-hand-side operator of the Fokker-Planck equation instead of replacing the usual time derivative on the left-hand side by a fractional time derivative [6]. In the case of Lévy flights this corresponds to representation

$$\frac{\partial P}{\partial t} = \left(\mathcal{D}_+^{1-\alpha}\left\{-D'(-\nabla^2)^{\sigma/2}P + D\nabla^2\right\}\right)(t)\,. \tag{6}$$

instead of the simple-minded (5).

The Green function of the fractional differential equation

$$(\mathcal{D}_+^{\alpha}\Delta_1)(t) + D'(\nabla^2)^{\sigma/2}\Delta_1 = 0$$

given by the derivative of the Mittag-Leffler function $E_\alpha(z)$ (see, e.g. [7]) as

$$\Delta_1(t,\mathbf{k}) = \int\limits_{s_0-i\infty}^{s_0+i\infty} \frac{e^{st}\,ds}{s^\alpha + D'k^\sigma} = \theta(t)\alpha t^{\alpha-1}E'_\alpha\left(-D'k^\sigma t^\alpha\right) \tag{7}$$

is singular in the limit $t \to 0$ so that a proper initial condition cannot be set and the normalization of the density is not conserved. On the contrary, the propagator Δ_2 of the problem (6) is the Green function of the fractional differential equation

$$\frac{\partial \Delta_2}{\partial t} + D'(\nabla^2)^{\sigma/2}\left(\mathcal{D}_+^{1-\alpha}\Delta_2\right)(t) = 0$$

given by the the Mittag-Leffler function as

$$\Delta_2(t,\mathbf{k}) = \int\limits_{s_0-i\infty}^{s_0+i\infty} \frac{e^{st}s^{\alpha-1}\,ds}{s^\alpha + D'k^\sigma} = \theta(t)E_\alpha\left(-D'k^\sigma t^\alpha\right) \tag{8}$$

has a finite limit at $t \to 0$: $\Delta_2(t,\mathbf{k}) \to 1$ and also obeys the normalization condition $\Delta_2(t,0) = 1$. Therefore, the physically correct setup of the problem is given by (6).

3 Kinetic equation

It is instructive to start the construction of the field theory of fluctuations from the (deterministic) nonlinear kinetic equation for the expected value of the fluctuating quantity. The consistent form of the kinetic equation with fractional time derivatives is of the form

$$\frac{\partial \varphi}{\partial t} = \mathcal{D}_+^{1-\alpha}V(\varphi) = -K\mathcal{D}_+^{1-\alpha}\varphi + \mathcal{D}_+^{1-\alpha}U(\varphi) + \mathcal{D}_+^{1-\alpha}\tilde{A}\,, \tag{9}$$

where $\tilde{A}$ is – for the time being – a fixed external field.

The nonlinear differential equation (9) allows for the tree-graph solution explicitly depending on initial conditions and coefficient functions and implicitly on boundary conditions through the Green function of the linear problem. All these conditions may contain randomness, whose effect on the solution is conveniently calculated with the use of the compact functional form of the tree-graph solution.

Let $\varphi[\tilde{A}]$ be solution of the generic fractional kinetic equation (9). Define generating function of solutions of the kinetic equation

$$G(A) = e^{A\varphi[\tilde{A}]} . \tag{10}$$

The linear part of the right-hand side of (9) yields the Mittag-Leffler propagator

$$\Delta_{12}^{ML}(t,\mathbf{k}) = \theta(t)E_\alpha\left[-\left(Dk^2 + D_\sigma k^\sigma\right)t^\alpha\right] . \tag{11}$$

Perturbation expansion for the generating functional is conveniently expressed in the functional-differential form as the S-matrix functional [8]

$$G(A) = \exp\left(\frac{\delta}{\delta\varphi}\Delta_{12}^{ML'}\frac{\delta}{\delta\tilde{\varphi}}\right)\exp\left[\tilde{\varphi}\mathcal{D}_+^{1-\alpha}U(\varphi) + \tilde{\varphi}\mathcal{D}_+^{1-\alpha}\tilde{A} + A\varphi\right]\Big|_{\tilde{\varphi}=\varphi=0}, \tag{12}$$

where $\Delta_{12}^{ML'}(t,\mathbf{x};t,\mathbf{x}') \equiv 0$. It should be noted that this condition is only a shorthand for the normal form of the interaction functional, the latter meaning that closed loops of single propagators are absent in the graphical representation of perturbation expansion (see [9] for details). It should be emphasized that (12) is a functional representation of the solution of a deterministic nonlinear equation. In the next section it will be used to construct the solution of a stochastic differential equation by simply averaging over the probability distribution of the external field. The iterative solution of the equation is unambiguous. The ostensible ambiguity which arises, when representation (12) is derived, is the same both in the deterministic and in the stochastic case [10] and thus is a property of the functional representation and not a consequence of the stochastic nature of the problem.

Renormalization is almost inevitable step in the analysis of a field theory. Representation (12) has the drawback that the interaction functional is nonlocal in time, in which situation the standard procedures of the field-theoretic renormalization are inapplicable.

This situation may be cured with the use of integration by parts in the interaction term

$$\tilde{\varphi}\mathcal{D}_+^{1-\alpha}U(\varphi) = U(\varphi)\mathcal{D}_-^{1-\alpha}\tilde{\varphi} = \int dt U(\varphi(t))\left(-\frac{1}{\Gamma(\alpha)}\frac{\mathrm{d}}{\mathrm{d}t}\int_t^\infty du\frac{\tilde{\varphi}(u)}{(t-u)^{1-\alpha}}\right). \tag{13}$$

In terms of the new variable $\tilde{\phi} = \mathcal{D}_-^{1-\alpha}\tilde{\varphi}$ the generating function

$$G(A) = \exp\left(\frac{\delta}{\delta\varphi}\Delta_{12}^{\alpha}{}'\frac{\delta}{\delta\tilde{\phi}}\right)\exp\left[\tilde{\phi}U(\varphi) + \tilde{A}\tilde{\phi} + A\varphi\right]\Big|_{\tilde{\phi}=\varphi=0} \tag{14}$$

contains a local interaction functional and the usual rules of renormalization apply. The presence of fractional derivatives brings about some new features, however, which will be discussed a bit later.

4 Langevin equation

Fractional Langevin equation is the fractional kinetic equation

$$\frac{\partial\varphi}{\partial t} = \mathcal{D}_+^{1-\alpha}V(\varphi) = -K\mathcal{D}_+^{1-\alpha}\varphi + \mathcal{D}_+^{1-\alpha}U(\varphi) + \mathcal{D}_+^{1-\alpha}f , \tag{15}$$

with the white-in-time Gaussian noise

$$\langle f(t,\mathbf{x})f(t',\mathbf{x}')\rangle = \delta(t-t')D(\mathbf{x}-\mathbf{x}') , \quad \langle f\rangle = 0 . \tag{16}$$

The generating function of solutions of this equation with fixed f is (14) with the substitution $\tilde{A} \to f$. Integration of the generating functional over noise yields the functional-differential representation for the generating function of the Langevin problem (15), (16)

$$G(A) = \exp\left(\frac{\delta}{\delta\varphi}\Delta_{12}^{\alpha}{}' \frac{\delta}{\delta\tilde{\phi}}\right)\exp\left[\tilde{\phi}U(\varphi) + \frac{1}{2}\tilde{\phi}D\tilde{\phi} + A\varphi\right]\Bigg|_{\tilde{\phi}=\varphi=0}. \qquad (17)$$

With the use of the identity

$$\exp\left(\frac{\delta}{\delta\varphi}\Delta_{12}^{\alpha}{}' \frac{\delta}{\delta\tilde{\phi}}\right) = \int \mathcal{D}\phi \int \mathcal{D}\tilde{\varphi}\, \exp\left[\tilde{\varphi}\left(-\mathcal{D}_{+}^{\alpha} - K\right)\phi + \phi\frac{\delta}{\delta\varphi} + \tilde{\varphi}\frac{\delta}{\delta\tilde{\phi}}\right].$$

representation (17) gives rise to the functional integral with the de Dominicis-Janssen dynamic action of quite standard form, let alone the fractional time derivative:

$$G(A) = \int \mathcal{D}\phi \int \mathcal{D}\tilde{\varphi}\, \exp\left[\tilde{\varphi}\left(-\mathcal{D}_{+}^{\alpha} - K\right)\phi + \tilde{\varphi}U(\phi) + \frac{1}{2}\tilde{\varphi}D\tilde{\varphi} + A\phi\right].$$

5 Renormalization

As spatial fluctuations in the fractional diffusion equations give rise to the Laplace operator, the usual time derivative is often generated by fluctuations as well in field-theoretic models generated by Langevin equations. Thus, the generic propagator in the Langevin problem is [3]

$$\Delta_{12}^{\alpha}(s, \mathbf{k}) = \frac{1}{c_1 s + c_\alpha s^\alpha + Dk^2 + D_\sigma k^\sigma}. \qquad (18)$$

Identification of canonical scaling dimensions and power counting in renormalization requires homogeneity of the propagator. Therefore, two terms in the denominator of the generic propagator (18) should be singled out to the free-field part of the dynamic action, which gives rise to the canonical dimensions of fields and the bare propagator of renormalized perturbation theory. The remaining two terms are then prescribed to the interaction and the canonical dimensions of their coefficient assume values corresponding to this choice. At first sight it would seem that the generic propagator (18) allows for several choices for the bare propagator. Analysis of similar situations in the field theory of static critical phenomena has revealed, however, that the consistent way to proceed is to separate the fractional terms to interaction [11, 12]

$$S_0 = \tilde{\varphi}\left(-c_1\partial_t + D\nabla^2\right)\phi + \frac{1}{2}\tilde{\varphi}D\tilde{\varphi}, \qquad (19)$$

$$S_I = \tilde{\varphi}\left(-c_\alpha \mathcal{D}_{+}^{\alpha} - D_\sigma(-\nabla^2)^{\sigma/2}\right)\phi + \tilde{\varphi}U(\phi). \qquad (20)$$

The other way round produces renormalized theory with divergences in the limit $\sigma \to 2$, $\alpha \to 1$. It should be noted, however, that there are cases, when the usual differential operators are not generated by renormalization (e.g. the diffusion-limited coagulation model $A + A \to A$) and the propagator must be used in the Mittag-Leffler form (11) [13].

It is a fundamental rule of the theory of renormalization that nonlocal terms of the dynamic action are not renormalized at all (see, e.g., [2]). This property has been the source of a great deal of confusion in the stochastic field theory. A detailed analysis of this issue has been given on the example of renormalization of the stochastic Navier-Stokes equation in [14, 15]. The crucial consequence in the

RG analysis of the fractional Langevin problem is that the fractional diffusion constant and the coefficient of the fractional time derivative are not renormalized and therefore the corresponding terms of the dynamic actio are not changed due to fluctuations. However, the fractional terms contribute to the renormalization of the local terms of the action. This gives rise to the phenomenon of generation terms: contributions corresponding to the local terms of the dynamic action are produced by renormalization (i.e. large-scale fluctuations) even if they were absent in the original action. Relevance of these contributions cannot be determined *a priori*: renormalization brings about changes in scaling dimensions of local terms and the assessment of relevance of fractional terms is possible only, when the effect of fluctuations to scaling of local terms has been calculated.

Splitting of the generic propagator (18) to the bare propagator and interaction quadratic in fields according to (19), (20) generates Feynman graphs with effective propagators with the structure

$$\frac{(c_\alpha s^\alpha + D_\sigma k^\sigma)^n}{(c_1 s + Dk^2)^{n+1}}\,. \tag{21}$$

Renormalization is carried out in the logarithmic theory. To assess the interplay of local and fractional terms deviations of the order of fractional derivatives from the integer order of the ordinary derivatives are assumed small. In practical calculations it is almost always the case that deviation from the critical dimension $\varepsilon = d_c - d$ is used as a regulator of UV-divergences of the field theory. In this case small $1-\alpha$, $2-\sigma$ are analytic regulators and analytic renormalization is called for in a consistent renormalization procedure.

In the usual models of critical dynamics there is a single regulator ε and the scheme of minimal subtractions (MS) is effective and popular (see, e.g. [2]). In case of several regulators the ray scheme is widely used: in this scheme all regulators are assumed to be of the same order. Therefore, just one regulator remains independent and the MS scheme is carried with respect to this independent regulator. Although popular, this approach is dubious to say the least.

The point is that in genuine analytic renormalization all singularities in regulators are removed from the graphs of perturbation theory.

All singularities are removed in the analytic renormalization: therefore, to be sure that renormalization is carried out consistently, the starting point should be the analytically regularized model logarithmic with respect to dimensional regulator as well. This is sometimes called the principle of maximum divergences. As in all renormalization schemes, there is a lot of freedom in the choice of any concrete version of the analytic renormalization. In particular, if the model is analytically regularized, renormalization may be carried out with the use of a scheme based on the subtraction of necessary coefficients of the Taylor expansion at a given value of external momenta of a divergent graph (the standard Bogolyubov-Parasyuk R operation).

It is a fundamental result of the renormalization theory that the Bogolyubov-Parasyuk R operation produces Green functions analytic in regulators [16]. The MS ray scheme does not share this property. A divergent (sub)graph gives rise to a factor of the form (n, m, l – integers) [16]

$$\frac{1}{n\varepsilon + m(2-\sigma) + 2l(1-\alpha)} = \frac{1}{\varepsilon}\,\frac{1}{(n + m\zeta + 2l\xi)}\,. \tag{22}$$

Contribution of a graph to renormalization constant is a product of these factors over all subgraphs multiplied by a function analytic in ε, $2-\sigma$ and $1-\alpha$ at the origin.

In the ray scheme the ratios ζ and ξ in (22) are finite and divergences are poles in ε. In the MS scheme only contributions singular in ε are extracted to the renormalization constants. In such an approach the ratio of two different expressions (22) is a finite quantity. Therefore, in the MS scheme applied with respect to ε not all factors $(n + m\zeta + 2l\xi)^{-1}$ are removed. The result is that the coefficient functions of the renormalization group are meromorphic functions in ε, $2-\sigma$ and $1-\alpha$, not analytic as they should be! This fact does not show at one-loop approximation, in which most of the results have been calculated. However, explicit two-loop results in the stochastic Navier-Stokes problem have substantiated the picture described here [15]. Moreover, in this problem it may be seen from the explicit pole structure similar to (22) that calculations in the MS ray scheme in high orders inevitably lead to contributions diverging at the physical values of the regulators. From here it follows that the MS ray scheme is at least dubious, if not completely inconsistent beyond the one-loop approximation. On the other hand, pure analytic renormalization is not well suited for calculation of renormalization constans essential in the usual renormalization-group analysis. Moreover, subtraction procedures based on analytic properties of coefficient functions of Feynman graphs are different in dimensional and analytic renormalization [16]. Therefore, in the case of combined dimensional and analytic regularization, which is the most relevant case in critical dynamics, the most straightforward procedure appears to be the subtraction scheme based on Taylor expansion at some normalization point.

6 Conclusion

In conclusion, construction of a renormalized field theory for the solution of a stochastic differential equation (Langevin equation) has been reviewed and analyzed. It is rather obvious that the result is stochastic field theory with fractional derivatives in the dynamic actions. In the analysis of renormalization of the model the important statement is that renormalization produces local counterparts of fractional derivatives with nontrivial scaling dimensions and these generation terms play an important rôle in the renormalization and the subsequent asymptotic analysis of the model. Another important property is that the fractional differential operators are not renormalized. Relevance of fractional and the corresponding usual derivatives to the asymptotic behaviour may be assessed only upon a renormalization group analysis of the renormalized model, in which the anomalous dimensions of the local differential operators are calculated. In has been pointed out that the consistent construction of the renormalization is produced directly, when the fractional differential operators are prescribed to the interaction in the perturbation theory. A renormalization procedure, which takes into account the interplay of fractional and usual differential operators inevitably leads to multi-parameter analytic regularization, possibly encompassing dimensional regularization as well. It has been pointed out that the popular in such cases ray scheme with minimal subtractions is not consistent beyond the one-loop order. It is conjectured that in multiloop calculations dimensional-analytic regularization with normalization-point subtractions is the consistent and convenient method to use.

References

[1] С.Г. Самко, А.А. Килбас, О.И. Маричев, Интегралы и производные дробного порядка и некоторые их приложения (Наука и техника, Минск, 1987); S.G. Samko, A.A. Kilbas, O.I. Marichev, *Fractional Integrals and Derivatives. Theory and Applications* (Gordon and Breach, Amsterdam, 1993)

[2] A.N. Vasil'ev, *The Field Theoretic Renormalization Group in Critical Behavior Theory and Stochastic Dynamics* (Chapman & Hall/CRC, Boca Raton, 2004)

[3] J. Adamek, M. Keller, A. Senftleben, H. Hinrichsen, J. Stat. Mech. P09002 (2005).

[4] P.C. Martin, E.D. Siggia, H.A. Rose, Phys. Rev. A **8**, 423 (1973); C. De Dominicis, J. Phys. (Paris) **37**, Suppl C1, 247 (1976); H.K. Janssen, Z. Phys. B **23**, 377 (1976)

[5] H. Hinrichsen, J. Stat. Mech. P07006 (2007)

[6] R. Metzler, E. Barkai, J. Klafter, Phys. Rev. Lett. **82**, 3563 (1999).
[7] A. Erdélyi (ed.), *Higher transcendental functions, volume III* (McGraw-Hill, New York, 1981)
[8] M. Hnatič, J. Honkonen, T. Lučivjanský, Acta Physica Slovaca **66**, 69 (2016)
[9] A.N. Vasiliev, *Functional Methods in Quantum Field Theory and Statistical Physics* (Gordon and Breach, Amsterdam, 1998)
[10] J. Honkonen, Theor. Math. Phys. **169**, 1460 (2011)
[11] J. Honkonen, M.Yu. Nalimov, J. Phys. A: Math. Gen. **22**, 751 (1989)
[12] J. Honkonen, J. Phys. A: Math. Gen. **23**, 825 (1990)
[13] J. Honkonen, *Fractional derivatives in stochastic reaction-diffusion problems*. The 15th Small Triangle Meeting on theoretical physics (Institute of Experimental Physics SAS, Košice, 2014) 74-79
[14] L.Ts. Adzhemyan, J. Honkonen, M.V. Kompaniets, A.N. Vasil'ev, Phys. Rev. E **68**, 055302(R) (2003)
[15] L.Ts. Adzhemyan, J. Honkonen, M.V. Kompaniets, A.N. Vasil'ev, Phys. Rev. E **71**, 036305 (2005)
[16] О.И. Завьялов, Перенормированные диаграммы Фейнмана (Наука, Москва, 1979); O.I. Zavialov, *Renormalized Quantum Field Theory* (Kluwer, Dordrecht, 1990)]

EPJ Web of Conferences **173**, 01006 (2018) https://doi.org/10.1051/epjconf/201817301006
Mathematical Modeling and Computational Physics 2017

Generation of Quantum Correlations in Bipartite Gaussian Open Quantum Systems

Aurelian Isar[1,*]

[1] *National Institute of Physics and Nuclear Engineering, Bucharest-Magurele, P.O. Box MG-6, Romania*

Abstract.
We describe the generation of quantum correlations (entanglement, discord and steering) in a system composed of two coupled non-resonant bosonic modes immersed in a common thermal reservoir, in the framework of the theory of open systems. We show that for separable initial squeezed thermal states entanglement generation may take place, for definite values of squeezing parameter, average photon numbers, temperature of the thermal bath, dissipation constant and strength of interaction between the two bosonic modes. We also show that for initial uni-modal squeezed states Gaussian discord can be generated for all non-zero values of the strength of interaction between the modes. Likewise, for an initial separable state, a generation of Gaussian steering may take place temporarily, for definite values of the parameters characterizing the initial state and the thermal environment, and the strength of coupling between the two modes.

1 Introduction

Characterizing and quantifying quantum correlations represent a key subject in the theory of quantum information [1–3]. In particular, quantum correlations in states of continuous variable systems play a crucial role in applications of quantum information processing and transmission [4]. In this respect, quantum entanglement is considered to be a strong physical resource for quantum information tasks and protocols [5]. However, not all non-classical properties of quantum correlations are described by entanglement. In this sense, quantum discord was introduced [6, 7] as a measure of quantum correlations which includes entanglement and which can also be present in separable states. Steering is also a type of quantum nonlocality first identified in the Einstein-Podolsky-Rosen paper [8], which is distinct from both nonseparability and Bell nonlocality.

In order to obtain a realistic description of quantum processes it is necessary to take decoherence and dissipation into consideration. Recently we studied, in the framework of the theory of open systems, the dynamics of quantum correlations of two, both uncoupled and respectively coupled, bosonic modes embedded in a common thermal bath, for initial Gaussian states of the system [9–15].

In the present work we analyze the possibility to generate quantum correlations in a system of two coupled bosonic modes interacting with a common thermal reservoir. The initial state of the subsystem is taken of Gaussian form and the evolution under the quantum dynamical semigroup assures the preservation in time of the Gaussian form of the state. We calculate the logarithmic

* e-mail: isar@theory.nipne.ro

negativity and show that for separable initial squeezed thermal states entanglement generation may take place, for definite values of squeezing parameter, average photon numbers, temperature of the thermal bath, dissipation constant and strength of interaction between the two bosonic modes. After its generation one can observe temporary suppressions and revivals of the entanglement. We also show that for initial uni-modal squeezed states the generation of Gaussian quantum discord takes place for all non-zero values of the strength of interaction between the bosonic modes. Finally, we illustrate that, for an initial separable squeezed thermal state, a generation of Gaussian quantum steering may take place temporarily, for definite values characterizing the initial state and the thermal environment, and the strength of interaction between the two modes. The thermal noise and dissipation destroy the steerability between the two parts and, unlike Gaussian quantum discord, which is decreasing asymptotically in time, the Gaussian quantum steering vanishes suddenly in a finite time.

The paper is organized as follows. In Sec. 2 we write the Markovian master equation for the density operator of the open system interacting with a general environment and solve the evolution equation for the covariance matrix of the state of the bimodal bosonic system. Then we describe in Sec. 3 the dynamics of quantum entanglement, Gaussian quantum discord and Gaussian quantum steering for the considered open system. A summary is given in Sec. 4.

2 Master equation for two bosonic modes interacting with the environment

In order to study the dynamics of the subsystem consisting of two coupled bosonic modes (harmonic oscillators) in weak interaction with a thermal reservoir, we use the axiomatic formalism based on completely positive quantum dynamical semigroups. In this framework the Markovian irreversible time evolution of an open system is described by the following Gorini-Kossakowski-Sudarshan-Lindblad master equation for the density operator $\rho(t)$ [16–19]:

$$\frac{d\rho(t)}{dt} = -\frac{i}{\hbar}[H,\rho(t)] + \frac{1}{2\hbar}\sum_j (2B_j\rho(t)B_j^\dagger - \{\rho(t), B_j^\dagger B_j\}_+), \tag{1}$$

where H denotes the Hamiltonian of the open system and the operators $B_j, B_j^\dagger$ are defined on the Hilbert space of H, and describe the interaction of the subsystem with a general environment.

The Hamiltonian of two coupled in coordinates non-resonant harmonic oscillators of identical mass m and frequencies ω_1 and ω_2 is given by

$$H = \frac{1}{2m}(p_x^2 + p_y^2) + \frac{m}{2}(\omega_1^2 x^2 + \omega_2^2 y^2) + qxy, \tag{2}$$

where x, y are the coordinates and p_x, p_y are the momenta of the two modes, q is the coupling parameter, and we take the operators B_j as polynomials of first degree in the canonical variables of coordinates and momenta.

The equations of motion for the quantum correlations of the canonical observables are the following (T denotes the transposed matrix) [19]:

$$\frac{d\sigma(t)}{dt} = Z\sigma(t) + \sigma(t)Z^{\mathrm{T}} + 2D, \quad Z = \begin{pmatrix} -\lambda & 1/m & 0 & 0 \\ -m\omega_1^2 & -\lambda & -q & 0 \\ 0 & 0 & -\lambda & 1/m \\ -q & 0 & -m\omega_2^2 & -\lambda \end{pmatrix}, \tag{3}$$

with real diffusion matrix D and dissipation coefficient λ. We introduced the 4×4 bimodal covariance matrix σ, with the elements defined as $\sigma_{ij} = \mathrm{Tr}[(A_iA_j + A_jA_i)\rho]/2, i, j = 1, \ldots, 4$, with

$\mathbf{A} = \{x, p_x, y, p_y\}$, which fully characterize any Gaussian state of a bimodal system (up to local displacements).

The time-dependent solution of Eq. (3) is given by $\sigma(t) = S(t)[\sigma(0) - \sigma(\infty)]S^{\mathrm{T}}(t) + \sigma(\infty)$ [19], where $S(t) \equiv \exp(Zt)$. The values at infinity are obtained from the equation $Z\sigma(\infty) + \sigma(\infty)Z^{\mathrm{T}} = -2D$. The covariance matrix, which is a real, symmetric and positive matrix entirely specifying a two-mode Gaussian state, has the following block structure, where 2×2 Hermitian matrices A and B are the covariance matrices for the single modes, and C contains the cross-correlations between the modes:

$$\sigma(t) = \begin{pmatrix} A & C \\ C^{\mathrm{T}} & B \end{pmatrix}. \tag{4}$$

3 Dynamics of quantum correlations

3.1 Generation of quantum entanglement

We use the logarithmic negativity as a measure to quantify the degree of entanglement of the two-mode Gaussian states: $E = \max\{0, -\log_2 2\tilde{n}_-\}$, where $\tilde{n}_-$ is the smallest of the two symplectic eigenvalues of the partial transpose $\tilde{\sigma}$ of the two-mode covariance matrix σ [20]: $2\tilde{n}_\mp^2 = \tilde{\Delta} \mp \sqrt{\tilde{\Delta}^2 - 4\det\sigma}$ and $\tilde{\Delta} = \det A + \det B - 2\det C$. We obtain [21, 22]:

$$E(t) = \max\{0, -\frac{1}{2}\log_2[4f(t)]\}, \tag{5}$$

where

$$f(t) = \frac{1}{2}(\det A + \det B) - \det C - \left(\left[\frac{1}{2}(\det A + \det B) - \det C\right]^2 - \det\sigma(t)\right)^{1/2}. \tag{6}$$

E determines the strength of entanglement for $E(t) > 0$, and if $E(t) = 0$, then the state is separable. In Refs. [23–28] we described the time evolution of the entanglement for a system of two uncoupled bosonic modes interacting with a common environment.

We take an initial two-mode squeezed thermal state, with the covariance matrix given by [29]

$$\sigma_{st}(0) = \begin{pmatrix} a & 0 & c & 0 \\ 0 & a & 0 & -c \\ c & 0 & b & 0 \\ 0 & -c & 0 & b \end{pmatrix}, \tag{7}$$

where

$$\begin{aligned} a &= (n_1 + \frac{1}{2})\cosh^2 r + (n_2 + \frac{1}{2})\sinh^2 r, \\ b &= (n_1 + \frac{1}{2})\sinh^2 r + (n_2 + \frac{1}{2})\cosh^2 r, \\ c &= (n_1 + n_2 + 1)\sinh r\cosh r. \end{aligned} \tag{8}$$

Here n_1, n_2 denote the mean thermal photon numbers associated with the two modes and r is the squeezing parameter. For $n_1 = 0$ and $n_2 = 0$, (7) becomes the covariance matrix of the two-mode squeezed vacuum state. If the squeezing parameter r satisfies the inequality $r > r_e$ [29], where

$$\cosh^2 r_e = \frac{(n_1 + 1)(n_2 + 1)}{n_1 + n_2 + 1}, \tag{9}$$

then the two-mode squeezed thermal state is an entangled state.

We suppose that the only non-zero quantum diffusion coefficients have the following form (we put $\hbar = 1$) [18]:

$$m^2\omega_1^2 D_{xx} = D_{p_xp_x} = \frac{m\omega_1\lambda}{2}\coth\frac{\omega_1}{2kT}, \quad m^2\omega_2^2 D_{yy} = D_{p_yp_y} = \frac{m\omega_2\lambda}{2}\coth\frac{\omega_2}{2kT}, \tag{10}$$

where k is Boltzmann constant and T the temperature of the thermal reservoir.

In Ref. [30] we described the dynamics of the quantum entanglement of a subsystem composed of two coupled bosonic modes interacting with a common thermal reservoir and have shown that, for a separable initial squeezed thermal state entanglement generation may take place, for definite values of the squeezing parameter, average photon numbers, temperature of the thermal bath, dissipation coefficient and strength of interaction between the two modes.

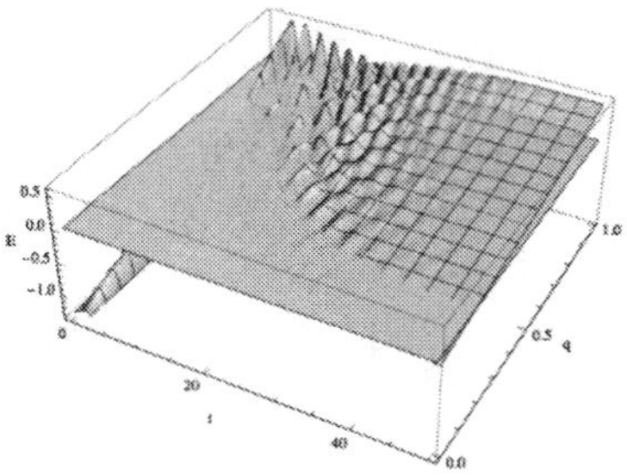

Figure 1. Logarithmic negativity E versus time t and interaction strength q for a separable initial squeezed thermal state with squeezing parameter $r = 0.3$, average photon numbers $n_1 = n_2 = 1$, dissipation constant $\lambda = 0.08$, and $\coth(\omega/2kT) = 1.1$ ($\omega_1 = \omega_2 \equiv \omega = 1, m = \hbar = 1$).

The evolution of separable initial squeezed thermal states with the covariance matrix given by Eq. (7) is illustrated in Fig. 1, where we represent the dependence of the logarithmic negativity $E(t)$ on time t and interaction strength q. One can observe that entanglement generation may take place, for definite values of the strength of interaction between the two modes, of squeezing parameter, average photon numbers, temperature of the thermal bath and dissipation constant. After its generation one can observe temporary suppressions and revivals of the entanglement.

Independent of the initial state, in the limit of large times the system evolves asymptotically to an equilibrium state which may be entangled or separable. The direct interaction between the two modes favours the generation or the preservation in time of the created entanglement, while the temperature of the thermal bath acts towards preventing the generation of entanglement, or suppressing it, once it was created. It is the competition between these two factors which determines the final state of being separable or entangled [30].

3.2 Generation of Gaussian discord

Quantum discord has been defined as the difference between two quantum analogues of classically equivalent expressions of the mutual information, which is a measure of total correlations in a quantum state. It was introduced [6, 7, 31] as a measure of all quantum correlations in a bipartite state, including entanglement. For pure entangled states quantum discord coincides with the entropy of entanglement. Quantum discord can have non-zero values also for some mixed separable states.

For bipartite continuous variable systems, closed formulas of the Gaussian quantum discord have been obtained for bipartite thermal squeezed states [32] and for all two-mode Gaussian states [33].

The Gaussian quantum discord of a general two-mode Gaussian state ρ_{12} can be defined as the quantum discord where the conditional entropy is restricted to generalized Gaussian positive operator valued measurements (POVM) on the mode 2. In terms of symplectic invariants it is given by (the symmetry between the two modes 1 and 2 is broken) [33] $D = f(\sqrt{\beta}) - f(\nu_-) - f(\nu_+) + f(\sqrt{\varepsilon})$, where

$$f(y) = \frac{y+1}{2}\log\frac{y+1}{2} - \frac{y-1}{2}\log\frac{y-1}{2}, \tag{11}$$

$$\varepsilon = \begin{cases} \dfrac{2\gamma^2 + (\beta-1)(\delta-\alpha) + 2|\gamma|\sqrt{\gamma^2 + (\beta-1)(\delta-\alpha)}}{(\beta-1)^2}, \\ \quad \text{if } (\delta-\alpha\beta)^2 \le (\beta+1)\gamma^2(\alpha+\delta) \\ \dfrac{\alpha\beta - \gamma^2 + \delta - \sqrt{\gamma^4 + (\delta-\alpha\beta)^2 - 2\gamma^2(\delta+\alpha\beta)}}{2\beta}, \\ \quad \text{otherwise,} \end{cases} \tag{12}$$

$\alpha = 4\det A, \beta = 4\det B, \gamma = 4\det C, \delta = 16\det\sigma$, and $\nu_\mp$ are the symplectic eigenvalues of the state, given by $2\nu_\mp^2 = \Delta \mp \sqrt{\Delta^2 - 4\delta}$, where $\Delta = \alpha + \beta + 2\gamma$.

In order to prove the possible generation of the Gaussian quantum discord, we have to consider an initial state of the form of a product state, which is the only kind of states which do not contain quantum discord [3]. The uni-modal squeezed states represent an interesting example of such product states with zero discord, described by the covariance matrix

$$\sigma(0) = \frac{1}{2}\begin{pmatrix} \cosh 2r & \sinh 2r & 0 & 0 \\ \sinh 2r & \cosh 2r & 0 & 0 \\ 0 & 0 & \cosh 2r & \sinh 2r \\ 0 & 0 & \sinh 2r & \cosh 2r \end{pmatrix}, \tag{13}$$

with squeezing parameter r.

In Ref. [34] we have shown that, for such an initial uni-modal squeezed states, the generation of Gaussian quantum discord takes place, immediately after the initial moment of time, for all nonzero values of the strength of interaction between the coupled bosonic modes. This phenomenon of discord generation is illustrated in Fig 2, where it is represented the dependence of the Gaussian quantum discord D on time t and squeezing parameter r for an initial uni-modal squeezed state, for the case of non-resonant modes. We notice that after reaching some maximum value, the generated Gaussian discord non-monotonically decreases in time and tends asymptotically for large times to some definite non-zero value. The non-zero asymptotic value of the Gaussian discord is the result of the fact that the two bosonic modes are coupled. In the absence of their coupling, the discord tends asymptotically for large times to zero, corresponding to an asymptotic product state.

3.3 Generation of Gaussian steering

To infer the steerability between two parties is equivalent with verifying the shared entanglement distribution by an untrusted party, by performing local measurements and classical communications [35]. It has been shown in Refs. [35, 36] that the steerability $A \to B$ is present if and only if the following relation does not hold:

$$\sigma + \frac{i}{2}0_A \oplus \Omega_B \ge 0, \quad \Omega = \begin{pmatrix} 0 & 1 \\ -1 & 0 \end{pmatrix}. \tag{14}$$

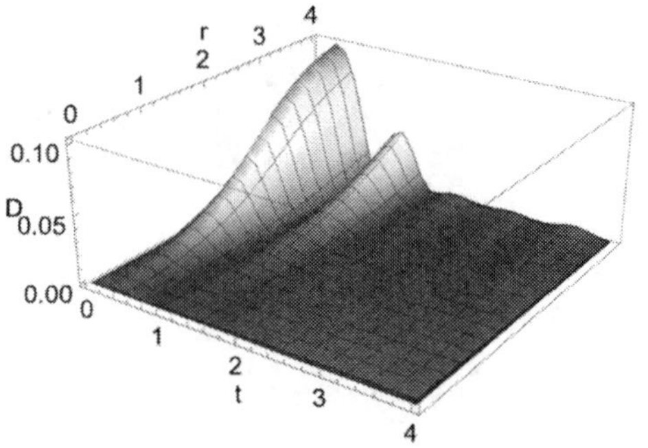

Figure 2. Gaussian quantum discord D versus time t and squeezing parameter r for an initial uni-modal squeezed state, non-resonant modes with $\omega_1 = 1$ and $\omega_2 = 4$, temperature $T = 0$ ($\coth(\omega/2kT) = 1$), strength of interaction between modes $q = 0.5$ and dissipation coefficient $\lambda = 1$.

A measure has been proposed of how much a state with covariance matrix σ is $A \to B$ steerable with Gaussian measurements, by quantifying the amount by which the condition (14) is violated, as follows [37]: $G^{A\to B}(\sigma) = \max\{0, -\ln 2\nu^B\}$, where ν^B is the symplectic eigenvalue of the Schur complement of A in covariance matrix σ. This quantity vanishes if and only if the state σ is not steerable by Gaussian measurements and is invariant under local symplectic transformations. Steering is in general asymmetric with respect to the interchange between steerable parties. The general quantity proposed in Refs. [37, 38] has a particularly simple form when the steered party has one mode:

$$G^{A\to B}(\sigma) = \max\{0, \frac{1}{2}\ln\frac{\det A}{4\det\sigma}\} = \max\{0, S(A) - S(\sigma)\}, \tag{15}$$

where S is the Renyi-2 entropy.

In Ref. [39] we considered the system of two uncoupled bosonic modes interacting with a common thermal reservoir and described the behaviour of the Gaussian quantum steering, when the initial state of the system is an entangled squeezed thermal state. We have shown that the suppression of the Gaussian steering takes place in a finite time, for all temperatures of the thermal reservoir and all values of the squeezing parameter, this behaviour being similar to the well-known phenomenon of entanglement sudden death. This kind of evolution of Gaussian steering and entanglement is in contrast with the dynamics of quantum discord, which decreases to zero asymptotically in time for uncoupled modes.

However, if we take a separable initial squeezed thermal state, with the covariance matrix of the form (7), then it is possible to generate Gaussian steering, for definite values of the squeezing parameter r and strength of interaction q between the coupled modes. In Fig. 3 we illustrate the generation of the Gaussian $A \to B$ quantum steering G for an initial separable squeezed thermal state, for a zero temperature of the thermal bath, as a function of time t and squeezing parameter r. We notice that after its generation, the Gaussian steering is suppressed in a finite time. There are also possible temporary revivals and suppressions of the Gaussian steering. Compared to the Gaussian quantum discord, which is decreasing asymptotically in time, the Gaussian quantum steering suffers a sudden death behaviour like quantum entanglement. The thermal noise and dissipation destroy Gaussian steering in a finite time. For large times the asymptotic value of Eq. (15) has only negative values for non-zero temperatures and is 0 for $T = 0$. Therefore for large times the Gaussian quantum steering is completely missing, independent of the fact that the two bosonic modes are coupled or not.

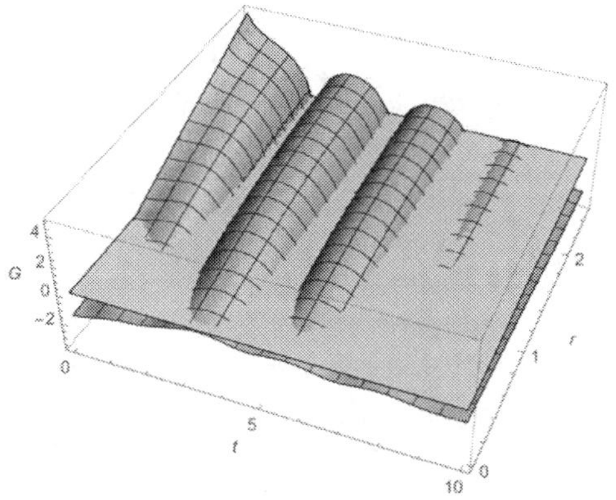

Figure 3. Gaussian quantum steering G versus time t and squeezing parameter r for a separable initial squeezed thermal state with $n_1 = 0.5, n_2 = 1$, in a thermal environment with temperature $T = 0$, dissipation coefficient $\lambda = 0.03$, strength of interaction between modes $q = 0.9$ and $\omega_1 = \omega_2 = 1$.

4 Summary

We investigated the generation of quantum correlations in a system composed of two coupled bosonic modes, embedded in a common thermal bath, in terms of the covariance matrix for Gaussian input states. The main conclusion is that the generation of quantum correlations strongly depends on the initial states of the subsystem, frequencies of the modes, parameters characterizing the thermal reservoir and the intensity of the coupling between the two modes. The competition between the direct interaction between the two bosonic modes and the influence of the thermal environment is decisively determining the possible generation of the quantum correlations and their time evolution.

For an initial separable state, for definite values of the mentioned parameters and for sufficiently large values of the interaction strength between the modes, this interaction makes possible the generation of the entanglement. After its generation, temporary suppressions and revivals of entanglement may also happen, and in the limit of large times the system evolves asymptotically to an equilibrium state which, depending on the values of the strength of interaction between the two bosonic modes, on temperature and dissipation coefficient, may be separable or entangled. Generation of Gaussian quantum discord in the case of an initial uni-modal squeezed state always happens, for all values of the strength of interaction between the two modes. Compared to the entanglement, Gaussian discord keeps for all finite times a non-zero value, including for asymptotic large times. Likewise, for an initial separable state, a generation of Gaussian quantum steering may take place temporarily, for definite values characterizing the initial state and the thermal environment, and the strength of interaction between the modes. In the limit of large times, the Gaussian steering is always zero.

In the context of the debate relative to the physical interpretation of quantum correlations, the present results, in particular the possibility of generating and preserving them in thermal environments for enough long times, might be useful in controlling quantum correlations in open systems and also for applications in the protocols of quantum information processing and communication.

Acknowledgments

The author acknowledges the financial support received from the Romanian Ministry of Research and Inovation, through the Project PN 16 42 01 01/2016.

References

[1] M. A. Nielsen and I. L. Chuang, *Quantum Computation and Quantum Information* (Cambridge University Press, Cambridge, 2000).
[2] C. Weedbrook, S. Pirandola, R. Garcia-Patron, N. J. Cerf, T. C. Ralph, J. H. Shapiro and S. Lloyd, Rev. Mod. Phys. **84**, 621 (2012).
[3] G. Adesso, T. R. Bromley and M. Cianciaruso, J. Phys. A: Math. Theor. **49**, 473001 (2016).
[4] S. L. Braunstein and P. van Loock, Rev. Mod. Phys. **77**, 513 (2005).
[5] R. Horodecki, P. Horodecki, M. Horodecki and K. Horodecki, Rev. Mod. Phys. **81**, 865 (2009).
[6] W. H. Zurek, Annalen der Physik (Leipzig) **9**, 853 (2000).
[7] H. Ollivier and W. H. Zurek, Phys. Rev. Lett. **88**, 017901 (2001).
[8] A. Einstein, B. Podolski and N. Rosen, Phys. Rev. **47**, 777 (1935).
[9] A. Isar, Eur. J. Phys. Spec. Topics **160**, 225 (2008).
[10] A. Isar, J. Russ. Laser Res. **30**, 458 (2009).
[11] A. Isar, Phys. Scr. **82**, 038116 (2010).
[12] T. Mihaescu and A. Isar, Rom. J. Phys. **60**, 853 (2015).
[13] S. Suciu and A. Isar, Rom. J. Phys. **60**, 859 (2015).
[14] H. A. Boura and A. Isar, Rom. J. Phys. **60**, 1278 (2015).
[15] I. Ghiu and A. Isar, Rom. J. Phys. **61**, 768 (2016).
[16] V. Gorini, A. Kossakowski and E. C. G. Sudarshan, J. Math. Phys. **17**, 821 (1976).
[17] G. Lindblad, Commun. Math. Phys. **48**, 119 (1976).
[18] A. Isar, A. Sandulescu, H. Scutaru, E. Stefanescu and W. Scheid, Int. J. Mod. Phys. E **3**, 635 (1994).
[19] A. Sandulescu, H. Scutaru and W. Scheid, J. Phys. A: Math. Gen. **20**, 2121 (1987).
[20] G. Adesso, A. Serafini and F. Illuminati, Phys. Rev. A **70**, 022318 (2004).
[21] A. Isar, Int. J. Quantum Inf. **6**, 689 (2008).
[22] A. Isar, Open Sys. Inf. Dynamics **16**, 205 (2009).
[23] A. Isar, J. Russ. Laser Res. **28**, 439 (2007)
[24] A. Isar, Phys. Scr. T **135**, 014033 (2009)
[25] A. Isar, J. Russ. Laser Res. **31**, 182 (2010).
[26] A. Isar, Phys. Scr. T **140**, 014023 (2010).
[27] A. Isar, Phys. Scr. T **147**, 014015 (2012).
[28] A. Isar, Phys. Scripta T **153**, 014035 (2013).
[29] P. Marian, T. A. Marian and H. Scutaru, Phys. Rev. A **68**, 062309 (2003).
[30] A. Isar, Open Sys. Information Dyn. **23**, 1650007 (2016).
[31] L. Henderson and V. Vedral, J. Phys. A: Math. Gen. **34**, 6899 (2001).
[32] P. Giorda and M. G. A. Paris, Phys. Rev. Lett. **105**, 020503 (2010).
[33] G. Adesso and A. Datta, Phys. Rev. Lett. **105**, 030501 (2010).
[34] A. Isar and T. Mihaescu, Eur. Phys. J. D **71**, 144 (2017).
[35] H. M. Wiseman, S. J. Jones and A. C. Doherty, Phys. Rev. Lett. **98**, 140402 (2007).
[36] S. J. Jones, H. M. Wiseman and A. C. Doherty, Phys. Rev. A **76**, 052116 (2007).
[37] I. Kogias, A. R. Lee, S. Ragy and G. Adesso, Phys. Rev. Lett. **114**, 060403 (2015).
[38] I. Kogias, P. Skrzypczyk, D. Cavalcanti, A. Acin and G. Adesso, Phys. Rev. Lett. **115**, 210401 (2015).
[39] T. Mihaescu and A. Isar, Rom. J. Phys. **62**, 107 (2017).

Modeling Quantum Behavior in the Framework of Permutation Groups

Vladimir Kornyak[1,*]

[1] *Laboratory of Information Technologies, Joint Institute for Nuclear Research, 141980 Dubna, Moscow Region, Russia*

Abstract. Quantum-mechanical concepts can be formulated in constructive finite terms without loss of their empirical content if we replace a general unitary group by a unitary representation of a finite group. Any linear representation of a finite group can be realized as a subrepresentation of a permutation representation. Thus, quantum-mechanical problems can be expressed in terms of permutation groups. This approach allows us to clarify the meaning of a number of physical concepts. Combining methods of computational group theory with Monte Carlo simulation we study a model based on representations of permutation groups.

1 Introduction

Since the time of Newton, differential calculus demonstrates high efficiency in describing physical phenomena. However, the infinitesimal analysis introduces infinities in the physical theories. This is often considered as a serious conceptual flaw: recall, for example, Dirac's frequently quoted claim that the most important challenge in physics is "to get rid of infinity". Moreover, differential calculus, being, in fact, a kind of approximation, may lead to descriptive losses in some problems – an illustrative example is given below in Sec. 3.1. In the paper, we describe a constructive version of the quantum formalism that does not involve any concepts associated with actual infinities.

The main part of the paper starts with Sec. 2, which contains a summary of the basic concepts of the standard quantum mechanics with emphasis on the aspects important for our purposes.

Sec. 3 describes a constructive modification of the quantum formalism. We start with replacing a continuous group of symmetries of quantum states by a finite group. The natural consequence of this replacement is unitarity, since any linear representation of a finite group is unitary. Further, any finite group is naturally associated with some cyclotomic field. Generally, a cyclotomic field is a dense subfield of the field of complex numbers. This can be regarded as an explanation of the presence of complex numbers in the quantum formalism. Any linear representation of a finite group over the associated cyclotomic field can be obtained from a permutation action of the group on vectors with natural components by projecting into a suitable invariant subspace. All this allows us to reproduce all the elements of the quantum formalism in invariant subspaces of the permutation representations.

In Sec. 4 we consider a model of quantum evolution inspired by the quantum Zeno effect – the most convincing manifestation of the role of observation in the dynamics of quantum systems. The model represents the quantum evolution as a sequence of observations with unitary transitions between them. The standard quantum mechanics assumes a single deterministic unitary transition between observations. In our model we generalize this assumption. We treat a unitary transition as a kind

*e-mail: vkornyak@gmail.com

of gauge connection – a way of identifying indistinguishable entities at different times. *A priori*, any unitary transformation can be used as a data identification rule. So, we assume that all unitary transformations participate in transitions between observations with appropriate weights. We call a unitary evolution *dominant* if it provides the maximum transition probability.[1] The Monte Carlo simulation shows a sharp dominance of such evolutions over other evolutions. To compare with a continuous description, we present also the Lagrangian of the continuum approximation of the model.

2 Formalism of quantum mechanics

Here is a brief outline of the basic concepts of the quantum mechanics. We divide these concepts into three categories: *states*, *observations and measurements*, and *time evolution*.

2.1 States

A ***pure quantum state*** is a ray in a Hilbert space $\mathcal{H}$ over the complex field $\mathbb{C}$, i.e., an *equivalence class* of vectors $|\psi\rangle \in \mathcal{H}$ with respect to the *equivalence relation* $|\psi\rangle \sim a\,|\psi\rangle$, where $a \in \mathbb{C},\ a \neq 0$. We can reduce the equivalence classes by *normalization*: $|\psi\rangle \sim \mathrm{e}^{\mathrm{i}\alpha}\,|\psi\rangle\,, \|\psi\| = 1, \alpha \in \mathbb{R}$. Finally, we can eliminate the phase "degree of freedom" α by transition to the rank one *projector* $\Pi_\psi = |\psi\rangle\langle\psi|$, which is a special case of a *density matrix*.

A ***mixed quantum state*** is described by a *general density matrix* ρ characterized by three properties: (a) $\rho = \rho^\dagger$, (b) $\langle\psi\,|\rho|\,\psi\rangle \geq 0$ for any $|\psi\rangle \in \mathcal{H}$, (c) $\mathrm{tr}\rho = 1$. In fact, any mixed state is a weighted mixture of pure states, i.e., its density matrix can be represented as a weighted sum of rank one projectors. We will denote the set of all density matrices by $\mathcal{D}(\mathcal{H})$.

The Hilbert space of a ***composite system***, $XY = X \times Y$, is the tensor product of the Hilbert spaces for its constituents: $\mathcal{H}_{XY} = \mathcal{H}_X \otimes \mathcal{H}_Y$. The ***states of a composite system***, $\mathcal{D}(\mathcal{H}_{XY})$, are classified into two types: *separable* and *entangled* states. The set of ***separable*** states, $\mathcal{D}_\mathrm{S}(\mathcal{H}_{XY})$, consists of the states $\rho_{XY} \in \mathcal{D}(\mathcal{H}_{XY})$ that can be represented as weighted sums of the tensor products of states of the constituents: $\rho_{XY} = \sum_k w_k \rho_X^k \otimes \rho_Y^k,\ \ w_k \geq 0,\ \ \sum_k w_k = 1$. The set of ***entangled*** states, $\mathcal{D}_\mathrm{E}(\mathcal{H}_{XY})$, is by definition the complement of $\mathcal{D}_\mathrm{S}(\mathcal{H}_{XY})$ in the set of all states: $\mathcal{D}_\mathrm{E}(\mathcal{H}_{XY}) = \mathcal{D}(\mathcal{H}_{XY}) \setminus \mathcal{D}_\mathrm{S}(\mathcal{H}_{XY})$.

2.2 Observations and measurements

The terms 'observation' and 'measurement' are often used as synonyms. However, it makes sense to separate these concepts: we treat observation as a more general concept which does not imply, in contrast to measurement, obtaining numerical information.

An ***observation*** is the detection ("click of detector") of a system found in the state ρ in the subspace $\mathcal{S} \leq \mathcal{H}$. The mathematical abstraction of the "detector in the subspace" $\mathcal{S}$ of a Hilbert space is the operator of projection, $\Pi_\mathcal{S}$, into this subspace. The result of quantum observation is random and its statistics is described by a probability measure defined on subspaces of the Hilbert space. Any such measure $\mu(\cdot)$ must be additive on any set of mutually orthogonal subspaces of a Hilbert space: if, e.g., $\mathcal{A}$ and $\mathcal{B}$ are mutually orthogonal subspaces, then $\mu(\mathrm{span}(\mathcal{A},\mathcal{B})) = \mu(\mathcal{A}) + \mu(\mathcal{B})$. Gleason proved [1] that, except the case $\dim\mathcal{H} = 2$, the only such measures have the form $\mu_\rho(\mathcal{S}) = \mathrm{tr}(\rho\Pi_\mathcal{S})$, where ρ is an arbitrary density matrix. If, in particular, ρ describes a pure state, $\rho = |\psi\rangle\langle\psi|$, and $\mathcal{S}$ is one-dimensional, $\mathcal{S} = \mathrm{span}(|\varphi\rangle)$, we come to the familiar Born rule: $\mathrm{tr}(\rho\Pi_\mathcal{S}) = \mathbf{P}_\mathrm{Born} = |\langle\varphi \mid \psi\rangle|^2$.

A ***measurement*** is a special case of observation, when the partition of a Hilbert space into mutually orthogonal subspaces is provided by a Hermitian operator A. Any such operator can be written as $A = \sum_k a_k \Pi_{e_k}$, where $a_1, a_2, \ldots \in \mathbb{R}$ is the *spectrum* of A, and $e_1, e_2, \ldots$ is an orthonormal basis of *eigenvectors* of A. "Click of the detector" Π_{e_k} is interpreted as that the *eigenvalue* a_k is the result of the measurement. The mean for multiple measurements tends to the *expectation value* of A in the state ρ: $\langle A\rangle_\rho = \mathrm{tr}(\rho A)$.

[1] In fact, the principle of least action in physical theories implies the selection of dominant evolutions among all possible ("virtual") evolutions. The apparent determinism of these evolutions can be explained by the sharpness of their dominance.

2.3 Time evolution

The ***time evolution*** of a quantum system is a unitary transformation of data between observations. For a density matrix, the unitary evolution takes the form

$$\rho_{t'} = U_{t't}\rho_t U_{t't}^{\dagger}, \tag{1}$$

where ρ_t is the state *after* observation at the time t, $\rho_{t'}$ is the state *before* observation at the time t', and $U_{t't}$ is the unitary transition between the observation times t and t'. In the standard quantum formalism, the time is considered as a continuous parameter, and the relation (1) becomes the *von Neumann equation* in the infinitesimal limit. The evolution of a pure state can be written as $|\psi_{t'}\rangle = U_{t't}|\psi_t\rangle$, and the corresponding infinitesimal limit is the *Schrödinger equation*. To emphasize the role of the observation in the quantum physics, we note that the unitary evolution is simply a change of coordinates in the Hilbert space and it is not sufficient to describe observable physical phenomena.

2.4 Emergence of geometry within large Hilbert space via entanglement

The quantum-mechanical theory does not need the geometric space as a fundamental concept – everything can be formulated using only the Hilbert space formalism. In this view, the observed geometry must emerge as an approximation. The currently popular idea [2–4] of the emergence of geometry within a Hilbert space is based on the notion of entanglement. Briefly, the scheme of extracting geometric manifold from the entanglement structure of a quantum state ρ in a Hilbert space $\mathcal{H}$ is as follows:

- The Hilbert space decomposes into a large number of tensor factors: $\mathcal{H} = \otimes_x \mathcal{H}_x,\ x \in X$. Each factor is treated as a point (or bulk) of geometric space to be built. A graph G – called *tensor network* – with vertices $x \in X$ and edges $\{x, y\} \in X \times X$ is introduced.
- The edges of G are assigned *weights* based on a *measure of entanglement*, a function that vanishes on separable states and is positive on entangled states. A typical such measure is the *mutual information*: $I(\rho_{xy}) = S(\rho_x) + S(\rho_y) - S(\rho_{xy})$, where ρ_x denotes the result of taking traces of ρ over all tensor factors except for the x-th (and similarly for ρ_y, ρ_{xy}); $S(a) = -\operatorname{tr}(a \log a)$ is the *von Neumann entropy*. The graph G is supplied with a metric derived from the weights of the edges.
- Finally, the graph G is approximately isometrically embedded in a smooth metric manifold of an as small as possible dimension using algorithms like the *multidimensional scaling* (MDS).

3 Constructive modification of the quantum formalism

David Hilbert, a prominent advocate of the free use of the concept of infinity in mathematics, wrote the following about the relation of the infinite to the reality: "*Our principal result is that the infinite is nowhere to be found in reality. It neither exists in nature nor provides a legitimate basis for rational thought – a remarkable harmony between being and thought.*" Adopting this view, we reformulate the quantum formalism in constructive finite terms without distorting its empirical content [5–7].

3.1 Losses due to continuum and differential calculus

Differential calculus (including differential equations, differential geometry, etc.) forms the basis of the mathematical methods in physics. The applicability of the differential calculus is based on the assumption that any relevant function can be approximated by linear relations at small scales. This assumption simplifies many problems in physics and mathematics, but at the cost of loss of completeness.

As an example, consider the problem of classifying simple groups. The concept of group is an abstraction of the properties of *permutations* (also called *one-to-one mappings* or *bijections*) of a set. Namely, an abstract group is a set with an *associative* operation, an *identity* element, and an *invertibility* for each element. There are two most common additional assumptions that make the notion of a group more meaningful: (a) *the group is a differentiable manifold* – such a group is called *Lie group*; (b) *the group is finite*. The empirical physics is insensitive to the assumption (b) – ultimately, any empirical description is reduced to a finite set of data. On the contrary, the assumption (a) implies severe constraints on the possible physical models.

The problem of the classification of the simple groups[2] under the assumption (a) turned out to be rather easy and was solved by two people (Killing and Cartan) in a few years. The result is four infinite series: A_n, B_n, C_n, D_n; and five exceptional groups: E_6, E_7, E_8, F_4, G_2.

The solution of the classification problem under the assumption (b) required the efforts of about a hundred people for over a hundred years [8]. But the result – *"the enormous theorem"* – turned out to be much richer. The list of finite simple groups contains 16 + 1 + 1 infinite series:

- *groups of Lie type*:
  $A_n(q)$, $B_n(q)$, $C_n(q)$, $D_n(q)$, $E_6(q)$, $E_7(q)$, $E_8(q)$, $F_4(q)$, $G_2(q)$,
  ${}^2A_n\left(q^2\right)$, ${}^2B_n\left(2^{2n+1}\right)$, ${}^2D_n\left(q^2\right)$, ${}^3D_4\left(q^3\right)$, ${}^2E_6\left(q^2\right)$, ${}^2F_4\left(2^{2n+1}\right)$, ${}^2G_2\left(3^{2n+1}\right)$;
- *cyclic groups of prime order*, $\mathbb{Z}_p$;
- *alternating groups*, A_n, $n \geq 5$;

and 26 *sporadic groups*: M_{11}, M_{12}, M_{22}, M_{23}, M_{24}, J_1, J_2, J_3, J_4, Co_1, Co_2, Co_3, Fi_{22}, Fi_{23}, Fi_{24}, HS, McL, He, Ru, Suz, $O'N$, HN, Ly, Th, B, M.

Note that finite groups have an advantage over the Lie groups in the sense that in the empirical applications any Lie group can be modeled by some finite group, but not vice versa.

3.2 Replacing a unitary group by a finite group

The main non-constructive element of the standard quantum formalism is the unitary group $\mathsf{U}(n)$, a set of cardinality of the continuum.

Formally, the group $\mathsf{U}(n)$ can be replaced by some finite group which is empirically equivalent to $\mathsf{U}(n)$ as follows. From the theory of quantum computing it is known that $\mathsf{U}(n)$ contains a dense finitely generated – and, hence, countable – matrix subgroup $\mathsf{U}_*(n)$. The group $\mathsf{U}_*(n)$ is *residually finite*, i.e., it has a reach set of non-trivial homomorphisms to finite groups.

In essence, it is more natural to assume that, at the fundamental level, there are finite symmetry groups and that the $\mathsf{U}(n)$'s are just continuum approximations of their unitary representations.

The following properties of the finite groups are important for our purposes:

- any finite group is a subgroup of a *symmetric group*,
- any linear representation of a finite group is *unitary*,
- any linear representation is a subrepresentation of some *permutation representation*.

3.3 "Physical" numbers

The basic number system in quantum formalism is the complex field $\mathbb{C}$. This non-constructive field can be obtained as a metric completion of many algebraic extensions of the rational numbers. We consider here constructive numbers that are closely related to the finite groups and are based on two primitives with clear intuitive meanings:

1. *natural numbers* ("counters"): $\mathbb{N} = \{0, 1, \ldots\}$;
2. kth *roots of unity*[3] ("algebraic form of the idea of k-periodicity"): $\mathsf{r}_k \mid \mathsf{r}_k^k = 1$.

These basic concepts are sufficient to represent all physically meaningful numbers.

[2] *The simple groups*, i.e., groups that do not contain nontrivial *normal subgroups*, are "building blocks" for all other groups.
[3] There are k different kth roots of unity. A kth root of unity is called *primitive* if $\mathsf{r}_k^m \neq 1$ for any m, $0 < m < k$.

We start by introducing $\mathbb{N}[\mathrm{r}_k]$, the extension of the *semiring* $\mathbb{N}$ by primitive kth root of unity. $\mathbb{N}[\mathrm{r}_k]$ is a *ring* if $k \geq 2$. This construction allows, in particular, to add *negative numbers* to the naturals: $\mathbb{Z} = \mathbb{N}[\mathrm{r}_2]$ is the extension of $\mathbb{N}$ by the primitive square root of unity. Further, by a standard mathematical procedure, we obtain the kth *cyclotomic field* $\mathbb{Q}(\mathrm{r}_k)$ as the fraction field of the ring $\mathbb{N}[\mathrm{r}_k]$. If $k \geq 3$, then the field $\mathbb{Q}(\mathrm{r}_k)$ is a *dense subfield* of $\mathbb{C}$, i.e., (constructive) cyclotomic fields are empirically indistinguishable from the (non-constructive) complex field. Note that $\mathbb{Q} \cong \mathbb{Q}(\mathrm{r}_2)$.

The importance of the cyclotomic numbers for the constructive quantum mechanics is explained by the following. Let us recall some terms [9]. The *exponent* of a group G is the least common multiple of the orders of its elements. A *splitting field* for a group G is a field that allows to split completely any linear representation of G into irreducible components. A *minimal splitting field* is a splitting field that does not contain proper splitting subfields. Although a minimal splitting field for a given group G may be non-unique, any minimal splitting field is a subfield of some cyclotomic field $\mathbb{Q}(\mathrm{r}_k)$, where k is a divisor of the exponent of G. Thus, to work with any unitary representation of G it is sufficient to use the kth cyclotomic field, where k is related to the structure of G.

3.4 Constructive representations of a finite group

Let a group G act by permutations on a set Ω, $|\Omega| = \mathsf{N}$. If we assume that the elements of Ω are "types" of some discrete entities ("ontological entities", "elements of reality"), then the collections of these entities can be described as elements of the module $H = \mathbb{N}^{\mathsf{N}}$ over the semiring $\mathbb{N}$ with the basis Ω. The decomposition of the action of G on the module H into irreducible components reflects the structure of the invariants of the action. In order for the decomposition to be complete, it is necessary to extend the semiring $\mathbb{N}$ to a splitting field, e.g., to a cyclotomic field $\mathbb{Q}(\mathrm{r}_k)$, where k is a suitable divisor of the exponent of G. With such an extension of the scalars, the module H is transformed into the Hilbert space $\mathcal{H}$ over $\mathbb{Q}(\mathrm{r}_k)$. This construction, with a suitable choice of the permutation domain Ω, allows us to obtain *any representation* of the group G in some invariant subspace of the Hilbert space $\mathcal{H}$. We obtain "quantum mechanics" within an invariant subspace if, in addition to unitary evolutions, projective measurements are also restricted to this subspace.

The above is illustrated in Figure 1 by the example of the natural action of the symmetric group S_N on the set $\Omega = \{e_1, \ldots, e_\mathsf{N}\}$. Note, that any symmetric group is a *rational-representation* group, i.e., the field of rational numbers $\mathbb{Q}$ is a splitting field for S_N.

4 Modeling quantum evolution

The fundamental discrete time $\mathcal{T}$ is represented by an ordered sequence of integers: $\mathcal{T} = \mathbb{N}$ or $\mathcal{T} = \mathbb{Z}$. We define a finite sequence of "instants of observations" as a subsequence of $\mathcal{T}$:

$$[t_0, t_1, \ldots, t_{k-1}, t_k, \ldots, t_n]. \tag{2}$$

The data of the model of the quantum evolution includes the sequence of the length $n+1$ for the states

$$[\rho_0, \rho_1, \ldots, \rho_{k-1}, \rho_k, \ldots, \rho_n] \tag{3}$$

and the sequence of the length n for unitary transitions between observations

$$[U_1, \ldots, U_k, \ldots, U_n]. \tag{4}$$

The standard quantum mechanics presupposes a single unitary evolution, U_k, between observations at the times t_{k-1} and t_k. The *single-step transition probability* takes the form

$$\mathbf{P}_k = \mathrm{tr}\left(U_k \rho_{k-1} U_k^\dagger \rho_k\right). \tag{5}$$

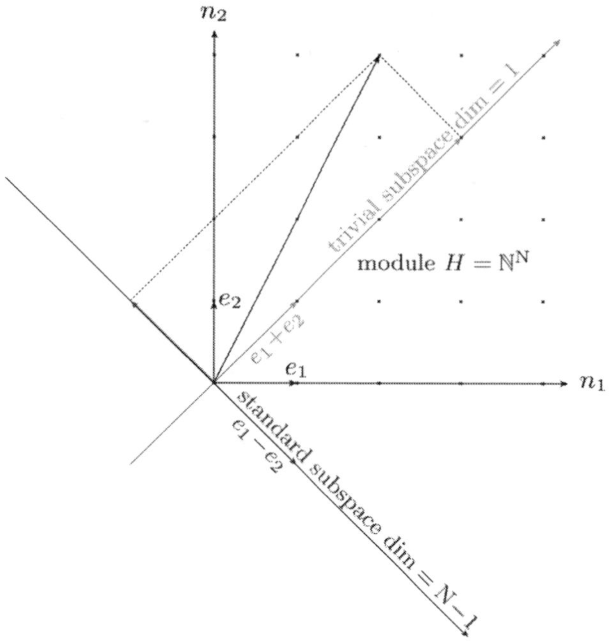

Figure 1. Natural representation of S_N decomposes into two irreducibles: 1D *trivial* and (N − 1)D *standard* representations.
Canonical bases:
in the trivial subspace
$e_1 + e_2 + \cdots + e_N$
in the standard subspace
$e_1 - e_2$
$e_2 - e_3$
$\vdots$
$e_{N-1} - e_N$

The evolution can be expressed via the Hamiltonian: $U_k = e^{-iH(t_k - t_{k-1})}$. In physical theories, Hamiltonians are usually derived from *the principle of least action*, which, like any extremal principle, implies the selection of a small subset of dominant elements in a large set of candidates. Thus it is natural to assume that, in fact, all unitary evolutions take part in the transition between observations with their weights, but only the dominant evolutions are manifested in observations. Therefore, in our model, we use the following modification of the single-step transition probability

$$\mathbf{P}_k = \sum_{m=1}^{M} w_{km} \operatorname{tr}\left(U_{k,m}\rho_{k-1}U_{k,m}^{\dagger}\rho_k\right), \tag{6}$$

where $U_{k,m} = \mathrm{U}(g_m)$, $g_m \in \mathrm{G}$; $\mathrm{G} = \{g_1, \ldots, g_M\}$ is a finite group; U is a unitary representation of G; w_{km} is the weight of the mth group element at the kth transition.

The operators $U_{k,m}$ that maximize $\operatorname{tr}\left(U_{k,m}\rho_{k-1}U_{k,m}^{\dagger}\rho_k\right)$ will be called *dominant evolutions*.

The *single-step entropy* is defined as $\Delta\mathbf{S}_k = -\log \mathbf{P}_k$. (7)

The continuum approximation of (7) leads to a *Lagrangian* $\mathcal{L}$. Taking the logarithm of the probability of the whole trajectory, $\mathbf{P}_{0\to n} = \prod_{k=1}^{n} \mathbf{P}_k$, we arrive at the *entropy of the trajectory* $\mathbf{S}_{0\to n} = \sum_{k=1}^{n} \Delta\mathbf{S}_k$, the continuum approximation of which is the *action* $\mathcal{S} = \int \mathcal{L} dt$.

4.1 Continuum approximation of a discrete model

A continuum approximation of the above model requires the following simplifying assumptions:

- The sequence (2) should be replaced by a continuous time interval $[t_0, t_n] \subseteq \mathbb{R}$.
- The sequences (3) and (4) are to be replaced by continuous functions of time, $\rho(t)$ and $U(t)$.
- The relation $\operatorname{tr}\left(\rho^2\right) = 1$ is necessary to ensure the continuity of the probability. This relation holds only for pure states $\rho = |\psi\rangle\langle\psi|$. So, we will consider ψ instead of ρ.
- Assuming that U belongs to a unitary representation of a Lie group, we use the Lie algebra approximation, $U \approx \mathbb{1} + iA$, where $A = A(t)$ is a function the values of which are Hermitian matrices.
- We introduce derivatives and use the linear approximations $\Delta A \approx \dot{A}\Delta t$ and $\Delta\psi \approx \dot{\psi}\Delta t$.

Applying these assumptions and approximations to the single-step entropy (7) and taking the infinitesimal limit we obtain the Lagrangian:

$$\mathcal{L} = \underbrace{\left\langle\psi\left|\dot{A}^2\right|\psi\right\rangle - \left\langle\psi\left|\dot{A}\right|\psi\right\rangle^2}_{\text{dispersion of } \dot{A} \text{ in the state } \psi} - \mathrm{i}\left(\left\langle\dot{\psi}\left|\dot{A}\right|\psi\right\rangle - \left\langle\psi\left|\dot{A}\right|\dot{\psi}\right\rangle + 2\left\langle\psi\left|\dot{A}\right|\psi\right\rangle\left\langle\psi\mid\dot{\psi}\right\rangle\right) - \left\langle\psi\mid\dot{\psi}\right\rangle^2.$$

4.2 Dominant unitary evolutions in a symmetric group

The dominant evolutions among the states represented by the vectors from the module $H = \mathbb{N}^{\mathsf{N}}$ for the group S_{N} can be computed as follows.

Let $|n\rangle = \begin{pmatrix} n_1 \\ \vdots \\ n_{\mathsf{N}} \end{pmatrix}$, $|m\rangle = \begin{pmatrix} m_1 \\ \vdots \\ m_{\mathsf{N}} \end{pmatrix}$, $|1\rangle = \begin{pmatrix} 1 \\ \vdots \\ 1 \end{pmatrix}$ be N-dimensional vectors with natural components.

The Born probabilities for the pair $|n\rangle$ and $|m\rangle$ are

$$\mathbf{P}_{\mathrm{nat}}\left(|n\rangle, |m\rangle\right) = \frac{\langle n \mid m\rangle^2}{\langle n \mid n\rangle\langle m \mid m\rangle} \quad \text{– natural representation,}$$

$$\mathbf{P}_{\mathrm{std}}\left(|n\rangle, |m\rangle\right) = \frac{\left(\langle n \mid m\rangle - \frac{1}{\mathsf{N}}\langle n \mid 1\rangle\langle 1 \mid m\rangle\right)^2}{\left(\langle n \mid n\rangle - \frac{1}{\mathsf{N}}\langle n \mid 1\rangle\langle 1 \mid n\rangle\right)\left(\langle m \mid m\rangle - \frac{1}{\mathsf{N}}\langle m \mid 1\rangle\langle 1 \mid m\rangle\right)} \quad \text{– standard representation.} \tag{8}$$

Let R_a denote the permutation (as well as its representation), that sorts the components of the vector $|a\rangle$ in some order. It is not hard to show that the unitary operator $U = R_m^{-1}R_n$ maximizes the probability $\mathbf{P}_*(U|n\rangle, |m\rangle)$, where the permutations R_n and R_m sort the vectors $|n\rangle$ and $|m\rangle$ *identically* in the case of a natural representation, and either *identically* or *oppositely*, depending on the value of the numerator in (8), in the case of a standard representation.

4.3 Energy of permutation

The Planck formula, $E = h\nu$, relates the energy to the frequency. This relation is reproduced by the quantum-mechanical definition of the energy as an eigenvalue of the Hamiltonian, $H = \mathrm{i}\hbar\ln U$, associated with a unitary transformation. Consider the energy spectrum of a unitary operator defined by a permutation. Let p be a permutation of the cycle type $\left\{\boldsymbol{\ell}_1^{m_1}, \ldots, \boldsymbol{\ell}_k^{m_k}, \ldots, \boldsymbol{\ell}_K^{m_K}\right\}$, where $\boldsymbol{\ell}_k$ and m_k represent lengths and multiplicities of cycles in the decomposition of p into disjoint cycles. A short calculation shows that the Hamiltonian of the permutation p has the following diagonal form

$$H_p = \begin{pmatrix} \mathbb{1}_{m_1}\otimes H_{\boldsymbol{\ell}_1} & & \\ & \ddots & \\ & & \mathbb{1}_{m_K}\otimes H_{\boldsymbol{\ell}_K} \end{pmatrix}, \quad \text{where} \quad H_{\boldsymbol{\ell}_k} = \frac{1}{\boldsymbol{\ell}_k}\begin{pmatrix} 0 & & & \\ & 1 & & \\ & & \ddots & \\ & & & \boldsymbol{\ell}_k - 1 \end{pmatrix}.$$

We shall call the least nonzero energy of a permutation the *base energy*:

$$\boldsymbol{\varepsilon} = \frac{1}{\max\left(\boldsymbol{\ell}_1, \ldots, \boldsymbol{\ell}_K\right)}. \tag{9}$$

The simulation shows that the base (*"ground state"*, *"zero-point"*, *"vacuum"*) energy is statistically more significant than other energy levels.

4.4 Monte Carlo simulation of dominant evolutions

Figure 2 shows several dominant evolutions for the standard representation of the groups S_{100} and S_{2000}. Each graph represents the time dependencies of the Born probabilities for the dominant evolutions between four randomly generated pairs of natural vectors. The dominant evolutions are marked by labeling their peaks with their base energies: $\boldsymbol{\varepsilon} \in \left\{\frac{1}{61}, \frac{1}{69}, \frac{1}{72}, \frac{1}{98}\right\}$ and $\boldsymbol{\varepsilon} \in \left\{\frac{1}{1416}, \frac{1}{1789}, \frac{1}{1939}, \frac{1}{1972}\right\}$ for S_{100} and S_{2000}, respectively. We see that with increasing the group size, non-dominant evolutions become almost invisible against the sharp peaks of the dominant evolutions.

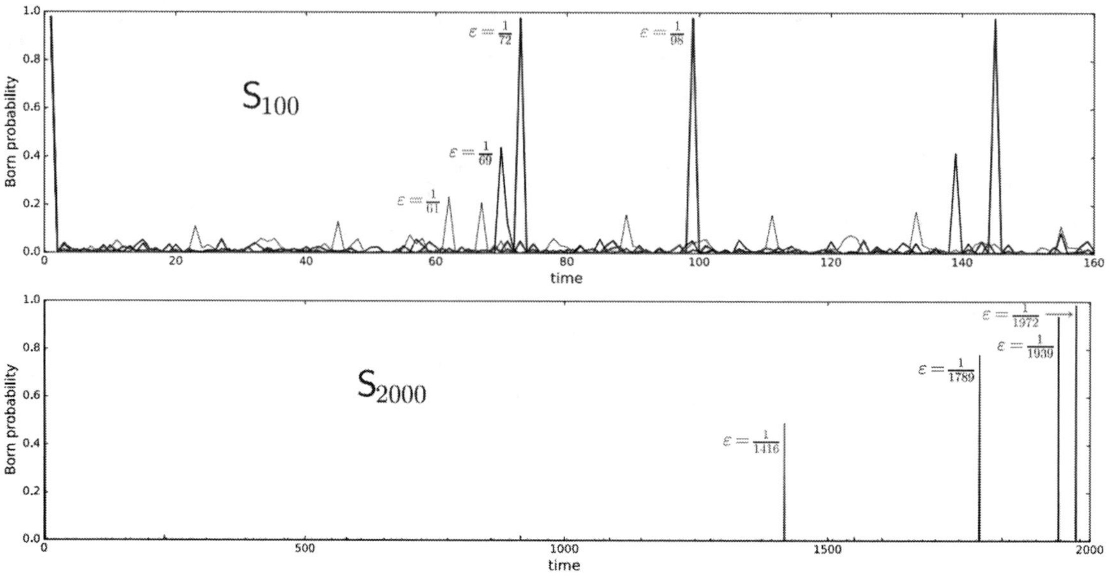

Figure 2. Dominant evolutions between randomly generated states. Born probability vs time

5 Summary

1. A constructive version of the quantum formalism can be formulated in terms of projections of permutations of finite sets onto invariant subspaces.
2. The quantum randomness is a consequence of the fundamental impossibility of tracing the individuality of indistinguishable entities in their evolution.
3. The natural number systems for the quantum formalism are the cyclotomic fields and the field of complex numbers is just their non-constructive metric completion.
4. The observable behavior of a quantum system is determined by the dominants among all possible quantum evolutions.
5. The principle of least action is a continuum approximation of the principle of selection of the most probable trajectories.

References

[1] A.M. Gleason, Indiana Univ. Math. J. **6**, 885–893 (1957)
[2] M. Van Raamsdonk, Gen. Rel. Grav. **42**, 2323–2329 (2010)
[3] J. Maldacena and L. Susskind, Fortschr. Phys. **61** 781–811 (2013)
[4] C. Cao, S.M. Carroll, and S. Michalakis, Phys. Rev. D **95**, 024031 (2017)
[5] V.V. Kornyak, EPJ Web of Conferences **108**, 01007 (2016)
[6] V.V. Kornyak, Mathematical Modelling and Geometry **3**, No 1, 1–24 (2015)
[7] V.V. Kornyak, Phys. Part. Nucl. **44**, No 1, 47–91 (2013)
[8] R. Solomon, Bull. Amer. Math. Soc. (N.S.) **38**, No 3, 315–352 (2001)
[9] Charles W. Curtis, Irving Reiner, *Representation Theory of Finite Groups and Associative Algebras* (John Wiley & Sons, 1962). Reedition by AMS Bookstore, 2006.

EPJ Web of Conferences **173**, 01008 (2018) https://doi.org/10.1051/epjconf/201817301008
Mathematical Modeling and Computational Physics 2017

Mathematical Modeling of Resonant Processes in Confined Geometry of Atomic and Atom-Ion Traps

Vladimir S. Melezhik[1,2,⋆]

[1] *Bogoliubov Laboratory of Theoretical Physics, Joint Institute for Nuclear Research, Dubna, Moscow Region, 141980, Russian Federation*
[2] *Peoples' Friendship University of Russia (RUDN University), 6 Miklukho-Maklaya St., Moscow, 117198, Russian Federation*

Abstract. We discuss computational aspects of the developed mathematical models for resonant processes in confined geometry of atomic and atom-ion traps. The main attention is paid to formulation in the nondirect product discrete-variable representation (npDVR) of the multichannel scattering problem with nonseparable angular part in confining traps as the boundary-value problem. Computational efficiency of this approach is demonstrated in application to atomic and atom-ion confinement-induced resonances we predicted recently.

1 Introduction

During the last three decades one can observe an impressive development of the physics of ultracold atoms [1] and cold ions [2]. Different aspects of this investigations attract big interest from the side of theoreticians and experimentalists. Thus, experimentalists have got a chance to work here with deterministically prepared quantum systems [3] with precise control of interparticle interaction, particle states and particle number [2, 3]. It opens a possibility for quantum simulation with fully controlled few-body systems in ultracold quantum gases [4–6]. The control of interparticle interaction is performed in quantum gases with the help of magnetic Feshbach resonances [1] which, in confined geometry of atomic traps, transform into so-cold confinement-induced resonances (CIRs) [7–13].

The mathematical modeling of resonant processes in the confined geometry of atomic traps is a special computational problem due to the strong nonseparability of the angular part of the system wave-function, what is the consequence of the strong particle-trap interaction. Therefore, the conventional theory is no longer valid here and the development of the low-dimensional theory, including the influence of the confinement, is needed. In our works we have developed computational methods [9, 10, 14, 15] for pair collisions in tight atomic waveguides and have found several novel effects in its applications: the CIRs in multimode regimes including effects of transverse excitations and deexcitations [10], the so-called dual CIR yielding a complete suppression of quantum scattering [9], and resonant molecule formation with transferring energy relies to center-of-mass excitation while forming molecules [16]. The last effect was recently confirmed in the Heidelberg experiment [3]. Our calculations have also been used for planning and interpretation of the Innsbruck experiment on

⋆e-mail: melezhik@theor.jinr.ru

investigation of CIRs in ultracold Cs gas [8]. We have also calculated the Feshbach resonance shifts and widths induced by atomic waveguides [11, 12] and predicted dipolar CIRs [13].

Recently, hybrid systems of laser-cooled trapped ions and ultracold atoms combined in a single experimental setup have emerged as a new platform for fundamental research in quantum physics. These experiments aim to inherit the most important advantages of such hybrid atom-ion systems and to show new emerging features providing a platform for both fundamental research in quantum physics and upcoming quantum technologies [17].

Here we discuss computational aspects of the methods we developed for low-dimensional ultracold processes with more extended presentation of results obtained recently for hybrid atom-ion confined scattering in waveguide-like traps. Specifically, this concerns the prediction of the atom-ion CIRs [18]. In the section 2 we formulate a typical scattering problem which models a resonant scattering in confined geometry of a waveguide-like trap. By using as an example the confined resonant scattering, we give a brief discussion of a nondirect product discrete variable representation (npDVR) in application to coupled 3D Schrödinger equations in the section 3. The formulation of the scattering problem as a boundary-value problem in the npDVR is also given here. In the section 4 applications of these methods to the quantitative description of atomic and atom-ion CIRs are discussed. Concluding remarks are given in the last section.

2 Resonant Scattering in Atomic and Atom-Ion Traps

The mathematical modeling of resonant scattering in confined geometry of atomic and atom-ion traps requires the development of efficient computational methods for integration of the system of 3D Schrödinger-like equations

$$\left(\left[-\frac{\hbar^2}{2\mu}\Delta_{\mathbf{r}} + \frac{1}{2}\mu(\omega_x^2 x^2 + \omega_y^2 y^2)\right]\hat{I} + \hat{V}(r)\right)|\psi(\mathbf{r})\rangle = E|\psi(\mathbf{r})\rangle, \tag{1}$$

strongly coupled by the matrix of effective interatomic potential $\hat{V}(r)$ ($\hat{I}$ is unit matrix here) and the derivation of the wave-function

$$|\psi(\mathbf{r})\rangle = \sum_{\alpha}\psi_\alpha(\mathbf{r})|\alpha\rangle\,, \qquad \alpha = \{e, c = 1, \dots\},$$

satisfying the scattering asymptotic

$$\psi_e(\mathbf{r}) = \left(e^{ik_0 z} + f_e e^{ik_0|z|}\right)\Phi_0(x, y)\,, \qquad \psi_c(\mathbf{r}) \to 0, \tag{2}$$

at $|z| \to +\infty$ for the fixed collision energy E, where

$$\mathbf{r} = (x, y, z) = (\rho\cos\phi, \rho\sin\phi, z) = (r\sin\theta\cos\phi, r\sin\theta\sin\phi, r\cos\theta)$$

is the relative variable between the colliding atoms [9–13] or the atom variable relative the heavy ion [18]. Here, $f_e(E)$ is the desired scattering amplitude, $\Phi_0(x, y)$ is the wave function of the ground-state of the 2D harmonic oscillator $\frac{1}{2}\mu(\omega_x^2 x^2 + \omega_y^2 y^2)$ describing the interaction of the atoms with the confining trap and $k_0 = \sqrt{2\mu(E - \hbar\omega_\perp)}/\hbar = \sqrt{2\mu E_\parallel}/\hbar \to 0$ is the relative momentum of two atoms in the open channel e, where $\omega_\perp = (\omega_x + \omega_y)/2$ and μ is the reduced mass of two colliding atoms [9–13] or the mass of atom colliding with heavy ion [18].

In the case of the atom-atom interaction a four-channel potential $\hat{V}(r)$ describing three lowest magnetic Feshbach resonances in Cs [19] (with $c = 1, 2, 3$) was used in works [11, 12]. Some of these resonances have been used in experimental investigation of the Cs confinement-induces resonances in anisotropic waveguide-like traps [8]. In the case of atom-ion confined scattering the problem (1, 2) was reduced to one channel scattering ($\alpha = e$) with scalar atom-ion potential $\hat{V}(r) \to V(r) \sim C_4/r^4$ [18].

Note, that in the presence of an anisotropic harmonic trap ($\omega_x \neq \omega_y$) the problem (1, 2) becomes non-separable in the 3D space $\{\rho, z, \phi\}$. To resolve the problem in the case of atom-atom confined scattering [11, 12] we have to integrate the system of four coupled 3D Schrödinger-like equations (1). In an isotropic trap ($\omega_x = \omega_y$) the problem admits separation of the angle ϕ and reduction to 2D case. In free space ($\omega_x = \omega_y = 0$) the angular part separates and the problem reduces to four coupled radial equations considered in [19]. The choice of the potential $\hat{V}(r)$ in the tensorial form ($\alpha = \{e, c\}$) with the diagonal terms $V_{\alpha\alpha}(r)$ as square-wells enabled the derivation [19] of an analytic model for describing Feshbach resonances in free space ($\omega_x = \omega_y = 0$). However, to develop an efficient computational scheme for the integration of the problem (1, 2) in a confining trap ($\omega_x \neq 0, \omega_y \neq 0$) is a challenging task due to sharp jumps at the edge of the interaction in the diagonal terms $V_{\alpha\alpha}(r)$ chosen in the form of square-well potential [11, 12]. In the case of atom-ion confined scattering the tensorial structure of the potential $\hat{V}(r)$ is absent but the need arises to get accurate solution for the long-range scalar atom-ion interaction $V(r) \sim C_4/r^4$ [18].

3 Method

The key feature of the system of equations (1) is the strong nonseparability of the angular part of the desired wave-function $|\psi(\mathbf{r})\rangle$ due to the strong coupling of the partial waves defined by the atom-trap interaction $\hat{W}(\mathbf{r}) = \frac{1}{2}\mu(\omega_x^2 x^2 + \omega_y^2 y^2)$. To solve the difficulty, the approximation of the angular part is done within the nondirect product discrete-variable representation (npDVR) which was suggested [20–23] and developed in our works [9–13, 16, 22–31]. It was shown that 2D npDVR is much more efficient computationally as compared to the conventional partial wave analysis due to its fast convergence and flexibility: there is no need for laborious calculation of the matrix elements under the change of the form of the interactions because any local interaction is a diagonal one in the npDVR.

In npDVR we have formulated the multichannel scattering problem with nonseparable angular part as a boundary-value problem and successfully applied this scheme to scattering in confined geometry of atomic and hybrid atom-ion traps [10–13, 18]. Here we discuss computational aspects of this approach and recent results obtained in quantitative analysis of the atom-ion CIRs.

We have also developed the efficient splitting-up-method in npDVR in application for the time-dependent few-dimensional Schrödinger equation [9, 16, 22–29]. Detailed description of this method one can find in our previous paper published in the EPJ (Web of conferences) [15].

3.1 Nondirect Product Discrete-Variable Representation

Different kinds of the 1D DVR or Lagrange-meshes are widely used in quantum mechanics computations [32–35] due to the simplicity (DVR leads to diagonal form for the interaction and compact form for the kinetic energy operator) and efficiency (fast convergence and stability) of this approach. In the work [20] we have theoretically estimated and confirmed by numerical experiment that the error very quickly decreases $\sim 1/(N+1)!$ with an increase in the number of DVR basis functions N. It ensures fast convergence of the 1D DVR. However, an extension of this representation to the 2D case (two angles θ and ϕ of the unit sphere) is a nontrivial problem. Actually, the simple idea to construct the 2D DVR as a direct product of two 1D DVRs leads to essential complication of the matrix of the angular part of the kinetic energy operator [33, 36]. As a result, the advantages of the 1D DVR, its simplicity and efficiency, are lost [33]. An alternative way to construct 2D DVR on an unit sphere is to use the spherical harmonics defined on a two-dimensional grid. However, in this case it becomes not possible to satisfy the orthogonality conditions for all the elements of the fixed set of this basis on the chosen grid [36, 37]. To overcome this difficulty we have suggested [22, 23] to use the basis of the orthogonalized combinations of the spherical harmonics on the two-dimensional grid over θ and ϕ variables. It turned out that this idea was very efficient for the time-dependent, as well as, stationary Schrödinger equations with two [10–13, 18], three [22–28] and four [9, 16, 29, 30] nonseparable spatial variables. Our idea how to construct the 2D npDVR [22, 23] was also successfully extended for computing vibrational levels of four-atom molecules [36] where it has got the title *non-direct product* DVR (npDVR). An alternative variant of 2D npDVR with non-direct product angular grid coinciding on the unit sphere with the nodes of the Lebedev quadratures was suggested in [37].

By applying the 2D npDVR for Eq. (1) we get the system of $4N$ Schrödinger-like coupled equations with respect to the unknown vector $\mathbf{u}(r) = \{\sqrt{\lambda_j}u_j^\alpha(r)\}$ (here $j = 1, \ldots, N$; $\alpha = e, c$)

$$\left(\hat{T}(r) + \hat{W}(r) + \hat{V}(r)\right)\mathbf{u}(r) = E\mathbf{u}(r) \tag{3}$$

where

$$T_{jj'}^{\alpha\alpha'} = -\frac{\hbar^2}{2\mu}\left[\delta_{jj'}\frac{d^2}{dr^2} - \frac{1}{r^2}\sum_{\nu=1}^{N} Y_{j\nu}l(l+1)(\mathbf{Y}^{-1})_{\nu j'}\right]\delta_{\alpha\alpha'}, \tag{4}$$

$$W_{jj'}^{\alpha\alpha'} = \frac{1}{2}\mu\left(\omega_x^2 x_j^2 + \omega_y^2 y_j^2\right)\delta_{jj'}\delta_{\alpha\alpha'}, \tag{5}$$

$$V_{jj'}^{\alpha\alpha'} = V_{\alpha\alpha'}(r)\,\delta_{jj'}\,. \tag{6}$$

Here $x_j = r\sin\theta_j\cos\phi_j$ and $y_j = r\sin\theta_j\sin\phi_j$ and the elements $u_j^\alpha(r)$ of the vector $\mathbf{u}(r)$ coincide with the values $r\psi_\alpha(r, \Omega_j)$ of desired wave function on N angular grid points $\Omega_j = \{\theta_j, \phi_j\}$. The matrix $Y_{j\nu}$ and its inverse $(Y^{-1})_{\nu j}$ are defined on the grid Ω_j as described in [9, 10, 15, 24] (here one can also find the procedure of construction of the 2D angular grid Ω_j and the definition of the weights λ_j).

Thus, by using the 2D npDVR we transform the initial scattering problem (1, 2) to the system of the Schrödinger-like coupled equations (3) with the following asymptotic

$$u_j^e(r) = r\left(e^{ik_0 z_j} + f_e e^{ik_0|z_j|}\right)\Phi_0(x_j, y_j)\,,\;\; u_j^c(r) = 0 \tag{7}$$

at $|z_j| = |r\cos\theta_j| \to +\infty$ and

$$u_j^\alpha(r) = 0, \tag{8}$$

at $r = 0$, which follows from the finiteness of the desired wave function $\psi_\alpha(\mathbf{r})$ ($\alpha = e, c$) at $r \to 0$.

3.2 Boundary conditions: reduction scattering problem to boundary-value problem

The problem (3-8) is defined on the semi-axis $r \in [0, +\infty)$ which is replaced by the interval $[0, r_m]$. The right edge of the interval r_m and any neighbour point r_{m-1} to the left of the edge must be chosen such that some of the points $z_j^m = r_m\cos\theta_j$ and $z_j^{m-1} = r_{m-1}\cos\theta_j$ (let say for $j_\theta \le j_m$ and for $j_\theta \ge N_\theta - j_m + 1$, here j_m is a constant) fall into the asymptotic region $|z_j^m| = |r_m\cos\theta_j| \to +\infty$ where the wave function $u_j^e(r)$ satisfies the asymptotic formulae (7) (see Fig. 1).

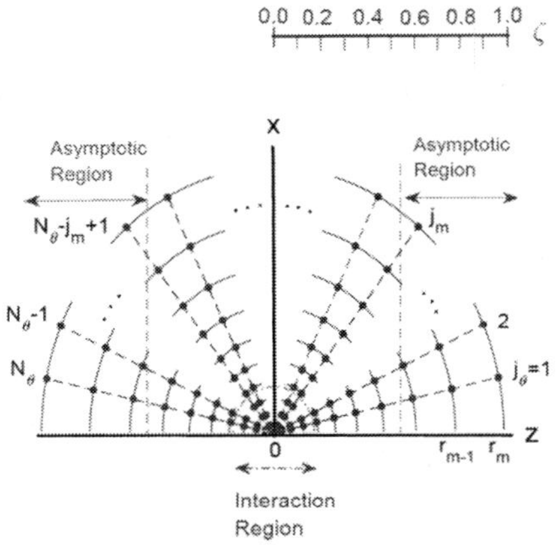

Figure 1. The schematic figure of the grid points in $x - z$ plane. The up horizontal axis shows the corresponding values of ζ.

By using two equations (7) in the points r_m and r_{m-1} we eliminate the unknown amplitude f_e and construct the two-point boundary condition at the right edge of the interval $[0, r_m]$

$$\begin{aligned} u_j^e(r_m) - u_j^e(r_{m-1}) e^{ik_0(|z_j^m|-|z_j^{m-1}|)} \frac{r_m \Phi_0(x_j^m, y_j^m)}{r_{m-1}\Phi_0(x_j^{m-1}, y_j^{m-1})} &= r_m\{e^{ik_0 z_j^m} - e^{ik_0(z_j^{m-1}-|z_j^{m-1}|+|z_j^m|)}\}\Phi_0(x_j^m, y_j^m) \\ &\quad (\text{if } j_\theta \le j_m \text{ or } j_\theta \ge N_\theta - j_m + 1), \\ u_j^e(r_m) &= 0 \quad (\text{if } j_m < j_\theta < N_\theta - j_m + 1), \\ u_j^c(r_m) &= 0, \end{aligned} \tag{9}$$

where $x_j^m = r_m \sin\theta_j \cos\phi_j$ and $y_j^m = r_m \sin\theta_j \sin\phi_j$.

Thus, the initial scattering problem (1, 2) is reduced to the boundary-value problem for the system of ordinary differential equations of second order (3) with the boundary conditions (8) and (9) which do not contain the unknown scattering amplitude f_e. After integration of the problem (3), (8) and (9) one can find the scattering amplitude f_e by mapping the calculated wave function $\psi_\alpha(r, \Omega_j)$ at the points $|z_j^m| = |r_m \cos\theta_j| \to +\infty$ with the asymptotic boundary condition (7). We solve the system of Eqs. (3) on a quasi-uniform radial grid [22, 27]

$$r_n = r_m \frac{e^{\gamma\zeta_n} - 1}{e^\gamma - 1}, \qquad n = 1, 2, \dots, N_r \tag{10}$$

of N_r grid points $\{r_n\}$ defined by mapping $r_n \in [0, r_m](r_m \to +\infty)$ onto the uniform grid $\zeta_n \in [0, 1]$ with the equidistant distribution $\zeta_n - \zeta_{n-1} = 1/N_r$. One can achieve a suitable distribution of the grid points for a specific interatomic and confining potential by varying N_r and the parameter $\gamma > 0$.

To integrate the boundary-value problem (3), (8) and (9) one can apply efficient algorithms. By using high order (six-order) finite-difference approximation for the radial derivatives on quasi-uniform grid we arrive at a system of algebraic equations with block-band structure which can be solved by the LU decomposition [38] or the sweep (also known as the Thomas algorithm [39]) method [40], which are very efficient for speed and memory.

4 Results

The high computational efficiency of the mathematical model (3,8) and (9) was demonstrated in application to the resonant scattering in confined geometry of waveguide-like atomic traps [10–12]. In Fig. 2 we present calculated atomic probability density distribution $|\psi_e(x,z)|^2$ in the vicinity of d-wave Cs Feshbach resonance (occurs at $B_0 = 47.8$ G [19]) for $B = 47.9$ G in the trap with $\omega_\perp = \omega_x = \omega_y = 14$ kHz [8]. This resonance is developed as a peak at the region $r \lesssim \bar{a}$ of the action of the square-well potential between the atoms in the plot of probability density distribution with a rapid oscillating behaviour over r (which is not clear in the figure), due to the constitution of quasi-molecular states. In the asymptotic region $|z| \to +\infty$ we observe slow oscillations of the channel wave function $\psi_e(x,z)$ (2) with the period of oscillations $2\pi/k_0 \to \infty$ defined by the very small colliding energy $k_0 = \sqrt{2\mu E_{\parallel}}/\hbar \to 0$. Fig. 3 illustrates the convergence of the method with respect

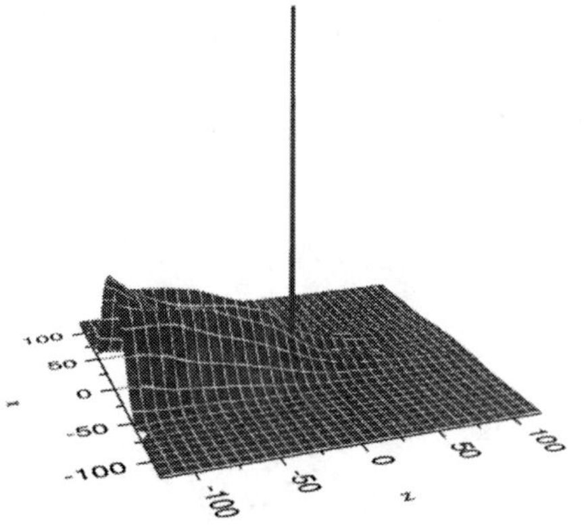

Figure 2. The calculated probability density distribution $|\psi_e(x,z)|^2$ in the entrance channel in the vicinity of d-wave Cs Feshbach resonance at $B = 47.9$ G. Variables are given in the units $\bar{a}$, $N = 30$ and $N_r = 1.0 \times 10^6$.

to the number of grid points N over the angular variable on a typical example of calculating the T coefficient $T = |1 + f_e|^2$ near s-wave Cs Feshbach resonance (which occurs at $B_2 = -11.16$ G [19]) for $B = 14.14$ G. It demonstrates rather fast convergence of the 2D npDVR over angular grid points N despite the strong coupling over angular variable provided by the atom-trap interaction potential $1/2\mu\omega_{\perp}^2 r^2 \sin^2\theta$ in a wide area of variation of the interatomic distance r. It is shown that for getting absolute accuracy on the level 10^{-3} it is enough to keep about $N = 10$ of equations in the system (3). Further computations with increasing N demonstrate monotonic convergence to more accurate T coefficient. To reach the level of accuracy $\simeq 10^{-4}$ it needs to increase N up to $\simeq 30$.

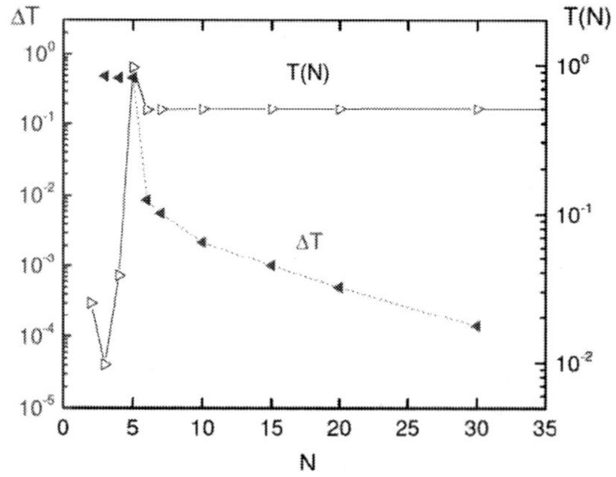

Figure 3. The dependence on the number of angular grid points N of the transmission coefficient $T(N)$ and absolute error $\Delta T(N) = |T(N) - T(N = 40)|$ in $T(N)$ due to npDVR. This results have been obtained for $B = 14.14$ G, $N_r = 1.0 \times 10^6$ and $r_m = 100\bar{a}$.

With this method the atom-ion CIRs were predicted and quantitatively investigated [18]. In Fig. 4 we present calculated transmission coefficients $T(R^*/a_{\perp}, a_{\perp}/a_{3D})$ for the atom-ion scattering in confined geometry of hybrid atom-ion trap $\omega_{\perp} = \omega_x = \omega_y$ as a function of ratios of the range of atom-ion interaction R^* to the width of atomic trap $a_{\perp} = \sqrt{\hbar/(\mu\omega_{\perp})}$ and $a_{\perp}$ to the atom-ion scattering length a_{3D} in free space. The minima in the coefficient T indicate the positions of the atom-ion CIRs. Before this investigation the CIR were known only for confined atomic scattering at the point $a_{\perp}/a_{3D} = 1.46$ for $R^*/a_{\perp} \to 0$ (see arrow in Fig. 4) [7–13]. Our investigation [18] has predicted conditions for CIR appearance in the confined atom-ion scattering. It was shown that due to longer tail of the atom-ion interaction with respect to atom-atom interaction, the limitation $R^*/a_{\perp} \to 0$ characterizing the atomic

CIR is overcame in hybrid atom-ion confined systems. Due to the importance of the CIRs for controlling the interparticle interactions further investigations are in demand. Specifically, the effect of the ion-micromotion induced heating is most actual unsolved problem. It asks for additional development of efficient computational methods [2]. Our computational schemes are look very promising in this respect.

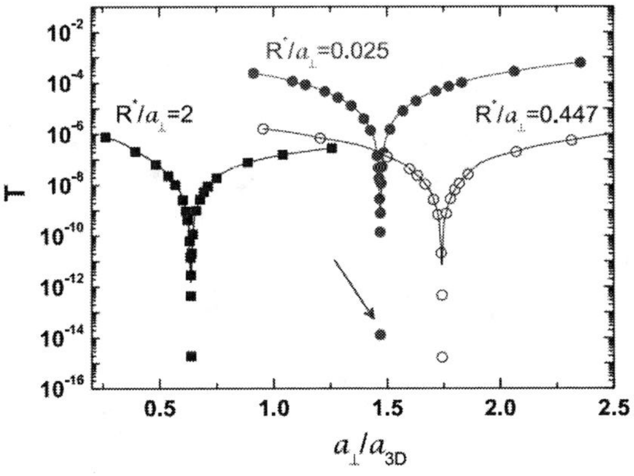

Figure 4. The transmission coefficients $T(R^*/a_\perp, a_\perp/a_{3D})$ calculated for the confined pair Li-Yb$^+$ collisions for different values $R^*/a_\perp$ and $a_\perp/a_{3D}$.

5 Conclusion

In this lecture we have considered computational aspects of the mathematical models we developed recently for ultracold two-body resonant processes in atomic and hybrid atom-ion traps. The main attention is paid to formulation in the npDVR of the multichannel scattering problem with nonseparable angular part in the confining traps as a boundary-value problem. The computational efficiency of this approach is demonstrated in application to atomic and atom-ion confinement-induced resonances we predicted recently.

The computational efficiency of our mathematical models achieved in quantitative analysis of different resonant processes and the supposed transparent procedure for its parallelization make the developed techniques very promising in application to other challenging problems of low-dimensional few-body physics. Note here the computationally demanding, and actual in physics of quantum gases and ions, problem of the impact of the ion-micromotion induced heating on the surrounding atomic gas in hybrid atom-ion ultracold systems [2, 18].

Acknowledgement

The work was financially supported by the Ministry of Education and Science of the Russian Federation (Agreement No. 02.a003.21.0008).

References

[1] C. Chin, R. Grimm, P. S. Julienne, and E. Tiesinga, Rev. Mod. Phys. **82**, 1225–1286 (2010)

[2] D. Leibfried, R. Blatt, C. Monroe, and D. Wineland, Rev. Mod. Phys. **75**, 281–324 (2003)

[3] S. Sala, G. Zürn, T. Lompe, A. N. Wenz, S. Murmann,F. Serwane, S. Jochim, A. Saenz, Phys. Rev. Lett. **110**, 203202–1–5 (2013)

[4] I. Bloch, J. Dalibard, and S. Nascimbene, Nature Physics **8**, 267–276 (2012)

[5] I. M. Georgescu, S. Ashhab, and F. Nori, Rev. Mod. Phys. **86**, 153-186 (2014)

[6] M. Dalmonte and S. Montangero, Contem. Phys. **57**, 388–412 (2016)

[7] M. Olshanii, Phys. Rev. Lett. **81**, 938–941 (1998)
[8] E. Haller, M. J. Mark, R. Hart, J. G. Danzl, L. Reichsöllner, V. Melezhik, P. Schmelcher, and H. C. Nägerl, Phys. Rev.Lett. **104**, 153203–1–4 (2010)
[9] V. S. Melezhik, J. I. Kim, and P. Schmelcher, Phys. Rev. A **76**, 053611–1–15 (2007)
[10] S. Saeidian, V. S. Melezhik, and P. Schmelcher, Phys. Rev. A **77**, 042721–1–15 (2008)
[11] S. Saeidian, V. S. Melezhik, and P. Schmelcher, Phys. Rev. A **86**, 62713–1–9 (2012)
[12] S. Saeidian, V. S. Melezhik, and P. Schmelcher, Journal of Physics B **48(15)**, 155301–1–10 (2015)
[13] P. Giannakeas, V. S. Melezhik, and P. Schmelcher, Phys. Rev. Lett. **111**, 183201–1–5 (2013)
[14] V. S. Melezhik, *Lecture Notes in Computer Science 7125* (Springer, Heidelberg Dordrecht London New-York, 2012) pp. 94–107.
[15] V. S. Melezhik, EPJ Web of Conf. **108**, 01008–1–10 (2016)
[16] V. S. Melezhik and P. Schmelcher, New Journal of Physics **11**, 073031–1–10 (2009)
[17] A. Härter and J. H. Denschlag, Contemp. Phys. **55**,33–45 (2014)
[18] V. Melezhik and A. Negretti, Phys. Rev. A **94**, 022704–1–8 (2016)
[19] A. D. Lange, K. Pilch, A. Prantner, F. Ferlaino, B. Engesser, H. C. Nägerl, R. Grimm, and C. Chin, Phys. Rev. A **79**, 013622-1–9 (2009)
[20] V. S. Melezhik, J. Comp. Phys. **92**, 67–81 (1991)
[21] V. S. Melezhik, Phys. Rev. A **48**, 4528–4538 (1993)
[22] V. S. Melezhik, Phys. Lett. A **230**, 203–208 (1997)
[23] V. S. Melezhik, *Atoms and Molecules in Strong External Fields* (Plenum, New-York and London, 1998) pp. 89–94.
[24] V. S. Melezhik and D. Baye, Phys. Rev. C **59**, 3232–3239 (1999)
[25] V. S. Melezhik and P. Schmelcher, Phys. Rev. Lett. **84**, 1870–1873 (2000)
[26] V. S. Melezhik and D. Baye, Phys. Rev. C **64**, 054612-1-11 (2001)
[27] P. Capel, D. Baye, and V. S. Melezhik, Phys. Rev. C **68**, 014612-1–13 (2003)
[28] V. S. Melezhik, J. S. Cohen, and Chi-Yu Hu, Phys. Rev. A **69**, 032709-1–13 (2004)
[29] J. I. Kim, V. S. Melezhik, and P. Schmelcher, Phys. Rev. Lett. **97**, 193203-1–4 (2006)
[30] J. I. Kim, V. S. Melezhik, and P. Schmelcher, Prog. Theor. Phys. Suppl. **166**, 159–168 (2007)
[31] E. A. Koval, O. A. Koval, and V. S. Melezhik, Phys. Rev. A **89**, 052710-1–9 (2014)
[32] J. V. Lill, G. A. Parker, and J. C. Light, Chem. Phys. Lett. **89**, 483–489 (1982)
[33] J. C. Light and T. Carrington,Jr., Adv. Chem. Phys. **114**, 263–310 (2000)
[34] D. Baye and P.-H. Heenen, J. Phys. A **19**, 2041–2059 (1986)
[35] D. Baye, Phys. Rep. **565**, 1–107 (2015)
[36] X-C. Wang and T. Carrington,Jr., J. Chem. Phys. **128**, 194109-1–8 (2008)
[37] D. J. Haxton, J. Phys. B **40**, 4443–4450 (2007)
[38] W. H. Press, S.A. Teukolsky, W.T. Vetterling, and B.P. Flannery, *Numerical Recipes* (Cambridge University Press, Cambridge, 1992)
[39] G. N. Bruce et al., Petrol. Trans. AIME **198**, 79 (1953)
[40] I. M. Gelfand and S. V. Fomin, *Calculus of Variations* (Dover Publications, New York, 2000)

EPJ Web of Conferences **173**, 01009 (2018) https://doi.org/10.1051/epjconf/201817301009
Mathematical Modeling and Computational Physics 2017

Two-Stage Approach to Image Classification by Deep Neural Networks

Gennady Ososkov[1,*] and *Pavel* Goncharov[2,**]

[1] *Laboratory of Information Technologies, Joint Institute for Nuclear Research, 141980 Dubna, Moscow region*
[2] *Sukhoi State Technical University of Gomel, Gomel, Belarus*

Abstract. The paper demonstrates the advantages of the deep learning networks over the ordinary neural networks on their comparative applications to image classifying. An autoassociative neural network is used as a standalone autoencoder for prior extraction of the most informative features of the input data for neural networks to be compared further as classifiers. The main efforts to deal with deep learning networks are spent for a quite painstaking work of optimizing the structures of those networks and their components, as activation functions, weights, as well as the procedures of minimizing their loss function to improve their performances and speed up their learning time. It is also shown that the deep autoencoders develop the remarkable ability for denoising images after being specially trained. Convolutional Neural Networks are also used to solve a quite actual problem of protein genetics on the example of the durum wheat classification. Results of our comparative study demonstrate the undoubted advantage of the deep networks, as well as the denoising power of the autoencoders. In our work we use both GPU and cloud services to speed up the calculations.

1 Introduction

The physicists have accumulated a solid experience on the use of artificial neural networks (ANN) to the solution of many high energy and nuclear physics (HENP) problems [1]. An important role in this experience plays the remarkable privilege to generate training samples of any arbitrary length needed for sufficient ANN training by Monte Carlo simulations of advanced theoretical models. However such comfortable circumstances largely distracted the physicists from the new deep learning neural network (NN) paradigm proposed in the newcoming challenging "Big Data" era asking for intensive data management. Since the problems to be solved are now getting more and more complicated, the ANN architecture needs to become deeper by adding more hidden layers for better parameterization and more adequate to problem modelling, but the ANNs mainly used by physicists are now obsolete and quite rarely can show the needed efficiency.

It is worth mentioning the series of researches devoted to the deep learning concepts published during the last decade [2, 3] in which the authors dealing with neural networks were inspired by the knowledge about the deep hierarchical structure of the mammals visual cortex [4]. There were

*e-mail: ososkov@jinr.ru
**e-mail: kaliostrogoblin3@gmail.com

proposed two ways of making deeper the neural network hierarchy. The first proposal was to add more hidden layers to a multilayer perceptron with its supervised learning paradigm [5], the second to change this paradigm by including into the neural network a hierarchy of additional layers but with different, unsupervised, learning. The second way implements an analog of the animal visual cortex and was proposed by G. Hinton [2] as a combination of several stacked layers of recurrent NN of Boltzmann machine type realizing unsupervised representations of input data with the last supervised layer completing a classification task. This type of deep networks was named Deep Belief Networks (DBN) referring one to the stochastic NNs and the probability distribution of their neurons. These stacked layers are restricted Boltzmann machines (RBM) [3] that can learn themselves a probability distribution over its set of inputs. The quite slow evolution of RBM layers and Hinton's trick to speed it up by the so-called "contrastive divergence" approach are described in [2, 6]. Nevertheless, due to the notable DBN disadvantages, there is now another type of deep NNs which is much wider applicable for the image classification. These are the Convolutional neural networks (CNN), which feed-forward many layer NN with the neurons in a layer connected only to a small region of the layer before it. The main CNN idea following the vision processing in living organisms is to convert every input image by scanning its small sub-regions and filtering them on a stack of convolutional and pooling layers into a set of elementary sub-images which correspond well to the original image, but are not sensitive to different orientations, scales or lighting. Then being trained by a labeled sample with the back-prop method, CNN obtains all weights of convolutional filters and constructs a library of image elements linked to different image classes (see exhaustive CNN description in [7]).

One more significant deep learning approach is to build out a deep autoassociative neural network of an autoencoder type. Autoencoder algorithms consist of training a neural network to produce an output that is identical to the input, but having fewer nodes in the middle hidden layer than in the input forming a "bottleneck". In this case one has built a tool for compressing data to a smaller number of the most informative features [8].

There are many other types of deep learning ANNs with quite different structures such as the Long Short Term Memory (LSTM), the Gated Recurrent Unit (GRU) networks, the Reinforcement Learning (RL) networks (see, for example, J. Schmidhuber's survey [9]. However we are not going to survey deep learning networks. Our goal is to illustrate the advantages of the deep learning approaches over the ordinary neural network on their comparative applications to image classifying. Besides we demonstrate the power of CNN on the example of its application to an actual problem of protein genetics.

2 Deep Learning study at the JINR

In the Laboratory of Information Technologies of JINR four neural network based programs for the image recognition were developed:

1. An autoassociative NN realizing the nonlinear principal component analysis (NLPCA) intended for the significant compression of the input data;

Three NN involve a special optimization of their structures and parameters as image classifiers:

2. Perceptron with one hidden layer;
3. Deep Belief Network;
4. Deep Neural Network (DNN) with initialization of normalized weights.

A study was recently fulfilled to compare the efficiency of these classifiers by testing them on two well-known datasets:

– the MNIST handwritten digit database [10] (70,000 handwritten digit images);

– the FERET face database [11] (40 people with 10 photos for each).

2.1 Data compression by autoencoders

An autoencoder (AE) accomplishes unsupervised learning for data compression to a smaller number of the most informative features by realizing the nonlinear principal component analysis (NLPCA). It was proposed by M. Kramer [8] as an autoassociative neural network with three hidden layers including the bottleneck layer in the middle. AE training is performed with sigmoidal activations by self-supervised backpropagation minimizing the sum of the squared errors (SSE) as the loss function.

We prefer to use the AE as a standalone program due to the following reasons:

– it gives us the same set of compressed data as the input vectors of the three neural networks to be compared further as classifiers;

– AE realized as an autoassociative neural network needs to optimize its structure, tune parameters of the activation function, choose a proper normalization of input data and obtain a fast convergence of the training process;

– AE provides highly effective data compression, which is getting especially important in the Big Data era.

In our study we considerably modified the Kramer scheme in order to improve its efficiency and speed up the training time by solving convergence problems. Our AE improvement consists in:

1. replacing sigmoidal activations by Leaky Rectified Linear (LReLU) activation [12]

$$f(x) = \begin{cases} x, & \text{if } x \geq 0, \\ \frac{x}{a}, & \text{otherwise,} \end{cases}$$

   with a carefully chosen specific parameter a different for each data set;

2. centering the training data by subtracting mean data value from each sample and then specific input normalization (MinMax was chosen);

3. applying tied weights;

4. stochastic gradient descent (SGD) with Adam optimization [13].

The idea of tied weights means the encode and decode weight matrices are simply transposes of each other, what gives less parameters to be optimized and stored and, therefore, faster training.

We use the Adaptive Moment Estimation (Adam) method for efficient SGD which requires first-order gradients only. Adam computes adaptive learning rates for each parameter and also keeps an exponentially decaying average of the past gradients, similar to the momentum of inertia [13].

2.2 Deep and shallow Neural Networks

Deep belief network. As already mentioned above, the deep belief networks (DBN) were proposed as probabilistic generative models which are composed of multiple layers of stochastic, latent variables [2]. DBN runs a pretraining procedure, using latent restricted Boltzmann machines (RBM) layers with a final layer of variables that performs discriminative fine-tuning of the desired outputs by backpropagating the error derivatives. Since there is a rich literature devoted to the detailed DBN

https://doi.org/10.1051/epjconf/201817301009

description and for the sake of brevity we shall not enter into the details on how it works, the reader may find them, e.g., in [2, 3].

Deep neural network. As mentioned above, the training time of the DBNs is always too long, but it is possible to use another type of classifiers the convergence of which is much faster, namely Deep Neural Networks (DNN) with the initialization performed by normalized weights [14]

$$W \sim U\left[-\frac{\sqrt{6}}{\sqrt{n_j + n_{j+1}}}, \frac{\sqrt{6}}{\sqrt{n_j + n_{j+1}}}\right], \tag{1}$$

where $U[-a, a]$ is the uniform distribution in the interval $(-a, a)$ and the n is the size of the previous layer (the number of W columns).

The DNN works much faster than DBN, because it avoids the pretraining procedure for the RBM hidden layers.

Shallow neural network.[1] For our comparative study we use also a feed-forward perceptron with one hidden layer and sigmoid activations as shallow neural network estimator. The input of the perceptron is an array of eigenvectors of dimension 64, which is received from the latent representation of the input images of the bottleneck layer of the autoencoder.

2.3 Selection of optimal hyperparameters for the network models

Improving the autoencoder. We need to choose first the optimal number of hidden neurons for the best input data recovery by our autoencoder. After calculating the dependences of the classification efficiency and the training time on different numbers of neurons in the bottleneck layer we conclude that the optimal number is 64. It guarantees 95% classification efficiency, while a larger number of neurons would result in unnecessary model complexity and cause overfitting.

Then we test how the applications of the tied weights and SGD Adam methods [13] would influence the convergence time. Test results are shown in fig. 1. The prefix "Tied/Untied" means the type of network weights: tied or separate weights, respectively. "GD" stands for the gradient descent abbreviation. As we can see in fig. 1, the Adam optimization method significantly reduces the convergence time, while the use of tied weights although do not benefit much for speeding up, reduce twice the number of network parameters what is important.

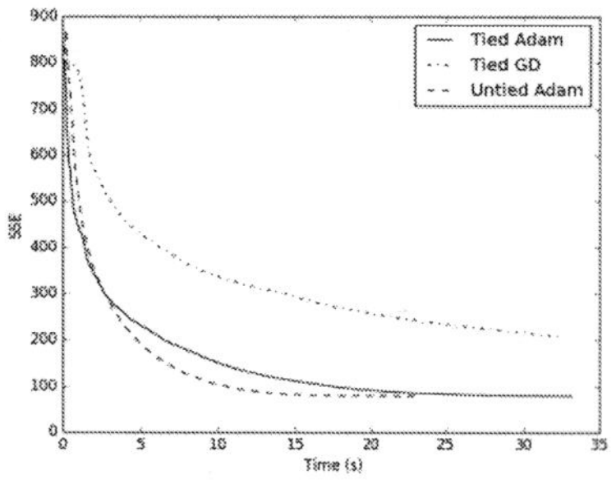

Figure 1. Decrease of the AE loss function in time in terms of the optimization method and the type of weights

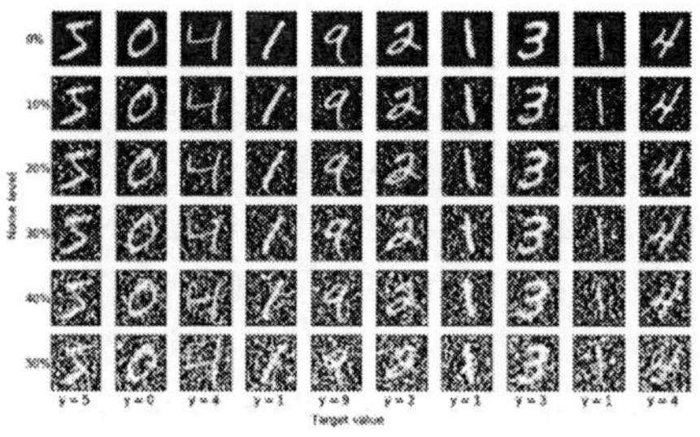

Figure 2. Examples of noisy images with different percentage of noise

[1]The concepts of shallow and deep NN are widely used in the literature (see, for example https://arxiv.org/pdf/1608.03287.pdf).

From similar considerations, we find that the number of neurons in both the mapping-demapping layers should not exceed 256. Eventually, the dimensions of the 5 autoencoder layers are chosen as **(m,256,64,256,m)**, where **m** is the length of the input vector.

In our further work with shallow, deep and deep belief neural networks we also use SGD Adam optimization.

Optimizations of deep classifiers and obtained results. We use grid search cross validation ([15]) for tuning DNN and DBN architectures. The training database is split in two parts: 80% as train set and another 20% as cross-validation set. Then we divide the cross-validation data into 3-fold batches, calculate the score during the fit of an estimator on each fold and take an average score value. After fitting the estimator on a parameter grid, we choose the parameters such as to maximize the cross-validation score, for which the network classification efficiency is selected.

All this gives two hidden layers for both the DNN and DBN while the optimal numbers of the hidden neurons per 1st and 2d layers are 300 and 100 respectively.

2.4 Results of testing by three classifiers

The images of both data sets rescaled by the autoencoder to 64-dimensional eigenvectors were classified by three of our classifiers with results presented in table 1.

The results look quite satisfactory. The deep learning techniques give the 100% classification efficiency on the faces database and the 92–98 percent efficiency on the handwritten digits dataset. Although it is not a problem for the humans to distinguish various faces, the DBN can make better predictions, even in the case of digits, than the humans can.

Table 1. Comparison of the different estimators

	Perceptron	DNN	DBN
MNIST	0.7977	0.9285	0.9839
FERET	0.8750	1.0000	1.0000

2.5 Autoencoder application for image denoising

Generally, the real image examples are not so clean – they can have a lot of noise. We test a remarkable feature of our autoencoder model to recover the input images with different percent of noise imposed on them. A denoising procedure takes noisy data as input and outputs this into data, from which the noise is removed.

Our classifiers are trained on non-corrupted images. The denoising autoencoder network (DAEN) is trained with 20% noise added to input images. Then the classifiers are tested on six datasets with noise levels in the range from 0% to 50% (see fig. 2) compressed previously by DAEN.

As the results in table 2 show, the efficiency keeps well even on the data with 30% noise level. Deep Belief Network performs better classification results again, but under noise increasing the DNN outperforms the DBN. Both deep classifier efficiencies are notably higher than those of the perceptron with one hidden layer.

3 Convolutional Neural Networks for image recognition

The CNN architecture involves a sequence of layers the each layer of which transforms one volume of activations to another through a filter which is a differentiable function.

https://doi.org/10.1051/epjconf/201817301009

Table 2. Comparison of classifying efficiency for noisy images in the case of applying LReLU activations and MinMax normalization of input

	FERET			MNIST		
Noise(%)	Perceptron	Deep NN	Deep Belief	Perceptron	Deep NN	Deep Belief
0	0.9166	0.9867	1.0000	0.7449	0.9182	0.9861
10	0.9083	0.9867	0.9833	0.6901	0.8779	0.9644
20	0.9000	0.9750	0.9666	0.6115	0.7832	0.8794
30	0.8750	0.9583	0.9583	0.5001	0.6443	0.6027
40	0.8500	0.9667	0.9499	0.4100	0.5202	0.3863
50	0.8000	0.9583	0.9333	0.3036	0.3973	0.2283

There are three main types of layers to build CNN architectures: the Convolutional layer, the Pooling (subsampling) Layer, and the Fully-Connected Layer (just MLP with backprop). There are also RELU (rectified linear unit) layers performing the max(0,x). Each Layer accepts an input 3D volume (x,y,RGB color) and transforms it to an output 3D volume. To construct all filters of convolutional layers our CNN must be trained by a labeled sample with the back-prop method [7].

We describe now the CNN use for the solution of an actual problem of the protein genetics, namely, the durum wheat classification by the labeled sample of electrophoretic spectra of gliadin (alcohol soluble protein) [16]. Each electrophoretogram becames after its digitalization a densitogram spectrum consisting of about 4000 pixels, as it shown in fig. 3. It can be considered as a simple genetic formula, which allows for a qualified expert to classify any spectrum to its corresponding protein.

The real size of the training sample was 3225 EF-densitograms for 50 sorts preliminarily classified by experts for each of wheat sorts. Having 4000 input values means that one would have a neural network with more than 10 millions of weights or equations to solve by the error back propagation method! There were several attempts to extract the most informative features from input data to reduce the data cardinally [16, 17]. Although it succeeded to reduce the input by more than an order of magnitude, the classification by multilayer perceptrons with one hidden layer failed to classify with an admissible efficiency more than 10 sorts of wheat [17].

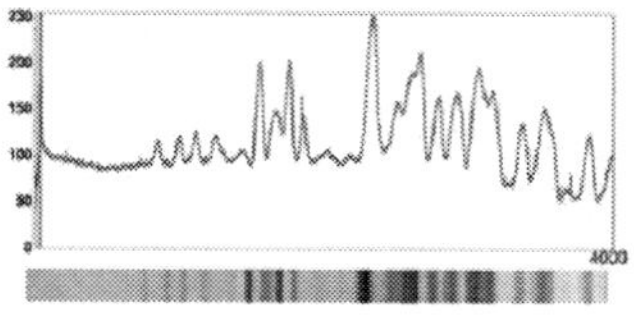

Figure 3. Example of the electrophoretogram strip (below) with its dencitogram (above)

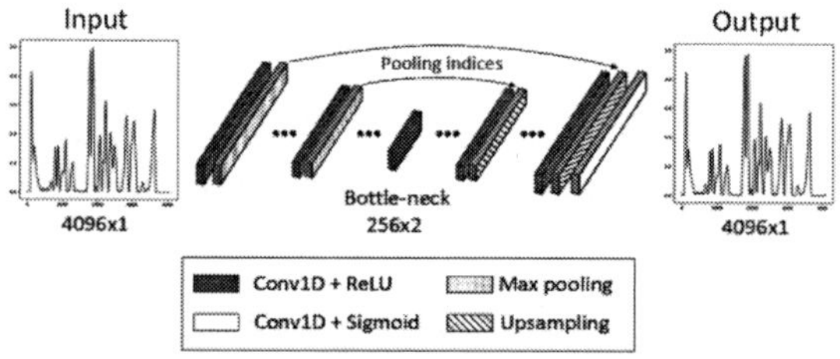

Figure 4. Convolutional Autoencoder scheme

To solve the problem of classifying 50 wheat sorts we proposed to use a two-step algorithm based on data-compression by an autoencoder at the first step followed then at the second step by applying either DNN or convolutional NN. To make a justified choice of AE we elaborate also a convolutional one (CAE) (see fig. 4).

In our study we have had to develop two versions of autoencoders and two versions of classifiers in order to compare them by the higher efficiency criterion with reasonable time consuming. We applied the improvements elaborated above for AE such as the tied weights and stochastic gradient descent (SGD) with Adam optimization. The total number of CAE parameters is 315 what is 5 orders of magnitude less than for DNN and gives a significant gain in the memory need to store AE as well as in training speed. The parameters of the deep and convolutional NN are shown in table 3.

Table 3. Parameters of the deep and convolutional NN

Parameter	CNN	DNN
Model size (MB)	3	66
Total epochs number	150	400
Epoch time (s)	1	0.2
Training time (s)	69	84

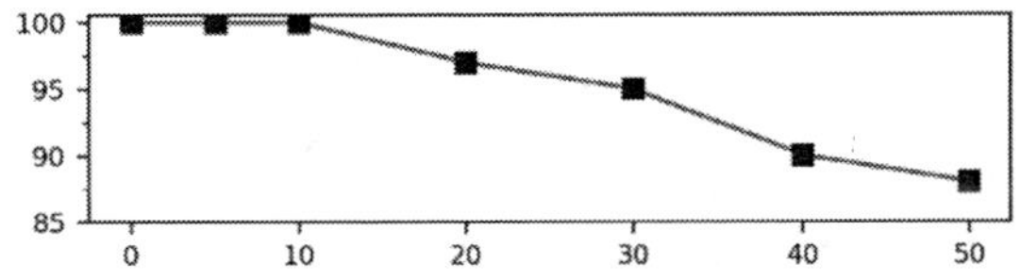

Figure 5. Classification efficiency vs sort numbers (the same for CNN and DNN)

The results of the protein classification efficiency are the same for both the CNN and DNN (fig. 5). One can see the advantages of the convolutional network as compared to DNN. With the same efficiency CNN takes 22 times less memory, while its training time is faster. These results demonstrate the overwhelming superiority of the deep approach in the solution of the protein classification problem.

4 Deep training with GPU acceleration in Cloud

In our study we speed up significantly our deep network calculations by using the multicore computational systems and multiprocessor graphics cards or GPUs via the cloud service facilities provided at the JINR heterogeneous computing cluster HybriLIT [18]. Although our classification models were developed on a notebook Dell Vostro 5480 (CPU: Intel Core i3-4500U @ 1.70 GHz (4 cores); RAM: 4096 MB; Intel HD Graphics 4400) we use HybriLIT cloud sevice and two free libraries to carry out deep training with GPU acceleration in the cloud with Nvidia Tesla K80: TensorFlow [19] and Keras [20].

These services greatly facilitate the deep learning performance. For the solution of a same problem, one training epoch for the convolutional autoencoder takes on the notebook 70–80 sec, but operating in the HybriLIT using TensorFlow and Keras it takes only 1 sec for one training epoch.

5 Conclusion

In our work, we propose a deep learning model for the image classification, which consists of two neural networks: an autoencoder for preliminary input compression and an estimator for classifying images. A comparative study of three neural classifiers: a perceptron with one hidden layer, DBN and DNN was accomplished on images of two famous benchmark data sets MNIST and FERET being previously compressed by the autoencoder.

A comparative study of three neural classifiers: a perceptron with one hidden layer, DBN and DNN was accomplished on images of both data sets being previously compressed by the autoencoder. The comparison of the investigated estimators shows that DBN provides the highest classification efficiency, but DNN accomplishes a faster training time having competitive efficiency. Both the DBN and DNN perform much better than the perceptron with one hidden layer.

It is important to note that the main efforts to deal with the deep learning networks were spent to quite a painstaking work for optimizing the structures of those networks and their components, as activation functions, weights, as well as the procedures of minimizing their loss function to improve their performances. We have to make a special study to find the optimal numbers of layers and hidden neurons, to apply tied weights and SGD Adam method, to choose the most suitable activation functions and weight normalization in order to improve the performance of all used neural nets and speed up their learning time. Additionally we observe such a remarkable feature of the deep autoencoders as their ability for denoising images after being trained on an image set corrupted in a special way. The denoising results of such autoencoder classification efficiency on FERET images are really striking: images corrupted with 50 percent of noise were classified with efficiency better than 90 percent.

Besides we extend our two stage deep learning model for the image classification to Convolutional Neural Networks. It was used to solve a quite actual problem of protein genetics on the example of the durum wheat classification with quite successful results.

In our study we speed up significantly our deep network calculations by using Tensorflow and Keras libraries and multicore computational systems and multiprocessor graphics cards or GPUs via the cloud service facilities provided at the JINR heterogeneous computing cluster HybriLIT.

References

[1] I. Kisel, V. Neskoromnyi, and G. Ososkov, Phys. Part. Nucl. **24**, 6, 657–676 (1993)

[2] G. E. Hinton and R. Salakhutdinov, Science **313**, 504–507 (2006)

[3] Y. Bengio, *Learning Deep Architectures for AI* (Now Publishers Inc, 2009) p. 144

[4] P. Lennie, Current biology **13**, (6), 493–497 (2003)

[5] Y. LeCun, L. Bottou, G. Orr, and K.-R. Müller, in *Neural Networks: Tricks of the Trade*, 9–50 (1998)

[6] M. Carreira-Perpinan, Aistats **10**, 33–40 (2005)

[7] *Web CNN description facility*, http://cs231n.github.io/convolutional-networks

[8] M. Kramer, AIChE Journal **37**, 2, 161–310 (1991)

[9] J. Schmidhuber, Neural Networks **61**, 85–117 (2015)

[10] *MNIST*, https://www.nist.gov/sites/default/files/documents/srd/nistsd19.pdf

[11] A. Phillips, H. Moon, P. Rauss, and S. Rizvi, IEEE Transactions on Pattern Analysis and Machine Intelligence **22**, 10 (2000)

[12] B. Xu, N. Wang, T. Chen, and M. Li, https://arxiv.org/abs/1505.00853

[13] D. Kingma and J. L. Ba, *Adam*, https://arxiv.org/abs/1412.6980

[14] X. Glorot and Y. Bengio, Aistats **9**, 249–256 (2010)

[15] M. Stone, Journal of the royal statistical society. Series B (Methodological), 111–147 (1974)

[16] A. Ruanet, A. Kudriavtsev, and S. Dadashev, Russian Journal of Genetics **37**, 10, 1207–1209 (2001)

[17] G.A. Ososkov and D.A. Baranov, Bulletin of PFUR Series Mathematics. Information Sciences. Physics. **3**, 142–148 (2010)

[18] *Web-portal of HybriLIT JINR computation facility*, http://hybrilit.jinr.ru

[19] *Speed up training with GPU-accelerated TensorFlow*, http://www.nvidia.com/object/gpu-accelerated-applications-tensorflow-benchmarks.html

[20] *Keras Python Deep Learning library*, https://keras.io/getting-started/sequential-model-guide/

Multiscale Multilevel Approach to Solution of Nanotechnology Problems

Sergey Polyakov[1,2,⋆] and *Viktoriia* Podryga[1,3,⋆⋆]

[1] *Keldysh Institute of Applied Mathematics of RAS, 4 Miusskaya square, 125047, Moscow, Russia*
[2] *National Research Nuclear University MEPhI (Moscow Engineering Physics Institute), 31, Kashirskoe shosse, 115409, Moscow, Russia*
[3] *National Research Center "Kurchatov Institute", 1, Akademika Kurchatova pl., 123182, Moscow, Russia*

Abstract. The paper is devoted to a multiscale multilevel approach for the solution of nanotechnology problems on supercomputer systems. The approach uses the combination of continuum mechanics models and the Newton dynamics for individual particles. This combination includes three scale levels: macroscopic, mesoscopic and microscopic. For gas–metal technical systems the following models are used. The quasihydrodynamic system of equations is used as a mathematical model at the macrolevel for gas and solid states. The system of Newton equations is used as a mathematical model at the meso- and microlevels; it is written for nanoparticles of the medium and larger particles moving in the medium. The numerical implementation of the approach is based on the method of splitting into physical processes. The quasihydrodynamic equations are solved by the finite volume method on grids of different types. The Newton equations of motion are solved by Verlet integration in each cell of the grid independently or in groups of connected cells. In the framework of the general methodology, four classes of algorithms and methods of their parallelization are provided. The parallelization uses the principles of geometric parallelism and the efficient partitioning of the computational domain. A special dynamic algorithm is used for load balancing the solvers. The testing of the developed approach was made by the example of the nitrogen outflow from a balloon with high pressure to a vacuum chamber through a micronozzle and a microchannel. The obtained results confirm the high efficiency of the developed methodology.

Introduction

The modern computing allows modeling very large and complex systems and processes. Computer modeling has become one of the most effective tools in many branches of science and production. The present work is devoted to the development and application of computer methods in the field of studying the complex gasdynamic processes in technical micro- and nanosystems. A feature of mathematical problems in this field is the simultaneous study of processes at many scales, including micro- and nanoscale. One of the modern and actively developing approaches to solving such problems is the multiscale approach which combines the methods of the continuum mechanics (MCM) and the Newton dynamics for individual particles. This allows efficient computer simulation of expensive and difficult to implement physical experiments.

⋆e-mail: polyakov@imamod.ru
⋆⋆e-mail: pvictoria@list.ru

In the present paper, an aspect of modeling technical microsystems is considered, which is related to computing the parameters of gas microflows under technical vacuum conditions. For a correct description of such processes, it is necessary to know in detail the properties of the real gases and reproduce them in a numerical experiment. Many properties of the gases are well studied experimentally in certain ranges of temperatures and pressures and are described in the literature. However, there are properties which can only be predicted theoretically on the basis of the kinetic theory of gases [1]. In this case, the obtained theoretical data correspond to a very limited range of temperatures and pressures and can differ substantially from the real properties of the gas. One of the ways to obtain the missing information about the properties of the gas medium in a given range of temperatures and pressures is the molecular dynamic simulation [2, 3].

A multiscale two-level approach to calculating the real gas flows in the microchannels of technical systems, applied in a wide range of Knudsen numbers, was presented in [4–6]. At the molecular level it allows to determine the parameters of the equation of state of a real gas [7], to calculate the kinetic properties of the gas [8], to determine the type of the boundary conditions at the walls of the microchannel [5, 6]. This approach is based on a combination of MCM and Newton dynamics for individual particles and includes two levels of modeling: macroscopic and microscopic. As a model at the macrolevel, the system of quasigasdynamic (QGD) [9] equations is used, as a model at the microlevel, the method of molecular dynamics (MD) [2, 3] is used. The implementation of this approach is based on splitting the problem into physical processes. At the macrolevel of detail the description of multicomponent gas media flows takes place. At the microlevel interactions are calculated for: 1) gas molecules between themselves (forming the equation of state of the mixture, determining the transport coefficients and realizing the mixing of the components); 2) gas molecules and atoms of solid surfaces (describing phenomena in boundary layers).

In what follows, a multiscale multilevel approach to solving nanotechnology problems with the help of supercomputer computing systems is described in some detail. This unification covers three scale levels: macroscopic, mesoscopic and microscopic. To realize it at the mesoscopic level the calculation of the motions of large particles in the gas flow is added, and at the microlevel the interactions of the gas molecules with the particle surface are calculated. The consideration of the mesoscopic level allows to track the detailed behavior of the individual large particles, for example, during their deposition on a substrate in order to create a given spatial nanostructure.

1 Mathematical formulation

The mathematical model includes three components corresponding to the three scale levels.

At the macroscopic level the QGD equations are used. In the case of a gas mixture, the system of QGD equations is written for each gas separately [9]. These equations are written below in the three-dimensional case for a mixture of gases in invariant form with respect to the coordinate system together with the equations of constraints and the equations of state:

$$\frac{\partial \rho_l}{\partial t} + \operatorname{div} \mathbf{W}_l^{(\rho)} = 0, \tag{1}$$

$$\frac{\partial}{\partial t}(\rho_l u_{l,k}) + \operatorname{div} \mathbf{W}_l^{(\rho u_k)} = S_l^{(\rho u_k)}, \tag{2}$$

$$\frac{\partial E_l}{\partial t} + \operatorname{div} \mathbf{W}_l^{(E)} = S_l^{(E)}, \tag{3}$$

$$\begin{cases} E_l = \rho_l\left(\frac{1}{2}|\mathbf{u}_l|^2 + \varepsilon_l\right), \quad H_l = \dfrac{E_l + p_l}{\rho_l}, \quad p_l = Z_l\rho_l R_l T_l, \quad \varepsilon_l = c_{V,l}T_l, \\ \gamma_l = \dfrac{c_{p,l}}{c_{V,l}}, \quad \mathrm{Pr}_l = \dfrac{\mu_l c_{p,l}}{\chi_l}, \quad \mathrm{Sc}_l = \dfrac{\mu_l}{\rho_l D_l}, \quad \mathrm{Ma}_l = \dfrac{|\mathbf{u}_l|}{a_l}, \quad \mathrm{Re}_l = \dfrac{\rho_l|\mathbf{u}_l|\lambda_l}{\mu_l}. \end{cases} \tag{4}$$

Here all variables with the index l relate to a gas of the type l, each component has its own concentration n_l, mass density $\rho_l = m_l n_l$ (m_l is the mass of a molecule of the gas l). Each gas is also characterized by its temperature T_l and macroscopic velocity $\mathbf{u}_l$. Other parameters of the mixture components: p_l are partial pressures of gases in the mixture; E_l, H_l and ε_l are total energy densities, enthalpies, and internal energies of the mixture components; $\mu_l = \mu_l(T_l)$, $D_l = D_l(T_l)$ and $\chi_l = \chi_l(T_l)$ are kinetic coefficients of mixture components, namely, coefficients of dynamic viscosity, diffusion and thermal conductivity. Variables $Z_l = Z_l(T_l, \rho_l)$, $\gamma_l = \gamma_l(T_l, \rho_l)$, $c_{V,l} = c_{V,l}(T_l)$, $c_{p,l} = c_{p,l}(T_l)$ and $R_l = k_B/m_l$ are compressibility coefficients, adiabatic indices, specific heat capacities and individual gas constants of the mixture components (k_B is the Boltzmann constant); Pr_l, Sc_l, Ma_l and Re_l denote Prandtl, Schmidt, Mach, and Reynolds numbers for the mixture components; λ_l are mean free paths; vectors $\mathbf{W}_l^{(\rho)}$, $\mathbf{W}_l^{(\rho u_k)}$ and $\mathbf{W}_l^{(E)}$ coincide, up to a sign, with the fluxes of mass density, the momentum density of the corresponding components, and energy density. The exchange terms $S_l^{(\rho u_k)}$ and $S_l^{(E)}$ take into account the redistribution of momentum and energy between the components of the mixture.

The system of equations (1)–(4) is closed by the corresponding initial and boundary conditions. The initial conditions are taken in accordance with the equilibrium state of the gas medium in the absence of interaction with external factors. The boundary conditions on solid surfaces are given in the form of mass density, momentum density and energy density fluxes across the boundary, calculated from empirical formulas or calculated using MD methods [4–6]. On the free surfaces of the computational domain, the so-called "soft" boundary conditions [9] are enforced.

At the microlevel, the evolution of the system under investigation is described by Newton's equations for a system of particles. The system of the equations of motion for particles of type l, which can be a gas, or a metal, has the following form

$$m_l \frac{d\mathbf{v}_{l,i}}{dt} = \mathbf{F}_{l,i}, \qquad \mathbf{v}_{l,i} = \frac{d\mathbf{r}_{l,i}}{dt}, \qquad i = 1, \ldots, N_l, \tag{5}$$

where i is particle index, l is particle type, N_l is total number of particles of type l. A particle of type l with index i has its own mass m_l, position vector $\mathbf{r}_{l,i} = \left(r_{x,l,i}, r_{y,l,i}, r_{z,l,i}\right)$, velocity vector $\mathbf{v}_{l,i} = \left(v_{x,l,i}, v_{y,l,i}, v_{z,l,i}\right)$ and total force $\mathbf{F}_{l,i} = \left(F_{x,l,i}, F_{y,l,i}, F_{z,l,i}\right)$ acting on it.

The forces acting on the i-th particle are calculated as a sum of its interactions with the surrounding particles, depending on the potential energy and the component responsible for the external action. The potential energy of the system is represented as a sum of partial energies, the calculation of which takes place according to the formula of the chosen interaction potential.

$$\mathbf{F}_{l,i} = -\frac{\partial U(\mathbf{r}_{l,1}, \ldots, \mathbf{r}_{l,N_l}, \ldots)}{\partial \mathbf{r}_{l,i}} + \mathbf{F}_{l,i}^{ext}, \qquad U = \sum U_{ll'}, \qquad i = 1, \ldots, N_l, \tag{6}$$

where U is total potential energy, $U_{ll'}$ is interaction potential for particles of type l with particles of type l', $\mathbf{F}_{l,i}^{\text{ext}}$ is the force of interaction with the external environment. The choice of the interaction potential is based on comparing the mechanical properties of the computer potential model and the real material.

The initial conditions at the microlevel are determined by the equilibrium or quasiequilibrium thermodynamic state of the particle system at a given temperature, pressure, and average momentum. The boundary conditions at the molecular level depend on the simulated situation.

At the mesoscopic level, in the presence of finely dispersed solid impurities in the gas mixture of the same chemical composition, a diffusion approximation is used in which the principal unknown quantity is the particle concentration C. For it, the following nonlinear convection-diffusion equation is used:

$$\frac{\partial C}{\partial t} = \operatorname{div}\left(D_C \nabla C - \mu_C \mathbf{F}_C C\right) + (\mathbf{u}, \nabla C). \tag{7}$$

Here D_C is the interdiffusion coefficient of the gas suspension that can be calculated in various ways [10, 11]. The coefficient μ_C describes the mobility of the impurity particles; in the general case it is determined in a complex manner and depends on the shape of the mesoparticles, the temperature and the viscosity of the medium. In this paper the well-known Stokes formula was used to determine the mobility coefficient, which is valid for spherical particles. The term $\mathbf{F}_C$ is the total force acting on the particles and calculated at the molecular level, $\mathbf{u}$ is average velocity of the gas mixture flow. The initial conditions for (7) consist in assigning a constant and, as a rule, a very small concentration of particles in the entire free space. The boundary conditions at solid walls are determined by the chosen interaction model (adhesion, reflection, sliding, etc.). The boundary conditions at the entrance to the medium consist in the assignment of the particle flux. The "soft" boundary conditions are used at the output.

If necessary, to trace the movements of the individual large particles (substantially exceeding the dimensions of the molecules of the gaseous medium), equations similar to (5), and (6) are used together with the equation (7).

2 Numerical approach

The numerical approach to the solution is based on the problem splitting into physical processes. In this case, the equations of the quasihydrodynamics (1)–(3) and the equation (7) are solved by the method of finite volumes on grids of various types. Newton dynamics equations are solved according to the Verlet integration [2] either in each grid cell independently or in groups of coupled cells.

The calculation of the macroparameters according to the QGD equations (1)–(3) and the equation (7) is carried out over a temporal grid using an explicit numerical algorithm which is based on the finite volume method on grids of arbitrary type. For the convenience of solving the problem in areas of complex geometry, hybrid block meshes consisting of cells of various shapes and sizes can be used. In the two-dimensional case it is proposed to use quadrangles and triangles, in the three-dimensional case — polyhedrons with a number of vertices from four to eight. All parameters of the gas components (density, pressure, temperature, velocity vector components, etc.) refer to the mass centers of the cells. Flux variables refer to the centers of the faces of the cells. Spatial approximations of the basic terms of the QGD equations are performed by standard methods. The chosen computing scheme on time is explicit and two-stage (predictor–corrector).

The system of MD equations is used in additional calculations independently, or as a subgrid algorithm applied within each control volume. At the microscopic level, at every step of the simulation, a system of ordinary differential equations is solved, corresponding to the Newton second law and describing the motion of particles of the molecular dynamic system. To integrate the equations of motion, the Verlet integration [2] is used.

To carry out a correct calculation of the QGD, the model is supplemented by real gas equations of state, transport coefficients and other accompanying parameters (enthalpies, average mean free paths, etc.), as well as real boundary conditions. In the case of a mixture of gases, it is necessary to add the corresponding exchange terms to the equations for the momentum and energy of each component. The calculation of these dependencies, coefficients, and conditions is performed using the MD methods.

The stability condition of the splitting method by physical processes includes standard conditions for the stability of an explicit finite-difference scheme ($\Delta t_1 \approx h^2$, where Δt_1 is time step, h is space step) at the macrolevel and the conditions for the stability of the Verlet schemes ($\Delta t_2 \approx 1/\max|F_2|$, $\Delta t_3 \approx 1/\max|F_3|$, where Δt_2, Δt_3 are time steps, $\max|F_2|$, $\max|F_3|$ are maximum absolute values of the forces of interaction between the particles at a considered step in time) at the meso- and microlevels.

To verify the proposed numerical methodology, a verification procedure was used, based on performing calculations with grinding the mesh at the macrolevel, and on performing calculations for systems with different number of particles and increasing the number of particles at the microlevel.

3 Algorithms and program implementation

The problem modeling on the basis of the multiscale approach under consideration with three levels of detail is carried out with the help of special algorithms that in general, depending on the degree of microlevel use, are divided into four classes.

Algorithms of class 1 are of interest for the study of the properties of gas media and the properties of solid surfaces with which the gas medium contacts in technical applications. As a numerical implementation of the approach in this case, the Velocity Verlet scheme acts. With the help of 1 class algorithms, a database of molecular dynamic calculations (DMDC) is accumulated for the properties of gases and solid materials, which can be used in the framework of other algorithms.

Algorithms of class 2 assume the solution of problems only at macro- and mesolevels based on QGD equations (1)–(3) and the convection-diffusion equation (7). In this case, the properties of the gas mixture components (the equations of state by pressure and energy, the kinetic coefficients — viscosity and thermal conductivity, the exchange terms in the equations for momentum and energy, the parameters of the boundary conditions) and the kinetic coefficients for equation (7) are determined from the above-mentioned DMDC accumulated in advance for the desired temperature and pressure range.

Algorithms of class 3 imply the simultaneous use of equations (1)–(3), equations (7) and Newton equations of mechanics (5), (6) in the calculations of QGD. Algorithms of class 3 are realized in the framework of the method of splitting into physical processes. It is assumed here that in the gaseous medium and at its boundaries it is possible to confine ourselves to a local consideration of the processes of interacting the gases of the mixture with each other, with the surfaces of large particles and with solid walls.

Algorithms of class 4 also assume the simultaneous use in the calculations of equations (1)–(3), (7) and (5), (6). The difference of this case from the previous one is that in some areas of the environment (usually at solid surfaces and in zones of a strong drop of the gas parameters), molecular dynamic calculations are carried out continuously without going to the macrolevel. In the same areas, the principle of locality of molecular interactions is not used, that is, in the general case, the algorithms of class 4 are non-local at the molecular level.

The parallelization is implemented based on the principles of geometric parallelism and rational partitioning of the computational domain. To balance the task loading, a dynamic algorithm is used. The code realization is made using MPI and OpenMP programming.

4 Calculation results

The code validation is carried out on the example of the problem of the nitrogen outflow from a balloon with high pressure to a vacuum chamber through a micronozzle and a microchannel. The computational geometry is shown in Fig. 1. A symmetric domain is selected, but the expected solution may not be symmetric.

The cylindrical micronozzle has a cross section of rectangular shape, the diameter $D_0 \approx 6.2\,\mu$m and length $L_0 = 6D_0 \approx 37.2\,\mu$m. The computational domain sizes of the balloon equate $D_1 = 6D_0$ and $L_1 = 10L_0$. The diameter and length of the microchannel equate $D_2 = 6D_0$ and $L_2 = 50L_0$. The nozzle and the microchannel connect the balloon with nitrogen and the open space of the vacuum

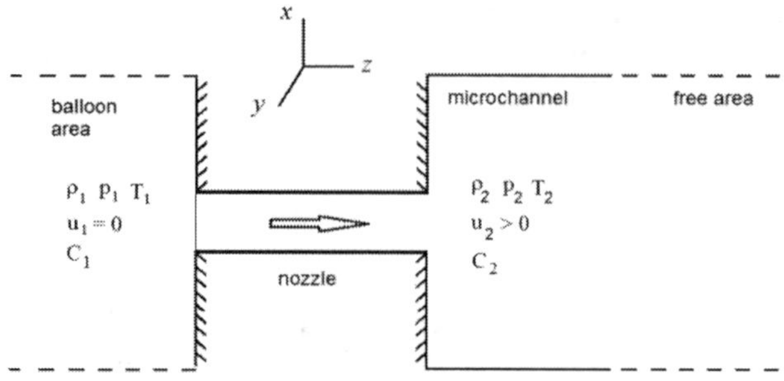

Figure 1. Computational domain

chamber. We assumed that all solid surfaces of the microsystem are covered with a layer of nickel. Copper clusters of a regular cubic form with a linear dimension of 10.83 nm ($30 \times 30 \times 30$ elementary cells with the edge length of the FCC lattice 0.361 nm) were considered as large particles.

At the initial moment the gas is at rest: $u_1 = u_2 = 0$. In this case, in the balloon the gas is under standard normal conditions: $T_1 = 295.15$ K, $p_1 = 101325$ Pa; in the nozzle, the microchannel and the vacuum chamber, the gas is at the same temperature, but at a lower pressure: $T_2 = T_1$, $p_2 = 10^{-5} p_1$.

The nozzle is blocked on the left by a partition which opens instantly at the beginning of the calculation. The inner surface of the nozzle is considered to be perfectly smooth and thermally insulated.

The calculations were performed using an algorithm of class 2 and were aimed at studying the flow parameters at the initial stage of flow acceleration. They showed that the internal energy of the gas in the balloon is, for a certain period of time, gradually converted into kinetic energy of the gas in the nozzle and in the channel. At the same time, a thermal shock wave passes through the gas medium. Subsequently, the gas temperature returns to the normal and the profiles of all gas macroparameters are aligned.

As an illustration of the above the Fig. 2 shows the distributions of the density, concentration of molecules, temperature, and local Mach number in the gas along the symmetry axis of the computational domain for a number of instants of time. An abbreviation a.u. means arbitrary units, here the spatial coordinates were normalized to the diameter of the micronozzle and the gas macroparameters were normalized to their values under normal conditions. Two-dimensional distributions of the concentration of gas molecules and the velocity modulus are shown in Fig. 3. They show the general picture of the flow near the nozzle and the entrance to the microchannel at the time $t = 13.26\,\mu$s, when the gas acceleration has already stopped, and a regular flow has been established.

In Fig. 4 the trajectories of copper clusters calculated in the case of triggering of single particles are shown. Clusters were placed in the middle of the calculated area of the balloon at zero speed. The flow of gas carried through the nozzle into the chamber accelerated them to a maximum speed equal to the average flow velocity. The calculated trajectories as a whole correspond to theoretical estimates. However, gravity was not taken into account. An even more accurate picture in the future will be obtained when this force is taken into account as well as the surface structure of the clusters and walls of the nozzle and microchannel.

Conclusion

A multiscale multilevel approach to the solution of nanotechnology problems by supercomputer systems is presented. The solution combines continuum mechanics models and Newton dynamics for individual particles at three scale levels: macroscopic, mesoscopic and microscopic. The description of the gas-metal technical systems as a mathematical model at the macrolevel is done using the quasihydrodynamic equations for gas and solid states. As a mathematical model at the meso- and

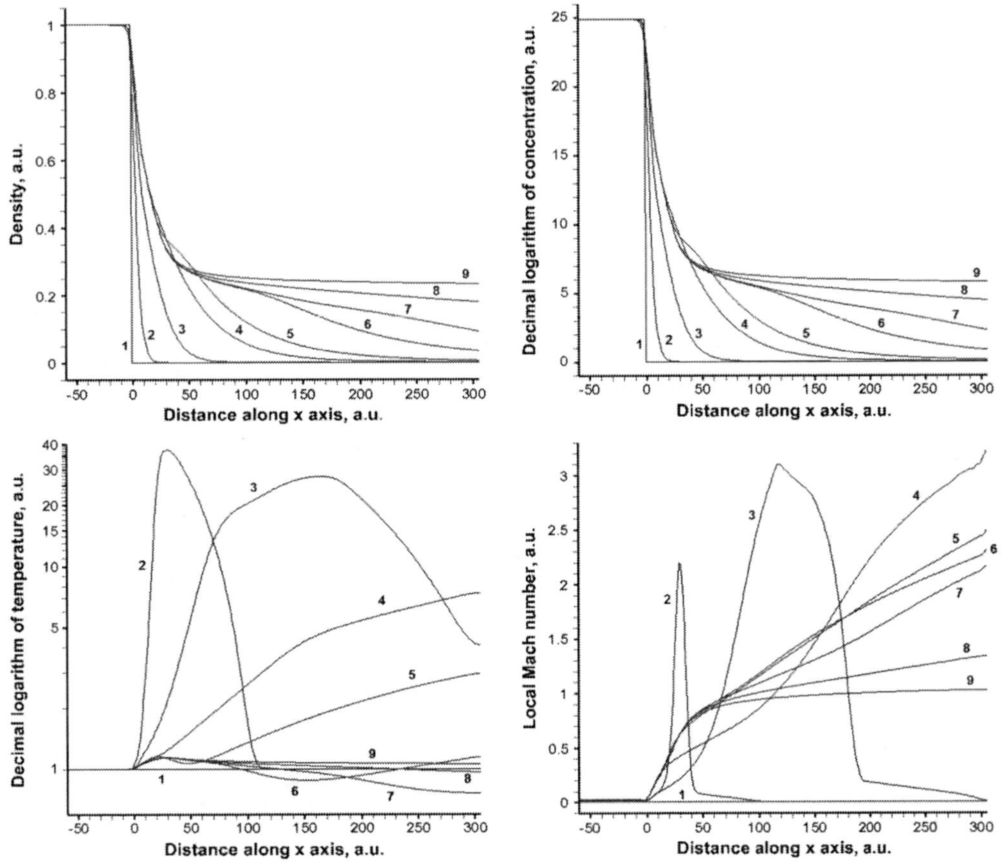

Figure 2. Distributions of density, concentration, temperature and local Mach number along the symmetry axis (x coordinate) of the computational domain. The curves labeled 1 to 9 correspond to the time moments t = 0, 0.0147, 0.1474, 0.7369, 1.474, 2.947, 4.421, 11.79, 73.69 μs

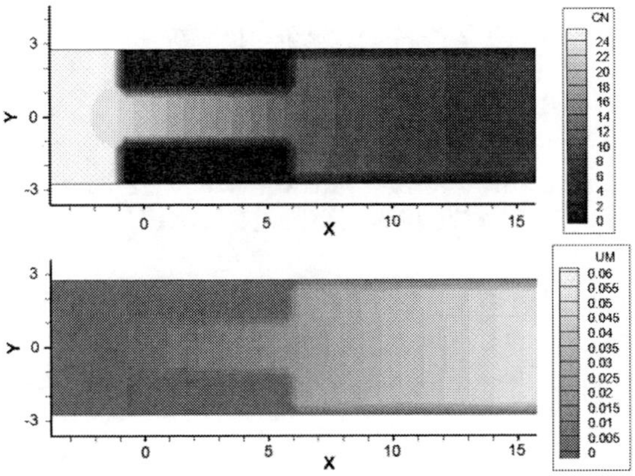

Figure 3. Two-dimensional distributions of gas concentration and velocity modulus near nozzle area at the time moment 13.26 μs

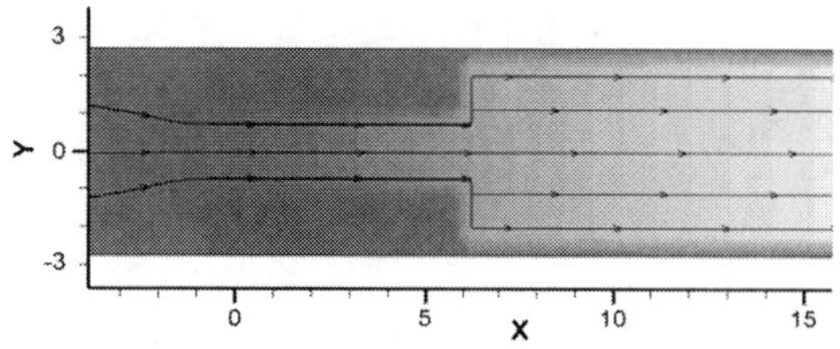

Figure 4. Illustration of coarse particles moving: streamlines of particles

microlevels, the systems of Newton equations written for nanoparticles of the medium and large particles moving in the medium are used. The numerical implementation of the solution involves the problem splitting into physical processes, the method of grids at the macrolevel and the Verlet integration at the meso- and microlevels. Four classes of algorithms and their parallel implementation are discussed. The implementation uses the principles of geometric parallelism, rational partitioning the computational domain, load balancing of the tasks. A special dynamic algorithm is used for load balancing the solvers. The algorithm validation is done on the example of the problem of the nitrogen outflow containing clusters of copper into vacuum. The obtained preliminary results correspond to theoretical estimates and confirm the high efficiency of the developed methodology.

Acknowledgments

This work was supported by the Russian Science Foundation (project No. 17-71-10045).

References

[1] J.O. Hirschfelder, C.F. Curtis, and R.B. Bird, *Molecular Theory of Gases and Liquids* (Wiley, New York, 1954)

[2] J.M. Haile, *Molecular Dynamics Simulations. Elementary Methods* (John Wiley & Sons, Inc., New York, 1992)

[3] D.C. Rapaport, *The Art of Molecular Dynamics Simulation* (Second edition, Cambridge University Press, Cambridge, 2004)

[4] V.O. Podryga, Doklady Mathematics **94** (1), 458–460 (2016)

[5] V.O. Podryga, IOP Conference Series: Materials Science and Engineering **158**, 012078 (2016)

[6] V.O. Podryga, Yu.N. Karamzin, T.A. Kudryashova, and S.V. Polyakov, *Proceedings of the VII European Congress on Computational Methods in Applied Sciences and Engineering (ECCOMAS Congress 2016)* **2** (ECCOMAS Congress 2016, Crete Island, Greece, June 5–10, 2016) 2331–2345

[7] V. Podryga and S. Polyakov, *Proceedings of the Fourth International Conference on Particle-Based Methods-Fundamentals and Applications (PARTICLES 2015)* (CIMNE, Barcelona, Spain, September 28–30, 2015) 779–788

[8] V.O. Podryga, Lecture Notes in Computer Science **10187**, 542–549 (2017)

[9] T.G. Elizarova, *Quasi-Gas Dynamic Equations* (Springer-Verlag, Berlin, 2009)

[10] V.Ya. Rudyak, A.A. Belkin, and S.L. Krasnolutskii, Thermophysics and Aeromechanic **12**(4), 489–507 (2005)

[11] V.Ya. Rudyak and S.L. Krasnolutskii, Technical Physics. The Russian Journal of Applied Physics **47**(7), 807–813 (2002)

EPJ Web of Conferences 173, 01011 (2018) https://doi.org/10.1051/epjconf/201817301011
Mathematical Modeling and Computational Physics 2017

A Generalized Technique in Numerical Integration

Hassan Safouhi[1,⋆]

[1] *Mathematical Division, Campus Saint-Jean*
University of Alberta
8406, 91 Street, Edmonton, Alberta T6C 4G9, Canada

Abstract. Integration by parts is one of the most popular techniques in the analysis of integrals and is one of the simplest methods to generate asymptotic expansions of integral representations. The product of the technique is usually a divergent series formed from evaluating boundary terms; however, sometimes the remaining integral is also evaluated. Due to the successive differentiation and anti-differentiation required to form the series or the remaining integral, the technique is difficult to apply to problems more complicated than the simplest. In this contribution, we explore a generalized and formalized integration by parts to create equivalent representations to some challenging integrals. As a demonstrative archetype, we examine Bessel integrals, Fresnel integrals and Airy functions.

1 Introduction

It is well known that in applied mathematics and in the numerical treatment of scientific problems, slowly convergent integrals occur abundantly. They are produced by approximation procedures depending on a parameter, iterative methods, quadrature schemes, perturbation techniques and reliable evaluation of functions defined by integrals. These slowly convergent integrals present severe numerical and computational difficulties. Traditional quadrature rules and summation techniques have failed to provide accurate approximations to such integrals. Numerous methods and techniques were developed for improving convergence of these challenging integrals [1–18] and extremely efficient methods were introduced such as numerical steepest descent, Filon-type and Levin-type methods. Unfortunately, their application to complicated integrals is extremely challenging. Examples of such integrals are the twisted tail, Bessel integrals, Sommerfeld integrals and the so-called molecular multicenter integrals involved in molecular structure calculations.

The technique of integration by parts has frequently been used to create divergent asymptotic series representations of integrals. As an example, we consider the Euler series arising from integrating the Euler integral by parts:

$$\int_x^\infty \frac{e^{-t}}{t}\mathrm{d}t \sim \frac{e^{-x}}{x}\sum_{l=0}^{\infty}(-1)^l \frac{l!}{x^l}, \quad x \to \infty. \tag{1}$$

⋆e-mail: hsafouhi@ualberta.ca

In this example, the remaining integral after n integrations by parts:

$$(-1)^n n! \int_x^\infty \frac{e^{-t}}{t^{n+1}} \mathrm{d}t, \tag{2}$$

is often discarded and the divergent series has either been summed straightforwardly or summed through the use of sequence transformations.

Another example of the use of integration by parts in numerical integration arises in [19], applied to the oscillatory spherical Bessel integral function involved in molecular integrals. Through a reformalized integration by parts with respect to $x\,\mathrm{d}x$, we transformed the initial spherical Bessel integral into an integral involving the simple sine function:

$$\int_0^\infty x^{n_x} \frac{\hat{k}_\nu[R_2\gamma(s,x)]}{[\gamma(s,x)]^{n_\gamma}} j_\lambda(vx)\,\mathrm{d}x = \int_0^\infty \left[\left(\frac{\mathrm{d}}{x\,\mathrm{d}x}\right)^\lambda \left(x^{n_x+\lambda-1} \frac{\hat{k}_\nu[R_2\gamma(s,x)]}{[\gamma(s,x)]^{n_\gamma}}\right)\right] \sin(x)\,\mathrm{d}x. \tag{3}$$

From a numerical point of view, the transformed sine integral function is more favorable than the initial Bessel integral function. The above integral transformation, which we caled the S transformation, was successfully applied to all molecular integrals leading to an unprecedented accuracy and efficiency [19, 20]. However, this transformation requires the boundary terms to vanish at both limits of integration and was only applied to spherical Bessel integrals [21]. In the present work, we generalize this approach to a larger class of integrals. This generalization does not require the boundary terms to vanish at both limits of integration. The contribution that will be introduced is a reformalized integration by parts with respect to $w(x)\,\mathrm{d}x$ for some weight function $w(x) = x^\mu$ for $\mu \in \mathbb{C}$ whose choice depends on the integrand. For the successive differentiation $\left(\frac{d}{x^\mu\, dx}\right)^k$ for $k = 0, 1, 2, \ldots$ required by this generalization, we will apply the Slevinsky-Safouhi formulae (SSF) [22].

We applied successfully the general reformalization of integration by parts to three well known challenging integrals, namely Bessel integrals, Fresnel integrals and Airy functions. The integral representations obtained for these integrals have better convergence properties and more favourable asymptotic behaviours.

For the computation of the integral representations, we propose a robust algorithm which will be referred to as the staircase algorithm. The algorithm uses both the boundary terms and the transformed integrals and progressively descends the integrand to an asymptotically favourable situation by applying the reformalized integration by parts at each iteration.

2 Slevinsky-Safouhi formulae (SSF) for differentiation

For more details on SSF and their applications, we refer the reader to [22, 23]. Applications of SSF in numerical integration of challenging integrals are presented in [24, 25].

Theorem 2.1 (The SSF 1 [22]) *Let $G(x) \in C^k$ with the term $\left(\frac{\mathrm{d}}{x^m\mathrm{d}x}\right)^k (x^{-n}G(x))$ well-defined and easy to compute for $m, n \in \mathbb{C}$. For $\mu, \nu \in \mathbb{C}$, the term $\left(\frac{\mathrm{d}}{x^\mu\mathrm{d}x}\right)^k (x^{-\nu}G(x))$ is given by:*

$$\left(\frac{\mathrm{d}}{x^\mu\mathrm{d}x}\right)^k (x^{-\nu}G(x)) = \sum_{i=0}^k A_k^i\, x^{n-\nu+i(m+1)-k(\mu+1)} \left(\frac{\mathrm{d}}{x^m\mathrm{d}x}\right)^i (x^{-n}G(x)), \tag{4}$$

with coefficients:

$$A_k^i = \begin{cases} 1 & \text{for} \quad i = k \\ (n - \nu - (k-1)(\mu+1))A_{k-1}^0 & \text{for} \quad i = 0,\ k > 0 \\ (n - \nu + i(m+1) - (k-1)(\mu+1))A_{k-1}^i + A_{k-1}^{i-1} & \text{for} \quad 0 < i < k. \end{cases} \tag{5}$$

Moreover, for $m \neq -1$, these coefficients have the explicit expression:

$$A_k^i = \sum_{j=0}^{i} \frac{(-1)^{i-j}\,(n - \nu + j(m+1) - (k-1)(\mu+1))_{k,\mu+1}}{(m+1)^i\, j!\,(i-j)!}, \tag{6}$$

where $(x)_{n,k}$ is the Pochhammer k-symbol [26] and can be computed as $\prod_{l=0}^{n-1}(x + kl)$.

The SSF 2 corresponds to the case $(\mu, \nu, m, n) = (0, 0, 1, 0)$ which is the most popular.

3 Reformalized integration by parts

Definition 3.1 *Let the following operator be known as integration of a function $f(x)$ by $w(x)\mathrm{d}x$, for some weight function $w(x)$:*

$$\int f(x)\, w(x)\mathrm{d}x. \tag{7}$$

Let us begin with the general integral $\int f(x)\mathrm{d}x$ for which the general antiderivative is $F(x)$. By performing the substitution $\varpi(x) \leftrightarrow x$, we arrive at $\int f(\varpi(x))\mathrm{d}\varpi(x)$. The differential element $\mathrm{d}\varpi(x)$ can be written as $\varpi'(x)\mathrm{d}x$ and by denoting $w(x) = \varpi'(x)$, we assert that this becomes *integration by* $w(x)\mathrm{d}x$, with a weight function $w(x)$.

The integral $\int f(x)\mathrm{d}x$ generally does not have a simple antiderivative; however, the integral $\int f(\varpi(x))\varpi'(x)\mathrm{d}x$ most certainly has an antiderivative with a simple representation with respect to the first integral, notably $F(\varpi(x))$.

Integration by parts is one of the main methods used in the formation of asymptotic series expansions of a given integral. When integrating by parts, we generally begin with:

$$g(x)\frac{\mathrm{d}}{\mathrm{d}x}h(x) = \frac{\mathrm{d}}{\mathrm{d}x}(g(x)h(x)) - h(x)\frac{\mathrm{d}}{\mathrm{d}x}g(x). \tag{8}$$

When integrating by parts by $x\mathrm{d}x$ as in [19], we begin with:

$$g(x)\frac{\mathrm{d}}{x\mathrm{d}x}h(x) = \frac{\mathrm{d}}{x\mathrm{d}x}(g(x)h(x)) - h(x)\frac{\mathrm{d}}{x\mathrm{d}x}g(x). \tag{9}$$

Our method, which formalizes this process is suitable for integrals of composite functions, like the aforementioned $f(\varpi(x))$.

In order to expand integrals of composite functions in asymptotic series, we must incorporate this formalism with integration by parts by $w(x)$ with $w(x) = x^\mu$ for $\mu \in \mathbb{Z}$. Naturally, then, we begin with:

$$g(x)\frac{\mathrm{d}}{x^\mu\mathrm{d}x}h(x) = \frac{\mathrm{d}}{x^\mu\mathrm{d}x}(g(x)h(x)) - h(x)\frac{\mathrm{d}}{x^\mu\mathrm{d}x}g(x). \tag{10}$$

By using the weight function $w(x) = x^\mu$, we must continue by integrating by $x^\mu \mathrm{d}x$. To clarify the following process, we make use of these functionals of $g(x)$ and $h(x)$:

$$G_l(x) = (-1)^l \left(\frac{\mathrm{d}}{x^\mu \mathrm{d}x}\right)^l g(x) \qquad \text{and} \qquad H_l(x) = \left(\frac{\mathrm{d}}{x^\mu \mathrm{d}x}\right)^{-l} h(x). \tag{11}$$

The computation of the functionals $G_l(x)$ is obtained using SSF.

With these functionals, we can rewrite (10) as:

$$G_0(x)H_0(x) = \frac{\mathrm{d}}{x^\mu \mathrm{d}x}(G_0(x)H_1(x)) + G_1(x)H_1(x), \tag{12}$$

and therefore with integration by $x^\mu \mathrm{d}x$, the first term on the right hand side becomes a boundary term and the second term on the right hand side becomes the transformed integral.

Let $f(x) \in C^m[a,b]$ be integrable on $[a,b]$ and be of the general form:

$$f(x) = G_0(x)H_0(x)\, x^\mu, \tag{13}$$

where the functionals $G_l(x)$ and $H_l(x)$ are defined as in (11).

Let us consider the integral:

$$\int_a^b f(x)\mathrm{d}x = \int_a^b G_0(x)H_0(x)x^\mu \mathrm{d}x. \tag{14}$$

By integrating the right hand side by parts, we obtain:

$$\int_a^b G_0(x)H_0(x)x^\mu \mathrm{d}x = G_0(x)H_1(x)\Big|_a^b + \int_a^b G_1(x)H_1(x)x^\mu \mathrm{d}x. \tag{15}$$

And, by integrating by parts another $n-1$ times, we obtain the equivalent representation:

$$\int_a^b f(x)\mathrm{d}x = \sum_{l=0}^{n-1} G_l(x)H_{l+1}(x)\Big|_a^b + \int_a^b G_n(x)H_n(x)x^\mu \mathrm{d}x. \tag{16}$$

Naturally, then, the question arises in how to use the equivalent representations. Indeed, one could simply neglect the remaining integral as $n \to \infty$ and form a series from boundary terms; or, one could evaluate the boundary terms up to m and continue to numerically evaluate the new integral. To these two proposed methods, there exist advantages and disadvantages. The boundary terms, for example, usually form divergent series, which are generally challenging to sum; however, that the integral becomes a simply series can lead to reduced calculation times. The new integral, on the other hand, may be asymptotically favourable; however, the computation of the new integrand may become challenging after a certain order.

The next algorithm we propose is actually a blend of the two extremes outlined above. We numerically integrate the integral to some point $x = x_0$, $a < x_0 < b$. At such a point, we now apply the first iteration of the reformalized integration by parts by $x^\mu dx$ to the remaining integral. We continue by evaluating the boundary terms at $x = b$ and $x = x_0$ and we numerically integrate the new integral to another point $x = x_1$, $x_0 < x_1 < b$. We continue in this manner until the desired accuracy is attained. This blended algorithm will be aptly known as the staircase algorithm, as it progressively descends the integrand into an evermore asymptotically favourable situation.

This methodology is concisely summarized in:

Algorithm 3.2 (The Staircase Algorithm) *The approximations to the integral* $\int_a^b f(x)\mathrm{d}x$ *take the following form. For* $a < x_0 < b$, *initialize:*

$$S_0 = \int_a^{x_0} G_0(x)H_0(x)x^\mu \mathrm{d}x, \tag{17}$$

For $l = 1, 2, \ldots, n$, *and for the sequence* $\{x_l\}_{l=1}^n$ *that satisfies* $a < x_{l-1} < x_l < b$ *define:*

$$S_l = S_{l-1} + G_l(x)H_{l+1}(x)\Big|_{x_{l-1}}^{b} + \int_{x_{l-1}}^{x_l} G_l(x)H_l(x)x^\mu \mathrm{d}x, \tag{18}$$

and the approximations to the integral $\int_a^b f(x)\mathrm{d}x$ *form the sequence* $\{S_l\}_{l=0}^n$.

3.1 Bessel Integrals

The integral that follows appeared in Numerical Recipes as a computational challenge and an example of application of sequence transformations:

$$\mathcal{I}_1 = \int_0^\infty \frac{x}{x^2+1} J_0(x)\,\mathrm{d}x = \int_0^\infty \frac{1}{x^2+1} J_0(x)\,x\,\mathrm{d}x. \tag{19}$$

We begin by noting that the integrand $f(x)$ satisfies $f(x) \in C^\infty(-\infty, +\infty)$. This integral is already in an appropriate form for the reformalized integration by parts, since we can subdivide the integrand as the product of:

$$G_0(x) = \frac{1}{x^2+1}, \quad H_0(x) = J_0(x), \quad \text{and} \quad w(x) = x. \tag{20}$$

Starting with these root forms, we can explicitly develop the functionals $G_l(x)$ and $H_l(x)$ as:

$$G_l(x) = \frac{2^l\, l!}{(x^2+1)^{l+1}} \quad \text{and} \quad H_l(x) = x^l J_l(x), \tag{21}$$

through Bessel function properties. The integral then has the equivalent representation:

$$\mathcal{I}_1 = 2^n\, n! \int_0^\infty \frac{x^{n+1}}{(x^2+1)^{n+1}} J_n(x)\,\mathrm{d}x. \tag{22}$$

In (22), all the boundary terms vanish.

In Numerical Recipes, the value of the integral is determined through Cauchy's residue theorem to be $\mathcal{I}_1 = K_0(1)$. Although the equivalent representation does not give this formula, it follows from (22) that:

$$2^n\,\Gamma(n+1) \int_0^\infty \frac{x^{n+1}}{(x^2+1)^{n+1}} J_n(x)\,\mathrm{d}x = K_0(1). \tag{23}$$

That is, each and every one of these integrals for parameterized n is also given explicitly by the Bessel function $K_0(1)$.

3.2 Fresnel Integrals

Let us consider the Fresnel-type integral:

$$\mathcal{I}_2(a,\upsilon) = \int_a^\infty \sin(\upsilon\, x^2)\mathrm{d}x. \tag{24}$$

We begin by noting that the integrand $f(x)$ satisfies $f(x) \in C^\infty(-\infty,+\infty)$. By identifying the functions:

$$G_0(x) = \frac{1}{x}, \quad H_0(x) = \sin(\upsilon\, x^2), \quad \text{and} \quad w(x) = x, \tag{25}$$

the functionals $G_l(x)$ and $H_l(x)$ can be readily determined as:

$$G_l(x) = \frac{(2l)!}{2^l\, l!\, x^{2l+1}} \quad \text{and} \quad H_l(x) = \frac{\sin(\upsilon\, x^2 - l\pi/2)}{(2\,\upsilon)^l}. \tag{26}$$

It then follows straightforwardly that (24) has the equivalent representations:

$$\mathcal{I}_2(a,\upsilon) = \frac{(2n)!}{(4\,\upsilon)^n\, n!}\int_a^\infty \frac{\sin(\upsilon\, x^2 - n\pi/2)}{x^{2n}}\mathrm{d}x - \sum_{l=0}^{n-1} \frac{2\,(2l)!}{(4\,\upsilon)^{l+1}\, l!}\frac{\sin(\upsilon\, x^2 - (l+1)\pi/2)}{x^{2l+1}}\Bigg|_{x=a}, \tag{27}$$

through application of the reformalized integration by parts..

Similarly, we obtain the equivalent representation for $\tilde{\mathcal{I}}_2(a,\upsilon) = \int_a^\infty \cos(\upsilon\, x^2)\mathrm{d}x$ which is given by:

$$\tilde{\mathcal{I}}_2(a,\upsilon) = \frac{(2n)!}{(4\,\upsilon)^n\, n!}\int_a^\infty \frac{\cos(\upsilon\, x^2 - n\pi/2)}{x^{2n}}\mathrm{d}x - \sum_{l=0}^{n-1} \frac{2\,(2l)!}{(4\,\upsilon)^{l+1}\, l!}\frac{\cos(\upsilon\, x^2 - (l+1)\pi/2)}{x^{2l+1}}\Bigg|_{x=a}. \tag{28}$$

3.3 Airy Functions

The Airy functions are given by:

$$\begin{aligned} \mathcal{I}_3(a,z) &= \int_a^\infty \cos\left(\frac{x^3}{3} + zx\right)\mathrm{d}x \\ &= \int_a^\infty \left[\cos(zx)\cos\left(\frac{x^3}{3}\right) - \sin(zx)\sin\left(\frac{x^3}{3}\right)\right]\mathrm{d}x. \end{aligned} \tag{29}$$

We begin by noting that the integrand $f(x)$ satisfies $f(x) \in C^\infty(-\infty,+\infty)$. By identifying the functions:

$$G_0(x) = x^{-2}\,{}^{\cos}_{\sin}(zx), \quad H_0(x) = {}^{\cos}_{\sin}\left(\frac{x^3}{3}\right), \quad \text{and} \quad w(x) = x, \tag{30}$$

the functionals $G_l(x)$ and $H_l(x)$ can be readily determined as:

$$G_l(x) = \left(\frac{-\mathrm{d}}{x^2\,\mathrm{d}x}\right)^l x^{-2}\,{}^{\cos}_{\sin}(zx) \quad \text{and} \quad H_l(x) = {}^{\cos}_{\sin}\left(\frac{x^3}{3} - \frac{l\pi}{2}\right). \tag{31}$$

The integral then has the equivalent representation:

$$\begin{aligned} \mathcal{I}_3(a,z) &= \sum_{l=0}^{n-1}\frac{(-1)^{l+1}}{x^{3l+2}}\sum_{i=0}^{l} A_l^i(zx)^i\cos\left(\frac{x^3}{3} + zx - \frac{(l+1-i)\pi}{2}\right)\Bigg|_a \\ &+ (-1)^n\int_a^\infty \sum_{i=0}^{n}\frac{A_n^i(zx)^i}{x^{3n}}\cos\left(\frac{x^3}{3} + zx - \frac{(n-i)\pi}{2}\right)\mathrm{d}x. \end{aligned} \tag{32}$$

4 Numerical Results and Discussion

As previously alluded to, a numerical evaluation of the integrals $\mathcal{I}_1$, $\mathcal{I}_2(a,v)$, $\tilde{\mathcal{I}}_2(a,v)$ and $\mathcal{I}_3(a,z)$ can be achieved through a multiplicity of methods. Historically, the popular candidates were the summation of the generally divergent boundary terms, although the transformed integral has also been used in cases where the boundary terms vanished. In this work, we implemented the Staircase algorithm, as the newest method to evaluate these integrals and serves as an original contribution to the overall methodology we will be developing.

The selection of an appropriate sequence $\{x_l\}_{l=0}^{n}$ for the integration subintervals in the Staircase algorithm is critical. This sequence should leave the integral subintervals to be numerically integrable – the points cannot be spaced so far apart that a quadrature routine is incapable of providing the solution. Conversely, the sequence should not be so squished that the sum of the boundary terms and the integral subintervals introduces a reduction in accuracy from the addition of a very large number and a very large negative number. Generally speaking, the dependence on l of the sequence $\{x_l\}_{l=0}^{n}$ resides in the integrand. For the cases at hand, the integrands are oscillatory, and thus if the difference $x_l - x_{l-1}$ is on the same order of magnitude as the difference between successive zeros of the integrand, both of the aforementioned scenarios can be averted.

As can be seen from the numerical table, the algorithm is capable of reaching high accuracy. This is mainly due to the fact that reformalized integration by parts progressively descends the integrand to an asymptotically favourable situation.

5 Numerical tables

Table 1. Numerical Results of Staircase Algorithm applied to Bessel integrals $\mathcal{I}_1$ with $x_l = 2\pi(l+1)$, Fresnel integrals with $\mathcal{I}_2(0,1)$. $x_l = \sqrt{2\pi(l+1)}$, and Airy functions $\mathcal{I}_4(0,1)$ with $x_l = \sqrt[3]{6\pi(l+1)}$.

l	S_l for $\mathcal{I}_1$	S_l for $\mathcal{I}_2(0,1)$	S_l for $\mathcal{I}_4(0,1)$
0	.414 193 276 771 795	.629 878 864 869 732	.432 683 511 614 577
1	.421 696 746 593 657	.627 294 419 199 049	.425 050 712 855 878
2	.421 072 353 906 909	.626 651 451 723 302	.425 018 211 967 205
3	.421 020 653 966 770	.626 655 488 072 699	.425 032 964 456 715
4	.421 023 974 243 269	.626 657 083 038 776	.425 033 674 972 581
5	.421 024 464 857 404	.626 657 073 115 469	.425 033 663 440 207
6	.421 024 443 256 394	.626 657 068 616 796	.425 033 661 201 175
7	.421 024 438 053 642	.626 657 068 644 466	.425 033 661 169 297
8	.421 024 438 183 915	.626 657 068 657 872	.425 033 661 174 750
9	.421 024 438 241 771	.626 657 068 657 790	.425 033 661 174 971
10	.421 024 438 241 364	.626 657 068 657 749	.425 033 661 174 961
11	.421 024 438 240 707	.626 657 068 657 749	.425 033 661 174 960
12	.421 024 438 240 700	.626 657 068 657 750	
13	.421 024 438 240 708		
Exact Values	.421 024 438 240 708	.626 657 068 657 750	.425 033 661 174 960

Acknowledgement

The author acknowledges the financial support for this research by the Natural Sciences and Engineering Research Council of Canada (NSERC) – Grant RGPIN-2016-04317.

References

[1] C. Brezinski, M. Redivo-Zaglia, *Extrapolation Methods: Theory and Practice* (Edition North-Holland, Amsterdam, 1991)
[2] P. Rabinowitz, Numer. Algo. **3**, 17 (1992)
[3] J. Lyness, ANZIAM J. **42**, 181 (2000)
[4] A. Sidi, *Practical Extrapolation Methods: Theory and Applications* (Cambridge U. P., Cambridge, 2003)
[5] J. McWilliams, J. Balusek, H. Gray, Comm. Stat. Simu. Comput. **33**, 1 (2004)
[6] D. Levin, A. Sidi, Appl. Math. Comput. **9**, 175 (1981)
[7] H. Gray, S. Wang, SIAM J. Numer. Anal. **29**, 271 (1992)
[8] R. Slevinsky, H. Safouhi, Numer. Algor. **48**, 301 (2008)
[9] R. Slevinsky, H. Safouhi, Int. J. Quantum Chem. **109**, 1741 (2009)
[10] D. Huybrechs, S. Vandewalle, SIAM J. Numer. Anal. **44**, 1026 (2006)
[11] D. Levin, J. Comput. Appl. Math. **67**, 95 (1996)
[12] S. Olver, IMA J. Num. Anal. **26**, 213 (2006)
[13] J. Lyness, J. Lottes, BIT Numer. Math. **49**, 397 (2009)
[14] J. Lyness, G. Hines, Math. Comp. **12**, 24 (1986)
[15] K. Kim, R. Cools, L. Ixaru, Appl. Numer. Math. **46**, 59 (2003)
[16] G. Evans, K. Chung, J. Webster, App. Numer. Math. **34**, 85 (2000)
[17] G. Adam, A. Nobile, IMA J. Numer. Anal. **11**, 271 (1991)
[18] G. Adam, S. Adam, N.M. Plakida, Comput. Phys. Comm. **154**, 49 (2003)
[19] L. Berlu, H. Safouhi, J. Phys. A: Math. Gen. **36**, 11791 (2003)
[20] L. Berlu, H. Safouhi, J. Phys. A: Math. Gen. **37**, 3393 (2004)
[21] H. Safouhi, J. Math. Chem. **29**, 213 (2001)
[22] R. Slevinsky, H. Safouhi, J. Comput. App. Math. **233**, 405 (2009)
[23] R. Slevinsky, H. Safouhi, Numerical Algorithm **66**, 457 (2014)
[24] P. Gaudreau, R. Slevinsky, H. Safouhi, SIAM J. Sci. Comput. **34**, B65 (2012)
[25] R. Slevinsky, H. Safouhi, Numer. Algor. **60**, 315 (2012)
[26] R. Díaz, E. Pariguan, Divulgaciones Matematica **15**, 179 (2007)

New Possibilities and Applications of the Least Squares Collocation Method

Vasiliy Shapeev[1,2,⋆], *Vasiliy* Belyaev[1,⋆⋆], *Sergey* Golushko[2,3,⋆⋆⋆], and *Semyon* Idimeshev[3,⋆⋆⋆⋆]

[1] *Institute of Theoretical and Applied Mechanics of SB RAS*
[2] *Novosibirsk State University*
[3] *Institute of Computational Technologies of SB RAS*

Abstract. Least squares collocation method (LSC) is a versatile numerical method for solving boundary value problems for PDE. The present article demonstrates the abilities of LSC to solve various problems – in particular, calculations of bending of isotropic irregular shaped plates and multi-layered anisotropic plates. In order to achieve higher accuracy, new versions of the method utilize high-degree polynomial spaces. The numerical experiments demonstrate high accuracy of the solutions.

1 Introduction

The least squares collocation method (LSC), as the name suggests, is a combination of two methods. In LSC, the initial differential problem

$$Lu(x) = f(x), \qquad x \in \Omega \tag{1}$$

with boundary conditions

$$lu(x) = g(x), \qquad x \in \partial\Omega \tag{2}$$

is projected by collocation into a finite-dimensional functional space. Different polynomial spaces are often used as functional spaces. Here, L and l denote differential operators, $u = (u_1, u_2, \dots, u_m)$ — unknown vector function, $x = (x_1, x_2, \dots, x_n)$, $x_1, x_2, \dots, x_n$ — independent variables, f and g — given vector functions.

The approximate solution $u_a(x)$ of the problem (1), (2) is represented as a linear combination of basis elements $\phi_i(x)$ of selected functional space with unknown coefficients c_i:

$$u_a(x) = \sum_{i=1}^{N} c_i \phi_i(x). \tag{3}$$

⋆e-mail: vshapeev@gmail.com
⋆⋆e-mail: belyaevasily@mail.ru
⋆⋆⋆e-mail: s.k.golushko@gmail.com
⋆⋆⋆⋆e-mail: idimeshev@gmail.com

If the initial problem is nonlinear, we can utilize linearization and include nonlinearity iterations in algorithm. Therefore, (1) and (2) are considered as linear onwards.

The domain Ω is covered by a grid, with a local solution required for each grid cell. Similarly to collocation method, in order to define a local solution, collocation equations are obtained by substituting (3) into (1) at given points $x_c \in \Omega$. Boundary conditions are obtained by substituting (3) into (1) at given points $x_b \in \partial\Omega$. Additionally, LSC requires matching conditions (MC) between solutions in neighbor cells. Matching conditions must not contradict the definition of problem solution. Choice of MC type is up to the mathematician. An example of MC are continuity requirements for analytical relations $l_m(u) = 0$ defined at selected points x_m on common sides of neighbor cells. Thus, the system of linear algebraic equations (SLAE) for unknown coefficients $c = (c_1, \dots, c_N)$ is constructed of collocation equations $Lu_a(x_c) = f(x_c)$, boundary conditions $lu_a(x_b) = g(x_b)$, and matching conditions. In order to define a local solution in each cell, an overdetermined (local) SLAE $A_e c = b_e$ with a rectangular full-rank $M \times N$ matrix A_e, $(M > N)$ must be formed. The set of all local SLAE produces the global SLAE $Ac = b$. The matrix of this global SLAE is coarse, therefore, algorithms that take matrix coarseness into consideration are best suited for its solution. In practice, for the global SLAE solution, it is convenient to use an iteration process based on direct methods for the local SLAE. In order to do this, all cells of the domain are sequentially processed in one "global" iteration, and in each cell, local SLAE are solved with direct methods. During this process, coefficients c in neighbor cells can be taken from the previous global iteration (Gauss-Jacobi process) or current iteration (Gauss-Seidel process). Implemented versions of LSC are further detailed in [1, 2, 7–10].

Several methods can be used to solve the corresponding linear least squares problem. Within the normal equations method, we define a vector $\check{c}$ from the set of all possible pseudosolutions of the overdetermined system $A_e c = b_e$ as a solution of the system $A_e^T A_e \check{c} = A_e^T b$ with normal (symmetrical and positive-definite) matrix $A_1 = A_e^T A_e$. The solution obtained in this way minimizes the residue functional $\Phi = \sum_{j=1}^{M} (r_j)^2$ on the set of all possible pseudosolutions of $A_e c = b_e$. Here, the residue $r = (r_1, \dots, r_M) \equiv A_e c - b_e$. The same solution $\check{c}$ of system $A_e c = b_e$ can also be found by QR decomposition of the matrix A_e. In order to achieve this, we use conventional methods to find an orthogonal matrix Q such that the first N rows of the matrix $R_e = QA_e$ would form the upper triangular matrix R; the solution is then found by solving the first N equations of the system $R_e c = Qb_e$. If ν is the matrix condition number, the following relations hold: $\nu(A_1) = \nu^2(A_e)$ and $\nu(QA_e) = \nu(A_e)$. Thus, if the SLAE is ill-conditioned, the second method is favorable. This circumstance was systematically taken into account during the development of LSC versions for solving various types of problems.

Presently, different LSC versions are implemented for various PDE system types, with different approximation orders, with regular and irregular grids including irregular domains, using multi-grid complexes, preconditioners, Krylov subspaces, utilizing varied polynomial spaces, different matching conditions. We simultaneously investigated method properties and applicability for problems with singularities, comparing numerical results with those published by other authors and found in benchmark problems. LSC has been successfully used in solving problems of viscous flow in cavities, calculating 3D thermophysical processes in laser welding of plates made of homogeneous and heterogeneous metals (including multi-layer composite insets), calculating stress in isotropic and anisotropic plates of various shapes, plates and rectangular beams made of composite materials, and so on. We briefly explain some of the aforementioned and new results.

2 *p*- and *hp*- versions of LSC method

Each cell of the grid Ω_i, $(i = 1, \ldots, K)$ has local coordinates for convenience, e.g. in case of a 2D domain:

$$\xi_1 = \frac{x_1 - x_1^*}{h}, \qquad \xi_2 = \frac{x_2 - x_2^*}{h}.$$

For a grid with rectangular cells (x_1^*, x_2^*) is cell center; $h = (h_1\, h_2)^{1/2}$ – cell dimensions along x_1, x_2 axes, respectively. We will use the term h-version of LSC if the grid has a great number of cells and the solution representation for (3) has a small number of basis elements. Conversely, a p-version of LSC has a single cell with possibly many basis elements in the representation (3). Initially, the researchers used and studied h-versions of LSC for solving mathematical physics equations and Navier-Stokes equations; these versions had small degree of basis polynomials [1], but sufficient for the approximation of (1), (2). However, achieving good solution accuracy using h-version with low approximation order requires substantially finer grids and respectively large SLAE. The inevitable rounding error accumulation significantly limits the potential of h-versions of LSC. p-versions of the method (pseudospectral versions) were later examined [7]: the domain consisted of one cell, and various high-degree polynomials were used to achieve acceptable solution accuracy. Numerical experiments showed that using p-versions may yield acceptable solution accuracy for applied problems, e.g. layered fluid flow, isotropic and anisotropic plate strength. However, if solution smoothness is limited, usage of p-version may be inefficient. Variants of hp-versions of LSC are most practical when used with moderate values of basis polynomial degrees and a moderate number of grid cells. They are more flexible and efficient than h- and p-versions.

Let us demonstrate the above by solving the Dirichlet problem for the Poisson equation:

$$\frac{\partial^2 u(x_1,x_2)}{\partial x_1^2} + \frac{\partial^2 u(x_1,x_2)}{\partial x_2^2} = \frac{\partial^2 u_{ex}(x_1,x_2)}{\partial x_1^2} + \frac{\partial^2 u_{ex}(x_1,x_2)}{\partial x_2^2}, \tag{4}$$

$$u(x_1,x_2) = u_{ex}(x_1 x_2), \quad (x_1,x_2) \in \partial\Omega, \qquad \Omega = [-1;1] \times [-1;1].$$

The exact solution is a product of functions from Runge's example [3]:

$$u_{ex}(x_1,x_2) = \frac{1}{(1 + 25\,x_1^2)}\frac{1}{(1 + 25\,x_2^2)}. \tag{5}$$

The approximate solution $u_a(x_1,x_2)$ has a form

$$u_a(\xi_1,\xi_2) = \sum_{i=0}^{N_1-1}\sum_{j=0}^{N_2-1} c_{ij} T_i(\xi_1) T_j(\xi_2), \qquad (\xi_1,\xi_2) \in [-1;1] \times [-1,1], \tag{6}$$

where $T_i(\xi)$ – i-th degree Chebyshev polynomials. The roots of polynomials serve as coordinates for collocation, matching, and boundary points. Usage of p-version (single cell) with $N_1 = N_2 = 100$ allowed to achieve a solution in C-norm with error $E_c = 1.96 \cdot 10^{-7}$, whereas hp-version with only four cells and $N_1 = N_2 = 25$ led to error $E_c = 3.10 \cdot 10^{-14}$. The advantage of the hp-version over the h-version was noticed while attempting to improve solution accuracy of the benchmark problem of viscous flow inside a square cavity with Reynolds number within the interval [1000; 7500]. Numerical experiments conducted on hp-version with a monomial basis (max degree 6) and Re=1000 yielded solution accuracy of $\approx 10^{-8}$. This was determined after a large number of numerical experiments and comparison with results of high-accuracy solutions published by other researchers.

3 Application of LSC method to calculation of stress and displacement fields in composite plates

Stress and displacement fields calculation in composite constructions and elements is an important problem for various industries. Numerical modeling in this field of knowledge often allows to substitute physical experiments, thus accelerating and cheapening the development of new materials and constructions. Researching strength properties of multi-layered composite plates under load often requires stress and displacement fields calculation. Assume that the plate layers consist of materials with different strength properties, and these materials are in elastic state. Such calculation can be performed by means of 3D elasticity theory. This approach requires solving a complex resource-intensive problem, i.e. solving a system of 3D equations for every layer while considering interaction conditions between layers. Solving this kind of problem requires significant computational power, which considerably lowers the effectiveness of numerical modeling. To solve this problem with a desktop computer, instead of 3D elasticity theory, we suggest using its 2D approximations. We use three composite plate theories: classical laminated plate theory (CLPT) [4], first shear deformation theory (FSDT) [4], and Grigolyuk-Chulkov theory (GCT) [5]. It is necessary to define their accuracy if possible and evaluate their applicability area. Therefore, the numerical method must not add any error to PDE solution greater than the error of the observed models. In order to evaluate the applicability of each model, results were compared to those derived from 3D elasticity theory. In particular, for bending problem of three-layered plate under uniform load [6]. A substantial number of calculations has been performed. Results of one of them are demonstrated in table 3.

Table 1. Values of bending functions in the center of a three-layered plate, calculated using 4 models: 3D elasticity theory, classical laminated plate theory (CLPT), first shear deformation theory (FSDT), Grigolyuk-Chulkov theory (GCT). Considered plates have varied thickness h/a, h – thickness, a – characteristic size in plate plane. The three right columns show percentage difference of bending values in each plate theory model relative to 3D elasticity theory model.

h/a	3D Elasticity	CLPT	FSDT	GCT	CLPT(%)	FSDT(%)	GCT(%)
0.25	2.618(−5)	4.680(−6)	2.200(−5)	2.544(−5)	82.12	15.95	2.81
0.1	1.333(−4)	7.310(−5)	1.139(−4)	1.306(−4)	45.16	14.55	2.02
0.05	7.078(−4)	5.850(−4)	6.710(−4)	7.015(−4)	17.35	5.20	0.89
0.02	9.427(−3)	9.140(−3)	9.230(−3)	9.421(−3)	3.05	2.09	0.07
0.01	7.368(−2)	7.310(−2)	7.341(−2)	7.360(−2)	0.79	0.37	0.11

4 Solving boundary value problems for PDE in irregular domains with high-order accuracy

Lately, mathematical modeling specialists have concluded that development of new methods with high-order accuracy is required, due to the necessity of modeling complex physical processes. In particular, high accuracy numerical methods are required for solving problems in irregular domains, since most simulated physical processes take place in such domains. Development of these methods based on conventional methods is impossible or significantly difficult. For instance, in finite difference methods, a boundary condition related to boundary points is assigned to nearest grid nodes not located on boundary. An error of order $O(h)$ is thus brought into the finite difference solution. In many cases, the approximation order of the solution is rarely higher than 1. With relative ease, the LSC method allows an approximation of the problem (1), (2) with high-order accuracy in variously shaped domains. Its evident property is that the approximate solution inside each grid cell is not strictly dependent on the cell shape. This allows to select a single layer of variously shaped cells near domain boundary and approximate the boundary conditions at points x_b which are located exactly on the boundary $\partial\Omega$. The above feature gives an opportunity to avoid the irremovable error related to shifting points away from boundary. Obviously, working with variously shaped cells complicates the algorithm for the problem (1), (2) approximation. Different implemented versions of these algorithms have relatively varied solution accuracy for the same domain. Since it is not possible to describe them in detail here, we only point out their characteristic features which are also used in other methods. For example, it became apparent that using a regular grid within the domain Ω is advisable in versions with rectangular cells implemented in [9, 10]. In this case, the transition from cell to cell during calculations is done by simply shifting the index (cell number) by 1. Therefore, there is no need to store neighbor position for every cell in computer memory or use search algorithms based on graph theory. The latter complicates working program algorithms and increases solution time, as specialists point out. Let the cell intersected by the domain boundary $\partial\Omega$ be called boundary cell; its part within Ω is called the irregular cell (IC), and its part outside Ω is the extraboundary part.

Table 2. $N_1 \times N_2$ — grid size, N_{iter} — number of iterations

$N_1 \times N_2$	E_c	N_{iter}	$\log_2 \frac{E_{N_1/2,N_2/2}}{E_{N_1,N_2}}$
10×10	$8.19 \cdot 10^{-9}$	63	—
20×20	$3.41 \cdot 10^{-10}$	120	4.58
40×40	$1.75 \cdot 10^{-11}$	321	4.28
80×80	$3.92 \cdot 10^{-13}$	729	4.48

LSC algorithms that utilize extraboundary parts, i.e. collocation points within them (type *I* algorithms), are simpler than algorithm versions which use only irregular cells (type *II* algorithms). The technique of grid refinement in multiple numerical experiments in various domains shows that like in other methods, using elongated cells and small cells next to larger cells decreases solution accuracy. A formal check for cell size was included in automated grid building programs. Small IC in type II

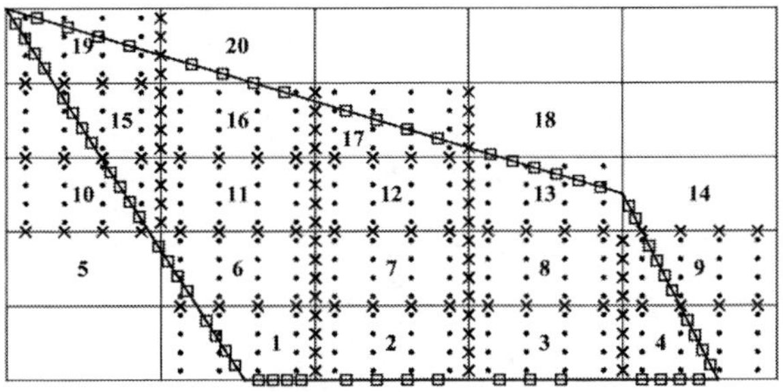

Figure 1. The numerical domain (grid size is 5 × 5). The symbol • denotes the collocation points; × — the matching points; □ — the points for recording the boundary conditions. The triangle IC 5 joins with the cell 6, the IC 14 and 18 join with the IC 13, the IC 20 joins with the cell 16.

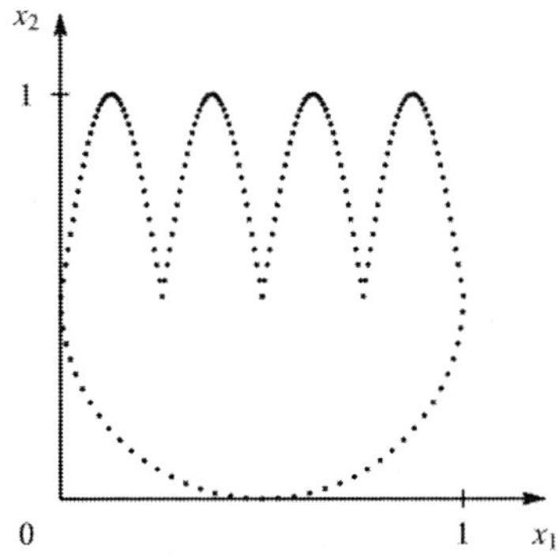

Figure 2. The boundary of domain defined by a discrete set of points. In this case, 5 bisplines have been used to describe the boundary of domain.

algorithms were automatically joined to nearest regular or irregular cells. A unified solution was then found for such joined cells. Mixed type algorithms which are more flexible than type I and II have also been implemented. A sample grid built with mixed algorithm is displayed in figure 1. Presently, three types of algorithms have been implemented which use 2nd, 4th, 8th degree polynomial spaces for solving the Dirichlet problem for the Poisson equation over arbitrary triangular and convex quadrilateral domains (figure 1), and a domain with curvilinear boundary comprised of smooth curve pieces [9, 10]. A curvilinear boundary domain is represented either as a set of joined analytic-form curve pieces, or a discrete set of points located on the boundary (figure 2). In the latter case, the program builds a continuous spline $(x_1(t), x_2(t))$ if boundary is smooth, or a piece-wise spline if boundary has breakpoints.

An example of discretely defined domain boundary can be seen in figure 2. We used it to verify convergence of numerical solution of the Dirichlet problem for the Poisson equation. Functions without singularities were used as boundary conditions in this example. Numerical experiments performed on a sequence of refining grids demonstrated high convergence order of numerical solution for problems without singularities. The regular cell size was formally used as cell size within the sequence. Tests show that the solution of a well-conditioned differential problem with high-degree polynomials has a high convergence order, and its accuracy on relatively coarse grids is close to the accuracy of the computer number representation. Results for solving the Dirichlet problem for the Poisson equation with test solution $u(x_1, x_2) = e^{x_1} + e^{x_2}$ over the quadrilateral domain with vertices $(0, 0)$, $(1, 0)$, $(0.8, 0.3)$, $(-0.3, 0.9)$ are shown in table 2. LSC method with 4th degree polynomial space was used in this case. The table demonstrates a convergence order not less than 4.

For stress and displacement fields calculation of variously shaped isotropic plates, we implemented mixed type algorithms using 4th degree polynomial space for solving the biharmonic equation in irregular domains. Plate deflection value is defined by the solution of the biharmonic equation

$$\frac{\partial^4 w(x_1, x_2)}{\partial x_1^4} + 2\frac{\partial^4 w(x_1, x_2)}{\partial x_1^2 \partial x_2^2} + \frac{\partial^4 w(x_1, x_2)}{\partial x_2^4} = \frac{q(x_1, x_2)}{D}, \qquad (x_1, x_2) \in \Omega. \tag{7}$$

In (7), $w(x_1, x_2)$ denotes the deflection of plate midplane, $q(x_1, x_2)$ – the transverse load, $D = Eh^3/12(1 - \nu^2)$ – the plate rigidity under bending, h – the plate thickness, E, ν – Young's module and Poisson ratio of isotropic plate material, domain Ω – the midplane. The following boundary conditions may be specified for plate edges:

$$w = 0, \qquad \frac{\partial w}{\partial n} = 0 \qquad \text{— clamped edge;} \tag{8}$$

$$w = 0, \qquad M_n w = 0 \qquad \text{— simply-supported edge;} \tag{9}$$

$$M_n w = 0, \qquad V_n w = 0 \qquad \text{— free edge.} \tag{10}$$

Here, n is the external normal to domain boundary $\partial\Omega$, M_n – the second order differential operator, V_n – the third order differential operator [12].

Table 3. Results of numerical experiments

$\Vert w_{N_1,N_2} - w_{N_1/2,N_2/2} \Vert_c$	Norm value	$\log_2 \frac{\Vert w_{N_1,N_2} - w_{N_1/2,N_2/2}\Vert_c}{\Vert w_{N_1/2,N_2/2} - w_{N_1/4,N_2/4}\Vert_c}$
$\Vert w_{20,20} - w_{10,10} \Vert_c$	$6.33 \cdot 10^{-4}$	3.16
$\Vert w_{40,40} - w_{20,20} \Vert_c$	$1.03 \cdot 10^{-4}$	2.61
$\Vert w_{80,80} - w_{40,40} \Vert_c$	$2.41 \cdot 10^{-5}$	2.09

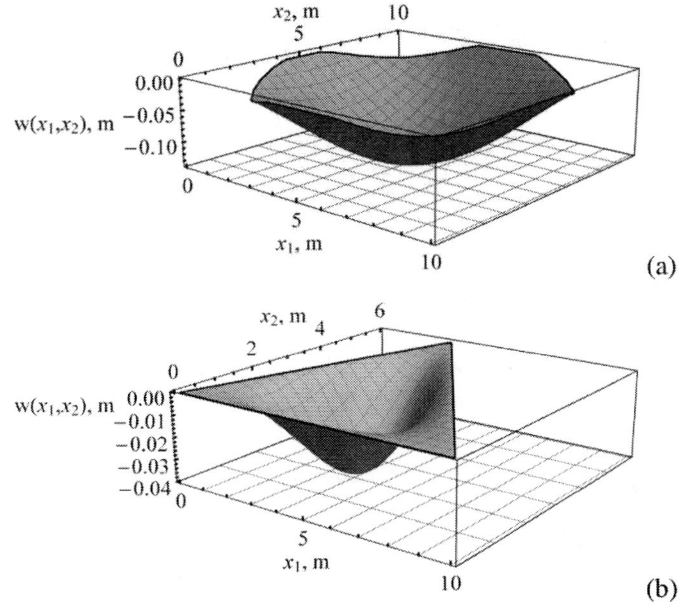

Figure 3. Deflection w of the plates under the action of transverse load: (a) plate with curvilinear boundary; (b) triangular plate, uniform load q = 1 MPa (edges are clamped). Here, $h = 0.1$ m, $E = 200$ GPa, $\nu = 0.28$. The image is scaled along the vertical axis for better visibility.

For $q = 10^5 \sin(\pi x_1 \backslash d_1) \sin(\pi x_2 \backslash d_2)$, (7) has the exact solution

$$w_e(x_1, x_2) = \frac{q d_1^4 d_2^4}{\pi^4 D \left(d_1^2 + d_2^2\right)^2}$$

($d_1 = d_2 = 10$ are the plate sizes). The boundary comprised of various curve pieces was used for testing purposes. The equation (7) is solved within this domain with corresponding right part and boundary values (9), in which instead of second equation we use $M_n w = M_n w_e$. The numerical solution of this test problem over a 64×64 grid reaches accuracy of $\approx 10^{-5}$ (figure 3a) for deflection value. Let us examine the problem with no known analytical solution. Assume the isotropic triangular plate with vertices (0, 0), (10, 0), (3, 6) is under the uniform transverse load $q = 1$ MPa. All plate edges are clamped (figure 3b). Results of this numerical experiment are shown in table 3.

5 Conclusion

Only several examples of LSC method calculations for PDE have been demonstrated in this article. This numerical method can be effectively applied to solving a broad variety of practically important problems due to its high versatility.

References

[1] A. G. Sleptsov, Modelirovanie v mekhanike **5 (22)**, 2, 101–126 (1991)

[2] V. I. Isaev and V. P. Shapeev, Comput. Math. and Math. Phys. **50**, 1758–1770 (2010)
[3] W. Cheney, *A Course in Approximation Theory* (W. Light Brooks, Cole, 2000) 360 pp.
[4] J.N. Reddy, *Mechanics of Laminated Composite Plates and Shells* (CRC Press, 2004) 856 pp.
[5] E.I. Grigoluk and T.M. Kulikov, *Multilayered Reinforced Shells. Design of Pneumatic Tires* (Mashinostroyenie, Moscow, 1988) 287 pp.
[6] N. Pagano, Composite Materials **4**, 20–34 (1970)
[7] S.K. Golushko, S.V. Idimeshev, and V.P. Shapeev, Computational technologies **19**, 5, 24–36 (2014)
[8] S.K. Golushko at al., in *MIT-2016*, 299–311 (Proceedings, Vrnjacka Banja, Serbia – Budva, 2017)
[9] V.P. Shapeev and V.A Belyaev, Computational technologies **21**, 5, 95–110 (2016)
[10] V.A. Belyaev and V.P. Shapeev, Computational technologies **22**, 4, 21 (2017)
[11] J.W. Demmel, *Applied numerical linear algebra* (Philadelphia, PA, USA, 1997) 416 pp.
[12] S.P. Timoshenko and S. Woinowsky-Krieger, *Theory of plates and shells, 2 ed.* (McGraw-Hill, New York, 1959) 580 pp.

Directed Percolation: Calculation of Feynman Diagrams in the Three-Loop Approximation

Loran Ts. Adzhemyan[1], *Michal* Hnatič[2,3,4], *Mikhail V.* Kompaniets[1], *Tomáš* Lučivjanský[4,5], and *Lukáš* Mižišin[2,3,⋆]

[1] *Saint Petersburg State University, St. Petersburg, Russia*
[2] *Bogoliubov Laboratory of Theoretical Physics, JINR, Dubna, Russia*
[3] *Slovak academy of Sciences, Institute of Experimental Physics, Košice, Slovak Republic*
[4] *Faculty of Sciences, P. J. Šafárik University in Košice, Košice, Slovak Republic*
[5] *Peoples' Friendship University of Russia (RUDN University), Moscow, Russia*

Abstract. The directed bond percolation process is an important model in statistical physics. By now its universal properties are known only up to the second-order of the perturbation theory. Here, our aim is to put forward a numerical technique with anomalous dimensions of directed percolation to higher orders of perturbation theory and is focused on the most complicated Feynman diagrams with problems in calculation. The anomalous dimensions are computed up to three-loop order in $\varepsilon = 4 - d$.

1 Introduction

The directed bond percolation process (DP) constitutes one of the simplest non-equilibrium systems in statistical physics [1–3]. It is one of the basic examples of an absorbing phase transition between an absorbing and active state. The DP universal class is important for the non-equilibrium physics and it may be considered as important as the Ising model for the static equilibrium systems.

An important method for an analysis of the second order phase transition is the renormalization group (RG) approach [4] and the ε-expansion where ε is the deviation from the upper critical dimension $d_c = 4$. In this work, the third-order pertubative theory is calculated numerically in the null-momentum (NM) subtraction scheme. By means of the scheme the renormalized Green function is obtained by $\mathcal{R}$ operation [4, 5] on Green function. Within this procedure anomalous dimensions γ are expressed in terms of irreducible renormalized Feynman diagrams, and thus the calculation of renormalization constants can be entirely skipped [6, 7]. The universal quantities are calculable in the form of the ε-expansion and they are independent of the choice of the renormalization scheme [4].

2 Model and Method

A field theoretical formulation of the percolation process [1–3] is based on the action functional,

$$S_0 = \tilde{\psi}\left(-\partial_t + D_0\partial^2 - D_0\tau_0\right)\psi + \frac{D_0\lambda_0}{2}\left[\tilde{\psi}^2\psi - \tilde{\psi}\psi^2\right], \tag{1}$$

⋆e-mail: lukas.mizisin@gmail.com

where ∂^2 is the Laplace operator, ∂_t is differentiation with respect to time, $\psi > 0$ is the coarse-grain density of percolating particles, $\tilde{\psi}$ is an auxiliary (Martin-Siggia-Rose) response field [9], D_0 is a diffusion coefficient, λ_0 is a positive coupling constant and τ_0 is a deviation from the threshold value (an analog of critical temperature in static models [4]). The model is studied near its upper critical dimension $d_c = 4$ using expansion in the formally small parameter $\varepsilon = d_c - d = 4 - d$.

The renormalized action functional for the DP process (1) can be written in the form

$$S_R = \tilde{\psi}\left(-Z_1\partial_t + Z_2 D\partial^2 - Z_3 D\tau\right)\psi + \frac{Z_4 D\lambda\mu^{\varepsilon/2}}{2}\left[\tilde{\psi}^2\psi - \tilde{\psi}\psi^2\right]. \tag{2}$$

On the other hand, the action functional S_R can also be obtained by the standard procedure [4] of multiplicative renormalization of all fields and parameters

$$\psi_0 = \psi Z_\psi, \qquad \tilde{\psi}_0 = \tilde{\psi} Z_{\tilde{\psi}}, \qquad D_0 = DZ_D, \qquad \lambda_0 = \lambda\mu^{\varepsilon/2}Z_\lambda, \qquad \tau_0 = \tau Z_\tau + \tau_c,$$

supplemented by similar relations between renormalization constants for terms in (2). From a direct comparison we can easily express RG constants for the model parameters in (3) through $Z_i, i = 1, \ldots, 4$.

In this work, a renormalization scheme at the normalization point (NP), $p = 0$, $\omega = 0$ and $\tau = \mu^2$ is considered. The counterterms are then specified at the normalization point, and for numerical calculation it is advantageous to express the renormalization constants in terms of normalized Green functions

$$\bar{\Gamma}_1 = \partial_{i\omega}\Gamma_{\tilde{\psi}\psi}\Big|_{\omega=0}^{p=0}, \quad \bar{\Gamma}_2 = \frac{-1}{2D}\partial_p^2\Gamma_{\tilde{\psi}\psi}\Big|_{\omega=0}^{p=0}, \quad \bar{\Gamma}_3 = \frac{\Gamma_{\tilde{\psi}\psi}\big|_{\tau=0} - \Gamma_{\tilde{\psi}\psi}}{D\tau}\Big|_{\omega=0}^{p=0}, \quad \bar{\Gamma}_4 = \frac{\Gamma_{\tilde{\psi}\tilde{\psi}\psi} - \Gamma_{\tilde{\psi}\psi\psi}}{D\lambda\mu^{\varepsilon/2}}\Big|_{\omega=0}^{p=0}, \tag{3}$$

that satisfy the conditions $\bar{\Gamma}_i^R|_{\tau=\mu^2} = 1$, $i = 1, 2, 3, 4$. The renormalization constant is defined in this way and does not depend on τ, as in the minimal subtraction (MS) scheme: The renormalization group equations are the same as in the MS scheme:

$$(\mu\partial_\mu + \beta_u\partial_u - \tau\gamma_\tau\partial_\tau - D\gamma_D\partial_D)\Gamma_i^R = (n_\psi\gamma_\psi + n_{\tilde{\psi}}\gamma_{\tilde{\psi}})\Gamma_i^R, \tag{4}$$

where μ is the renormalization mass, n_ψ and $n_{\tilde{\psi}}$ are the numbers of the corresponding fields in Green function, $\gamma_x = \mu\partial_\mu \log Z_x$ is anomalous dimension and $\beta_u = \mu\partial_\mu u = u(-\epsilon - \gamma_u)$ is describing a flow of the charge $u = \lambda^2$. By applying these equations, one can obtain relations for the normalized functions

$$(\mu\partial_\mu + \beta_u\partial_u - \tau\gamma_\tau\partial_\tau - D\gamma_D\partial_D)\bar{\Gamma}_i^R = \gamma_i\bar{\Gamma}_i^R, \tag{5}$$

where the relations among the anomalous dimensions fields, parameters and γ_i have the forms:

$$\gamma_1 = 2\gamma_\psi \qquad \gamma_2 = 2\gamma_\psi + \gamma_D \qquad \gamma_3 = 2\gamma_\psi + \gamma_D + \gamma_\tau \qquad \gamma_4 = 3\gamma_\psi + \gamma_D + \gamma_\lambda\,. \tag{6}$$

Taking into account the renormalization scheme, the anomalous dimension can be presented in terms of renormalized derivatives of the one-particle irreducible Green function $\bar{\Gamma}_i$ at the normalization point [6–8]. The renormalization procedure is implemented on the diagram in terms of the $\mathcal{R}$ operation. Let us define functions f_i by means of the $\mathcal{R}$ operation

$$f_i \equiv \mathcal{R}[-\tilde{\tau}\partial_{\tilde{\tau}}\bar{\Gamma}_i(\tilde{\tau})]\Big|_{\tilde{\tau}=1}, \tag{7}$$

The relation for anomalous dimensions in terms of the quantities f_i with respect to $\tilde{\tau}$ in the NP is

$$\gamma_i = \frac{2f_i}{1 + f_2}, \qquad i = 1, 2, 4. \tag{8}$$

The main advantage of these considerations is that the $\mathcal{R}$ operation can be expressed in terms of the subtracting operation $1-\mathcal{K}_i$ [4], which eliminates all divergences from the Feynman graphs $\mathcal{R}\Gamma = \prod(1-\mathcal{K}_i)\Gamma$, where the product runs over all relevant subgraphs of a given Feynman diagram, including the diagram as a whole. Then we can obtain the representation for the $\mathcal{R}$ operation [6, 8] in the null-momentum (NM) scheme

$$\mathcal{R}\chi = \prod_i \frac{1}{n_i!} \int_0^1 da_i\,(1-a_i)^{n_i} \partial_{a_i}^{n_i+1} \chi(\{a\}), \tag{9}$$

where the product is taken over all one-irreducible subgraphs χ_i (including the diagram χ as a whole) with canonical dimensions $n_i \geq 0$ and a_i are parameters that stretch the moments flowing into the i-th subgraph inside the given graph.

3 Contribution of Feynman diagrams

Calculation of contribution to graphs was easily performed up to the two-loop approximation [10] in the time-momentum representation. In the three-loop, most Feynman diagrams [1, 10] are evaluated in the same way. Exceptions are 3 diagrams from $\Gamma_{\tilde{\psi}\psi}$ and 9 diagrams from $\Gamma_{\tilde{\psi}\psi\psi}$ (Fig. 1). All these diagrams have one common feature, they contain two 1-loop subgraphs.

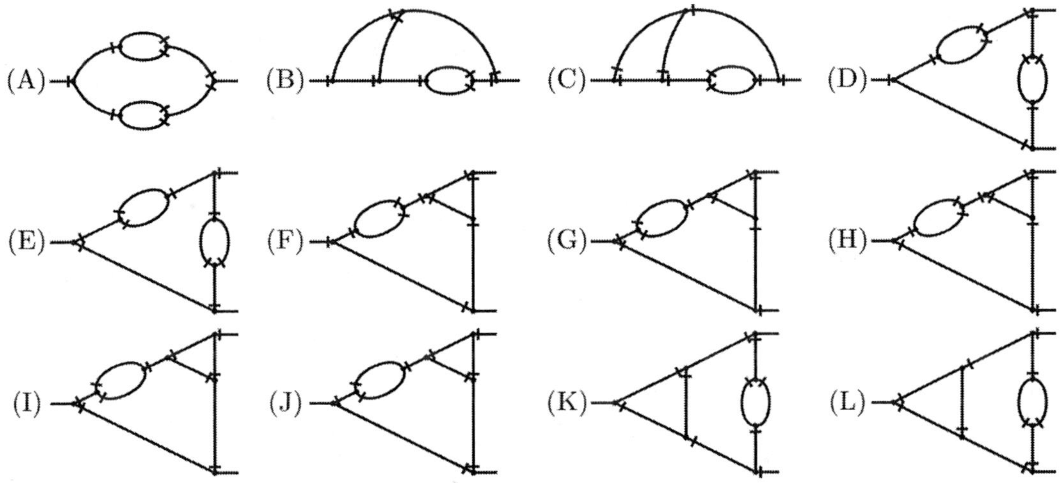

Figure 1. Three-loop Feynman diagrams which can not be calculated in the time-momentum representation. The diagrams (A) – (C) belong to $\Gamma_{\tilde{\psi}\psi}$ and the diagrams (D) – (L) are part of $\Gamma_{\tilde{\psi}\psi\psi}$.

One of the possible ways to solve this problem was to calculate the contribution of the 1-loop diagram $\Gamma_{\tilde{\psi}\psi}$ and obtain prescription when the divergent part of diagram has the form

$$\text{quadratic:} \quad P_{R_2} = \frac{\Gamma(\varepsilon)\Gamma(2-\varepsilon)}{4}\frac{1}{\varepsilon-1}\left[\left(\frac{-i\omega}{2}+\frac{p^2}{4}+1\right)^{1-\varepsilon} - 1 - (1-\varepsilon)\left(\frac{-i\omega}{2}+\frac{p^2}{4}\right)\right],$$

$$\text{logarithmic:} \quad \dot{P}_{R_0} = \frac{\Gamma(\varepsilon)\Gamma(2-\varepsilon)}{4}\left[\left(\frac{-i\omega}{2}+\frac{p^2}{4}+1\right)^{-\varepsilon} - 1\right], \tag{10}$$

where $\tau = 1$, $\Gamma(\cdot)$ denotes Gamma functions, p and ω denote the external momentum and frequency with respect to the subgraph. The diffusion constant D is excluded. In the case of 1-loop subgraph $\Gamma_{\tilde{\psi}\psi\psi}$ part a logarithmic divergence is obtained

$$T_{R_0} = \frac{\Gamma(\varepsilon)\Gamma(2-\varepsilon)}{8}\left[\int_0^1 da\left(\frac{-i\omega a}{2} + \frac{p^2 a}{4} + 1 - \frac{a^2p^2}{4}\right)^{-\varepsilon} - 1\right]. \tag{11}$$

The final result of Feynman diagrams (Fig. 1) can be obtained using the relation for the 1-loop subgraphs. In the first step of calculation for the diagram, the integrand is expressed in the frequency-momentum representation. In the second step the $\mathcal{R}$ operation from (7) is applied and the final symbolic integrand for numerical evaluation is found. In this integrand the divergent 1-loop contribution is replaced with respect to the previous relations. The last step is the numerical calculation where the integration is taken with respect to two or three parameters.

Now the final result of all diagrams for $\Gamma_{\tilde{\psi}\psi}$ can be derived. Using relations (8) one arrives to the following expressions for anomalous dimension in three-loop approximation

$$\gamma_1 = u\left(-\frac{1}{8} + \frac{\varepsilon}{16} - \frac{\pi^2\varepsilon^2}{192}\right) + u^2(0.026645 - 0.042537\varepsilon) - u^3 0.0190775,$$

$$\gamma_2 = u\left(-\frac{1}{16} + \frac{\varepsilon}{32} - \frac{\pi^2\varepsilon^2}{384}\right) + u^2(0.010607 - 0.016903\varepsilon) - u^3 0.006826.$$

Our results from the two-loop approximation with the new way of calculation is in good agreement with previous calculation [10] and also agrees very well with analytic calculation [1] in the ε-expansion. The calculation is in progress for 150 Feynman diagrams $\Gamma_{\tilde{\psi}\psi\psi}$ (in order to obtain γ_3, γ_4).

Acknowledgement

This research was supported by the Ministry of Education and Science of the Russian Federation (Agreement No. 02.a03.21.0008) and the Ministry of Education, Science, Research, and Sport of the Slovak Republic (VEGA Grant No. 1/0345/17).

References

[1] H. K. Janssen and U. C. Täuber, *Ann. Phys.* **315**, 147–192, (2004)
[2] J. L. Cardy and R. L. Sugar, J. Phys. A **13**, L423–L427, (1980)
[3] H. Hinrichsen, Adv. Phys. **49**, 815–958 (2000)
[4] A. N. Vasilev, *The Field Theoretic Renormalization Group in Critical Behavior Theory and Stochastic Dynamics* ([in Russian], PIYaF, St. Petersburg, 1998; English trans., Chapman and Hall/CRC, Boca Raton, Fla, 2004)
[5] J. C. Collins, *Renormalization: an introduction to renormalization, the renormalization group and the operator-product expansion* (Cambridge university press, 1984)
[6] L. Ts. Adzhemyan, M. V. Kompaniets, *Theor. Math. Phys.* **169**, 1450-1459 (2011)
[7] L. Ts. Adzhemyan, M. V. Kompaniets, S. V. Novikov, V. K. Sazonov, *Theor. Math. Phys.* 175(3), 719–728 (2013)
[8] L. Ts. Adzhemyan, M. V. Kompaniets, *Journal of Physics Conference Series* **523**, 012049, (2014)
[9] P. C. Martin, E. D. Siggia, H. A. Rose, *Physical Review A*, **8** (1), 423, (1973)
[10] L. Ts. Adzhemyan, et al., *EPJ Web of Conferences.* **108** (EDP Sciences, 2016)

Heat Exchange Between Electrons and Phonons in Nanosystems at Sub-Kelvin Temperatures

Dragoş-Victor Anghel[1,*] and *Sergiu* Cojocaru[1,**]

[1] *Institutul National de C&D pentru Fizica si Inginerie Nucleara – Horia Hulubei, 077125 Magurele, Romania*

Abstract. Ultra-sensitive nanoscopic detectors for electromagnetic radiation consist of thin metallic films deposited on dielectric membranes. The metallic films, of thickness d of the order of 10 nm, form the thermal sensing element (TSE), which absorbs the incident radiation and measures its power flux or the energies of individual photons. To achieve the sensitivity required for astronomical observations, the TSE works at temperatures of the order of 0.1 K. The dielectric membranes are used as support and for thermal insulation of the TSE and are of thickness $L - d$ of the order of 100 nm (L being the total thickness of the system). In such conditions, the phonon gas in the detector assumes a quasi-two-dimensional distribution, whereas quantization of the electrons wavenumbers in the direction perpendicular to the film surfaces leads to the formation of quasi two-dimensional electronic sub-bands. The heat exchange between electrons and phonons has an important contribution to the performance of the device and is dominated by the interaction between the electrons and the antisymmetric acoustic phonons.

High sensitivity electromagnetic radiation detectors for space applications are nanometer-size devices which work at sub-Kelvin temperatures [1, 2]. Such a detector consists of a thermal sensing element (TSE) which is deposited on a supporting membrane. The TSE is formed of one or several metallic layers and has the role of absorbing and measuring the incident electromagnetic radiation. The supporting membrane is dielectric and provides the thermal insulation of the TSE from the bulk material [1–4]. To ensure the level of sensitivity required by the astrophysical observations, the thickness of the TSE should be of the order of 10 nm, the thickness of the supporting membrane is of the order of 100 nm, whereas the working temperature of the device may be about 100 mK [5, 6]. The layers of materials that form the detector are shown schematically in Fig. 1.

The TSE may consist of a normal metal strip coupled to a superconducting antenna [2, 3, 5–9]. When the incident radiation is absorbed, its energy is dissipated into the normal metal strip. At the detector's working temperature, the electrons are weakly coupled to the lattice and therefore we can associate an effective temperature to each of these subsystems: T_e will denote the temperature of the electrons and T_{ph} will denote the temperature of the lattice (phonons).

Being at different temperatures, a net heat power P flows between the electron system and the phonon system, due to the electron-phonon interaction. This interaction is described by the deformation potential Hamiltonian [10–12]

$$\hat{H}_{def} = \frac{2}{3} E_F \int_{V_{el}=A\times d} d^3\mathbf{r}\, \Psi^\dagger(\mathbf{r})\, \Psi(\mathbf{r}) \nabla \cdot \mathbf{u}(\mathbf{r}), \tag{1}$$

*e-mail: dragos@theory.nipne.ro
**e-mail: scojocaru@theory.nipne.ro

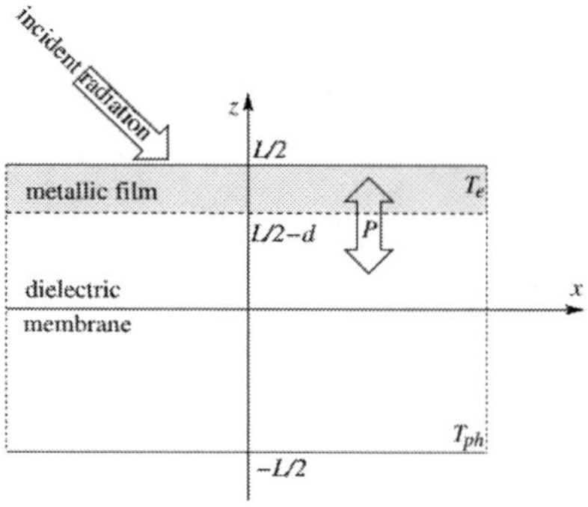

Figure 1. The schematic representation of the metallic film (the layer between $z = L/2 - d$ and $L/2$), on top of the supporting dielectric membrane (between $z = -L/2$ and $x = L/2 - d$). The incident radiation is absorbed by the metallic film, increasing the temperature of the electrons T_e above the temperature of the lattice T_{ph}. The electron-phonon heat exchange is P. The upper and lower surfaces of the film and membrane are parallel to the (xy) plane.

where E_F is the Fermi energy of the electrons, A is the area of the metallic film, d is the thickness (see Fig. 1), $\Psi^\dagger(\mathbf{r})$ and $\Psi(\mathbf{r})$ are the electron field creation and annihilation operators, respectively, whereas $\mathbf{u}$ is the phonon field. The volume of the metallic film is $V_{el} \equiv Ad$ and we do not take into account any other interaction in the system (like, for example, electron-electron, phonon-phonon or interaction with impurities). For three-dimensional (3D) systems, the heat power $P^{3D} \propto T_e^5 - T_{ph}^5$ [13]. If the phonons are confined to a lower dimensionality s, then [12, 14, 15]

$$P^{sD} \propto T_e^{s+2} - T_{ph}^{s+2}. \tag{2}$$

In general, we write $P^{sD} \equiv P^{(0)} - P^{(1)}$, where $P^{(0)}$ is defined as the heat power from electrons to phonons and $P^{(1)}$ is the heat power from phonon to electrons, so that, from Eq. (2), we have $P^{(0)} \propto T_e^{s+2}$ and $P^{(1)} \propto T_{ph}^{s+2}$.

In the detector systems investigated by us, the supporting membrane is made of silicon nitride (SiN_x) and the metallic film is copper (Cu). The phonon modes and the thermal properties of nanoscopic SiN_x membranes have been investigated, for example, in [16–18] and it was noticed that in a temperature range below 1 K a dimensionality crossover of the phonon modes takes place, namely a quasi-two-dimensional (2D) phonon gas distribution transforms – as the temperature increases – into a 3D distribution. For example, the phonon gas in a 100 nm thick membrane is well approximated by a 2D distribution at temperatures below 200 mK and therefore one would expect $s = 2$ and the power flux proportional to $T_e^4 - T_{ph}^4$ in this temperature range. If one takes into account the discretization of the electrons' wavevectors in the direction perpendicular to the surface of the metallic layer (we shall call this the *quantum well description*–QW), together with the confinement of the phonon modes, the form (2) changes. The quasi-2D phonon modes in the system can be divided into three categories [19]: horizontal shear (h), symmetric (s) and antisymmetric (a) modes. From these three types of phonons, only the s and a modes interact with the electrons and therefore contribute to the heat flux. The heat flux due to the s modes is $P^{2D}_{QWs} \propto T_e^4 - T_{ph}^4$ and the heat flux due to the a modes is $P^{2D}_{QWa} \propto T_e^{3.5} - T_{ph}^{3.5}$; the total flux is $P^{2D}_{QW} = P^{2D}_{QWs} + P^{2D}_{QWa}$. In the low temperature limit (T_e and $T_{ph} \to 0$), the antisymmetric modes have the dominant contribution to the heat exchange, as can be seen in Fig. 2.

In conclusion, the heat exchange between electrons and phonons in the layered structure investigated is dominated by the interaction between the electrons and the antisymmetric phonon modes, which leads to a temperature dependence $P^{2D}_{QW} \propto T_e^{3.5} - T_{ph}^{3.5}$ that does not support the conjecture (2).

Acknowledgement

Financial support from the Romnian ANCS project PN 16420101 and from the Romania-JINR collaboration projects, JINR Order 220/10.04.2017, positions 18, 19, 21, 22 are gratefully acknowledged.

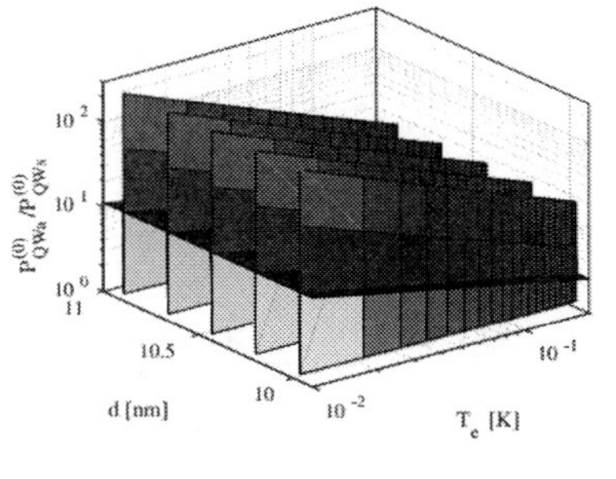

Figure 2. The ratios $P^{(0)}_{QWa}/P^{(0)}_{QWs}$ for $L = 100$ nm. In the temperature range of interest ($T_e < 0.2$ K), the dominant contribution comes from the antisymmetric phonons. The same is true for $P^{(1)}_{QWa}/P^{(1)}_{QWs}$. The sharp crests that appear parallel to the T_e axis are exceptions to the simple power-law dependence of P^{2D}_{QW} and are not discussed here (see [11] for details).

References

[1] M.J. Myers, W. Holzapfel, A.T. Lee, R. O'Brient, P.L. Richards, H.T. Tran, P. Ade, G. Engargiola, A. Smith, H. Spieler, Appl. Phys. Lett. **86**, 114103 (2005)
[2] F. Giazotto, T.T. Heikkilä, A. Luukanen, A.M. Savin, J.P. Pekola, Rev. Mod. Phys. **78**, 217 (2006)
[3] D.V. Anghel, L. Kuzmin, Appl. Phys. Lett. **82**, 293 (2003)
[4] K.L. Denis, N.T. Cao, D.T. Chuss, J. Eimer, J.R. Hinderks, W. Hsieh, S.H. Moseley, T.R. Stevenson, D.J. Talley, K. U.-yen et al., AIP Conf. Proc. **1185**, 371 (2009)
[5] D.V. Anghel, A. Luukanen, J.P. Pekola, Appl. Phys. Lett. **78**, 556 (2001)
[6] D.V. Anghel, J.P. Pekola, J. Low Temp. Phys. **123**, 197 (2001)
[7] M. Nahum, J.M. Martinis, Appl. Phys. Lett. **63**, 3075 (1993)
[8] M. Nahum, P.L. Richards, C.A. Mears, IEEE Trans. Appl. Superc. **3**, 2124 (1993)
[9] L.S. Kuzmin, Physica B p. 2129 (2000)
[10] J.M. Ziman, *Electrons and Phonons* (Harcourt College Publishers, 1976) 566 pp.
[11] D. Anghel, S. Cojocaru, Solid State Commun. **227**, 56 (2016)
[12] S. Cojocaru, D.V. Anghel, Phys. Rev. B **93**, 115405 (2016)
[13] F.C. Wellstood, C. Urbina, J. Clarke, Phys. Rev. B **49**, 5942 (1994)
[14] J.K. Viljas, T.T. Heikkilä, Phys. Rev. B **81**, 245404 (2010)
[15] F.W.J. Hekking, A.O. Niskanen, J.P. Pekola, Phys. Rev. B **77**, 033401 (2008)
[16] D.V. Anghel, J.P. Pekola, M.M. Leivo, J.K. Suoknuuti, M. Manninen, Phys. Rev. Lett. **81**, 2958 (1998)
[17] T. Kühn, D.V. Anghel, J.P. Pekola, M. Manninen, Y.M. Galperin, Phys. Rev. B **70**, 125425 (2004)
[18] T. Kühn, D.V. Anghel, Y.M. Galperin, M. Manninen, Phys. Rev. B **76**, 165425 (2007), cond-mat/07051936
[19] B.A. Auld, *Acoustic Fields and Waves in Solids, 2nd Ed.* (Robert E. Krieger Publishing Company, 1990)

Functional Integral Approach to the Solution of a System of Stochastic Differential Equations

Edik Ayryan[1,2,⋆], *Alexander* Egorov[3,⋆⋆], *Dmitri* Kulyabov[1,2,⋆⋆⋆], *Victor* Malyutin[3,⋆⋆⋆⋆], and *Leonid* Sevastianov[2,4,†]

[1] *Laboratory of Information Technologies, Joint Institute for Nuclear Research, Dubna, Russia*
[2] *Peoples' Friendship University of Russia (RUDN University), Moscow, Russia*
[3] *Institute of Mathematics, National Academy of Sciences of Belarus, Minsk, Belarus*
[4] *Bogoliubov Laboratory of Theoretical Physics, Joint Institute for Nuclear Research, Russia*

Abstract. A new method for the evaluation of the characteristics of the solution of a system of stochastic differential equations is presented. This method is based on the representation of a probability density function p through a functional integral. The functional integral representation is obtained by means of the Onsager-Machlup functional technique for a special case when the diffusion matrix for the SDE system defines a Riemannian space with zero curvature.

1 Introduction

Stochastic differential equations (SDE) are very popular in physics, chemistry, biology and etc. [1, 2]. There are many publications devoted to different aspects of SDE theory [1–3]. Except for some very special cases, SDE cannot be solved analytically and numerical methods are needed. The most common approach to the numerical solution of SDE is based on the discretization of a time interval and numerical simulation of SDE solutions at discrete times [4–7]. In this work a new method for evaluating of the characteristics of a SDE solution is presented. This method is based on the representation of the probability density function (PDF) p through a functional integral. The Onsager-Machlup functional technique is used to get the functional integral representation [8–11]. For one dimension case this approach was considered in [12]. In this work this approach is given for a system of SDE. Here we consider the Onsager-Machlup functional technique only for the flat space when the diffusion matrix for a SDE system defines a Riemannian space with zero curvature.

In section 2 the representation of PDF p through a functional integral is discussed. In section 3 the method of approximate evaluation of functional integrals is presented. Using the minimal action principle [13], we can distinguish among all trajectories the classical trajectory for which the action S takes the extremal value. The classical trajectory is found as the solution of the multidimensional Euler-Lagrange equation. Further, we use the expansion of the action with respect to the classical

⋆e-mail: ayrjan@jinr.ru
⋆⋆e-mail: egorov@im.bas-net.by
⋆⋆⋆e-mail: kulyabov_ds@rudn.university
⋆⋆⋆⋆e-mail: malyutin@im.bas-net.by
†e-mail: sevastianov_la@rudn.university

trajectory to compute the integral. In section 4 approximate and exact values for the expectations of some functionals of solution of specific system of SDE are presented.

2 Representation of characteristics through functional integral

We consider the SDE system (the Ito interpretation)

$$\begin{cases} dx_1(t) = a_1(\vec{x}, t)\, dt + \sum_{j=1}^{d} g_{1j}(\vec{x}, t)\, dw_j(t), \\ \dots \\ dx_d(t) = a_d(\vec{x}, t)\, dt + \sum_{j=1}^{d} g_{dj}(\vec{x}, t)\, dw_j(t), \end{cases} \tag{1}$$

where the solution $\vec{x}(t)$ satisfies the initial condition $\vec{x}(t_0) = \vec{x}_0$, $\vec{w}(t)$ is a d-dimensional Wiener process.

By means of the Onsager-Machlup functional technique the PDF can be represented in a form of the functional integral

$$p(\vec{x}, t, \vec{x}_0, t_0) = \int D[\vec{x}] \exp\Big\{ - \int_{t_0}^{t} L_0(\vec{x}(\tau), \dot{\vec{x}}(\tau))\, d\tau \Big\}, \tag{2}$$

$$D[\vec{x}] = \lim_{N\to\infty} \prod_{j=1}^{N-1} \prod_{k=1}^{d} dx_{kj} \prod_{j=1}^{N} \frac{1}{(2\pi\Delta t)^{d/2}\, \sqrt{\det G(\vec{x}_{j-1}, t_{j-1})}}, \qquad x_{kN} = x_k(t),$$

$$L_0(\vec{x}(\tau), \dot{\vec{x}}(\tau)) = \frac{1}{2} \sum_{k,j=1}^{d} G_{kj}^{-1}(\vec{x}(\tau), \tau) \times [\dot{x}_k - A_k(\vec{x}(\tau), \tau)][\dot{x}_j - A_j(\vec{x}(\tau), \tau)],$$

$$A_k(\vec{x}(\tau), \tau) = a_k(\vec{x}(\tau), \tau) - \frac{1}{2} \sum_{i,j=1}^{d} g_{ij}(\vec{x}(\tau), \tau) \frac{\partial}{\partial x_i} g_{kj}(\vec{x}(\tau), \tau),$$

where G denotes a matrix of elements

$$G_{ij}(\vec{x}(\tau), \tau) = \sum_{k=1}^{d} g_{ik}(\vec{x}(\tau), \tau) g_{jk}(\vec{x}(\tau), \tau).$$

3 Evaluation of functional integrals

The quantity $L_0(\vec{x}(\tau), \dot{\vec{x}}(\tau))$ in (2) can be considered as the Lagrangian of the system, and the quantity $S = \int_{t_0}^{t} L_0(\vec{x}(\tau), \dot{\vec{x}}(\tau))\, d\tau$ can be considered as its action. Using the minimal action principle [13], we can distinguish among all trajectories the classical trajectory $\vec{x}_{cl}$ for which the action S takes the extremal value. The classical trajectory is found as the solution of the multidimensional Euler-Lagrange equation

$$\begin{cases} \dfrac{d}{dt}\left(\dfrac{\partial L_0}{\partial \dot{x}_1}\right) - \dfrac{\partial L_0}{\partial x_1} = 0, \\ \vdots \\ \dfrac{d}{dt}\left(\dfrac{\partial L_0}{\partial \dot{x}_d}\right) - \dfrac{\partial L_0}{\partial x_d} = 0. \end{cases} \tag{3}$$

Further, to compute the integral, we can use the decomposition of the action with respect to the classical trajectory $\vec{x}_{cl}$

$$S[\vec{x}(\tau)] \approx S[\vec{x}_{cl}(\tau)] + \frac{1}{2}\delta^2 S[\vec{x}_{cl}(\tau)].$$

The second order variation $\delta^2 S[\vec{x}_{cl}(\tau)]$ can be written in the form

$$\delta^2 S[\vec{x}_{cl}(\tau)] = \int_{t_0}^{t} \sum_{i,j=1}^{d} \delta x_i \Lambda_{ij} \delta x_j \, d\tau,$$

where $\vec{x} = \vec{x}_{cl} + \vec{\delta x}$,

$$\Lambda_{ij} = \left[\frac{\partial^2 L}{\partial x_i \partial x_j}\right]_{x_{cl}} + \left[\frac{\partial^2 L}{\partial x_i \partial \dot{x}_j}\right]_{x_{cl}} \frac{d}{dt} - \frac{d}{dt}\left[\frac{\partial^2 L}{\partial \dot{x}_i \partial x_j}\right]_{x_{cl}} - \frac{d}{dt}\left[\frac{\partial^2 L}{\partial \dot{x}_i \partial \dot{x}_j}\right]_{x_{cl}} \frac{d}{dt}.$$

After these transformations the formula (2) is written in the form

$$p(\vec{x}, t, \vec{x}_0, t_0) = \exp\big\{ - S[\vec{x}_{cl}(\tau)]\big\} \int D[\vec{\delta x}] \exp\Big\{ - \frac{1}{2} \int_{t_0}^{t} \sum_{i,j=1}^{d} \delta x_i \Lambda_{ij} \delta x_j \, d\tau\Big\}.$$

By evaluating the functional integral from the exponential function with the quadratic form in exponent we obtain result for the PDF p.

4 Numerical results

In this section we consider an application of the proposed method for the evaluation of the expectation of some functionals of the SDE system solutions

$$\begin{cases} dx_1(t) = (a_1 x_1 + c_1 \sqrt{x_1 x_2} + b_1)\, dt + \sigma_1 \sqrt{x_1}\, dw_1(t), \\ dx_2(t) = (a_2 x_2 + c_2 \sqrt{x_1 x_2} + b_2)\, dt + \sigma_2 \sqrt{x_2}\, dw_2(t), \end{cases} \quad (4)$$

$$x_1(t_0) = x_{10}, \qquad x_2(t_0) = x_{20}.$$

The approximate and exact values of the expectation of some functionals of the SDE system solution (4) for concrete coefficients and initial conditions are presented in Tables 1 and 2. The form of the functionals is chosen in such way that we can calculate the exact values of the expectations. To compute the function $\vec{x}_{cl}$ approximately we use the partitioning of the interval $[t_0, t]$ into n parts.

Table 1. Approximate and exact values for expectations for $a_1 = a_2 = 2$, $c_1 = c_2 = -2$, $\sigma_1 = \sigma_2 = 1$, $b_1 = b_2 = 1/4$, $t_0 = 0$, $t = 1$, $x_{10} = x_{20} = 1/4$

	$E[\sqrt{x_1}]$	$E[\sqrt{x_2}]$	$E[(x_1 - x_2)]$
Approximate values $n = 40$	0.4651	0.4651	0.00003
Approximate values $n = 120$	0.4837	0.4837	0.00002
Exact values	0.5	0.5	0

5 Conclusion

A new method for evaluating of the characteristics of a SDE solution is proposed. From results presented in Tables 1 and 2 one can see that the increase of the parameter n leads to the decreasing differences between exact and approximate values.

Table 2. Approximate and exact values for expectations for $a_1 = a_2 = 2$, $c_1 = c_2 = -2$, $\sigma_1 = 9$, $\sigma_2 = 10$, $b_1 = 81/4$, $b_2 = 100/4$, $t_0 = 0$, $t = 1$, $x_{10} = x_{20} = 4$

	$E[\sqrt{x_1}]$	$E[\sqrt{x_2}]$	$E[(x_1 - x_2)]$
Approximate values $n = 40$	6.037	10.72	−11.17
Approximate values $n = 120$	5.962	11.51	−13.53
Exact values	5.806	13.19	−15.17

Acknowledgements

This work was financially supported by the Ministry of Education and Science of the Russian Federation (the Agreement number 02.a03.21.0008) and by the Belarussian Republic's Fund for Basic Research (project F16D-002).

References

[1] C.W. Gardiner, *Handbook of Stochastic Methods: For Physics, Chemistry, and the Natural Sciences* (Springer-Verlag; 2 Sub edition, January 1986)
[2] N.G. Van Kampen, *Stochastic Processes in Physics and Chemistry* (North-Holland, Amsterdam, 2007)
[3] I.I. Gikhman and A. V. Skorokhod, *Stochastic Differential Equations* ("Naukova Dumka", Kiev, 1968)
[4] D. F. Kuznetsov, *Numerical Integration of Stochastic Differential Equations* (Polytechnical University Publishing House, Saint Petersburg, 2001)
[5] P. E. Kloeden and E. Platen, *Numerical Solution of Stochastic Differential Equations* (Springer, Berlin, 1999)
[6] P. E. Kloeden, E. Platen, and H. Schurz, *Numerical Solution of Stochastic Differential Equations Through Computer Experiments* (Springer, Berlin, 1994)
[7] G. N. Milstein, *The Numerical Integration Solution of Stochastic Differential Equations*, (Kluwer Academic, Dordrecht, 1995)
[8] L. Onsager and S. Machlup, Phys. Rev. **91**, 1505 (1953)
[9] F. Langouche, D. Roekaerts, and E. Tirapegui, *Functional Integration and Semi-Classical Expansions* (D. Reidel, Dordrecht, 1982)
[10] H.S. Wio, *Application of Path Integration to Stochastic Process: An Introduction* (World Scientific, Singapore, 2013)
[11] E. Bennati, M. Rosa-Clot, and S. Taddei, Int. J. Theor. Appl. Finan. **02**, 381 (1999)
[12] E. A. Ayryan, A. D. Egorov, D. S. Kulyabov, V. B. Malyutin, and L. A. Sevastianov, Matematicheskoe Modelirovanie **28**, 11, 113–125, (2016)
[13] R. P. Feynman and A. R. Hibbs, *Quantum Mechanics and Path Integrals* (McGraw-Hill, New York, 1965)

Interpolation Environment of Tensor Mathematics at the Corpuscular Stage of Computational Experiments in Hydromechanics

Alexander Bogdanov[1,⋆], Alexander Degtyarev[1,⋆⋆], Vasily Khramushin[1,⋆⋆⋆], and Yulia Shichkina[1,2,⋆⋆⋆⋆]

[1] *St. Petersburg State University, Universitetskii prospect 35,198504 Saint-Petersburg, Russia*
[2] *St. Petersburg State Electrotechnical University, Popova ulitca 5, 197376 Saint-Petersburg, Russia*

Abstract. Stages of direct computational experiments in hydromechanics based on tensor mathematics tools are represented by conditionally independent mathematical models for calculations separation in accordance with physical processes. Continual stage of numerical modeling is constructed on a small time interval in a stationary grid space. Here coordination of continuity conditions and energy conservation is carried out. Then, at the subsequent corpuscular stage of the computational experiment, kinematic parameters of mass centers and surface stresses at the boundaries of the grid cells are used in modeling of free unsteady motions of volume cells that are considered as independent particles. These particles can be subject to vortex and discontinuous interactions, when restructuring of free boundaries and internal rheological states has place. Transition from one stage to another is provided by interpolation operations of tensor mathematics. Such interpolation environment formalizes the use of physical laws for mechanics of continuous media modeling, provides control of rheological state and conditions for existence of discontinuous solutions: rigid and free boundaries, vortex layers, their turbulent or empirical generalizations.

1 Introduction

The paper discusses some generalizations and justifications for the practical application of continual-corpuscular models of the computational hydromechanics. Tensor mathematics in the simulation of processes in continuous medium mechanics synthesizes two-stage computational experiments with the help of the principal (linear) components of the hydromechanics laws in the vicinity of adjacent fluid particles at controlled time intervals [1]. The use of the classical apparatus of the tensor calculus allows one to formalize the key approximations of hydromechanics in the three-dimensional space and the absolute time. This creates the necessary tools and ample mechanisms for direct algorithmic modeling of various spatio-temporal processes in the nearest environment of a computational cell.

⋆e-mail: a.v.bogdanov@spbu.ru
⋆⋆e-mail: a.degtyarev@spbu.ru
⋆⋆⋆e-mail: v.khram@gmail.com
⋆⋆⋆⋆e-mail: strange.y@mail.ru

An intuitive-natural variant of a direct computational experiment constructing is represented by two stages of spatial integration of the first order. These stages connect the differentials of kinematic and rheological processes into the general solution of mechanics equations by means of extrapolation of the second order in time.

At the **first**, continual stage, the traditional kinematic increments (differentials) within the original numerical cells-corpuscles are determined. These are accelerations in the motion-rotation-deformation arising under the action of volume / mass forces and surface stresses on the boundaries between adjacent nodes of elementary numerical cells.

At the **second**, corpuscular stage, the updated, nonstationary physical processes are constructed over a number of formally independent corpuscular particles, the mechanics of interaction of which directly affects the change in the rheological properties of the continuous medium. After this, the local surface stresses are rebuilt in the original nodes of the grid calculation area.

Then the **essence of the computational experiment** is reduced to the following [2, 3]. (1) Free polarized corpuscles in the initial stress field accelerate the kinematic flows of a continuous medium. There is a substitution of the emerging depressions by the anisotropic material substance (a functionally generalized physical field). (2) At the corpuscular stage of the calculations, physical fields are updated in empirical gradient, direct linear and divergent dependences for volumetric and surface stresses on liquid particles, taking into account the current rheological state of the spatial flows of the continuous medium.

The flows and transformation of the material medium are uniformly formalized in tensor form on the continuum and corpuscular stages of the computational experiment.

2 Direct continual-corpuscular computational experiment

A computational experiment in hydromechanics is designed on the basis of explicit schemes on two stages of physical processes modeling and 3D tensor mathematics. A specially synthesized functional-algorithmic apparatus with a strict and definite representation of mechanics and physics laws in architecture and programming languages of modern computers is used [4].

Additive growth of physical quantities occurs according to a single scalar argument in time. All spatial operations are based on successive products defining mechanics laws as interpolation dependencies between the internal Cartesian and transformed Euclidean bases of the rheological characteristics of the continuous medium. The syntax of the order of multiplicands is the following:

$$\text{“vectors} \bullet (\text{rheology} \bullet \text{geometry} \bullet \text{derivatives}) \bullet \text{scalars”},$$

where rheological and geometric tensors always exist, while derivatives (local velocities and strains) can be degenerate. Vector quantities are used in the reconstruction of physical and geometric bases in mechanics laws. Scalars are invariant under spatial transformations. In particular, they are used in the traditional differentiation / integration of hydromechanical equations in time.

Correct work with information about current state of simulated continuous medium and controlled execution of mechanics laws are provided by the following key rules of tensor mathematics:

I – numerical objects are defined in dimensional form;

II – all computational operations are context-sensitive, i.e. they have **first order** approximation in space and time;

III – duality of absolute and local reference frames, i.e. separation into stages of conditionally independent physical processes.

In what follows, we use uppercase letters for physical quantities in the global reference frame, and lowercase for quantities in connected local bases, which are essentially difference differentials. Notations are described in details in [5, 6].

Now we describe the mathematical and algorithmic features of the build up of a computational experiment.

1. In the computational experiment, the **first continual or kinematic** stage is built on the initial nodes of the grid region according to the specified mass inertial forces and stresses on the conjugate faces of the computational cells. The tensor models of the rheological state of the simulated fluid participate in multiplications. These models include the prehistory of the hydrodynamic processes for relaxation matching according to the "forecast and correction" scheme in the equations of motion. The result of the calculations is the updated space of increments in the intensity of the continuous medium flows. It fixes the displacements of the mass centers of the numerical cells and the current convective velocities in conjugate nodes of the computational domain.

The generalized laws of the mechanics apply to smooth spatial functions: momentum vectors $\vec{v} \cdot \overset{>}{\mathbf{m}} = \vec{f} \cdot t$ [kg · m/s] and tensors of intensities of the internal flows in the tensor form $\overset{\vee}{\boldsymbol{\rho}} \cdot \overset{\wedge}{\mathbf{v}} = \overset{\vee}{\mathbf{r}} \cdot \overset{\wedge}{\mathbf{f}} \cdot t = \overset{>}{\mathbf{f}} \cdot t$ [kg/s].

2. At the second corpuscular stage the mechanics of the interaction of the virtual numerical objects which inherit the specific physical (rheological) properties of the continuous medium is considered:

Elasticity → displacement – cell compression;

Viscosity → radiation – diffusion rate;

Inertia → flow shift – acceleration of corpuscles.

The reconstruction of the kinematics of the flows affects the distribution of the mass forces and surface stresses. Kinematic displacements of mobile and deformed liquid particles are added to harmonize the laws of hydromechanics and rheology of the internal state of the particles at the corpuscular stage. An algorithmic sequence is formed: {new synthesized bases → local and convective velocities → stresses}

At the corpuscular stage, the calculation nodes are shifted to the centers of the virtual numeric cells. This requires algorithmic methods for quick selection of values in conjugate interpolation spaces $\{{}^{\circ}\overset{\wedge}{\mathbf{r}} = {}^{\circ}_{+}\vec{R}_i - {}^{\circ}_{\circ}\vec{R}_i\}$ [m^3], where new coordinates of the centers of mass are reconstructed using the differences $\{{}^{+}\overset{\wedge}{\mathbf{r}} = {}^{+}_{+}\vec{R}_i - {}^{\circ}_{\circ}\vec{R}_i\}$ [m^3].

In the case of small deformations inherent in models of the theory of elasticity, the local displacement tensor is formed by summation over the entire time period. In fluid mechanics, the local velocity and displacement $\{\overset{\wedge}{\mathbf{v}} \cdot t = {}^{+}\overset{\wedge}{\mathbf{r}} - {}^{\circ}\overset{\wedge}{\mathbf{r}}\}$ [m^3] are supplemented by velocity (acceleration) increments of the calculated time interval t [s]. They are calculated using the updated velocities of the numeric cells $\{{}^{+}\overset{\wedge}{\boldsymbol{\omega}} = {}^{+}_{+}\vec{V}_i - {}^{\circ}_{\circ}\vec{V}_i\}$ [m^3/s], or they are selected from the first stage in the case of small deformations $\overset{\wedge}{\boldsymbol{\omega}} = \vec{\omega}^i = \Delta\vec{V}_i =_{+} \vec{V}_i -_{\circ} \vec{V}_i$ [m^3/s].

The rheology of the continuous environment is proved in commensurable scales. It is formalized by simple expressions: $\{{}^{+}\overset{>}{\mathbf{m}} = {}^{\circ}\overset{>}{\mathbf{m}} + f(\overset{\vee}{\boldsymbol{\rho}} \cdot \overset{\wedge}{\mathbf{v}}) \cdot t\}$ [kg], where $\overset{>}{\mathbf{m}}$ [kg] is the inertial mass of the liquid particle; $f(\overset{\vee}{\boldsymbol{\rho}} \cdot \overset{\wedge}{\mathbf{v}})$ [kg/s] is actually the law of mechanics as tensor function of the product of internal state (density of liquid particle $\overset{\vee}{\boldsymbol{\rho}}$ [kg/m^3)]) under the influence of the local speeds tensor $\overset{\wedge}{\mathbf{v}}$ [m^3/s].

In the space of the trivectors of the local velocities, deformation displacements relative to moving corpuscles are specified: $\overset{\wedge}{\boldsymbol{\varepsilon}} \approx \overset{\wedge}{\mathbf{v}} \cdot t$ [m^3]. Also the tensor of tension is determined: $\overset{<}{\boldsymbol{\sigma}} = (\overset{\wedge}{\boldsymbol{\varepsilon}} \cdot \overset{\vee}{\boldsymbol{\kappa}} + \overset{\wedge}{\boldsymbol{\omega}} \cdot \overset{\vee}{\boldsymbol{\eta}}) \times \lambda$ [kg m^2/c^2], where $\overset{\vee}{\boldsymbol{\kappa}}$ [N/m^3] is the modulus of elastic deformation; $\overset{\vee}{\boldsymbol{\eta}}$ [N s/m^3] is the tensor of speed viscosity; λ [m] is the scale coefficient characterizing the distance between the interacting liquid particles.

The tensors of the stress $\overset{<}{\boldsymbol{\sigma}}$ [N m], local velocities $\overset{\wedge}{\boldsymbol{\omega}}$ [m^3/s] and deformations $\overset{\wedge}{\boldsymbol{\varepsilon}}$ [m^3] can be zero or degenerate. It is excluded the calculation of their inversions. Change of bases is performed only in operations of product with the form tensors, such as $\overset{>}{\mathbf{v}} = \overset{\vee}{\mathbf{r}} \cdot \overset{\wedge}{\mathbf{v}}$ [1/s] or $\overset{<}{\mathbf{v}} = \overset{\wedge}{\mathbf{r}} \cdot \overset{\vee}{\mathbf{v}}$ [1/s] without affecting the time count in the denominator.

3 Conclusion

The described procedure of the computational experiment takes into account all rheological properties of the medium. The constructed scheme ensures unambiguous interpretation of physical processes by means of linear interpolation, which is performed by tensor operators in conjugate coordinate systems [5]. The solution is carried out in three stages:

1. integral accumulation of the amount of motion and deformation of particles of liquid;
2. local displacements and deformations of liquid particles over the calculated time step;
3. increment of velocities for local displacements and deformations of virtual numerical cells.

The continuum (Eulerian) and corpuscular (Lagrange) stages are connected by the iterative matching of the kinematic and rheological characteristics of the physical field in explicit numerical schemes. In this situation, we obtain non-stationary computational processes according to mathematical models for *viscous, elastic and compressible liquids*.

Scalar temperature distribution, variation of elastic gas density related with sound velocity, and other characteristics can be taken into account in applied problems in the calculation of internal flows of moving particles. Due to the use of explicit numerical schemes, it is possible to introduce empirical refinements during calculations. This synthesizes an extended solution of the applied problem of aerodynamic mechanics. In such problems tensor spaces are represented by linear interpolation dependencies in space and time. This applies to smooth distributions of scalar, vector and tensor - anisotropic spatial functions. Such functions are valid in the kinematic calculations of the interaction of particles, when we have similar stages in the synthesis of vector gradients and the analysis of resulting divergences for local density changes. In the tensor formulation, the modeling algorithms correspond to the properties of real liquids. Virtual computing processes are interpreted by inertial external and internal interactions of particle-cells.

Acknowledgement

The research is partly supported by Russian Foundation for Basic Research (project N 16-07-00886, 16-07-01113, 17-29-04288). The paper has been prepared within the scope of the state project "Initiative scientific project" of the main part of the state plan of the Ministry of Education and Science of Russian Federation (task N 2.6553.2017/8.9 BCH Basic Part).

References

[1] A. B. Degtyarev and V. N. Khramushin, Mathematical Modeling **26**, 11, 4–17 (2014)
[2] A. Bogdanov, A. Degtyarev, and V. Korkhov, EPJ Web of Conferences **108**, 01002 (2016)
[3] A. Bogdanov and V. Khramushin, EPJ Web of Conferences **108**, 02013 (2016)
[4] A. Degtyarev, V. Khramushin, and Y. Shichkina, AIP Conference Proceedings **1863**, 110006 (2017)
[5] A. Degtyarev and V. Khramushin, EPJ Web of Conferences **108**, 02018 (2016)
[6] V. N. Khramushin, *3D Tensor Mathematics for the Computational Fluid Mechanics Experience* (FEB RAS, Vladivostok, 2005)

EPJ Web of Conferences **173**, 02005 (2018) https://doi.org/10.1051/epjconf/201817302005
Mathematical Modeling and Computational Physics 2017

On the Relativistic Separable Functions for the Breakup Reactions

Serge G. Bondarenko[1,⋆], *Valery V.* Burov[1], and *Elena P.* Rogochaya[1]

[1] *Joint Institute for Nuclear Research, 141980 Dubna, Moscow region, Russia*

Abstract. In the paper the so-called modified Yamaguchi function for the Bethe-Salpeter equation with a separable kernel is discussed. The type of the functions is defined by the analytic stucture of the hadron current with breakup - the reactions with interacting nucleon-nucleon pair in the final state (electro-, photo-, and nucleon-disintegration of the deuteron).

1 Introduction

One of the most consistent nucleon-nucleon (NN) interaction theories is based on the solution of the Bethe-Salpeter (BS) equation [1]. In this case, we have to deal with a nontrivial integral equation for the bound state (deuteron) or for the interacting unbound NN pair.

The approximations based on a kernel with particle exchange are difficult to solve. An effective and solvable approach to the exact solution of the BS equation uses the separable *Ansatz* for the interaction kernel in the BS equation [2]. In this case one can transform an initial integral equation into a system of linear equations. The parameters of the kernel are obtained by fitting the phase shifts, inelasticity and low-energy parameters for the respective partial-wave states.

The first separable parameterizations were worked out within nonrelativistic models. The separable functions (called form factors) of the interaction kernel used in these models had no poles on the real axis in the relative energy complex plane [3, 4]. However, such poles appeared when the interaction kernel was relativistically generalized.

In some cases they do not prevent to perform the calculations (for example in elastic reactions). However, at high energies, one have to deal with several thresholds corresponding to the production of one, two and more mesons of different types, which is clearly not feasible to deal with. A more practical approach is to employ a phenomenological covariant separable kernel, which do not exhibit the meson-production thresholds and can even be constructed in a singularity-free fashion, using separable form factors and Wick-rotation prescription as it is done in the present paper. Thus, an accurate description of the on-shell nucleon-nucleon data is possible up to quite high energies. One then can hope that the obtained separable interaction kernels have also a reasonable off-shell behavior, so that they can be applied to other reactions as well.

⋆e-mail: bondarenko@jinr.ru

2 Bethe-Salpeter formalism

We start with the partial-wave decomposed Bethe-Salpeter equation for the nucleon-nucleon scattering matrix T (in the rest frame of the two-nucleon system):

$$\begin{aligned} t_{L'L}(p'_0, p', p_0, p; s) &= v_{L'L}(p'_0, p', p_0, p; s) \\ &+ \frac{i}{4\pi^3} \sum_{L''} \int dk_0 \int k^2\, dk \frac{v_{L'L''}(p'_0, p', k_0, k; s) t_{L''L}(k_0, k, p_0, p; s)}{(\sqrt{s}/2 - E_k + i\epsilon)^2 - k_0^2}. \end{aligned} \tag{1}$$

Here t is the partial-wave decomposed T matrix and v is the kernel of the NN interaction, $E_k = \sqrt{k^2 + m^2}$. There is only one term in the sum for the singlet (uncoupled triplet) case ($L = J$) and there are two terms for the coupled triplet case ($L = J \mp 1$). We introduce the square of the total momentum $s = P^2 = (p_1 + p_2)^2$ and the relative momentum $p = (p_1 - p_2)/2$ [$p' = (p'_1 - p'_2)/2$] (for details, see reference [2]).

Assuming the separable form (rank I) for the partial-wave decomposed kernels of NN interactions:

$$v_{L'L}(p'_0, p', p_0, p; s) = \lambda g^{[L']}(p'_0, p') g^{[L]}(p_0, p), \tag{2}$$

we can solve Eq. (1) and write for the T matrix:

$$t_{L'L}(p'_0, p', p_0, p; s) = \tau(s) g^{[L']}(p'_0, p') g^{[L]}(p_0, p), \tag{3}$$

with the function $\tau(s)$ being given by

$$\tau(s) = 1/(\lambda^{-1} + h(s)). \tag{4}$$

The function $h(s)$ has the following form:

$$h(s) = \sum_L h_L(s) = -\frac{i}{4\pi^3} \int dp_0 \int p^2\, dp \sum_L \frac{[g^{[L]}(p_0, p)]^2}{(\sqrt{s}/2 - E_p + i\epsilon)^2 - p_0^2}. \tag{5}$$

The simplest separable function $g(p_0, p)$ which can be used is a covariant generalization of the non-relativistic *Yamaguchi*-type [5, 6] function:

$$g(p_0, p) = \frac{1}{p_0^2 - p^2 - \beta^2 + i\epsilon}, \tag{6}$$

where β is a parameter.

2.1 Modified Yamaguchi-type functions

Let us consider the integral $h(s)$ (Eq. 5). Taking into account the pole structure of the propagators:

$$p_0^{(1,2)} = \pm\sqrt{s}/2 \mp E_p \pm i\epsilon \tag{7}$$

and of the g functions:

$$p_0^{(3,4)} = \mp E_\beta \pm i\epsilon \tag{8}$$

and using the Cauchy theorem, the $h(s)$ function can be written as follows:

$$\frac{1}{2\pi^2} \int p^2 dp \frac{1}{(s/4 - \sqrt{s}E_p + m^2 - \beta^2)^2} \frac{1}{\sqrt{s} - 2E_p + i\epsilon}. \tag{9}$$

To calculate this integral one should analyze the factor $f = (s/4 - \sqrt{s}E_p + m^2 - \beta^2)$ in the denominator as a function of s:

- if $2(m-\beta) < \sqrt{s} < 2(m+\beta)$ then always $f < 0$ and the function $1/f^n$ *is integrable* for any integer n and any E_p;
- for a bound state $\sqrt{s} = M_d = (2m - \epsilon_D)$. Since for minimal $\beta_{\min} = 0.2$ GeV always $\beta_{\min} > \epsilon_D/2$ then the function $1/f^n$ *is integrable* for any integer n and any E_p;
- if $\sqrt{s} < 2(m-\beta)$ or $\sqrt{s} > 2(m+\beta)$ then f can be positive and negative and $1/f^n$ *is non-integrable* for even n at any E_p.

The critical value $s^c = 4(m+\beta)^2$ corresponds to the laboratory kinetic energy of np-pair $T^c_{lab} = 4\beta + 2\beta^2/m \simeq 4\beta$. If $\beta_{\min} = 0.2$ GeV then $T^{\min}_{lab} = 0.8$ GeV.

So, if we consider breakup processes of the deuteron such as photo-, electro- and nucleon-breakup Yamaguchi-functions can be used only if the laboratory kinetic energy of the NN-pair is less than $T^{\min}_{lab}$. To avoid this restriction we suggest to use Yamaguchi-type functions modified in the following way:

$$g_Y(p_0,p) = 1/(p_0^2 - p^2 - \beta^2) \longrightarrow g_{MY}(p_0,p) = 1/((p_0^2 - p^2 - \beta^2)^2 + \alpha^4),$$

here Y stands for the Yamaguchi and MY – for Modified Yamaguchi functions.

To work with the modified Yamaguchi-type functions the procedure of p_0 integration should be modified, too. This procedure is worthy of a special discussion. The poles of the $h(s)$ integral with the modified Yamaguchi-type functions are:

$$p_0^{(3,4)} = \pm\sqrt{p^2 + \beta^2 + i\alpha^2},$$
$$p_0^{(5,6)} = \pm\sqrt{p^2 + \beta^2 - i\alpha^2}. \qquad (10)$$

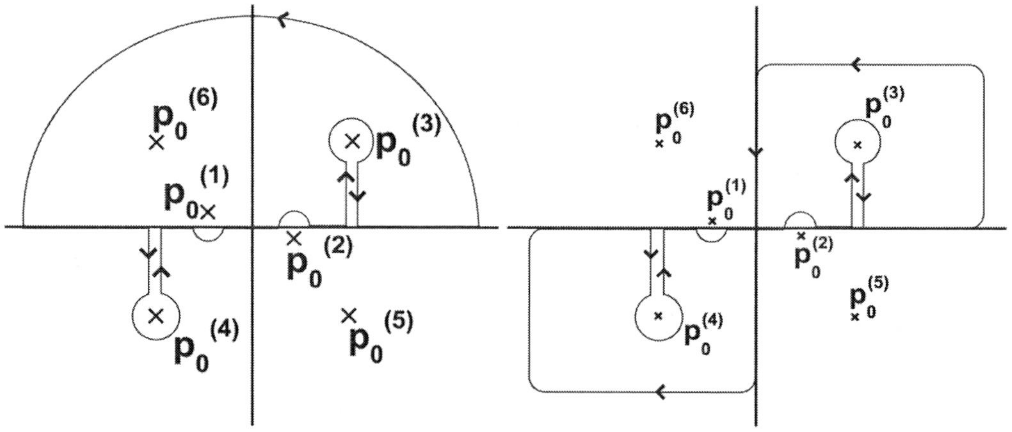

Figure 1. (a) Contour for the integration over p_0 according to the Cauchy theorem. (b) Contour for the integration over p_0: the Wick rotation.

All poles and the contour of integration are pictured in Fig. 1(a,b). The idea how to choose the contour appeared owing to [7, 8]. It is:

1. the contour must envelope the poles of g form factors which will be inside the standard contour in the $\alpha \to 0$ limit. "Standard" means the contour used in the quantum field theory calculations

with a propagator which has poles only on the real axis in the p_0 complex plane; one of them is circled from below and the other – from above. So, the path of integration is defined by an appropriate contour for the propagator;

2. the calculation over the presented path leads to a pure real contribution from the form factor poles and, therefore, to the unitary S matrix (or the corresponding unitarity condition for the T matrix). We also obtain a correct transition to ordinary form factors of type $g \sim 1/(p_0^2 - p^2 - \beta^2)^2$ in the $\alpha \to 0$ limit.

In general, the modified Yamaguchi-type functions can be written as:

$$g_i^{[a]}(p_0, p) = \frac{(p_{ci} - p_0^2 + p^2)^{n_i}(p_0^2 - p^2)^{m_i}}{((p_0^2 - p^2 - \beta_{1i}^2)^2 + \alpha_{1i}^4)^{k_i}((p_0^2 - p^2 - \beta_{1i}^2)^2 + \alpha_{1i}^4)^{l_i}}, \tag{11}$$

where the parameters – n_i, m_i, k_i, l_i (integer), $p_{ci}, \beta_{1i}, \beta_{2i}, \alpha_{1i}, \alpha_{2i}$ (real) depend on the channel $[a]$ under consideration. Such g form factors are used to describe neutron-proton scattering observables (phase shifts, inelasticities, low-energy parameters and deuteron characteristics) for the total angular momentum $J = 0, 3$ in a wide energy range (see [9]–[13]).

Acknowledgement

One of the authors (S.G. Bondarenko) would like to thank the organizers of the International Conference "Mathematical Modeling and Computational Physics, 2017" (JINR, Dubna, July 3–7, 2017) for the invitation and opportunity to present our results.

This work was partially supported by the Russian Foundation for Basic Research grant № 16-02-00898.

References

[1] E.E. Salpeter and H.A. Bethe, Phys. Rev. **84**, 1232 (1951)

[2] S.G. Bondarenko, V.V. Burov, A.V. Molochkov, G.I. Smirnov, and H. Toki, Prog. Part. Nucl. Phys. **48**, 449 (2002)

[3] L. Mathelitsch, W. Plessas, and W. Schweiger, Phys. Rev. **C26**, 65 (1982)

[4] J. Haidenbauer and W. Plessas, Phys. Rev. **C30**, 1822 (1984)

[5] Y. Yamaguchi, Phys. Rev. **95**, 1628 (1954)

[6] Y. Yamaguchi, Y. Yamaguchi, Phys. Rev. **95**, 1635 (1954)

[7] R.E. Cutkosky, P.V. Landshoff, D.I. Olive, and J.C. Polkinghorne, Nucl. Phys. **B12**, 281 (1969)

[8] T.D. Lee and G.C. Wick, Nucl. Phys. **B9**, 209 (1969)

[9] S.G. Bondarenko, V.V. Burov, W.-Y. Pauchy Hwang, and E.P. Rogochaya, Nucl. Phys. **A832**, 233 (2010)

[10] S.G. Bondarenko, V.V. Burov, W.-Y. Pauchy Hwang, and E.P. Rogochaya, Nucl. Phys. **A848**, 75 (2010)

[11] S.G. Bondarenko, V.V. Burov, and E.P. Rogochaya, Phys. Lett. **B705**, 264-268 (2011); Nucl. Phys. Proc. Suppl. **219–220**, 126–129 (2011)

[12] S.G. Bondarenko, V.V. Burov, and E.P. Rogochaya, Nucl. Phys. Proc. Suppl. **245**, 291–297 (2013)

[13] S.G. Bondarenko, V.V. Burov, S.E. Kemelzhanova, E.P. Rogochaya, and N. Sagimbaeva, *Proceedings of the 12th International Workshop "Relativistic Nuclear Physics from Hundreds of MeV to TeV"* (Slovak Republic, Stara Lesna, June 16–20, 2014)

Two-Parameter Harmonic Oscillator Expansion Method for Calculating Non-Relativistic Ground State Energy of Coulomb Three-Particle Systems with Two Identical Particles

Algirdas Deveikis[1,⋆]

[1] *Department of Applied Informatics, Vytautas Magnus University, Vileikos 8, LT-44404, Kaunas, Lithuania*

Abstract. The variational method in oscillator representation with individual parameters for each Jacobi coordinate is applied to the non-relativistic calculation of the ground state energy of a number of three-particle Coulomb systems, consisting of two identical particles and a different one. The accuracy and convergence rate of the calculations in the constructed oscillator basis are studied up to a total of 28 oscillator quanta. The results are compared with those of the traditional approach using only one such nonlinear variational parameter. The method with individual parameters for Jacobi coordinates is found to possess a number of advantages as compared to the traditional approach.

1 Introduction

The variational method in the oscillator representation has a number of significant advantages [1] and is widely used in quantum physics for calculating the properties of quantum systems. However, for long-range potentials of Coulomb type the method of harmonic oscillator expansion is known to show a relatively low rate of convergence. This circumstance considerably complicates the application of this method to the calculation of systems with realistic potentials. In Ref. [2], a new variational method is proposed for calculating the energy spectrum of three-particle Coulomb systems, consisting of two identical particles and a different one. In this method, individual parameters for each Jacobi coordinate are introduced instead of only one such parameter in the traditional approach. In the referred paper, this method was applied to calculate the ground state energy of a number of systems in the oscillator basis with zero number of oscillator quanta. In Ref. [3] the method using the oscillator basis with individual parameters for Jacobi coordinates was developed for the non-relativistic calculation of the ground state energy of a Coulomb system consisting of three different particles. For a nonzero number of oscillator quanta it showed much higher accuracy and convergence rate than the calculation with one nonlinear variational parameter. Keeping in mind the advantages of introducing individual parameters for Jacobi coordinates in the case of systems of three different particles, in the present work we apply the method of [2] to the calculation of the ground state energy for a number of Coulomb systems, comprising two identical particles and a different one at nonzero numbers of oscillator quanta and report results of the comparison of its accuracy and convergence rate with those of the traditional calculation. Thanks to the availability of practically "exact" solutions for three-particle Coulomb

⋆e-mail: algirdas.deveikis@vdu.lt

systems [4–9], it appears possible to perform an actually rigorous evaluation of the advantage of the method [2] over the traditional approach.

2 Hamiltonian and harmonic oscillator expansion method

The non-relativistic intrinsic Hamiltonian for the three-particle bound-state system with two identical particles and a different one coupled by Coulomb interaction can be written as [2]

$$\frac{\hat{H}_{\rm intr.}(\rho_1,\rho_2)}{\nu_2 c^2} = -\frac{1}{2}\nu_{12}\frac{1}{\gamma_1^2}\Delta_{\rho_1} - \frac{1}{2}\frac{1}{\gamma_2^2}\Delta_{\rho_2} + \frac{2Z_1Z_2\alpha}{\left|\frac{1}{\nu_{12}}\gamma_1\rho_1 - \frac{1}{2}\gamma_2\rho_2\right|} + \frac{Z_2^2\alpha}{\gamma_2\rho_2}, \tag{1}$$

where ρ_1 and ρ_2 are the translation-invariant Jacobi coordinates, γ_1 and γ_2 are the individual parameters (oscillator sizes) for each Jacobi coordinate, in contrast to one such parameter used in the traditional approach, ν_1 and ν_2 are the reduced masses for the introduced Jacobi coordinates, and $\nu_{12} = \nu_1/\nu_2$. The construction of the translation-invariant and asymmetric wave function of the system and the derivation of the expression for the matrix elements of the Hamiltonian (1) using the Racah technique and Talmi-Moshinsky transformations are thoroughly discussed in Ref. [2]. We should note that, due to the Coulomb character of the interaction between the particles, the calculated energy of the ground state is independent of the spin of the particles. The process of calculating the ground state energy of the considered system consists in a sequence of diagonalizations of the Hamiltonian (1) matrix aimed at finding the values of the parameters γ_1 and γ_2 that minimize the lowest eigenvalue. The approach proposed in Ref. [2] is reduced to the traditional approach by the simple substitution $\gamma = \gamma_1 = \gamma_2$. Then the minimization of the lowest eigenvalue of the matrix of the appropriate Hamiltonian is performed with respect to the unique parameter γ. The dimension of the matrix is determined by the basis cut-off parameter $N_{\rm max}$ that allows the inclusion of all the states of the system with the total number of oscillator quanta not exceeding this value.

3 Results and discussion

The ground state energies of a number of three-particle Coulomb systems involving up to 28 oscillator quanta have been computed by means of the present method. The algorithm developed in the present work was tested by constructing the variational surfaces of the ground state energy of the considered systems. The characteristic shape of the resulting surfaces is shown in Fig. 1 on the example of the e^+ee system. The study of the form of the variational surfaces shows that the considered method is very suitable for the numerical minimization with respect to the nonlinear parameters γ_1 and γ_2, since the obtained surfaces exhibit a global minimum only and no local minima. The results of the calculation of the ground state energy of the considered three-particle systems are summarized in Table 1. The first two columns present the notation of the calculated systems (here the commonly used notation of particles is used) and the ratio of the masses of one of the identical particles and the different one, respectively. The next two columns denoted as $H(\gamma)$ report the minimizing value of the parameter γ and the energy of the ground state of the systems calculated using the traditional approach. The next three columns denoted by $H(\gamma_1,\gamma_2)$ present the minimizing values of the parameters γ_1 and γ_2, as well as the ground state energy of the systems calculated using the individual parameters for each Jacobi coordinate. The last column presents the "exact" values of the energy and the corresponding references. The dash in the first row of Table 1 for the results of calculation using the traditional approach means that for the *epp* system the convergence rate of the one-parameter computations was so low that a negative value of the ground state energy could not be obtained up to 28 oscillator quanta.

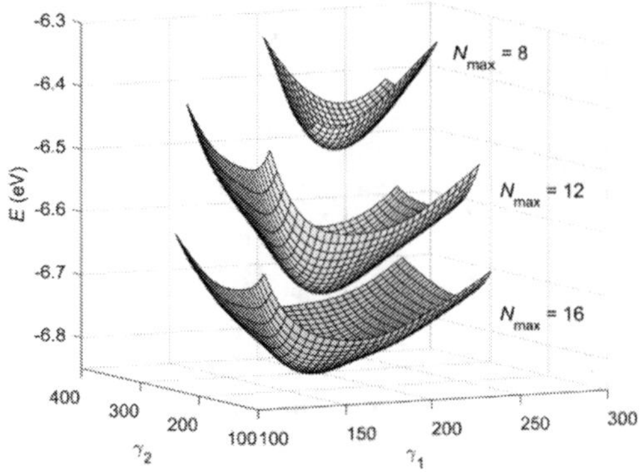

Figure 1. Variational surfaces of the ground state energy of the $e^{+}ee$ system at $N_{\max} = 8, 12, 16$ plotted against the nonlinear parameters γ_1 and γ_2.

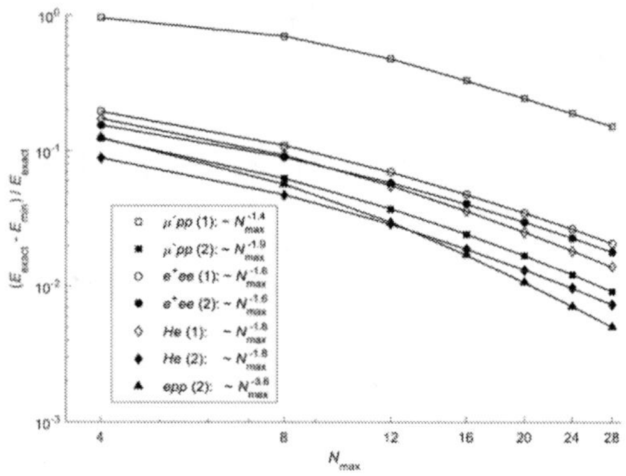

Figure 2. Comparison of the convergence rate for the methods with one general parameter and with individual parameters for Jacobi coordinates in the calculation of three-particle systems $\mu^{-}pp$, $e^{+}ee$, He, *epp*.

Results illustrating the computation accuracy and the convergence rate of the traditional method with a single nonlinear parameter and of the method with two nonlinear parameters are reported in Fig. 2 for particular systems. To guide the eye, the symbols which illustrate the dependence on $N_{\max}$ of the outputs of a same method are connected with straight lines. The significantly better performance of the present method is obvious. Following [10], the convergence rate is calculated as the power q of the number of oscillator quanta in the expression

$$E_{\min} = E_{\text{exact}} + P N_{\max}^{-q} , \tag{2}$$

where E_{exact} is the result of the "exact" calculation, P is a constant, $E_{\min}$ is the value of the ground state energy of the system, calculated for the given $N_{\max}$. The value of the power q was determined as the slope of the straight line drawn through the last two calculated points in logarithmic scale. The obtained results allow the following conclusions. First, for systems of atomic type (He) the method with two parameters demonstrates much higher precision of the calculations as compared to the single-parameter method, although the convergence rates of both methods in this case are practically the same. Second, for systems with close masses the effect of introducing individual parameters for each Jacobi coordinate is small. The example of the $e^{+}ee$ system shows only a minor increase of the accuracy of calculation in the case of using individual parameters as compared to the traditional approach. The convergence rate in this case is similar for both methods. Third, as shown by the example of the $\mu^{-}pp$ system, with increasing the molecular character of the system, the introduction of individual parameters for Jacobi coordinates substantially improves the accuracy and the convergence rate of the calculations as compared to the traditional approach.

4 Summary and conclusions

The harmonic oscillator expansion method using individual parameters for each Jacobi coordinate shows considerable advantage over the traditional approach using only one such parameter in the systems with the masses of two identical particles strongly differing from that of the third particle, as

Table 1. Ground state energies $E_{\min}$ (in eV) for Coulomb three-particle systems with two identical particles calculated for the 28 harmonic oscillator excitation quanta ($N_{\max} = 28$).

System	m_2/m_1	$H(\gamma)$		$H(\gamma_1, \gamma_2)$			E_{exact}
		γ	$E_{\min}$	γ_1	γ_2	$E_{\min}$	
epp	1.8362×10^3	–	–	97.7	8.70×10^4	−16.167	−16.249 [4, 5]
$\mu^- pp$	8.8802	400	−2359.3	118	8.33×10^2	−2756.1	−2781.6 [4, 6]
$K^- pp$	1.9006	222	−8845.6	148	325	−8962.3	−9104.3 [4]
$e^+ ee$	1	198	−6.9803	168	237	−7.0008	−7.1295 [4]
$p\mu^-\mu^-$	1.1261×10^{-1}	184	−2585.7	227	138	−2604.5	−2654.9 [4, 7]
$\alpha\mu^-\mu^-$	2.8347×10^{-2}	73.8	−15658	93.7	49.3	−15748	−15865 [4, 7]
$\mu^+ ee$	4.8363×10^{-3}	185	−13.854	244	124	−14.028	−14.288 [4, 8]
$K^+ ee$	1.0351×10^{-3}	185	−13.906	247	124	−14.085	−14.345 [4]
pee	5.4462×10^{-4}	185	−13.913	246	124	−14.092	−14.353 [4, 7]
He	1.3709×10^{-4}	74.2	−77.900	95.4	48.0	−78.423	−79.003 [4, 9]
Li^+	7.8208×10^{-5}	46.6	−195.77	59.6	30.0	−196.89	−198.08 [4]

well as for small and moderate numbers of oscillator quanta. The latter fact is of particular importance for practical calculations, since in the variational methods the computing time and the computational complexity, as a rule, grow exponentially with the increase of the number of oscillator quanta. We should emphasize the essential advantage of the considered method in the calculation of molecular systems, for which, in the case of strongly expressed molecular character such as *epp*, the traditional approach is practically inapplicable.

The introduction of a larger number of nonlinear variational parameters complicates the computation procedure. However, the improvement of calculation accuracy allows the solution of an eigenvalue problem with reduced cut-off $N_{\max}$ and, therefore, the computational burden and computing time are reduced, too. The developed method is not restricted to Coulomb systems only. The Coulomb potential can be replaced with or used in combination with practically any pair potential. The developed variational approach with the oscillator basis having individual parameters for each Jacobi coordinate opens new promising potentialities for the solution of cluster models in quantum mechanics.

References

[1] G.P. Kamuntavičius, J. Math. Phys. **55**, 042103 (2014)
[2] A. Deveikis, J. Mod. Phys. **7**, 290 (2016)
[3] A. Deveikis, Lith. J. Phys. **57**, 66 (2017)
[4] E.Z. Liverts and N. Barnea, Comput. Phys. Commun. **184**, 2596 (2013)
[5] M.M. Cassar and G.W.F. Drake, J. Phys. B **37**, 2485 (2004)
[6] S. Kilic, J.P. Karr, and L. Hilico, Phys. Rev. A **70**, 042506 (2004)
[7] L.U. Ancarani, K.V. Rodriguez, and G. Gasaneo, EPJ Web Conf. **3**, 02009 (2010)
[8] A.M. Frolov, J. Phys. B: At. Mol. Opt. Phys. **50**, 105102 (2017)
[9] V.I. Korobov, Phys. Rev. A **66**, 024501 (2002)
[10] L.M. Delves, Adv. Nucl. Phys. **5**, 1 (1972)

EPJ Web of Conferences **173**, 02007 (2018) https://doi.org/10.1051/epjconf/201817302007
Mathematical Modeling and Computational Physics 2017

Quasi-Vector Model of Propagation of Polarized Light in a Thin-Film Waveguide Lens

Dmitriy Divakov[1,*], *Mikhail* Malykh[1], *Leonid* Sevastianov[1,2], *Anton* Sevastianov[1], and *Edik* Ayryan[2]

[1] *Peoples' Friendship University of Russia (RUDN University), 117198, Moscow, Russia*
[2] *Joint Institute for Nuclear Research, 141980, Dubna, Moscow region, Russia*

Abstract. Maxwell equations describe the propagation with diffraction of waveguide modes through a thin-film waveguide lens. If the radius of the thin-film lens is large, then the thickness of the lens varies slowly in the yz plane. For this case we propose the model, which is based on the assumption of a small change in the electromagnetic field in a direction y. Under this assumption the vector diffraction problem is reduced to a number of scalar diffraction problems. The solutions demonstrate the vector nature of the electromagnetic field, which allows us to call the proposed model a quasi-vector model.

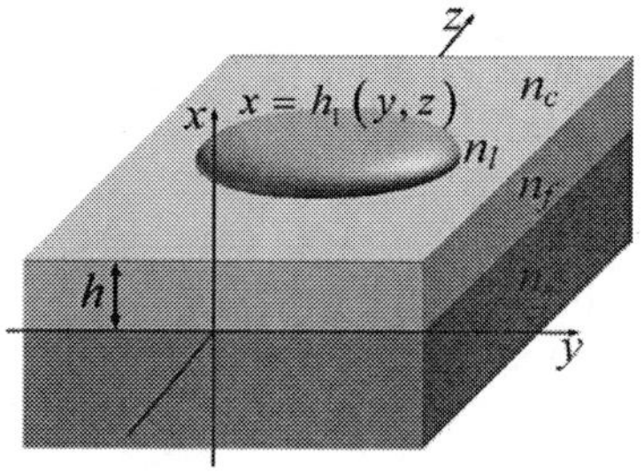

Figure 1. Thin-film waveguide lens of radius R_l with varible height $h_l(y, z)$

1 Introduction

The first and the second Maxwell equations in the component representation in the Cartesian coordinate system for time-harmonic fields have the following form [1]:

$$\frac{\partial H_z}{\partial y} - \frac{\partial H_y}{\partial z} = -ik_0\varepsilon E_x, \qquad \frac{\partial H_x}{\partial z} - \frac{\partial H_z}{\partial x} = -ik_0\varepsilon E_y, \qquad \frac{\partial H_y}{\partial x} - \frac{\partial H_x}{\partial y} = -ik_0\varepsilon E_z, \tag{1}$$

$$\frac{\partial E_z}{\partial y} - \frac{\partial E_y}{\partial z} = ik_0 H_x, \qquad \frac{\partial E_x}{\partial z} - \frac{\partial E_z}{\partial x} = ik_0 H_y, \qquad \frac{\partial E_y}{\partial x} - \frac{\partial E_x}{\partial y} = ik_0 H_z. \tag{2}$$

*e-mail: divakov_dv@rudn.university

2 Zeroth order approximation

If the waveguide lens has a radius $R_l \gg \lambda$, where λ is the wavelength, then $|\partial h_1/\partial y| \ll 1$ and assuming in this case that the electromagnetic field also varies slowly along y, we introduce a small parameter defined as $\delta = \max\{\,|\partial E_\alpha/\partial y|,\ |\partial H_\alpha/\partial y|,\ \ \alpha = x, y, z\,\}$. In the zeroth order approximation over the small parameter δ the equations (1), (2) take the following form [2, 3]:

$$\frac{\partial H_x^{(0)}}{\partial z} - \frac{\partial H_z^{(0)}}{\partial x} = -ik_0 n^2 E_y^{(0)}, \qquad ik_0 H_x^{(0)} = -\frac{\partial E_y^{(0)}}{\partial z}, \qquad ik_0 H_z^{(0)} = \frac{\partial E_y^{(0)}}{\partial x}, \tag{3}$$

$$\frac{\partial E_x^{(0)}}{\partial z} - \frac{\partial E_z^{(0)}}{\partial x} = ik_0 H_y^{(0)}, \qquad ik_0 n^2 E_x^{(0)} = \frac{\partial H_y^{(0)}}{\partial z}, \qquad -ik_0 n^2 E_z^{(0)} = \frac{\partial H_y^{(0)}}{\partial x}, \tag{4}$$

where $n(x, y, z)$ is the refractive index of the waveguide. The system of equations (1), (2) is now represented as two independent subsystems: the subsystem (3) for TE-mode and (4) for TM-mode, each subsystem can be represented as one equation of the second order for the leading component (E_y for TE-mode and H_y for TM-mode) and two equations relating the remaining components through the leading one [1]. We consider the problem of waveguide diffraction of one TE_0-mode incident on irregularity [2]. The formulation of the problem in the zeroth order approximation has the form [3]:

$$\begin{cases} \left(\Delta_{x,z} + k_0^2 n^2(x, y, z)\right) E_y^{(0)} = 0, \\ E_y^{(0)}\Big|_{z<-R_l} = A(y)\, e^{ik_0\beta_0 z}\varphi_0(x) + \sum_{j=1}^{N} R_j^{\mathrm{TE}}(y)\, e^{-ik_0\beta_j z}\varphi_j(x), \\ E_y^{(0)}\Big|_{z>R_l} = \sum_{j=1}^{N} T_j^{\mathrm{TE}}(y)\, e^{ik_0\beta_j z}\varphi_j(x), \qquad E_y^{(0)} \underset{x\to\pm\infty}{\longrightarrow} 0, \end{cases} \tag{5}$$

where $\Delta_{x,z} = \partial^2/\partial x^2 + \partial^2/\partial z^2$, $R_j^{\mathrm{TE}}(y)$ and $T_j^{\mathrm{TE}}(y)$ are reflection and transmission coefficients of the component E_y (complex values, as in [4]), $A(y)$ determines the amplitude of the waveguide mode incident on the lens, R_l is radius of the waveguide lens. The eigenvalues β_j and eigenfunctions $\varphi_j(x)$ determine the TE-modes of the regular waveguide ($z > R_l,\ z < -R_l$). The components $H_x^{(0)}$ and $H_z^{(0)}$ are expressed in terms of $E_y^{(0)}$ by means of (3). In addition, there is no TM-mode in the incident radiation, which leads us in the zeroth order approximation to $H_y^{(0)} = E_x^{(0)} = E_z^{(0)} \equiv 0$, therefore the reflection and transmission coefficients of the components $E_x^{(0)}$ and $E_z^{(0)}$ are zeros.

We seek the approximate solution $\tilde{E}_y^{(0)}$ of the problem (5) in the form: $\tilde{E}_y^{(0)} = \sum_{j=1}^{N} V_j(y, z)\,\varphi_j(x)$ where $V_j(y, z)$ are the desired functions [2, 3, 5]. We substitute this solution into (5) and perform the Galerkin method procedure [2]. As a result we obtain the mixed boundary value problem with respect to the desired Kantorovich coefficient functions $V_j(y, z)$ [3, 5, 6]:

$$\begin{cases} \vec{v}''_{zz} + \mathbf{Q}(y, z)\,\vec{v} = \vec{0}, \\ \left(\vec{v}'_z + ik_0\mathbf{D}\vec{v}\right)_{z=-R_l} = 2ik_0\mathbf{D}\vec{a}_{n_0}, \qquad \left(\vec{v}'_z - ik_0\mathbf{D}\vec{v}\right)_{z=R_l} = \vec{0}. \end{cases} \tag{6}$$

For each $y = y_m$, the problem (6) is a boundary value problem for a system of ordinary differential equations, in spite of the fact that the coefficient functions $V_j(y, z)$ depend on two variables. Differentiation is carried out only on the fast variable z and the dependence on the slow variable y is parametric.

Solving (5), we define the leading component $E_y^{(0)}$ and the components $H_x^{(0)}$ and $H_z^{(0)}$ of the TE-mode by the formula (3). The remaining components corresponding to the TM-mode will be zero. The zeroth order approximation, therefore, does not describe the process of hybridization of modes and we now proceed to the first order approximation.

3 First order approximation

We get the Maxwell equations in the first order approximation: $\vec{E} = \vec{E}^{(0)} + \vec{E}^{(1)}$, $\vec{H} = \vec{H}^{(0)} + \vec{H}^{(1)}$ considering that $\partial E_\alpha^{(1)}/\partial y$, $\partial H_\alpha^{(1)}/\partial y$ for $\alpha = x, y, z$ have the second order of smallness. We obtain the following equations for the TE-mode:

$$\left(\Delta_{x,z} + k_0^2 n^2 (x, y, z)\right) E_y^{(1)} = 0, \qquad H_x^{(1)} = -\frac{1}{ik_0}\frac{\partial E_y^{(1)}}{\partial z}, \qquad H_z^{(1)} = \frac{1}{ik_0}\frac{\partial E_y^{(1)}}{\partial x}. \tag{7}$$

The formulation of the diffraction problem for the TE-mode in the first order approximation has the following form:

$$\begin{cases} \left(\Delta_{x,z} + k_0^2 n^2 (x, y, z)\right) E_y^{(1)} = 0, \\ E_y^{(1)}\Big|_{z<-R_l} = \sum_{j=1}^{N} R_j^{(1)}(y)\, e^{-ik_0\beta_j z}\varphi_j(x), \quad E_y^{(1)}\Big|_{z>R_l} = \sum_{j=1}^{N} T_j^{(1)}(y)\, e^{ik_0\beta_j z}\varphi_j(x), \quad E_y^{(1)} \underset{x\to\pm\infty}{\to} 0. \end{cases} \tag{8}$$

In the boundary conditions of the problem (8), which determine the reflected and transmitted fields, the incident field is absent, since it is completely taken into account in the zeroth order approximation, and in the first approximation gives only the first-order corrections to the reflection and transmission coefficients. Applying the incomplete Galerkin method to the diffraction problem (8), we obtain a boundary value problem for a homogeneous system of second-order differential equations with homogeneous boundary conditions of the third kind, which has only a trivial solution [2]. Thus, the correction of the first order $E_y^{(1)}$ vanishes identically and also $H_x^{(1)} = H_z^{(1)} = 0$. The first-order correction $H_y^{(1)}$ is also zero. Corrections of the first order of smallness to the remaining components of the field are equal, respectively to:

$$E_x^{(1)} = \frac{-1}{ik_0 n^2}\frac{\partial H_z^{(0)}}{\partial y}, \qquad E_z^{(1)} = \frac{1}{ik_0 n^2}\frac{\partial H_x^{(0)}}{\partial y}. \tag{9}$$

It follows from (9) that the TE-mode incident on the irregularity excites the components of the TM-mode of the first order of smallness, that is, the solution obtained in the first approximation describes the hybridization of waveguide modes, which demonstrates the vector nature of the model.

4 Numerical experiment

We solve numerically the problem of waveguide diffraction of monochromatic light in the waveguide lens with the following input data: $n_c = 1.0$, $n_s = 1.47$, $n_f = 1.565$ and $n_l = 1.9$. The thickness of the waveguide layer is 2λ where $\lambda = 0.55\mu$ is the wavelength. The waveguide mode TE_0 with a unit amplitude ($A(y) = 1$) incidents on a waveguide lens with a radius $R_l = 20\lambda$. We consider the transmission coefficient of the component $E_y^{(0)}$ of the second waveguide mode, which was excited by the energy redistribution between modes during the propagation of the mode TE_0. The order of smallness of $T_2^{\mathrm{TE}}(y)$, as follows from the Figure 2, corresponds to the assumption that the value of $\partial E_y^{(0)}/\partial y$ is small. The remaining components of the field have the transmission coefficients similar of the same order of smallness.

The transmission coefficients of the components $E_x^{(1)}$ and $E_z^{(1)}$ (see Fig. 3) are by an order smaller than the coefficient of transmission of $E_y^{(0)}$, which corresponds to the first approximation of the problem under consideration. The corrections of the first order of smallness to the remaining components of the field are zeros.

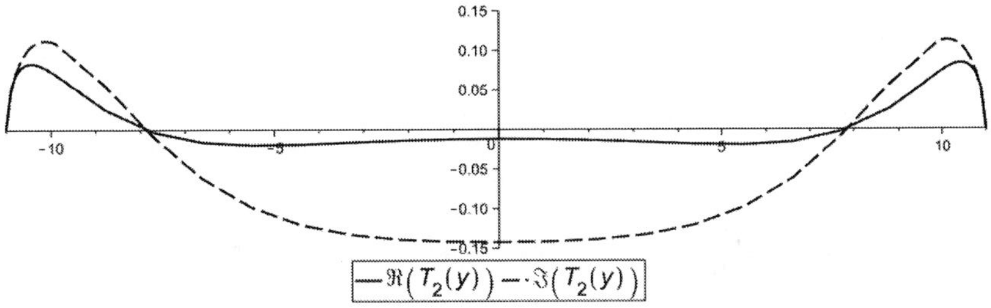

Figure 2. Transmission coefficient $T_2^{\mathrm{TE}}(y)$ of the component $E_y^{(0)}$ of the mode TE_1

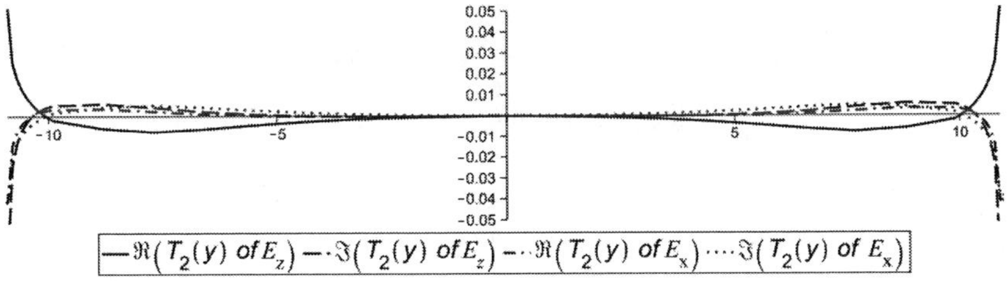

Figure 3. Transmission coefficient $T_2^{\mathrm{TM}}(y)$ of the components $E_z^{(1)}$ and $E_x^{(1)}$ of the mode TM_1

5 Conclusion

The paper describes a quasi-vector model of propagation of waveguide modes through a thin-film waveguide lens of large radius. Under the assumption of a small change in the electromagnetic field with respect to y, one can construct a perturbation theory with respect to the corresponding small parameter. The proposed model allows us to formulate in the zeroth order approximation a series of scalar diffraction problems the TE-mode, the solutions of m-th problem describe the TE-mode evolution in the longitudinal section $y = y_m$. In the first approximation, these solutions are used to calculate the corrections $E_x^{(1)}$, $E_z^{(1)}$ perturbating the components of the TE- and TM-modes. Thus, as a result of solving a series of scalar problems, we obtain a solution that possesses the properties of the solution of the vector problem $\vec{E} = \left(E_x^{(1)}, E_y^{(0)}, E_z^{(1)}\right)$, $\vec{H} = \left(H_x^{(0)}, 0, H_z^{(0)}\right)$.

Acknowledgements

The publication was prepared with the support of the "RUDN University Program 5-100". This research was supported by RFBF grants No 15-07-08795, No 16-07-00556.

References

[1] E. Ayryan et al., Proc. of SPIE **10337**, 103370I (2017)
[2] A.G. Sveshnikov, Comp. Math and Math. Phys. **2**, 1, 186–190 (1963)
[3] M.D. Malykh et al., RUDN Journal of Math., Inf. Sci. and Phys. **25**:1, 56–68 (2017)
[4] A.G. Sveshnikov and A.S. Il'inskii, Zh. vych. mat. **3**:3, 478–488 (1963)
[5] L.V. Kantorovich and V.I. Krylov, *Approximate Methods of Higher Analysis* (Wiley, New-York, 1964)
[6] D.V. Divakov, L.A. Sevastianov, and N.E. Nikolaev, EPJ Web of Conf. **108**, 02020 (2016)

EPJ Web of Conferences **173**, 02008 (2018) https://doi.org/10.1051/epjconf/201817302008
Mathematical Modeling and Computational Physics 2017

Splitting of Potential Curves in the Two-Coulomb-Centre Problem

Michal Hnatič[1,2,3,*], *Viktor M.* Khmara[3,**], *Volodymyr Yu.* Lazur[4,***], and *Oleksandr K.* Reity[4,****]

[1] *Department of Theoretical Physics, SAS, Institute of Experimental Physics, Watsonova 47, 040 01 Košice, Slovakia*
[2] *Peoples' Friendship University of Russia (RUDN University), 6 Miklukho-Maklaya St, Moscow, 117198, Russian Federation*
[3] *Faculty of Science, P. J. Šafárik University, Šrobárová 2, 041 54 Košice, Slovakia*
[4] *Department of Theoretical Physics, Uzhhorod National University, Voloshyna 54, 88000 Uzhhorod, Ukraine*

Abstract. The quasiclassical expression for the exchange interaction ΔE of potential curves at the points of their quasicrossing in the two-Coulomb-centre problem is found. It can be used for the calculation of cross sections of charge exchange processes between hydrogen or hydrogen-like atoms and bare nuclei.

1 Introduction

In order to solve many problems arising in the theory of slow atomic collisions, e.g. for the calculation of the matrix element $\Delta(R)$ of the exchange interaction between a hydrogen atom (or H-like ion) and a bare nucleus, it is necessary to know the two-centre radial and angular Coulomb spheroidal wave functions (CSWF) [1]. In the general non-resonant case ($Z_1 \neq Z_2$), the exchange matrix element $\Delta(R)$ was determined formerly [2, 3] under the condition requiring that the two-Coulomb-centre spheroidal wave function reach the one-centre parabolic wave function when an electron approaches one of the nuclei. The correct result for the exchange energy splitting $\Delta E = 2\Delta(R)$ at the quasicrossing points can actually be obtained only (see [4]) when the wave functions of the zeroth order approximation are considered in the spheroidal system of coordinates. The fact is that the exchange matrix element $\Delta(R)$ is defined by the asymptotic region of electron coordinates where the one-centre parabolic and spheroidal wave functions of a hydrogen atom differ substantially from each other. To be more specific, at large distances from the nucleus, a set of several Coulomb parabolic wave functions makes a contribution to the asymptotic behaviour of CSWF. This circumstance makes it difficult to apply the comparison equation method [1, 5] to determine the mentioned asymptotic behaviour (see, for instance, [4]).

In this paper, we provide a combined approach to the solution of the quantum two-Coulomb-centre Z_1eZ_2 problem. The slightly modified perturbation theory has been used to determine the

* e-mail: hnatic@saske.sk
** e-mail: viktor.khmara@student.upjs.sk
*** e-mail: volodymyr.lazur@uzhnu.edu.ua
**** e-mail: okreity@gmail.com

local behaviour of the solutions of the Z_1eZ_2 problem near a certain nucleus [6, 7]. We propose to employ the quasiclassical approach (the WKB method) to construct the asymptotic expansions of the angular CSWF in the internuclear region. In contrast to the perturbation theory, the quasiclassical approximation is not restricted by the smallness of the interaction and, hence, has a wider applicability domain, allowing the qualitative description for the behaviour and properties of quantum mechanical systems. Using the obtained quasiclassical CSWF and the Firsov surface integral method [8], we calculate the first three terms of the asymptotic behaviour of the exchange interaction potential of an ion with an atom for the general nonresonance case.

2 The wave functions of the two-Coulomb-centre problem

The Schrödinger equation of the Z_1eZ_2 problem after separation of the variables in the prolate spheroidal coordinates (ξ, η, ϕ), using the new variables $\mu = (\xi - 1)R/2$, $\nu = (1 + \eta)R/2$ and replacing the wave function by the product

$$\Psi(\vec{r}, R) = \frac{U(\xi, R)}{\sqrt{\xi^2 - 1}} \frac{V(\eta, R)}{\sqrt{1 - \eta^2}} \frac{e^{\pm im\phi}}{\sqrt{2\pi}} = \frac{\psi(\xi, \eta, R)}{\sqrt{(\xi^2 - 1)(1 - \eta^2)}} \frac{e^{\pm im\phi}}{\sqrt{2\pi}}, \tag{1}$$

results in the following system of differential equations ($\gamma = \sqrt{-2E}$):

$$U''(\mu) - \left[\gamma^2 - \frac{Z_1 + Z_2 + \lambda_\xi/R}{\mu} - \frac{Z_1 + Z_2 - \lambda_\xi/R}{R + \mu} + \frac{R^2(m^2 - 1)}{4\mu^2(R + \mu)^2}\right] U(\mu) = 0, \tag{2}$$

$$V''(\nu) - \left[\gamma^2 - \frac{Z_1 - Z_2 - \lambda_\eta/R}{\nu} + \frac{Z_1 - Z_2 + \lambda_\eta/R}{R - \nu} + \frac{R^2(m^2 - 1)}{4\nu^2(R - \nu)^2}\right] V(\nu) = 0 \tag{3}$$

with boundary conditions $U(0) = U(\infty) = 0$ and $V(0) = V(R) = 0$.

The solutions of the equations (2), (3) for $\mu \ll R$ (in the area near the internuclear axis) and $\nu \ll R$ (close to the Z_1 centre) can be obtained by means of the perturbation theory. The details of such a calculation can be found in [6, 7], here we give only the final result:

$$\psi^{\text{pert}}(\mu, \nu, R) = C(R) U^{\text{pert}}(\mu) V^{\text{pert}}(\nu), \tag{4}$$

$$U^{\text{pert}} = f_{n_1}^{(0)}(\mu) + \sum_{p=1}^{3} \sum_{k=-p}^{p} c_{n_1+k}^{(p)} f_{n_1+k}^{(0)}(\mu), \quad V^{\text{pert}} = f_{n_2}^{(0)}(\nu) + \sum_{p=1}^{3} \sum_{k=-p}^{p} c_{n_2+k}^{(p)} f_{n_2+k}^{(0)}(\nu), \tag{5}$$

$$f_{n_i}^{(0)}(x) = \left[\frac{(n_i + m)!}{n_i!(m!)^2(2n_i + m + 1)}\right]^{1/2} (2\gamma x)^{(m+1)/2} e^{-\gamma x} F(-n_i, m + 1, 2\gamma x), \tag{6}$$

where $F(-n_i, m + 1, 2\gamma x)$ is the confluent hypergeometric function and the coefficients $C(R)$ and $c_{n_i+k}^{(p)}$ $(p = 1, 2, 3)$ were found in [6, 7].

In the under-barrier region, the wave function of an electron moving in the field of well-separated nuclei (i.e., when R is large) is mostly concentrated near the internuclear axis, exponentially vanishing with the increase of μ. Considering this fact, the quasiradial wave function $U^{\text{pert}}(\mu)$ seems to be fairly accurate here. An entirely different situation occurs for the quasiangular function: when an electron moves from the first nucleus to the second one (and ν changes from 0 to R), the influence of Z_1 nucleus is gradually replaced by the influence of Z_2. The perturbation theory is not sufficient in this area of the under-barrier electron motion. Therefore, we propose to use the quasiangular WKB-function [6, 7]

$$V^{\text{quas}} = \frac{C_0}{\sqrt{q}} \exp\left[-\int_{\nu_2}^{\nu} q(\nu')d\nu' + S_1 + S_2 + O(R^{-4})\right], \; q(\nu) = \sqrt{\gamma^2 - \frac{2\tilde{Z}_1}{\nu} - \frac{2\tilde{Z}_2}{R - \nu} + \frac{(m^2 - 1)}{4\nu^2(1 - \nu/R)^2}}, \tag{7}$$

where the quasiclassical corrections S_1 and S_2 are determined by the formulae

$$S_1 = -\frac{\tilde{Z}_1}{4\gamma^3\nu^2}\left(1+\frac{17\tilde{Z}_1}{6\gamma^2\nu}\right)+\frac{\tilde{Z}_1\tilde{Z}_2}{2\gamma^5R^3}\ln\frac{\nu}{R-\nu}+\frac{m^2-1}{16\gamma^3}\left(\frac{1}{\nu^3}+\frac{1}{\nu^2R}-\frac{1}{R(R-\nu)^2}-\frac{1}{(R-\nu)^3}\right)$$
$$+\frac{\tilde{Z}_1\tilde{Z}_2}{4\gamma^5R}\left(\frac{3}{(R-\nu)^2}-\frac{3}{\nu^2}+\frac{1}{R}\left[\frac{1}{R-\nu}-\frac{1}{\nu}\right]\right)+\frac{\tilde{Z}_2}{4\gamma^3(R-\nu)^2}\left(1+\frac{17\tilde{Z}_2}{6\gamma^2(R-\nu)}\right)+C_1, \quad (8)$$

$$S_2 = \frac{\tilde{Z}_1}{4\gamma^4\nu^3}+\frac{\tilde{Z}_2}{4\gamma^4(R-\nu)^3}+C_2, \quad \tilde{Z}_{1,2}=\left[\pm(Z_1-Z_2)-\lambda_2/R\right]/2. \quad (9)$$

The constants C_0, C_1 and C_2 were found by the exact matching of V^{quas} (7) with V^{pert} from (5). Their explicit expressions are given in [7]. Note also that the expression for C_0 contains a misprint: the additional factor $\sqrt{\gamma}\,[n_2!(n_2+m)!(2n_2+m+1)]^{-1/2}$ was omitted there.

Considering the quasiangular function (7), the final expression for the wave function centred on the atom (e, Z_1) is of the form

$$\psi(\mu,\nu,\phi) = C(R)\,\frac{R^2}{4}\,\frac{U^{\text{pert}}(\mu)}{\sqrt{\mu(R+\mu)}}\,\frac{V^{\text{quas}}(\nu)}{\sqrt{\nu(R-\nu)}}\,\frac{e^{\pm im\phi}}{\sqrt{2\pi}}. \quad (10)$$

3 Splitting of the potential curves at the quasicrossing points

The calculation of the exchange splitting $\Delta E(R)$ of the potential curves at the quasicrossing points can be carried out using the well-known surface integral formula [8]

$$\Delta E = \oint_S \left(\Psi_I^*\vec{\nabla}\Psi_{II}-\Psi_{II}^*\vec{\nabla}\Psi_I\right)\cdot d\vec{S}. \quad (11)$$

Here Ψ_I and Ψ_{II} are the electron wave functions of the quasimolecule (Z_1, e, Z_2). In the separated atoms limit ($R\to\infty$) those turn into the wave function of the hydrogen-like atom (e, Z_1) and of the ion (e, Z_2), respectively; $d\vec{S} = \vec{n}\,dS$; S is the surface enclosing the half-space containing nuclei Z_1; $\vec{n}$ is the surface vector normal to S. As an integration surface in (11) we have chosen the paraboloid of revolution $\eta = \text{const}$. Omitting the details of this calculation, we provide only the resulting formula for ΔE:

$$\Delta E = \frac{2\gamma^2(-1)^{n_2+n_2'}(2\gamma R)^{n_2+n_2'+m+1}\,e^{-\gamma R}}{\left[n\,n'\,n_2!\,(n_2+m)!\,n_2'!\,(n_2'+m)!\right]^{1/2}}\left\{1-\frac{1}{2\gamma R}\left[\frac{A_2^2+A_2'^2}{4}+A_2A_2'+\frac{1-m^2}{2}\right]\right.$$
$$-\frac{A_2+A_2'}{2\gamma R}-\frac{A_1}{2\gamma^2R}\left(\frac{Z_1}{n}+\frac{Z_2}{n'}\right)+\frac{[A_2^2+A_2'^2+4A_2A_2'+2(1-m^2)]^2}{128\gamma^2R^2}+\frac{A_1(3A_1+1-m^2)}{4\gamma^2R^2}$$
$$\left.+\frac{A_2^3+A_2'^3+(A_2A_2'-4A_1+2m^2-6)(A_2+A_2')}{32\gamma^2R^2}+\frac{(A_1-1)(A_2+A_2')^2+2A_2A_2'(A_1-2)}{8\gamma^2R^2}\right\}. \quad (12)$$

Here $A_i = 2n_i+m+1$, $A_i' = 2n_i'+m+1$. In the resonance case ZeZ, formula (12) gives the difference between the energies of gerade and ungerade states in agreement with the results of [3, 9]. It should be noted also that the formulae for H_2^+ in [1, 3] contain a mistake in the terms of the order of R^{-2}.

Concerning the splitting of the potential curves obtained by means of the comparison equation method, it should be noted that various formulae for the splitting of potential curves at the quasicrossing points are available. These can be written compactly as follows:

$$\Delta E = T_{\{\text{KS},\text{P},\text{PKS}\}}\,\delta(n_2, n_2', m, p), \quad (13)$$

$$T_{\text{KS}} = 2\frac{(Z_2-Z_1)^2}{(n_2'-n_2)^3}, \qquad T_{\text{P}} = \frac{\partial E_1}{\partial n_2}+\frac{\partial E_2}{\partial n_2'}, \qquad T_{\text{PKS}} = \frac{\partial E_1/\partial n_2}{\sqrt{1+\frac{\partial\beta}{\partial E_1}\frac{\partial E_1}{\partial n_2}}}+\frac{\partial E_2/\partial n_2'}{\sqrt{1-\frac{\partial\beta}{\partial E_2}\frac{\partial E_2}{\partial n_2'}}}, \quad (14)$$

where δ is determined by formula (4.36) from [1]. Komarov and Slavyanov [2] first proposed to determine T by differentiating the expression $E = -(Z_2 - Z_1)^2/(2\beta^2)$ with respect to the indices, giving $T = T_{\mathrm{KS}}$. Power [3] criticized this expression and noted that it does not give a correct result for the splitting in the limiting case $Z_1 = Z_2$. Thus, Power proposed to differentiate the half-sum of the eZ_1- and eZ_2-energies with respect to the indices and take $T = T_{\mathrm{P}}$ (this result is presented in [1]). Finally, Ponomarev [10] noticed that the dependence of $\beta(E)$ on the energy should be considered when differentiating the eZ_1- and eZ_2-energies leading to $T = T_{\mathrm{PKS}}$. The numerical values of pre-exponential factor, given by these formulae, differ from each other.

It is of interest to estimate the limits of applicability and practical accuracy of our asymptotic formula (12) for ΔE by its comparison with the results of numerical integration of a $Z_1 e Z_2$ problem. In Table 1 we compare the values of ΔE in the $Z_1 e Z_2$ system ($Z_1 = 1$, $4 \leqslant Z_2 \leqslant 8$) provided by our formula (12) with the analytical results ΔE_{PKS} [1] and ΔE_B [11] as well as with the numerical calculations ΔE_{num} [12] (Nlm are spherical quantum numbers in the limit of the united atom ($R = 0$)). The comparison shows that the values of ΔE are quite close to the exact ones. The proximity of these results convincingly demonstrates the usefulness of the WKB method developed here.

Table 1. Adiabatic energy splittings ΔE at the quasicrossing points R_c in the system (p, e, Z_2)

Z_2	(Nlm)–$(N'l'm')$	R_c	ΔE (12)	ΔE_{PKS} [1]	ΔE_B [11]	ΔE_{num} [12]
4	$(4,3,0)$–$(3,2,0)$	7.76	6.66×10^{-2}	6.56×10^{-2}	—	6.94×10^{-2}
5	$(5,4,0)$–$(4,3,0)$	12.9	4.07×10^{-3}	6.09×10^{-3}	4.16×10^{-3}	4.25×10^{-3}
6	$(6,5,0)$–$(5,4,0)$	21.4	2.40×10^{-5}	3.37×10^{-5}	2.41×10^{-5}	—
7	$(7,6,0)$–$(6,5,0)$	31.9	2.06×10^{-8}	2.44×10^{-8}	2.14×10^{-8}	—
8	$(8,7,0)$–$(7,6,0)$	44.3	3.04×10^{-12}	4.51×10^{-12}	2.88×10^{-12}	—

Acknowledgements. This work is supported by the Ministry of Education and Science of the Russian Federation (Agreement No. 02.a03.21.0008) and the Ministry of Education, Science, Research, and Sport of the Slovak Republic (VEGA Grant No. 1/0345/17).

References

[1] I.V. Komarov, L.I. Ponomarev, and S.Yu. Slavyanov, *Spheroidal and Coulomb Spheroidal Functions* (Moscow: Nauka, 1976) [in Russian]

[2] I.V. Komarov, S.Yu. Slavyanov, J. Phys. B: At. Mol. Phys. **1**, 1066 (1968)

[3] J.D. Power, Phil. Trans. Roy. Soc. London Ser. A **274**, 663 (1973)

[4] T.M. Kereselidze, M.I. Chibisov, in *Physics of Electronic and Atomic Collisions* (LFTI, Leningrad, 1989) p. 160

[5] T.P. Grozdanov, R.K. Janev, V.Yu. Lazur, Phys. Rev. A **32**, 3425 (1985)

[6] M. Hnatič, V.M. Khmara, V.Yu. Lazur, O.K. Reity, EPJ Web of Conf. **108**, 02028 (2016)

[7] M. Hnatič, V.M. Khmara, V.Yu. Lazur, O.K. Reity, Theor. Math. Phys. **190**, 345 (2017)

[8] O.B. Firsov, Zh. Eksp. Teor. Fiz. **21**, 1001 (1951)

[9] P.J. Greenland, J. Phys. B: At. Mol. Phys. **11**, 3563 (1978)

[10] L.I. Ponomarev, Sov. Phys. JETP **28**, 971 (1969)

[11] A.A. Bogush and V.S. Otchik, J. Phys. A: Math. Gen. **30**, 559 (1997)

[12] I.V. Komarov and N.F. Truskova, JINR Report No. P4-11445 (1978)

EPJ Web of Conferences **173**, 02009 (2018) https://doi.org/10.1051/epjconf/201817302009
Mathematical Modeling and Computational Physics 2017

Diffusion Processes in the A-Model of Vector Admixture: Turbulent Prandtl Number

Eva Jurčišinová[1,2,⋆], *Marián* Jurčišin[1,2,3,⋆⋆], and *Richard* Remecky[1,2,⋆⋆⋆]

[1] *Institute of Experimental Physics, SAS, Košice, Slovakia*
[2] *Bogoliubov Laboratory of Theoretical Physics, JINR, Dubna, Russian Federation*
[3] *Dep. of Theoretical Physics and Astrophysics, Faculty of Science, P.J. Šafárik University, Košice, Slovakia*

Abstract. Using analytical approach of the field theoretic renormalization-group technique in two-loop approximation we model a fully developed turbulent system with vector characteristics driven by stochastic Navier-Stokes equation. The behaviour of the turbulent Prandtl number $\mathrm{Pr}_{\mathcal{A},t}$ is investigated as a function of parameter $\mathcal{A}$ and spatial dimension $d > 2$ for three cases, namely, kinematic MHD turbulence ($\mathcal{A} = 1$), the admixture of a vector impurity by the Navier-Stokes turbulent flow ($\mathcal{A} = 0$) and the model of linearized Navier-Stokes equation ($\mathcal{A} = -1$). It is shown that for $\mathcal{A} = -1$ the turbulent Prandtl number is given already in the one-loop approximation and does not depend on d while turbulent Prandt numbers in first two cases show very similar behaviour as functions of dimension d in the two-loop approximation.

1 Introduction

The problems related to the behavior of various admixtures in turbulent environments are among the most studied in the framework of the turbulent fluid dynamics. Typical examples are scalar and vector impurities in given turbulent systems, e.g., the temperature in the problem of thermal diffusion or the problem of an impurity with internal structure in turbulent environments [1, 2], as well as the weak magnetic field in a turbulent plasma (electrically conductive turbulent fluid) described by the turbulent magnetohydrodynamics (MHD) [3, 4].

An effective approach for theoretical analysis of diffusion processes in fully developed turbulent environments is the field theoretic renormalization group (RG) technique (see, e.g., Ref. [5] and references cited therein). During the last few decades various field theoretical models of passive advection were developed for the analysis of the intermittency and anomalous scaling behavior of various correlation functions, for the investigation of deviations from the simple scaling predictions of the classical phenomenological Kolmogorov-Obukhov theory [6]. Properties of the anomalous scaling were reported in all these investigations.

On the other hand, one of the most important characteristics of the diffusion processes in fluids is the corresponding Prandtl number. The Prandtl number is defined as the ratio of the coefficient

⋆e-mail: jurcisine@saske.sk
⋆⋆e-mail: jurcisin@saske.sk
⋆⋆⋆e-mail: remecky@saske.sk

of kinematic viscosity to the coefficient of diffusivity of the system. The Prandtl numbers usually strongly depend on the microscopic structure of the fluid when the value of the Reynolds number (Re) of the flow is low. The situation changes in the state of the so-called fully developed turbulence, i.e., where the fluid motion is characterized by very high Reynolds number (in the ideal case Re $\to \infty$), when the Prandtl numbers reach universal values which are known as effective or turbulent Prandtl numbers.

In the present paper we investigate the turbulent vector Prandtl number in the framework of the general $\mathcal{A}$ model [7–10] of a passive vector quantity advected by the turbulent velocity field driven by the stochastic Navier-Stokes equation. This model describes three physically important cases of passive vector admixture, namely, the passive advection of the magnetic field in a conductive turbulent environment in the framework of the kinematic MHD turbulence with $\mathcal{A} = 1$ [8, 11, 12], the passive admixture of a vector impurity by the Navier-Stokes turbulent flow with $\mathcal{A} = 0$ [13, 14], and the model of linearized Navier-Stokes equation with $\mathcal{A} = -1$.

2 Mathematical Model of the Turbulent System

In what follows, we consider the advection of a passive vector field $\mathbf{b} \equiv \mathbf{b}(x)$, where $x \equiv (t, \mathbf{x})$, driven by the incompressible Navier-Stokes velocity field $\mathbf{v}(x)$ described by the following system of stochastic equations:

$$\partial_t \mathbf{b} + (\mathbf{v} \cdot \partial)\mathbf{b} = u_0 \nu_0 \Delta \mathbf{b} + \mathcal{A}(\mathbf{b} \cdot \partial)\mathbf{v} - \partial Q + \mathbf{f}^{\mathbf{b}}, \tag{1}$$

$$\partial_t \mathbf{v} + (\mathbf{v} \cdot \partial)\mathbf{v} = \nu_0 \Delta \mathbf{v} - \partial \mathcal{P} + \mathbf{f}^{\mathbf{v}}, \tag{2}$$

where ν_0 is viscosity coefficient, u_0 is dimensionless reciprocal general vector Prandtl number, $\mathcal{P}$ and Q represent corresponding pressures, and the statistics of the random forces $\mathbf{f}^{\mathbf{b}}$ and $\mathbf{f}^{\mathbf{v}}$ are supposed Gaussian with zero means and correlation functions

$$D^b_{ij}(x; x') \equiv \langle f^b_i(x) f^b_j(x')\rangle = \delta(t - t')\, C_{ij}(|\mathbf{x} - \mathbf{x}'|/L), \tag{3}$$

$$D^v_{ij}(x, x') \equiv \langle f^v_i(x) f^v_j(x')\rangle = \delta(t - t') \int \frac{d^d\mathbf{k}}{(2\pi)^d} g_0 \nu_0^3 k^{4-d-2\varepsilon} P_{ij}(\mathbf{k})\, e^{i\mathbf{k}(\mathbf{x}-\mathbf{x}')}, \tag{4}$$

where the explicit form of the function C_{ij} is not important, $P_{ij}(\mathbf{k}) = \delta_{ij} - k_i k_j / k^2$ is the ordinary transverse tensor projector, g_0 plays the role of the coupling constant of the model, and the physical value of formally small parameter $0 < \varepsilon \leq 2$ is $\varepsilon = 2$.

The stochastic model given by Eqs. (1)-(4) is equivalent to the field theoretic model of the double set of fields $\Phi = \{\mathbf{v}, \mathbf{b}, \mathbf{v}', \mathbf{b}'\}$ with the action functional given as follows

$$\begin{aligned} S(\Phi) = {} & \frac{1}{2} \int dt_1\, d^d\mathbf{x}_1\, dt_2\, d^d\mathbf{x}_2\, [v'_i(x_1) D^v_{ij}(x_1; x_2) v'_j(x_2) + b'_i(x_1) D^b_{ij}(x_1; x_2) b'_j(x_2)] \\ & + \int dt\, d^d\mathbf{x}\, \{\mathbf{v}'[-\partial_t \mathbf{v} + \nu_0 \Delta \mathbf{v} - (\mathbf{v} \cdot \partial)\mathbf{v}]\} \\ & + \int dt\, d^d\mathbf{x}\, \{\mathbf{b}'[-\partial_t \mathbf{b} + \nu_0 u_0 \Delta \mathbf{b} - (\mathbf{v} \cdot \partial)\mathbf{b} + \mathcal{A}(\mathbf{b} \cdot \partial)\mathbf{v}]\}, \end{aligned} \tag{5}$$

The field theoretic model (5) corresponds to the standard Feynman diagrammatic perturbation theory which is used for the analysis of the original stochastic model. Note that the formulation of the stochastic problem given by Eqs. (1)–(4) through the field theoretic model with the action functional Eq. (5) allows one to use the well-defined field theoretic tools, e.g., the RG technique, to analyze the problem. The RG analysis of the field theoretic model described by the action functional (5) is based on analysis of UV divergences of the model. The detailed analysis is given in [15].

3 Prandtl Number in the Two-Loop Approximation

We are interested in the IR asymptotic scaling behavior of the correlation functions of the model deep inside the inertial interval. This behavior is driven by the corresponding IR stable fixed point of the RG equations. Coordinates of the fixed point (g_*, u_*) deep inside the inertial interval are determined by the requirement of vanishing of the corresponding β functions (see Ref. [15] for details).

The nontrivial IR stable fixed point, i.e., the IR stable fixed point with $g_* \neq 0$ and $u_* \neq 0$, in the two-loop approximation of the present model reads

$$g_* = g_*^{(1)}\varepsilon + g_*^{(2)}\varepsilon^2 + O(\varepsilon^3), \qquad u_* = u_*^{(1)} + u_*^{(2)}\varepsilon + O(\varepsilon^2), \tag{6}$$

where

$$g_*^{(1)} = \frac{(2\pi)^{2d}}{S_d}\frac{8(d+2)}{3(d-1)}, \qquad g_*^{(2)} = \frac{(2\pi)^{2d}}{S_d}\frac{8(d+2)}{3(d-1)}\lambda, \tag{7}$$

and

$$u_*^{(1)} = \frac{1}{3a_2}\left[-2a_2 - \frac{2^{1/3}b_1}{(b_2+b_3)^{1/3}} + \frac{(b_2+b_3)^{1/3}}{2^{1/3}}\right], u_*^{(2)} = \frac{2(d+2)}{d[1+2u_*^{(1)}]}\left[\lambda - \frac{128(d+2)^2}{3(d-1)^2}\mathcal{B}(u_*^{(1)})\right], \tag{8}$$

with the explicit forms of the coefficients $\mathcal{B}(u_*^{(1)})$, λ, and a_i and b_i for $i = 1, 2$ given in [15].

Using the same procedure as in Ref. [16] the corresponding two-loop formula for the turbulent inverse vector Prandtl number u_{eff} in the framework of the general $\mathcal{A}$ model of passively advected vector quantity can be derived and written in the similar form as in Ref. [16], namely,

$$u_{\text{eff}} = u_*^{(1)}\left(1 + \varepsilon\left\{\frac{1+u_*^{(1)}}{1+2u_*^{(1)}}\left[\lambda - \frac{128(d+2)^2}{3(d-1)^2}\mathcal{B}(u_*^{(1)})\right] + \frac{(2\pi)^d}{S_d}\frac{8(d+2)}{3(d-1)}\left[a_v - a_b(u_*^{(1)})\right]\right\}\right), \tag{9}$$

where a_v and $a_b(u_*^{(1)})$ are given by the corresponding expansion to the leading order in ε of the scaling functions of response functions $\langle bb'\rangle$ and $\langle vv'\rangle$ of the advected vector and velocity field, respectively. For more details and the explicit expressions for the aforementioned quantities see [15].

The turbulent vector Prandtl number in the two-loop approximation is given as the inverse value of the effective inverse Prantl number, i.e.,

$$\text{Pr}_{\mathcal{A},t} = 1/u_{\text{eff}}. \tag{10}$$

4 Discussion

In the present work we devoted special attention to the three important and physically interesting cases of the general model of vector advection, namely, to the model of the linearized Navier-Stokes equation $\mathcal{A} = -1$, to the passive admixture of vector impurity by Navier-Stokes turbulent flow $\mathcal{A} = 0$ and to the model of passive advection of the magnetic field in the framework of kinematic MHD turbulence $\mathcal{A} = 1$.

It is shown that the turbulent Prandtl number $\text{Pr}_{l,t}$ in the model of linearized Navier-Stokes equation ($\mathcal{A} = -1$) is exactly given already in the one-loop approximation, $\text{Pr}_{l,t}^{(1)} = \text{Pr}_{l,t}$, see Fig. 1. Higher loop approximations vanish and the numerical value is equal to 1 for all values of the spatial dimension $d > 2$. The nontrivial conclusion is that in this model, coefficients of turbulent diffusivity and the turbulent viscosity are numerically equal.

On the other hand, the two-loop turbulent Prandtl numbers $\text{Pr}_{m,t}$ ($\mathcal{A} = 1$) and $\text{Pr}_{v,t}$ ($\mathcal{A} = 0$) show very similar behaviours as functions of the spatial dimension d, see Fig. 1. They are both increasing functions of d and both go to 1 in the limit $d \to \infty$.

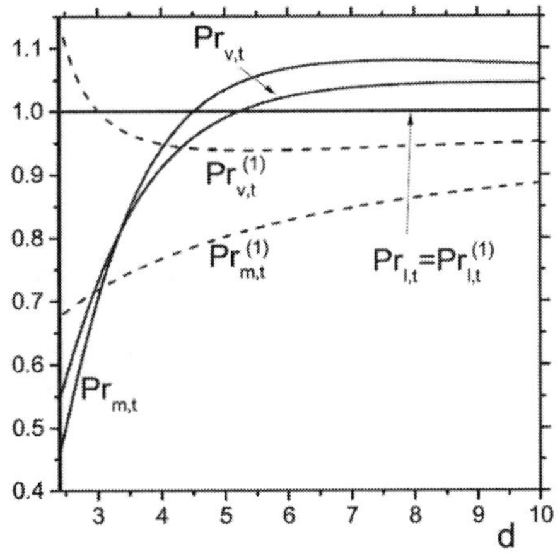

Figure 1. The behavior of the two-loop turbulent Prandtl numbers in the general $\mathcal{A}$ model for three special cases, namely for $\mathcal{A} = -1$ ($Pr_{l,t}$), $\mathcal{A} = 0$ ($Pr_{v,t}$) and for $\mathcal{A} = 1$ ($Pr_{m,t}$). The corresponding one-loop curves $Pr^{(1)}_{...}$ are also shown for comparison.

Acknowledgements

The work was supported by the VEGA grant No. 2/0065/17.

References

[1] A.S. Monin and A.M. Yaglom, *Statistical Fluid Mechanics, Vol. 2* (Cambridge, MA, MIT Press, 1975)
[2] W.D. McComb, *The Physics of Fluid Turbulence* (Oxford, Clarendon, 1990)
[3] A. Yoshizawa, S.I. Itoh, and K. Itoh, *Plasma and Fluid Turbulence: Theory and Modelling* (Bristol, IoP, 2003)
[4] D. Biskamp, *Magnetohydrodynamic Turbulence* (Cambridge, Cambridge University Press, 2003)
[5] A.N. Vasil'ev, *The Field Theoretic Renormalization Group in Critical Behavior Theory and Stochastic Dynamics* (Chapman & Hall/CRC, CRC Press, 2004)
[6] A.N. Kolmogorov, Dokl. Akad. Nauk SSSR **30**, 301 (1941)
[7] L.Ts. Adzhemyan, N.V. Antonov, A. Mazzino, P. Muratore-Ginanneschi, and A.V. Runov, Europhys. Lett. **55**, 801 (2001)
[8] N.V. Antonov, M. Hnatich, J. Honkonen, and M. Jurčišin, Phys. Rev. E **68**, 046306 (2003)
[9] H. Arponen, Phys. Rev. E **79**, 056303 (2009)
[10] N.V. Antonov and N.M. Gulitskiy, Theor. Math. Phys. **176**, 851 (2013)
[11] E. Jurčišinová, M. Jurčišin, and R. Remecký, Phys. Rev. E **84**, 046311 (2011)
[12] E. Jurčišinová, M. Jurčišin, R. Remecký, and P. Zalom, Phys. Rev. E **87**, 043010 (2013)
[13] E. Jurčišinová, M. Jurčišin, and R. Remecký, Phys. Rev. E **88**, 011002(R) (2013)
[14] E. Jurčišinová, M. Jurčišin, and P. Zalom, Phys. Rev. E **89**, 043023 (2014)
[15] E. Jurčišinová, M. Jurčišin, and R. Remecký, Phys. Rev. E **93**, 033106 (2016)
[16] L.Ts. Adzhemyan, J. Honkonen, T.L. Kim, and L. Sladkoff, Phys. Rev. E **71**, 056311 (2005)

On the Generation of Random Ensembles of Qubits and Qutrits

Computing Separability Probabilities for Fixed Rank States

Arsen Khvedelidze[1,2,3,4,⋆] and *Ilya* Rogojin[4,⋆⋆]

[1] *A.Razmadze Mathematical Institute, Iv.Javakhishvili Tbilisi State University, Tbilisi, Georgia*
[2] *Institute of Quantum Physics and Engineering Technologies, Georgian Technical University, Tbilisi, Georgia*
[3] *National Research Nuclear University, MEPhI (Moscow Engineering Physics Institute), Moscow, Russia*
[4] *Laboratory of Information Technologies, Joint Institute for Nuclear Research, Dubna, Russia*

Abstract. The generation of random mixed states is discussed, aiming for the computation of probabilistic characteristics of composite finite dimensional quantum systems. In particular, we consider the generation of random Hilbert-Schmidt and Bures ensembles of qubit and qutrit pairs and compute the corresponding probabilities to find a separable state among the states of a fixed rank.

1 Introduction

The easily stated problem of finding the probability of a binary composite quantum system being in a separable state turns out to be a challenging computational issue even for simple systems, such as qubit-qubit or qubit-qutrit pairs. Although, the issue of the "separability probability" was formulated almost twenty years ago, nowadays it still remains a subject of interest in the theory of quantum information and quantum communications. For references on relevant studies, rigorous notions and definitions we cite a comprehensive review [1] and a recent paper by P. Slater [2]. Based on the classical theorems of the geometric probability theory [3, 4] attributed to J. Sylvester, the computation of the quantum "separability probability" can be reduced to the evaluation of the relative volume of the separable states with respect to the volume of the whole state space. In order to avoid these subtle and cumbersome numerical calculations of multidimensional integrals (for a generic 2-qubit case the integrals are over the semi-algebraic domain of a 15-dimensional Euclidean space), the Monte-Carlo ideology with a specific algorithm for the generation of random variables has been used (cf. [5]–[8] and references therein). Our report aims to present several results on the numeric studies of the separability probabilities for random qubit-qubit and qubit-qutrit Hilbert-Schmidt and Bures ensembles. In particular, we discuss the dependence of the "separability probability" on the Bloch vectors of each subsystem for the random states of all the possible ranks.

2 The separability probability

There are two equivalent ways to define and compute the separability probability of quantum states. The first is to follow the geometric probability theory, considering the state space of a composite

⋆e-mail: akhved@jinr.ru
⋆⋆e-mail: virus-atl@inbox.ru

quantum system as the Riemannian space and defining the separability probability as the two volumes ratio:

$$\mathcal{P}_{\text{sep}} = \frac{\text{Vol (Separable states)}}{\text{Vol (All states)}} \,. \tag{1}$$

The second way consists in generating the ensemble of random states distributed in correspondence with the volume measure in (1) and evaluating the separability probability:

$$\mathcal{P}_{\text{sep}} = \frac{\text{Number of separable states in ensemble}}{\text{Total number of states in ensemble}} \,. \tag{2}$$

Representation (1) is well appropriate for theoretical studies, but attempting to use it for a practical evaluation leads to big computational difficulties (see e.g., [2] and references therein). In its turn, the equation (2) allows to use highly effective numerical computational techniques. Below, based on representation (2), the results of the studies on the separability probability of the qubit-qubit and qubit-qutrit systems will be stated.

3 Generating the Hilbert-Schmidt and Bures random states

Keeping in mind the calculations of the separability probability of states with different ranks we briefly describe the algorithm for the generation of density matrices from the Hilbert-Schmidt and the Bures ensembles of a n-level quantum system.

- The first step in the construction of both ensembles is the generation of $n \times n$ matrices Z from the complex Ginibre ensemble, i.e. the family of matrices in which each entry is an independent complex Gaussian random variable of mean zero and variance one;
- Using Z and its complex conjugate $Z^\dagger$ one can represent the density matrices ϱ_{HS} describing states from the Hilbert-Schmidt ensemble in the following form:

$$\varrho_{\text{HS}} = \frac{ZZ^\dagger}{\operatorname{tr}(ZZ^\dagger)} \,; \tag{3}$$

- The density matrices ϱ_{B} from the Bures ensemble are generated using the Ginibre matrices Z and matrices $U \in SU(n)$ distributed over the $SU(n)$ group manifold according to the Haar measure,

$$\varrho_{\text{B}} = \frac{(I+U)ZZ^\dagger\left(I+U^\dagger\right)}{\operatorname{tr}\left((I+U)ZZ^\dagger\,(I+U^\dagger)\right)} \,. \tag{4}$$

Since we are interested in the study of the entanglement of states with a fixed rank of density matrices, the above algorithm requires further specification. For states of a rank lower than the maximal one we proceed as follows.

•Rank-3 states• We start with the generation of complex rank-3 Ginibre matrices. It is known that any rank-3 complex 4×4 matrix admits the following representation:

$$Z = P_Z \left(\begin{array}{ccc|c} & & & x_1 \\ & A & & x_2 \\ & & & x_3 \\ \hline y_1 & y_2 & y_3 & D \end{array}\right) Q_Z \,, \tag{5}$$

where P_Z and Q_Z denote the permutation matrices and A is a regular element of the Ginibre ensemble of 3×3 matrices. Additionally assuming that the 3-tuples $Y = (y_1, y_2, y_3)$ and $X = (x_1, x_2, x_3)$ are

Table 1. The separability probability for the maximal rank-4 states in qubit-qubit and qubit-qutrit systems

States	System	Probability
HS ensemble	$2 \otimes 2$	0.2424
	$2 \otimes 3$	0.0270
Bures ensemble	$2 \otimes 2$	0.0733
	$2 \otimes 3$	0.0014

Table 2. The separability probability for non-maximal rank qubit-qubit states

States	Rank	Probability
HS ensemble	3	0.1652
	2	0
	1	0
Bures ensemble	3	0.0494
	2	0
	1	0

composed of standard normal random complex variables, while the entry D is $D = YA^{-1}X$, we arrive at the presentation of the rank-3 elements from the Ginibre ensemble of complex 4×4 matrices.
•**Rank-2 states**• Similarly, the 4×4 Ginibre matrix of rank-2 can be written as,

$$Z = P_Z \left(\begin{array}{c|c} A & B \\ \hline C & D \end{array} \right) Q_Z \,, \tag{6}$$

where A, B and C are 2×2 complex Ginibre matrices, while the 2×2 matrix D reads, $D = CA^{-1}B$.
•**Rank-1 states**• Finally, the 4×4 complex Ginibre matrix Z of rank-1 admits the representation:

$$Z = P_Z \left(\begin{array}{c|ccc} a & y_1 & y_2 & y_3 \\ \hline x_1 & & & \\ x_2 & & D & \\ x_3 & & & \end{array} \right) Q_Z \,, \tag{7}$$

where a, x_1, x_2, x_3 and y_1, y_2, y_3 are standard normal random complex variables, while D is:

$$D = \frac{1}{a} \begin{pmatrix} x_1 y_1 & x_1 y_2 & x_1 y_3 \\ x_2 y_1 & x_2 y_2 & x_2 y_3 \\ x_3 y_1 & x_3 y_2 & x_3 y_3 \end{pmatrix} .$$

Now, using the generic Ginibre matrix and the representations (5)–(7) for non-maximal rank matrices we can build both the Hilbert-Schmidt and the Bures states of the required rank following either equation (3) or equation (4).

4 Computing the separability probability

The computation of the sought-for separability probability (2) requires testing the generated states on their separability. With this goal in mind, the Peres-Horodecki criterion [1] has been used in our calculations.

4.1 Results of the computations

In table 1 the results of the calculations of the separability probability of the generic (rank-4) states from the Hilbert-Schmidt and Bures ensembles of qubit-qubit and qubit-qutrit pairs are collected. The results of the computations of the separability probabilities of non-maximal rank states of a qubit-qubit system are given for both ensembles in table 2. It is worth noticing the zero separability probability for the rank-2 and rank-1 states. This result is consistent with the known statement, that if a density matrix of 2-qubits ϱ is such that $\text{rank}(\varrho) < d_A = \text{rank}(\varrho_A)$,[1] then ϱ is not separable [9]. Indeed, the above mentioned inequality holds true, since during the generation of rank-2 and rank-1 almost all reduced matrices are not singular.

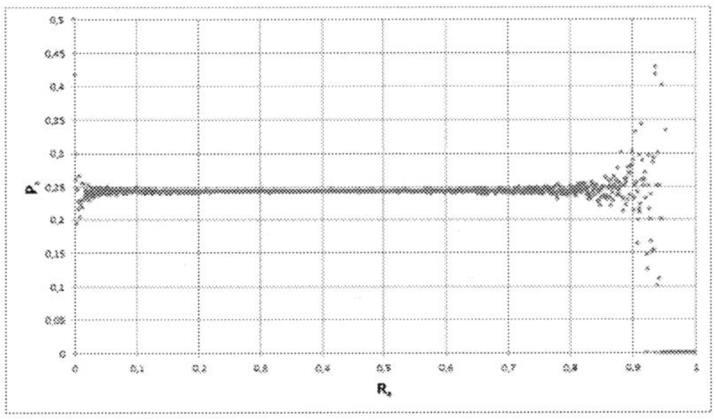

(a) Maximal rank states

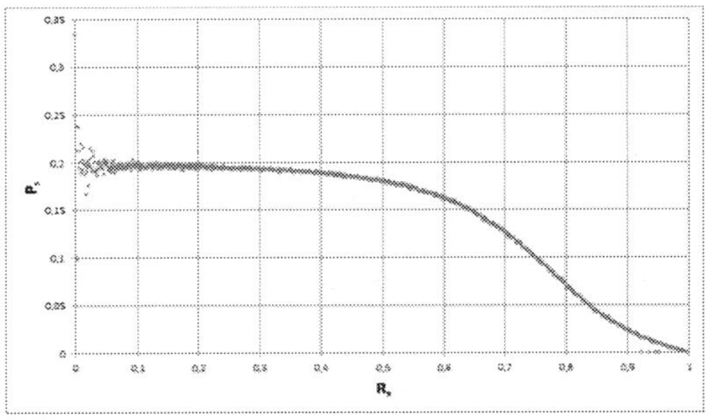

(b) Rank-3 states

Figure 1. The separability probability for the Hilbert-Schmidt qubit-qubit states as a function of the Bloch radius of the first qubit

5 Concluding remarks

Using the above described algorithms as tools for the generation of ensembles of random states of different ranks, various characteristics of entanglement in composite quantum systems can be studied. As an example, in figures 1a and 1b, we illustrate the separability probability of 2-qubit systems as a function of the Bloch radius of the constituent qubit for the states of maximal and sub-maximal ranks. Our analysis shows that the distribution of the separability probability with respect to the Bloch radius of the qubit is uniform for maximal rank states (see figure 1a), while for the rank-3 states the deviation from the total separability probability is depicted in figure 1b.

References

[1] I. Bengtsson and K. Zyczkowski, *Geometry of Quantum States: An Introduction to Quantum Entanglement* (Cambridge University Press, Cambridge, 2006)
[2] P.B. Slater, arXiv:1701.01973v7, (2017)
[3] D.A. Klain and G.C. Rota, *Introduction to Geometric Probability* (Cambridge University Press, Cambridge, 1997)
[4] L.A. Santalo, *Integral Geometry and Geometric Probability* (Cambridge University Press, Cambridge, 2004)
[5] S.L. Braunstein, Phys. Lett. **A 219**, 169, (1996)
[6] K. Zyczkowski and H-J. Sommers, J. Phys. A: Math. Theor. **34**, 7111, (2001)
[7] V.A. Osipov, H.J. Sommers, and K. Zyczkowski, J. Phys. A: Math. Theor. **43**, 055302, (2010)
[8] A. Khvedelidze and I. Rogojin, J. of Math. Sciences **209**, 6, 988–1004, (2015)
[9] M. Ruskai and E.M. Werner, J. Phys. A. Math. Theor. **42**, 095303, (2009)

[1] Here, ϱ_A denotes the reduced density matrix obtained by taking the partial trace with respect to the second qubit.

EPJ Web of Conferences **173**, 02011 (2018) https://doi.org/10.1051/epjconf/201817302011
Mathematical Modeling and Computational Physics 2017

On the Stratifications of 2-Qubits X-State Space

Arsen Khvedelidze[1,2,3,4,⋆] and *Astghik* Torosyan[4,⋆⋆]

[1] *A. Razmadze Mathematical Institute, Iv.Javakhishvili Tbilisi State University, Tbilisi, Georgia*
[2] *Institute of Quantum Physics and Engineering Technologies, Georgian Technical University, Tbilisi, Georgia*
[3] *National Research Nuclear University, MEPhI (Moscow Engineering Physics Institute), Moscow, Russia*
[4] *Laboratory of Information Technologies, Joint Institute for Nuclear Research, Dubna, Russia*

Abstract. The 7-dimensional family $\mathfrak{P}_X$ of the so-called mixed X-states of 2-qubits is considered. Two versions of stratifications of $\mathfrak{P}_X$, i.e., its decomposition into strata according to orbit types of the adjoint actions of two groups, are described. The first stratification is due to the action of global unitary group $G_X \subset SU(4)$, while the second one corresponds to the action of the local unitary group $LG_X \subset G_X$. The equations and inequalities in the invariants of the corresponding groups, determining each stratification component, are given.

1 Introduction

The understanding of a symmetry that a physical system possesses, as well as this symmetry breaking pattern allows us to explain uniquely a wide variety of phenomena in many areas of physics, including elementary particle physics and condensed matter physics [1]. The mathematical formulation of symmetries related to the Lie group action consists of the detection of the stratification of the representation space of the corresponding symmetry group. Dealing with closed quantum systems the symmetries are realized by the unitary group actions and the quantum state space plays the role of the symmetry group representation space. Below, having in mind these observations, we will outline examples of the stratifications occurring for a quantum system composed of a pair of 2-level systems, two qubits. We will analyze symmetries associated with two subgroups of the special unitary group $SU(4)$. More precisely, we will consider the 7-dimensional subspace $\mathfrak{P}_X$ of a generic 2-qubit state space, the family of X-states (for definition see [2], [3] and references therein) and reveal two types of its partition into sets of points having the same symmetry type. The primary stratification originates from the action of the invariance group of X-states, named the global unitary group $G_x \subset SU(4)$, whereas the secondary one is due to the action of the so-called local group $LG_x \subset G_X$ of the X-states.

2 X-states and their symmetries

The mixed 2-qubit X-states can be defined based on the purely algebraic consideration. The idea is to fix the subalgebra $\mathfrak{g}_X := \mathfrak{su}(2) \oplus \mathfrak{su}(2) \oplus \mathfrak{u}(1) \in \mathfrak{su}(4)$ of the algebra $\mathfrak{su}(4)$ and define the density matrix

⋆e-mail: akhved@jinr.ru
⋆⋆e-mail: astghik@jinr.ru

of the X-states as

$$\varrho_X = \frac{1}{4}\left(I + i\mathfrak{g}_X\right). \tag{1}$$

In order to coordinatize the X-state space we use the tensorial basis for the $\mathfrak{su}(4)$ algebra, $\sigma_{\mu\nu} = \sigma_\mu \otimes \sigma_\nu$, $\mu, \nu = 0, 1, 2, 3$. It consists of all possible tensor products of two copies of Pauli matrices and a unit 2×2 matrix, $\sigma_\mu = (I, \sigma_1, \sigma_2, \sigma_3)$, which we order as follows (see details in [3]):

$$\lambda_1, \dots, \lambda_{15} = \frac{i}{2}\left(\sigma_{x0}, \sigma_{y0}, \sigma_{z0}, \sigma_{0x}, \sigma_{0y}, \sigma_{0z}, \sigma_{xx}, \sigma_{xy}, \sigma_{xz}, \sigma_{yx}, \sigma_{yy}, \sigma_{yz}, \sigma_{zx}, \sigma_{zy}, \sigma_{zz}\right). \tag{2}$$

In this basis the 7-dimensional subalgebra $\mathfrak{g}_X$ is generated by the subset $\alpha_X = (\lambda_3, \lambda_6, \lambda_7, \lambda_8, \lambda_{10}, -\lambda_{11}, \lambda_{15})$, and thus the unit norm X-state density matrix is given by the decomposition:

$$\varrho_X = \frac{1}{4}\left(I + 2i \sum_{\lambda_k \in \alpha_X} h_k \lambda_k\right). \tag{3}$$

The real coefficients h_k are subject to the polynomial inequalities ensuring the semi-positivity of the density matrix, $\varrho_X \geq 0$:

$$\mathfrak{P}_X = \left\{h_i \in \mathbb{R}^7 \mid (h_3 \pm h_6)^2 + (h_8 \pm h_{10})^2 + (h_7 \pm h_{11})^2 \leq (1 \pm h_{15})^2\right\}. \tag{4}$$

Using the definition (1) one can conclude that the X–state space $\mathfrak{P}_X$ is invariant under the 7-parameter group, $G_X := \exp(\mathfrak{g}_X) \in SU(4)$:

$$g\varrho_X g^\dagger \in \mathfrak{P}_X, \qquad \forall g \in G_X. \tag{5}$$

Group G_X plays the same role for the X-states as the special unitary group $SU(4)$ plays for a generic 4-level quantum system, and thus is termed the *global unitary group* of the X–states. According to [3], group G_X admits the representation:

$$G_X = P_\pi \left(\begin{array}{c|c} e^{-i\omega_{15}} SU(2) & 0 \\ \hline 0 & e^{i\omega_{15}} SU(2)' \end{array}\right) P_\pi, \quad \text{with} \quad P_\pi = \begin{pmatrix} 1 & 0 & 0 & 0 \\ 0 & 0 & 0 & 1 \\ 0 & 0 & 1 & 0 \\ 0 & 1 & 0 & 0 \end{pmatrix}. \tag{6}$$

Correspondingly, the *local unitary group* of the X-states is

$$LG_X = P_\pi \exp\left(\imath \frac{\phi_1}{2}\right) \otimes \exp\left(\imath \frac{\phi_2}{2}\right) P_\pi \subset G_X. \tag{7}$$

3 Global orbits and state space decomposition

Now we give a classification of the global G_X-orbits according to their dimensionality and isotropy group. Every density matrix ϱ_X can be diagonalized by some element of the global G_X group. In other words, all global G_X-orbits can be generated from the density matrices, whose eigenvalues form the partially ordered simplex $\underline{\Delta}_3$, depicted on Figure 1.

The tangent space to the G_X-orbits is spanned by the subset of linearly independent vectors, built from the vectors: $t_k = [\lambda_k, \varrho_X]$, $\lambda_k \in \alpha_X$. The number of independent vectors t_k determines the dimensionality of the G_X-orbits and is given by the rank of the 7×7 Gram matrix:

$$\mathcal{G}_{kl} = \frac{1}{2}\operatorname{Tr}(t_k t_l). \tag{8}$$

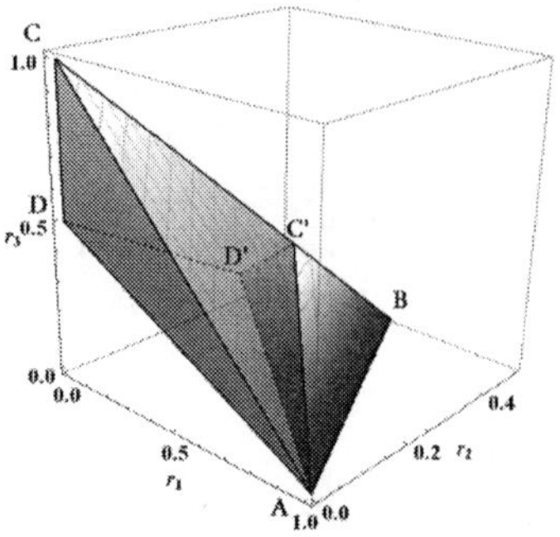

Figure 1. The tetrahedron $ABCD$ describes the partially ordered simplex $\underline{\Delta}_3 := \{ \sum_{i=1}^{4} r_i = 1\,,\ \{1 \geq r_1 \geq r_2 \geq 0\} \cup \{1 \geq r_3 \geq r_4 \geq 0\}\}$ of the density matrix eigenvalues, while the tetrahedron $ABC'D'$ inside it corresponds to a 3D simplex with the following complete order: $\{ \sum_{i=1}^{4} r_i = 1\,,\ 1 \geq r_1 \geq r_2 \geq r_3 \geq r_4 \geq 0 \}$.

The Gram matrix (8) has three zero eigenvalues and two double multiplicity eigenvalues:

$$\mu_\pm \;=\; -\frac{1}{8}\left((h_3 \pm h_6)^2 + (h_8 \pm h_{10})^2 + (h_7 \pm h_{11})^2\right). \tag{9}$$

Correspondingly, the G_X-orbits have dimensionality of either 4, 2 or 0. The orbits of maximal dimensionality, $\dim(O)_{\text{Gen}} = 4$, are characterized by non-vanishing $\mu_\pm \neq 0$ and consist of the set of density matrices with a generic spectrum, $\sigma(\varrho_X) = (r_1, r_2, r_3, r_4)$. If the density matrices obey the equations

$$h_6 = \pm h_3\,, \quad h_{10} = \pm h_8\,, \quad h_{11} = \pm h_7\,, \tag{10}$$

they belong to the so-called degenerate orbits, $\dim(O)_\pm = 2$. The latter are generated from matrices which have a double degenerate spectrum of the form, $\sigma(\varrho_X) = (p, p, r_3, r_4)$ and $\sigma(\varrho_X) = (r_1, r_2, q, q)$ respectively. Finally, there is a single orbit $\dim(O)_0 = 0$, corresponding to the maximally mixed state $\varrho_X = \frac{1}{4}I$.

Considering the diagonal representative of the generic G_X-orbit one can be convinced that its *isotropy group* is

$$H = P_\pi \left(\begin{array}{c|c} e^{i\omega}\exp i\frac{\gamma_1}{2}\sigma_3 & 0 \\ \hline 0 & e^{-i\omega}\exp i\frac{\gamma_2}{2}\sigma_3 \end{array}\right) P_\pi\,, \tag{11}$$

while for a diagonal representative with a double degenerate spectrum the isotropy group is given by one of two groups:

$$H_+ = P_\pi \left(\begin{array}{c|c} e^{i\omega}SU(2) & 0 \\ \hline 0 & e^{-i\omega}\exp i\frac{\gamma_2}{2}\sigma_3 \end{array}\right) P_\pi\,, \quad H_- = P_\pi \left(\begin{array}{c|c} e^{i\omega}\exp i\frac{\gamma_1}{2}\sigma_3 & 0 \\ \hline 0 & e^{-i\omega}SU(2)' \end{array}\right) P_\pi\,.$$

For the single zero dimensional orbit the isotropy group H_0 coincides with the whole invariance group, $H_0 = G_X$. Therefore, the isotropy group of any element of G_X-orbits belongs to one of these conjugacy classes: $[H]$, $[H_\pm]$ or $[H_0]$. Moreover, a straightforward analysis shows that $[H_+] = [H_-]$. Hence, any point $\varrho \in \mathfrak{P}_X$ belongs to one of three above-mentioned types of G_X-orbits[1], denoted afterwards as $[H_t]\,, t = 1, 2, 3$. For a given H_t, the associated *stratum* $\mathfrak{P}_{[H_t]}$, defined as the set of all points whose stabilizer is conjugate to H_t:

$$\mathfrak{P}_{[H_t]} := \{y \in \mathfrak{P}_X |\ \text{isotropy group of } y \text{ is conjugate to } H_t\}$$

determines the sought-for decomposition of the state space $\mathfrak{P}_X$ into strata according to the orbit types:

$$\mathfrak{P}_X = \bigcup_{\text{orbit types}} \mathfrak{P}_{[H_t]}\,. \tag{12}$$

[1]The *orbit type* $[\varrho]$ of a point $\varrho \in \mathfrak{P}_X$ is given by the conjugacy class of the isotropy group of point ϱ, i.e., $[\varrho] = [G_{\varrho_X}]$.

The strata $\mathfrak{P}_{(H_i)}$ are determined by this set of equations and inequalities:

$$(1)\quad \mathfrak{P}_{[H]} := \{h_i \in \mathfrak{P}_X \mid \mu_+ > 0,\, \mu_- > 0\}\,, \tag{13}$$

$$(2)\quad \mathfrak{P}_{[H_+]} \cup \mathfrak{P}_{[H_-]} := \{h_i \in \mathfrak{P}_X \mid \mu_+ = 0,\, \mu_- > 0\} \cup \{h_i \in \mathfrak{P}_X \mid \mu_+ > 0,\, \mu_- = 0\}\,, \tag{14}$$

$$(3)\quad \mathfrak{P}_{[H_0]} := \{h_i \in \mathfrak{P}_X \mid \mu_+ = 0,\, \mu_- = 0\}\,. \tag{15}$$

4 Local orbits and state space decomposition

Analogously, one can build up the X-state space decomposition associated with the local group LG_X action. For this action the dimensionality of LG_X-orbits is given by the rank of the corresponding 2×2 Gram matrix constructed out of vectors t_3 and t_6. Since its eigenvalues read:

$$\mu_1 = -\frac{1}{8}\left((h_8 + h_{10})^2 + (h_7 + h_{11})^2\right)\,, \qquad \mu_2 = -\frac{1}{8}\left((h_8 - h_{10})^2 + (h_7 - h_{11})^2\right)\,, \tag{16}$$

the LG_X-orbits are either generic ones with the dimensionality of $\dim(O_L)_{\mathrm{Gen}} = 2$, or degenerate $\dim(O_L)_\pm = 1$, or exceptional ones, $\dim(O_L)_0 = 0$. The LG_X-orbits can be collected into the strata according to their orbit type. There are three types of strata associated with the "local" isotropy subgroups of $LH \in LG_X$. Correspondingly, one can define the following "local" strata of state space:

- the generic stratum, $\mathfrak{P}^L_{[I]}$, which has a trivial isotropy type, $[I]$, and is represented by the inequalities: $\mathfrak{P}^L_{[I]} := \{h_i \in \mathfrak{P}_X \mid \mu_1 > 0,\, \mu_2 > 0\}\,,$
- the degenerate stratum, $\mathfrak{P}^L_{[H^\pm_L]}$, collection of the orbits whose type is $[H^\pm_L]$, with the subgroup either $H^+_L = I \times \exp(iu\sigma_3)$, or $H^-_L = \exp(iv\sigma_3) \times I$. The stratum defining equations read respectively:

$$h_{10} = \pm h_8\,, \qquad h_{11} = \pm h_7\,, \tag{17}$$

- the exceptional stratum, $\mathfrak{P}_{[LG_X]}$ of the type $[LG_X]$, determined by the equations: $h_{11} = h_{10} = h_8 = h_7 = 0$.

Therefore, the local group action prescribes the following stratification of 2-qubit X–state space:

$$\mathfrak{P}_X = \mathfrak{P}_{[I]} \cup \mathfrak{P}_{[H^+_L]} \cup \mathfrak{P}_{[H^-_L]} \cup \mathfrak{P}_{[LG_X]}\,. \tag{18}$$

5 Concluding remarks

In the present article two stratifications of quantum state space associated with the global and local unitary symmetries of two qubits in the X-states are described. Stratifications encode information on system's physical characteristics. The first one, due to the global unitary symmetry, is related to the properties of a system as a whole, while the second stratification, according to the local symmetries, comprises information on the entanglement, cf. [4]. In an upcoming publication, based on the introduced stratifications of state space, we plan to analyze an interplay between these two symmetries and particularly to determine the entanglement/separability characteristics of every stratum.

References

[1] L. Michel, et.al., Physics Reports **341**, 7, (2001)
[2] T. Yu and J.H. Eberly, Quantum Inf. Comput. **7**, 459–468, (2007)
[3] A. Khvedelidze and A. Torosyan, Journal of Mathematical Sciences **224**, 349–359, (2017)
[4] X. Chen, Z.-C. Gu, and X.-G. Wen, Phys. Rev. **B 82**, 155138, (2010)

EPJ Web of Conferences **173**, 02012 (2018) https://doi.org/10.1051/epjconf/201817302012
Mathematical Modeling and Computational Physics 2017

Quark-Parton Model and Relativistic Quantum Mechanics

Boris Kostenko[1,⋆]

[1] *Joint Institute for Nuclear Research, Dubna, Russian Federation*

Abstract. An attempt to treat the asymptotic freedom and the quark confinement as a self-consistent problem in the framework of relativistic quantum mechanics is realized. It is shown that the confinement of quarks induces a change of their helicities together with a simultaneous alteration of orbital momenta, so that the total angular momentum of each quark is conserved. This observation may cast light on the so-called proton spin puzzle after some additional numerical estimations.

1 Introduction

The Dirac field of a free quark polarized at rest along z-axis may be written in the form

$$\psi = \sqrt{\frac{E+m}{2m}}\left(1, 0, 0, \frac{p}{E+m}\right)^T$$

in a reference frame moving along the direction of polarization. Using the probabilistic quantum mechanical interpretation [1] of the separate components of ψ, it is easy to check that the electric charge and the z-component of the quark spins are conserved after transformation into the moving frame. Of course, the quarks in a nucleon are not free and that can change drastically the picture. In particular, virtual excited states of quarks may appear in the bound state. Short-living in the rest frame, they become very long-living in the high-speed frame. We may even consider them as real excitations after transformation to the infinite momentum frame which underlies the parton picture of the nucleon structure. Just as the excited states of the hydrogen atom are described taking into account corresponding physical symmetries, the almost real excitations of a nucleon may be naturally treated in the same way. In contrast to the hydrogen atom which is characterized by spherical symmetry, the states of quarks in a relativistic nucleon should be axially symmetric relative to the direction of its motion. This fact brings into play the projection of the orbital angular momentum L_z^i of the quarks on the symmetry axis. Indeed, if we admit the possibility of the observation of excited states of the nucleon (e.g., in the deep-inelastic lepton scattering on it) we should expect that the total helicity of the quarks $\sum S_z^i$ in the nucleon is not equal to the projection of nucleon's spin along the direction of movement. Only the total angular momentum, $\sum(S_z^i + L_z^i)$, may be constant. Below we shall consider these general arguments in more detail and try to estimate a possible value of the effect.

⋆e-mail: bkostenko@jinr.ru

2 Free partons in the nucleon

Following the parton model which considers the quarks as free particles and taking into account the axial symmetry of their state, we write the Dirac equation for free fermions in cylindrical coordinates,

$$\left[i\left(\gamma^r\frac{\partial}{\partial r}+\gamma^\varphi\frac{\partial}{r\partial\varphi}+\gamma^z\frac{\partial}{\partial z}\right)+\gamma^0E-m\right]\psi(r,\varphi,z;E)=0.$$

Here[1]

$$\gamma^r=\begin{pmatrix}0&\sigma^r\\-\sigma^r&0\end{pmatrix},\quad \gamma^\varphi=\begin{pmatrix}0&\sigma^\varphi\\-\sigma^\varphi&0\end{pmatrix},\quad \gamma^z=\begin{pmatrix}0&\sigma^z\\-\sigma^z&0\end{pmatrix},\quad \gamma^0=\begin{pmatrix}I&0\\0&-I\end{pmatrix},$$

and

$$\sigma^r=\begin{pmatrix}0&e^{-i\varphi}\\e^{i\varphi}&0\end{pmatrix},\quad \sigma^\varphi=\begin{pmatrix}0&-ie^{-i\varphi}\\ie^{i\varphi}&0\end{pmatrix},\quad \sigma^z=\begin{pmatrix}1&0\\0&-1\end{pmatrix},\quad I=\begin{pmatrix}1&0\\0&1\end{pmatrix}.$$

The substitution

$$\psi=\begin{pmatrix}q(r,\varphi)\\\bar q(r,\varphi)\end{pmatrix}e^{ip_zz}$$

separates the wave functions of the quarks and antiquarks and reduces the Dirac equation to the system of two partial differential equations,

$$\begin{cases}(E-m)q+\left[i\left(\sigma^r\frac{\partial}{\partial r}+\frac{\sigma^\varphi}{r}\frac{\partial}{\partial\varphi}-\sigma^zp_z\right)\right]\bar q=0,\\[2mm](E+m)\bar q+\left[i\left(\sigma^r\frac{\partial}{\partial r}+\frac{\sigma^\varphi}{r}\frac{\partial}{\partial\varphi}-\sigma^zp_z\right)\right]q=0.\end{cases}$$

From the second equation of the system, we get $\bar q$ in terms of q and substitute the result into the first one. Thereby we obtain

$$\left(\frac{\partial^2}{\partial r^2}+\frac{1}{r}\frac{\partial}{\partial r}+\frac{1}{r^2}\frac{\partial^2}{\partial\varphi^2}\right)q+p_\perp^2q=0,\tag{1}$$

where $p_\perp=\sqrt{E^2-(p_z^2+m^2)}$ is the transverse momentum of the quark. The separation of the variables by means of the substitution $q(r,\varphi)=u(r)e^{in\varphi}$, where n is an integer due to the single-valuedness of $q(r,\varphi)$, leads to the Bessel equation,

$$\frac{\partial^2u}{\partial r^2}+\frac{1}{r}\frac{\partial u}{\partial r}-\frac{n^2}{r^2}u+p_\perp^2u=0.$$

Its solutions which are finite at $r=0$ are the Bessel functions of the first kind, $u(r)\sim J_n(p_\perp r)$. The corresponding wave functions of the quarks in the transverse plane, $q(r,\varphi)=u(r)e^{in\varphi}$, are eigenstates of the z-component of the orbital momentum operator $\hat L_z=-i\partial/\partial\varphi$. In principle, the wave functions of this type can describe the orbital momentum contribution to the total helicity of the proton. However, they cannot describe the localization of the quarks inside the nucleon because $J_n(p_\perp r)$ have oscillating asymptotics damped as $1/\sqrt{p_\perp r}$ at long distances. Since $\int_0^{r_N}u^2(r)r\,dr<\int_{r_N}^\infty u^2(r)r\,dr=\infty$ these solutions describe "wave functions" of quarks localized mainly outside the nucleon.

[1] The expressions for γ-matrices in cylindrical coordinates can be found taking into account the vectorial form of their transformations, $\gamma^s=\gamma^i\partial x^s/\partial x^i$.

3 Confined partons in the nucleon

Here we consider a model describing the quark confinement within an external cylindrically symmetric static potential $V(r)$. Now the Dirac equation will involve the expression $E - V(r)$ instead of E. Then the equation (1) will undergo a change too, and instead of it we find

$$\left(\frac{\partial^2}{\partial r^2} + \frac{1}{r}\frac{\partial}{\partial r} + \frac{1}{r^2}\frac{\partial^2}{\partial \varphi^2} + p_\perp^2\right) q(r,\varphi) + i\sigma^r \frac{-dV/dr}{E - V + m}\left[i\left(\sigma^r \frac{\partial}{\partial r} + \frac{\sigma^\varphi}{r}\frac{\partial}{\partial \varphi}\right) - \sigma^z p_z\right] q(r,\varphi) = 0.$$

As compared to (1), an additional term proportional to the force $-dV/dr$ appears. The key observation explaining the mechanism of the quark spin flipping is as follows. The additional term contains the matrix

$$\sigma^r \sigma^z = -i\sigma^\varphi,$$

which operates on the quark state with the positive helicity in the following way

$$-i\sigma^\varphi \begin{pmatrix} 1 \\ 0 \end{pmatrix} = \begin{pmatrix} 0 \\ e^{i\varphi} \end{pmatrix}.$$

Thus, due to it, the projection of the spin changes its sign, the value of the orbital momentum increases by the unit and the total angular momentum remains constant,

$$(S_z + L_z)\, q = \frac{1}{2} q, \quad (S_z + L_z)\, q' = \frac{1}{2} q', \quad q = (1,0)^T, \quad q' = -i\sigma^\varphi q.$$

Now the separation of the r and φ variables is done through the substitution

$$q(r,\varphi) = \begin{pmatrix} u_+(r) e^{in\varphi} \\ -iu_-(r) e^{i(n+1)\varphi} \end{pmatrix},$$

where u_+ and u_- are the wave functions of the quarks with positive and negative helicities. For the case $n = 0$, which indicates that there are no orbital excitations in the region of the asymptotic freedom and all of them appear only due to the confinement force, we arrive at the system

$$\begin{cases} \left(\frac{\partial^2}{\partial r^2} + \frac{1}{r}\frac{\partial}{\partial r} + p_\perp^2\right) u_+(r) - \frac{dV/dr}{E - V + m}\left[\frac{\partial u_+(r)}{\partial r} + p_z u_-(r)\right] = 0, \\ \left(\frac{\partial^2}{\partial r^2} + \frac{1}{r}\frac{\partial}{\partial r} + p_\perp^2 - \frac{1}{r^2}\right) u_-(r) - \frac{dV/dr}{E - V + m}\left[\left(\frac{\partial}{\partial r} + \frac{1}{r}\right) u_-(r) + p_z u_+(r)\right] = 0. \end{cases} \tag{2}$$

We have assumed for $V(r)$ a realistic confinement potential found in the framework of lattice QCD in [2]. Following the parton model, we suppose that quarks are free in the vicinity of the nucleon center, $r < 0.7$ fm, and that the confinement force is manifested at larger distances in the region of the so-called gluon Y-string. Now the mathematical problem may be formulated as a Cauchy one for the system (2) with the initial conditions at $r_0 = 0.7$ fm

$$u_+(r_0) = J_0(p_\perp r_0), \quad u'_+ = J_0'(p_\perp r_0), \quad u_-(r_0) = 0, \quad u'_- = 0.$$

The Runge-Kutta-Fehlberg algorithm recommended by MAPLE (`rkf45 procedure`) has not succeeded in finding a numerical solution compatible with the confinement conditions $u_+, u_-|_{r\to\infty} \to 0$. The following method has been applied instead. The confinement potential is written in the ladder-shaped form, so that $dV/dr = \Delta V(r_i)\delta(r - r_i)$. Everywhere, apart from points r_i, the system (2) results in two Bessel equations for u_+ and u_- separately. Their solutions for $p_\perp^2 > 0$ are $u_+(r) = A_1 J_0(p_\perp r) + A_2 Y_0(p_\perp r)$, $u_-(r) = B_1 J_1(p_\perp r) + B_2 Y_1(p_\perp r)$, where J_i and Y_i are Bessel functions of the 1st and 2nd kind, correspondingly. Similarly, for $p_\perp^2 < 0$ we have solutions expressed in terms of modified Bessel functions of the 1st and 2nd kind, $u_+(r) = C_1 I_0(|p_\perp| r) + C_2 K_0(|p_\perp| r)$, $u_-(r) = D_1 I_1(|p_\perp| r) + D_2 K_1(|p_\perp| r)$.

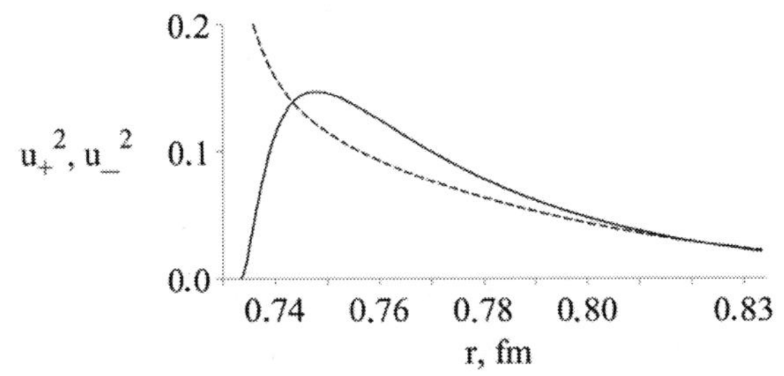

Figure 1. Quarks' wave functions squared corresponding to positive and negative helicities in the region of sub-barrier penetration (dashed and solid lines, respectively). They were calculated for $A_2, B_2, C_1, D_1 = 0$ (see text). The total polarization of the quarks is negative and equal to −0.0144 in the interval of radii shown in the figure.

4 Computing algorithm and results

The idea of our numerical solution is to join the above quoted Bessel functions by imposing on them matching conditions at points r_i to satisfy the system (2). The conditions turn out to be

$$\begin{cases} \Delta u'_+(r_{i+1}) = \dfrac{\Delta V(r_i)}{E - V(r_i) + m}\,(u'_+(r_i) + p_z u_-(r_i)), \\ \Delta u'_-(r_{i+1}) = \dfrac{\Delta V(r_i)}{E - V(r_i) + m}\left(u'_-(r_i) + \dfrac{u_-(r_i)}{r_i} + p_z u_+(r_i)\right) \end{cases} \tag{3}$$

and can be found from (2) after neglecting the contribution of all finite terms at $r = r_i$. If all $A_i, B_i \neq 0$ we may impose some additional conditions. For example, for $u_+(r)$ one of the additional conditions in the region $p_\perp^2 > 0$ may be taken in a form

$$A_{1i}J_0(p_{\perp i}r_i) + A_{2i}Y_0(p_{\perp i}r_i) = A_{1i-1}J_0(p_{\perp i-1}r_{i-1}) + A_{2i-1}Y_0(p_{\perp i-1}r_{i-1}).$$

Similar condition for $u_-(r)$ may be obtained by substitutions $A \to B$ and $J_0 \to J_1$, $Y_0 \to Y_1$. Analogues conditions may be written in the region $p_\perp^2 < 0$. This mathematical trick leads to *constant* solutions despite $K_i(p_\perp r), Y_i(|p_\perp| r) \to \infty$ at $|p_\perp| \to 0$. They can describe behavior of the wave functions of quarks in the neighborhood of the point r_{stop} at which $p_\perp(r) = 0$ and, thus, connect the classically accessible region (CAR) with the region of subbarrier penetration (SPR).

If one of the coefficients A_i and one of B_i is equal to zero the functions u_+ and u_- themselves should undergo breaks. In this case we have two equations (2) for the determination of 2 nonzero coefficients. The same condition holds for the coefficients C_i and D_i. The solutions of this type should be chosen according to the general Galerkin prescription to fulfil the necessary boundary conditions (of the finiteness here) at $r > r_{\text{stop}}$.

Our results were obtained for $p_z = 27.6$ GeV in agreement with the HERMES experiment [3], $m = 0$ according to the parton model, $p_\perp(r = 0.7) = 2$ GeV (may be adjusted). The calculations have shown that the confinement potential cannot stimulate significantly the quark helicity flipping in CAR, but the effect is striking in SPR (see Fig. 1). For an exact numerical estimation of the effect, it is necessary to solve a problem of the smooth matching the `rcf45` solutions at a point near r_{stop} in SPR where $u_+(r)$ still drops with the solutions of the Galerkin type described above.

Acknowledgement

The author thanks Gh. Adam, N.V. Makhaldiani and A.B. Pestov for numerous valuable remarks.

References

[1] J.D. Bjorken and S.D. Drell, *Relativistic Quantum Mechanics* (McGraw-Hill, New York, 1964)
[2] T.T. Takahashi, H. Matsufuru, Y. Nemoto, and H. Suganuma, Phys. Rev. Lett. **86**, 18 (2001)
[3] A. Airapetian et al., the HERMES Collaboration, Phys. Rev. **D75**, 12007 (2007)

Fractional Langevin Equation Model for Characterization of Anomalous Brownian Motion from NMR Signals

Vladimír Lisý[1,2,⋆] and Jana Tóthová[2,⋆⋆]

[1]*Laboratory of Radiation Biology, Joint Institute for Nuclear Research, 141980 Dubna, Moscow Region, Russia*
[2]*Department of Physics, Faculty of Electrical Engineering and Informatics, Technical University of Košice, Park Komenského 2, 042 00 Košice, Slovakia*

Abstract. Nuclear magnetic resonance is often used to study random motion of spins in different systems. In the long-time limit the current mathematical description of the experiments allows proper interpretation of measurements of normal and anomalous diffusion. The shorter-time dynamics is however correctly considered only in a few works that do not go beyond the standard Langevin theory of the Brownian motion (BM). In the present work, the attenuation function $S(t)$ for an ensemble of spins in a magnetic-field gradient, expressed in a form applicable for any kind of stationary stochastic dynamics of spins with or without a memory, is calculated in the frame of the model of fractional BM. The solution of the model for particles trapped in a harmonic potential is obtained in a simple way and used for the calculation of $S(t)$. In the limit of free particles coupled to a fractal heat bath, the results compare favorably with experiments acquired in human neuronal tissues.

1 Introduction

The present work connects two fields: the Brownian motion (BM) of particles in soft condensed materials and Nuclear Magnetic Resonance (NMR) as a tool of its observation. NMR has proven to be a very effective non-invasive method of studying molecular self-diffusion and diffusion in various materials [1]. Brownian particles in some media often show interesting behaviors, such as a dynamics that significantly differs from that predicted by the standard Langevin theory [2]. It seems natural to apply NMR methods to follow such unusual BM. However, as discussed in [3–5], the mathematical description of suitable NMR experiments in the literature is valid only for long times when the particles are in diffusion regime (normal or anomalous). The only exception is the memoryless Langevin model for which correct interpretations of the NMR experiments have been proposed [6]. Possible memory effects in the shorter-time particle dynamics are ignored or incorrectly taken into account. The aim of our recent papers [3–5, 7] was to overcome this limitation and to calculate the all-time attenuation functions of the NMR signals from spin-bearing particles resulting from the BM with memory. We have considered the Brownian particles in Maxwell fluids [3, 7] and the BM with hydrodynamic memory [4]. Here we focus on a popular model of fractional BM [8–11]. This model emerges naturally,

⋆e-mail: lisy@jinr.ru
⋆⋆e-mail: jana.tothova@tuke.sk

e.g., in viscoelastic media, and interpolates between the standard Langevin equation (LE) and a model with constant memory [11]. We obtain the solution of the generalized Langevin equation (GLE) for this model in an exceedingly simple way. The Brownian particles can be free or trapped in a harmonic potential. The results for the mean square displacement (MSD) of the particles are then applied to calculate the attenuation function $S(t)$ for an ensemble of spins in a magnetic-field gradient. $S(t)$ is used in the form obtained by the method of accumulation of phases in the frame rotating with the resonance frequency [3] and applicable for any time and kind of the stochastic motion od spins. Although our work was mainly aimed to contribute to adequate interpretations of the NMR experiments on the BM in complex fluids, in the limit of free particles coupled to a fractal heat bath our results correct the description and give favorable comparison with experiments acquired in human neuronal tissues [12].

2 Harmonically bounded particle described by the fractional LE

If a particle is moving in a trap modeled by a Hookean potential with elastic constant k, the GLE reads

$$M\dot{v}(t) + \int_0^t G(t-t')\,v(t')\,dt' + k\int_0^t v(t')\,dt' = f(t), \tag{1}$$

where M is the mass of the particle, $x(t)$ is its position and $v(t) = \dot{x}(t)$ its velocity. We will consider the memory kernel $G(t) = \gamma_\varepsilon \varepsilon t^{\varepsilon-1}$, $t > 0$, $1 > \varepsilon > 0$. Within the linear response theory the second fluctuation-dissipation theorem [13] relates $G(t)$ to the random force $f(t)$, $\langle f(t)f(0)\rangle = k_B T G(t)$, where k_B is the Boltzmann constant and T is the temperature. When $\varepsilon \to 0$, $G(t) \to 2\gamma\delta(t)$ and Eq. (1) becomes the standard memoryless LE with the white noise force and a purely viscous Stokes friction force $-\gamma v(t)$. To our opinion, the simplest method of solving linear GLE equations, which goes back to the old work [14], is as follows [15–19]. If we are interested in finding the MSD of the particle, $X(t) = \langle [x(t) - x(0)]^2\rangle$, we have merely to replace $x(t)$ in (1) by $X(t)$ and to substitute the stochastic force driving the particle with $2k_BT$. With the initial conditions $X(0) = \dot{X}(0) = 0$ the new equation is immediately solved by using the Laplace transformation,

$$\tilde{X}(s) = \int_0^\infty X(t)e^{-st}dt = \frac{2k_BT}{Ms}\left[s^2 + \frac{\gamma_\varepsilon}{M}\Gamma(\varepsilon+1)\,s^{1-\varepsilon} + \frac{k}{M}\right]^{-1}, \tag{2}$$

where $\Gamma(z)$ is the gamma function. Thus, in this way the exact solution of (1), first obtained in [10], can be easily got in a few steps. The limiting cases of short and long times correspond to large and small s, respectively, so that if $\varepsilon < 1$ then at $s \to 0$ one finds $\tilde{X}(s) \simeq 2k_BT/ks$ and $X(t) \simeq 2k_BT/k$, if $t \to \infty$. At $s \to \infty$, $X(s) \simeq 2k_BT/Ms^3$ and $X(t) \simeq k_BTt^2/M$. Corrections to these expressions can be obtained from the exact representation of (2) in the time domain through the Mittag-Leffler functions $E^{(k)}_{\alpha,\beta}(y)$ [10],

$$X(t) = \frac{2k_BT}{M}\sum_{k=0}^{\infty}\frac{(-1)^k}{k!}(\omega t)^{2k}t^2E^{(k)}_{1+\varepsilon,3+k-\varepsilon k}\left(-\varepsilon\gamma_\varepsilon\Gamma(\varepsilon)t^{1+\varepsilon}/M\right), \tag{3}$$

where $E^{(k)}_{\alpha,\beta}(y) = \left(d^k/dy^k\right)\sum_{j=0}^{\infty} y^j\Gamma^{-1}(\alpha j + \beta)$, $\alpha > 0$, $\beta > 0$. It should be noted that the short-time expansion of this solution for $X(t)$, as given in [10], Eq. (41), does not correspond to the solution of the standard LE (Eq. (1) in the limit $\varepsilon \to 0$ [20]). If $\omega^2 = k/M$ and $\omega_M = \gamma/M$, in [10], Eq. (41), the t^4 term should be $(1/12)(\omega_M^2 - \omega^2)t^4$, *i.e*, the term containing ω_M is missing. Another way to show that the expansions in [10] should be completed is to consider the limit of a free Brownian particle (when $\omega = 0$). The velocity autocorrelation function in this case must be $\langle v(0)v(t) = (k_BT/M)\exp(-\omega_M t)$, which does not agree with Eq. (42) [10]. The limit of (3) for a free particle was used in [12] to calculate the NMR signals from human neuronal tissues, however, with an incorrect spin-echo attenuation $S(t)$ (see [3]). The basic formulas used to calculate $S(t)$ are presented in the following section.

3 Attenuation of the NMR signals due to Brownian motion

Below we will consider experiments with the NMR signals from a liquid or gaseous system with spin-bearing Brownian particles placed in a strong magnetic field B_0 along the axis x. The magnetization of spins is modulated by a field gradient $g(t)$. General formulas for the attenuation function $S(t)$ of the observed NMR signal that are applicable to any kind of stochastic motion of spins, including their dynamics with memory, have been obtained in [3, 7]. In the evaluation of $S(t)$ it was only assumed that the studied random processes are stationary in the sense that the autocorrelation function of a fluctuating dynamical variable x at times t and t' depends only on the time difference, $t - t'$: $\langle x(t)x(t')\rangle = \langle x(t-t')x(0)\rangle$ [1]. One more assumption is that the distribution of the random variables is Gaussian, or that the accumulated phase ϕ in the frame rotating with the resonance frequency $\gamma_n B_0$, where γ_n is the nuclear gyromagnetic ratio, is small. In these cases the nuclear induction signal observed in the presence of a steady magnetic-field gradient after the $\pi/2$ rf pulse applied at time $t = 0$, $S(t) = \langle \exp[i\phi(t)]\rangle = \exp[-\langle\phi^2(t)/2\rangle]$ [1], is $S(t) = \exp[-(1/2)\gamma_n^2 g^2 \int_0^t t' X(t')dt']$. To our knowledge, this simple formula for a constant gradient g was for the first time proposed in [7]. In the pulsed-gradient spin echo experiment [1] between two gradient pulses, the duration of each being δ, a π rf pulse is applied. Let the first gradient pulse begins at time t_0 after the $\pi/2$ rf pulse and the second one at time $t = \Delta$. The result for $t < \tau$ (up to the second rf pulse) is the same as for a steady gradient (with $t = \delta$). Due to stationarity, after the second rf and gradient pulses the result of the calculation of the echo signal observed at $t = 2\tau$ does not depend on t_0 and can be evaluated from [3]

$$S(\delta,\Delta) = \exp\left\{-(1/2)\gamma_n^2 g^2 \left[\int_0^\delta dt' \int_0^\delta dt'' X(t''-t'+\Delta) - 2\int_0^\delta dt' (\delta - t') X(t')\right]\right\}. \quad (4)$$

One can check that this equation gives the known results for particles in diffusion regime, when $X(t) = Ct^\alpha$ (C is a temperature-dependent parameter, $\alpha = 1$ corresponds to normal diffusion with $C = 2D$ and the diffusion coefficient $D = k_B T/\gamma$, $\alpha < 1$ to sub-diffusion, and $\alpha > 1$ to super-diffusion [15]. In our study the solution (3) was used to obtain the NMR signal decay from Eq. (4). The function $S(t)$ has been analyzed in detail for different situations (free and trapped particles, weak and strong traps, short and long times). As an important example that generalizes the known results for standard Langevin and diffusion models (normal and anomalous) we show that in the long-time approximation ($\gamma_\varepsilon\Gamma(\varepsilon+1)\,t^{\varepsilon+1}/M \gg 1$) and weak traps ($M\omega^2 t^{1-\varepsilon} \ll \gamma_\varepsilon\Gamma(1+\varepsilon)$) the decay of the NMR signal at any time after the second gradient pulse is given by the mean of the square of the accumulated phase

$$\left\langle \phi^2(\delta,\Delta)\right\rangle = \frac{2k_B T\gamma_n^2 g^2}{\gamma_\varepsilon\Gamma(1+\varepsilon)\Gamma(2-\varepsilon)}\left[\psi_{1-\varepsilon} - \frac{M\omega^2\Gamma(2-\varepsilon)}{\gamma_\varepsilon\Gamma(1+\varepsilon)\Gamma(3-2\varepsilon)}\psi_{2-2\varepsilon} + \cdots\right]. \quad (5)$$

Here, $\psi_\alpha = [(\Delta+\delta)^{\alpha+2} + (\Delta-\delta)^{\alpha+2} - 2\Delta^{\alpha+2} - 2\delta^{\alpha+2}][(\alpha+1)(\alpha+2)]^{-1}$. When $\varepsilon = 0$, $\langle\phi^2(\delta,\Delta)\rangle = 2k_B T\gamma_n^2 g^2\gamma^{-1}[\psi_1 - M\omega^2\psi_2/2\gamma + \cdots]$ with $\psi_1 = \delta^2(\Delta - \delta/3)$ and $\psi_2 = (\delta\Delta)^2$. The function $S(t)$ calculated from the first term in (5) at this limit can be compared with the results [12] of the attenuation of steady-gradient ($\Delta = \delta = t$) spin-echo signal from water molecules anomalously diffusing in a human brain tissue. We have obtained a very good agreement with these measurements but, since in [12] an erroneous formula for $S(t)$ was used (see [3]), the parameters ε and the generalized diffusion coefficient $D_\varepsilon = k_B T/\gamma_\varepsilon$ extracted from the experiments are very different. For example, instead of $\varepsilon = 0.31$ determined in [12] (Table 1), the best correspondence with the measurements is obtained for ε about 0.42. Even if ε was correctly extracted from the experiments, D_ε found in [12] would have to be divided by $2(2^{1-\varepsilon} - 1)/(2-\varepsilon)$, which, for $\varepsilon \in [0, 1]$ changes from 0 to 1 (e.g., if $\varepsilon = 0.5$, see Table II [12], D_ε should be about 1.8 times larger).

4 Conclusion

The phenomenological model of fractional Brownian motion is one of the most popular and best justified models in the soft condensed matter physics. It generalizes other models of the stochastic dynamics described by the generalized Langevin equation. On the other hand, various NMR methods serve as probes for accurate characterization of the random motion of particles in different systems, including the biological ones due to the non-invasive character of these measurements. In spite of the long time since the invention of such methods as the NMR spin echo, its description in the literature is limited to long times when the spin-bearing particles are in the diffusion regime, or to the stochastic motion without memory. Nowadays it is well known that in the description of the behavior of many systems this simplification fails. In our work we have proposed an interpretation of the NMR experiments that could provide a characterization of the Brownian motion in systems described by the fractional generalized Langevin model. The new formulas for the attenuation function of the NMR signals can be directly used to describe the behavior of spins exhibiting any kind of the stochastic dynamics not only in the diffusion limit, but for all times. Although the presented approach assumes Brownian particles much larger than the surrounding molecules, surprisingly it leads to a very good agreement with experiments on diffusion of water molecules in neuronal tissues.

Acknowledgement

This work was supported by the Ministry of Education and Science of the Slovak Republic through grant VEGA No. 1/ 0348/15.

References

[1] P. T. Callaghan, *Translational Dynamics and Magnetic Resonance: Principles of Pulsed Gradient Spin Echo NMR* (Oxford University Press, Oxford, 2014)
[2] W. T. Coffey, Yu. P. Kalmykov, J. T. Waldron, *The Langevin Equation* (World Scientific, New Jersey *e.a.*, 2005)
[3] V. Lisý, J. Tóthová, J. Mag. Res. **276**, 1 (2017)
[4] V. Lisý, J. Tóthová, J. Mol. Liq. **234**, 182 (2017)
[5] J. Tóthová, V. Lisý, Acta Phys. Pol. **131**, 1111 (2017)
[6] J. Stepišnik, Physica B **198**, 299 (1994)
[7] V. Lisý, J. Tóthová, Phys. Rev. Lett. **117**, 249701 (2016)
[8] E. Lutz, Phys. Rev. E **64**, 051106 (2001)
[9] A. D. Viñales, M. A. Despósito, Phys. Rev. E **73**, 016111 (2006)
[10] M. A. Despósito, A. D. Viñales, Phys. Rev. E **80**, 021111 (2009)
[11] I. Goychuk, Phys. Rev. E **80**, 046125 (2009)
[12] J. M. Cooke *et al.*, Phys. Rev. E **80**, 061102 (2009)
[13] R. Kubo, Rep. Prog. Phys. **29**, 255 (1966)
[14] V. Vladimirsky, Zh. Eksp. Teor. Fiz. **12**, 199 (1942)
[15] J. Tóthová *et al.*, Eur. J. Phys. **32**, 645 (2011)
[16] J. Tóthová *et al.*, Eur. J. Phys. **32**, L47 (2011)
[17] J. Tóthová, L. Glod, V. Lisý, Int. J. Thermophys. **34**, 629 (2013)
[18] J. Tóthová, V. Lisý, Acta Phys. Slovaca **65**, 1 (2015)
[19] J. Tóthová, V. Lisý, Phys. Lett. A **380**, 2561 (2016)
[20] L. Glod *et al.*, J. Electr. Eng. **63**, 53 (2012)

EPJ Web of Conferences **173**, 02014 (2018) https://doi.org/10.1051/epjconf/201817302014
Mathematical Modeling and Computational Physics 2017

Diffraction of Electromagnetic Waves on a Waveguide Joint

Mikhail Malykh[1], *Leonid* Sevastianov[1], *Anastasiya* Tyutyunnik[1,*], and *Nikolai* Nikolaev[2]

[1] *Department of Applied Probability and Informatics, Peoples' Friendship University of Russia (RUDN University)*
[2] *Department of Applied Physics, Peoples' Friendship University of Russia (RUDN University)*

Abstract. In general, the investigation of the electromagnetic field in an inhomogeneous waveguide doesn't reduce to the study of two independent boundary value problems for the Helmholtz equation. We show how to rewrite the Helmholtz equations in the "Hamiltonian form" to express the connection between these two problems explicitly. The problem of finding monochromatic waves in an arbitrary waveguide is reduced to an infinite system of ordinary differential equations in a properly constructed Hilbert space. The calculations are performed in the computer algebra system Sage.

1 Introduction

The mathematical theory of the waveguide systems, based on the works of A.N. Tikhonov, A.A. Samarsky [1] and A.G. Sveshnikov [2], created in the middle of the last century. It has become an important chapter of the computational electrodynamics. The solution of waveguide diffraction problems is necessary for the design of integrated optical and fiber optic devices. The linear dimensions of these devices along different directions may differ by several orders of magnitude, for example, the radius of the Lüneburg thin-film waveguide lens exceeds its thickness by more than 104 times. Such a difference in size leads to insurmountable problems in the discretization of continuous models. The aim of our research is to overcome these difficulties. For example, the height of the Lüneburg waveguide lens is less than its diameter by 10^4 times, which inevitably leads to specific computational peculiarities [3, 4]. Therefore the development of modern program packages for the calculation of fields in the closed waveguides with extremal parameters remains an important challenge for the computational electrodynamics.

2 Reduction of the system of Maxwell's equations to two scalar problems

The investigation of the electromagnetic field in a regular waveguide filled with a homogeneous substance is reduced to the investigation of two independent boundary value problems for the Helmholtz equation [5]. In the case of a waveguide filled with an inhomogeneous substance, a relationship arises between the modes of these two problems, which in the numerical experiments can not always be fully taken into account. The first papers on the mathematical theory of the waveguide [6] had in mind the generalization of the Fourier method, therefore the question of reducing the system of Maxwell's

*e-mail: tyutyunnik_aa@pfur.ru

equations to two scalar equations was in the shadow of the theorem on the completeness of the system of normal modes. Meanwhile, writing this pair is not at all difficult, if we go over to the "Hamiltonian form" of the Helmholtz equations.

Consider a waveguide of a constant simply connected section S with ideally conducting walls. We will not make any starting assumptions with respect to the filling ϵ, μ of this waveguide. The axis Oz is directed along the cylinder axis, the normal to ∂S will be denoted as $\vec{n}$, the tangent vector perpendicular to $\vec{e}_z$ as $\vec{\tau}$. For a basis we take the system of Maxwell's equations, from which we exclude E_z and H_z. For brevity's sake let us assume $\nabla = (\partial_x, \partial_y)^T$ and $\nabla' = (-\partial_y, \partial_x)^T$. By excluding the longitudinal components from Maxwell's equations, we get

$$\begin{cases} \vec{e}_z \times \partial_z \vec{H}_\perp + \nabla \dfrac{1}{ik\mu} (\mathrm{curl}\, E_\perp)_z \times \vec{e}_z + ik\epsilon \vec{E}_\perp = 0, \\ \vec{e}_z \times \partial_z \vec{E}_\perp - \nabla \dfrac{1}{ik\epsilon} (\mathrm{curl}\, H_\perp)_z \times \vec{e}_z - ik\mu \vec{H}_\perp = 0. \end{cases} \tag{1}$$

At the boundary, the conditions of ideal conductivity are satisfied to

$$\vec{E}_\perp \times \vec{n} = 0, \qquad H_\perp \cdot \vec{n} = 0, \qquad E_z = 0. \tag{2}$$

A two-dimensional analogue of the Helmholtz decomposition makes it possible, without loss of generality, to seek a solution of the system (1) in the form

$$\vec{E}_\perp = \nabla u_e + \nabla' v_e, \qquad \vec{H}_\perp = \nabla v_h + \nabla' u_h.$$

The four scalar functions introduced here will be called potentials, and we'll always assume that they satisfy the boundary conditions $u_e = u_h = n \cdot \nabla v_e = n \cdot \nabla v_h = 0$ at the boundary of the waveguide. Substituting these expressions into (1), we obtain

$$\begin{cases} \nabla' \Big(\partial_z v_h - \dfrac{1}{ik\mu} \Delta v_e \Big) + ik\epsilon \nabla' v_e - \nabla \partial_z u_h + ik\epsilon \nabla u_e = 0, \\ \nabla' \Big(\partial_z u_e + \dfrac{1}{ik\epsilon} \Delta u_h \Big) - ik\mu \nabla' u_h - \nabla \partial_z v_e - ik\mu \nabla v_h = 0. \end{cases} \tag{3}$$

The first two of the three boundary conditions (2) are automatically satisfied. With the usual choice of basis, the third condition follows from the Dirichlet conditions imposed on the function u_h. If ϵ and μ are constants, then the system (3) can be rewritten as a pair of uncoupled Helmholtz equations with respect to v_e, u_h.

The presentation of a whole series of problems in the mathematical theory of waveguides can be carried out uniformly if we consider the field as a function of z with values in a suitably chosen Hilbert space [7]. In other words, we will use the weak formulation of the problem of finding the potentials in the general case.

We will consider the solution

$$w = (v_h, v_e, u_e, u_h)$$

of the system (3) as a map of $z \in \mathbb{R}$ to a Hilbert space

$$\mathfrak{H} = \Big\{ \, v_h, v_e, u_e, u_h \in W_2^2(S) \colon u_e, u_h, n \cdot \nabla v_e, n \cdot \nabla v_h = 0 \text{ on } \partial S \, \Big\}.$$

In this space the equations of the system (3) can be written as follows

$$\partial_z b(\tilde{w}, w) + ika(\tilde{w}, w) + \frac{1}{ik} c(\tilde{w}, w) = 0, \qquad \forall \tilde{w} \in \mathfrak{H}.$$

Here the forms a, b, c are bounded in $\mathfrak{H}$, therefore, by the Riesz theorem, the above condition can be rewritten in terms of the operators generated by these bilinear forms. Thus any electromagnetic field in the waveguide satisfies the system of ordinary differential equations

$$B\frac{dw}{dz} + ikAw + \frac{1}{ik}Cw = 0 \tag{4}$$

in the Hilbert space $\mathfrak{H}$. Since the embedding of the Sobolev spaces is compact, the operators A and B are compact, and B is also self-adjoint.

3 Normal modes of a waveguide

Similar to the finite-dimensional case, any solution of the system (4) is a superposition of normal modes, that is, solutions that depend on z as $e^{ik\beta z}$. To calculate these modes we obtain an eigenvalue problem with respect to the spectral parameter β:

$$\beta Bw = Aw - \frac{1}{k^2}Cw. \tag{5}$$

The discreteness of the spectrum and the completeness of the system of normal waves are obtained as a corollary of the theorem of M. V. Keldysh about the operator pencil [8].

The derivation of the numerical solution of this eigenvalue problem is divided into two steps:

- discretization, that is, truncation of the infinite matrices (the Galerkin method),
- solving the resulting algebraic eigenvalue problem.

In the standard approach to calculate the modes, we need to select some components of the vectors $\vec{E}$ and $\vec{H}$ and write an approximate problem for them using the finite element method or the finite difference method [9]. Our computing experiments have shown that the system (4) is more convenient for applying the Galerkin method because the truncation of operators A, B and C in this basis give us at once the algebraic eigenvalues problem in the form

$$\beta\hat{B}w = \hat{A}w - \frac{1}{k^2}\hat{C}w. \tag{6}$$

The work proposed by us contains, in our opinion, an elegant and effective generalization of the incomplete Galerkin method [10]. In practice we calculate the matrix elements numerically with small errors but preserve all symmetry of the waveguide. As the 2nd step, the standard approach suggests using the Krylov Subspace Methods [11], but in modern computer algebra system Sage [12] we may work over the field of algebraic numbers without any numerical errors.

This approach to computing eigenmodes was tested on several waveguides [13]. Our computing experiments have shown that in such a way we can calculate all the running modes (that is, modes with real eigenvalues β) and the small locked modes with correct multiplicities. The last circumstance is unexpected and indicates new application for symbolical calculations in numerical analysis.

4 The joint of two waveguides

Let the filling ϵ, μ of the waveguide have a jump at $z = 0$, so this plane is the joint of two waveguides. Any solution of the differential equation (4) at $z < 0$ and at $z > 0$ can be represented as the superposition of normal modes. The problem about the diffraction of the wave

$$\sum_{\beta>0} F_n w_n^{(1)} e^{ik\beta_n^{(1)} z}$$

at the joint $z = 0$ is reduced to the following: for the given waveguide, the functions $\epsilon_{1,2}(x, y)$ and $\mu_{1,2}(x, y)$ and the numbers $F_1, \ldots,$ find numbers R_n, T_n such that

$$\sum_{\beta>0} F_n w_n^{(1)} e^{ik\beta_n^{(1)} z} + \sum_{\Re\beta<0} R_n w_n^{(1)} e^{ik\beta_n^{(1)} z} = \sum_{\Re\beta>0} T_n w_n^{(2)} e^{ik\beta_n^{(2)} z}$$

at $z = 0$. To solve this problem we can calculate the normal waves as above and solve a linear system with respect to R and T using the standard routine in computer algebra system (CAS).

5 Conclusion

We list the main results of our work.

- We can introduce four potentials, by means of which the system of Maxwell's equations in the waveguide reduces to an infinite system of linear ODEs and is convenient for numerical analysis by the Galerkin method.
- To compute the normal waves we can use CAS and save multiplicities and other qualitative properties of eigenvalues.
- The problem about diffraction on joint is reduced to a system of linear algebraic equations which can also be solved symbolically.

Acknowledgments

The calculations in this article are performed in the computer algebra system Sage [12]. The publication was prepared with the support of the "RUDN University Program 5-100". The work was partially supported by RFBF grants No 15-07-08795, No 16-07-00556.

References

[1] A.A. Samarskiy, *Izbrannyye trudy* (Moscow: MAX Press, 2003) 531 pp.

[2] I. E. Mogilevskii, A. G. Sveshnikov , *Matematicheskiye zadachi teorii difraktsii* (Moscow: Faculty of Physics MSU, 2010) 197 pp.

[3] A.A. Egorov, L.A. Sevastyanov, Kvantovaya Elektronika **39** (6), 566–574 (2009)

[4] A.A. Egorov et al., Kvantovaya Elektronika **40** (9), 830–836 (2010)

[5] W.C. Chew, *Lectures on theory of microwave and optical waveguides* (http://wcchew.ece.illinois.edu/chew/course/tgwAll20121211.pdf, 2012) 362 pp.

[6] A.A. Samarskiy, A.N. Tikhonov, Zhurnal tehnicheskoy fiziki **18** (7), 959–970 (1948)

[7] A.N. Bogolyubov, M.D. Malykh, Computations mathematics and mathematical physics **43** (4), 560–563 (2003)

[8] A.N. Bogolyubov, A.L. Delitsyn, A.G. Sveshnikov, Computational mathematics and mathematical physics **38** (11), 1815–1823 (1998)

[9] K.S. Chiang, Optical and Quantum Electronics **26**, S113–S134 (1994)

[10] A.N. Bogolyubov, A.I. Erokhin, I.E. Mogilevskii, Computational Mathematics and Mathematical Physics **52** (6), 932–936 (2012)

[11] Y. Saad, *Numerical methods for large eigenvalue problems* (Philadelphia, PA : Society for Industrial and Applied Mathematics, 2011) 276 pp.

[12] *SageMath*, http://www.sagemath.org

[13] M. D. Malykh, L. A. Sevastianov, A. A. Tiutiunnik, N. E. Nikolaev, Journal of Electromagnetic Waves and Applications, 1–13 (2017)

EPJ Web of Conferences **173**, 02015 (2018) https://doi.org/10.1051/epjconf/201817302015
Mathematical Modeling and Computational Physics 2017

Minimum Period of Rotation of Millisecond Pulsars and Pulsar Matter Equations of State

Sergey Mikheev[1,*] and *Victor* Tsvetkov[1,**]

[1] *Tver State University*

Abstract. Based on the findings of our previous studies of fast-rotating Newtonian polytropes, we found the relation between the minimum pulsar rotation period, the value of pulsar central density, and the polytropic index. From this relation we come to the conclusion that the value of minimum central density of a pulsar with a peak period is $2.5088 \cdot 10^{14}$ g/cm^3.

1 Introduction

The astrophysical data is indicative of the presence of a minimum period of rotation of the millisecond pulsars. Now this period is 1.396648 ms. The indicated value is reached long before the pulsar destruction under the influence of centrifugal forces. This phenomenon is easily explained by the fast growth of the angular momentum losses near the bifurcation point of the pulsar configuration due to intense pulsar gravitational radiation.

On the 11th of February 2016 the experimental discovery of gravitational waves by collaborations LIGO and VIRGO was announced [1].

We suggest the possibility of detection of gravitational waves generated by rapidly rotating magnetized gravitating configurations (pulsars) [2].

2 The Polytrope Equation of State of a Pulsar

For designation of gravitating configurations with the equations of state for matter in the form of polytropes the term Newtonian polytrope is generally accepted [3].

As shown in [4], both the ideal Fermi gas equation of state and the realistic Bethe-Johnson and Raid equations can be approximated by the polytrope with the corresponding index n.

Assume that the pulsar configuration density is given by:

$$\rho = \sum_{a+b+c=0}^{a+b+c=P} \rho_{abc} x^a y^b z^c. \tag{1}$$

The parameter X determining the intensity of pulsar gravitational radiation is given by:

$$X = \rho_{[20]0} = \rho_{200} - \rho_{020}. \tag{2}$$

*e-mail: Mikheev.SA@tversu.ru
**e-mail: Tsvetkov.VP@tversu.ru

In our papers [5, 6] the equation for the definition of the X-values was proposed:

$$A(\varepsilon, n)X + b_k(n)X^3 = \eta_m, \tag{3}$$

where $\varepsilon = \omega^2/(4\pi G\rho_0)$, $\omega = 2\pi/T$, T is the pulsar rotation period, ρ_0 is the pulsar center density, $\eta_m = B_{0in}^2 \sin^2\alpha/(32\pi^2 G\rho_0^2 a^2)$, a is the equatorial radius of the configuration, G is the gravitation constant, B_0 is the effective value of intensity of internal magnetic field at the center of configuration.

According to the mathematical catastrophe theory, the equation (3) describes a classic case of catastrophe A_3 iff $\eta_m = 0$.

Let the critical points ε_k be defined by the statement:

$$A(\varepsilon_k, n) = 0. \tag{4}$$

Then $\varepsilon_k = \varepsilon_k(n)$.

At the critical points, $X = X_k = \sqrt[3]{\eta_m/b_k}$, and away from the critical points, $X = \eta_m/A(\varepsilon, n)$. For real pulsars, $\eta_m \sim 10^{-9}$–10^{-12}. Therefore X_k and X can differ by 6–8 orders of magnitude.

The parameter X determines the intensity of pulsar gravitational radiation, and correspondingly at the critical points the intensity will be higher by many orders of magnitude in comparison with its value away from the critical points.

The calculation of $\varepsilon_k(n)$ was carried out using a program package developed and implemented by us [5, 6].

The $\varepsilon_k(n)$ plot is given in Fig. 1.

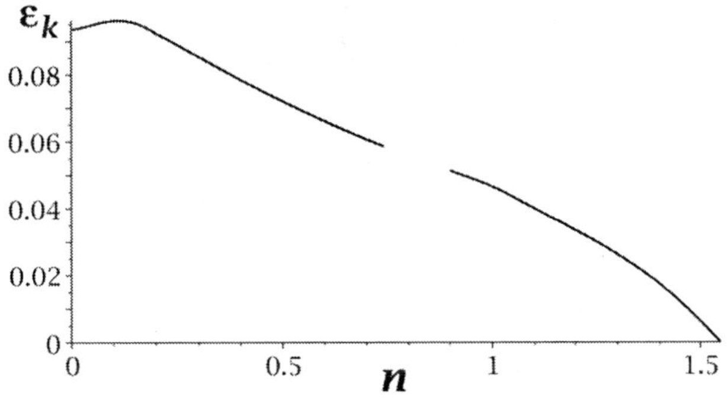

Figure 1. The plot of the function $\varepsilon_k(n)$

From Fig. 1 we can see that there are no critical points at polytropic index values $0.74 \leq n \leq 0.904$ and $n \geq 1.55$, and therefore, the gravitational radiation of pulsars with these state equations will be weaker by many orders of magnitude.

The intense gravitational radiation of a pulsar leads to the loss of its angular momentum M according to the following equation:

$$\dot{M} = -\omega^{-1} I_g, \tag{5}$$

where I_g is the intensity of its gravitational radiation.

The growth of the rotational speed of a newly born neutron star is responsible for the rapid decrease in its radius at the moment of its birth, and in a dual system – for twisting as well the accretion of matter from the second component. According to the equation (5), the growth of the angular speed of the pulsar rotation at the critical point ω_k will be stopped due to the fast growth of the gravitational radiation intensity near this point. Thus, the millisecond pulsar is born with the lowest value of the rotation period. From (5) the following relation arises:

$$T > T_k = \sqrt{\pi/(G\rho_0\varepsilon_k(n))}. \tag{6}$$

From (6) the function $\rho_0 = \rho_0(n, T_k)$ follows which can be represented graphically in Fig. 2.

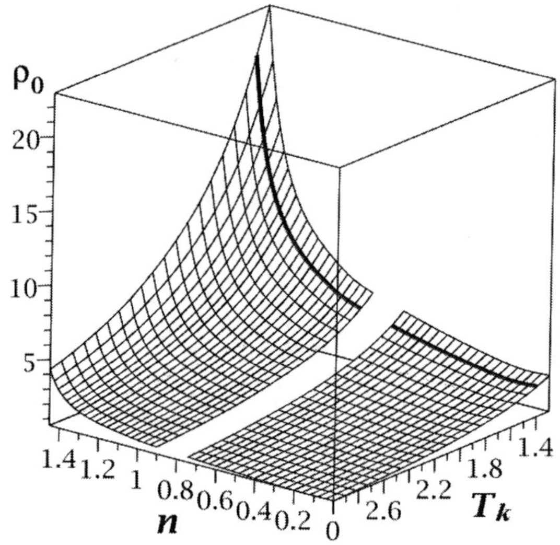

Figure 2. The plot of the function $\rho_0(n, T_k)$. The thick line shows the curve ρ_0 $(n, T_k = 1.396648)$; ρ_0 is given in 10^{14} g/cm^3; T_k is given in *msec*.

The region of discontinuity of the function $\varepsilon_k(n)$ in the interval of $0.74 \leq n \leq 0.904$ leads to a doubly connected surface $\rho_0(n, T_k)$ in Fig. 2. The domain $n \geq 1.55$ does not represent a significant interest for the problem under consideration and is not mapped in Fig. 2.

Fig. 2 displays the presence of ranges of the polytropic index n for which there is no the limitation (6) for the period of the pulsar rotation. At the present time there are all reasons to take the value of 1.396648 ms for the minimum pulsar rotation period. If pulsars with a substantially smaller period would be found, their equations of state would be close to the polytrope with an index of either $0.74 \leq n \leq 0.904$ or $n \geq 1.55$.

For other polytropic index values, Fig. 2 imposes the constraint: $\rho_0 \geq 2.5088 \cdot 10^{14}$ g/cm^3.

Let us consider the possibility of detection of gravitational waves generated by rapidly rotating magnetized gravitating configurations (pulsars) [2].

The possibility of detecting this radiation would allow unique inferences on the equations of state of the pulsar overdense nuclear matter. The pulsar gravitational radiation is generated at doubled rotation speed of pulsars and based on the asymmetry parameter of the matter density distribution relative to the X-ray pulsar's rotation axis related to its magnetic field and magnetic axis directed at an angle to the rotation axis.

In principle, in order to obtain any data on object locations, it is necessary to use observation data obtained both by gravitational wave detectors and conventional radiotelescopes. The most precise object direction sensing is provided by its electromagnetic radiation. High hopes follow from the possibility to set up signal match experiments in which gravitational waves will be detected together with electromagnetic ones [2].

First of all, such objects include supernovas which trigger bursts of electromagnetic radiation. If their shapes differ from the axially symmetric one under collapsing, they will also emit gravitational waves. Accordingly, with the help of gravitational waves we can "see" the instant of supernova explosion whereas in the optical channel this information is delayed for several hours required for the explosion development whereafter the instant of supernova explosion can be observed as a stellar flare [2].

If gravitational waves followed by a flare would be detected, this would allow to confirm with a high probability that the detected radiation is really associated with supernova explosion. Besides, such a delay on the path of 50–100 mln light years long would prove that gravitational waves propagate with the speed of light.

3 Conclusions

From our report, it follows that the minimum pulsar rotation period of the order of 1.3 ms observable by experience, indirectly shows that the millisecond pulsar configurations have critical points and there is intense gravitational radiation near these points.

In this paper we have discussed the possibility of detecting gravitational waves generated by rapidly rotating magnetized gravitating configurations (pulsars).

References

[1] B.P. Abbott et al., Phys. Rev. Let. **116, 061102**, 1–16 (2016)
[2] K.S. Torn, Priroda **#4 (860)**, 34–45 (1987)
[3] S.L. Shapiro and S.A. Teukolsky, *Black Holes, White Dwarfs and Neutron Stars: The physics of compact objects* (New York: Wiley, 1983), 645 pp.
[4] S.A. Mikheev and V.P. Tsvetkov, Physics of Particles and Nuclei Letters **7**, 1, 1–4 (2010)
[5] S.A. Mikheev and V.P. Tsvetkov, Physics of Particles and Nuclei Letters **5**, 4, 405–412 (2008)
[6] S.A. Mikheev and V.P. Tsvetkov, Physics of Particles and Nuclei Letters **10**, 3, 234–242 (2013)

EPJ Web of Conferences **173**, 02016 (2018) https://doi.org/10.1051/epjconf/201817302016
Mathematical Modeling and Computational Physics 2017

Near-Perfect Matchings on Cylinders $C_m \times P_n$ of Odd Order

Sergey N. Perepechko[1,⋆]

[1] *Petrozavodsk State University, Russia*

Abstract. A close relationship was established between the number of perfect and near-perfect matchings on cylinders $C_m \times P_n$. Generating functions are obtained for the number of near-perfect matchings in these graphs for fixed odd $m \leqslant 13$. A conjecture is put forth on the properties of the denominators of generating functions for arbitrary odd m.

1 Introduction

One of the classical lattice models of statistical mechanics is the dimer problem. In this model, the rigid dimers (diatomic molecules) are treated as the edges of a graph together with the vertices incident to it. Each lattice node cannot be occupied by more than one atom, so the edges must be non-adjacent.

In the dimer model, close-packed lattice coverings containing the maximum possible number of dimers are studied. The combinatorial interpretation of the model leads to the need to enumerate the maximum matchings in certain set of lattice graphs.

For many kinds of lattices of even order, used in applications, the maximum matchings are perfect matchings. Near-perfect matchings are a natural analogue of perfect matchings in graphs of odd order. A near-perfect matching is one in which exactly one vertex is unmatched. A vertex of the graph, which is not saturated by a matching, will be called *vacancy*.

A significant part of the work related to the enumeration of matchings deals with perfect matchings. At the same time, the quantitative characteristics of other kinds of matchings require further study. In this paper cylinders $C_m \times P_n$ are considered when both parameters m and n are odd. Among the works of the past years devoted to the study of such graphs, the paper [1] is of the greatest interest. This paper gives a closed-form expression for the number of near-perfect matchings when the vacancy is on the boundary of a cylinder. In [2] Kong briefly mentions the results of numerical calculations on cylinders for a fixed value of the parameter m. The author reports about the similarity of asymptotic expansions for the total number of near-perfect matching on cylinders and rectangular lattices $P_m \times P_n$, while omitting the details of the computations.

In our work we use a universal method for enumerating matchings. This allows us to avoid many of the restrictions inherent to the Pfaffian technique, and to study in detail the dependence of the number of near-perfect matchings on the vacancy location. The results of investigation clearly demonstrate that parity considerations are just as important for solving this problem as for counting perfect matchings.

Extending the results of numerical calculations by Kong, we reveal a deep connection between the number of near-perfect and perfect matchings on the cylinder with the same value of m. In fact, for

⋆e-mail: persn@newmail.ru

fixed m and changing n, the number of near-perfect matchings and the number of perfect matchings are solutions of the same recurrence relation if the vacancy location does not change. Based on this result, we can conjecture the form of the denominator of the generating function for the total number of near-perfect matchings on the cylinder for any odd m.

2 Basic notation

Consider the graph $G_{m,n} = C_m \times P_n$. This graph can be viewed as n copies of C_m placed sequentially with edges joining the corresponding vertices of C_m. We number all the cycles C_m by integers from 1 to n. The ordinal number put in parentheses will be indicated as superscript as shown at the left of Figure 1. The union of graphs $\cup_{k=1}^{n} C_m^{(k)}$ forms a spanning subgraph of $G_{m,n}$.

Let $\mathcal{V}$ be the set of vertices of $G_{m,n}$ and $K_m^{v}(n)$, $(v \in \mathcal{V})$ – the number of near-perfect matchings on the cylinder when the vacancy is fixed at the node v. We denote by $\mathcal{V}_k$ the set of vertices of the cycle $C_m^{(k)}$. Consider two graphs $G_{m,n} - v_k'$ and $G_{m,n} - v_k''$, where $v_k', v_k'' \in \mathcal{V}_k$, $v_k' \neq v_k''$. Since these graphs are isomorphic, then $K_m^{v_k'}(n) = K_m^{v_k''}(n)$. Let's introduce the notation $\hat{K}_m^{(k)}(n) = K_m^{v_k}(n)$, $(v_k \in \mathcal{V}_k)$. The set of values of $\hat{K}_m^{(k)}(n)$, $(k = 1, 2, \ldots, n)$ will be called the profile of near-perfect matchings on cylinder $G_{m,n}$. This function satisfies the simple symmetry relation $\hat{K}_m^{(k)}(n) = \hat{K}_m^{(n+1-k)}(n)$, $(k = 1, 2, \ldots, \lfloor n/2 \rfloor)$.

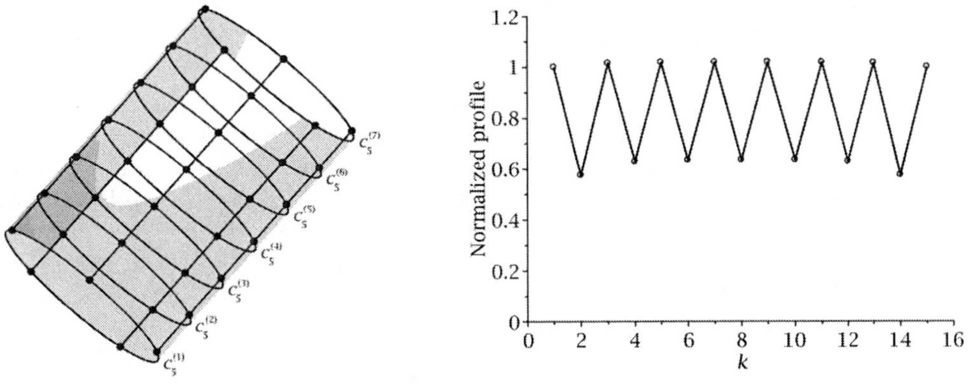

Figure 1. Cylinder $C_5 \times P_7$ and normalized profile of near-perfect matchings on $C_5 \times P_{15}$

The exponential dependence of the number of near-perfect matchings on both parameters of the graph makes it difficult to compare different profiles. It is convenient to introduce *normalized* profiles by dividing all $\hat{K}_m^{(k)}(n)$ by $\hat{K}_m^{(1)}(n)$. An example of such a profile for a small cylinder is shown at the right in Figure 1. The main advantage of the normalized profiles is their weak dependence on the parameters m and n.

For the existence of a near-perfect matching in the graph, its order must be odd, so in the future the parameter m will always be considered odd. The different kinds of maximum matchings discussed in this paper will be distinguished by superscripts as follows:

B – vacancy occurs at the boundary,
$\hat{K}_m^{B}(n) = \hat{K}_m^{(1)}(2n+1)$ – the number of near-perfect matchings on $G_{m,2n+1}$ graph with one fixed vacant node on the boundary. $G_m^{B}(z) = \sum_{n=0}^{\infty} \hat{K}_m^{B}(n)\, z^n$ – generating function for the sequence $\{\hat{K}_m^{B}(n)\}$.

P – perfect matchings,
$K_m^{P}(n)$ – the number of perfect matchings on the $G_{m,2n}$ graph. $G_m^{P}(z) = \sum_{n=0}^{\infty} K_m^{P}(n)\, z^n$ – generating function for the sequence $\{K_m^{P}(n)\}$.

N – near-perfect matchings,
$\hat{K}_m^N(n) = m \cdot \sum_{k=1}^{2n+1} \hat{K}_m^{(k)}(2n+1)$ – the total number of near-perfect matchings on $G_{m,2n+1}$ graph.
$G_m^N(z) = \sum_{n=0}^{\infty} \hat{K}_m^N(n)\, z^n$ – generating function for the sequence $\{\hat{K}_m^N(n)\}$.

3 Counting matchings

The method proposed by Wilf was used to count matchings [3]. This method allows us to reduce the problem of counting perfect matchings in a graph of order n to evaluating the coefficients of a certain polynomial in n variables. It is attractive due to its universality and ease of implementation in computer algebra systems. In our implementation of the algorithm we used a 64-bit version of Maple 17. Due to the extensive set of library functions built into this CAS, the program code turned out to be very compact.

To generate the graphs studied, the `GraphTheory` package was used. The same package was used to perform all operations on graphs. Calling routine `CartesianProduct(PathGraph(n), CycleGraph(m))` with odd m and n created a cylinder of odd order. To calculate the profile of near-perfect matchings, one vertex belonging to the cycle $C_m^{(k)}$ was selected for each value of k from 1 to $\lceil n/2 \rceil$. This vertex, together with the edges incident to it, was removed from the cylinder. Then, to evaluate $\hat{K}_m^{(k)}(n)$, the algorithm by Wilf was applied to the resulting even-order graph. The second half of the profile was reconstructed on the basis of symmetry considerations. To accelerate the computations, the values of $\hat{K}_m^{(k)}(n)$ for different k were calculated in parallel. To do this, the tools of the `Threads` package were used.

After calculating the profile of near-perfect matchings on the graph $G_{m,n}$, we can find the value of $\hat{K}_m^N(\lfloor n/2 \rfloor)$, which was of primary interest in the paper by Kong. Then the parameter n was increased by 2 and all calculations were repeated on the graph $G_{m,n+2}$. Multiple repetition of the calculations made it possible to obtain initial segments of the sequences $\{\hat{K}_m^{(k)}(n)\}$ and $\{\hat{K}_m^N(n)\}$. In those cases when these segments turned out to be sufficiently long, it was possible to derive recurrence relations. For example, the sequence $\{\hat{K}_3^B(n)\}$ satisfies a second-order relation $\hat{K}_3^B(n+1) = 5\hat{K}_3^B(n) - \hat{K}_3^B(n-1)$. However, $\{K_3^P(n)\}$ also satisfies the same recurrence relation, which indicates close connections between perfect and near-perfect matchings.

For Cartesian product graphs and fixed values of m, the method by Wilf has properties which are typical for algorithms from the complexity class **FPT**. The resources required to compute $\hat{K}_m^{(k)}(n)$ depend exponentially on m and are polynomial in n. In turn, the upper bound of the order of the recurrence relations satisfied by the sequence $\{K_m^P(n)\}$ is equal to $2^{\lfloor m/2 \rfloor}$, and this boundary as a rule is achieved [4]. To derive such relation, the initial segment must contain more than $2^{\lfloor m/2 \rfloor + 1}$ elements. Under such circumstances, obtaining analytical expressions is possible only for small m. At moderate values of m, we must limit ourselves to the numerical values of the number of near-perfect matchings.

All calculations were carried out on a personal computer with an Intel Core i7-980 3.33 GHz processor and 24 GB RAM. The computer was running a 64-bit version of Windows 7. Calculations were performed in the 64-bit version of Maple 17. Available facilities make it possible to find linear recurrence relations and generating functions $G_m^B(z), G_m^N(z)$ on cylinders for $m \leqslant 13$. Profiles of near-perfect matchings were evaluated on $G_{m,n}$ graphs for $15 \leqslant m \leqslant 19$ and $n \leqslant n^*$, where $n^* = 135$ for $m = 15, 17$, and $n^* = 85$ for $m = 19$.

4 Main results

In this section, we will discuss, on simple examples, those properties of near-perfect matchings that hold for all the studied values of m. The generating functions $G_m^P(z)$ will be considered known, since they were found in [4] for $m \leqslant 30$.

Let's compare $G_m^B(z)$ and $G_m^P(z)$ for the simplest case $m = 3$

$$G_3^B(z) = \frac{1}{1 - 5z + z^2}, \qquad G_3^P(z) = \frac{1 - z}{1 - 5z + z^2}.$$

The equality of the denominators $G_3^B(z)$ and $G_3^P(z)$ means that $\hat{K}_3^B(n)$ can be represented as a linear combination of $K_3^P(n)$. This property makes it much easier to obtain the asymptotic behavior of $\hat{K}_3^B(n)$ for large n

$$\hat{K}_3^B(n) = \frac{1}{3}\left(K_3^P(n+1) - K_3^P(n)\right), \qquad \lim_{n\to\infty} \hat{K}_3^B(n)/K_3^P(n) = \frac{\sqrt{3}+\sqrt{7}}{2\sqrt{3}}.$$

The quantity on the right-hand side of the last equality is closely related to the asymptotics of the number of perfect matchings in $G_{3,2n}$ graphs. This suggests that, for other odd m, the values of $\hat{K}_m^B(n)$ and $K_m^P(n)$ can also be calculated from the same formulas. As a result of simplifying expression (2) of [1], we conclude that to evaluate the $\hat{K}_m^B(n)$, we can substitute the odd values m and n in the formula (2) from [4], and then divide the obtained value by $\sqrt{m}$.

To compare with the results of Kong's numerical computations, it is necessary to clarify the properties of the functions $G_m^N(z)$. Their most important feature is that the denominators of $G_m^N(z)$ are the squares of the denominators of $G_m^P(z)$. In this case, as shown below, $\hat{K}_m^N(n)$ is also a linear combination of $K_m^P(n)$, but with coefficients linearly dependent on n

$$G_3^N(z) = 3\frac{1 + 2z - z^2}{(1 - 5z + z^2)^2}, \qquad \hat{K}_3^N(n) = \frac{27n + 17}{21}K_3^P(n+1) - \frac{15n+5}{21}K_3^P(n).$$

For small m, we can even write out an explicit expression directly through the parameters of the graph

$$\hat{K}_3^N(n) = \frac{1}{7}\left((2+\sqrt{21})n + 2 + \frac{11}{\sqrt{21}}\right)\left(\frac{\sqrt{7}+\sqrt{3}}{2}\right)^{2n+2} + \frac{1}{7}\left((2-\sqrt{21})n + 2 - \frac{11}{\sqrt{21}}\right)\left(\frac{\sqrt{7}-\sqrt{3}}{2}\right)^{2n+2}.$$

An immediate consequence of the above formula is that the asymptotic expansions of the free energy per lattice site contain logarithmic corrections for any odd m.

With increasing m, the expressions for $G_m^N(z)$ become more complicated, but all the regularities formulated in this section remain valid. For example, for $m = 5$, the denominator of $G_5^N(z)$ is also the square of the denominator of $G_5^P(z)$

$$G_5^P(z) = \frac{(1-z)(1-7z+z^2)}{1 - 19z + 41z^2 - 19z^3 + z^4}, \qquad G_5^N(z) = 5\frac{1 + 10z - 56z^2 + 84z^3 - 24z^4 - 10z^5 + z^6}{(1 - 19z + 41z^2 - 19z^3 + z^4)^2}.$$

We have no reason to doubt the correctness of this property for *all* odd m. However, since verification was performed in a limited range of m, it has to be considered as a conjecture.

Conjecture. For *all* odd values of m the denominator of $G_m^N(z)$ is always the square of the denominator of $G_m^P(z)$.

In conclusion, it should be noted that numerous data that have been omitted due to a lack of space can be obtained from the author on request.

References

[1] F.Y. Wu, W.-J. Tzeng, and N. Sh. Izmailian, Physical Review **E83**, 011106 (2011)
[2] Y. Kong, Physical Review **E74**, 011102 (2006)
[3] H.S. Wilf, Journal of Combinatorial Theory **4**, 246 (1968)
[4] A.M. Karavaev and S.N. Perepechko, Matematicheskoe Modelirovanie **26**:11, 18 (2014)

EPJ Web of Conferences **173**, 02017 (2018) https://doi.org/10.1051/epjconf/201817302017
Mathematical Modeling and Computational Physics 2017

Generalized Darcy's Law in Filtration Theory

Yuri P. Rybakov[1,⋆] and *Natalya V.* Semenova[1,⋆⋆]

[1] *RUDN University, Moscow, Russian Federation*

Abstract. We study the hydrodynamics of flow in a porous medium modeling the grain filling in filters. Using the lattice approximation, we derive the structure of the current in porous media and obtain the transverse diffusion coefficient D which proves to be proportional to the diameter d of the grains as constituents of the medium. We consider the axially-symmetric stationary flow in a cylindrical filter and show that the vertical velocity takes its maximal value at the wall, this effect being known as the "near-wall" one. We analyze the solution to the Euler equation with the modified Darcy force, which depends not only on the velocity but also on the gradient of the pressure included in the Darcy coefficient. Finally, within the scope of the perturbation method, we derive the main filtration equation and discuss the influence of modifying the Darcy's law on the efficiency of the filtration process.

1 Introduction. Lattice approximation for the flow in porous media

The main difficulty in the filtration process seems to be the so-called "near-wall" effect, that is an anomalously large value of the flow velocity near the wall due to larger values of the gaps between the wall and the grains, the effectiveness of the filtration being decreasing [1]. In order to take this effect into account, let us first consider the discrete variant of the mass conservation equation and number the lattice vertexes by the indexes i, j (transverse to the flow) and k (along the flow), the corresponding Cartesian coordinates being x, y and z, respectively. Let us denote the local vertical stream of the fluid with the unit density by $G_{ijk} = S_0 u_{ijk}$, where u_{ijk} is the velocity of the flow and S_0 is the area of the gap between the grains. Therefore, the local mass conservation law reads

$$G_{ijk} = rG_{ijk-1} + p\left(G_{i-1jk-1} + G_{i+1jk-1}\right) + q\left(G_{ij-1k-1} + G_{ij+1k-1}\right), \tag{1}$$

where the branching coefficients r, p, q are introduced and it is supposed that $p = q$ due to the x, y symmetry of the flow. Inserting (1) into the mass conservation equation

$$\sum_{ij} G_{ijk} = \sum_{ij} G_{ijk-1}, \tag{2}$$

one immediately gets the following constraint on the branching coefficients: $r + 4p = 1$. Identifying now the lattice spacing with the diameter d of the grain, it can be proved through (1), (2) that in the

⋆e-mail: rybakov_yup@pfur.ru
⋆⋆e-mail: dobroe_slovo@inbox.ru

continuous limit the following differential equation holds:

$$\partial_z (ru_z) = d\,\partial_x (p\,\partial_x u_z) + d\,\partial_y \left(p\,\partial_y u_z\right). \tag{3}$$

One can rewrite the equation (3) in the form of the stationary conservation law: $\mathrm{div}\vec{j} = 0$, where the components of the current $\vec{j}$ in cylindrical coordinates ρ, z read: $j_\rho = u_\rho - D\,\partial_\rho u_z$, $j_z = ru_z$, and the transverse diffusion coefficient is introduced: $D = p\,d$. It is worth-while to stress that the effect of the transverse diffusion in porous media is widely discussed in literature [2, 3].

2 "Near-wall" effect and generalized Darcy's law

The well-known Darcy's law of filtration was found empirically in 1856 by the French engineer H. Darcy [4] who specialized in constructing fountains. According to this law, the velocity of the fluid flow in a porous medium appears to be proportional to the pressure gradient and depends on the structure of the grains as constituent elements of the medium. Later various modifications of the Darcy law were proposed, with the correction terms containing some degrees of the velocity, which was considered as a formal small parameter [5]. As an alternative to such an activity, we suggest a dynamical approach based on the structure of the fundamental Euler equation, which contains the velocity $\vec{u}$, the pressure P, the gravity acceleration $\vec{g}$ and the Darcy force:

$$(\vec{u}\nabla)\vec{u} + \nabla P = \vec{f} = \vec{g} - k_D\,\vec{u}. \tag{4}$$

We suggest a generalization of the Darcy law by including in k_D the natural invariant $I = (\vec{u}\nabla)P$ in the simplest linear form: $k_D = k_0 + k^*\,I$. We intend to analyze the dependence of the filtration process on the coefficient k^*. First of all, we write down the mass balance equation and the Euler equation (4) in cylindrical coordinates, using the denotations $u_\rho = w$, $u_z = u$:

$$\rho r \partial_z u + \partial_\rho \left[\rho\left(w - D\,\partial_\rho u\right)\right] = 0, \tag{5}$$

$$\left(w\,\partial_\rho + u\,\partial_z\right)u + \partial_z P = g - (k_0 + k^* I)\,u, \tag{6}$$

$$\left(w\,\partial_\rho + u\,\partial_z\right)w + \partial_\rho P = -(k_0 + k^* I)\,w, \tag{7}$$

where the invariant I reads $I = \left(w\,\partial_\rho + u\,\partial_z\right)P$. Multiplying the equations (6) and (7) by u and w respectively and adding the results, one finds for the invariant I the expression:

$$I = \left[1 + k^*\left(u^2 + w^2\right)\right]^{-1}\left[g\,u - k_0 u^2 - \frac{1}{2}\left(w\,\partial_\rho + u\,\partial_z\right)\left(u^2 + w^2\right)\right]. \tag{8}$$

Using (8), one can eliminate the pressure P between the equations (6) and (7), the resulting equation for u and w being:

$$\partial_\rho\left[\left(w\,\partial_\rho + u\,\partial_z\right)u + (k_0 + k^*\,I)\,u\right] = \partial_z\left[\left(w\,\partial_\rho + u\,\partial_z\right)w + (k_0 + k^*\,I)\,w\right]. \tag{9}$$

We define the stream function Ψ through the substitution: $\rho\,r\,u = -\partial_\rho\Psi$, $\rho\left(w - D\,\partial_\rho u\right) = \partial_z\Psi$, the equation (5) being satisfied identically. We now separate the constant part u_0 of the velocity, with the stream function $\psi_0 = \psi_0(\rho)$. To explain the "near-wall" effect, one can use the perturbation method by putting $\Psi = \psi_0 + \psi$, $\quad u = u_0 + \tilde{u}$, where it is supposed that $\psi \ll \psi_0$, $\tilde{u} \ll u_0$, $w \ll u_0$. Within the perturbation strategy, one can simplify the invariant I by putting $I = I_0 + \tilde{I}$, where

$$I_0 = \left(1 + k^*\,u_0^2\right)^{-1}\left(g\,u_0 - k_0\,u_0^2\right), \qquad \tilde{I} = \left(1 + k^*\,u_0^2\right)^{-1}\left[(g - 2\,k_0\,u_0)\,\tilde{u} - u_0^2\,\partial_z\tilde{u}\right], \tag{10}$$

and the smallness of the parameter $\epsilon = k^* u_0^2 \ll 1$ is taken into account. Inserting (10) into (9), one can represent the latter equation in the form:

$$u_0 \left(\partial^2_{\rho z} \tilde{u} + k^* \partial_\rho \tilde{I} \right) + (k_0 + k^* I_0)\, \partial_\rho \tilde{u} = u_0\, \partial_z^2 w + (k_0 + k^* I_0)\, \partial_z w, \quad \tilde{u} = -\frac{1}{\rho r} \partial_\rho \psi, \quad w = \frac{1}{\rho} \partial_z \psi + D \partial_\rho \tilde{u}. \tag{11}$$

Taking into account that $D(\rho) = p(\rho)d$, $r(\rho) = 1 - 4\,p(\rho)$, one can solve (11) by separation of variables: $\psi = R(\rho)Z(z)$, $Z = \exp(-\lambda z)$, $\lambda = \text{const}$. The resulting equation for $R(\rho)$ is:

$$\left(\frac{1}{\rho\, r} R' \right)' \left[\lambda u_0 - k_0 - k^* u_0 \left(2\,g - 3\,k_0 u_0 + \lambda u_0^2 \right) \right] +$$
$$+ \lambda \left[\frac{\lambda}{\rho} R + D \left(\frac{1}{\rho\, r} R' \right)' \right] \left[\lambda u_0 - k_0 - k^* u_0 \left(g - k_0 u_0 \right) \right] = 0. \tag{12}$$

In view of the fact that the branching coefficient $p(\rho)$ vanishes at the wall of the filter, i.e. at $\rho = a$, one gets $p(a) = 0$ and derives from the boundary condition $w(0) = w(a) = 0$ that for the function $R(\rho)$: $R(0) = R(a) = 0$.

We are now in a position to remark that in the model in question, beside the parameter ϵ, another small parameter can be introduced: $\mu = d/a \ll 1$. Therefore, we can use the perturbation decomposition:

$$R = R_0 + \epsilon R_1 + \mu R_2 + \cdots, \qquad \lambda = \lambda_0 + \epsilon \lambda_1 + \mu \lambda_2 + \cdots,$$

which permits us to deduce from (12) the following equation for R_0:

$$\left(\frac{1}{\rho\, r} R_0' \right)' + \frac{\lambda_0^2}{\rho} R_0 = 0. \tag{13}$$

It is reasonable to consider first the domain $0 \le \rho < a - d$, for which $r(\rho) = r_0 = \text{const}$. Using the substitution: $R_0 = \rho K_-(x)$, $x = \rho\, \lambda_0 \sqrt{r_0}$, one reduces (13) to the Bessel equation, with the solution $K_-(x) = A\, J_1(x)$, $A = \text{const}$. On the other hand, in the very small domain $a \ge \rho > a - d$, where $R_0(a) = 0$, one can put $r \approx 1$, $R_0 = \rho K_+(\rho \lambda_0)$ and use the approximation: $K_+ \approx B\, \lambda_0 (\rho - a)$, $B = \text{const}$. Matching the functions K_- and K_+, together with their first derivatives, at the point $\rho = a - d$, one gets the following algebraic system:

$$A\, J_1 \left[(a-d) \lambda_0 \sqrt{r_0} \right] = -B\, \lambda_0 d; \qquad A\, J_1' \left[(a-d) \lambda_0 \sqrt{r_0} \right] \sqrt{r_0} = B. \tag{14}$$

As can be easily seen, the system (14) is equivalent to the condition:

$$J_1 \left[(a-d) \lambda_0 \sqrt{r_0} \right] = -J_1' \left[(a-d) \lambda_0 \sqrt{r_0} \right] \sqrt{r_0}\, \lambda_0 d. \tag{15}$$

Taking into account that $a \gg d$, one concludes that the condition (15) yields the following equation for λ_0:

$$J_1 \left(a\, \lambda_0 \sqrt{r_0} \right) = 0. \tag{16}$$

Denoting the n-th nontrivial zero of the Bessel function $J_1(x)$ as x_n, one can represent the general solution to the equation (11) as follows:

$$\psi = \sum_{n=1}^{\infty} A_n \rho J_1 \left(\rho\, \lambda_0^n \sqrt{r_0} \right) \exp\left(-\lambda_0^n z \right), \tag{17}$$

where $\lambda_0^n = x_n/a\sqrt{r_0}$. In particular, for the first mode one finds the root $x_1 \approx 3.8317$. As for the amplitudes A_n in (17), they can be obtained from the initial velocity profile (at $z = 0$). It is worthwhile to underline that the solution (17) permits one to explain the aforementioned "near-wall" effect. In fact, if one takes into account that the main contribution to the velocity profile is given by the first mode, one easily finds from the well-known identity for the Bessel functions: $x\,J_0(x) = J_1(x) + x\,J_1'(x)$ that the vertical velocity near the wall of the filter proves to be proportional to $J_0(x_1)$, i. e. it takes its maximal value at the wall. Let us now estimate the constant part u_0 of the velocity. To this aim, we consider the equation (6) at $\rho = 0$:

$$u\partial_z u - g + k_0 u + \partial_z P\left(1 + k^*\,u^2\right) = 0 \tag{18}$$

and integrate its left-hand side over z in the interval $z \in [0, \ell]$, where ℓ stands for the height of the filter. The result of integration can be approximately represented as follows:

$$\triangle P\,(1 + k^* u_0^2) + (g - k_0 u_0)\ell = 0, \tag{19}$$

where $\triangle P = P(z = 0) - P(z = \ell)$, and the smallness of the parameters ϵ, μ is taken into account.

3 Conclusion

We analyzed for the first time the origin of the "near-wall" effect in the filtration practice and showed its connection with the transverse diffusion coefficient vanishing at the wall of the filter. The other problem concerned the generalization of the Darcy's law, which was motivated by the idea that the Darcy coefficient k_D should have a dynamical origin. To this end, we inserted in k_D the natural dynamical invariant I in the simplest linear form $k_D = k_0 + k^*\,I$ and showed that the main filtration relation (19) proved to be modified via deforming the pressure: $P \to P\left(1 + k^*\,u_0^2\right)$. As a result, the filtration velocity u_0 should be also modified.

Acknowledgement

This work was financially supported by the Ministry of Education and Science of the Russian Federation (the Agreement number 02.a03.21.0008).

References

[1] V.N. Lukerchenko, D.N. Maslov, Yu.P. Rybakov, T.M. Shabalina, G.N. Shikin, and V.A. Yanushkevich, *Mathematical Models of Nonlinear Phenomena, Processes and Systems: From Molecular Scale to Planetary Atmosphere. Chapter 14.* (Editors: A.B. Nadykto, L.A. Uvarova; Nova Science Publishers, Hauppauge, New York, 2013) Pp. 217–224
[2] K. Vafai, *Handbook of Porous Media* (CRC Press, Boca Raton, FL, 2015) 959 pp.
[3] F.A.L. Dullien, *Porous Media: Fluid Transport and Pore Structure* (Academic Press, San Diego, 2012) 574 pp.
[4] A.E. Sheidegger, *The Physics of Flow Through Porous Media* (MacMillan, New York, 1960) 256 pp.
[5] P.Ya. Polubarinova – Kochina, *Theory of Ground Water Movement* (Princeton University Press, Princeton, 1960) 664 pp.

EPJ Web of Conferences **173**, 02018 (2018) https://doi.org/10.1051/epjconf/201817302018
Mathematical Modeling and Computational Physics 2017

Nonlinear Spinor Field in Non-Diagonal Bianchi Type Space-Time

Bijan Saha[1,*]

[1] *Laboratory of Information Technologies, Joint Institute for Nuclear Research, 141980 Dubna, Moscow region, Russia*

Abstract. Within the scope of the non-diagonal Bianchi cosmological models we have studied the role of the spinor field in the evolution of the Universe. In the non-diagonal Bianchi models the spinor field distribution along the main axis is anisotropic and does not vanish in the absence of the spinor field nonlinearity. Hence within these models perfect fluid, dark energy etc. cannot be simulated by the spinor field nonlinearity. The equation for volume scale V in the case of non-diagonal Bianchi models contains a term with first derivative of V explicitly and does not allow exact solution by quadratures. Like the diagonal models the non-diagonal Bianchi space-time becomes locally rotationally symmetric even in the presence of a spinor field. It was found that depending on the sign of the coupling constant the model allows either an open Universe that rapidly grows up or a close Universe that ends in a Big Crunch singularity.

1 Introduction

Since the first nonlinear generalization of the Dirac equation by Ivanenko [1], the nonlinear Dirac equation (NLD) has emerged as a practical model in many physical systems: (i) extended particles; (ii) the gap solitons in nonlinear optics; (iii) Light solitons in waveguide arrays and experimental realization of an optical analog for relativistic quantum mechanics; (iv) Bose-Einstein condensate in honeycomb optical lattices; (v) Phenomenological models in quantum chromodynamics; (vi) Cosmology; (vii) Chiral models of Skyrme and Faddeev; (vii) Chiral models of Graphene etc. (cf. [2] and references therein).

We study the role of the spinor field in the evolution of the Universe. A key feature of the nonlinear spinor field is that under a suitable choice of the nonlinearity it can (i) eliminate the problem of the initial singularity giving rise to a regular solution; (ii) accelerate the isotropization process of an initially anisotropic space-time and (iii) generate late time acceleration of the expansion of the Universe (cf. [3] and references therein).

Moreover, thanks to its flexibility a nonlinear spinor field can simulate different characteristics of the matter from a perfect fluid to the dark energy and describe the different stages of the evolution of the Universe.

Some recent studies suggest that flexible though it is, the presence of non-diagonal components of the energy-momentum tensor of the spinor field together with Fierz identity impose very severe

*e-mail: bijan@jinr.ru

restrictions on the geometry of the Universe as well as on the spinor field, thus justifying our previous claim that the spinor field is very sensitive to the gravitational one. But those studied were performed within the scope of diagonal metrics. The purpose of this report is to extend that study to non-diagonal cases and clarify the role of spinor field in the evolution of the Universe.

2 Basic equations and their solutions

We choose the gravitational field in the form

$$ds^2 = dt^2 - a_1^2(t)\,dx_1^2 - [h^2(x_3)\,a_1^2(t) + f^2(x_3)\,a_2^2(t)]\,dx_2^2 - a_3^2\,dx_3^2 + 2a_1^2(t)\,h(x_3)\,dx_1\,dx_2, \qquad (1)$$

where $a_1(t)$, $a_2(t)$, $a_3(t)$ are functions of time and $f(x_3)$ and $h(x_3)$ are some functions of x_3. Depending on the value of δ the metric (1) describes Bianchi type-*II* (*BII*), Bianchi type-*VIII* (*BVIII*) or Bianchi type-*IX* (*BIX*) model, respectively:

$$\delta = -\frac{f''}{f} = \begin{cases} 0, & BII, \\ -1, & BVIII, \\ 1, & BIX. \end{cases} \qquad (2)$$

The nonlinear spinor field is given by [4]

$$L_{\rm sp} = \frac{\imath}{2}\left[\bar\psi\gamma^\mu\nabla_\mu\psi - \nabla_\mu\bar\psi\gamma^\mu\psi\right] - m_{\rm sp}\bar\psi\psi - F, \qquad (3)$$

where $F = F(K)$ is the self-interaction (nonlinear term) with K taking one of the following expressions $\{I,\ J,\ I+J,\ I-J\}$. Here $I = S^2 = (\bar\psi\psi)^2$ and $J = P^2 = \left(\imath\bar\psi\gamma^5\psi\right)^2$ are invariants constructed from bilinear spinor forms. ∇_μ is the covariant derivative of the spinor field such that $\nabla_\nu\psi = \partial_\nu\psi - \Gamma_\nu\psi$, $\nabla_\nu\bar\psi = \partial_\nu\bar\psi + \bar\psi\Gamma_\nu$, and Γ_ν is the spinor affine connection. On account of the spinor field equations the energy-momentum tensor corresponding to (3) takes the form

$$\begin{aligned} T_\mu^{\ \rho} &= \frac{\imath}{4} g^{\rho\nu}(\bar\psi\gamma_\mu\partial_\nu\psi + \bar\psi\gamma_\nu\partial_\mu\psi - \partial_\mu\bar\psi\gamma_\nu\psi - \partial_\nu\bar\psi\gamma_\mu\psi) \\ &- \frac{\imath}{4} g^{\rho\nu}\bar\psi(\gamma_\mu\Gamma_\nu + \Gamma_\nu\gamma_\mu + \gamma_\nu\Gamma_\mu + \Gamma_\mu\gamma_\nu)\psi - \delta_\mu^\rho(2KF_K - F). \end{aligned} \qquad (4)$$

In order to find the metric functions we have to write the Einstein equations corresponding to (1) and (3) and well as the spinor field equations. We also impose proportionality conditions relating the shear σ and expansion θ, that yields

$$a_i = X_i V^{Y_i}, \qquad \prod_{i=1}^{3} X_i = 1, \qquad \sum_{i=1}^{3} Y_i = 0, \qquad (5)$$

with V being the volume scale defined as $V = a_1 a_2 a_3$.

From the spinor field equations one obtains the system of equations for the bilinear spinor forms that in our case gives [5]

$$S = V_0/V, \qquad V_0 = \text{const}. \qquad (6)$$

Further, choosing the spinor field nonlinearity in the form

$$F = \sum_k \lambda_k I^{n_k} = \sum_k \lambda_k S^{2n_k} = \lambda_0 S^{2n_0} + \lambda_1 S^{2n_1} + \lambda_2 S^{2n_2}, \qquad (7)$$

on account of the non-diagonal components from the diagonal Einstein equations and the equations for bilinear spinor forms one finds [5]

$$\dot{V} = Y, \tag{8a}$$

$$\dot{Y} = \frac{3\kappa}{2}\sqrt{\frac{q_2^3}{q_3}}\frac{h'}{f'}V^{3q_1/2}\Phi_1(V, A_0^3, Y) + \Phi_2(V, A_0^3, Y), \tag{8b}$$

$$\dot{A}_0^3 = \Phi_1(V, A_0^3, Y), \tag{8c}$$

where $A_0^3 = VA^3$ with A^3 being the component of the pseudo-vector $A^\mu = \left(\bar{\psi}\gamma^5\gamma^\mu\psi\right)$ and

$$\Phi_1(V, A_0^3, Y) = -\frac{3q_1}{4q_2}\,A_0^3\,V^{-(q_1+4/3)}Y + \frac{b}{\kappa}\sqrt{q_2^5 q_3}V^{5q_1/2-2/3},$$

$$\Phi_2(V, A_0^3, Y) = \frac{1}{4}q_2^4 V^{4q_1+1/3}\left(\frac{h'}{f}\right)^2 + \frac{1}{2}q_2 q_3 V^{q_1+1/3}\frac{f''}{f}$$
$$+\frac{3\kappa}{2}\left[\left(m_{\rm sp} + \lambda_0\right) + 2\lambda_1(1-n_1)V^{1-2n_1} + 2\lambda_2(1-n_2)V^{1-2n_2}\right].$$

Depending on the concrete model we define $f(x_3)$ from (2), whereas the function $h(x_3)$ is defined from

$$h'' - \frac{f'}{f}h' = bf, \qquad b = \text{const}. \tag{9}$$

For the *BII* model we have $f(x_3) = px_3 + q$ and $h(x_3) = c_1(px_3^2/2 + qx_3) + c_2$ for $b = 0$ and $h(x_3) = (1/3)bpx_3^3 + x_3^2/2(c_1p + bq) + c_1qx_3 + c_2$ for $b \neq 0$,with p, q, c_1, c_2 being some arbitrary constants.

For the *BVIII* model we have $f(x_3) = \sinh(x_3)$ and $h(x_3) = \cosh(x_3)$ for $b = 0$ and for $b \neq 0$ $h(x_3) = b\,(x_3\cosh(x_3) - \sinh(x_3)) + c_1\cosh(x_3) + c_2$,with c_1, c_2 being some arbitrary constants.

For the *BIX* model we have $f(x_3) = \sin(x_3)$ and $h(x_3) = \cos(x_3)$ for $b = 0$ and for $b \neq 0$ $h(x_3) = b\,(\sin(x_3) - x_3\cos(x_3)) - c_1\cos(x_3) + c_2$.

In what follows we solve the system (8) numerically. We set $n_k = n_0 : 1 - 2n_0 = 0$ which gives $n_0 = 1/2$. In this case the corresponding term can be added with the mass term. We assume that $q_1 > 0 : 4q_1 + 1/3 > 0$. For the nonlinear term to be dominant at large time, we set $n_k = n_1 : 1 - 2n_1 > 4q_1 + 1/3$, i.e., $n_1 < 1/3 - 2q_1$. And finally, for the nonlinear term to be dominant at the early stage we set $n_k = n_2 : 1 - 2n_2 < 0$, i.e., $n_2 > 1/2$. Since we are interested in a qualitative picture of the evolution, let us set $\kappa = 1$, $V_0 = 1$, $q_2 = 1$, $q_3 = 1$. The initial values are taken as $V(0) = 0.01$, $Y(0) = 0.1$, $A_0^3(0) = 1$. We have also set $t \in [0, 2]$ with step size 0.001 and $x_3 = 0.2k \in [0, 1]$ with a step $k = 0, 1, \ldots, 5$.

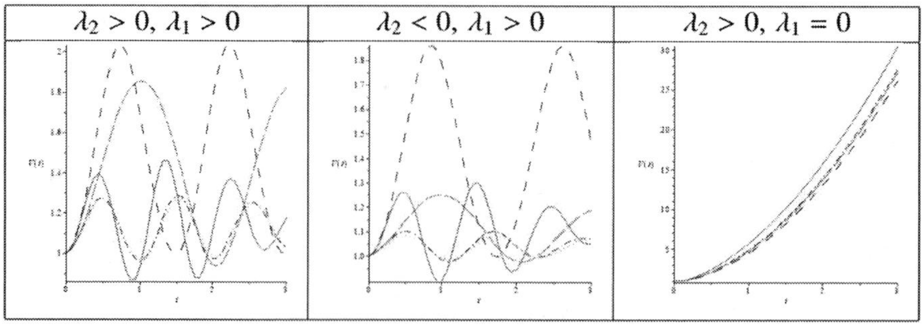

Figure 1. Evolution of V in case *BIX* space-time with nontrivial b

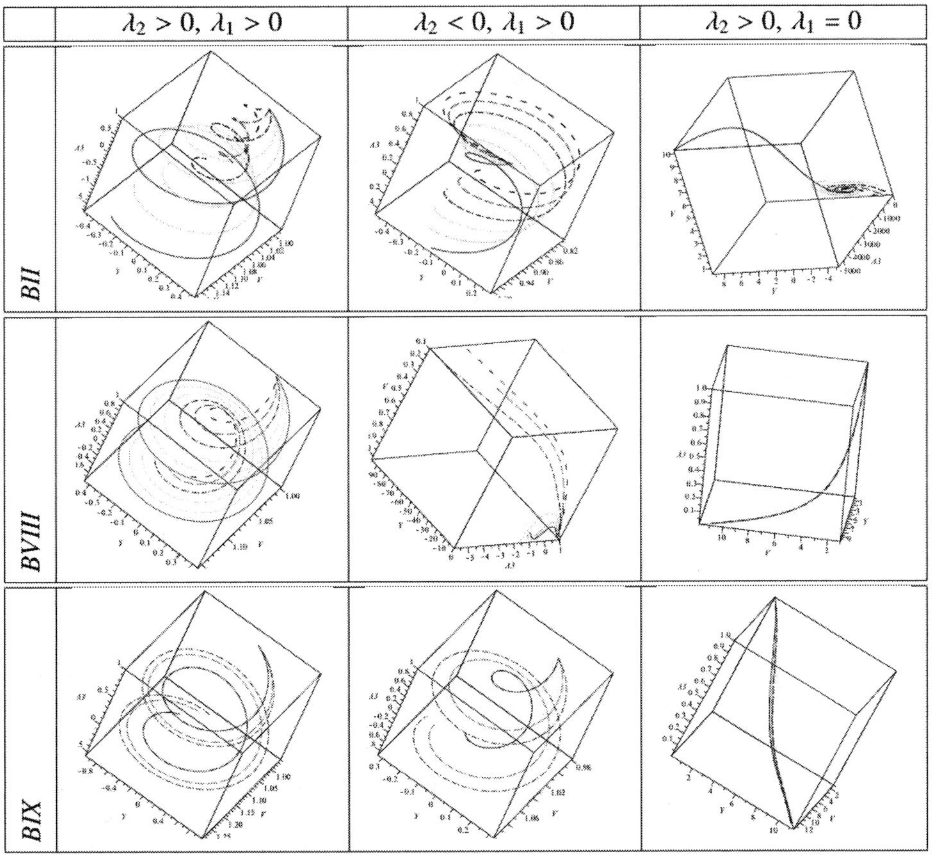

Figure 2. Phase diagram of $[V, \dot{V}, A_0^3]$ in case of nontrivial b for $[\lambda_2 > 0,\ \lambda_1 > 0]$, $[\lambda_2 < 0,\ \lambda_1 > 0]$ and $[\lambda_2 > 0,\ \lambda_1 = 0]$ for Bianchi type – *II*, *VIII* and *IX*

In figure 1 the evolution of V corresponding to these cases is given for *BIX* model as in the cases we have the qualitatively same picture. In figure 2 we have plotted the phase diagram of $[V, \dot{V}, A_0^3]$ for $[\lambda_2 > 0,\ \lambda_1 > 0]$, $[\lambda_2 < 0,\ \lambda_1 > 0]$ and $[\lambda_2 > 0,\ \lambda_1 = 0]$ for Bianchi type – *II*, *VIII* and *IX*, respectively. In the figures solid, dot, dash, dash-dot,long-dash and space-dash correspond to $k = 0, 1, 2, 3, 4, 5$.

Acknowledgments

This work is supported in part by a joint Romanian-LIT, JINR, Dubna Research Project, theme no. 05-6-1119-2014/2016.

References

[1] D. Iwanenko, Physikalische Zeitschrift der Sowjetunion **13**, 141–150 (1938)
[2] B. Saha, Eur. Phys. J. Plus **131**, 242–17 (2016)
[3] B. Saha, Astrophys. Space Sci. **331**, 243 (2011)
[4] B. Saha, Phys. Rev. D **64**, 123501 (2001)
[5] B. Saha, ArXiV. 1705.07773 [gr-qc] (2017)

EPJ Web of Conferences **173**, 02019 (2018) https://doi.org/10.1051/epjconf/201817302019
Mathematical Modeling and Computational Physics 2017

Generalized System of Riccati-Type Equations

Robert M. Yamaleev[1,⋆]

[1]*Laboratory of Information Technologies, Joint Institute for Nuclear Research, 6, Joliot Curie St., 141980 Dubna, Moscow Region, Russia*

Abstract. A new system of generalized Riccati-type equations is derived. An interconnection between the solutions of n-th order differential equations and the solutions of a generalized system of Riccati-type equations is established. Inverse mapping from the solutions of generalized Riccati-type equations onto the linearly independent solutions of the n-th order differential equation is constructed.

1 Introduction

The structure of the standard Riccati equation is defined in terms of a first order derivative and a second order polynomial. The Riccati equation is associated foremost with differential equation and the Möbius transformation [1]. Analogously, the generalized system of n-th order Riccati-type equations is also associated with the n-th order differential equations. In the present paper this problem is formulated as follows.

Consider the evolution equation with respect to a parameter t generated by the finite dimensional operator H

$$\frac{d}{dt}\Psi(t) = H\Psi(t), \quad \Psi(0) = \Psi_0, \tag{1}$$

the direct closed-form solution of which is given by the formula

$$\Psi(t) = \exp(tH)\,\Psi_0.$$

The finite dimensional operator H is represented by an $n \times n$ matrix which obeys the characteristic polynomial equation

$$f(H) = 0. \tag{2}$$

As a matter of convenience let us suppose that *the characteristic polynomial* coincides with the *minimal polynomial*

$$f(X) = X^n + \sum_{k=1}^{n}(-1)^k a_k X^{n-k}, \qquad a_k \in C. \tag{3}$$

Let E be a *companion matrix* of the polynomial $f(X)$ associated with operator H. The companion matrix obeys the same characteristic equation, namely,

$$f(E) = 0. \tag{4}$$

⋆e-mail: yamaleev@jinr.ru

Besides the evolution equation generated by the operator H one may consider an evolution generated by the polynomial $f(X)$. This evolution is described by the n-th order Riccati equation

$$\frac{d}{dt}U = f(U). \tag{5}$$

The aim of the present contribution is to establish a mapping between solutions of the equations (1) and (5).

2 Generalized trigonometric functions as solutions of high-order Riccati equation

In the same way as the usual complex algebra induces the trigonometry, the *general complex algebra* GC_n induces representations of the set of generalized trigonometric functions [2], [3]. A matrix representation of the GC_n algebra is given by the companion matrix. The *companion matrix* E is the representation of the equivalent class of all $n \times n$ matrices with trace a_1, determinant a_n and the sum of corresponding minors $a_i, 2, \ldots, n-1$. The explicit form of this matrix is defined as follows

$$E = \begin{pmatrix} 0 & 0 & 0 & 0 & 0 & (-1)^{n+1}a_n \\ 1 & 0 & 0 & 0 & 0 & (-1)^n a_{n-1} \\ 0 & 1 & 0 & 0 & 0 & (-1)^{n-1}a_{n-2} \\ \vdots & \vdots & \ddots & \ddots & \vdots & \vdots \\ 0 & 0 & \cdots & 1 & 0 & -a_2 \\ 0 & 0 & \cdots & 0 & 1 & a_1 \end{pmatrix}. \tag{6}$$

An analogy of the Euler formula is defined by the series

$$\exp\left(\sum_{k=1}^{n-1} E^k \phi_k\right) = Q(E), \tag{7}$$

where the polynomial $Q(U)$ denotes an $(n-1)$-degree polynomial of the form

$$Q(U) = g_0(\phi) + U g_1(\phi) + U^2 g_2(\phi) + \cdots + U^{n-1} g_{n-1}(\phi). \tag{8}$$

The parameter ϕ stands for the set of $(n-1)$ parameters $\phi := (\phi_1, \phi_2, \phi_3, \ldots, \phi_{n-1})$.

The structure of the set of differential equations for generalized trigonometric functions $g_0(\phi)$, $g_1(\phi)$, $g_2(\phi)$, …, $g_{n-1}(\phi)$ is governed by the matrix E and its degrees $E^k, k = 1, \ldots, n-1$ formulated in the standard way:

$$\frac{\partial}{\partial \phi_k}\mathbf{v}^g(\phi) = E^k \mathbf{v}^g(\phi), \qquad \phi = (\phi_1, \phi_2, \ldots, \phi_{n-1}), \quad k = 1, \ldots, n-1; \tag{9}$$

where $\mathbf{v}^g(\phi)$ means a vector of components

$$\mathbf{v}^g = [g_0, g_1, g_2, \ldots, g_{n-1}]^T. \tag{10}$$

As proved in [4], the differential equations (2.9) are reduced to an n-th order Riccati equation

$$\frac{d}{d\phi_{n-1}}U = f(U), \tag{11}$$

under the set of constraints

$$g_k(\phi) = 0, \qquad k = 2, 3, \ldots, n-1. \tag{12}$$

In this approach the solution of the n-th order Riccati equation is defined as a fraction of two trigonometric functions

$$U(\phi_{n-1}) = -\frac{g_0(\phi_{n-1})}{g_1(\phi_{n-1})}, \tag{13}$$

where ϕ_{n-1} depends of $(n-2)$ parameters $\phi_{n-1}(\phi_1, \phi_2, \ldots, \phi_{n-2})$, this dependence in a implicit way is defined by the constraints (12). The transformation of the linear system of evolution equations into the canonical form of the n-th order Riccati equation requires the $n-2$ constraints (12). Under these constraints the polynomial $Q(U)$ of order $(n-1)$ is reduced to a linear function of the form

$$Q(U) = g_0 + U g_1. \tag{14}$$

Then, the solution of the equation $Q(U) = 0$ turns out to the solution to the n-th order Riccati equation (11). This observation prompts us the idea to seek differential equations for the roots of the polynomial $Q(U)$ of order $(n-1)$. As a result, we get a system of Riccati-type equations for the functions

$$u_k = u_k(\phi), \quad k = 1, 2, 3, \ldots, n-1, \qquad \phi = (\phi_1, \phi_2, \ldots, \phi_{n-1}), \tag{15}$$

where u_k are roots of the polynomial $Q(u_k)$.

3 System of Riccati-type equations

Consider the polynomial of the $(n-1)$-th order,

$$Q(U) = \sum_{j=0}^{n-1} U^j g_j(\phi), \tag{16}$$

with roots $u_k(\phi)$, $k = 1, 2, 3, \ldots, n-1$; $\phi = (\phi_1, \phi_2, \ldots, \phi_{n-1})$. The coefficients of the polynomial $g_j(\phi)$, $j = 0, 1, 2, \ldots, n-1$ are solutions of the system of evolution equations

$$\partial_i\, g_j = \sum_{m=1}^{n} (E^i)_j^m g_{m-1}, \qquad i = 1, \ldots, n-1. \tag{17}$$

The main result is given by the following

Theorem. *The functions $u_k(\phi)$, $k = 1, \ldots, n-1$ obey the following system of nonlinear equations*

$$F(u_m) \sum_{k=1}^{n-p} a_{n-k-p} \partial_k\, u_m = A_p\, f(u_m), \qquad m = 1, \ldots, n-1. \tag{18}$$

where $F(u_m)$ is the $(n-2)$-degree truncated polynomial

$$F(u_m) = \left.\frac{dQ(U)}{dU}\right|_{U=u_m} = u_m^{n-2} + \sum_{k=0}^{n-3} u_m^k A_k(m) = \prod_{k=1,\, k\neq m}^{n-1} (u_m - u_k), \tag{19}$$

and $A_p(m)$ is the p-th coefficient of the polynomial $F(u_m)$. □

The explicit form of the system of equations (19) is presented as follows

$$F(U)\,(\partial_{n-1} - a_1\partial_{n-2} + a_2\partial_{n-3} - a_3\partial_{n-4} + \cdots + (-1)^{n-1}a_{n-2}\partial_1)U = A_{n-1}\,f(U),$$

$$\cdots$$

$$F(U)\,(\partial_k - a_1\partial_{k-1} + a_2\partial_{k-2} - a_3\partial_{k-3} + \cdots + (-1)^k a_{k-1}\partial_1)U = A_k\,f(U),$$

$$\cdots$$

$$F(U)\,(\partial_5 - a_1\partial_4 + a_2\partial_3 - a_3\partial_2 + a_4\partial_1)U = A_5\,f(U),$$
$$F(U)\,(\partial_4 - a_1\partial_3 + a_2\partial_2 - a_3\partial_1)U = A_4\,f(U),$$
$$F(U)\,(\partial_3 - a_1\partial_2 + a_2\partial_1)U = A_3\,f(U),$$
$$F(U)(\partial_2 - a_1\partial_1)U = A_2\,f(U),$$
$$F(U)\partial_1 U = A_1\,f(U). \tag{20}$$

If the basic polynomial $f(X)$ is a polynomial of the n-th order then the functions u_k, $k = 1, \ldots, n-1$ are roots of the following polynomial of the order $(n-1)$:

$$Q(U) = g_{n-1}U^n + \cdots + g_3U^3 + g_2U^2 + g_1U + g_0. \tag{21}$$

Each of the roots u_k obeys the system of differential equations (20), where

$$F(U) = A_1U^{n-2} + A_2U^{n-3} + \cdots + A_{n-2}U + A_{n-1}, \tag{22}$$

and

$$A_1 = 1, \qquad -A_2 = V + W + Y + Z, \qquad A_3 = VW + VY + VZ + WY + WZ + YZ + \cdots. \tag{23}$$

4 Conclusion

We have derived the system of Riccati-type equations from the linear system of evolution equations. The method can be applied in the problem of transformation of the n-th order differential equation into a Riccati-type equation. It is expected, the present method will be useful in the theory of finite dimensional quantum mechanics.

5 Acknowledgments

The author is greatly indebted to Professor Giuseppe Dattoli for his attentive reading of this work and generous encouragements. Author thanks Dr. Adan Rodriguez-Dominguez for fruitful discussions.

References

[1] A. Rodriguez-Dominguez and A. Martinez-Gonzalez, J. Nonlin. Math. Phys. **53 (4)**, 265–280 (2012)
[2] D. Babusci, G. Dattoli, E. Di Palma, and E. Sabia, Adv. Appl. Cliff. Al. **72 (2)**, 271–281 (2012)
[3] R.M. Yamaleev, J. Adv. Appl. Cliff. Al. **15 (2)**, 123–150 (2005)
[4] R.M. Yamaleev, J. Math. Anal. Appl. **420 (1)**, 12, 334–347 (2014)

EPJ Web of Conferences **173**, 03001 (2018) https://doi.org/10.1051/epjconf/201817303001
Mathematical Modeling and Computational Physics 2017

Linear Approximation of Volume Integral Equations for the Problem of Magnetostatics

Pavel Akishin[1,*] and *Andrey* Sapozhnikov[1]

[1] *Joint Institute for Nuclear Research, Dubna, Russia*

Abstract. The volume integral equation method is considered for magnetic systems. New modeling results are reported.

1 Introduction

Issues concerning the volume integral equation method for the calculations of magnetic systems are considered in the present report. We focus the following discussion on three new points:

1. Short description of the generators of a finite element based 3D grid enabling numerical solution.
2. Matrix element calculation algorithm.
3. Numerical results obtained for quadrupole magnet agree with those obtained by the famous code **TOSCA** [1].

Other issues will be discussed in a forthcoming regular paper.

2 Volume integral equation method

Let $\bar{B}(\bar{a})$, $\bar{H}(\bar{a})$, $\bar{M}(\bar{a})$ be the induction, the intensity and the magnetization of the magnetic field at the point $\bar{a}$. The values $\bar{B}$, $\bar{H}$, $\bar{M}$ are connected with the following nonlinear ratios:

$$\bar{H}(\bar{a}) = \frac{\bar{B}(\bar{a})}{\mu\left(|\bar{B}(\bar{a})|\right)\mu_0}, \qquad \bar{M}(\bar{a}) = \frac{\bar{B}(\bar{a})}{\mu_0} - H(\bar{a}), \tag{1}$$

where μ_0 is the absolute magnetic permeability of the vacuum, $\mu(x)$ is the magnetic permeability.

The following integral equation takes place:

$$\bar{H}(\bar{a}) = \bar{H}^S(\bar{a}) + \frac{1}{4\pi}\nabla_{\bar{a}} \int\limits_G \left(\bar{M}(\bar{x}), \nabla_a \frac{1}{|\bar{x}-\bar{a}|}\right) dv_{\bar{x}}, \tag{2}$$

where $\bar{H}^S(\bar{a})$ is the magnetic field from current winding, G is the area filled by iron. The field $\bar{H}^S(\bar{a})$ can be found according to the Biot-Savart law:

$$\bar{H}^S(\bar{a}) = \frac{1}{4\pi}\,\mathrm{Rot}_{\bar{a}} \int\limits_{R^3} \frac{\bar{J}(\bar{x})}{|\bar{x}-\bar{a}|}\, dv_{\bar{x}}, \tag{3}$$

*e-mail: akishin@jinr.ru

where $\bar{J}(\bar{x})$ is the current density at $\bar{x}$.

The difficulty of applying the integral approach is connected with the singularity of the kernel of the integral equations. This is the reason why only a piecewise approximation of unknown parameters within element division area is used in the famous code GFUN3D [2]. The alternative for the collocation method is an integration over dividing elements. It allows to use higher order approximation for the unknown variables. The most convenient mathematical approach for constructing such type of approximations is the finite elements method (FEM).

Let us divide the area G into tetrahedrons $\{G_i\}$. We suppose that the fragmentation $G = \bigcup_{i=1}^{N} G_i$ satisfies the requirements of FEM. Let us assume $\{\bar{P}_k, k = 1, \ldots, L\}$ is the set of all vertices in all tetrahedrons $\{G_i\}$. Let us introduce the notation $\bar{H}_k = \bar{H}(\bar{P}_k)$, $\bar{M}_k = \bar{M}(\bar{P}_k)$, $\bar{B}_k = \bar{B}(\bar{P}_k)$. We denote $f_k(\bar{x})$ as a node function, associated with the vertex $\bar{P}_k$. The functions $\{f_k(\bar{x})\}$ on each tetrahedron are linear functions, equal to 1 at the vertex $\bar{P}_k$ and to 0 at any other vertices. Using these notations we define the linear approximations for the vectors $\bar{B}(\bar{a})$, $\bar{H}(\bar{a})$, $\bar{M}(\bar{a})$:

$$\hat{B}(\bar{a}) = \sum_{k=1}^{L} f_k(\bar{a})\bar{B}_k, \qquad \hat{M}(\bar{a}) = \sum_{k=1}^{L} f_k(\bar{a})\bar{M}_k, \qquad \hat{H}(\bar{a}) = \sum_{k=1}^{L} f_k(\bar{a})\bar{H}_k. \tag{4}$$

We define a discretized formulation of the magnetostatics problem, using the finite element linear approximation within the elements of the division:

$$\sum_{j=1}^{L} \int_G f_i(\bar{a}) f_j(\bar{a}) \bar{H}_j \, dv_{\bar{a}} = \int_G f_i(\bar{a}) \bar{H}^S(\bar{a}) \, dv_{\bar{a}} + \sum_{j=1}^{L} \int_G f_i(\bar{a}) \frac{\nabla_a}{4\pi} \left[\int_G f_j(\bar{x}) \left(\bar{M}_j, \nabla_a \frac{1}{|\bar{x}-\bar{a}|} \right) dv_{\bar{x}} \right] dv_{\bar{a}}, \ i = \overline{1, L}. \tag{5}$$

Let the matrices $[C]$ and $[A]$ be matrices of $[3L \times 3L]$ dimension, consisting of diagonal matrices $[C_{ij}]$ and non-diagonal matrices $[A_{ij}]$ of $[3 \times 3]$ dimension:

$$[C] = \begin{bmatrix} [C_{11}] & \cdots & [C_{1L}] \\ \vdots & \ddots & \vdots \\ [C_{L1}] & \cdots & [C_{LL}] \end{bmatrix}, \qquad [A] = \begin{bmatrix} [A_{11}] & \cdots & [A_{1L}] \\ \vdots & \ddots & \vdots \\ [A_{L1}] & \cdots & [A_{LL}] \end{bmatrix}.$$

The matrices $[C_{ij}]$ are the following diagonal matrices and the matrices $[A_{ij}]$ satisfy the following relation for any constant vector $\bar{M}$:

$$[C_{ij}] = [\mathbf{E}] \int_G f_i(\bar{a}) f_j(\bar{a}) \, dv_{\bar{a}}, \qquad [A_{ij}]\bar{M} = \int_G f_i(\bar{a}) \, dv_{\bar{a}} \frac{\nabla_a}{4\pi} \left[\int_G f_j(\bar{x}) \left(\bar{M}, \nabla \bar{a} \frac{1}{|\bar{x}-\bar{a}|} \right) dv_{\bar{x}} \right]. \tag{6}$$

We use the following notations:

$$\hat{B} = \left(\bar{B}_1, \ldots, \bar{B}_L\right)^T, \qquad \hat{M}(\hat{B}) = \mu_0 \left(\bar{M}(\bar{B}_1), \ldots, \bar{M}(\bar{B}_L)\right)^T,$$

$$\hat{H}^S = \mu_0 \left(\int_G f_1(\bar{a}) \bar{H}^S(\bar{a}) \, dv_{\bar{a}}, \ldots, \int_G f_L(\bar{a}) \bar{H}^S(\bar{a}) \, dv_{\bar{a}} \right)^T.$$

Taking into account (1) the system (5) can be written as:

$$[C]\hat{B} = \hat{H}^S + ([A] + [C]) \hat{M}(\hat{B}). \tag{7}$$

Using node functions of higher order similarly to (5) it is possible to formulate a discretization with quadratic, cubic or higher approximation of variables within the element.

3 Generation of the finite element mesh

To build a discretization for the integral equations (5), the region of calculations should be divided into tetrahedrons satisfying the requirements of FEM. Depending on the task, there are different requirements for the mesh elements. In subregions where the solution changes faster, more detailed discretization is needed and, as a result, the element size must be smaller. And vice versa, within the regions of slow solution changing, detailed division leads to a huge number of elements, thus complicating solving the final discretized system of equations. Details of the multidimensional finite element mesh generating are discussed in [3], where the 3D mesh generator **3dfemmesh** oriented on modeling electromagnetic fields in large-scale electrophysical machines has been proposed. The algorithm used is based on the representation of the problem domain as a combination of standard macroblocks with initial generation of a two-dimensional mesh on their boundary followed by generation of a three-dimensional mesh in each block individually. The program has a graphical interface for the data entry and a visual assessment of the quality of the partition, it calculates a number of criteria for evaluating the quality of the resulting mesh.

4 Matrix elements calculation

The problem of defining matrix coefficients of the discretized equations can be reduced to calculating sixfold, singular in general, integrals from (6) by two tetrahedrons. The simplest way to evaluate these integrals done by a cubature. It might happen that the integrand function shows an isolated singularity or, even worse, the function being integrated is singular at every point of the volume under integration. In these cases the use of a cubature formula is not possible and we need suitable procedures for evaluating the matrix coefficients. The method which allows to decrease the time necessary for the calculation of the coefficients of the discretizated systems is described below. We start from the remark that the matrix coefficients entering (6) can be represented generically as integrals:

$$J^{n,j,l}_{m,i,k} = \int\limits_{G_m}\int\limits_{G_n}\left[f_i(\bar{x})\, f_j(\bar{a})\, \frac{\partial^2}{\partial x_k \partial a_l}\frac{1}{|\bar{x}-\bar{a}|}\right] dv_{\bar{x}}\, dv_{\bar{a}}.$$

Taking into account that $\{f_p(\bar{x})\}$ are linear functions and, as a consequence, the function gradient vector is constant inside each tetrahedron, such volume integrals can be reduced to surface integrals:

$$J^{n,j,l}_{m,i,k} = \oint\limits_{\partial G_m}\oint\limits_{\partial G_n} \frac{f_i(\bar{x}) f_j(\bar{a}) (d\bar{S}_x, \bar{e}_k)(d\bar{S}_a, \bar{e}_l)}{|\bar{x}-\bar{a}|} - 0.5\frac{\partial f^m_i(\bar{x})}{\partial x_k}\frac{\partial f^n_j(\bar{a})}{\partial a_l}\oint\limits_{\partial G_m}\oint\limits_{\partial G_n}\frac{(d\bar{S}_x, d\bar{S}_a)}{|\bar{x}-\bar{a}|} - \quad (8)$$

$$-0.5\frac{\partial f^m_i(\bar{x})}{\partial x_k}\oint\limits_{\partial G_m}\oint\limits_{\partial G_n} f_j(\bar{a})(dS_a, \bar{e}_l)\frac{((\bar{x}-\bar{a}), d\bar{S}_x)}{|\bar{x}-\bar{a}|} - 0.5\frac{\partial f^n_j(\bar{a})}{\partial a_l}\oint\limits_{\partial G_m}\oint\limits_{\partial G_n} f_i(\bar{x})(dS_x, \bar{e}_k)\frac{((\bar{a}-\bar{x}), d\bar{S}_a)}{|\bar{a}-\bar{x}|}.$$

The notation $\partial f^m_i(\bar{x})/\partial x_k$ means that the derivative is calculated on the tetrahedron G_m. It is important to note that the region G consists of a union of tetrahedrons. Then the borders $\{\partial G\}$ are triangles. Thus, calculating the expressions (8) reduces a 6D integral to the sum of 4D integrals over two triangles. There are four types of positional relationships of the triangles in space: triangles do not cross, triangles have one mutual vertex, triangles have two mutual vertices, and triangles are congruent. For the first type use of a cubature formula is possible. Meanwhile, in the other cases, the cubature is unreliable due to the singularity of the expressions being integrated. Let us note that the expression under integration in (8) can be represented as a sum of homogeneous functions. The method of integration from [4] allows to reduce the singular integrals of homogeneous functions to superpositions of regular integrals of lower order, the calculation of which can be reliably done by cubature.

5 Iterative method for solving the nonlinear discretized system

To achieve the requested approximation accuracy in practice it is necessary to split the region G into smaller elements, which leads to huge dimension rise of the nonlinear discretized systems of the equations. The use of methods for handling high order matrices is extremely difficult. So, for solving discretizated systems of equations (7) a simple iterative process is used:

$$[C]\hat{B}_{k+1} = (\hat{H}^S + ([A] + [C])\,\hat{M}(\hat{B}_k)), \qquad \hat{B}_0 = \bar{0}, \qquad k = 1, 2, \ldots \tag{9}$$

This process ends when the equations residuals become less than a previously defined value ε. For solving linear systems of equations $[C]\bar{x} = \bar{y}$, use is made of the incomplete Cholesky expansion method in combination with the conjugate gradient method [5].

6 Magnet systems modeling

The above mentioned method of volume integral equations was tested for different magnet systems modeling. Here we give a single instance: the computer model of the BOOSTER quadrupole magnet of the **NICA** project with the help of the generator **3dfemmesh** is given Fig. 1a. The yoke field saturation is presented in Fig. 1b. The results shown in Fig. 1c point to the very good agreement of the present method and of the famous code **TOSCA** [1] based on solving partial differential equations.

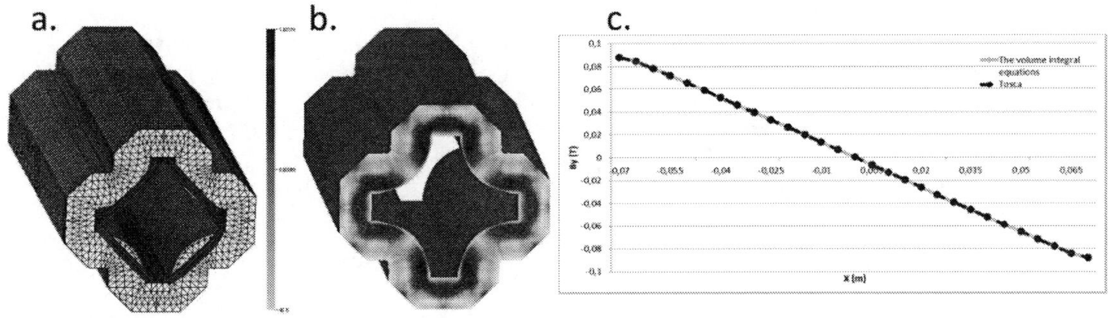

Figure 1. a. – FEM discretization of BOOSTER quadrupole magnet volume; b. – yoke field saturation; c. – comparison of the outputs of the present method and TOSCA

References

[1] J. Simkin and C.W. Trowbridge, IEE Proc. **127** B, no. 6, 368–374 (1980)
[2] A.G. Armstrong et al., *GFUN3D User Guide*, RL-76-029/A, (1976)
[3] P.G. Akishin and A.A. Sapozhnikov, Communication JINR 11-2015-58, Dubna, (2015) http://wwwinfo.jinr.ru/programs/jinrlib/3dfemmesh/
[4] P.G. Akishin, Communication JINR 11-83-558, Dubna, (1983)
[5] J.A. Meijerink and H.A. van der Vorst, Math. Comput. **31**, 148–162 (1977)

EPJ Web of Conferences **173**, 03002 (2018) https://doi.org/10.1051/epjconf/201817303002
Mathematical Modeling and Computational Physics 2017

Modeling Static Electric Field Effect on Nematic Liquid Crystal Director Orientation in Side-Electrode Cell

Alexander Ayriyan[1,*], *Edik* Ayryan[1,2], *Alexandre* Egorov[3,**], *Maria* Dencheva-Zarkova[4], *Georgi* Hadjichristov[4], *Yordan* Marinov[4], *Igor* Maslyanitsyn[3], *Alexander* Petrov[4], *Lidia* Popova[4], *Vladimir* Shigorin[3], *Alfredo* Strigazzi[5,6], and *Sofia* Torgova[7]

[1] *Joint Institute for Nuclear Research, Dubna, Russia*
[2] *Peoples' Friendship University of Russia (RUDN University), Moscow, Russia*
[3] *Prokhorov General Physics Institute, Russian Academy of Sciences, Moscow, Russia*
[4] *Georgi Nadjakov Institute of Solid State Physics, Bulgarian Academy of Sciences, Sofia, Bulgaria*
[5] *Physics Institute of Condensed Matter and Complex Systems, CNISM, Politecnico di Torino, Torino, Italy*
[6] *National Research Nuclear University MEPhI (Moscow Engineering Physics Institute), Moscow, Russia*
[7] *Lebedev Physical Institute, Russian Academy of Sciences, Moscow, Russia*

Abstract. A two-dimensional model of Fredericks effect was used for the investigation of the static electric field influence on nematic liquid crystal director orientation in the side-electrode cell. The solutions of the equations describing the model were obtained by finite-difference methods. Fredericks transition threshold for the central part of the cell, as well as dependencies of the distribution of the director orientation patterns on the electric field and location, were obtained. The numerical results are found to agree qualitatively with the experiment. Further investigations are needed to elucidate completely the Fredericks effect.

1 Introduction

A change of the preferential orientation of liquid crystal (LC) molecules (director), in an electric or magnetic field exceeding a threshold value, results in significant change of LC properties. This effect (known as orientation Fredericks transition) is used in numerous applications and extensively studied [1]. LC cells with plane parallel glass platelets with thin LC layer between them and side electrodes are used in numerous experiments devoted to electro-optical properties of LC (see e.g. [2, 3]).

Such cells were used for studies of second harmonic generation in samples of nematic liquid crystal (NLC) 4-n-pentyl-4'-cyanobiphenyl (5CB) [3]. Domain formation with quite different director orientations has been found to occur when a high pulse-periodic voltage was applied to the electrodes. This is to some extent inconsistent with the common point of view that changes in the NLC director orientation are smooth. The discovered effect can be used for a number of applications and therefore, its investigation is of great interest.

*e-mail: ayriyan@jinr.ru
**e-mail: yegorov@kapella.gpi.ru

An investigation whether the origin of the observed effect originates in the electric field magnitude or in the pulse-periodic nature of the field is important. Moreover, it is interesting to study the space-modulated LC structure formed by the field. In this paper, a two-dimensional mathematical model for the experimentally observed domains is proposed. The mathematical simulation of the director orientation in static electric fields in the side-electrode cell has been carried out. The Fredericks transition threshold for the central part of the cell, as well as dependencies of the distribution of the director orientation patterns on the electric field magnitude and location were obtained.

2 Mathematical formulation of the problem

The design of the LC cell is presented in fig. 1. The cell walls are glass platelets. The surfaces of the platelets are covered with surfactant providing LC homeotropic orientation, when long 5CB molecular axes are parallel to the Y-axis. A static electric field is directed along to the X-axis. It is supposed that only longitudinal and transverse bends of the director appear in the field and the change in the director orientation (director distortion) is defined by the angle θ between the director and the Y-axis at the equilibrium of the electric field and elasticity torques. Therefore, the angle θ in our case is a function of the x and y coordinates.

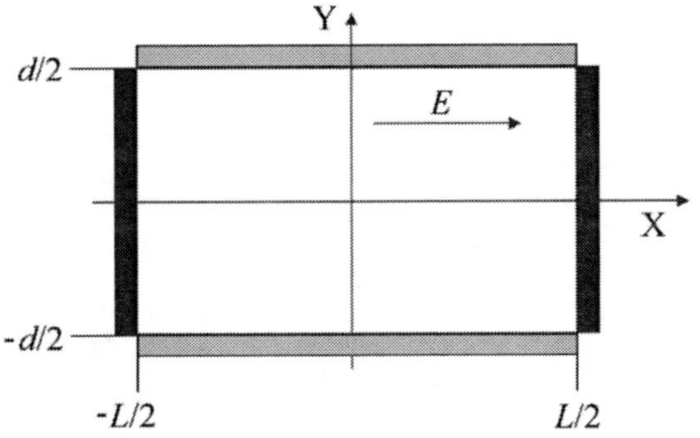

Figure 1. Cross-section of the cell used in the calculations. The electrodes are in black, glass platelets are in grey. The thickness of LC layer is d and the inter-electrode distance is L.

The standard procedure of free energy density minimization togather with qualitative approximations for the simplification of the solution result in the nonliner differential equation,

$$4K\left(\theta_{xx} + \theta_{yy}\right) + 0.5\varepsilon_0\Delta\varepsilon E^2 \sin(2\theta) = 0. \tag{1}$$

Here $K = 0.5\,(K_{11} + K_{33})$, with a splay elastic constant $K_{11} = 7 \cdot 10^{-12}$ N, a bend elastic constant $K_{33} = 10^{-13}$ N [4], the dielectric constant of the vacuum $\varepsilon_0 = 8.854 \cdot 10^{-12}$ F/m, the dielectric anisotropy $\Delta\varepsilon$ [4], $\theta_{\alpha\alpha} \equiv \partial^2\theta/\partial\alpha^2$ with $\alpha = x, y$.

The definition domain is $\Omega = \{(x, y) \mid -L/2 \le x \le L/2,\ -d/2 \le y \le d/2\}$, with $L = 2 \cdot 10^{-3}$ m and $d = 10^{-4}$ m. The boundary conditions are

$$\theta(x, -d/2) = 0, \qquad \theta(x, d/2) = 0, \qquad \theta(-L/2, y) = \pi/2, \qquad \theta(L/2, y) = \pi/2.$$

For convenience, the eq. (1) was converted to:

$$\theta_{xx} + \theta_{yy} + qE^2 \sin(2\theta) = 0, \tag{2}$$

where $q = 1.822882353$ V^{-2}.

Well-known finite–difference methods were used for the solution of the eq. (2) [5]. The programs for numerical solution of the two-dimensional parabolic partial differential equation were developed both in FORTRAN and C.

3 Results of Calculation and Discussion

Calculated dependencies of LC director orientation on the electric field magnitude and location are shown in fig. 2. Fig. 2a unveils the existence of a threshold value of $1.6 \cdot 10^3$ V/m of the electric field above which a sharp director rotation occurs. This value can be taken for the Fredericks effect threshold for two-dimensional deformation of LC director in static electric field. A director rotation angle $\theta = 45°$ is reached at the field value $E \sim 2 \cdot 10^4$ V/m. At $E \geq 5 \cdot 10^4$ V/m the director is predominantly oriented along the field.

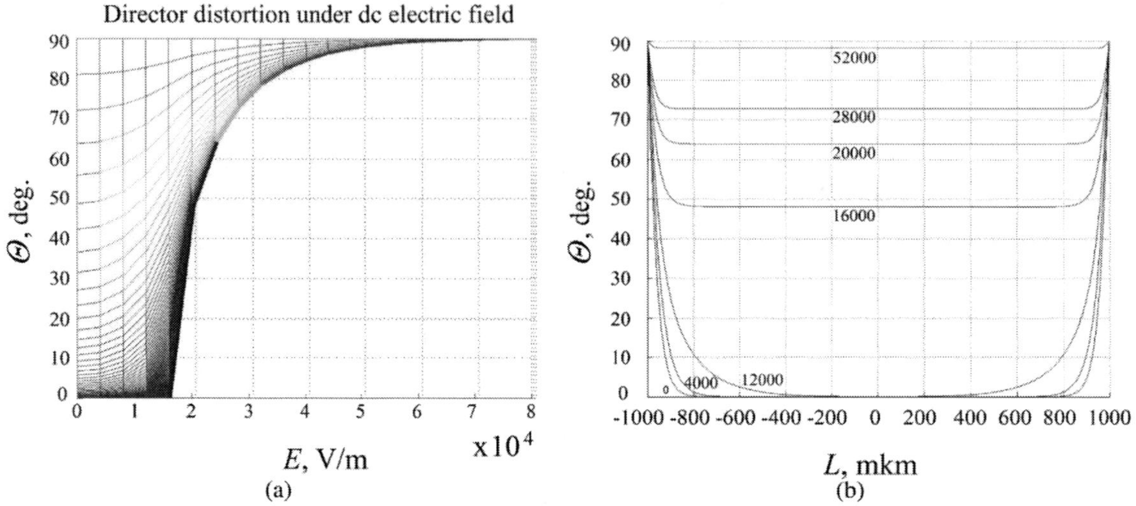

Figure 2. (a) Dependence of NLC director orientation (angle θ) on the static electric field at different positions between the electrodes (oblique lines). (b) Dependence of the director orientation (angle θ) on the position between the electrodes at different electric field values (V/m).

Fig. 2b points to the fact that a gradual change of the director orientation occures in regions close to the electrodes. The summed widths of these regions do not exceed 10% of the inter-electrode distance, they are of the order of the LC layer thickness (d = 100 mkm) irrespective of the field values. At the same time, in the central region the magnitude of the angle θ is almost constant at each given E, it increases with the field magnitude and reaches 90° at the electrodes.

The size of the above-mentioned regions is important for the comparison with the experiment. Experimental images of LC cell under the influence of the pulse-periodic electric field are shown in fig. 3 [3]. Side dark domains are the electrodes (the inter-electrode distance of $2 \cdot 10^{-3}$ m corresponds to the respective parameter of the model), central dark domain with the varying width is the region of LC undistorted (initial) homeotropic orientation and two light domains with varying width are the regions of changed (distorted) director orientation. The left image (fig. 3a) was taken in the absence of the external field and the right one (fig. 3b) after the field was applied for 16 min.

The width of the light stripes in fig. 3a does not exceed 10% of the total inter-electrode distance. This is in good agreement with the calculation results at $E = 0$. However in the fig. 3b this width increases and reaches magnitudes from 15% to 30% of the distance. One can use some value of the angle θ as a boundary point, dividing "dark" and "light" regions according to their values of the angle θ (e.g., at $\theta < 10°$ – region is dark, at $\theta > 10°$ – region is light) to separate all points in the image into dark and light domains. Then in the interval $(0.4–1.6) \cdot 10^4$ V/m the obtained angle patterns, formally, will have a qualitative agreement with the experiment. However a number of questions associated with the contrast of dark and light domains in fig. 3 remains open.

One can see that the transition from one domain to another is sharp and all parts of light domain correspond to similar director orientation. At the same time it can be concluded from fig. 2b that in the above-mentioned interval of the field the solution demonstrates:

1) smooth change of the angle around any value of θ, and

2) strong changes of the angle θ (from 10° to 90°) take place in near-electrode regions.

This indicates that the used two-dimensional model of LC cell and static electric field cannot completely account for the observed experimental effect, which is likely caused by the pulse-periodic nature of the applied experimental field. Further studies for the elucidation of this point are needed.

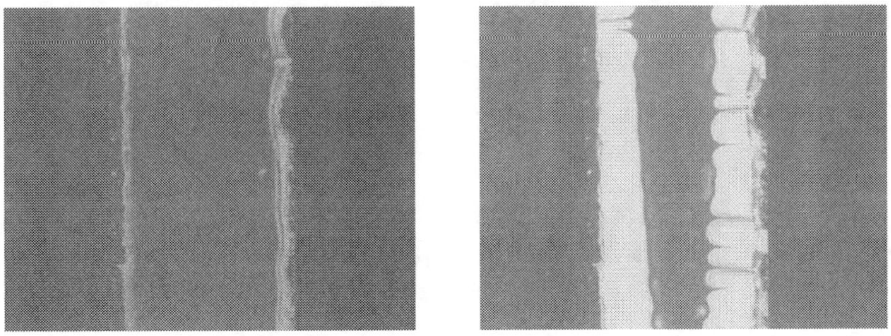

(a) field off (b) field on

Figure 3. Changes of the director orientation in a cell with initially homeotropic oriented NLC, subjected to a pulse-periodic electric field [4].

4 Summary

A two-dimensional model of the Fredericks effect in side-electrode cell was used for the investigation of the NLC director distribution. Fredericks transition threshold for the central part of the cell as well as dependencies of the distribution of the director orientation on the location and electric field magnitude were obtained. The results of the calculation were compared with the experiment.

The calculations have shown that the used model of LC cell with side electrodes results in correct homeotropic NLC director orientation in the absence of an electric field. At the same time, the introduction of static electric field to the model cannot completely explain the observed experimental effect, which is, likely, due to the pulse-periodic nature of the applied experimental field.

Acknowledgements. E. A. Ayryan appreciates financial support of the Ministry of Education and Science of the Russian Federation (the Agreement number 02.a03.21.0008). A. S. Ayriyan thanks the JINR grant (project no. 17-602-01).

References

[1] I. Ch. Khoo, *Liquid Crystals* (Wiley Interscience, New York, 2007) 384 pp.

[2] Y. G. Marinov, G. B. Hadjichristov, A. G. Petrov, S. Sridevi, U. S. Hiremath, C. V. Yelamaggad, and S. K. Prasad, Journal of Physics CS **253**, 012060 (2010)

[3] A. S. Ayriyan, E. A. Ayrjan, A. A. Egorov, G. B. Hadjichristov, Y. G. Marinov, I. A. Maslyanitsyn, A. G. Petrov, J. Pribis, L. Popova, V. D. Shigorin, A. Strigazzi, and S. I. Torgova, Physics of Wave Phenomena **24** (4), 259 (2016)

[4] A. Bogi and S. Faetti, Liq. Cryst. **28** (5), 729 (2001)

[5] A. A. Samarsky, *The Theory of Difference Schemes* (Marcel Dekker Inc., New York, 2001) 788 pp.

Model of the Phase Transition Mimicking the Pasta Phase in Cold and Dense Quark-Hadron Matter

Alexander Ayriyan[1,⋆] and *Hovik* Grigorian[1,2,⋆⋆]

[1] *Joint Institute for Nuclear Research, Joliot-Curie 6, 141980 Dubna, Russia*
[2] *Yerevan State University, Alek Manyukyan 1, 0025 Yerevan, Republic of Armenia*

Abstract. A simple mixed phase model mimicking so-called "pasta" phases in the quark-hadron phase transition is developed and applied to static neutron stars for the case of DD2 type hadronic and NJL type quark matter models. The influence of the mixed phase on the mass-radius relation of the compact stars is investigated. Model parameters are chosen such that the results are in agreement with the mass-radius constraints.

1 Introduction and Motivation

The possible hadron-quark phase transition modelling is usually made with the use of the so-called Maxwell construction, where the two phases are assumed to be separated. However, due to the finite surface tension effects, a mixed phase with non-trivial structure (the so-called "pasta" phase) could be thermodynamically preferred [1–6].

A simple model of phase transition construction mimicking the "pasta" phase is proposed. The model is parameterized by the use of an additional pressure corresponding to the impact of the structural effects in the mixed phase to the critical pressure of the Maxwell construction. The construction ignoring the details of the mixed phase structure is simple for calculations and therefore it can be easily used for investigations of the influence of the mixed phase on the mechanical characteristics of the compact stars [5].

The question of the existence of the mixed phase is actual in the investigations of the hybrid compact star structures and characteristics, which could be compared to the available observational data [7, 8]. In the sequel an application of this model to the structure and mass-radius relation of the static neutron stars is presented using the hadronic matter model with density-dependent meson-nucleon coupling (DD2) [9], and the Nambu-Jona-Lasinio model (NJL) with 8-quark interaction [10] for quark matter. The method could be extended to the finite temperature case along the whole first oder transition border of the QCD phase diagram and could be used in the simulations of heavy ion collisions as well (e.g. [11, 12]). The possibility of such extension is also discussed.

2 The model

We assume that the hadronic and quark matter phases are described by equation of state (EoS) denoted $P_H(\mu)$ and $P_Q(\mu)$ respectively which describe the dependence of pressure on chemical potential at $T = 0$ case relevant for the netrons star (NS) modeling.

⋆e-mail: ayriyan@jinr.ru
⋆⋆e-mail: hovik.grigorian@gmail.com

It is assumed that a first order Maxwell phase transition exists between the given two phases with a critical chemical potential μ_c satisfying the following relation:

$$P_Q(\mu_c) = P_H(\mu_c) = P_c. \tag{1}$$

The Maxwell construction can be modified if the assumption is fulfilled that close to the phase transition point the EoS of both phases are changing due to their interactions (see fig. 1), so that the effective mixed phase EoS can be described as

$$P_M(\mu) = \alpha(\mu - \mu_c)^p + \beta(\mu - \mu_c)^q + P_c + \Delta P, \tag{2}$$

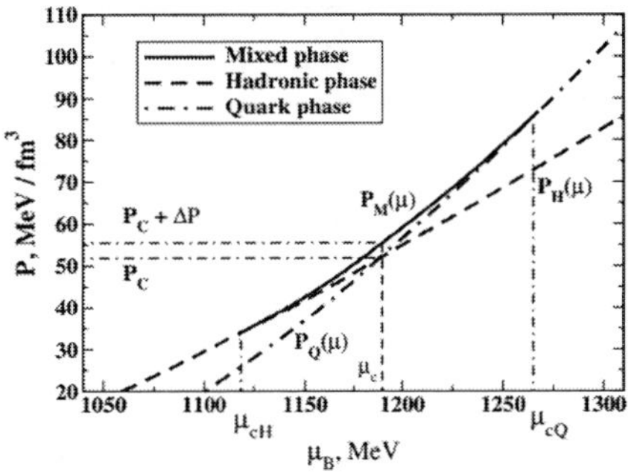

Figure 1. A model of the equation of state with mixed phase based on the Maxwell construction.

where $p > q$ are natural numbers for instance. Here the additional pressure at μ_c is considered to be the free parameter of the model:

$$P_M(\mu_c) = P_H(\mu_c) + \Delta P. \tag{3}$$

This research is focused on the quadratic form of the EoS of the mixed phase ($p = 2$ and $q = 1$):

$$P_M(\mu) = \alpha(\mu - \mu_c)^2 + \beta(\mu - \mu_c) + P_c + \Delta P. \tag{4}$$

The transition from the H-phase to the M-phase is smooth without jump in the density $n(\mu) = dP(\mu)/d\mu$. Thus there are the new critical chemical potentials: μ_{cH} for the transition from the H-phase to the M-phase and μ_{cQ} for the transition from the M-phase to the Q-phase. As a result, there are four unknowns including the coefficients α and β from Eq. (2). The phase transition conditions are

$$\begin{cases} P_M(\mu_{cH}) = P_H(\mu_{cH}), \\ P_M(\mu_{cQ}) = P_Q(\mu_{cQ}), \end{cases} \qquad \begin{cases} n_M(\mu_{cH}) = n_H(\mu_{cH}), \\ n_M(\mu_{cQ}) = n_Q(\mu_{cQ}). \end{cases} \tag{5}$$

The quantities α and β are expressed from the set of equations for the pressure

$$\begin{cases} \alpha = \dfrac{1}{\mu_{cQ} - \mu_{cH}} \left(\dfrac{P_Q(\mu_{cQ}) - (\Delta P + P_c)}{\mu_{cQ} - \mu_c} - \dfrac{P_H(\mu_{cH}) - (\Delta P + P_c)}{\mu_{cH} - \mu_c} \right), \\ \beta = \dfrac{P_H(\mu_{cH}) - (\Delta P + P_c)}{\mu_{cH} - \mu_c} + \dfrac{P_Q(\mu_{cQ}) - (\Delta P + P_c)}{\mu_{cQ} - \mu_c} - \dfrac{P_Q(\mu_{cQ}) - P_H(\mu_{cH})}{\mu_{cQ} - \mu_{cH}}, \end{cases} \tag{6}$$

and are eliminated form the set of equations for the densities (eq. (5)). Solving the remaining equations for the densities one can find the values of the critical chemical potentials μ_{cH} and μ_{cQ}.

3 Results

For the illustration purposes, the described model is applied for the construction the mixed phase EoS (see fig. 1) between the hadronic and quark matter respectively given by the DD2 type model [9] and the NJL model with 8-quark interaction [10]. The results of the construction are shown in fig. 2. Here the varied pressure is given by the dimensionless parameter $\Delta_P = \Delta P/P_c$. The parameter Δ_P is varied within 10%. This completely covers the whole range of the additional pressure obtained by the microscopic models of the mixed phase [1–4, 6]. Note, that the P-ε dependence of the mixed phase is not a linear function by the construction, but is very close to it.

With the use of this set of the EoS models the mechanical characteristics of the compact stars are investigated. The mass and radius relationships of the static spherically symmetric neutron stars are

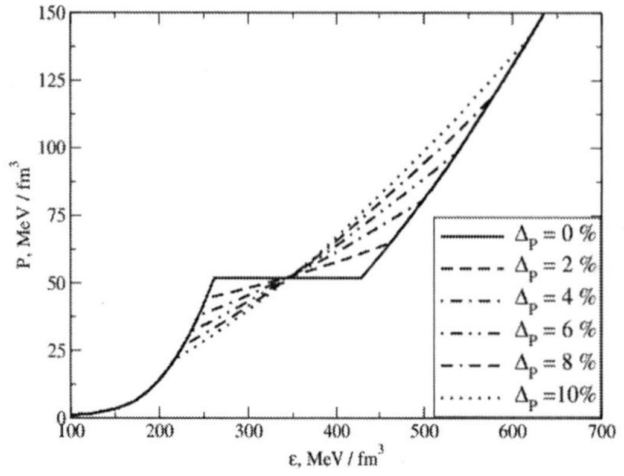

Figure 2. Influence of the additional pressure in the mixed phase on the phase transition.

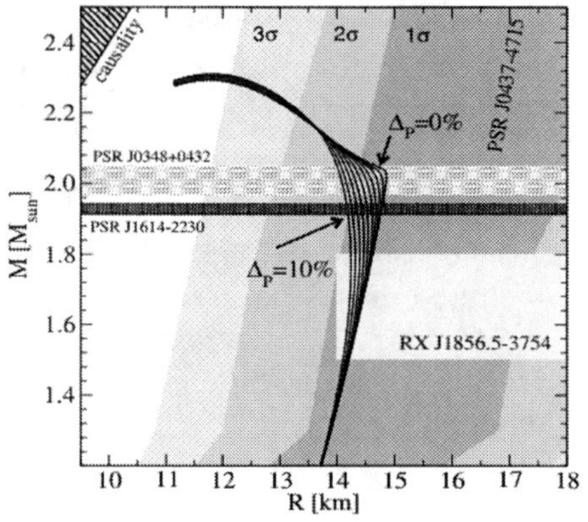

Figure 3. Influence of the mixed phase on the mass-radius relation.

obtained as solutions of the Tolman-Oppenheimer-Volkoff (TOV) equations [13, 14]:

$$\begin{cases} \dfrac{dP(r)}{dr} = -\dfrac{GM(r)\varepsilon(r)}{r^2} \dfrac{\left(1 + \frac{P(r)}{\varepsilon(r)}\right)\left(1 + \frac{4\pi r^3 P(r)}{M(r)}\right)}{\left(1 - \frac{2GM(r)}{r}\right)}, \\ \dfrac{dM(r)}{dr} = 4\pi r^2 \varepsilon(r), \end{cases} \tag{7}$$

where $\hbar = c = 1$, r is the coordinate distance from the center, $M(r)$, $\varepsilon(r)$ and $P(r)$ are the mass, the energy density and the pressure profiles correspondingly, and G is the gravitational constant. Each of these profiles, representing solutions of the TOV equations, depends on the value of the energy density in the star center $\varepsilon(0)$. The relation between the radius R where the pressure is $P(R) = 0$ and the total mas $M = M(R)$ are shown in the fig. 3 for different values of the parameter Δ_P.

The presence of the mixed phase changes the EoS softness and the maximum mass becomes larger when the maximum radius decreases. The heaviest known pulsars PSR J0348+0432 and PSR J1614−2230 have respectively the following masses $2.01^{+0.04}_{-0.04}$ $M_\odot$ [15] and $1.928^{+0.017}_{-0.017}$ $M_\odot$ [16, 17]. Therefore the maximum possible mass given by a model should exceed this limits. These constraints are shown in the fig. 3. It is demonstrated that all considered EoS models with the mixed phase satisfy this requirement. On this plot the constraints from the radius estimations of the two pulsars PSR J0437−4715 [18] and RX J1856.5−3754 [19] are also shown and the results are in agreement with this.

4 Conclusions and Discussion

A simple model of the hadron-quark phase transition mimicking the "pasta" structure is proposed. It is parameterized using the additional pressure due to the structure of the mixed phase. The model is based on the standard Maxwell construction of the first order phase transitions. It ignores the details of the mixed phase structure, at the same time it is easy to use for the investigation of the influence of the mixed phase on the mechanical characteristics of the compact star. The structure and mass-radius relations of the static neutron stars are presented for the case of the DD2 type hadronic vs NJL type quark matter EoS models. It is shown that the variation of the parameter Δ_P in the relevant range of the additional pressure, for the mixed phase, changes the maximum mass and the radius of the star configurations within limits imposed by the observational constraints.

The method could be extended to the finite temperature case along the whole first oder transition border of the QCD phase diagram as it is shown in the fig.4 assuming that the additional pressure decreases roportionally to the increases in the temperature. Thus the case when the additional pressure goes from the value Δ_P at ($\mu = \mu_c, T = 0$) down to zero at the Critical End Point (CEP) is demonstrated in fig. 4. The first order phase transition (the jump in density) disappears at the CEP, therefore the possible mixed phase is also disappearing. For the crossover transition one needs to apply another construction. The considered approach for the transition construction is independent of the internal details of the transition, therefore it could be applied both for the beta-equilibrated and for the symmetric nuclear matter. Thus it could be used in the simulations of heavy ion collisions.

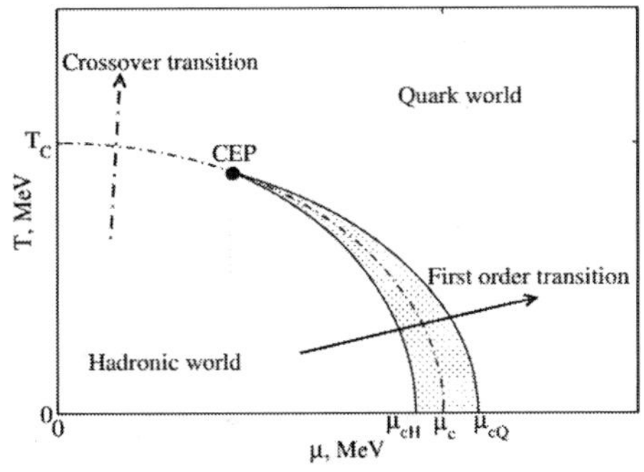

Figure 4. A simple schematic view to the QCD phase diagram. The grey area is the area of mixed phase of the first order hadron-quark phase transition.

Acknowledgements

The authors thank Prof. David Blaschke, Prof. Dmitry Voskresensky, Dr. David Edwin Alvarez Castillo, Dr. Vahagn Abgaryan and Konstantin Maslov for fruitful discussions. Prof. Stefan Typel and Dr. Sanjin Benic provided tables of the EoS models, special thanks to them. The research was carried out under financial support of the Russian Science Foundation (project No. 17-12-01427).

References

[1] D. N. Voskresensky et al., Nucl. Phys. A **723**, 291–339 (2003)
[2] T. Tatsumi et al., Nucl. Phys. A **718**, 359–362 (2003)
[3] T. Endo et al., Nucl. Phys. A **749**, 333–336 (2005)
[4] T. Endo et al., Prog. Theor. Phys. **115** (2), 337–353 (2006)
[5] T. Noda et al., Astrophys. J. **765**:1 (2013)
[6] N. Yasutake et al., Phys. Rev. C **89**, 065803 (2014)
[7] M. Bejger, P. Haensel, and J. L. Zdunik, Mon. Not. Roy. Astron. Soc. 359, 699–706 (2005)
[8] D. E. Alvarez-Castillo and D. Blaschke, Phys. Part. Nucl. **46**, 846–848 (2015)
[9] S. Typel, Eur. Phys. J. A **52**:16 (2016)
[10] S. Benic, Eur. Phys. J. A **50**:111 (2014)
[11] V. D. Toneev, Eur. Phys. J. WoC **7**, 02005 (2010)
[12] K. A. Bugaev et al., Phys. Part. Nucl. Lett. **12** (2), 238–245 (2015)
[13] R. C. Tolman, Phys. Rev. **55** (4), 364–373 (1939)
[14] J. R. Oppenheimer and G. M. Volkoff, Phys. Rev. **55** (4), 374–381 (1939)
[15] J. Antoniadis, P. C. C. Freire, N. Wex, et al., Science **340**, 6131 (2013)
[16] P. Demorest, T. Pennucci, S. Ransom, et al., Nature **467**, 1081–1083 (2010)
[17] E. Fonseca et al., Astrophys. J. **832**, 167 (2016).
[18] S. Bogdanov, Astrophys. J. **762**:96 (2013)
[19] H. Hambaryan, R. Neuhauser, V. Suleimanov, and K. Werner, J. Phys., Conf. Ser. **496**, 012015 (2014)

EPJ Web of Conferences **173**, 03004 (2018) https://doi.org/10.1051/epjconf/201817303004
Mathematical Modeling and Computational Physics 2017

Methods of Investigation of Equations that Describe Waves in Tubes with Elastic Walls and Application of the Theory of Reversible and Weak Dissipative Shocks

Igor Bakholdin[1,⋆]

[1] *Keldysh Institute of Applied Mathematics, Miusskaya sq. 4, 125047 Moscow, Russia*

Abstract. Various models of a tube with elastic walls are investigated: with controlled pressure, filled with incompressible fluid, filled with compressible gas. The non-linear theory of hyperelasticity is applied. The walls of a tube are described with complete membrane model. It is proposed to use linear model of plate in order to take the bending resistance of walls into account. The walls of the tube were treated previously as inviscid and incompressible. Compressibility of material of walls and viscosity of material, either gas or liquid are considered. Equations are solved numerically. Three-layer time and space centered reversible numerical scheme and similar two-layer space reversible numerical scheme with approximation of time derivatives by Runge-Kutta method are used. A method of correction of numerical schemes by inclusion of terms with high-order derivatives is developed. Simplified hyperbolic equations are derived.

1 Introduction

Models that describe waves in elastic tubes are of great interest for technical and biological applications. Here equations based on the hyperelastic model and the complete membrane model are analyzed. Such equations were derived in [1] and analytically investigated in [2] and [3]. The purpose of this work is the generalization of the models, the application of the theory of reversible and weakly dissipative shocks [4], the development of numerical methods.

The equations of wave motion of the walls of an elastic incompressible cylindrical tube with fixed internal and external pressure are given by [2]

$$\left(R\sigma_1 z'/\lambda_1^2\right)' - P^* r r' = \rho R\ddot{z}, \quad \left(R\sigma_1 r'/\lambda_1^2\right)' - \sigma_2/\lambda_2 + P^* r z' = \rho R\ddot{r}, \tag{1}$$

$$\lambda_1 = \sqrt{r'^2 + z'^2}, \quad \lambda_2 = r/R, \quad \lambda_3 = h/H, \quad \sigma_i = \lambda_i W_{\lambda_i} - p, \quad P^* = P/H.$$

Here the prime denotes the derivative with respect to the variable Z, which stands for Lagrange initial longitudinal spatial coordinate, the upper dots denote derivative with respect to the time. Unknowns z and r determine the surface of the tube in the cylindrical system of coordinates, the z axis of this system coincides with the central line of the tube. The parameter P stands for the difference between internal and external pressures, the unknown p is the pressure in the material. The parameter ρ is the

⋆e-mail: ibbakh@yandex.ru

density of material per square unit. The thickness of the wall is denoted as h. Here λ_i are the principal stretches and σ_i are the principal Cauchy stresses, W is the stress function. It is assumed that in the absence of any external forces $z = Z$, $r = R$, $h = H$. Due to the condition of incompressibility,

$$\lambda_1\lambda_2\lambda_3 = 1, \quad \sigma_i = \lambda_i\hat{W}_i, \quad \hat{W}_i = \hat{W}_{\lambda_i}, \quad \hat{W} = W(\lambda_1, \lambda_2, 1/(\lambda_1\lambda_2)).$$

The stress function used in this paper for numerical calculations corresponds to the Gent material

$$W = -\tilde{\mu}J_m \ln\left[1 - (\lambda_1^2 + \lambda_2^2 + \lambda_3^2 - 3)/J_m\right].$$

It is considered that this stress function is applicable for rubber.

For the case of a fluid-filled tube with arbitrary pressure we must include the equations [3]

$$\dot{r}z' - r'\dot{z} + vr' + \frac{1}{2}rv' = 0, \tag{2}$$

$$\rho_f(\dot{v}z' - v'\dot{z} + vv') + P' = 0. \tag{3}$$

Here v is the fluid velocity, ρ_f is the fluid density.

2 Main generalizations and simplification of the model

The analysis shows that for the case of the gas-filled tube the equation (2) must be replaced by

$$(\dot{\rho}_f z' - \rho_f'\dot{z})r^2 + 2\rho_f r(\dot{r}z' - r'\dot{z}) + (\rho_f v r^2)' = 0. \tag{4}$$

In the barotropic case only one additional equation of state is required. An additional equation for energy is required for the complete gas dynamics model.

The viscosity of the fluid or gas may be taken into account if the additional term $c_{v2}v''$ is included in (3). The viscosity of the material of the walls is enforced within the Kelvin-Voight model [5]

$$\sigma_i \to \sigma_i + \sigma_{vi}, \quad \sigma_{vi} = \nu_{vi}\dot{\lambda}_i, \quad i = 1, 2, 3.$$

It can be shown that for some simplified version of the equations this model leads to the inclusion of the term $c_{v2}v''$ in the equation (3) also.

The compressibility of the material of the wall may be taken into account if a stress function (potential) without p is used. A two dimensional potential (potential with hat) can be derived from the equation $\sigma_3 = p_e$, where p_e is the external pressure.

In order to take into account the meridional bending resistance of the thin walls additional "bending resistance pressure" must be formally included in (1) but not in (3) [6, 7]

$$P \to P + P_b, \quad P_b = -\cos\alpha\left[\frac{\partial^2}{\partial x^2}\left(D\frac{\partial^2}{\partial x^2}r\right) - \tan\alpha\frac{\partial^2}{\partial x^2}\left(D\frac{\partial^2}{\partial x^2}z\right)\right], \quad \frac{\partial}{\partial x} = \frac{\cos\alpha}{z'}\frac{\partial}{\partial Z}, \quad \tan\alpha = \frac{r'}{z'}.$$

Here α is the angle of inclination of the wall of the tube. This approach is based on the Germain-Lagrange formula. The bending resistance coefficient D must be calculated for the anisotropic linearized state

$$D = \{W_1 + \lambda_1(W_{11} - W_{13}) + [W_3 + \lambda_3(W_{33} - W_{31})]\lambda_3/\lambda_1\}h^3/12.$$

If the geometric nonlinearity is not taken into account then $P_b = P_{bs} = -Dr''''$ and the bending pressure is included only in the equation for r. Only the term $-br''''$ is added to the left part in this case. A problem is not well-posed if $\sigma_1 < 0$ and initial equations are used. The inclusion of the bending resistance enforces well-posedness.

A non-differential term related with the latitudinal (circular) bending resistance [8] may be included in pressure formula. But if the wall of the tube is weakly stretched (this is a typical case when inclusion of meridional bending resistance is required) or if h/R is close to zero this term vanishes.

The term simplified equations is used in the theory of shocks to denote hyperbolic equations describing waves with slow variations. The simplified form of the controlled pressure equations is

$$\dot{u} = q'; \quad \rho R\dot{q} = \left(R\sigma_1 u/\lambda_1^2\right)' - P^* r', \quad \lambda_1 = u = z', \quad P^* = \sigma_2/(\lambda_2 r u).$$

No simplification is required for the equations (2), (3), (4). These equations describe non-wave zones for Riemann problem solutions. Calculations of solutions of these equations require the use of special methods of calculations such as Godunov method in order to obtain generalized solutions with discontinuities.

3 Numerical methods and results of calculations

Standard three-layer time and space centered scheme and space centered two-layer perdictor-corrector schemes with approximation of time derivatives by Runge-Kutta method were used for all cases. These schemes were used for all models analysed in the theory of reversible and weakly dissipative shocks. The schemes are explicit usually. Similar implicit numerical schemes are used for the approximation of complex time-space derivatives if it is necessary.

For the case of fluid filled tube the equations were transformed to the form with excluded pressure

$$\left[\rho_f^* z' - \left(\frac{r'\rho R^2}{rz'^2\xi}\right)'\right]\dot{v} - \left[\frac{r'\rho R^2}{rz'^2\xi} + \left(\frac{\rho R^2}{2z'^2\xi}\right)'\right]\dot{v}' - \frac{\rho R^2}{2z'^2\xi}\dot{v}'' = \rho_f^*(v'q - vv') + \left[-P_b^* + \frac{\mathcal{P}}{\xi}\right]' + c_{v2}v'' - \{c_{v4}v''''\},$$

$$\dot{r}z' - r'q + vr' + \frac{1}{2}rv' = 0 + \{c_{r6}r''''''z'\}, \quad \dot{z} = q - \{c_{z4}z''''\}, \quad \rho R\dot{q} = \left(R\sigma_1\frac{z'}{\lambda_1^2}\right)' - (P^* + P_b^*)rr',$$

$$P^* + P_b^* = \frac{P + P_b}{H} = \frac{-1}{\xi}\left[\frac{r'\rho R}{rz'^2}\dot{v} + \frac{\rho R}{2z'^2}\dot{v}' + \mathcal{P}\right], \qquad \xi = 1 + \frac{r'^2}{z'^2},$$

$$\mathcal{P} = \frac{1}{rz'}\left[\left(R\sigma_1\frac{r'}{\lambda_1^2}\right)' - \frac{r'}{z'}\left(R\sigma_1\frac{z'}{\lambda_1^2}\right)' - \frac{\sigma_2}{\lambda_2}\right] - \frac{\rho R}{rz'^2}(q-v)\left(\frac{z'q - vr' - \frac{1}{2}rv'}{z'}\right)' - \frac{\rho R}{rz'^3}\left(r'q - vr' - \frac{1}{2}rv'\right)\left(\frac{1}{2}v' - q'\right).$$

Here terms in braces may be included for effectiveness of calculations (see below).

For the case of controlled pressure the three-layer scheme was mainly used [6]. But for cases of fluid-filled and gas-filled tubes boundary numerical instability is observed. For fluid-filled tubes this instability may be corrected by inclusion of dissipative zones near boundaries. An additional term with second-order derivative must be included in the equation for v. If the space step is decreased then the coefficient before this term must be increased. Good results are achieved if this term is approximated by the DuFort-Frankel method. This is not convenient for calculations. That is why two-layer predictor-corrector scheme was used for the cases of fluid-filled and gas-filled tube. This scheme possesses non-correct scheme dissipation. The growth of disturbances is observed for long-time calculations. But nevertheless convergence is achieved under a stability condition that is more sharper than the natural Courant condition. This effect may be removed if additional terms with higher-order derivatives are included. The magnitudes of the coefficients of this additional terms are taken to be proportional to the square of the space step. Hence the order of approximation of the scheme is preserved. Such approach may be used non only for the case of equations of tube. Estimations of the required magnitudes of the coefficients in the general case are made. Note that for Runge-Kutta method half of time step predictor is used. Growth of disturbances may be excluded if the predictor is applied for the entire time step. But such numerical scheme possesses great numerical dissipation and its application for non-dissipative and weakly dissipative problems is not convenient.

Numerical solutions of the Riemann problem are found. Results of calculations correspond to the theory of reversible and weakly dissipative shocks. Typical types of shock structures [4] are found for all cases. Examples of graphs for the case of controlled pressure are given in [6]. Here graphs of solutions for fluid-filled tube are presented in fig. 1, $v_0 = 0.1$, $R = 1$, $\rho = 1$, $\rho_f^* = 1$, $\tilde{\mu} = 2$, $J_m = 30$, $z'_{-\infty} = 1.1$, $t = 1000$, $b = 0.1$. Here v_0 is initial velocity of fluid. Non-stationary chaotic structure of mainly longitudinal elastic waves moving to the left, solitary wave structure of mainly transversal elastic waves moving to the left, kink, solitary wave structure of mainly transversal elastic waves moving to the right and structure with emission of mainly longitudinal elastic waves moving to the right are noticed.

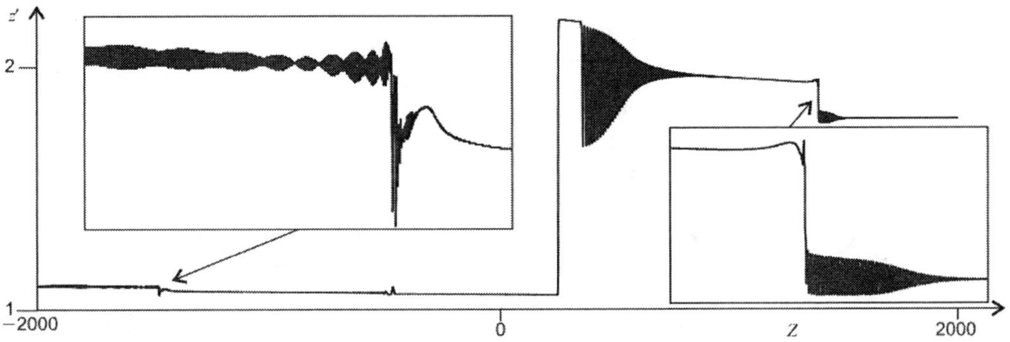

Figure 1. Solution of Riemann problem

The results of the investigation show that the theory of non-dissipative shocks is applicable for the equations of tubes with elastic walls. Effective numerical method of correction of the two-step scheme is developed. It can be used for reversible equations if the three-layer scheme is not applicable.

Acknowledgement

This work was supported by Russian Foundation for Basic Research, grant 15-01-04357a.

References

[1] M. Epstein and J. W. Johnston, Proc. Roy. Soc. London A **457**, 2009, 1195–1213 (2001)
[2] Y. B. Fu and S. P. Pearce, IMA J. Appl. Math. **75**, 4, 581–602 (2010)
[3] Y. B. Fu and A. Il'ichev, IMA J. Appl. Math. **75**, 2, 257–268 (2010)
[4] I. B. Bakholdin, J. Appl. Math. Mech. **78**, 6, 599–612 (2014)
[5] A. G. Kulikovskii and E. I. Sveshnikova, *Nonlinear Waves in Elastic Media* (CRC Press, Boca Raton, New York, London, Tokio, 1995)
[6] I. B. Bakholdin, J. Comp. Math. and Math. Phys. **55**, 11, 1884–1898 (2015)
[7] I. B. Bakholdin, Prikl. Mat. Mekh. **81**, 5, 593–609 (2017) (J. Appl. Math. Mech. in press)
[8] Y. B. Fu, J. L. Liu, and G. S. Francisco,J. of the Mechanics and Physics of Solids **90**, 45–60 (2016)

EPJ Web of Conferences **173**, 03005 (2018) https://doi.org/10.1051/epjconf/201817303005
Mathematical Modeling and Computational Physics 2017

Development of Continuum-Atomistic Approach for Modeling Metal Irradiation by Heavy Ions

Balt Batgerel[1,3], *Stefka* Dimova[2], *Igor* Puzynin[1], *Taisia* Puzynina[1], *Ivan* Hristov[2], *Radoslava* Hristova[2], *Zafar* Tukhliev[1,⋆], and *Zarif* Sharipov[1]

[1] *Joint Institute for Nuclear Research, 141980, Dubna, Russia*
[2] *Sofia University "St. Kliment Ohridski", Sofia, Bulgaria*
[3] *Mongolian State University of Science and Technology, Ulaanbaatar, Mongolia*

Abstract. Over the last several decades active research in the field of materials irradiation by high-energy heavy ions has been worked out. The experiments in this area are labor-consuming and expensive. Therefore the improvement of the existing mathematical models and the development of new ones based on the experimental data of interaction of high-energy heavy ions with materials are of interest. Presently, two approaches are used for studying these processes: a thermal spike model and molecular dynamics methods. The combination of these two approaches – the continuous-atomistic model – will give the opportunity to investigate more thoroughly the processes of irradiation of materials by high-energy heavy ions. To solve the equations of the continuous-atomistic model, a software package was developed and the block of molecular dynamics software was tested on the heterogeneous cluster HybriLIT.

1 Introduction

The continuous-atomistic approach for modeling the interaction of high-energy heavy ions (HEHI) with various materials combines the equation of thermal conductivity of the electron gas of the thermal spike (TS) model [1, 2] and the equations of motion of atoms of the molecular dynamics (MD) method [3]. This combination, in contrast to the MD method, allows one to consider processes over a wide range of energy of the incident heavy ion to the target (since inelastic energy losses predominate at high energies of the incident ion and the MD method does not work), and unlike the TS model, it allows to obtain more detailed information on the simulated system. A continuous-atomistic model (CAM) [4] was proposed in 2003 to describe the effect of short-pulsed lasers on metal targets. The use of combined methods requires the development of new numerical schemes, algorithms and program packages. Since the combined models consist of equations of two different classes, special attention is to be paid to the consistency of the numerical methods for both classes of equations. Modeling within the framework of the combined approach requires considerable computational resources, and thus it is necessary to develop new parallel algorithms for multiprocessor computer systems. To carry out the simulation, a software package was developed with the possibility of using it on high-performance computing systems [5].

⋆e-mail: zafar@jinr.ru

The aim of the present work is to simulate the irradiation of metal targets by HEHI in the framework of the CAM, in terms of the parameters of the source function and the electron-phonon interaction coefficient, as well as the development and refinement of the capabilities of the CAM to describe the available experimental data.

2 Formulation of the problem

The continuous-atomistic model [4], which describes the processes in a metal film under the action of a short-pulse laser, is taken as basis of the CAM. The model described below assumes a continuous equation for the thermal conductivity of the electronic subsystem of the TS model in Cartesian coordinates and equations of motion of the atoms within the MD method:

$$C_e(T_e)\frac{\partial T_e}{\partial t} = \mathrm{div}\,(\lambda_e \mathrm{grad}\,(T_e) - g(T_e)(T_e - T_l) + A_e(x, y, t), \tag{1}$$

$$m_i \frac{d^2\mathbf{r}_i}{dt^2} = \mathbf{F}_i + \zeta m_i \mathbf{v}_i^T, \tag{2}$$

where

$$\zeta = \frac{1}{n}\sum_{k=1}^{n} g(T_e)V_n(T_e^k - T_l) / \sum_i m_i(\mathbf{v}_i^T)^2,$$

$$A_e(x, y, t) = b\,S_{\mathrm{inel}}(0)\exp\left(-\frac{(t-t_0)^2}{2\sigma_t^2}\right)\exp\left(-\frac{\sqrt{(x-x_0)^2 + (y-y_0)^2}}{r_0}\right).$$

A detailed description of equation (1) is given in [2], and of equations 2 – in [4].

In this paper we consider, as an example, the irradiation of nickel by uranium ions of energy of 700 MeV for different values of the parameter r_0 of the source function $A_e(x, y, t)$ and of the electron-phonon interaction coefficient $g(T_e)$. The choice of the target and the incident ion is due to the fact that the authors of [2] performed a simulation of the same ion-target combination within the TS model. The use of the two-dimensional CAM is motivated by the fact that the processes are being modeled in a thin (1–10 nm thick) material.

3 Numerical methods

To solve the equation (1), the computational domain is discretized by a uniform grid over which finite-differences are used [6]. The Linked-List Cell MD Algorithm [7] is chosen to integrate the equations (2) of the particle motion. An important step is to calculate the temperature field of the irradiated material. To this aim, the computational domain is divided into superimposed cells so that in each cell the number of particles is 100–1000, and the temperature in each cell is calculated. In contradistinction to [4], this approach makes possible the increase of the density of points of the temperature field of the target.

In the computational schemes for solving equations (1)–(2) it is necessary to preserve:

1) the consistency of the steps in time;

2) the temperature relationship at the node points of the finite-difference method and in the cells of the calculating region of the MD method.

To solve the system (1)–(2), a software package TS-MD was implemented on a multiprocessor system with shared memory within the framework of the OpenMP standard. Results characterising the efficiency of the MD part of the TS-MD software are given in [5].

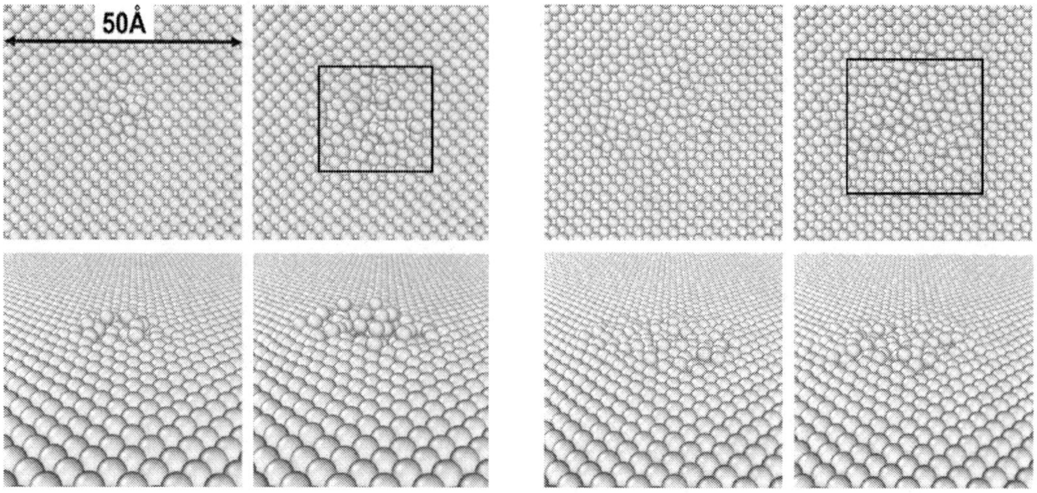

(a) (b)

Figure 1. The dynamics of the destruction of the crystal lattice for r_0 = 0.5 nm (a) and r_0 = 1 nm (b) at time instants t_1 = 100 fs (left) and t_2 = 300 fs (right). The upper pictures are top views and the lower pictures are views at an angle of 45 degrees.

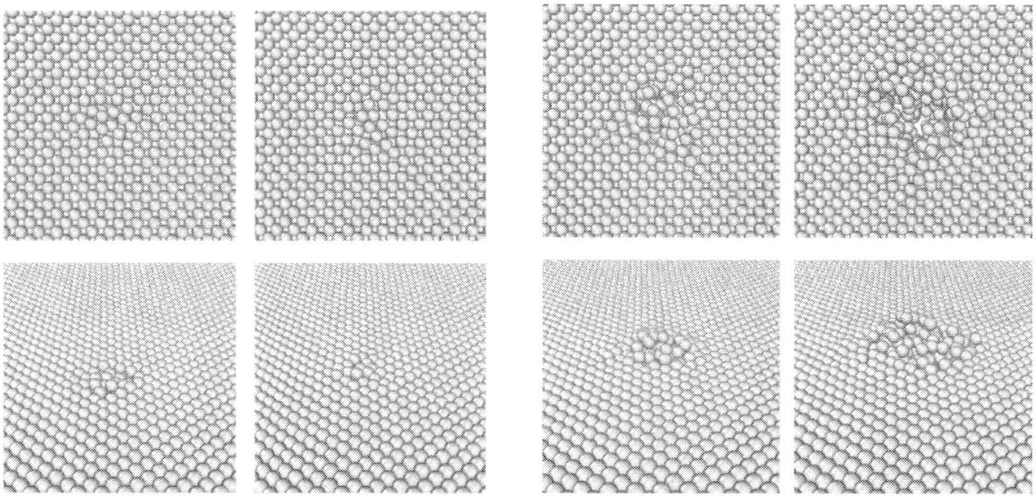

(a) (b)

Figure 2. Dynamics of the destruction of the crystal lattice for parameters $g = 0.5g_{Ni}$ (a) and $g = 2g_{Ni}$ (b) at instants t_1 = 100 fs (left) and t_2 = 300 fs (right). The upper pictures are top views and the bottom pictures are views at an angle of 45 degrees.

4 Results and discussion

To solve the equations of the CAM, a software package was developed the block of MD software of which was tested on the heterogeneous cluster HybriLIT [8].

The simulation of the process of interaction of uranium ions with a nickel target was carried out as follows. A nickel sample of size $35 \times 35 \times 1.1$ nm with 140700 particles was selected as a target. The physical parameters of equation (1) for nickel are taken from [2]. In the MD method, the embedded atom model (EAM) potential is used as the interparticle potential. The most important and interesting result in the interaction of HIHE with solids is the formation of spatial fractures of the crystal lattice-tracks [2]. We have investigated this process in two cases. In the first case simulation was made for two values of the parameter r_0 (0.5 nm, 1 nm) of the source function and for the electron-phonon interaction coefficient $g = g_{Ni}$. In Fig. 1 a) $r_0 = 0.5$, b) $r_0 = 1$ the results are given respectively at instants $t_1 = 100$ fs (left) and $t_2 = 300$ fs (right). The upper figures are top views and the lower figures are views at an angle of 45 degrees. In these figures, the square is the area of destruction of the crystal lattice. It can be seen that at smaller values of the parameter r_0, the crystal lattice is severely damaged in a small region. For larger values the damage area increases, but the nature of the fractures does not change significantly. These results show the dependence of the model on the parameter r_0. When this parameter is changed, the density of energy distribution over space changes. The parameter r_0 of the source function for HIHE is estimated to be within 1–2.5 nm [2]. In the second stage, we change the electron-phonon interaction coefficient: $g_1 = 0.5g_{Ni}$, $g_2 = 2g_{Ni}$. The results obtained are shown in Fig. 2 a), b), respectively, at time instants $t_1 = 100$ fs and $t_2 = 300$ fs. The results show that the increase of the electron-phonon interaction coefficient g results in the increase of the energy transfer to the lattice.

5 Conclusion

Results of modeling the interaction of a flux of uranium ions of energy 700 MeV with a nickel target in terms of the parameters of the source function and the electron-phonon interaction coefficient are obtained using the developed software package of the CAM. The results of modeling, let us infer the following conclusions:

1. Changing the parameter r_0 in the source function strongly affects the results through the energy dependence of the incident ion and the type of ion. This factor becomes significant when the ion penetrates in depth the target since the moving ion loses more than 70–80 % of its initial energy in the first half of the path.

2. The choice of the coefficient of electron-phonon interaction is determined by straightforward comparison with the experimental data, since it corresponds to the amount of transferred energy from electrons to the lattice.

Acknowledgements

The work was financially supported by the RFBR grant No. 17-01-00661-a and partially supported by a grant of the Plenipotentiary Representative of the Republic of Bulgaria at the JINR. The work of I. Hristov, R. Hristova, S. Dimova is partially supported by the NSF of BMSE under Grant DFNI-I02/9 and Grant DFNI-I02/8.

References

[1] I.M. Lifshits, M.I. Kaganov, and L.V. Tanatarov, Atomic Energy **6**, 391–402 (1959)
[2] I.V. Amirkhanov, A. Hofman, I.V. Puzynin et. al., PEPAN **37**, N.6, 837–866 (2006)
[3] H.T. Holmurodov, I.V.Puzynin et all., PEPAN. **34**, N.2, 472 (2003)
[4] D. Ivanov and L. Zhigilei, Phys. Rev. **B68**, 064114 (2003)
[5] S. Dimova, I. Puzynin, T. Puzynina et. al., CEUR Workshop Proceedings **1787**, 184 (2016)
[6] A.A. Samarsky and A.V. Gulin, *Numerical methods* (Nauka, Moscow, 1989) 432
[7] http://cacs.usc.edu/education/cs596/01-1LinkedListCell.pdf
[8] http://hybrilit.jinr.ru/
[9] A.V. Holian et al., Phys. Rev. **A43**, N.6, 2655 (1991)

Generation and Analysis of a New Implicit Difference Scheme for the Korteweg-de Vries Equation

Yuri Blinkov[1,*], *Vladimir* Gerdt[2,**], and *Konstantin* Marinov[3,***]

[1] *Saratov State University, Astrakhanskaya Street 83, 410012 Saratov, Russia*
[2] *Joint Institute for Nuclear Research, Joliot-Curie 6, 141980 Dubna, Russia*
[3] *International University "Dubna", Universitetskaya 19, 141980 Dubna, Russia*

Abstract. In this paper we apply our computer algebra based algorithmic approach to construct a new finite difference scheme for the two-parameter form of the Korteweg-de Vries equation. The approach combines the finite volume method, numerical integration and difference elimination. Being implicit, the obtained scheme is consistent and unconditionally stable. The modified equation for the scheme shows that its accuracy is of the second order in each of the mesh sizes. Using exact one-soliton solution, we compare the numerical behavior of the scheme with that of the other two schemes known in the literature and having the same order of accuracy. The comparison reveals numerical superiority of our scheme.

1 Introduction

In our previous paper [1] the general algorithmic approach suggested in [2] for generating difference schemes was worked out for evolution equations of the form

$$u_t = \partial_x (P + a\, u_{m-1}), \quad u_m := \partial_x^m u, \quad P \in \mathcal{R}, \quad \mathrm{ord}_{\partial_x}(P) < m-1, \quad m \geq 2, \quad a \in \mathbb{R}, \tag{1}$$

where $\mathcal{R}$ is the ordinary differential polynomial ring with undeterminate u and derivative ∂_x. We denote the differential order of P by $\mathrm{ord}_{\partial_x}(P)$. The (differential) coefficient field of $\mathcal{R}$ is assumed to be a field of rational functions in parameters (constants) the coefficients of which are rational numbers.

In the present paper we generate and study a new difference scheme for the two-parameter form of the Korteweg-de Vries (KdV) equation

$$u_t + (P + \beta u_{xx})_x = 0, \quad P = \frac{\alpha}{2} u^2, \quad \alpha, \beta \in \mathbb{R} \tag{2}$$

which belongs to the class (1). Eq. (2) was extensively studied in the literature both analytically and numerically (see monograph [3] and refereces therein). It is integrable by the inverse scattering transform method, possesses a rather wide class of exact solutions and infinitely many local conservation laws. Besides, various difference schemes have been derived for its numerical solution. The paper is organized as follows. In Sect. 2 we describe the generation of a new difference scheme by our method and prove its convergence. In Sect. 3 we compare the numerical behavior of the obtained scheme with

[*] e-mail: BlinkovUA@info.sgu.ru
[**] e-mail: gerdt@jinr.ru
[***] e-mail: marinov.kohctahtih@gmail.com

that of two other schemes, see [3], having the same order of accuracy. The exact one-soliton solution of (2) is used as a benchmark.

2 Generation of a difference scheme for the KdV equation

To generate a difference scheme for Eq. (2) we follow [1, 2] and use a regular grid with $t_{n+1} - t_n = \tau$, $x_{j+1} - x_j = h$. First, we convert Eq. (2) into the integral conservation law form

$$u_t + (P + \beta u_{xx})_x = 0 \implies \oint_\Gamma -(P + \beta u_{xx})\,dt + u\,dx = 0, \tag{3}$$

where Γ is the following rectangular integration contour ("control volume")

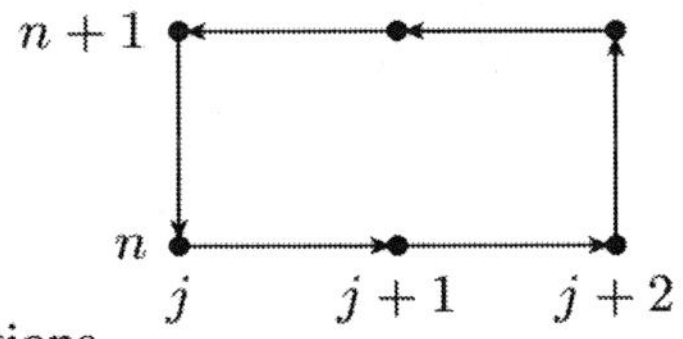

and add to Eq. (3) the integral relations

$$\int_{x_j}^{x_{j+2}} u_{xx}\,dx = u_x(t, x_{j+2}) - u_x(t, x_j)\,, \quad \int_{x_j}^{x_{j+1}} u_x\,dx = u(t, x_{j+1}) - u(t, x_j)\,. \tag{4}$$

Then, we evaluate the integrals (3) and (4) numerically. To evaluate (3), we use the trapezoidal rule for integration over t and the midpoint rule for integration over x. The left integral in (4) is evaluated by the trapezoidal rule and the right integral by the midpoint rule. This results in a system of difference equations for the grid functions $u_j^n, (u_x)_j^n, (u_{xx})_j^n$. The algebraic (difference) elimination of $(u_x)_j^n$ and $(u_{xx})_j^n$ from this system yields a difference scheme for Eq. (2). Such elimination can be done by using Maple (see [1]).

The derived difference scheme

$$\begin{gathered}\frac{u_j^{n+1} - u_j^n}{\tau} + \frac{\alpha\left((u^2)_{j+1}^{n+1} - (u^2)_{j-1}^{n+1}\right) + \alpha\left((u^2)_{j+1}^{n} - (u^2)_{j-1}^{n}\right)}{8h} + \\ \frac{\beta\left(u_{j+2}^{n+1} - 2u_{j+1}^{n+1} + 2u_{j-1}^{n+1} - u_{j-2}^{n+1}\right) + \beta\left(u_{j+2}^{n} - 2u_{j+1}^{n} + 2u_{j-1}^{n} - u_{j-2}^{n}\right)}{4h^3} = 0\end{gathered} \tag{5}$$

can be solved by the simple iteration method (cf. [1]).

Using the library SymPy [4] written in Python we computed the modified equation (or the first differential approximation [5]) of the scheme

$$\begin{aligned}&u_t + \alpha u u_1 + \beta u_3 + h^2\left[\alpha\left(\frac{1}{6}uu_3 + \frac{1}{2}u_1u_2\right) + \frac{\beta}{4}u_5\right] + \tau^2\left[\alpha^3\left(\frac{1}{12}u_3 + \frac{3}{4}u^2u_1u_2 + \frac{1}{2}uu_1^3\right) + \right.\\ &\alpha^2\beta\left(\frac{1}{4}u^2u_5 + \frac{7}{4}uu_1u_4 + \frac{11}{4}uu_2u_3 + \frac{9}{4}u_1^2u_3 + \frac{11}{4}u_1u_2^2\right) + \alpha\beta^2\left(\frac{1}{4}uu_7 + u_1u_6 + \right.\\ &\left.\left.\frac{9}{4}u_2u_5 + \frac{7}{2}u_3u_4\right) + \frac{\beta^3}{12}u_9\right] + O(\tau^4, \tau^2h^2, h^4) = 0\,, \quad u_k := \partial_x^k u\,.\end{aligned} \tag{6}$$

The scheme (5) is implicit, and hence it is unconditionally stable. Eq. (6) shows the consistency of (5) with (2). Therefore, (5) is convergent (cf. [6]).

3 Numerical behavior of the generated scheme

We compare the numerical behavior of scheme (5) with that of two other schemes from [3]. Similar to the scheme (5), these two schemes provide $O(\tau^2, h^2)$ approximation to Eq. (2).

Scheme I. The explicit scheme ([3], Eq. 1.80):

$$u_i^{n+1} = u_i^{n-1} - \frac{\alpha\tau}{h} u_i^n \left(u_{i+1}^n - u_{i-1}^n\right) - \frac{\beta\tau}{h^3}\left(u_{i+2}^n - 2u_{i+1}^n + 2u_{i-1}^n - u_{i-2}^n\right) \tag{7}$$

stable for

$$\tau \leq \frac{2h^3}{3\sqrt{3}\beta} \cong 0.384\frac{h^3}{\beta}\,.$$

The modified equation reads

$$u_t + \alpha u u_1 + \beta u_3 + h^2\left(\frac{\alpha}{6}uu_3 + \frac{\beta}{4}u_5\right) - \tau^2\Big[\alpha^3\Big(\frac{1}{6}u^3u_3 + \frac{3}{2}u^2u_1u_2 + uu_1^3\Big) + \alpha^2\beta\Big(\frac{1}{2}u^2u_5 + \frac{7}{2}uu_1u_4 + \frac{11}{2}uu_2u_3 + \frac{9}{2}u_1^2u_3 + \frac{11}{2}u_1u_2^2\Big) + \alpha\beta^2\left(\frac{1}{2}uu_7 + 2u_1u_6 + \frac{9}{2}u_2u_5 + 7u_3u_4\right) + \frac{\beta^3}{6}u_9\Big] + O(\tau^4, h^4) = 0\,. \tag{8}$$

Scheme II. The implicit scheme ([3], Eq. 1.96):

$$\frac{u_j^{n+1} - u_j^n}{\tau} + \frac{\alpha}{4h}\left[u_j^n\left(u_{j+1}^{n+1} - u_{j-1}^{n+1}\right) + u_j^{n+1}\left(u_{j+1}^n - u_{j-1}^n\right)\right] + \frac{\beta}{4h^3}\left(u_{j+2}^{n+1} - 2u_{j+1}^{n+1} + 2u_{j-1}^{n+1} - u_{j-2}^{n+1} + u_{j+2}^n - 2u_j^{n+1} + 2u_{j-1}^n - u_{j-2}^n\right) = 0 \tag{9}$$

the modified equation of which is given by

$$u_t + \alpha u u_1 + \beta u_3 + h^2\left(\frac{\alpha}{6}uu_3 + \frac{\beta}{4}u_5\right) + \tau^2\Big[\alpha^3\Big(\frac{1}{12}u^3u_3 + \frac{1}{4}u^2u_1u_2\Big) + \frac{\alpha^2\beta}{4}\Big(u^2u_5 + 5uu_1u_4 + 7u_1^2u_3 + 9uu_2u_3 + 11u_1u_2^2\Big) + \alpha\beta^2\Big(\frac{1}{4}uu_7 + u_1u_6 + \frac{9}{4}u_2u_5 + 3u_3u_4\Big) + \frac{\beta^3}{12}u_9\Big] + O(\tau^4, \tau^2h^2, h^4) = 0\,. \tag{10}$$

We performed numerical comparison of the difference schemes (5), (7) and (9) by using the Python-based library SciPy [7]. As a benchmark, we considered the initial value problem $u(0, x) = u_{\text{exact}}(0, x)$ for the exact one-soliton solution

$$u_{\text{exact}}(t, x) = \frac{2k_1^2}{\cosh\left(k_1(x - 4k_1^2 t)\right)^2} \tag{11}$$

to the KdV equation (2) with $\alpha = 6$, $\beta = 1$ and $k_1 = 0.4$. In so doing, for all schemes we fixed the spacial mesh step $h = 0.25$. Then, for the explicit scheme (7) we fixed $\tau = 0.37h^3/\beta$, and for the implicit schemes (5), (9) we considered the following two values of τ: $\tau = 0.1h$ and $\tau = h$. For these values of the mesh sizes we constructed a numerical solution in the interval $-50 \leq x \leq 50$ with periodic boundary conditions (cf. [3], p.49). The numerical accuracy was estimated by the Frobenius norm

$$\left\|u - u_{\text{exact}}\right\|_F =: \sqrt{\sum_n \sum_j \left|u_j^n - u_{\text{exact}}(t_n, x_j)\right|^2}\,. \tag{12}$$

The time dynamics of (12) for the schemes (5), (7) and (9) shown below (left pictures) demonstrates the superiority of our scheme (5) over the two other schemes. It is confirmed by the numerical values of the left-hand sides of the modified equations (6), (8) and (10) for the one-soliton solution (11) at $t = 0$ (right pictures) τ and the equality of the terms of order h^2 in Eq. (8) and Eq. (10).

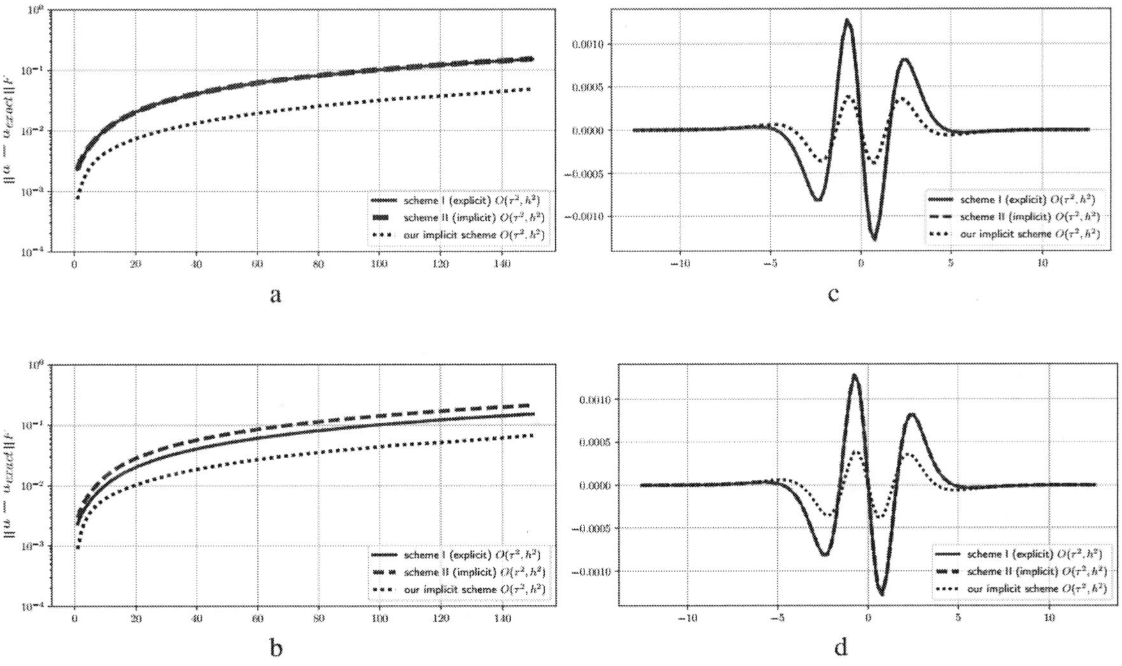

Figure 1. The time dynamics of descrepancy (12) for with $\tau = h/10$ (a) and $\tau = h$ (b) for the implicit schemes, and the values (c) and (d) of the left-hand sides of the modified equations for these values of τ, respectively.

4 Conclusion

By applying the modern methods of computer algebra related to differential and difference algebra, we generated and analysed the new implicit second order finite difference scheme (5) for the classical KdV equation. The symbolic analysis based on the construction of modified equations combined with numerical experiments demonstrated superiority of the scheme over the two other second order schemes known in the literature.

5 Acknowledgements

The work is supported in part by the Russian Foundation for Basic Research (grant 16-01-00080).

References

[1] V.P. Gerdt, Yu.A. Blinkov, and K.B. Marinov, Program. Comput. Soft. **43** (2), 84–89 (2017)
[2] Yu.A. Blinkov, V.P. Gerdt, and V.V. Mozzhilkin, SIGMA **2** (051) (2006) arXiv:math.RA/0605334
[3] V.Yu. Belashov and S.V. Vladimirov, *Solitary Waves in Dispersive Complex Media. Theory· Simulation · Applications* (Springer-Verlag, Berlin, 2005) 292 pp.
[4] `http://www.sympy.org/en/`
[5] Y.I. Shokin, *The Method of Differential Approximation* (Springer-Verlag, Berlin, 1983)
[6] E.E. Rosinger, *Nonlinear Equivalence, Reduction of PDEs to ODEs and Fast Convergent Numerical Methods* (Pitman, Boston, 1982)
[7] `http://scipy.org`

EPJ Web of Conferences **173**, 03007 (2018) https://doi.org/10.1051/epjconf/201817303007
Mathematical Modeling and Computational Physics 2017

Peculiarities of Matrix-Element Calculations with Few Coulomb Functions for Particles' Scattering Processes

Ochbadrakh Chuluunbaatar[1,2,⋆], *Konstantin* Kouzakov[3], and *Yuri* Popov[4,1]

[1] *Joint Institute for Nuclear Research, Dubna, Moscow region 141980, Russia*
[2] *Institue of Mathematics, National University of Mongolia, Ulan-Bator, Mongolia*
[3] *Department of Nuclear Physics and Quantum Theory of Collisions, Faculty of Physics, Lomonosov Moscow State University, Moscow 119991, Russia*
[4] *Skobeltsyn Institute of Nuclear Physics, Lomonosov Moscow State University, Moscow 119991, Russia*

Abstract. Specific features of the matrix-element calculations in the momentum-space representation are discussed for the single ionization of the He atom by fast proton impact in the case when the Coulomb interactions of all three charged fragments in the final state are taken into account. It is shown that a "soft" smoothing of the Coulomb singularity does not affect the accuracy of the calculations in a certain region of the smoothing parameter values.

1 Introduction

The computation of integrals containing Coulomb functions is typically encountered in the problems of scattering involving charged particles. A well-known example of such scattering problems is the ionization of an atomic target by electron or ion impact. In the last two decades, due to the remarkable progress of the cold target recoil ion momentum spectroscopy (COLTRIMS) [1], much interest has been directed toward the study of the single ionization of atomic targets by fast ions in the perturbative regime. Different theoretical approaches, in particular, those based on the plane wave first Born approximation (PWFBA), can be very robustly tested by comparing their predictions for the fully differential cross sections (FDCS) with the COLTRIMS data.

Currently there are two main theoretical approaches to the study of the ionization of a quantum target under the impact of a positive ion. These are the continuum distorted wave-eikonal initial state (CDW-EIS) method in the semiclassical and fully quantum formulations respectively [2, 3]. In the semiclassical approach, one treats the projectile ion as a classical particle moving along a straight-line trajectory with an impact parameter ρ and velocity v_p, which induces a time-dependent perturbation of the quantum target. In the quantum approach, one solves stationary quantum scattering problems for all particles involved in the ionization reaction.

It is to be mentioned that the CDW-EIS model involves three or more Coulomb functions or their asymptotic expressions in the matrix-element integration. They enter the descriptions of the interactions of the charged fragments both in the initial and in the final channels of the reaction. The matrix element itself can be written either in the real space or in the momentum space. There

⋆e-mail: chuka@jinr.ru

are calculations in real-space representation with three and even four Coulomb functions (see, for instance, [4]). However, due to the complexity of the involved integrations, only simple trial wave functions of the target are employed. In the momentum space one has less three integrations due to the conservation laws but, at the same time, the Coulomb functions in the momentum-space representation have singularities with which one has to cope with and to perform correctly the regularization of the integrals.

In this work, we present calculations of the FDCS beyond the PWFBA theory, in order to explain the shift of the measured binary and recoil peaks in the scattering plane by a few degrees towards the incident proton direction with respect to the PWFBA predictions [5]. To this purpose we examine the well-known 3C (or Brauner-Briggs-Klar) model [6] and three different helium ground-state wave functions. Atomic units (a.u., $\hbar = e = m_e = 1$) are used throughout unless otherwise specified.

2 Theory

Let us recall some well-known properties of a continuum Coulomb wave function. This function is a solution of the stationary Schrödinger equation and in the real-space representation it has the form

$$\tilde{\phi}^{\pm}(\vec{k},\vec{r};Z) = e^{-\pi\eta/2}\Gamma(1 \pm i\eta)e^{i\vec{k}\cdot\vec{r}}\,{}_1F_1(\mp i\eta, 1; \pm ikr - i\vec{k}\cdot\vec{r}), \qquad \tilde{\phi}^{-}(\vec{k},\vec{r};Z) = \tilde{\phi}^{+*}(-\vec{k},\vec{r};Z). \quad (1)$$

In (1), $\eta = Z/v$, $Z = Z_1Z_2$, where Z_1 and Z_2 are the charges of the colliding fragments, v, $\vec{k}$ and $\vec{r}$ are respectively the relative velocity, momentum and position of the colliding fragments, and ${}_1F_1$ is the confluent hypergeometric function.

To calculate the Fourier transform of the Coulomb function one usually introduces a smoothing parameter λ, for example, as follows:

$$\begin{aligned}\phi^{\pm}(\vec{k},\vec{p};Z) &= \lim_{\lambda\to+0}\int d^3r\, e^{-\lambda r - i\vec{p}\cdot\vec{r}}\,\tilde{\phi}^{\pm}(\vec{k},\vec{r};Z) = 8\pi e^{-\pi\eta/2}\Gamma(1\pm i\eta) \\ &\times \lim_{\lambda\to+0}\left[\lambda(1\pm i\eta)\frac{[(\lambda\mp ik)^2+p^2]^{\pm i\eta}}{[(\vec{p}-\vec{k})^2+\lambda^2]^{2\pm i\eta}} \mp i\eta(\lambda\mp ik)\frac{[(\lambda\mp ik)^2+p^2]^{-1\pm i\eta}}{[(\vec{p}-\vec{k})^2+\lambda^2]^{1\pm i\eta}}\right]. \quad (2)\end{aligned}$$

It is well known that in the limit $\lambda \to +0$ the integral of a product involving the first strongly singular term in the square brackets and any smooth function yields zero, and the second term is used as a wave function in the momentum space. The latter term is also singular that makes certain difficulties to deal with, a reason for which most authors prefer to work in the real-space representation.

Let us turn to this problem. For the sake of clarity, we start with some preliminary definitions. We designate the incident proton momentum by $\vec{p}_i$ and the final momentum by $\vec{p}_s$, the electron momentum by $\vec{k}_e$, the final ion momentum by $\vec{K}_{\text{ion}}$. The momentum transfer is $\vec{Q} = \vec{p}_i - \vec{p}_s = \vec{k}_e + \vec{K}_{\text{ion}}$. In the experiment for which the calculations have been carried out, $E_p = 1$ MeV, i.e. $v_p = \sqrt{2E_p/m_p} = 6.35$ a.u., $E_e = 6.5$ eV. From the energy conservation one obtains the z component of the momentum transfer as $Q_z = (-\varepsilon_0^{\text{He}} + \varepsilon_0^{\text{He}^+} + E_e)/v_p = 0.18$ a.u. (the z axis is directed along the initial proton momentum $\vec{p}_i$). The transverse component is $Q_\perp \approx m_p v_p \theta_s = 0.73$ a.u. (θ_s is the proton scattering angle).

The general expression for the triple differential cross section (TDCS) is

$$\frac{d^3\sigma}{dE_e\, d\Omega_e\, d\Omega_s} = k_e\frac{m_p^2}{(2\pi)^5}|T_{\text{fi}}|^2. \quad (3)$$

The final state of the discussed reaction contains three charged fragments: the proton, electron, and He^+ ion. Below we consider a correlated 3C final-state wave function [6]. In this case the amplitude

$T_{\rm fi}$ splits into the sum of three integrals, $T_{\rm fi} = A_1 + A_2 + A_3$, where

$$A_1 = 2\sqrt{2}\int \frac{d^3p}{(2\pi)^3}\,\phi^{-*}(\vec{v}_p - \vec{k}_e, \vec{p} + \vec{v}_p - \vec{k}_e; -1) I(\vec{p}_r, \vec{Q} + \vec{p}_r - \vec{p}; 1) G(\vec{k}_e, \vec{p}, 0), \tag{4}$$

$$A_2 = -\sqrt{2}\int \frac{d^3p}{(2\pi)^3}\, I(\vec{v}_p - \vec{k}_e, \vec{p} + \vec{v}_p - \vec{k}_e; -1)\phi^{-*}(\vec{p}_r, \vec{Q} + \vec{p}_r - \vec{p}; 1) G(\vec{k}_e, \vec{p}, 0), \tag{5}$$

$$A_3 = -4\pi\sqrt{2}\int \frac{d^3p_1}{(2\pi)^3}\frac{d^3p_2}{(2\pi)^3 p_2^2}\,\phi^{-*}(\vec{v}_p - \vec{k}_e, \vec{p}_1 + \vec{v}_p - \vec{k}_e; -1) \times \phi^{-*}(\vec{p}_r, \vec{Q} + \vec{p}_r - \vec{p}_1 - \vec{p}_2; 1) G(\vec{k}_e, \vec{p}_1, \vec{p}_2). \tag{6}$$

In (4)–(6)

$$\vec{p}_r = [(4m_p + 1)\vec{p}_s - m_p\vec{K}]/(5m_p + 1) \approx (4m_p/5)\vec{v}_p + (1/5)\vec{k}_e - \vec{Q},$$

and also we have used the following notations:

$$I(\vec{p}, \vec{q}; Z) = \lim_{\lambda\to+0}\int \frac{d^3r}{r} e^{-\lambda r}\tilde{\phi}^{-*}(\vec{p}, \vec{r}; Z) e^{i\vec{q}\cdot\vec{r}} = \lim_{\lambda\to+0} 4\pi\, e^{-\pi\eta/2}\Gamma(1 + i\eta)\frac{[q^2 - (p + i\lambda)^2]^{i\xi}}{[(\vec{q} - \vec{p})^2 + \lambda^2]^{(1+i\xi)}}, \tag{7}$$

and

$$G(\vec{k}, \vec{q}_1, \vec{q}_2) = \int d^3r_1 d^3r_2\, \tilde{\phi}^{-*}(\vec{k}, \vec{r}_1; -1) e^{i\vec{q}_1\cdot\vec{r}_1}\varphi_0^{\rm He^+}(\vec{r}_2) e^{i\vec{q}_2\cdot\vec{r}_2}\Phi_0^{\rm He}(\vec{r}_1, \vec{r}_2). \tag{8}$$

For most of the He ground-state models the function $G(\vec{k}, \vec{q}_1, \vec{q}_2)$ is an analytical function or an integral of much lower dimension.

Three initial trial helium wave functions $\Phi_0^{\rm He}$ are employed: a weakly correlated Roothaan-Hartree-Fock (RHF) [7] function, $\varepsilon_0^{\rm RHF} = -2.8617$ a.u., a simple Silverman-Platas-Matsen (SPM) function [8] of the configuration interaction family, $\varepsilon_0^{\rm SPM} = -2.8952$ a.u., and a strongly correlated function (CF) [9] which explicitly depends on the r_{12} distance between the electrons in helium, $\varepsilon_0^{\rm CF} = -2.903724$ a.u. The helium energy of the last function is very close to the experimental value, $\varepsilon_0^{\rm He} = -2.903724$ a.u.

3 Results and discussion

The purpose of the present study is to investigate the λ dependence of the TDCS. In this regard it should be noted that the integrand functions in (4) and (5) in the vicinity of the weak singular points $\vec{p} = 0$ and $\vec{p} - \vec{Q} = 0$ at $\lambda \to 0+$ have the behaviors

$$\frac{\lambda}{(\lambda^2 + p^2)^2}\frac{1}{[\lambda^2 + (\vec{p} - \vec{Q})^2]} \tag{9}$$

and

$$\int_{\Omega_0} d^3p \frac{\lambda}{(\lambda^2 + p^2)^2} = 2\pi\left(\arctan\left(\frac{P}{\lambda}\right) - \frac{\lambda P}{\lambda^2 + P^2}\right) \to \pi^2, \qquad p \le P \ll 1, \tag{10}$$

$$\int_{\Omega_Q} d^3p \frac{1}{(\lambda^2 + (\vec{p} - \vec{Q})^2)} = 4\pi\left(P - \lambda\arctan\left(\frac{P}{\lambda}\right)\right) \to 4\pi P, \qquad |\vec{p} - \vec{Q}| \le P \ll 1. \tag{11}$$

The integrand function of the term (6) in the vicinity of the weak singular points $\vec{p}_1 = 0$ and $\vec{p}_1 + \vec{p}_2 - \vec{Q} = 0$ at $\lambda \to 0+$ has the behavior

$$\frac{\lambda}{(\lambda^2 + p_1^2)^2}\frac{1}{p_2^2}\frac{\lambda}{(\lambda^2 + (\vec{p}_1 + \vec{p}_2 - \vec{Q})^2)^2}. \tag{12}$$

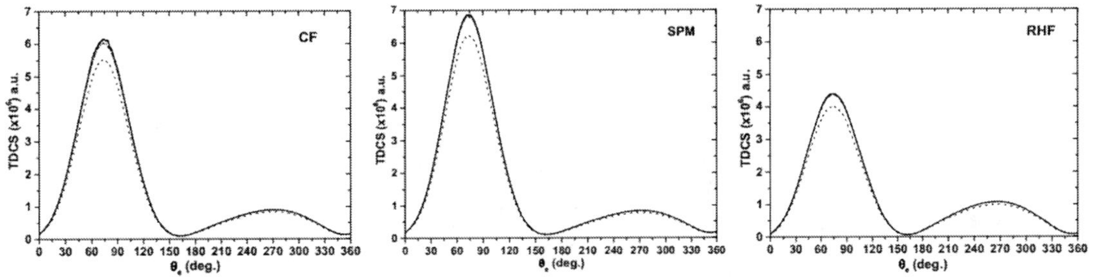

Figure 1. TDCS in coplanar geometry using different trial helium ground-state wave functions: CF (left), SPM (middle), and RHF (right). The solid line corresponds to $\lambda = 10^{-4}$, the dash-dotted line to $\lambda = 10^{-3}$, and the dashed line to $\lambda = 10^{-2}$.

It means that the integral (6) has a singular point $\vec{p}_2 = -\vec{p}_1 + \vec{Q}$ in each direction of $\vec{p}_1$ and is numerically unstable. After using special transformation for each trial function, RHF, SPM and CF, the term (6) has the same singularities as the terms (4) and (5). Finally, the 3D, 4D and 7D integrals (depending on the trial functions) are calculated numerically using the Fortran code Cuhre [10].

We use (2) and keep the parameter λ small but finite, namely in the range 10^{-4}–10^{-2}. Our numerical approach exhibits quite good agreement with different test analytical calculations using the function (1). However, when λ falls below 10^{-4}, problems with the accuracy arise. At the same time, within the domain indicated above we observe good accuracy and stability of calculations. Numerical results for TDCS using different trial helium ground-state wave functions are shown in Fig. 1. Thus, the present study demonstrates the existence of a finite domain of smoothing-parameter values where the numerical calculations exhibit very good convergence and stability.

Acknowledgments

The present research benefited from computational resources of the Central Information and Computer Complex and the HybriLIT heterogeneous computing cluster of the Joint Institute for Nuclear Research. Yu. P. is grateful to Russian Foundation for Basic Research (RFBR) for the financial support under the grant No. 16-02-00049-a. O. Ch. acknowledges support from the Hulubei-Meshcheryakov program JINR-Romania.

References

[1] J. Ullrich *et al*, J. Phys. B **30**, 2917 (1997); R. Dörner *et al*, Phys. Rep. **330**, 95 (2000); J. Ullrich *et al*, Rep. Prog. Phys. **66**, 1463 (2003)
[2] Dz. Belkic, R. Gayet, and A. Salin, Phys. Rep. **56**, 281 (1979)
[3] D. S. F. Crothers and L. J. Dube, Adv. At. Mol. Opt. Phys. **30**, 287 (1993)
[4] A. Mondal, Ch. Mandal, and M. Purkait, EPJD **70**, 16 (2016)
[5] H. Gassert et al., Phys. Rev. Lett. **116**, 073201 (2016)
[6] M. Brauner, J. S. Briggs, and H. Klar, J. Phys. B: At. Mol. Opt. Phys. **22**, 2265 (1989)
[7] E. Clementi and C. Roetti, At. Data Nucl. Data Tables **14**, 177 (1974)
[8] J. N. Silvermann, O. Platas, and F. A. Matsen, J. Chem. Phys. **32**, 1402 (1960)
[9] O. Chuluunbaatar *et al*, Phys. Rev. A **74**, 014703 (2006)
[10] http://www.feynarts.de/cuba/

EPJ Web of Conferences **173**, 03008 (2018) https://doi.org/10.1051/epjconf/201817303008
Mathematical Modeling and Computational Physics 2017

General Solution of the Rayleigh Equation for the Description of Bubble Oscillations Near a Wall

Ivan Garashchuk[1,*], *Dmitry* Sinelshchikov[1,**], and *Nikolay* Kudryashov[1,***]

[1] *NRNU MEPhI, 115409, Russia, Moscow, Kashirskoe shosse, 31*

Abstract. We consider a generalization of the Rayleigh equation for the description of the dynamics of a spherical gas bubble oscillating near an elastic or rigid wall. We show that in the non–dissipative case, i.e. neglecting the liquid viscosity and compressibility, it is possible to construct the general analytical solution of this equation. The corresponding general solution is expressed via the Weierstrass elliptic function. We analyze the dependence of this solution properties on the physical parameters.

1 Introduction

Understanding the dynamics of the spherical gas bubbles in a liquid is of high importance for various applications such as biomedical and industrial ones [1]. There are several models for the description of gas bubbles oscillations, which mostly generalize the Rayleigh equation (see, e.g. [1, 2]). Although, usually, these models are studied numerically, it has been recently shown [3–5] that it is possible to study such models analytically. For example, in [3, 4] the general solutions of the Rayleigh equation and its N-dimensional generalization for both an empty and a gas–filled bubble were found. In this work we consider another generalization of the Rayleigh equation, which describes oscillations of a spherical gas bubble near an elastic or rigid wall [6, 7]. This model has not been studied analytically yet and, therefore, the aim of this work is to find its general solution and investigate the influence of the physical parameters on the properties of this solution.

2 Main results

We consider the following equation

$$R_{tt} + \frac{3}{2} \cdot \frac{1+4AR}{1+3AR} \cdot \frac{R_t^2}{R} - \frac{\widetilde{P_0} R_0^{3\gamma} R^{-3\gamma-1}}{\rho_1(1+3AR)} + \frac{2\sigma}{\rho_1 R^2 (1+3AR)} + \frac{P}{\rho_1 R(1+3AR)} = 0, \tag{1}$$

which describes the motion of a spherical gas bubble near an elastic or rigid wall [6, 7], if one neglects the liquid viscosity and compressibility. Here t is the time, R is the radius of the bubble, γ is the polytropic exponent, ρ_1 is the liquid density, σ is the surface tension, $\widetilde{P_0} = P_0 + 2\sigma/R_0$, P_0 is the

*e-mail: ivan.mail4work@yandex.ru
**e-mail: disine@gmail.com
***e-mail: nakudr@gmail.com

ambient pressure of the gas inside the bubble, R_0 is the ambient radius of the bubble and P is the far–field pressure, which we assume to be constant. The parameter A is $1/(6d)$ for the rigid wall or

$$A = \frac{(\rho_1 - \beta)(\beta - \rho_3)}{6h(\rho_1 + \beta)(\rho_3 + \beta)} - \frac{\rho_1 - \beta}{6d(\rho_1 + \beta)} - \frac{\beta - \rho_3}{(\beta + \rho_3)6(d + h)} \tag{2}$$

for the elastic wall. Here ρ_2 is the wall density, ρ_3 is the density of the material surrounding the wall, $\beta = \rho_2 \nu/(1 - \nu)$ is a characteristic of the wall, ν is the Poisson ratio, h is the thickness of the wall given in mm and d denotes the distance between the wall and the center of the bubble.

For finding the general solution of (1) we use the approach that was proposed in [3, 4]. First, we introduce the dimensionless variables $u = R/R_0$, $\tau = t\omega_0$, where

$$\omega_0^2 = \frac{3\gamma\tilde{P}_0 - 2\sigma/R_0}{\rho_1 R_0^2(1 + 3AR_0)}\,.$$

Then, equation (1) can be rewritten as

$$u_{\tau\tau} + \frac{3}{2}\cdot\frac{1 + 4AR_0u}{1 + 3AR_0u}\cdot\frac{u_\tau^2}{u} - \frac{\alpha u^{-3\gamma-1}}{1 + 3AR_0u} + \frac{\delta}{u^2(1 + 3AR_0u)} + \frac{\epsilon}{u(1 + 3AR_0u)} = 0\,, \tag{3}$$

where

$$\alpha = \frac{\widetilde{P_0}}{\rho_1 R_0^2\omega_0^2}, \qquad \delta = \frac{2\sigma}{\rho_1 R_0^3\omega_0^2}, \qquad \epsilon = \frac{P}{\rho_1 R_0^2\omega_0^2}\,.$$

Note that equation (3) admits the following first integral for $\gamma \neq 1$

$$u^3(1 + 3AR_0u)u_\tau^2 + \frac{2}{3}\epsilon u^3 + \delta u^2 + \frac{2\alpha}{3(\gamma - 1)}u^{3(1-\gamma)} = C_1, \tag{4}$$

where C_1 is an integration constant. This constant can be considered as the total energy of the bubble and, therefore, is positive. Below we consider the case $\gamma = 4/3$, which corresponds to the behaviour of diatomic gases between isothermal and adiabatic [3]. Note that the case of $\gamma = 1$, which correspond to isothermal behaviour, is not considered due to appearance of a logarithmic term in the first integral of (3).

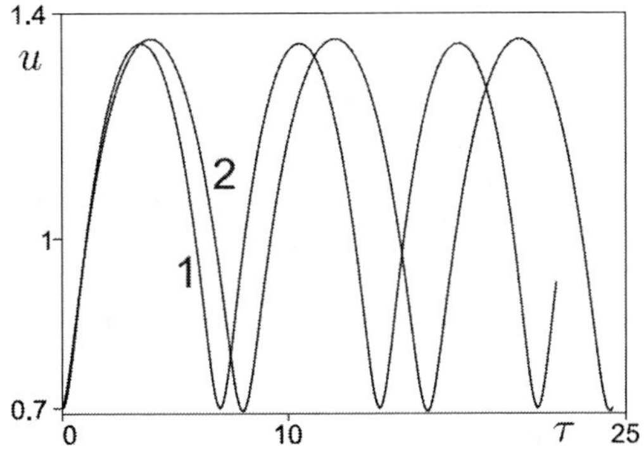

Figure 1. Solution (10) for the case of the elastic (curve 1) and rigid (curve 2) walls corresponding to the initial conditions $u(0) = 0.7$, $u_t(0) = 0$

Then, at $\gamma = 4/3$, with the help of the transformations

$$v = \frac{1}{u}, \qquad d\zeta = \frac{1}{u^2\sqrt{1 + 3AR_0u}}\,d\tau \tag{5}$$

from (3) we get

$$v_{\zeta\zeta} - \frac{3}{2}\frac{v_\zeta^2}{v} + \alpha v^3 - \delta - \frac{\epsilon}{v} = 0. \tag{6}$$

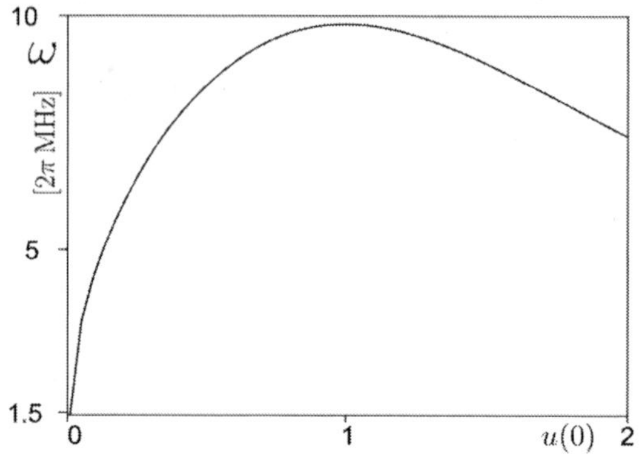

Figure 2. Dependence of the angular frequency of solution (10) on the initial strain

The general solution of (6) is obtained in terms of the Weierstrass elliptic function

$$v = \frac{8\mu^2\wp\{\zeta - \zeta_0, g_2, g_3\} + 8C_1\mu - \delta\mu^3 - 32\alpha}{4\mu^3\wp\{\zeta - \zeta_0, g_2, g_3\} - 2C_1\mu^2 + 16\alpha\mu}, \tag{7}$$

where ζ_0 is an arbitrary constant,

$$\begin{aligned} g_2 &= \frac{\delta\,\mu^4 C_1 + 64\,C_1\,\mu\,\alpha - 16\,\alpha\,\delta\,\mu^3 - 256\,\alpha^2}{4\mu^4} = \frac{\delta C_1}{4} + \frac{4\alpha\epsilon}{3}, \\ g_3 &= \frac{\left(4\,C_1 - \delta\,\mu^2\right)\left(C_1^2\mu - 4\alpha\,C_1 - \alpha\,\delta\,\mu^2\right)}{8\mu^4} = \frac{\epsilon C_1^2}{24} + \frac{\delta^2\alpha}{8}, \end{aligned} \tag{8}$$

and the parameter μ is a real solution of the following equation

$$\epsilon\mu^4 - 3(4C_1\mu - 16\alpha - \delta\mu^3) = 0. \tag{9}$$

Finally, inverting transformations (5) we can find the general solution of (3) as follows

$$u = \frac{4\mu^3\wp\{\zeta - \zeta_0, g_2, g_3\} - 2C_1\mu^2 + 16\alpha\mu}{8\mu^2\wp\{\zeta - \zeta_0, g_2, g_3\} + 8C_1\mu - \delta\mu^3 - 32\alpha}, \qquad \tau = \int_0^\zeta u^2\sqrt{1 + 3AR_0 u}\, d\xi. \tag{10}$$

Formula (10) gives us parametric representation of the general solution of (3), where ζ can be considered as a parameter.

It can be shown that, for the physically relevant values of the parameters, solution (10) is periodic with a real period. One can investigate the dependence of the period or angular frequency of this solution on the parameters. For all analytical calculations we use the same values of the physical parameters as in [6]. For the results presented in Figs. 1 and 2 we also assume that $d = R_0$, i.e. the bubble touches the wall. Notice that the typical range of bubbles oscillations frequency is $\nu = \omega/2\pi = 10^6 \div 10^7$ Hz.

Plots of solution (10) for both rigid and elastic wall cases are shown in Fig. 1. One can see that both the period and magnitude of the oscillations in the case of an elastic wall are slightly smaller

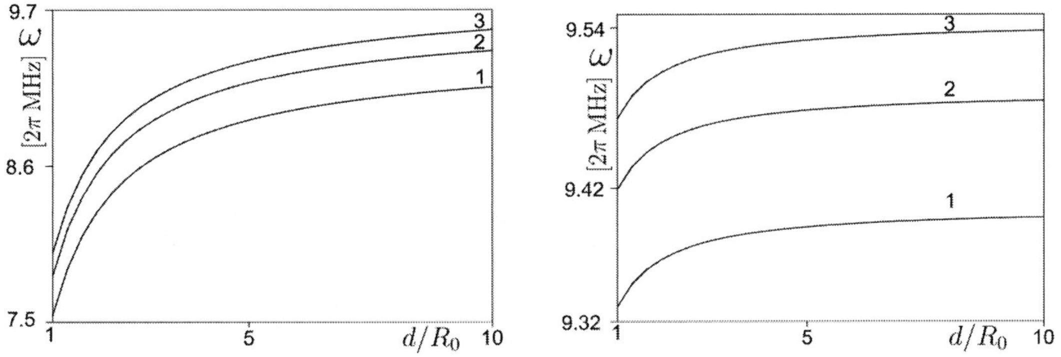

Figure 3. Dependence of angular frequency on the distance from the elastic wall for different values of the initial strain for the rigid wall curve 1: u(0)=0.7; curve 2: u(0)=0.8; curve 3: u(0)=0.9 (left figure) and for the elastic wall curve 1: u(0)=0.7; curve 2: u(0)=0.73; curve 3: u(0)=0.75 (right figure).

than those in the case of a rigid wall. In Fig. 2 the dependence of the dimensional angular frequency of solution (10) is plotted in terms of the initial strain, i.e. on the ratio $R(0)/R_0 = u(0)$. This figure shows us that the bubble oscillations are essentially nonlinear, since their frequency strongly depends on the initial values. In Fig. 3 the dependence of the angular frequency of the bubble oscillations is plotted in term of the distance from the bubble to the elastic or the rigid wall for different values of the initial strain. One can see that in the both cases the frequency of oscillations decreases when the bubble comes closer to the wall and also decreases with the initial strain. However, in the case of the elastic wall the oscillation frequency is smaller than that in the case of the rigid wall.

3 Conclusion

In this work we have considered a model for the description of the dynamics of a spherical gas bubble near an elastic or a rigid wall. We have shown that it is possible to construct the general analytical solution of this model in the non–dissipative case. We have found this solution in explicit form in the adiabatic case with the help of nonlocal transformations. Some physical properties of the obtained solution have been discussed.

Acknowledgement

This work is supported by Russian Science Foundation project 17-71-10241.

References

[1] C.E. Brennen, *Cavitation and bubble dynamics* (Cambridge University Press, 2013) 286 pp.
[2] M.S. Plesset and A. Prosperetti, Annual review of fluid mechanics **9**, 145–185 (1977)
[3] N.A. Kudryashov and D.I. Sinelshchikov, Journal of Physics A: Mathematical and Theoretical **47**, 405202 (2014)
[4] N.A. Kudryashov and D.I. Sinelshchikov, Physics Letters A **379**, 798–802 (2015)
[5] N.A. Kudryashov and D.I. Sinelshchikov, Regular and Chaotic Dynamics **20**, 486–496 (2015)
[6] A.A. Doinikov, L. Aired, and A. Bouakaz, Physics in Medicine and Biology **56**, 6951 (2011)
[7] A.A. Doinikov and A. Bouakaz, Journal of Applied Mechanics and Technical Physics **54**, 867–876 (2013)

EPJ Web of Conferences **173**, 03009 (2018) https://doi.org/10.1051/epjconf/201817303009
Mathematical Modeling and Computational Physics 2017

Interpolation Hermite Polynomials For Finite Element Method

Alexander Gusev[1,⋆], *Sergue* Vinitsky[1,2,⋆⋆], *Ochbadrakh* Chuluunbaatar[1,3], *Galmandakh* Chuluunbaatar[1], *Vladimir* Gerdt[1], *Vladimir* Derbov[4], *Andrzej* Góźdź[5], and *Pavel* Krassovitskiy[6]

[1] *Joint Institute for Nuclear Research, Dubna, Russia*
[2] *RUDN University, Moscow, Russia, 6 Miklukho-Maklaya st, Moscow, 117198*
[3] *Institute of Mathematics, National University of Mongolia, Ulaanbaatar, Mongolia*
[4] *N.G. Chernyshevsky Saratov National Research State University, Saratov, Russia*
[5] *Institute of Physics, University of M. Curie-Skłodowska, Lublin, Poland*
[6] *Institute of Nuclear Physics, Almaty, Kazakhstan*

Abstract. We describe a new algorithm for analytic calculation of high-order Hermite interpolation polynomials of the simplex and give their classification. A typical example of triangle element, to be built in high accuracy finite element schemes, is given.

1 Introduction

For more than half a century, the finite element method (FEM) has won universal recognition as an efficient method for solving the most diverse problems of mathematical physics and engineering. In the multidimensional case the finite element grids of various shapes are used. The problem of constructing high-order interpolation polynomials for FEM has a simple solution only for simplex finite elements, such as the well known Lagrange interpolation polynomials (LIPs) [1]. Meanwhile, in the case of a d-dimensional simplex domain, the LIPs of the order $p' \geq 1$ are often sought by compiling and solving systems of $(d + p')!/d!/p'!$ linear algebraic equations [2].

However, there are problems in which values of directional derivatives of the solutions are also necessary. They are of particular importance when high smoothness between the elements is required, or when highly accurate values of the gradient of the solution are necessary. The construction of such basis functions, referred to as Hermite interpolation polynomials (HIPs), is not possible on arbitrary meshes. This is one of the most important and difficult problems in the FEM and its applications in different fields, solved to date explicitly only for certain particular cases [2, 3].

In this paper we report a new algorithm for the calculation of HIPs of the order $p' = \kappa_{\max}(p+1)-1$ providing continuity of the approximating piecewise polynomial functions and of their directional derivatives up to an order κ' along the normals to the boundaries of the simplex finite elements in the Euclidean space $\mathcal{R}^d$, which reduces to solving systems consisting of

$$\frac{(d+p')!}{d!\,p'!} - \frac{(d+p)!}{d!\,p!} \cdot \frac{(d+\kappa^{\max}-1)!}{d!\,(\kappa^{\max}-1)!}$$

⋆e-mail: gooseff@jinr.ru
⋆⋆e-mail: vinitsky@theor.jinr.ru

linear algebraic equations, where p and $\kappa^{\max}$ will be specified below.

2 Algorithm for Calculating the Hermite Interpolating Polynomials

The conventional FEM implementation for a problem defined in terms of the set of coordinates $z = (z_1, \ldots, z_d) \in \mathcal{R}^d$ performs all calculations in local (reference) coordinates $z' = (z'_1, \ldots, z'_d) \in \mathcal{R}^d$, in which the $d+1$ coordinates of the *simplex vertices* are [3]: $\hat{z}'_j = (\hat{z}'_{j1}, \ldots, \hat{z}'_{jd})$, $\hat{z}'_{jk} = \delta_{jk}$, $j = 0, \ldots, d$,

$$z_i = \hat{z}_{0i} + \sum_{j=1}^{d} \hat{J}_{ij} z'_j, \quad z'_i = \sum_{j=1}^{d} (\hat{J}^{-1})_{ij}(z_j - \hat{z}_{0j}), \quad \frac{\partial}{\partial z'_i} = \sum_{j=1}^{d} \hat{J}_{ji} \frac{\partial}{\partial z_j}, \quad \frac{\partial}{\partial z_i} = \sum_{j=1}^{d} (\hat{J}^{-1})_{ji} \frac{\partial}{\partial z'_j}, \tag{1}$$

where $\hat{J}_{ij} = \hat{z}_{ji} - \hat{z}_{0i}$, $i, j = 1, \ldots, d$, given by the corresponding $d+1$ physical coordinates $\hat{z}_j = (\hat{z}_{j1}, \ldots, \hat{z}_{jd})$.

In the local coordinates of the d-dimensional simplex Δ, the LIP $\varphi_r(z')$ of the order $p' = p$ which equates one at the node points $\xi'_r = (n_1/p, \ldots, n_d/p)$, $n_i \geq 0$, $n_1 + \cdots + n_d \leq p$, $r = 1, \ldots, (d+p)!/d!/p!$ and zero at the remaining node points $\xi'_{r'}$, i.e., $\varphi_r(\xi'_{r'}) = \delta_{rr'}$, is determined by the formula:

$$\varphi_r(z') = \left(\prod_{i=1}^{d} \prod_{n'_i=0}^{n_i-1} \frac{z'_i - n'_i/p}{n_i/p - n'_i/p} \right) \left(\prod_{n'_0=0}^{n_0-1} \frac{1 - z'_1 - \ldots - z'_d - n'_0/p}{n_0/p - n'_0/p} \right), \qquad n_0 = p - n_1 - \cdots - n_d. \tag{2}$$

Step 1. To construct the HIPs in the local coordinates z', we define the set of auxiliary polynomials $\varphi_r^{\kappa_1 \ldots \kappa_d}(z')$ referred to as AP1

$$\varphi_r^{\kappa_1 \ldots \kappa_d}(\xi'_{r'}) = \delta_{rr'} \delta_{\kappa_1 0} \ldots \delta_{\kappa_d 0}, \quad \left. \frac{\partial^{\mu_1 \ldots \mu_d} \varphi_r^{\kappa_1 \ldots \kappa_d}(z')}{\partial z_1'^{\mu_1} \ldots \partial z_d'^{\mu_d}} \right|_{z'=\xi'_{r'}} = \delta_{rr'} \delta_{\kappa_1 \mu_1} \ldots \delta_{\kappa_d \mu_d}, \tag{3}$$
$$0 \leq \kappa_1 + \kappa_2 + \cdots + \kappa_d \leq \kappa_{\max} - 1, \quad 0 \leq \mu_1 + \mu_2 + \cdots + \mu_d \leq \kappa_{\max} - 1.$$

Here in contrast to the LIPs, the values of the functions themselves, and of their derivatives up to the order $\kappa_{\max} - 1$ are specified at the node points ξ'_r. The explicit expressions of AP1 are given by

$$\varphi_r^{\kappa_1 \kappa_2 \ldots \kappa_d}(z') = w_r(z') \sum_{\mu_1 + \cdots + \mu_d = 0, 1, \ldots, \kappa^{\max}-1} a_r^{\kappa_1 \ldots \kappa_d, \mu_1 \ldots \mu_d} (z'_1 - \xi'_{r1})^{\mu_1} \times \cdots \times (z'_d - \xi'_{rd})^{\mu_d}, \tag{4}$$
$$w_r(z') = \left(\prod_{i=1}^{d} \prod_{n'_i=0}^{n_i-1} \frac{(z'_i - n'_i/p)^{\kappa^{\max}}}{(n_i/p - n'_i/p)^{\kappa^{\max}}} \right) \left(\prod_{n'_0=0}^{n_0-1} \frac{(1 - z'_1 - \ldots - z'_d - n'_0/p)^{\kappa^{\max}}}{(n_0/p - n'_0/p)^{\kappa^{\max}}} \right), \quad w_r(\xi'_r) = 1,$$

where the coefficients $a_r^{\kappa_1 \ldots \kappa_d, \mu_1 \ldots \mu_d}$ are calculated from recurrence relations obtained by the substitution of Eq. (4) into the conditions (3).

Step 2. To enforce a uniquely defined polynomial basis, two types of auxiliary polynomials $Q_s(z)$ denoted respectively AP2 and AP3 are defined. AP2 and AP3 are linearly independent of AP1 from Eq. (4) and satisfy the following conditions at the node points $\xi'_{r'}$ of AP1:

$$Q_s(\xi'_{r'}) = 0, \quad \left. \frac{\partial^{\kappa'_1 \kappa'_2 \ldots \kappa'_d} Q_s(z')}{\partial z_1'^{\mu_1} \partial z_2'^{\mu_2} \ldots \partial z_d'^{\mu_d}} \right|_{z'=\xi'_{r'}} = 0, \quad s = 1, \ldots, K, \quad K = \frac{(d+p')!}{d!p'!} - \frac{(d+p)!}{d!p!} \frac{(d+\kappa^{\max}-1)!}{d!(\kappa^{\max}-1)!}. \tag{5}$$

To provide the continuity of derivatives, the polynomials referred to as AP2 are asked to satisfy the condition

$$\left. \frac{\partial^k Q_s(z')}{\partial n_{i(s)}^k} \right|_{z'=\eta'_{s'}} = \delta_{ss'}, \quad s, s' = 1, \ldots, T_1(\kappa'), \quad k = k(s'), \tag{6}$$

Table 1. Characteristics of the HIP bases

$[p\kappa_{\max}\kappa']$	[120]	[131]	[141]	[231]	[152]	[162]	[241]	[173]
$p' = \kappa_{\max}(p+1) - 1$	3	5	7	8	9	11	11	13
$N_{1p'} = (p'+1)(p'+2)/2$	10	21	36	45	55	78	78	105
$K = p(p+1)\kappa_{\max}(\kappa_{\max}-1)/4$	1	3	6	9	10	15	9	21
$N(\mathrm{AP1}) = N_{\kappa_{\max}p'}$	9	18	30	36	45	63	60	84
$N(\mathrm{AP2}) = T_1(\kappa')$	0	3	3	6	9	9	6	18
$N(\mathrm{AP3}) = K - T_1(\kappa')$	1	0	3	3	1	6	12	3
$N_{\kappa_{\max}p'} = (p+1)(p+2)\kappa_{\max}(\kappa_{\max}+1)/4$								
Restriction of derivative order κ': $3p\kappa'(\kappa'+1)/2 \le K$								

where $\eta'_{s'} = (\eta'_{s'1}, \ldots, \eta'_{s'd})$ are conveniently chosen points lying on the faces of various dimensionalities (from 1 to $d-1$) of the d-dimensional simplex Δ and do not coincide with the nodal points of HIP ξ'_r, where Eq. (3) is valid, $\partial/\partial n_{i(s)}$ is the directional derivative along the vector n_i, normal to the corresponding i-th face of the d-dimensional simplex Δ_q at the point $\eta_{s'}$ in the physical frame, which is recalculated to the point $\eta'_{s'}$ of the face of the simplex Δ in the local frame using the relations (1). Calculating the number $T_1(\kappa)$ of independent parameters required to provide the continuity of derivatives to the order κ, we determine its maximal value κ' that can be obtained for the schemes with given p and $\kappa_{\max}$ and, correspondingly, the additional conditions (6).

$T_2{=}K{-}T_1(\kappa')$ parameters remain independent and, correspondingly, T_2 additional conditions are added, necessary for the unique determination of the polynomials referred to as AP3,

$$Q_s(\zeta'_{s'}){=}\delta_{ss'}, \quad s, s'{=}T_1(\kappa'){+}1, \ldots, K, \tag{7}$$

where $\zeta'_{s'} = (\zeta'_{s'1}, \ldots, \zeta'_{s'd}) \in \Delta$ are the chosen points belonging to the simplex without the boundary, but not coincident with the node points of AP1 ξ'_r.

The auxiliary polynomials AP2 and AP3 are given by the expression

$$Q_s(z') = \left(\prod_{t=0}^{d} z_t'^{k_t}\right) \sum_{j_1,\ldots,j_d} b_{j_1,\ldots,j_d;s} z_1'^{j_1} \ldots z_d'^{j_d}, \qquad z'_0 = 1 - z'_1 - \cdots - z'_d. \tag{8}$$

For AP2 $k_t = 1$ if the point η_s, at which the additional conditions (6) are specified, lies on the corresponding face of the simplex Δ, and $k_t = \kappa'$, otherwise, $t = 0, \ldots, d$. For AP3 $k_t = \kappa'$, $t = 0, \ldots, d$. The coefficients $b_{j_1,\ldots,j_d;s}$ are determined from the uniquely solvable system of linear equations, obtained as a result of the substitution of Eq. (8) into the conditions (5)–(7).

Step 3. As a result, we get the required set of basis HIPs $\varphi^\kappa_r(z') = \{\check{\varphi}^\kappa_r(z'), Q_s(z')\}$, $\kappa = \kappa_1, \ldots, \kappa_d$, composed of the polynomials $Q_s(z')$ of the type AP2 and AP3, and the polynomials $\check{\varphi}^\kappa_r(z')$

$$\check{\varphi}^\kappa_r(z'){=}\varphi^\kappa_r(z'){-}\sum_{s=1}^{K} c_{\kappa;r;s} Q_s(z'), \qquad c_{\kappa;r;s}{=}\begin{cases} \left.\dfrac{\partial^k \varphi^\kappa_r(z')}{\partial n^k_{i(s)}}\right|_{z'=\eta'_s}, & Q_s(z') \in \mathrm{AP2}, \\ \varphi^\kappa_r(\zeta_s), & Q_s(z') \in \mathrm{AP3}. \end{cases} \tag{9}$$

Step 4. The AP1 $\check{\varphi}^\kappa_r(z')$ from (9), where κ denotes the directional derivatives along the local coordinate axes, are recalculated using Eqs. (1) into $\check{\varphi}^{\check{\kappa}}_r(z')$ specified in the local coordinates, but now $\check{\kappa} = \check{\kappa}_1, \ldots, \check{\kappa}_d$ denotes the directional derivatives along the physical coordinate axes.

For example, at $d = 2$ the derivatives $\partial/\partial n_i$ along the direction n_i, perpendicular to the appropriate face $i = 0, 1, 2$ in the physical frame are expressed in terms of the partial derivatives $\partial/\partial z'_j$, $j = 1, 2$ in

Table 2. The 20 HIPs $p{=}1$, $\kappa_{\max}{=}4$, $\kappa'{=}1$, $p'{=}7$, remaining 16 HIPs are obtained by permutation $z_1 \leftrightarrow z_2$.

AP1 : $\xi_0{=}(0,0)$, $\xi_1{=}(1,0)$, $\xi_2{=}(0,1)$, $P_0(z_j) = (20z_j^3{-}70z_j^2{+}84z_j{-}35)$, $P_1(z_j) = (10z_j^2{-}24z_j{+}15)$			
$\varphi_2^{0,0}{=}{-}z_2^4P_0(z_2)$	$\varphi_2^{0,1}{=}z_2^4(z_2{-}1)P_1(z_2)$	$\varphi_2^{1,0}{=}z_1z_2^4P_1(z_2)$	$\varphi_0^{0,0}{=}{-}z_0^4P_0(z_0)$
$\varphi_2^{0,2}{=}{-}z_2^4(z_2{-}1)^2(4z_2{-}5)/2$	$\varphi_2^{1,1}{=}{-}z_1z_2^4(z_2{-}1)(4z_2{-}5)$	$\varphi_2^{2,0}{=}{-}z_1^2z_2^4(4z_2{-}5)/2$	$\varphi_0^{0,1}{=}z_0^4z_2P_1(z_0)$
$\varphi_2^{0,3}{=}z_2^4(z_2{-}1)^3/6$	$\varphi_2^{1,2}{=}z_1z_2^4(z_2{-}1)^2/2$	$\varphi_2^{2,1}{=}z_1^2z_2^4(z_2{-}1)/2$	$\varphi_2^{3,0}{=}z_1^3z_2^4/6$
$\varphi_0^{2,0}{=}{-}z_0^4z_1^2(4z_0{-}5)/2$	$\varphi_0^{1,1}{=}{-}z_0^4z_1z_2(4z_0{-}5)$	$\varphi_0^{0,3}{=}z_0^4z_2^3/6$	$\varphi_0^{1,2}{=}z_0^4z_1z_2^2/2$
AP2 : $\eta_1{=}(0,1/2), \eta_2{=}(1/2,0), \eta_3{=}(1/2,1/2)$		AP3 : $\zeta_4{=}(1/4,1/2), \zeta_5{=}(1/2,1/4), \zeta_6{=}(1/4,1/4)$	
$Q_1{=}8z_1z_2^2z_0^2(12z_1^2{-}7z_1{-}8z_1z_2{-}8z_2^2{+}8z_2)/f_{11}$		$Q_4{=}1024z_0^2z_1^2z_2^2(4z_2{-}1)$	
$Q_3{=}4z_1^2z_2^2z_0(12z_2^2{-}17z_2{+}5{-}17z_1{+}32z_1z_2{+}12z_1^2)/f_{01}$		$Q_6{=}1024z_0^2z_1^2z_2^2(4z_0{-}1)$	

the local frame of the triangle Δ, using the relations (1), as

$$\frac{\partial}{\partial n_i}{=}f_{i1}\frac{\partial}{\partial z_1'}{+}f_{i2}\frac{\partial}{\partial z_2'}, \quad i{=}1,2, \quad \frac{\partial}{\partial n_0}{=}(f_{01}{+}f_{02})\frac{\partial}{\partial z_1'}{+}(f_{01}{-}f_{02})\frac{\partial}{\partial z_2'}, \tag{10}$$

where $f_{ij} = f_{ij}(\hat{z}_0, \hat{z}_1, \hat{z}_2)$ are functions of the coordinates of the vertices $\hat{z}_0, \hat{z}_1, \hat{z}_2$ of the triangle Δ_q in the physical frame. The characteristics of the polynomial basis of HIPs on the *triangle element* Δ at $d = 2$, including known Argyris triangle [131] (see [3]), are presented in Table 1. Table 2 presents the results of executing the Algorithm for calculating the HIPs in the case ($p{=}1$, $\kappa_{\max}{=}4$, $\kappa'{=}1$, $p'{=}7$), AP1: $\varphi_r^\kappa(z')$, AP2 and AP3: $Q_s(z')$, and the corresponding coefficients $c_{\kappa;r;s}$ are calculated using Eqs. (9). The notations are as follows: ξ_r, η_s, ζ_s are the coordinates of the nodes, which are specified in the conditions (3), (6) or (7), respectively, $z_0 = 1 - z_1 - z_2$, the arguments of functions and the primes in the notations of independent variables are omitted. The explicit expressions for the HIPs from Table 1 were calculated too, but are not presented here because of the paper size limitations (they can be obtained under request to the authors or using the program TRIAHP implemented in Maple, which will be published in the library JINRLIB). The calculations were carried out using the Intel Pentium CPU 987, x64, 4 GB RAM, the Maple 16, during 6 seconds.

3 Conclusion

We presented a symbolic-numeric algorithm, implementable in any computer algebra system, in particular, the Maple system, for analytical calculation of the basis of Hermite interpolation polynomials of several variables of the simplex, which can be used to construct a FEM computational scheme of high-order accuracy. The efficiency of the FEM computational schemes using high-order accuracy LIPs and HIPs for benchmark calculations of exactly solvable problems for the triangle membrane and hypercube is shown in the forthcoming paper of this issue.

The work was partially supported by the RFBR (grants Nos. 16-01-00080 and 17-51-44003 Mong_a), the MES RK (Grant No. 0333/GF4), the Bogoliubov-Infeld program, the RUDN University Program 5-100 and grant of Plenipotentiary of the Republic of Kazakhstan in JINR.

References

[1] N.C. Bachvalov, N.P. Zhidkov, and G.M. Kobelkov, *Chislennye metody* (Nauka, Moscow, 1987), *in Russian.*

[2] A.R. Mitchell and R. Wait, *The Finite Element Method in Partial Differential Equations* (John Wiley & Sons, Chichester, 1977)

[3] P. Ciarlet, *The Finite Element Method for Elliptic Problems* (North-Holland, Amsterdam, 1978)

High-Accuracy Finite Element Method: Benchmark Calculations

Alexander Gusev[1,⋆], *Sergue* Vinitsky[1,2,⋆⋆], *Ochbadrakh* Chuluunbaatar[1,3], *Galmandakh* Chuluunbaatar[1], *Vladimir* Gerdt[1], *Vladimir* Derbov[4], *Andrzej* Góźdź[5], and *Pavel* Krassovitskiy[6]

[1]*Joint Institute for Nuclear Research, Dubna, Russia*
[2]*RUDN University, Moscow, Russia, 6 Miklukho-Maklaya st, Moscow, 117198*
[3]*Institute of Mathematics, National University of Mongolia, Ulaanbaatar, Mongolia*
[4]*N.G. Chernyshevsky Saratov National Research State University, Saratov, Russia*
[5]*Institute of Physics, University of M. Curie-Skłodowska, Lublin, Poland*
[6]*Institute of Nuclear Physics, Almaty, Kazakhstan*

Abstract. We describe a new high-accuracy finite element scheme with simplex elements for solving the elliptic boundary-value problems and show its efficiency on benchmark solutions of the Helmholtz equation for the triangle membrane and hypercube.

1 Introduction

In Ref. [1] we present a new algorithm for the calculation of high-order Lagrange and Hermite interpolation polynomials (LIP and HIP) of the simplex in analytical form, their classification and a typical example of the triangle element for high-accuracy finite element method (FEM) schemes. In this paper we show the efficiency of the schemes using high-order accuracy LIP and HIP for benchmark calculations of the exactly solvable problems for the triangle membrane and hypercube.

2 Formulation of the Problem

Consider the self-adjoint boundary-value problem (BVP) for the elliptic differential equation:

$$(D-E)\,\Phi(z) \equiv \left(-\frac{1}{g_0(z)}\sum_{ij=1}^{d}\frac{\partial}{\partial z_i}g_{ij}(z)\frac{\partial}{\partial z_j}+V(z)-E\right)\Phi(z)=0. \tag{1}$$

It is assumed that $g_0(z) > 0$, $g_{ji}(z) = g_{ij}(z)$ and $V(z)$ are real-valued functions, continuous together with their generalized derivatives to a given order in the domain $z \in \bar{\Omega} = \Omega \cup \partial\Omega \in \mathcal{R}^d$ with the piecewise continuous boundary $S = \partial\Omega$, which provides the existence of nontrivial solutions obeying the boundary conditions (BCs) of the first (I) or the second (II) kind

$$(\mathrm{I}): \quad \Phi(z)|_S = 0, \qquad (\mathrm{II}): \quad \left.\frac{\partial\Phi(z)}{\partial n_D}\right|_S = 0, \qquad \frac{\partial\Phi(z)}{\partial n_D} = \sum_{ij=1}^{d}(\hat{n},\hat{e}_i)g_{ij}(z)\frac{\partial\Phi(z)}{\partial z_j}, \tag{2}$$

⋆e-mail: gooseff@jinr.ru
⋆⋆e-mail: vinitsky@theor.jinr.ru

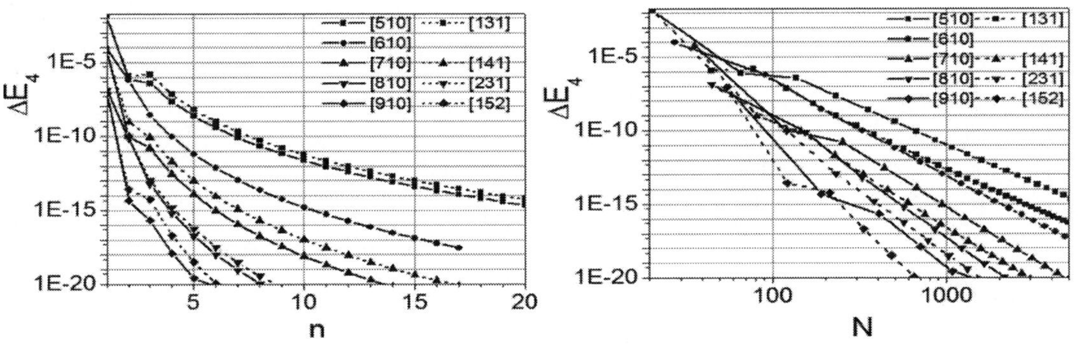

Figure 1. The error ΔE_4 of the eigenvalue E_4^h versus the number of elements n and the length of the vector N

where $\frac{\partial\Phi_m(z)}{\partial n_D}$ is the derivative along the conormal direction, $\hat{n}$ is the outer normal to the boundary of the domain $S = \partial\Omega$, $\hat{e}_i$ is the unit vector of $z = \sum_{i=1}^{d} \hat{e}_i z_i$, and $(\hat{n}, \hat{e}_i)$ is the scalar product in $\mathcal{R}^d$.

For a discrete spectrum problem, the functions $\Phi_m(z)$ from the Sobolev space $H_2^{s\geq 1}(\Omega)$, $\Phi_m(z) \in H_2^{s\geq 1}(\Omega)$, corresponding to the real eigenvalues E: $E_1 \leq E_2 \leq \cdots \leq E_m \leq \cdots$, obey the conditions

$$\langle \Phi_m(z)|\Phi_{m'}(z)\rangle = \int_\Omega dz\, g_0(z)\Phi_m(z)\Phi_{m'}(z) = \delta_{mm'}, \quad dz = dz_1 \cdots dz_d. \tag{3}$$

The FEM solution of the BVPs (1)–(3) reduces to the stationary points of the functional

$$\Pi(\Phi_m, E_m, z) = \int_\Omega dz \left[\sum_{ij=1}^{d} g_{ij}(z) \frac{\partial \Phi_m(z)}{\partial z_i} \frac{\partial \Phi_m(z)}{\partial z_j} + g_0(z)\Phi_m(z)(V(z) - E_m)\Phi_m(z) \right]. \tag{4}$$

The polyhedral domain $\Omega = \Omega_h(z) = \bigcup_{q=1}^{Q} \Delta_q$ is divided in finite elements in the form of simplexes Δ_q with HIPs or LIPs $\varphi_r^\kappa(z)$, calculated using the Algorithm of [1]. The piecewise polynomial functions $P_l(z)$ of the order p' with continuous derivatives to the order κ' are constructed by joining the polynomials $\varphi_r^\kappa(z)$ to the finite elements $\Delta_q \in \Omega_h(z)$. The expansion of the sought solution $\Phi_m(z)$ in the basis of piecewise polynomial functions $P_l(z)$, $\Phi_m^h(z) = \sum_{l=1}^{N} P_l(z)\Phi_{lm}^h$ and its substitution into the variational functional (4) leads to the generalized algebraic eigenvalue problem (AEP), $(\mathbf{A} - \mathbf{B}E_m^h)\mathbf{\Phi}_m^h = 0$, solved using the standard method of [2]. The elements of the symmetric matrices of stiffness $\mathbf{A}$ and mass $\mathbf{B}$ comprise integrals like Eq. (4), which are calculated on the elements Δ_q in the domain $\Omega = \Omega_h(z) = \bigcup_{q=1}^{Q} \Delta_q$, recalculated into the local coordinates on the master element Δ.

The deviation of the approximate solution E_m^h,$\Phi_m^h(z) \in \mathcal{H}_2^{\kappa'+1\geq 1}(\Omega_h)$ from the exact one E_m,$\Phi_m(z) \in \mathcal{H}_2^2(\Omega)$ is theoretically estimated as [2]

$$\Delta E_m = \left|E_m - E_m^h\right| \leq c_1 h^{2p'}, \qquad \left\|\Phi_m(z) - \Phi_m^h(z)\right\|_0 \leq c_2 h^{p'+1}, \tag{5}$$

where $\|\Phi_i(z)\|_0^2 = \int_\Omega g_0(z)dz\Phi_i(z)\Phi_i(z)$, h is the maximal size of the finite element Δ_q, p' is the order of the FEM scheme, m is the number of the eigenvalue, c_1 and c_2 are coefficients independent of h.

3 BVP for Helmholtz equation in an equilateral triangle

As an example, we consider the solution of the BVP (1)–(3) at $d = 2$, $g_0(z) = g_{ij}(z) = 1$, and $V(z) = 0$ for the Helmholtz equation (HEQ) under the BC (II) of the second kind (2) in the domain Ω

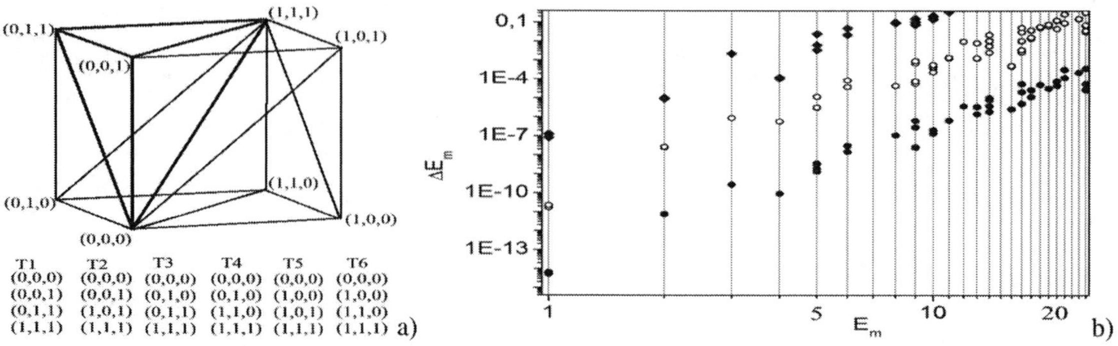

Figure 2. a) The division of the 3D cube into $3! = 6$ equal tetrahedrons (T1,...,T6). b) The error $\Delta E_m = E_m^h - E_m$ calculated by FEM using sixth-order LIPs versus the exact eigenvalue E_m. Squares: the cube having the edge π divided into 6 tetrahedrons. Circles: the cube with having the edge π divided into 2^3 cubes, each comprised of 6 tetrahedrons. Solid circles: the cube with the edge π divided into 4^3 cubes, each comprised of 6 tetrahedrons.

assumed to be an equilateral triangles with the side $4\pi/3$ partitioned into $Q = n^2$ equilateral triangles Δ_q of sides $h = 4\pi/3n$. The eigenvalues of this problem having degenerate spectrum are the integers $E_m = m_1^2 + m_2^2 + m_1 m_2 = 0, 1, 1, 3, 4, 4, 7, 7, \ldots,\ m_1, m_2 = 0, 1, 2, \ldots.$ Figure 1 shows the errors $\Delta E_m = |E_m^h - E_m|$ of the eigenvalue $E_4^h(z)$ depending on the number n (the number of elements being n^2) and on the length N of the vector $\mathbf{\Phi}_m^h$ of the AEP for the FEM schemes from the fifth to the ninth order of accuracy: using LIP with the labels $[p\kappa_{\max}\kappa'] = [510], \ldots, [910]$, and using HIP with the labels [131], [141], [231] and [152] from Table 1 of paper [1], conserving the continuity of the first and the second derivative of the approximate solution, respectively. As seen from Fig. 1, the errors of the eigenvalue $\Delta E_4^h(z)$ of the FEM schemes of the same order $p' = \kappa_{\max}(p+1) - 1$ are nearly similar and correspond to the theoretical estimates (5), but in the FEM schemes conserving the continuity of the first and the second derivatives of the approximate solution, the matrices of smaller dimension are used that correspond to the length of the vector N smaller by 1.5–2 times than in the schemes with LIP that conserve only the continuity of the functions themselves at the boundaries of the finite elements. Computations were carried out using a dual processor desktop 2 x Xeon 3.2 GHz, 4 GB RAM, Intel Fortran 77 with quadruple precision real*16. The computing time was 3 minutes.

4 BVP for Helmholtz equation in a d-dimensional hypercube

For benchmark calculations we use the BVP for the HEQ with the BC (II) in a d-dimensional hypercube with the edge length π. Since the variables are separated, the eigenvalues $E_m = E_{m_1,\ldots,m_d}$ are sums of squared integers, $E_m = E_{m_1,\ldots,m_d} = m_1^2 + \cdots + m_d^2$, $m_k = 0, 1, \ldots$, $k = 1, \ldots, d$.

Assertion (see also [3]). The hypercube is divided into $d!$ equal simplexes. The vertices of each simplex are located on broken lines, composed of d mutually perpendicular edges, and the extreme vertices of all polygons are located on one of the diagonals of the hypercube (for $d = 3$ see Fig. 2a).

Algorithm. *Input.* A single d-dimensional hypercube with vertices the coordinates of which are either 0 or 1 in the Euclidean space $\mathcal{R}^d$. The chosen diagonal of the hypercube connects the vertices with the coordinates $(0, \ldots, 0)$ and $(1, \ldots, 1)$.

Output. $z_k^{(i)} = (z_{k,1}^{(i)}, \ldots, z_{k,d}^{(i)})$ the coordinates of i-th simplex.

Local. The coordinates of the vertices of the polygonal line are $z_k = (z_{k,1}, \ldots, z_{k,d})$, $k = 0, \ldots, d$.

1. For all $i = (i_1, \ldots, i_d)$, the permutations of the numbers $(1, \ldots, d)$:

Table 1. The lower part of the exact spectrum E_m and the calculated spectrum E_m^h for the 6D hypercube.

E_m	E_m^h
0	0.183360983479286 e-10 ,
1	1.00023, 1.00034, 1.00034, 1.00034, 1.00034, 1.00034,
2	2.04760, 2.04760, 2.04760, 2.04760, 2.04760, 2.04760, 2.04760, 2.04760, 2.04760, 2.07391, 2.08478, 2.08478, 2.08478, 2.08478, 2.08478,
3	3.15060, 3.15196, 3.15196, 3.15196, 3.15196, 3.15196, 3.15780, 3.15780, 3.15780, 3.15780, 3.15780, 3.16319, 3.16319, 3.16319, 3.16319, 3.16319, 3.16319, 3.16319, 3.16319, 3.16319

1.1 For all $k = 0, \ldots, d$ and $s = 1, \ldots, d$: $z_{k,s}^{(i)} = \{1, i_s \leq k, ; 0, i_s > k\}$
1.2 if $\det(z_{k,s}^{(i)})_{k,s=1}^{d} = -1$ then $z_{k,d}^{(i)} \leftrightarrow z_{k,d-1}^{(i)}$.

3D HEQ for the cube. In Fig. 2b we show the error of eigenvalues of the 3D BVP for the HEQ at $d = 3$ with the BC (II) using the FEM scheme with 3D LIP of the order $p = p' = 6$. As can be seen from Fig. 2b, the errors of the eigenvalues lie on parallel lines in the double logarithmic scale which corresponds to the theoretical error estimates (5) for the eigenvalues depending on the maximal size of the finite element. In the case of the cube with the edge π divided into 4^3 cubes, each of them comprised by 6 tetrahedrons, the matrices **A** and **B** had the dimension 15625 × 15625. The matrices **A** and **B** were calculated analytically using Maple 2015, 2x 8-core Xeon E5-2667 v2 3.3 GHz, 512 GB RAM, GPU Tesla 2075, and the AEP was solved during 20 minutes using Intel Fortran.

6D HEQ for the hypercube. We solved HEQ at $d = 6$ with the BC (II) using FEM scheme with 6D LIP of the order $p = p' = 3$. The 6D hypercube having the edge π was divided into $n = d! = 6! = 720$ simplexes (the size of the finite element being equal to π). On each of them $N_1(p) = (p+d)!/(d!p!) =$ 84 third-order LIPs were used. The matrices **A** and **B** had the dimension 4096 × 4096. The lower part of the spectrum E_m is shown in Table 1. The errors of the second, the third and the fourth degenerate eigenvalues are equal to 0.0003, 0.05 and 0.15, respectively. Note, that applying the third-order scheme for solving the BVPs of smaller dimension d, we obtained errors of the same order. The calculation time was 9234.46 seconds using Maple 2015, 2x 8-core Xeon E5-2667 v2 3.3 GHz, 512 GB RAM, GPU Tesla 2075.

5 Conclusion

We demonstrated the efficiency of the schemes using the high-order accuracy LIP and HIP for benchmark calculations of the exactly solvable Helmholtz problems for the triangle and hypercube. The detailed description of the FEM schemes with multivariate LIP and HIP, using a more economical numerical calculation of FEM integrals by means of the cubature formulas, together with benchmark calculations of the parametric BVP for the two-center Coulomb problem, that provide accuracy of the order 10^{-12} for eigenvalues of a low part of the spectrum, will be presented elsewhere.

The work was partially supported by the RFBR (grants Nos. 16-01-00080 and 17-51-44003 Mong_a), the MES RK (Grant No. 0333/GF4), the Bogoliubov-Infeld program, the RUDN University Program 5-100 and grant of Plenipotentiary of the Republic of Kazakhstan in JINR.

References

[1] A. Gusev et al, *Interpolation Hermite Polynomials For Finite Element Method* (in this issue)
[2] K.J. Bathe, *Finite Element Procedures in Engineering Analysis* (Eng. Cliffs, NY, 1982)
[3] D. G. Mead, Proc. Amer. Math. Soc. **76**, 302–304 (1979)

Numerical Solution of the Time Dependent 3D Schrödinger Equation Describing Tunneling of Atoms from Anharmonic Traps

Ilyas Ishmukhamedov[1,2,3,⋆], *Altay* Ishmukhamedov[2,4,⋆⋆], and *Vladimir* Melezhik[1,5,⋆⋆⋆]

[1] *Bogoliubov Laboratory of Theoretical Physics, Joint Institute for Nuclear Research, 141980 Dubna, Moscow Region, Russian Federation*
[2] *Al-Farabi Kazakh National University, Almaty 050040, Republic of Kazakhstan*
[3] *Institute of Nuclear Physics, the Ministry of Energy of the Republic of Kazakhstan, Almaty 050032, Republic of Kazakhstan*
[4] *K.I. Satpaev Institute of Geological Sciences, Almaty 050010, Republic of Kazakhstan*
[5] *Peoples' Friendship University of Russia (RUDN University), 6 Miklukho-Maklaya St., 117198 Moscow, Russian Federation*

Abstract. We present an efficient numerical method for the integration of the 3D Schrödinger equation. A tunneling problem of two interacting bosonic atoms confined in a 1D anharmonic trap has been successfully solved by means of this method. We demonstrate fast convergence of the final results with respect to spatial and temporal grid steps. The computational scheme is based on the operator-splitting technique with the implicit Crank-Nicolson algorithm on spatial sixth-order finite-differences. The computational time is proportional to the number of spatial grid points.

1 Introduction

Pair correlation plays a major role in quantum tunneling and there are a only few works in the context of ultracold physics which accurately take into account the interatomic interaction [1–7]. In the present paper, we consider a tunneling of two interacting atoms from an anharmonic optical trap. We solve this problem by means of a numerical method which is based on the split-operator approximation [1–3, 8, 9], the Crank-Nicolson scheme and the sixth-order spatial finite-differences.

2 Method

We consider the time-dependent Schrödinger equation (throughout $\hbar = \omega = m = 1$ units are used)

$$i\frac{\partial\psi(x_1, x_2, t)}{\partial t} = H(x_1, x_2)\psi(x_1, x_2, t), \qquad (1)$$

with the Hamiltonian

$$H(x_1, x_2) = H_1(x_1) + H_2(x_2) + V^{(aa)}(x_1 - x_2), \qquad (2)$$

⋆e-mail: i.ishmukhamedov@mail.ru
⋆⋆e-mail: ishmukhamedov.altay@gmail.com
⋆⋆⋆e-mail: melezhik@theor.jinr.ru

where $V^{(\mathrm{aa})}$ describes the atom-atom interaction and $H_j(x_j)$ denote the single-atom Hamiltonians

$$H_j(x_j) = -\frac{1}{2}\frac{\partial^2}{\partial x_j^2} + V^{(\mathrm{at})}(x_j), \qquad j = 1,2, \tag{3}$$

in which $V^{(\mathrm{at})}$ describes the atom-trap interaction. The partial derivatives in (3) are approximated by sixth-order finite-differences.

To integrate (1) we use the split-operator method the basic ideas of which have been proposed in [8] and which has been developed in [2, 3, 9] in connection with the solution of confined ultracold atom-atom collisions in waveguide-like traps:

$$\begin{aligned}\psi(x_1, x_2, t + \Delta t) = \exp\left\{-i\frac{\Delta t}{2}V^{(\mathrm{aa})}(x_1 - x_2)\right\}\exp\{-i\Delta t H_1(x_1)\}\exp\{-i\Delta t H_2(x_2)\}\\ \times\exp\left\{-i\frac{\Delta t}{2}V^{(\mathrm{aa})}(x_1 - x_2)\right\}\psi(x_1, x_2, t).\end{aligned} \tag{4}$$

The actions of the operators $\exp\left\{-i\Delta t H_j(x_j)\right\}$ are approximated by the implicit Crank-Nicolson scheme, which maintains the accuracy order of the split-operator scheme (4) to $O(\Delta t^3)$.

When considering a problem of particle tunneling it is necessary to correctly model the outgoing wave function at $x_1, x_2 \to \pm\infty$. This can be done by introducing into the original Hamiltonian (2) a complex absorbing potential (CAP) in the form suggested in [10]:

$$W(x_j) = w_c(|x_j| - x_c)^2\theta(|x_j| - x_c), \qquad j = 1,2, \tag{5}$$

where $\theta(x)$ is the Heaviside step function. The values of interest of the parameters w_c and x_c are fixed after Eq. (8).

3 Results

The efficiency of our method is demonstrated in the solution of the topical problem of tunneling of cold atoms [1]. The atom-trap interaction is chosen of the standing-wave form:

$$V^{(\mathrm{at})}(x) = \begin{cases} 1/(2\beta^2)\cdot\sin^2(\beta x), & |x| \le \pi/\beta, \\ 0, & |x| > \pi/\beta, \end{cases} \tag{6}$$

where $\beta = 2\pi/\lambda \cdot \sqrt{\hbar/(m\omega)}$ (in standard units), λ and ω represent the trap wavelength and frequency correspondingly. For $\beta = 0.427471$, this models an actual atom-trap interaction [1, 11, 12] in modern experiments [13] on ultracold atoms (figure 1).

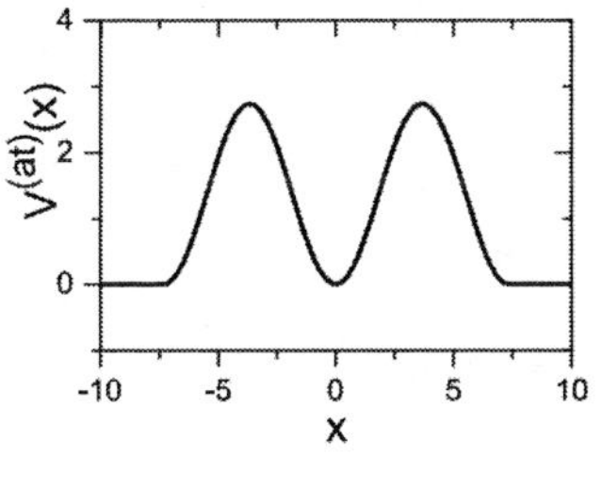

Figure 1. Trap potential $V^{(\mathrm{at})}(x)$.

The atom-atom interaction is taken of Gaussian form:

$$V^{(\mathrm{aa})}(x_1 - x_2) = -V_0\exp\left\{-\frac{(x_1 - x_2)^2}{2r_0^2}\right\}, \tag{7}$$

where the range of the interaction is fixed to $r_0 = 0.1$ [4, 14] and the interaction strength V_0 is varied.

We calculate the total probability $P(t)$ in the box $|x_1, x_2| \leq 20$. For a wide range of values V_0, $P(t)$ decays exponentially. This allows us to extract the tunneling rate γ as

$$\gamma = -\frac{1}{P(t)}\frac{dP(t)}{dt}. \tag{8}$$

We find that the parameters of the CAP (5), $w_c = -0.1$ and $x_c = 10$, lie in the domain where the tunneling rate γ does not change significantly under the variation of these parameters.

To solve (1), the value of the initial wave function is to be known. We fix it by replacing the trap (6) with the "closed" sextic (i.e., the power series expansion of (6) up to x_j^6) trapping potential [1, 15] and calculating the eigenstates of the doubly excited relative and center-of-mass motion states of the stationary Schrödinger equation by using the numerical scheme from [12].

Figure 2 shows the fast convergence of the errors

$$\text{error}(\delta) = \frac{|\gamma(\delta \to 0) - \gamma(\delta)|}{\gamma(\delta \to 0)}, \qquad \delta = \Delta x,\ \Delta t. \tag{9}$$

and tunneling rates γ with the spatial Δx and temporal Δt steps.

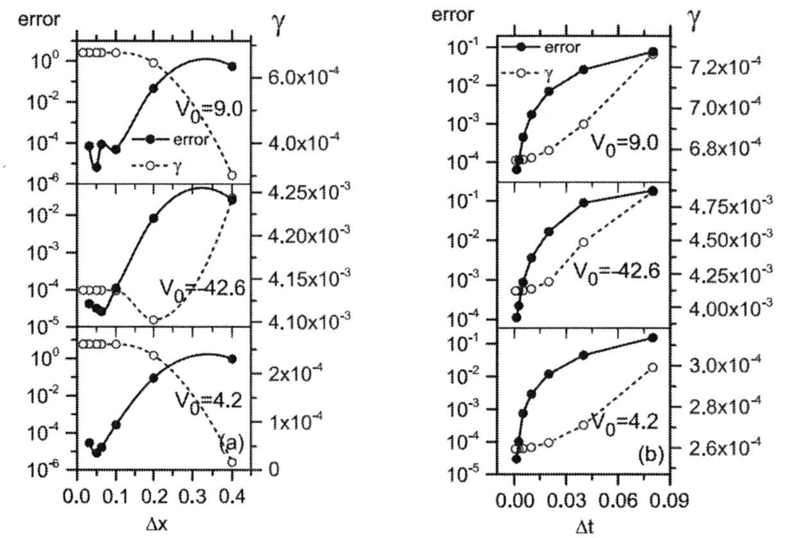

Figure 2. (a) Dependence of the relative errors of the computed γ values (closed circles, left side scale) and dependence of the tunneling rate γ (open circles, right side scale) on the spatial grid step Δx. (b) Dependence of the relative errors of the computed γ values (closed circles, left side scale) and dependence of the tunneling rate γ (open circles, right side scale) on the temporal grid step Δt. Outputs are reported for three values of the $V^{(aa)}$ potential: $V_0 = 9$ and -42.6 (doubly excited relative motion state at $t \leq 0$) and $V_0 = 4.2$ (doubly excited center-of-mass motion state at $t \leq 0$).

In figure 3 one can see that the computational time grows linearly with the number of spatial grid points N, according to the sweeping method theory [16].

4 Conclusion

We have solved a tunneling problem of two cold interacting atoms by using a numerical method. The interatomic coupling strength is varied in a wide range – from strong attraction to strong repulsion.

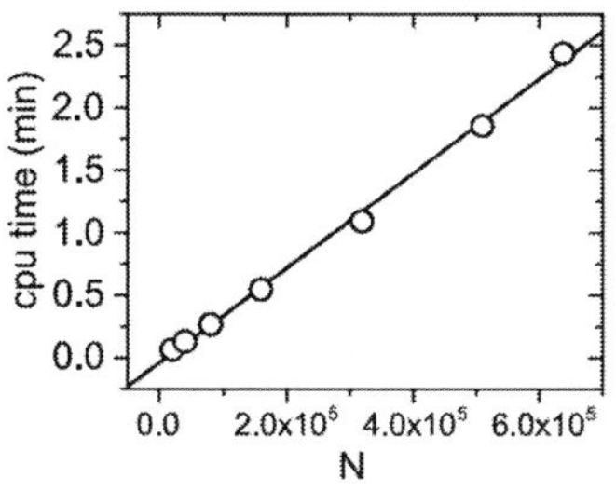

Figure 3. Dependence of the CPU time (in minutes) for the derivation of the tunneling rate γ (open circles) at $V_0 = 9$ (doubly excited relative motion state at $t \leq 0$) on the number of spatial grid points $N = N_1 \times N_2$ (N_j defines the grid along x_j). The solid line is a fit function. The computation has been performed until the temporal variable reaches the value $t = 10$ using the server at the Bogoliubov Laboratory of Theoretical Physics, JINR (i7a: 8-core Intel E5-2690 @2.9-3.8GHz, 128 GB ECC RAM, GPU GTX680).

The method is based on the split-operator approximation of the time-evolution operator and Crank-Nicolson approximation of the single-particle Hamiltonians. We have found that the error of the method reduces quite fast with reducing the step sizes of the spatial and temporal grids. The method can be extended to more complicated problems with more degrees of freedom.

Acknowledgement

The work was financially supported by the Ministry of Education and Science of the Russian Federation (Agreement No. 02.a003.21.0008) and by the Grant of the Plenipotentiary Representative of the Republic of Kazakhstan to JINR.

References

[1] I.S. Ishmukhamedov and V.S. Melezhik, Phys. Rev. A **95**, 062701 (2017)
[2] V.S. Melezhik, J.I. Kim, and P. Schmelcher, Phys. Rev. A **76**, 053611 (2007)
[3] V.S. Melezhik, EPJ Web Conf. **108**, 01008 (2016)
[4] S.E. Gharashi and D. Blume, Phys. Rev. A **92**, 033629 (2015).
[5] P. M. Krassovitskiy and F. M. Pen'kov, J. Phys. B: At. Mol. Opt. Phys. **47**, 225210 (2014)
[6] V. O. Nesterenko, A. N. Novikov, and E. Suraud, Laser Phys. **24**, 125501 (2014)
[7] A. U. J. Lode, A. I. Streltsov, O. E. Alon, H.-D. Meyer, and L. S. Cederbaum, J. Phys. B **42**, 044018 (2009)
[8] G.I. Marchuk, *Methods of Numerical Mathematics* (Springer-Verlag, New York, 1975) Sec. 4.3.3
[9] V.S. Melezhik, *Atoms and Molecules in Strong External Fields* (Plenum, New-York and London, 1998) pp. 89–94
[10] U. V. Riss and H.-D. Meyer, J. Phys. B **26**, 4503 (1993)
[11] S.-G. Peng, H. Hu, X.-J. Liu, and P.D. Drummond, Phys. Rev. A **84**, 043619 (2011)
[12] I.S. Ishmukhamedov, D.T. Aznabayev, and S.A. Zhaugasheva, Phys. Part. Nuclei Lett. **12**: 680 (2015)
[13] E. Haller, M. J. Mark, R. Hart, J. G. Danzl, L. Reichsollner, V. Melezhik, P. Schmelcher, and H.-C. Nägerl, Physical Review Letters, **104**, 153203, (2010)
[14] I.S. Ishmukhamedov, D.S. Valiolda, and S.A. Zhaugasheva, Phys. Part. Nuclei Lett. **11**: 238 (2014)
[15] S. Grishkevich and A. Saenz, Phys. Rev A **80**, 013403 (2009)
[16] I.M. Gelfand and S.V. Fomin, *Calculus of Variations*, (Dover Publications, New York, 2000)

EPJ Web of Conferences **173**, 03012 (2018) https://doi.org/10.1051/epjconf/201817303012
Mathematical Modeling and Computational Physics 2017

Optimal Approximation of Biquartic Polynomials by Bicubic Splines

Viliam Kačala[1,*] and *Csaba* Török[1,**]

[1] *Institute of Computer Science, Faculty of Science, P. J. Šafárik University in Košice, Slovakia*

Abstract. Recently an unexpected approximation property between polynomials of degree three and four was revealed within the framework of two-part approximation models in 2-norm, Chebyshev norm and Holladay seminorm. Namely, it was proved that if a two-component cubic Hermite spline's first derivative at the shared knot is computed from the first derivative of a quartic polynomial, then the spline is a clamped spline of class C^2 and also the best approximant to the polynomial.
Although it was known that a 2×2 component uniform bicubic Hermite spline is a clamped spline of class C^2 if the derivatives at the shared knots are given by the first derivatives of a biquartic polynomial, the optimality of such approximation remained an open question.
The goal of this paper is to resolve this problem. Unlike the spline curves, in the case of spline surfaces it is insufficient to suppose that the grid should be uniform and the spline derivatives computed from a biquartic polynomial. We show that the biquartic polynomial coefficients have to satisfy some additional constraints to achieve optimal approximation by bicubic splines.

1 Introduction

Lately a surprising property of polynomials within the framework of two-part models in case of function approximation has been found. Papers [1], [2] show that a two-component uniform cubic Hermite spline curve of class C^1 by its definition will be also of class C^2, if its coefficients are computed from a corresponding quartic polynomial. This relation was later confirmed for 2×2 component bicubic splines and biquartic polynomials [3]. However, unlike the two-component spline curve, the resulting four component spline surface is not automatically the best approximant to the biquartic polynomial.

The goal of this paper is to show that the coefficients of the biquartic polynomial have to satisfy some constraints to achieve optimal approximation by bicubic splines.

Let define the inputs, the given and computed coefficients to construct a 2×2 spline surface.

Consider a grid $[u_0, u_1, u_2] \times [v_0, v_1, v_2]$. According to [4], [5], a spline surface $S \equiv S(x, y)$ is defined by the given values

$$z_{i,j}, \quad i = 0, 1, 2, \quad j = 0, 1, 2$$

*e-mail: viliam.kacala@student.upjs.sk
**e-mail: csaba.torok@upjs.sk

at the equispaced grid-points and the given first order directional derivatives

$$d^x_{i,j}, \quad i = 0, 2, \quad j = 0, 1, 2$$

at the boundary verticals,

$$d^y_{i,j}, \quad i = 0, 1, 2, \quad j = 0, 2$$

at the boundary horizontals and the cross derivatives

$$d^{x,y}_{i,j}, \quad i = 0, 2, \quad j = 0, 2$$

at the four corners of the grid.

To compute a bicubic clamped spline surface S of class C^2 one needs to define each component of a quadruple $[z_{i,j}, d^x_{i,j}, d^y_{i,j}, d^{x,y}_{i,j}]$ at every grid-point $[u_i, v_j]$, with the properties

$$z_{i,j} = S(u_i, v_j), \quad d^x_{i,j} = \frac{\partial S(u_i, v_j)}{\partial x}, \quad d^y_{i,j} = \frac{\partial S(u_i, v_j)}{\partial y}, \quad d^{x,y}_{i,j} = \frac{\partial^2 S(u_i, v_j)}{\partial x \partial y},$$

such that the adjacent spline segments would be twice continuously differentiable. This can be achieved using either de Boor's algorithm [4], [6] or a newer, computationally more efficient approach based on an interrelation between bicubic splines and biquartic polynomials [3].

For the sake of completeness we will define the relationship between splines and biquartic polynomials.

Definition 1 *The Hermite spline*

$$S(x, y) = \begin{cases} S_{0,0}(x, y) & \textit{for } (x, y) \in [u_0, u_1] \times [v_0, v_1], \\ S_{0,1}(x, y) & \textit{for } (x, y) \in [u_0, u_1] \times [v_1, v_2], \\ S_{1,0}(x, y) & \textit{for } (x, y) \in [u_1, u_2] \times [v_0, v_1], \\ S_{1,1}(x, y) & \textit{for } (x, y) \in [u_1, u_2] \times [v_1, v_2] \end{cases}$$

is based on the biquartic polynomial

$$P(x, y) = \sum_{i=0}^{4} \sum_{j=0}^{4} a_{i,j} x^i y^j, \qquad a_{4,4} \neq 0,$$

if the coefficients of S are computed from P according to

$$S(u_i, v_j) = z_{i,j} = P(u_i, v_j), \quad i, j = 0, 1, 2, \qquad \frac{\partial S(u_i, v_j)}{\partial x} = d^x_{i,j} = \frac{\partial P(u_i, v_j)}{\partial x}, \quad i = 0, 2, \ j = 0, 1, 2,$$

$$\frac{\partial S(u_i, v_j)}{\partial y} = d^y_{i,j} = \frac{\partial P(u_i, v_j)}{\partial y}, \quad i = 0, 1, 2, \ j = 0, 2, \qquad \frac{\partial^2 S(u_i, v_j)}{\partial x \partial y} = d^{x,y}_{i,j} = \frac{\partial^2 P(u_i, v_j)}{\partial x \partial y}, \quad i, j = 0, 2.$$

It is proved in [3] that S will be of class C^2 if it is based on any biquartic polynomial and the grid $[u_0, u_1, u_2] \times [v_0, v_1, v_2]$ is uniform. The task is to find additional constraints under which the 2×2-component spline $S = \{S_{0,0}, S_{1,0}, S_{0,1}, S_{1,1}\}$ will be an optimal approximant to the biquartic polynomial as well.

2 Approximation optimality

The Holladay seminorm is given by the functional ${}^{x_0,x_1}_{y_0,y_1}\|P\|^2 = \int_{x_0}^{x_1}\int_{y_0}^{y_1}\left|\frac{\partial^4 P(x,y)}{\partial x^2 \partial y^2}\right|^2 dy\,dx$, see [7] and [8].

Theorem 1 *Consider a grid $[u_0, u_1, u_2] \times [v_0, v_1, v_2]$ with $u_0 < u_1 < u_2$, $v_0 < v_1 < v_2$ and a biquartic polynomial P. Then the 2×2-component Hermite spline $S = \{S_{0,0}, S_{1,0}, S_{0,1}, S_{1,1}\}$ will be of class C^2 and the best approximant to P in the Holladay seminorm, if S is based on P and*

$$u_1 = \frac{1}{2}(u_0 + u_2), \quad v_1 = \frac{1}{2}(v_0 + v_2), \quad 0 = a_{2,4} = a_{4,2} = a_{3,4} = a_{4,3}, \tag{1}$$

Proof:

We have only to prove that ${}^{u_0,u_2}_{v_0,v_2}\|P - S\|^2$ attains its minimum under the assumptions of the theorem. First we derive the stationary point and then we find the conditions under which the norm attains its minimum in the stationary point.

The approximation error of the four-component Hermite spline in Holladay seminorm is (see [3])

$$\begin{aligned} N = {}^{u_0,u_2}_{v_0,v_2}\|P - S\|^2 &= {}^{u_0,u_1}_{v_0,v_1}\|P - S_{0,0}\|^2 + {}^{u_0,u_1}_{v_1,v_2}\|P - S_{0,1}\|^2 + {}^{u_1,u_2}_{v_0,v_1}\|P - S_{1,0}\|^2 + {}^{u_1,u_2}_{v_1,v_2}\|P - S_{1,1}\|^2 \\ &= {}^{u_0,u_1}_{v_0,v_1}\|ZA(x, y, u_0, u_1, v_0, v_1)\|^2 + {}^{u_0,u_2}_{v_1,v_2}\|ZA(x, y, u_0, u_1, v_1, v_2)\|^2 \\ &+ {}^{u_1,u_2}_{v_0,v_1}\|ZA(x, y, u_1, u_2, v_0, v_1)\|^2 + {}^{u_1,u_2}_{v_1,v_2}\|ZA(x, y, u_1, u_2, v_1, v_2)\|^2, \end{aligned}$$

where, e.g.,

$$\begin{aligned} ZA(x, y, x_0, x_1, y_0, y_1) = {} & Z(x, x_0, x_1)(a_{4,0} + a_{4,1}y + a_{4,2}y^2 + a_{4,3}y^3) + \\ & Z(y, y_0, y_1)(a_{0,4} + a_{1,4}x + a_{2,4}x^2 + a_{3,4}x^3) + \\ & a_{4,4}[Z(x, x_0, x_1)\sigma(y, y_0, y_1) + Z(y, y_0, y_1)\sigma(x, x_0, x_1) + Z(x, x_0, x_1)Z(y, y_0, y_1)], \end{aligned}$$

and $Z(t, t_0, t_1) = (t - t_0)^2(t - t_1)^2$, $\quad \sigma(t, t_0, t_1) = (t(t_0 + t_1) - t_0t_1) \cdot (2t^2 - t(t_0 + t_1) + t_0t_1)$.

Now we compute the partial derivatives of N

$$N_{u_1} = \frac{\partial N(x, y, u_0, u_1, u_2, v_0, v_1, v_2)}{\partial u_1} = -16(u_0 - u_2)(u_2 + u_0 - 2u_1) \cdot (u_0^2 - 2u_0u_1 + 2u_1^2)(v_0 - v_2)R_x,$$

$$\begin{aligned} N_{v_1} &= \frac{\partial N(x, y, u_0, u_1, u_2, v_0, v_1, v_2)}{\partial v_1} = -16(v_0 - v_2)(v_2 + v_0 - 2v_1) \cdot (v_0^2 - 2v_0v_1 + 2v_1^2)(u_0 - u_2)R_y, \\ N_{u_1,v_1} &= \frac{\partial^2 N(x, y, u_0, u_1, u_2, v_0, v_1, v_2)}{\partial u_1 \partial v_1} \\ &= -16a_{5,5}^2(v_0 - v_2)(v_2 + v_0 - 2v_1)(v_0^2 - 2v_0v_1 + 2v_1^2)(u_0 - u_2)(u_2 + u_0 - 2u_1)(u_0^2 - 2u_0u_1 + 2u_1^2), \end{aligned}$$

where the expressions R_x and R_y depend on v_0, v_1, v_2, $a_{4,2}$, $a_{4,3}$, $a_{4,4}$ and u_0, u_1, u_2, $a_{2,4}$, $a_{3,4}$, $a_{4,4}$, respectively. From the first derivatives N_{u_1} and N_{v_1} we get that a stationary point located at

$$[u_1, v_1] = \left[\frac{1}{2}(u_2 + u_0), \frac{1}{2}(v_2 + v_0)\right].$$

Since N_{u_1} and N_{v_1} have no other real roots, $[u_1, v_1]$ denotes the location of a global extremum. To decide whether this is a point of minimum or maximum, the Hessian $H = N_{u_1,u_1}N_{v_1,v_1} - (N_{u_1,v_1})^2$ at point $[u_1, v_1]$ is to be computed. Since $N_{u_1,v_1} = 0$ at the stationary point, see above, after the evaluation of N_{u_1,u_1} and N_{v_1,v_1}, the substitution of u_1, v_1 and some algebraic transformations, the following expression is obtained for H,

$$H = \psi(u_0, u_2, a_{4,2}, a_{4,3}, a_{4,4})\,\psi(v_0, v_2, a_{2,4}, a_{3,4}, a_{4,4}),$$

where

$$\psi(t_0, t_2, b_0, b_1, b_2) = 16b_0^2 + \left(48(t_0 + t_2)b_1 + 64(t_0^3 + t_0t_2 + t_2^3)b_2\right)b_0 + 48(t_0^2 + t_0t_2 + t_2^2)\,b_1^2$$
$$+ 144(t_0^3 + t_0^2t_2 + t_0t_2^2 + t_2^3)b_2b_1 + (115t_0^4 + 116t_0^3t_2^1 + 114t_0^2t_2^2 + 116t_0^1t_2^3 + 115t_2^4)\,b_2^2.$$

Let $0 = a_{2,4} = a_{4,2} = a_{3,4} = a_{4,3}$. Then

$$H = \psi(u_0, u_2, 0, 0, a_{4,4})\,\psi(v_0, v_2, 0, 0, a_{4,4}),$$

where

$$\psi(t_0, t_2, 0, 0, b_2) = \left(115t_0^4 + 116t_0^3t_2^1 + 114t_0^2t_2^2 + 116t_0^1t_2^3 + 115t_2^4\right)b_2^2.$$

Since the quartic polynomial $\psi(t_a, t_c, 0, 0, a_{4,4})$ has no real root and is positive at least at one point, it results that $\psi(t_a, t_c, 0, 0, a_{4,4}) > 0$. Hence it follows immediatelly that the Hessian $H > 0$, which means that under the assumption (1) the approximation error in Holladay seminorm N has global minimum at the point $[u_1, v_1] = \left[\frac{1}{2}(u_2 + u_0), \frac{1}{2}(v_2 + v_0)\right]$. □

Remark 1 The conditions on $a_{i,j}$ in (1) can be expressed using function values and derivatives as well, however, not uniquely. It follows from the theorem that, in general a bicubic spline, cannot optimally approximate even the nearest higher degree quartic polynomial.

3 Summary

We derived the constraints under wich a 2×2-component interpolating bicubic spline of class C^2 optimally approximates a biquartic polynomial in Holladay seminorm.

4 Acknowledgement

This work was partially supported by projects Technicom ITMS 26220220182 and APVV-15-0091 Effective algorithms, automata and data structures.

References

[1] Cs. Török, Matematicheskoe modelirovanie **26**, 11, 33–36 (2014)

[2] J. Buša, L. Miňo, and Cs. Török, in *Proceedings of 1st year of the Spring School of Doctoral Students* (UPJŠ, Košice, Slovakia, 2015)

[3] L. Miňo, I. Szabó, and Cs. Török, Open Computer Science **6**, 1, 1–7 (2016)

[4] C. de Boor, Journal of Mathematics and Physics **41** (3), 212–218 (1962)

[5] E.V. Shikin and A.I. Plis, *Handbook on Splines for the User* (CRC Press, 1995) 141 pp.

[6] D. Salomon, *Curves and Surfaces for Computer Graphics* (Springer, 2006), 1–460

[7] *Advanced Course on FAIRSHAPE* (Editors J. Hoscheck and P. Kaklis, B. G. Teubner Stuttgart, 1996) 210 pp.

[8] J. Stoer and R. Bulirsch, *Introduction to numerical analysis* (Springer, 2002) 99 pp.

EPJ Web of Conferences **173**, 03013 (2018) https://doi.org/10.1051/epjconf/201817303013
Mathematical Modeling and Computational Physics 2017

Data Mining in Institutional Economics Tasks

Igor Kirilyuk[1,⋆], *Anna* Kuznetsova[2,⋆⋆], and *Oleg* Senko[3,⋆⋆⋆]

[1] *Institute of Economics of RAS, Moscow, Russia*
[2] *Emanuel Institute of Biochemical Physics, Moscow, Russia*
[3] *FRC "Informatics and Control" of RAS, Moscow, Russia*

Abstract. The paper discusses problems associated with the use of data mining tools to study discrepancies between countries with different types of institutional matrices by variety of potential explanatory variables: climate, economic or infrastructure indicators. An approach is presented which is based on the search of statistically valid regularities describing the dependence of the institutional type on a single variable or a pair of variables. Examples of regularities are given.

1 Introduction

The theory of institutional matrices proposed by S.G. Kirdina ([1, 2]) supposes that two types of basic public institutions (or two types of institutional matrices in other words) may be identified. The first type may be characterized by a redistributive economy, a unitary political system, and communitarian ideology. This basic type will be referred to as X type. The second type that will be referred to as Y type is complementary to the first type. It is characterized by the market economy, the federal political system and the ideology of individualism. The countries of the world may be divided in two groups: X countries with dominance of X type of the institutional matrices and Y countries with dominance of Y type of the institutional matrices. Our goal is to check the institutional matrices theory and also to describe the relationship between the economic development characteristics and the dominating institutional type. In this regard we try to find statistically significant regularities allowing to discriminate between two types of countries by macroeconomic variables, climate and infrastructure indicators. The presence of valid patterns in the data provides strong evidence of the correctness of the institutional matrices theory. In [3] the relationship between the dominating type of institutional matrices and the climate or infrastructure variables was studied. In [4] the optimal valid partitioning (OVP) method was applied to compare joint dynamics of macroeconomic variables in groups of X and Y countries. A specially developed technique for evaluating multiple testing effects was used. A statistically significant relationship was found to hold between the institutional type and mutual dynamics of the gross domestic product (GDP), domestic credit and the governmental spendings.

⋆e-mail: igokir@rambler.ru
⋆⋆e-mail: azfor@yandex.ru
⋆⋆⋆e-mail: senkoov@mail.ru

2 Statistical tools

Today different data mining tools are available that may be useful in tasks of valid regularities search. There are several peculiarities that influence the choice of analytical tools. We must compare rather small groups of countries where experts are sure about the dominant type of institutional matrices. The full number of such countries varies from 25 to 35. Climate indicators include average monthly temperatures and precipitation amounts, their average year minimum, maximum and variations. Infrastructure indicators include road and railway density, irrigation levels and others. Average values of the macroeconomic variables over some time periods are less informative than various parameters describing joint dynamics of macroeconomic variables. The number of such parameters may achieve several tens. Hence the total number of factors involved in analyses may significantly exceed the total number of countries. Isolated factors often are not sufficiently informative and so it is important to analyse their combinations. There is no evidence that the studied factors may be distributed normally. In our research we used the OVP method [3]. OVP technique is aimed to explore the dependence of a target value on the set of explanatory variables. The OVP method includes the search of one-dimensional and two-dimensional regularities. A one-dimensional regularity is described by a threshold for the corresponding factor that separates in the best way the compared groups. Two quadrants to the left and to the right of the threshold are formed. Example of one-dimensional regularity is given at figure 1.

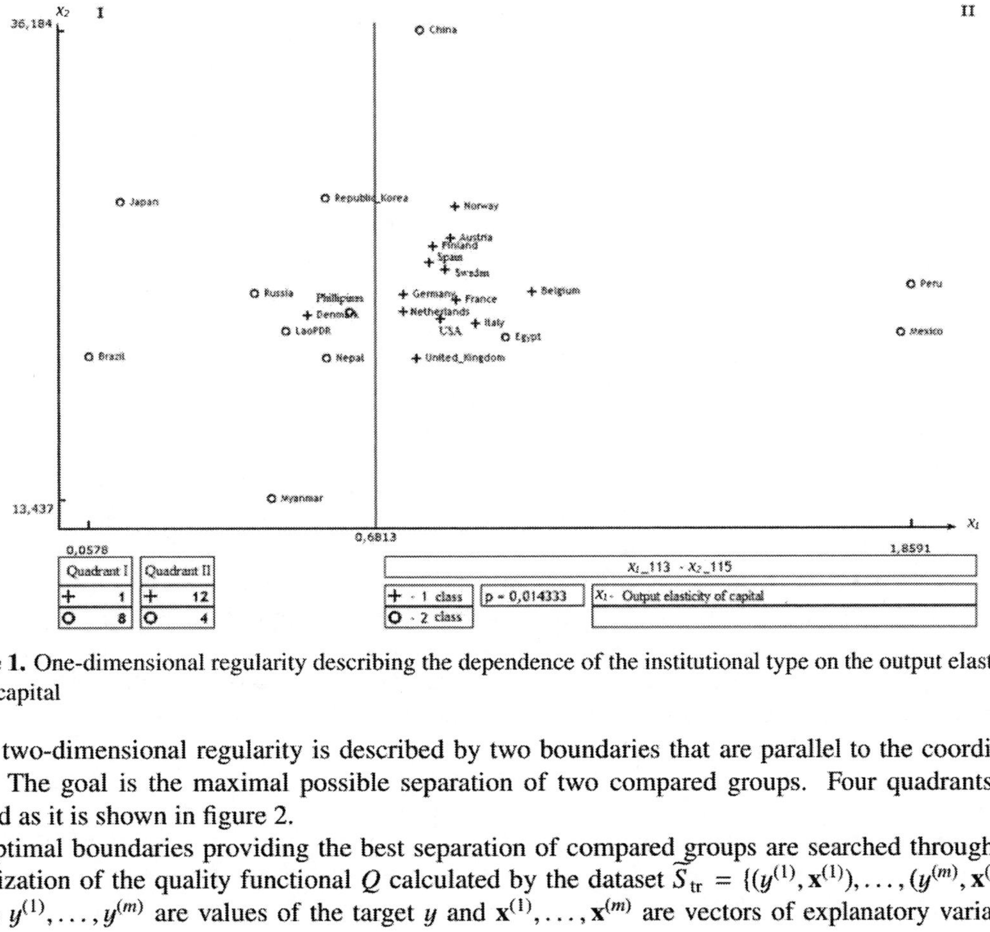

Figure 1. One-dimensional regularity describing the dependence of the institutional type on the output elasticity of the capital

A two-dimensional regularity is described by two boundaries that are parallel to the coordinate axes. The goal is the maximal possible separation of two compared groups. Four quadrants are formed as it is shown in figure 2.

Optimal boundaries providing the best separation of compared groups are searched through the optimization of the quality functional Q calculated by the dataset $\widetilde{S}_{\mathrm{tr}} = \{(y^{(1)}, \mathbf{x}^{(1)}), \ldots, (y^{(m)}, \mathbf{x}^{(m)})\}$, where $y^{(1)}, \ldots, y^{(m)}$ are values of the target y and $\mathbf{x}^{(1)}, \ldots, \mathbf{x}^{(m)}$ are vectors of explanatory variables

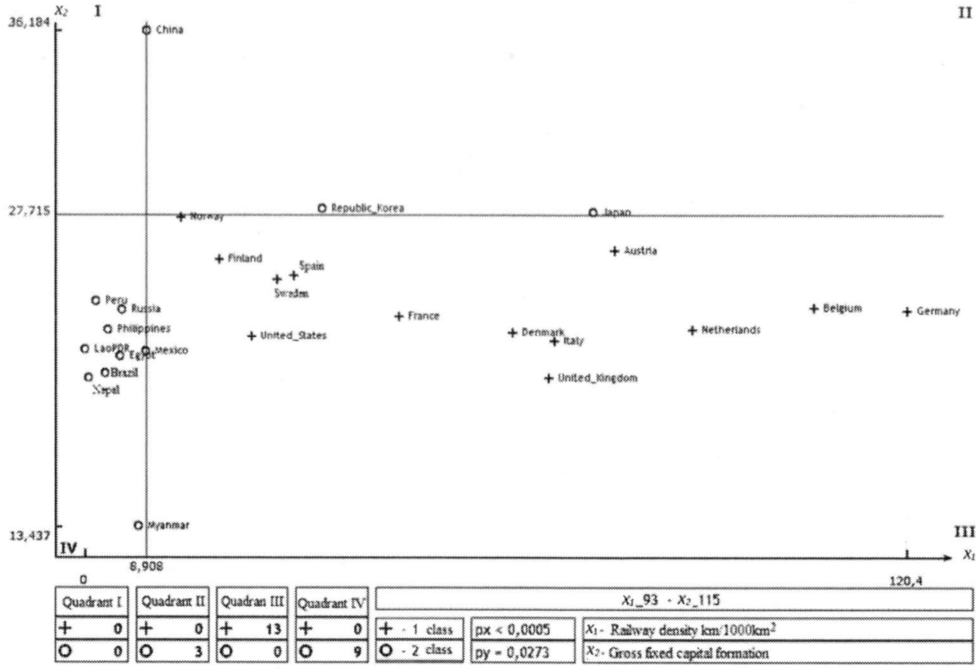

Figure 2. Two-dimensional regularity describing the dependence of the institutional type on the railway density and GFCF

$x_1, \ldots, x_n$ at objects from data set $\widetilde{S}_{\mathrm{tr}}$: $Q = \frac{1}{D_y}\sum_{i=1}^{r} m_i \cdot (\hat{y}_i - \hat{y}_0)^2$, where $\hat{y}_0$ denotes the average of the target values y at objects of $\widetilde{S}_{\mathrm{tr}}$, $\hat{y}_i$ denotes the average of the target values y at objects of $\widetilde{S}_{\mathrm{tr}}$ from the quadrant Q_i, m_i is the number of $\widetilde{S}_{\mathrm{tr}}$ objects from Q_i, r denotes the number of quadrants. The validity of the regularities is evaluated with the help of a random permutation test (RPT). The values of the functional Q on the data set $\widetilde{S}_{\mathrm{tr}}$ are compared with the values of the functional Q calculated on a set of random samples generated from $\widetilde{S}_{\mathrm{tr}}$. The positions of the values of the target y in the random sample are randomly permuted relatively to the fixed positions of the explanatory variables vectors. The proportion of random samples for which Q does not exceed the value calculated on $\widetilde{S}_{\mathrm{tr}}$ is used as the p-value estimate. It was shown [4] that the RPT tests fulfill the null hypothesis that the target y is independent on the corresponding explanatory variable.

In the case of two-dimensional regularities the combination of two null hypothesis is tested. Let the dependence of the target y on two explanatory variables x_1 and x_2 is studied. At the first step a first null hypothesis is tested: the dependence of y on x_1 and x_2 is exhaustively described by a simple one-dimensional regularity with the threshold for the variable x_2 coinciding with the corresponding border value in the tested two-dimensional regularity. To test this hypothesis comparison is made with the values of the functional Q calculated on a set of random samples generated from $\widetilde{S}_{\mathrm{tr}}$. The procedure coincides with the technique used when one-dimensional regularity is verified but permutations are allowed only inside two subsets of $\widetilde{S}_{\mathrm{tr}}$ with x_2 values to the left and to the right of the threshold. Thus a p-value evaluating the validity of x_1 contribution to the two-dimensional regularity is calculated. In the same way a p-value evaluating the validity of x_2 contribution to the two-dimensional regularity is calculated. The two-dimensional regularity is assumed to be valid only if contributions of both variables are valid at some fixed accuracy level. It must be noted that the RPT does not ask for a priori assumptions about the probability distributions. There is no any requirement on sample sizes.

The permutation test evaluates the validity of isolated regularities. But in tasks where the number of tested data patterns is large multiple testing problem exists: the probability that a really random patterns will be incidentally identified as a valid regularity is also large. For example in tasks with a large number of explanatory variables the probability that an empirical distribution for at least one variable occasionally corresponds to a valid one-dimensional regularity is high.

Usually a Bonferroni correction [4] is used to receive more realistic validity estimates: the calculated p-values are multiplied by the number of tested explanatory variables. The multiple testing problem is more difficult when two-dimensional regularities are searched because it is necessary to multiply the calculated p-values by a very large number of tested pairs of variables. The Bonferroni correction is too pessimistic when it is applied to two-dimensional regularities search. An alternative approach was discussed in [4]. Two dimensional regularities found in a true data set were compared with two-dimensional regularities found in random samples that were obtained from the true data set by target permutation. Such procedure is more correct than the inferences got from the standard Bonferroni procedure.

3 Results

In this paper we present new results of recent researches. A statistically significant ($p = 0.014$) relationship was found to hold between institutional type and the output elasticity of the capital (OEC) parameters in the Cobb-Douglas production function [5]. This regularity is shown in figure 1. It is seen that the OEC is less than the threshold 0.6813 for 8 X countries (Russia, Japan, Republic of Korea, Phillipines, Laos, Nepal, Brazil, Myanmar) and only for one Y country – Denmark. On the contrary, the OEC is larger than the threshold for 12 Y countries and only for 4 X countries. The OEC for Mexico and Peru are extremely high.

From the figure 2 a two dimensional significant regularity follows which describes the relationship between the institutional type and two variables: the gross fixed capital formation (GFCF) ($p < 0.0005$) and the railway density ($p = 0.0273$). The quadrant IV corresponding to low railway density and relatively low GFCF contains 9 X countries, the quadrant III corresponding to high railway density and relatively low GFCF contains all the 13 countries of Y type, the quadrant II corresponding to high railway density and high GFCF contain 3 countries of X type: Japan, China, Republic of Korea.

The existence of regularities shown in figures 1 and 2 show that the parameters of the production functions or GFCF may be related to the dominating type of the institutional matrices. Data from [6] was used.

Acknowledgement

The work was supported by RFBR grant 17-02-00207.

References

[1] S. Kirdina, Journal of Economic Issues **47**, 309–322 (2014)
[2] S. Kirdina-Chandler, Journal of Economic Issues **51**, 476–485 (2017)
[3] I. L. Kirilyuk, A. I. Volynsky, M. S. Kruglova, A. V. Kuznetsova, A. A. Rubinstein, and O. V. Senko, Computer Research and Modeling **51**, 923–939 (2015) [in Russian]
[4] I. L. Kirilyuk, A. V. Kuznetsova, O. V. Senko, and A. M. Morozov, Pattern recognition and image analysis **27**, 94–104 (2015)
[5] C. W. Cobb and P. H. Douglas, American Economic Review **51** (18), 139–165 (1928)
[6] R. C. Feenstra, R. Inklaar, and M. P. Timmer, American economic review **105** (10), 3150–3182 (2015)

EPJ Web of Conferences **173**, 03014 (2018) https://doi.org/10.1051/epjconf/201817303014
Mathematical Modeling and Computational Physics 2017

Extrapolation of Functions of Many Variables by Means of Metric Analysis

Alexandr Kryanev[1,2,⋆], *Victor* Ivanov[2,1,⋆⋆], *Anastasiya* Romanova[1,⋆⋆⋆], *Leonid* Sevastianov[3,2,⋆⋆⋆⋆], and *David* Udumyan[4,1,3,†]

[1] *National Research Nuclear University "MEPhI", Moscow, Russia*
[2] *Joint Institute for Nuclear Research (JINR), Dubna, Moscow region, Russia*
[3] *Peoples' Frendship University of Russia (PFUR University), Moscow, Russia*
[4] *University of Miami, USA*

Abstract. The paper considers a problem of extrapolating functions of several variables. It is assumed that the values of the function of m variables at a finite number of points in some domain D of the m-dimensional space are given. It is required to restore the value of the function at points outside the domain D. The paper proposes a fundamentally new method for functions of several variables extrapolation. In the presented paper, the method of extrapolating a function of many variables developed by us uses the interpolation scheme of metric analysis. To solve the extrapolation problem, a scheme based on metric analysis methods is proposed. This scheme consists of two stages. In the first stage, using the metric analysis, the function is interpolated to the points of the domain D belonging to the segment of the straight line connecting the center of the domain D with the point M, in which it is necessary to restore the value of the function. In the second stage, based on the auto regression model and metric analysis, the function values are predicted along the above straight-line segment beyond the domain D up to the point M. The presented numerical example demonstrates the efficiency of the method under consideration.

1 Introduction

One of the main problems of data processing in many areas is the problem of functions of several variables extrapolation. In the presented paper, the method of a function of many variables extrapolation, developed by us, uses the interpolation scheme of metric analysis. Below is a brief description of the metric analysis of interpolation of the values of functions of several variables and its application [3].

⋆e-mail: avkryanev@mephi.ru
⋆⋆e-mail: ivanov@jinr.ru
⋆⋆⋆e-mail: 2015anr@mail.ru
⋆⋆⋆⋆e-mail: sevastianov_la@rudn.university
†e-mail: mathudum@gmail.com

2 Extrapolation scheme

At the first and second stages of the scheme of extrapolation of functions of many variables presented in the article, the interpolation of metric analysis is used. The extrapolation scheme uses the interpolation method for the functional dependence:

$$Y = F(X_1, \ldots, X_m) = F(\vec{X}), \tag{1}$$

where the function $F(\vec{X})$ is unknown and is subject to recovery, either at one point $\vec{X}^*$ or in a set of given points on the basis of known values of the function $Y_k, k = 1, \ldots, n,$ in points $\vec{X}_k = (X_{k1}, \ldots, X_{km})^T$ [1].

According to the method of interpolation, based on metric analysis, interpolation values are found as solutions of problems of minimum of the measure of metric uncertainty with respect to the point $\vec{X}^* = (X_1^*, \ldots, X_m^*)^T$ [1]–[4]

$$\sigma^2_{ND}(Y^*; \vec{z}^*) = (W\vec{z}^*, \vec{z}^*), \tag{2}$$

where W is a matrix of the metric uncertainty and the interpolation value is determined by a linear combination

$$Y^* = \sum_{i=1}^{n} z_i^* Y_i, \qquad \sum_{i=1}^{n} z_i^* = 1, \tag{3}$$

and is given by

$$Y^* = \frac{(W^{-1}\vec{Y}, \vec{1})}{(W^{-1}\vec{1}, \vec{1})}. \tag{4}$$

The matrix of the metric uncertainty is defined by

$$W = \begin{pmatrix} \rho^2(\vec{X}_1, \vec{X}^*)_{\vec{w}} & (\vec{X}_1, \vec{X}_2)_{\vec{w}} & \ldots & (\vec{X}_1, \vec{X}_n)_{\vec{w}} \\ (\vec{X}_2, \vec{X}_1)_{\vec{w}} & \rho^2(\vec{X}_2, \vec{X}^*)_{\vec{w}} & \ldots & (\vec{X}_2, \vec{X}_n)_{\vec{w}} \\ \ldots & \ldots & \ldots & \ldots \\ (\vec{X}_n, \vec{X}_1)_{\vec{w}} & (\vec{X}_n, \vec{X}_2)_{\vec{w}} & \ldots & \rho^2(\vec{X}_n, \vec{X}^*)_{\vec{w}} \end{pmatrix}, \tag{5}$$

where $\rho^2(\vec{X}_i, \vec{X}^*)_{\vec{w}} = \sum\limits_{k=1}^{m} w_k (X_{ik} - X_k^*)^2$, $(X_i, X_j)_{\vec{w}} = \sum\limits_{k=1}^{m} w_k (X_{ik} - X_k^*) \cdot (X_{jk} - X_k^*)$, $i \neq j = 1, \ldots, n$, where w_k, $k = 1, \ldots, m$ are metric weights (see [3]).

Next, a metric analysis scheme is used to predict (extrapolate) the function of a single variable using an autoregressive model of metric analysis.

Let us consider a function $y = f(x)$ of one variable x with known values $Y_1 = f(x_1), \ldots, Y_n = f(x_n)$ for $x_1 < \cdots < x_n \in [x_1, x_n]$. It is required to find the extrapolated value Y_{n+1} for x_{n+1}.

The problem of finding the extrapolated value Y_{n+1} is reduced to the problem of interpolation of functions of several variables by means of a nonlinear autoregressive model [3, 4]:

$$\begin{array}{rcl} y(x_{l+1}) = Y_{l+1} &=& F(Y_1, \ldots, Y_l), \\ y(x_{l+2}) = Y_{l+2} &=& F(Y_2, \ldots, Y_{l+1}), \\ \ldots \quad \ldots \ldots \ldots & & \ldots\ldots \\ y(x_n) = Y_n &=& F(Y_{n-l}, \ldots, Y_{n-1}). \end{array} \tag{6}$$

Then the extrapolation of the function $y = f(x)$ is reduced to the interpolation of the function of l variables $Y = F(y_1, y_2, \ldots, y_l)$ with values in $n - l$ points

$$\begin{array}{rcl} \vec{X}_1 &=& (Y_1, \ldots, Y_l)^T, \\ \vec{X}_2 &=& (Y_2, \ldots, Y_{l+1})^T, \\ \ldots\ldots\ldots & & \\ \vec{X}_{n-l} &=& (Y_{n-l}, \ldots, Y_{n-1})^T. \end{array} \tag{7}$$

The extrapolated value $Y_{\text{ext}} = Y_{n+1}$ is defined as an interpolation value of the function $Y = F(y_1, y_2, \ldots, y_l)$ at the point $\vec{X}^*$:

$$Y_{n+1} = F(\vec{X}^*) = \frac{(W^{-1}\vec{1}, \vec{Y})}{(W^{-1}\vec{1}, \vec{1})}, \tag{8}$$

where $\vec{X}^* = (Y_{n-l+1}, \ldots, Y_n)^T$, W^{-1} is the inverse matrix for the $(n-l) \times (n-l)$ matrix of metric uncertainty, $\vec{Y} = (Y_{l+1}, \ldots, Y_n)^T$ is the $(n-l)$-dimensional vector of the values of the extrapolated function.

The number l determines the dimension of the space of vectors, and its value in the test was found as a solution of extremal problem [1, 2]

$$l^* = \text{argmin}\, ||\vec{Y}_{\text{re}} - \vec{Y}_{\text{ext}}||, \tag{9}$$

where minimization is done in l, $\vec{Y}_{\text{re}}$ is the vector of realized values and $\vec{Y}_{\text{ext}}$ is the vector of extrapolated values.

The extrapolation scheme for the function of several variables (1) at a given point $\vec{X}^* = (X_1^*, \ldots, X_m^*)^T$ consists of two stages.

At the first stage, the point $\vec{X}_0 = (X_{01}, \ldots, X_{0m})^T$ is selected inside the cluster of the realized values of the function $Y = F(\vec{X})$. Then the points $\vec{X}_0$ and $\vec{X}^*$ are connected by a straight line segment

$$(1-s) \cdot \vec{X}_0 + s \cdot \vec{X}^*, \quad 0 \leq s \leq 1, \tag{10}$$

which is divided into L equal segments with nodes

$$\vec{S}_k = (S_{k1}, \ldots, S_{km})^T, \quad k = 1, \ldots, L+1, \quad \vec{S}_{L+1} = \vec{X}^*. \tag{11}$$

In the points

$$\vec{S}_k = (S_{k1}, \ldots, S_{km})^T, \quad k = 1, \ldots, l, \quad l < L+1, \tag{12}$$

belonging to the cluster and rectilinear segment (10), the interpolation of the values of the function (1) is performed using the scheme (4)–(5) on the set of known values of the function Y_i, $i = 1, \ldots, n$ at the points $\vec{X}_i = (X_{i1}, \ldots, X_{im})^T$ belonging to the cluster.

At the second stage, the interpolated values, which were calculated at the first step $\vec{Y} = (Y_1, \ldots, Y_l)^T$ at the points (12), are successively extrapolated to the remaining nodes $\vec{S}_k = (S_{k1}, \ldots, S_{km})^T$, $k = l+1, \ldots, L+1$, where

$$f(s) = F((1-s) \cdot \vec{X}_0 + s \cdot \vec{X}^*), \quad 0 \leq s \leq 1. \tag{13}$$

By using the autoregressive scheme (6)–(9) at the nodes

$$s = s_k, \quad k = l+1, \ldots, L+1, \quad s_k = s_{k-1} + \Delta s, \quad \Delta s = \frac{||\vec{X}^* - \vec{X}_0||}{L}, \tag{14}$$

one can obtain the extrapolated value

$$Y_{\text{ext}} = F(\vec{X}^*) = f(s_{L+1}). \tag{15}$$

3 Numerical result

Example. Function $Y = F(x_1, \ldots, x_{10}) = \sum_{i=1}^{10} x_i^2 + \sum_{i=1}^{10} x_i + 6 \cdot \sin(2(x_1 + x_{10}))$ is considered.

Values of function are set in 8 points:

(0.1;0.6;0.9;0.1;0.1;0.7;0.6;0.1;0;0.9), (0.3;0.7;0.3;0.7;0.2;0.3;0.4;0.8;0.9;0.1), (0.5;0.4;0.1;0.4;0.6;0.1;0;0.3;0.6;0.4), (0.4;0.2;0.3;0.8;0.8;0.2;0.4;0.2;0.3;0.3), (0.2;0.7;0.4;0.3;0.3;0.8;1;0.5;1;0.2), (1;0;0.8;0.6;0.9;0.5;0.3;0.4;0.5;0.3), (0.3;1;0.8;0.5;1;0.9;0.5;0.4;0.3;1), (1;0.8;0.7;0.6;0.8;0.8;1;1;0.9;1)

It is necessary to find the the extrapolated value of Y at the point (1.3;1;0.6;0.8;1;0.8;1.5;1;1.4;1.3).

Extrapolation is made at the points of a straight line connecting the point of the center of domain (0.5;0.5;0.5;0.5;0.5;0.5;0.5;0.5;5;0.5) of the given 8 points and the point (1.3;1;0.6;0.8;1;0.8;1.5;1;1.4;1.3).

For the example in question m=10, n=8, l^*=4, L=20.

Table 1 lists the exact and extrapolated values of the function for k =13,...,21 (see (13)–(14)).

Table 1. Numerical example

k	13	14	15	16	17	18	19	20	21
Exact values	11.58	11.78	12.11	12.61	13.27	14.1	15.1	16.29	17.63
Extrapolated values	11.83	11.97	12.34	12.77	13.32	13.91	14.68	15.37	16.12

4 Conclusion

In this paper a fundamentally new method of the extrapolation for functions of several variables is proposed. The obtained numerical results of extrapolation of functions of several variables, taken from different areas, show that the presented scheme can provide a reliable accuracy of extrapolation results.

The publication was prepared under the support of the “PFUR University Program 5-100”.

References

[1] A.V. Kryanev, G.V. Lukin, and D.K. Udumyan, *Metric analysis and data processing* (M.: Ed. Science, 2012)

[2] A.V. Kryanev and D.K. Udumyan, Int. Journal of Mathematical Analysis **8** (45), 2221–2228 (2014)

[3] A.V. Kryanev and D. K. Udumyan, Int. Journal of Mathematical Analysis **8** (60), 2971–2978 (2014)

[4] A.V. Kryanev, D.K. Udumyan, G.V. Lukin, and V.V. Ivanov, Applied Mathematical Sciences **8** (22), 1053–1060 (2014)

EPJ Web of Conferences **173**, 03015 (2018) https://doi.org/10.1051/epjconf/201817303015
Mathematical Modeling and Computational Physics 2017

Shift of Shapiro Step in High Critical Temperature Superconductors

Kirill Kulikov[1,2,⋆], *Yury* Shukrinov[1,2], *Majed* Nashaat[1,3], and *Akinobu* Irie[4]

[1] *BLTP, JINR, Dubna, Moscow Region, 141980, Russia*
[2] *State University of "Dubna", Dubna, Moscow region, 141980 Russia*
[3] *Department of Physics, Cairo University, Cairo, Egypt*
[4] *Department of Electrical and Electronic Systems Engineering, Utsunomiya University, Japan*

Abstract. The influence of the charge imbalance effect on a system of intrinsic Josephson junctions of high temperature superconductors under external electromagnetic radiation is investigated. We find that the charge imbalance is responsible for the slope of the Shapiro step in the IV-characteristic. The nonperiodic boundary conditions shift the Shapiro step from the canonical position which is determined by the frequency of the external radiation. We also find how the system parameters influence the shift of the Shapiro step.

1 Introduction

The phase dynamics of the layered superconducting materials have attracted a great interest because of the rich and interesting physics on one side and application perspectives on the other side [1, 2]. In particular, the nonequilibrium effects created by stationary current injection in high-T_c materials have been studied very intensively during the last two decades [1–7]. However, the charge imbalance in the systematic perturbation theory was considered only indirectly as far as it is induced by fluctuations of the scalar potential [1, 2, 5]. In Ref. [8], it is taken into account as an independent degree of freedom and therefore the results are different from those of earlier treatments. In addition, due to the fact that the charge is not screened in the superconducting layers the system forms intrinsic Josephson junctions (IJJ) [9, 10]. *Such system cannot be at equilibrium at any value of the electrical current.* The influence of the charge coupling on the Josephson plasma oscillations was studied in [6, 9].

During the last few years, two theoretical models are widely used to describe the IJJ: the capacitively coupled Josephson junctions (CCJJ) model and the charge imbalance (CIB) model. In the CCJJ model a non-vanishing generalized scalar potential appears due to the breaking of the charge neutrality, but in CIB model it is related to the quasiparticle charge imbalance. Actually, the relaxation length of the charge imbalance in the layered system could be much larger than any other characteristic lengths. Therefore, both effects could exist in HTSC simultaneously because the thickness of the superconducting layers is smaller than the Debye length and thus obviously less than the relaxation length of the charge imbalance. In the present paper we study the nonequilibrium effects created by current injection in a stack of IJJ under external electromagnetic radiation.

⋆e-mail: kulikov@theor.jinr.ru

2 Model

A system of $N + 1$ superconducting layers (S-layers) presented in Fig. 1 is characterized by the order parameter $\Delta_l(t) = |\Delta| \exp(i\theta_l(t))$ and time-dependent phase $\theta_l(t)$. The thickness of the S-layer is

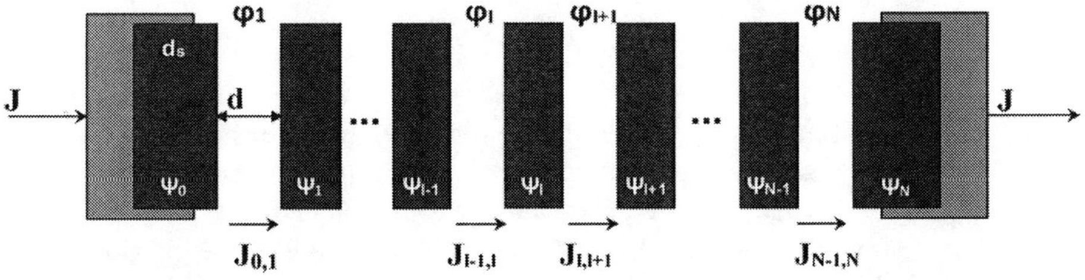

Figure 1. Layered system of $N + 1$ superconducting layers forms a stack of Josephson junctions. Since the 0-th and N-th layers are in contact with normal metal, their thicknesses d_s^0 and d_s^N are different from the thickness of the other S-layers d_s inside of the stack due to the proximity effect.

comparable with the Debay screening r_D length that leads to the generalized Josephson relation [11]. The total current density J_l through each S-layer is given as a sum of displacement, superconducting, quasiparticle, diffusion and nonequilibrium terms. Those equations together with the kinetic equations for the nonequilibrium potential $\Psi_l(t)$ describe the physics of IJJs in HTSC. In the dimensionless form the system of equations is

$$\dot{v}_l = \left[I - \sin\varphi_l - \beta\dot{\varphi}_l + A\sin\omega\tau + I_{\text{noise}} + \psi_l - \psi_{l-1}\right], \tag{1}$$

$$\dot{\varphi}_1 = v_1 - \alpha(v_2 - (1+\gamma)v_1) + \frac{\psi_1 - \psi_0}{\beta}, \tag{2}$$

$$\dot{\varphi}_l = (1 + 2\alpha)v_l - \alpha(v_{l-1} + v_{l+1}) + \frac{\psi_l - \psi_{l-1}}{\beta}, \tag{3}$$

$$\dot{\varphi}_N = v_N - \alpha(v_{N-1} - (1+\gamma)v_N) + \frac{\psi_N - \psi_{N-1}}{\beta}, \tag{4}$$

$$\zeta_0\dot{\psi}_0 = \eta_0\left(I + A\sin\omega\tau - \beta\dot{\varphi}_{0,1} + \psi_1 - \psi_0\right) - \psi_0, \tag{5}$$

$$\zeta_l\dot{\psi}_l = \eta_l(\beta[\dot{\varphi}_{l-1,l} - \dot{\varphi}_{l,l+1}] + \psi_{l-1} + \psi_{l+1} - 2\psi_l) - \psi_l,$$

$$\zeta_N\dot{\psi}_N = \eta_N\left(-I - A\sin\omega\tau + \beta\dot{\varphi}_{N-1,N} + \psi_{N-1} - \psi_N\right) - \psi_N, \tag{6}$$

where the dot shows the derivative with respect to $\tau = \omega_p t$, v_l between the layers $l-1$ and l, $v_l(t) \equiv v_{l,l-1}(t)$, $\varphi_l(t)$ is the phase difference across the layers $l-1$ and l, $\alpha = \epsilon\epsilon_o/2e^2N(0)d$ is the coupling parameter, ϵ is the dielectric constant, ϵ_o is the vacuum permittivity, d is the distance between the superconducting layers and $N(0)$ is the density of states, $I = J/J_c$ is the dimensionless current density, J_c is the critical current density, $\omega_p = \sqrt{2eJ_c/(\hbar C)}$ is the plasma frequency, C is the capacitance. Other dimensionless parameters are the dissipation parameter $\beta = \hbar\omega_p/(2eRI_c)$, R is the junction resistance, the normalized quasiparticle relaxation time $\zeta_l = \omega_p\tau_{qp}$ and the nonequilibrium parameter $\eta_l = 4\pi r_D^2\tau_{qp}/(d_s^l R)$, where d_s^i is the thickness of the S-layers, and τ_{qp} is the quasiparticle relaxation time. The parameter of the nonperiodic boundary conditions γ is $\gamma = d_s/d_s^0 = d_s/d_s^N$. The term $A\sin\omega\tau$ describes the effect of the external radiation with the amplitude A and frequency ω, which are normalized to J_c and ω_p, respectively. To reflect the experimental situation, we have added the noise I_{noise} in the bias current with the amplitude $\sim 10^{-8}$ which is produced by a random number generator and its amplitude is normalized to the critical current density value J_c.

This system of equations is solved numerically using the fourth order Runge-Kutta method. We assume that due to the proximity effect the thickness of the first and the last S-layers are larger than the middle one. Therefore, the nonequilibrium parameters depend on the parameter of boundary conditions γ, $\eta_{0,N} = \gamma\eta_l$, where $l = 1, 2, \ldots, N-1$. We consider the underdamped case with the McCumber parameter $\beta_c = 25$ or $\beta = 0.2$.

3 Results

In [11] we have shown that in the system of intrinsic Josephson junctions of high temperature superconductors under external electromagnetic radiation the charge imbalance is responsible for the slope in the Shapiro step in the IV-characteristic. The value of the slope increases with the nonequilibrium parameter. We demonstrate that the coupling between junctions leads to the distribution of the slope values along the stack. It was shown also that the nonperiodic boundary conditions shift the Shapiro step from the canonical position $V = N\omega$.

The simulated IV-characteristics of the JJs stack in the case without the charge imbalance $\eta = 0$ (dashed line) and at $\eta = 0.2$ (solid line) are presented in Fig. 2. The IV-curve without the charge

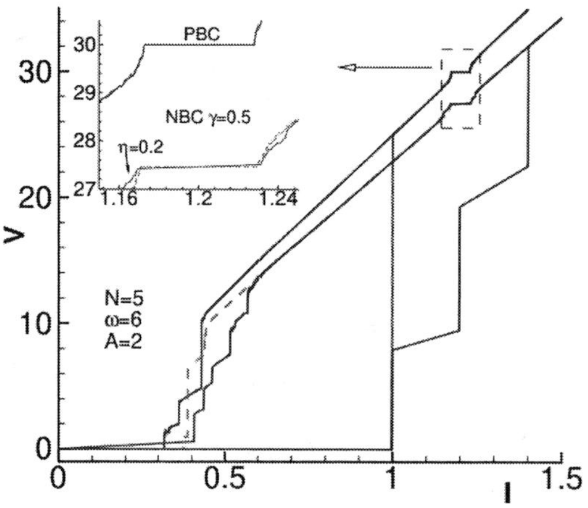

Figure 2. The IV-characteristics of JJ stacks without the charge imbalance $\eta = 0$ (dashed line) and at $\eta = 0.2$ (solid line). The results for the periodic boundary condition are shown for comparison. The enlarged parts of the IV-characteristics with the SS are shown in the inset.

imbalance at periodic boundary conditions is shown as well. The simulations have been made for a stack with five JJs, a coupling parameter $\alpha = 0.5$ and $\gamma = 0.5$. We see that the position of the SS (at periodic boundary condition (PBC)) corresponds to the canonical value of the SS voltage $V = 30$ in agreement with the value of the external frequency $\omega = 6$ and a number of junctions in the stack $N = 5$. The nonperiodic boundary conditions with $\gamma \neq 0$ shifts the outermost branch relatively to the curve at PBC, leading to a corresponding shift of the Shapiro steps. The charge imbalance manifests itself as appearance of the slope in the Shapiro step, which is clearly demonstrated in the inset for the case $\eta = 0.2$.

The influence of the coupling parameter on the shift of the Shapiro step is shown in Fig. 3(a). The increase of α leads to the increase of the shift value. The steps on the IV-characteristic at $\alpha = 0, 0.2, 0.6, 1$ are indicated by the large dashed rectangle. The IV-characteristics at large α demonstrate also additional steps to those occurring on an internal branch. These Shapiro steps are evidenced inside the inset to the small dashed rectangle.

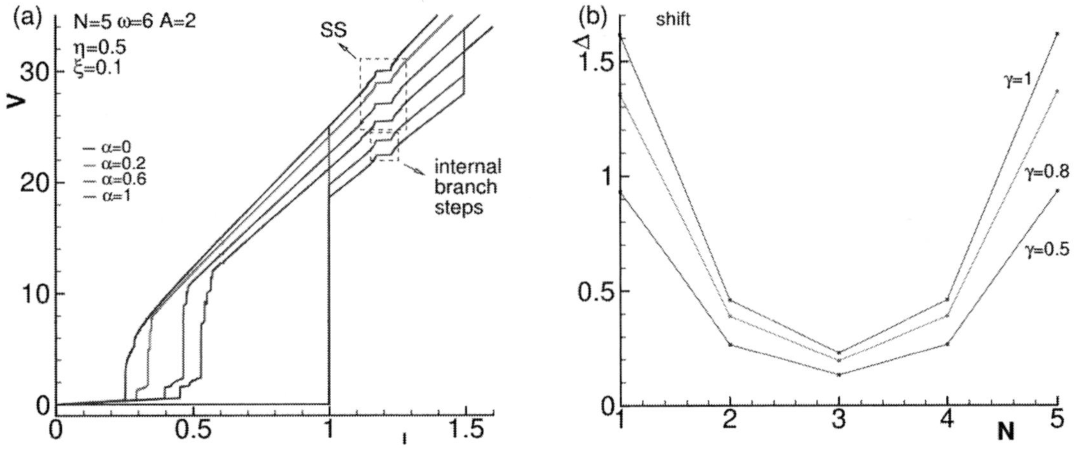

Figure 3. (a) IV-characteristics of the stack of $N = 5$ JJ at four coupling parameter values $\alpha = 0, 0.2, 0.6, 1$. (b) Distribution of the SS shift along the stack of $N = 5$ JJ at three parameter values $\gamma = 0.5, 0.8, 1$.

The shift of the Shapiro step depends on the value of γ and on the coupling parameter α. Fig. 3(b) shows the distribution of the shift of the step along the stack with parameters $\gamma = 0.5, 0.8, 1$. One can see that the maximum of the shift values occur at the first and last Josephson junctions. The thicknesses of the superconducting layers of those junctions are larger than the others. The distribution of the shift can be also seen in the middle layers due to the coupling between the Josephson junctions.

Thus, the Shapiro step shows a shift of its position from the canonical value $N\omega$, where N is the number of junctions in the stack and ω is the frequency of the external radiation. The value of this shift depends on the boundary conditions and the coupling between the Josephson junctions. Due to the coupling, the effect of the boundary conditions is extended to the neighboring junctions.

Acknowledgement

The reported study was funded by RFBR according to the research projects 15–29–01217 and 16–52–45011_ India.

References

[1] S. Artemenko and A. Kobelkov, Phys. Rev. Lett. **78**, 3551 (1997)
[2] C. Preis, C. Helm, J. Keller, A. Sergeev, and R. Kleiner, in *Superconducting Superlattices II: Native and Artificial* (Proceedings of SPIE Volume **3480**, 1998) p. 236
[3] S. E. Shafranjuk and M. Tachiki, Phys. Rev. B **59**, 14087 (1999)
[4] C. Helm, C. Preis, C. Walter, and J. Keller, Phys. Rev. B **62**, 6002 (2000)
[5] C. Helm, J. Keller, C. Preis, and A. Sergeev, Physica C **362**, 43 (2001)
[6] C. Helm, L. N. Bulaevskii, E. M. Chudnovsky, and M. P. Maley, Phys. Rev. Lett. **89**, 057003 (2002)
[7] L. N. Bulaevskii, C. Helm, A. R. Bishop, and M. P. Maley, Europhys Lett. **58**, 057003 (2002)
[8] J. Keller and D. A. Ryndyk, Phys. Rev. B **71**, 054507 (2005)
[9] T. Koyama and M. Tachiki, Phys. Rev. B **54**, 16183 (1996)
[10] D. A. Ryndyk, Phys. Rev. Lett. **80**, 3376 (1998)
[11] Yu. M. Shukrinov, M. Nashaat, K. V. Kulikov, R. Dawood, H. El Samman, and Th. M. El Sherbini, EPL **115**, 20003 (2016)

EPJ Web of Conferences **173**, 03016 (2018) https://doi.org/10.1051/epjconf/201817303016
Mathematical Modeling and Computational Physics 2017

Finite Difference Schemes as Algebraic Correspondences between Layers

Mikhail Malykh[1,*] and *Leonid* Sevastianov[1]

[1] *Department of Applied Probability and Informatics, Peoples' Friendship University of Russia (RUDN University)*

Abstract. For some differential equations, especially for Riccati equation, new finite difference schemes are suggested. These schemes define protective correspondences between the layers. Calculation using these schemes can be extended to the area beyond movable singularities of exact solution without any error accumulation.

Introduction

Modern development of computer science revived the old investigations on the solvability of differential equations in the finite terms. There are remarkable differential equations which can be integrated in computer algebra systems (CAS). In our work we want to talk about remarkable differential equations in another sense: for these equations there are finite difference schemes that correctly describe singularities of exact solutions.

1 Finite difference method and singularities

Consider an ordinary differential equation (ODE)

$$\frac{dy}{dx} = f(x, y), \qquad f \in \mathbb{Q}(x, y). \tag{1}$$

If a singular point of the solution to the Cauchy problem

$$\frac{dy}{dx} = f(x, y), \qquad y|_{x=a} = y_0$$

depends on the initial data then it is called movable singularity of ODE. By the Painlevé theorem these singularities are always algebraic, that is, the equation in the neighborhood of such singularity can be expanded into Puiseux series

$$y = \sum_{s=n}^{\infty} k_s (x - c)^{s/m},$$

where $m > 0$ and n are integers and $k_n \neq 0$ [1]. The power $p = n/m$ is called the order of the singularity.

*e-mail: malykh_md@rudn.university

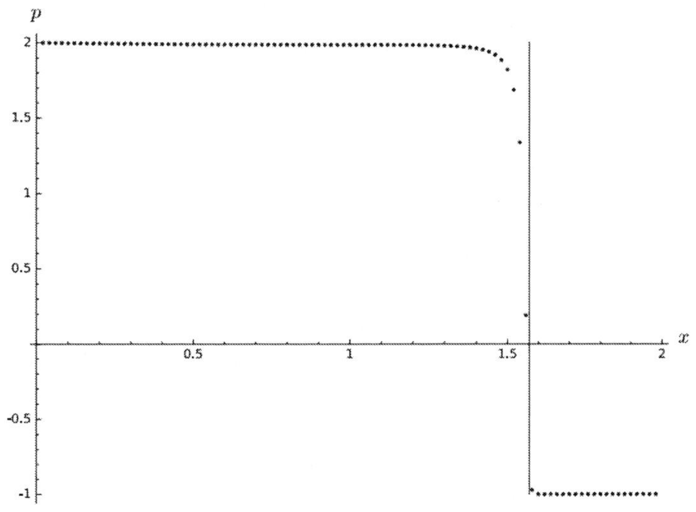

Figure 1. The graph of the effective order of the finite difference scheme of the example 1

In general we don't know the exact solution $y = \varphi(x)$ of the given Cauchy problem but we can calculate an approximate solution by finite difference approach. This suggests replacing the original differential equation with an algebraic equation (scheme) of the form

$$F(y, \hat{y}; x, \hat{x}) = 0$$

in commonly used notations [2]. This equation defines a correspondence between neighboring layers y and $\hat{y}$, which are usually investigated as points on two affine straight lines.

Problem 1 *Given a Cauchy problem and an interval $a < x < b$, we want to detect mobile singularities on this interval by analysis of one or several approximate solutions.*

Many authors think that this problem can't be solved because finite difference method describes mistakenly the solution in the neighborhood of a singularity. However a practical solution was given by E. Alshina et al. in 2005 [3], applications and development of these ideas see in [4–7]. There is such a scheme (ex. Complex Rosenbrock scheme, CROS), within which the approximate solution goes to a finite value when the exact solution has a pole. For CROS the ratio

$$\frac{y(x_n, \Delta x) - y(x_n, \Delta x/2)}{y(x_n, \Delta x/2) - y(x_n, \Delta x/2^2)}$$

is equal approximately to 2^2 in regular points and 2^p in the neighborhood of a singularity of order p. The logarithm of this ratio is called the effective order of scheme: its jump indicates the position of the singularity and its value after the jump is equal to the order of the singularity.

Example 1 *In fig. 1 we can see that the numerical solution of the Cauchy problem*

$$\frac{dy}{dx} = 1 + y^2, \qquad y|_{x=0} = 0$$

has a pole at $x \simeq 1.6$. It is true because the exact solution $y = \tan x$ has a pole at $x = \pi/2$. The calculations were made in our package for Sagemath [8].

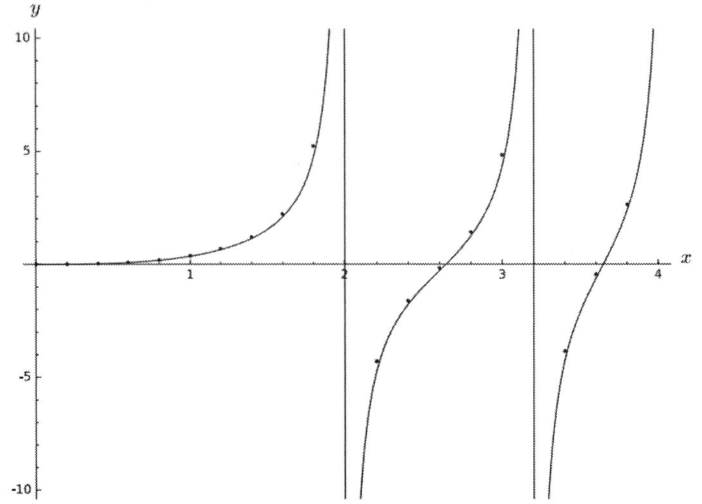

Figure 2. The graph of the solution for example 2: the solid line shows the graph of the exact solution, points mark the graph of the approximate solution

2 Riccati equation

All the commonly used schemes describe mistakenly the solutions of Cauchy problem in the neighborhood of singularities. However, the Riccati equation

$$\frac{dy}{dx} = p(x) + q(x)y + r(x)y^2 \tag{2}$$

can be approximated by the scheme

$$\frac{\Delta y}{\Delta x} = p + qy + ry\hat{y}. \tag{3}$$

The calculations by this scheme can be extended to the area beyond the pole without any error accumulation, a feature noticed in [9].

Example 2 *The initial value problem*

$$\frac{dy}{dx} = x^2 + y^2, \qquad y|_{x=0} = 0,$$

can be integrated in Bessel functions. We have calculated the exact solution in Maple and compared it to the approximate solution computed by our scheme (3) *in Sagemath. In fig. 2 we can see that points of the numerical solution give good approximation of the solution everywhere.*

Theorem 1 *Let p, q, r denote rational functions which have not any pole on the interval $[a, b]$, and let $y = g(x)$ be the exact solution of Cauchy problem for the Riccati equation* (2) *on this interval. For any $\delta > 0$ there exist the finite constants M_x and M_y such that any approximate solution $\{y_1, \dots, y_N\}$ found by the scheme* (3) *satisfies the estimate*

$$|y_n - g(x_n)| \leq M_y \Delta x,$$

if the step size Δx satisfies the condition $\Delta x \leq M_x$ and distance between x_n and the poles is larger than δ.

3 Finite difference method and algebraic geometry

When we write difference scheme we map differential equation to a pair (V, F), where 1) V is a layer, that is, an algebraic variety which can depend on x; 2) F is a difference scheme, that is, an algebraic correspondence between $V(x)$ and $V(\hat{x})$. We can define the notion of approximation and approximate solutions in the frame of the Weierstrass Vorbereitungssatz. The simplest, in the geometric sense, correspondences aren't used in the standard theory of the finite differences. In the simplest case the layer is a projective straight line $\mathbb{P}$ and the correspondence is a projective one-to-one transformation. Such schemes do not exist for all the differential equations (1) but only for the Riccati equation (2). One of these schemes is the scheme (3) the remarkable properties of which were described above.

Natural generalization of the projective one-to-one transformation is the n-to-n projective correspondence on algebraic varieties. We can prove that there is a projective n-to-n scheme for a given differential equation iff the exact general solution of the differential equation depends on constant algebraically. This class was investigated in early works by Painlevé [10, 11]. Thus, the classical transcendental functions are good functions not only from the point of view of the power series expansions. The evaluation of these functions by finite difference schemes can be extended to the area beyond singularities without any error accumulation.

Conclusions

Given the Riccati equation (2), the existence of a pole inside the investigated interval does not necessarily results in the accumulation of the discretization errors in a finite difference method. The reason is a bad choice of finite difference scheme. Projective n-to-n finite difference schemes always exist for the differential equations which can be integrated in classical (in Painlevé sense) transcendental functions.

Acknowledgments

The calculations in this article are performed in the computer algebra system Sage [12]. The publication was prepared with the support of the "RUDN University Program 5-100". The work was partially supported by RFBF grants No 15-07-08795, No 16-07-00556.

References

[1] W.W. Golubev, *Vorlesungen über Differentialgleichungen im Komplexen* (Berlin: VEB Deutscher Verlag der Wissenschaften, 1958)

[2] A.A. Samarskii, *The Theory of Difference Schemes* (NY, 2001)

[3] E.A. Al'shina, N.N. Kalitkin, and P.V. Koryakin, Comp. Math. and Math. Phys. **45** (10), 1769–1779 (2005)

[4] A.A. Belov, Doklady Mathematics **93** (3), 334–338 (2016)

[5] A.A. Belov, Computational Mathematics and Mathematical Physics **57** (1), 122–132 (2017)

[6] M.O. Korpusov, D.V. Lukyanenko, A.A. Panin, and E.V. Yushkov, Journal of Mathematical Analysis and Applications **442** (2), 451–468 (2016)

[7] M.O. Korpusov, D.V. Lukyanenko, A.A. Panin, and E.V. Yushkov, Mathematical Methods in the Applied Sciences **40** (7), 2336–2346 (2017)

[8] M.D. Malykh, Cros for Sagemath (2017). Available from: https://malykhmd.neocities.org

[9] M.D. Malykh, RUDN Journal of Mathematics, Information Sciences and Physics **3**, 5–9 (2015)

[10] H. Umemura, Nagoya Math. J. **119**, 1–80 (1990)

[11] M.D. Malykh, Journal of Mathematical Sciences **209** (6), 935–952 (2015)

[12] SageMath [Internet]. Available from: http://www.sagemath.org

EPJ Web of Conferences **173**, 03017 (2018) https://doi.org/10.1051/epjconf/201817303017
Mathematical Modeling and Computational Physics 2017

The Influence of the Relaxation Time on the Dynamic Hysteresis in Perovskite Solar Cells

Alexandra Palici[1], *George Alexandru* Nemnes[2,3,⋆], *Cristina* Besleaga[1], *Lucian* Pintilie[1], *Dragos-Victor* Anghel[3], *Ioana* Pintilie[1], and *Andrei* Manolescu[4]

[1] *National Institute of Materials Physics, Magurele 077125, Ilfov, Romania*
[2] *University of Bucharest, Faculty of Physics, Materials and Devices for Electronics and Optoelectronics Research Center, 077125 Magurele-Ilfov, Romania*
[3] *Horia Hulubei National Institute for Physics and Nuclear Engineering, 077126 Magurele-Ilfov, Romania*
[4] *School of Science and Engineering, Reykjavik University, Menntavegur 1, IS-101 Reykjavik, Iceland*

Abstract. We investigate the dynamic behavior of perovskite solar cells by focusing on the relaxation time τ, which corresponds to the slow de-polarization process from an initial bias pre-poled state. The dynamic electrical model (DEM) is employed for simulating the J-V characteristics for different bias scan rates and pre-poling conditions. Depending on the sign of the initial polarization normal or inverted hysteresis may be induced. For fixed pre-poling conditions, the relaxation time, in relation to the bias scan rate, governs the magnitude of the dynamic hysteresis. In the limit of large τ the hysteretic effects are vanishing for the typical range of bias scan rates considered, while for very small τ the hysteresis is significant only when it is comparable with the measurement time interval.

1 Introduction

The dynamic hysteresis [1] observed in perovskite solar cells (PSCs) has become a notorious hallmark. Yet recently an increasing number of studies reported PSCs with diminished hysteresis and relatively large power conversion efficiencies (PCEs). However, knowing that the dynamic effects strongly depend on solar cell pre-conditioning and measurement history, unambigous comparisons regarding the magnitude of the observed hysteresis are rather difficult. Presently, a rather large consensus [2] has been reached concerning the origin of the dynamic hysteresis, which points to the ion migration within the perovskite layer as the main reason for the slow process, characterized by a relaxation time τ which is typically in the order of seconds. In particular, iodine migration was already evidenced in several reports, as well as in our own samples [3].

In this context, we investigate the hysteretic behavior dependence on the relaxation time τ at different bias pre-poling conditions. We base our analysis on the dynamic electrical model (DEM) [4], which is able to accurately reproduce the experimental J-V characteristics. In particular, a unified picture comprising normal (NH) and inverted (IH) hysteresis in the same device structure subject to different bias pre-conditioning was recently given by DEM, validated by J-V measurements [5].

⋆e-mail: nemnes@solid.fizica.unibuc.ro

DEM provides a time dependent description of the J-V characteristics, by solving the coupled system of differential equations with given initial conditions as described in detail in Refs. [4, 5]:

$$-R_s C_0 \frac{\partial I}{\partial t} = I_s\left(e^{\frac{q(V+IR_s)}{nk_BT}} - 1\right) + \left(\frac{R_s}{R_{sh}} + 1\right)I + \frac{V}{R_{sh}} + C_0\frac{\partial V}{\partial t} + \mathcal{A}\frac{\partial P_{nl}}{\partial t} - I_{ph}, \quad (1)$$

$$\frac{\partial P_{nl}}{\partial t} = -\frac{P_{nl}(t) - P_{nl,\infty}(U_c(t))}{\tau}. \quad (2)$$

Thus, the time dependence of the current and non-linear polarization, $I(t)$ and $P_{nl}(t)$, are obtained for an input voltage scan $V(t)$.

The core assumption of DEM is enclosed in Eq. (2), which describes the time evolution of P_{nl} using a single relaxation time τ. A condition of the same type was used for the surface polarization voltage in the model proposed by Ravishankar et al. [6], where it was conjectured that the cation migration together with an accumulation of holes is responsible for the hysteretic effects. According to Eq. (2) the non-linear polarization P_{nl} follows, within the relaxation time τ, the equilibrium value $P_{nl,\infty} = (U_c/V_{oc})P_\infty$, which is a bias dependent quantity, where $U_c = V + IR_s$. The notations are the same as in Refs. [4, 5].

We assume the relaxation time is set by the morphology of the interfaces and layers of the PSC, i.e. it is a structure dependent quantity, although it may be further influenced by other factors like pre-poling conditions, temperature etc. DEM suitably describes the experimental J-V characteristics using a single relaxation time, although multiple polarization mechanisms with different relaxation times may be present, e.g. induced by the drifting of several ionic species. If these individual contributions can be differentiated, then Eq. (2) may be extended to accomodate a more detailed description. In the next section we discuss the dynamic hysteretic effects with the focus on changing τ, for two different initial polarizations, which correspond to normal and inverted hysteresis.

2 Dynamic hysteretic effects – the τ dependence

We simulate the dynamic J-V characteristics following the measurement protocol introduced in Ref. [5], i.e. the samples are pre-poled with an initial polarization P_0 and subsequently a reverse-forward (R-F) bias scan with constant rate $\alpha = \partial V/\partial t$ is performed. Figure 1 shows R-F scans corresponding to four different relaxation times, $\tau = 5, 10, 20, 40$ s, for an initial polarization $P_0 = 5P_\infty$ and the same set of bias scan rates, $\alpha = 5, 20, 100, 1000$ mV/s. Since $P_0 > 0$ for each relaxation time, normal hysteresis behavior is obtained for any τ as explained in Ref. [4]: the hysteresis is diminished for very small and very large scan rates, while the short-circuit current I_{sc} increases with α. However, modifying τ the overall picture regarding the dynamic hysteretic effects is changing.

In the limit of large relaxation times ($\tau = 40$ s) the hysteresis is diminished for all bias scan rates, while the increase of I_{sc} is minimal. Setting $\tau \to \infty$ in Eq. (2), the current corresponding to the variation of the non-linear polarization vanishes, i.e. $\partial P_{nl}/\partial t \to 0$, and, in the limit of small geometrical capacitance $C_0 \to 0$, the stationary case is recovered. Therefore, for a typical range of bias scan rates considered here and relatively large τ, the J-V characteristics are closing in to the stationary case, roughly corresponding to the 5 mV/s J-V characteristics.

Decreasing the relaxation time and keeping α fixed, I_{sc} increases. Furthermore, for fixed, finite value of τ, the hysteresis has a nonmonotonic dependence on α, reaching a maximum when the reverse/forward scan time V_{oc}/α becomes comparable with τ. This behavior can be explained as follows. For smaller τ values, $\partial P_{nl}/\partial t$ is getting larger at given α, enhancing the output current. Increasing α, the hysteresis reaches a maximum, after which is decreasing. For very fast bias scans, the non-linear polarization remains practically unchanged, $P_{nl} \approx P_0$.

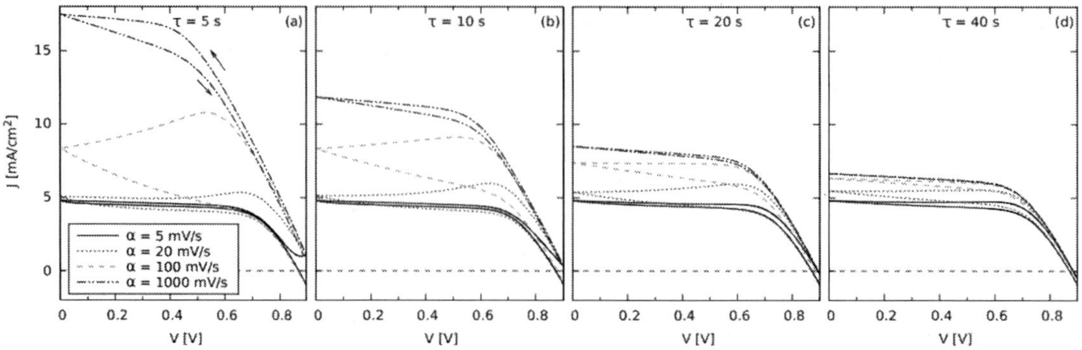

Figure 1. Normal hysteresis effects under changing the relaxation time τ: 5 s (a), 10 s (b), 20 s (c), 40 s (d). The bias scan rates are $\alpha = 5, 20, 100, 1000$ mV/s. The initial polarization $P_0 = 5P_\infty$. The DEM parameters are: the series resistance $R_s = 250\ \Omega$, the shunt resistance $R_{sh} = 20$ kΩ, the diode saturation current $I_s = 0.1$ pA, the diode ideality factor $n = 1.53$, the photogenerated current $I_{ph} = 0.43$ mA, the device area $\mathcal{A} = 0.125$ cm^2, the open-circuit voltage $V_{oc} = 0.9$ V (stationary case) and the open-circuit polarization $P_\infty = 12$ mC/cm^2. The arrows indicate the NH behavior.

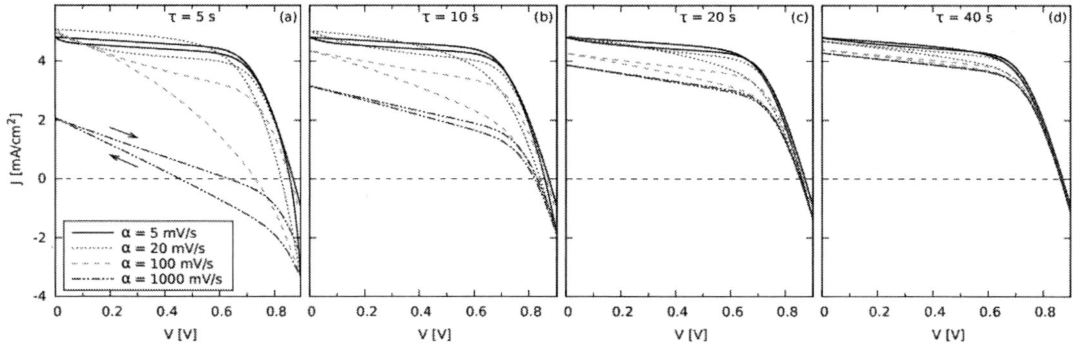

Figure 2. Inverted hysteresis effects obtained for the same set of parameters as in Fig. 1, apart from the initial polarization $P_0 = -P_\infty$. The IH is present for $V > V_c$, i.e. to the right of the crossing point, while NH is visible for $V < V_c$ in some instances.

A different situation occurs when the initial polarization is negative, $P_0 < 0$, and an inverted hysteresis is evidenced, i.e. the current in forward scan is larger than in reverse [5]. Assuming the same parameters of the DEM and changing only $P_0 = -P_\infty$, we obtain the J-V characteristics depicted in Fig. 2. With respect to τ variations, there is an overall similar behavior as in the NH case. However, under IH pre-poling conditions, the resulting I_{sc} decreases for higher bias scan rates, except for small τ values and slowest scan rates of 5 and 20 mV/s. To explain this, one should note that for these two scan rates, a relatively small $|P_0|$ induces a mixed hyteresis, characterized by a crossing point at bias V_c, which separates the IH behavior at large biases from the NH occuring at smaller ones, e.g. near short-circuit. For $\alpha = 5$ or 20 mV/s, P_0 decays significantly during the reverse scan particularly for small τ and subsequently P_{nl} follows $P_{nl,\infty}(U_c)$ as the bias is further decreased, resulting in NH behavior at small bias. By contrast, for $\alpha = 100$ or 1000 mV/s, P_{nl} changes are similar on both reverse and forward scans, while $\partial P_{nl}/\partial t$ becomes larger, so that eventually the hysteresis becomes completely inverted and I_{sc} decreases. Similarly, increasing τ we have $P_{nl} \approx P_0$ and the completely IH behavior is obtained at even lower scan rates, while the hysteresis is diminished.

From experimental point of view, achieving a certain initial polarization depends on the structure of the PCS, as well as on the pre-poling conditions, like poling bias V_{pol} and poling time t_{pol} [5]. In terms of relaxation times, it is more dificult to obtain the same P_0 as τ increases. For large τ values, the PSC is hysteresis-less, while for very short ones a fast enough scan rate can produce an apparently hysteresis-less behavior, although far away from the stationary case. Probing the dynamic behavior of the PSCs by different bias scan rates and poling conditions is recommended for a correct evaluation of the stationary J-V characteristics and can provide additional insight into the ion migration related degradation.

3 Conclusions

We investigated the hysteretic behavior of PSCs under the change of the relaxation time scale, which is a central quantity in the DEM. The two types of the hysteresis, normal or inverted, determined by the sign of the initial polarization are analyzed in this context, for a typical range of bias scan rates. In the case of large relaxation times the dynamic hysteresis is vanishing and the J-V characteristics become close to the stationary case even for the very large scan rates considered. In contrast, for smaller relaxation time scales, the short-circuit current changes significantly, either increasing for NH case or being reduced for IH case. The hysteresis is enhanced at intermediate scan rates, while at very fast and very slow scan rates it is diminished. It is therefore worth noting that measuring the J-V characteristics at rather high bias scan rates could lead to an apparent hysteresis-less behavior, with relatively large PCE, although far away from the stationary characteristics. In addition, the hysteretic behavior is consistently evaluated if the reverse and forward scans are correlated. Otherwise, independent scans having presumably different initial polarizations could yield a rather unpredictable and irreproducible results. The experimental determination of the specific relaxation times and the investigation of the dynamic effects, monitored over time, may also provide further insight into ion migration induced degradation of the PSCs.

Acknowledgement

This work was supported by Romania-JINR cooperation project JINR Order 220/10.04.2017, pp. 24 and pp. 94 and EEA Financial Mechanism 2009-2014 under project contract no. 8SEE/30.06.2014.

References

[1] H. J. Snaith, A. Abate, J. M. Ball et al., J. Phys. Chem. Lett. **5**, 1511 (2014)

[2] W. Tress, J. Phys. Chem. Lett. **8**, 3106 (2017)

[3] Cristina Besleaga, Laura Elena Abramiuc, Viorica Stancu, A. G. Tomulescu, M. Sima, Liliana Trinca, N. Plugaru, L. Pintilie, G. A. Nemnes, Mihaiela Iliescu, H. G. Svavarsson, A. Manolescu and Ioana Pintilie, J. Phys. Chem. Lett. **7**, 5168 (2016)

[4] G. A. Nemnes, Cristina Besleaga, A. G. Tomulescu, Ioana Pintilie, L. Pintilie, K. Torfason, A. Manolescu, Sol. Energy Mater. Sol. Cells **159**, 197 (2017)

[5] G. A. Nemnes, Cristina Besleaga, Viorica Stancu, Daniela Emilia Dogaru, Lucia Nicoleta Leonat, L. Pintilie, K. Torfason, M. Ilkov, A. Manolescu, Ioana Pintilie, J. Phys. Chem. C **121**, 11207 (2017)

[6] S. Ravishankar, O. Almora, C. Echeverria-Arrondo, E. Ghahremanirad, C. Aranda, A. Guerrero, F. Fabregat-Santiago, A. Zaban, G. Garcia-Belmonte, and J. Bisquert, J. Phys. Chem. Lett. **8**, 915 (2017)

EPJ Web of Conferences **173**, 03018 (2018) https://doi.org/10.1051/epjconf/201817303018
Mathematical Modeling and Computational Physics 2017

Optimization of the Magnetic Field Homogeneity Area for Solenoid Type Magnets

Eugene Perepelkin[1,⋆], *Rima* Polyakova[2,⋆⋆], *Aleksandr* Tarelkin[1], *Alexander* Kovalenko[2], *Pavel* Sysoev[1], *Marianne* Sadovnikova[1], and *Ivan* Yudin[2⋆⋆⋆]

[1] *Lomonosov Moscow State University, GSP-1, Leninskie Gory, Moscow, 119991, Russian Federation*
[2] *Joint Institute for Nuclear Research, 6 Joliot-Curie St., Dubna, Moscow region, 141980, Russian Federation*

Abstract. Homogeneous magnetic fields are important requisites in modern physics research. In this paper we discuss the problem of magnetic field homogeneity area maximization for solenoid magnets. We discuss A-model and B-model, which are basic types of solenoid magnets used to provide a homogeneous field, and methods for their optimization. We propose C-model which can be used for the NICA project. We have also carried out a cross-check of the C-model with the parameters stated for the CLEO II detector.

1 Introduction

Detectors such as CMS CERN, ATLAS CERN, MPD NICA and FCC in Berlin use solenoid magnets under severe tolerance limits on the homogeneity of the magnetic field. The main problem in constructing these systems is the maximization of the homogeneous field area with adequate accuracy inside the detector. In this work we discuss two basic solenoid magnets models, ways of optimizing of these models and our new model within the designated problem. For comparing the models we selected parameters which provide the same magnetic field at the center of each model $B_{\text{center}} = 0.5$ T. Computations were performed (using TOSCA software) by the finite element method on tetrahedral mesh, which had been specially generated taking into account the structural features of the detectors [1].

2 Ways to increase the magnetic field homogeneity area

2.1 A-model

This is the reference model for our study. We got geometry and dimensions of the model (Fig. 1) and made calculations in order to assess whether any part of it needs improvement, and, if so, to establish a benchmark for our work. The model parameters: current density $J = 1.89767173 \cdot 10^6$ A/m^2, coil cross section $S = 0.212 \cdot 4.2$ m^2, full current $I = 1.68968691 \cdot 10^6$ A.

⋆e-mail: pevgeny@mail.ru
⋆⋆e-mail: polykovarv@mail.ru
⋆⋆⋆e-mail: yudin@jinr.ru

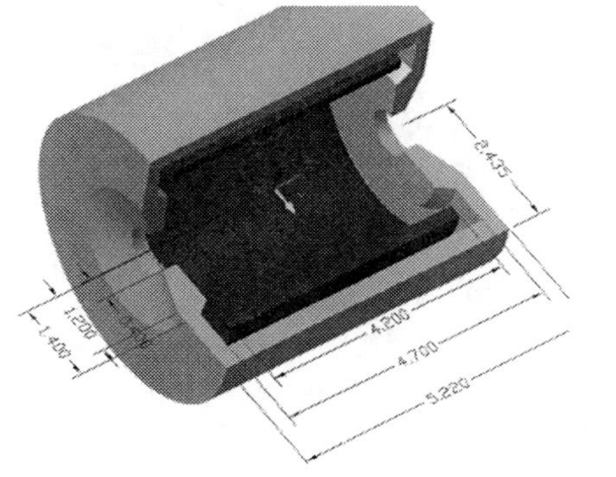

Figure 1. A-model geometry

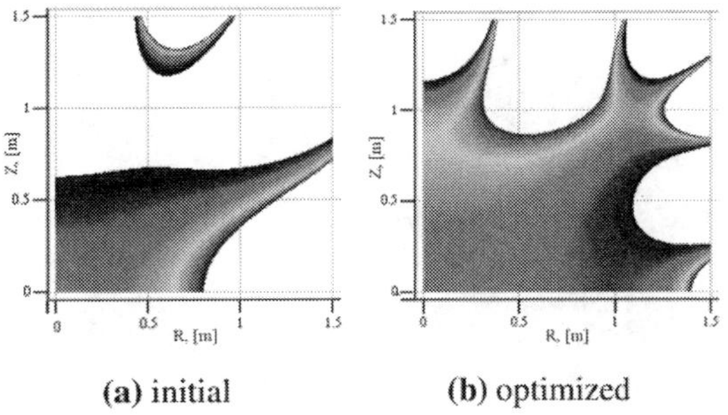

Figure 2. A-model area of field homogeneity

We pay particular attention to providing a broad area of magnetic field homogeneity with accuracy ±0.1%, a tolerance often required for real detectors. Figure 2a shows that the magnetic field homogeneity area isn't big enough (on this figure and further figures mapping field homogeneity, we represent the distribution of the homogeneous field on a given plane inside the detector given its symmetry; gray-scale gradient shows variation of the field within range 0.5±0.005 T). It occupied less than 1/4 of working space inside the detector. So, the model needs optimization.

2.2 A-model optimization

We used four coils (three double and one single) with different currents, instead of one single unified coil (Fig. 3). We supposed that the final field homogeneity map of all coils would be given by overlapping the contribution of each coil separately in accordance with Root Mean Square (Fig. 4). This assumption is valid for small magnetic fields. Figure 2b shows the area of field homogeneity

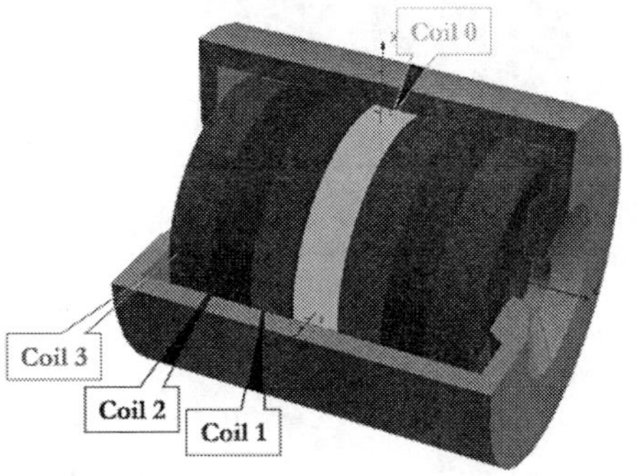

Figure 3. Optimized A-model geometry

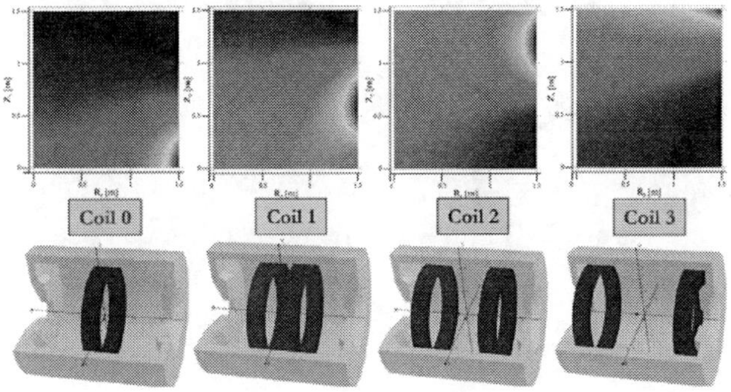

Figure 4. RMS optimization

after optimization with accuracy ±0.1%. The homogeneity area coverage increases 1.76 times in comparison with that of the initial model.

2.3 B-model

The B-model geometry is shown in Fig. 5. Figure 6a shows the field homogeneity map with accuracy ±0.1% for the B-model.

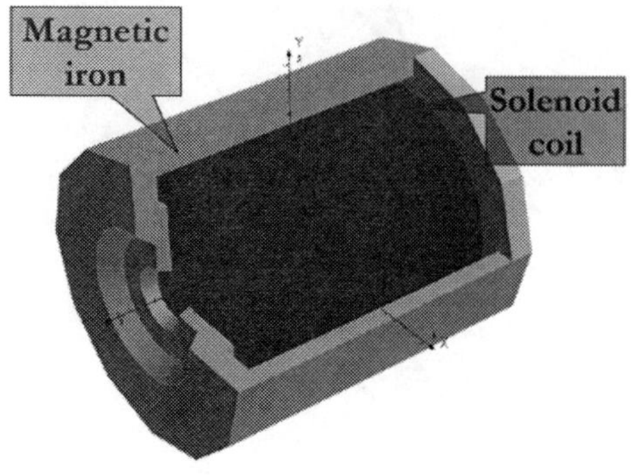

Figure 5. B-model geometry

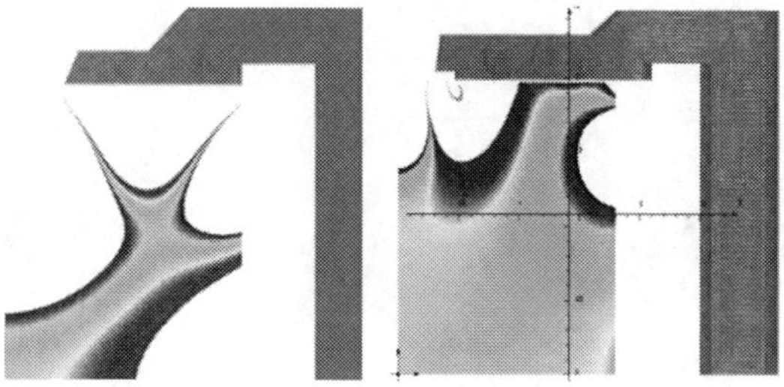

(a) initial **(b)** optimized

Figure 6. B-model area of field homogeneity

2.4 B-model optimization

It is possible to improve the characteristics of the B-model by adding two coils to each side of the detector (Fig. 7). We select the appropriate current density for each coil in order to provide enough homogeneity field coverage: $J_{\text{main}} = 1.047247 \cdot 10^7$ A/m^2, $J_{\text{add1}} = 6.293004 \cdot 10^6$ A/m^2 and $J_{\text{add2}} = 1.255567 \cdot 10^6$ A/m^2. Figure 6b shows the map of the field homogeneity for the optimized B-model with the same level of accuracy $\pm 0.1\%$.

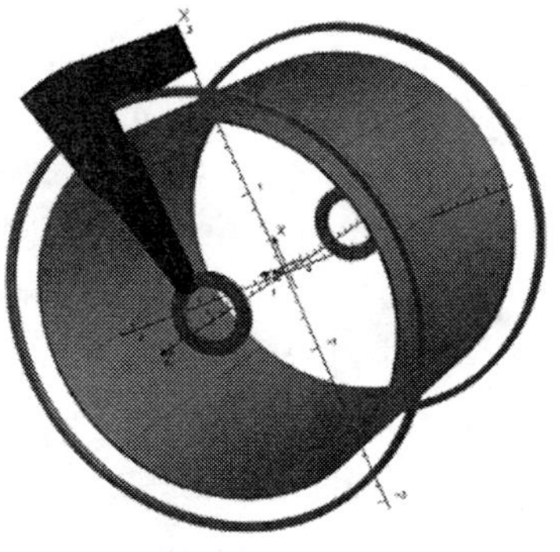

Figure 7. Optimized B-model geometry

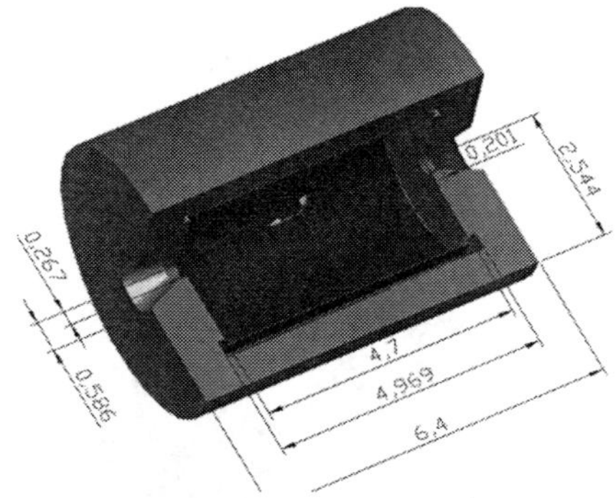

Figure 8. C-model geometry

2.5 C-model

We created the model with a single coil striving for construction simplicity. The dimensions and geometry of the model are given in Fig. 8. It is worth noting that the magnetic field distribution is sensitive to the size of the gap, or, in other words, how deeply the coil goes into the ferromagnetic on the edges of the model. The model parameters: current density $J = 9.956410099 \cdot 10^6$ A/m^2, coil cross section $S = 0.04 \cdot 4.7$ m^2, full current $I = 1.871805098 \cdot 10^6$ A. Figure 9 shows the area of field homogeneity obtained for the C-model.

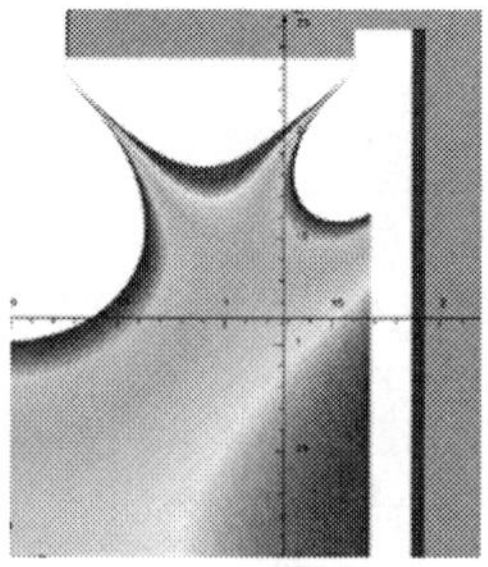

Figure 9. C-model area of field homogeneity

Table 1. Comparison of models

Model	Area, m^2	η, %	Num. coils
A	0.799	35.5	1
(optimization)	1.439	64	4
B	1.089	48.4	1
(optimization)	2.258	98.4	3
C	2.057	91.4	1

3 Comparison of models

The comparison of the main features of models, presented in Table 1, points to data obtained for a square area $S = 1.5 \cdot 1.5$ m^2 and corresponds with a homogeneity level of $\pm 0.1\%$. In all cases the models characteristics could be improved in different ways to get further optimization.

4 CLEO II cross-check

We conducted a cross-check of the C-model by creating a prototype satisfying the requirements for the CLEO II detector [2]. It is asked to provide a magnetic field 1.5 T with uniformity $\pm 0.2\%$ over 95% of the drift chamber volume.

The model geometry is represented in Figure 10. Figure 11 shows the map of field homogeneity obtained. The result of the simulation satisfies all the formulated requirements.

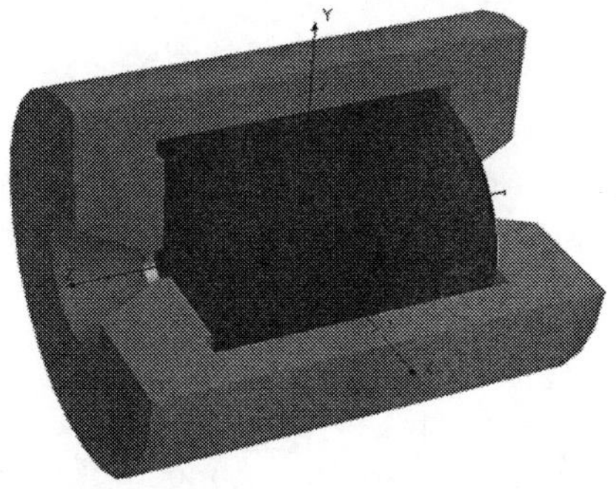

Figure 10. Model geometry for cross-check

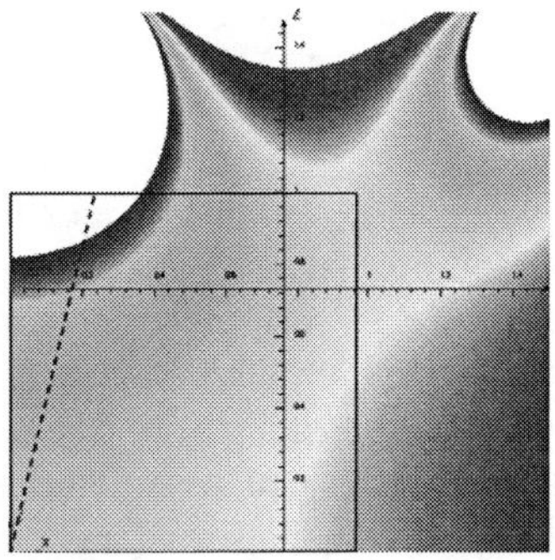

Figure 11. Map of field homogeneity for cross-check

5 Conclusion

1. Optimization algorithms for increasing the homogeneity area have been suggested.
2. A new model of the magnet detector has been proposed.
3. The new model has satisfied the requirements asked for the CLEO II detector.

References

[1] E.E. Perepelkin, R.V. Polyakova, A.D. Kovalenko, P.N. Sysoev, M.B. Sadovnikova, A.A. Tarelkin, and I.P. Yudin, RUDN Journal of MIPh. **25**, 3, 253–265 (2017)

[2] Kubota, Yuichi, et al., Nuclear Instruments and Methods in Physics Research Section A: Accelerators, Spectrometers, Detectors and Associated Equipment **320**, 1–2, 66–113 (1992)

EPJ Web of Conferences **173**, 03019 (2018) https://doi.org/10.1051/epjconf/201817303019
Mathematical Modeling and Computational Physics 2017

Estimation of Magnetic Field Growth and Construction of Adaptive Mesh in Corner Domain for the Magnetostatic Problem in Three-Dimensional Space

Eugene Perepelkin[1]* and *Aleksandr* Tarelkin[1]**

[1] *Lomonosov Moscow State University, GSP-1, Leninskie Gory, Moscow, 119991, Russian Federation*

Abstract. A magnetostatics problem arises when searching for the distribution of the magnetic field generated by magnet systems of many physics research facilities, e.g., accelerators. The domain in which the boundary-value problem is solved often has a piecewise smooth boundary. In this case, numerical calculations of the problem require consideration of the solution behavior in the corner domain. In this work we obtained an upper estimation of the magnetic field growth using integral formulation of the magnetostatic problem and propose a method for condensing the differential mesh near the corner domain of the vacuum in the three-dimensional space based on this estimation.

1 Introduction

Many physics research facilities use magnetic systems of various configurations, e.g., a system of spectrometric magnets. It is very important to know with a good accuracy the distribution of the magnetic field generated by this system. The problem is actually reduced to the formulation of a magnetostatics problem of finding the distribution of the magnetic field generated by the magnetic system under consideration. Since the magnetic system has a complicated configuration, the solution to the problem is usually sought using numerical methods. This paper discusses the case when the domain over which a boundary value magnetostatics problem is solved has piecewise smooth boundary. In this case, the solution to the problem or the derivative solutions can have a singularity, thus the numerical search for the solution requires the use of special methods.

2 Integral formulation of the magnetostatic problem

The integral formulation allows the magnetic field to be represented as

$$\vec{H}(s) = \vec{H}_C(s) - \nabla_s \int\limits_{\Omega_f} \left(\vec{M}(p), \nabla_p \psi(s,p)\right) dv_p, \tag{1}$$

where $\vec{H}_C$ is the field from the current sources, $\vec{M}$ is the ferromagnetic magnetization vector, the function $\Psi(s,p)$ is equal to $1/4\pi r_{sp}$ or $1/2\pi \cdot \ln r_{sp}$ for the three-dimensional and the two-dimensional case respectively, and Ω_f is the ferromagnetic domain. The magnetization vector is defined as $\vec{M}=$

*e-mail: pevgeny@mail.ru
**e-mail: tarelkin.aleksandr@physics.msu.ru

$\mu_0\chi(H)\vec{H} = \mu_0(\mu(H)-1)\vec{H}$, where μ_0 is a constant, $\chi(H)$ is the magnetic susceptibility, and $\mu(H)$ is the permeability of the ferromagnetic.

3 Estimation of magnetic field growth

The following theorem characterizes the magnetic field growth in the 3D space (the estimation for the 2D space was obtained in [1]):

Theorem

If the permeability function of the ferromagnetic has the representation $\mu(H) = 1 + A/H - B/H^2$ when $H \to \infty$, then the magnetic field in the corner domain outside a ferromagnetic satisfies the condition

$$H(s) \le C_0 \ln \frac{1}{r_s} + w(s), \tag{2}$$

where C_0 is a constant, $w(s)$ is a bounded function, and r_s is the distance to the corner.

Proof

At high fields ($H \to \infty$), the representation $\mu(H) = 1 + A/H - B/H^2$ when $H \to \infty$ is valid, where A and B are positive constants. Consequently, at $H \to \infty$, $M = |\vec{M}|$ is limited by a constant $M_0 = \mu_0 A$.

From (1) we obtain $\vec{H}(s) = \vec{H}_C(s) - \nabla_s \int_{\Omega_f} \left(\vec{M}(p), \vec{r}_{sp}/r_{sp}^3\right) dv_p$. Here the first term doesn't contribute directly to the effect of significant growth of the magnetic field near the corner point, and we therefore estimate the second term

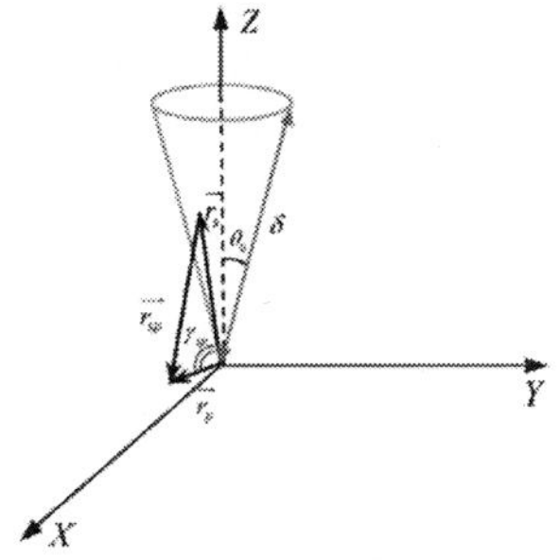

Figure 1. The conical volume

$$\left|\nabla_s \int_{\Omega_f} \left(\vec{M}(p), \frac{\vec{r}_{sp}}{r_{sp}^3}\right) dv_p\right| \le \left|\int_{\Omega_f} \frac{\partial}{\partial x_s} \frac{M^{(x)}(p)\bar{x} + M^{(y)}(p)\bar{y} + M^{(z)}(p)\bar{z}}{\bar{x}^2 + \bar{y}^2 + \bar{z}^2} dv_p\right| +$$

$$+\left|\int_{\Omega_f} \frac{\partial}{\partial y_s} \frac{M^{(x)}(p)\bar{x} + M^{(y)}(p)\bar{y} + M^{(z)}(p)\bar{z}}{\bar{x}^2 + \bar{y}^2 + \bar{z}^2} dv_p\right| + \left|\int_{\Omega_f} \frac{\partial}{\partial z_s} \frac{M^{(x)}(p)\bar{x} + M^{(y)}(p)\bar{y} + M^{(z)}(p)\bar{z}}{\bar{x}^2 + \bar{y}^2 + \bar{z}^2} dv_p\right| \le$$

$$\le 6M_0 \int_{\Omega_f} \frac{1}{r_{sp}^3} dv_p,$$

where $\bar{x} = x_s - x_p$, $\bar{y} = y_s - y_p$ and $\bar{z} = z_s - z_p$. We calculate the integral

$$\int_{\Omega_f} \frac{1}{r_{sp}^3} dv_p = \int_{v_\delta} \frac{1}{r_{sp}^3} dv_p + \int_{\Omega_f \setminus v_\delta} \frac{1}{r_{sp}^3} dv_p,$$

where $v_\delta = \Omega_f \cap S_\delta(Q)$ is the conical volume v_δ at the corner point (see Figure 1).

Let us consider the integral over the domain v_δ

$$\int_{\Omega_f} \frac{1}{r_{sp}^3} dv_p = \int_0^{\varphi_0} d\varphi_p \int_0^{\theta_0} \sin\theta_p \, d\theta_p \int_0^{\delta} \frac{r_p^2}{r_{sp}^3} dr_p = \int_0^{\varphi_0} d\varphi_p \int_0^{\theta_0} \sin\theta_p \, d\theta_p \int_0^{\delta} \frac{r_p^2}{\left(r_p^2 + r_s^2 - 2r_p r_s \cos\gamma_{sp}\right)^{3/2}} dr_p =$$

$$= \int_0^{\varphi_0} d\varphi_p \int_0^{\theta_0} \sin\theta_p \, d\theta_p \left(\int_0^1 \frac{t^2 dt}{\left(1 + t^2 - 2t\cos\gamma_{sp}\right)^{3/2}} + \int_1^{\delta/r_s} \frac{t^2 dt}{\left(1 + t^2 - 2t\cos\gamma_{sp}\right)^{3/2}} \right),$$

$t = r_p/r_s$ and $\cos\gamma_{sp} = \sin\theta_p \sin\theta_s \cos(\varphi_p - \varphi_s) + \cos\theta_p \cos\theta_s$. Then we use the expression for the generating function

$$\frac{1}{\sqrt{1 + t^2 - 2t\cos\varphi_{sp}}} = \begin{cases} \sum\limits_{m=0}^{+\infty} P_m\left(\cos\varphi_{sp}\right) t^m, & |t| < 1, \\ \sum\limits_{m=0}^{+\infty} P_m\left(\cos\varphi_{sp}\right) t^{-m-1}, & |t| < 1, \end{cases}$$

and obtain

$$\int_{\Omega_f} \frac{1}{r_{sp}^3} dv_p = \int_0^{\varphi_0} d\varphi_p \int_0^{\theta_0} \sin\theta_p \, d\theta_p \left[\int_0^1 t^2 dt \sum_{m,n,k=0}^{+\infty} P_m P_n P_k t^{m+n+k} + \int_1^{\delta/r_s} t^2 dt \sum_{m,n,k=0}^{+\infty} P_m P_n P_k t^{-(m+n+k+3)} \right] =$$

$$= \int_0^{\varphi_0} d\varphi_p \int_0^{\theta_0} \sin\theta_p \, d\theta_p \left[\sum_{m,n,k=0}^{+\infty} P_m P_n P_k \left. \frac{t^{m+n+k+3}}{m+n+k+3} \right|_0^1 + \sum_{m+n+k\neq 0}^{+\infty} P_m P_n P_k \left. \frac{t^{-(m+n+k)}}{-(m+n+k)} \right|_1^{\delta/r_s} \right] +$$

$$+ \int_0^{\varphi_0} d\varphi_p \int_0^{\theta_0} \sin\theta_p \, d\theta_p P_0^2 \ln t|_1^{\delta/r_s} = \omega_0 \ln\frac{\delta}{r_s} + \sum_{m,n,k=0}^{+\infty} \frac{\alpha_{m,n,k}}{m+n+k+3} + \sum_{m+n+k\neq 0}^{+\infty} \frac{\alpha_{m,n,k}}{m+n+k} \left[1 - \left(\frac{r_s}{\delta}\right)^{m+n+k} \right] =$$

$$= C_1 \ln\frac{1}{r_s} + w_1(s),$$

where $\alpha_{m,n,k} = \int_0^{\varphi_0} d\varphi_p \int_0^{\theta_0} \sin\theta_p P_m\left(\cos\gamma_{sp}\right) P_n\left(\cos\gamma_{sp}\right) P_k\left(\cos\gamma_{sp}\right) d\theta_p$, C_1 is a constant, and $w_1(s)$ is a bounded function. Thus, the validity of expression (2) is ascertained.

Remark
The estimation of field growth given above was obtained for an arbitrary form of the ferromagnetic. It is easy to select the constants δ, θ_0 and φ_0 for defining any model geometry.

4 Method of mesh condensing in the corner domain

Based on the estimation obtained we propose to condense the mesh approaching the corner point. Let us define the consistent step of the magnetic field value change as $\Delta h = H_0/(C_1 N)$, where H_0 is the approximate maximal value of the magnetic field in the domain under consideration (it can be obtained from pre-simulation with using uniform mesh), C_1 is a constant and N is the number of mesh nodes in a linear dimension. The main idea of the mesh construction is that each step of the mesh Δl_i corresponds with one and the same Δh [2].

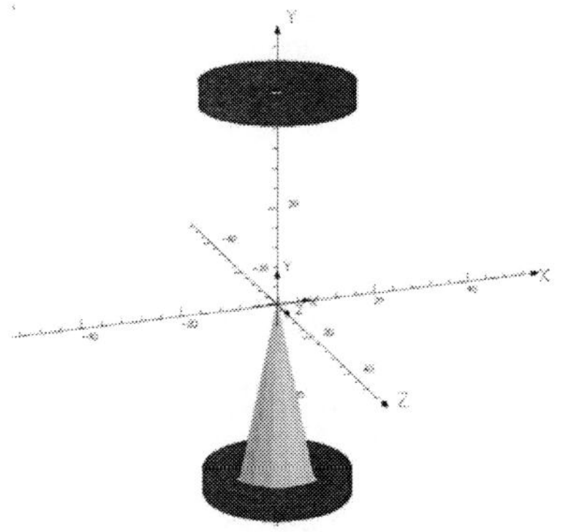

Figure 2. Model geometry

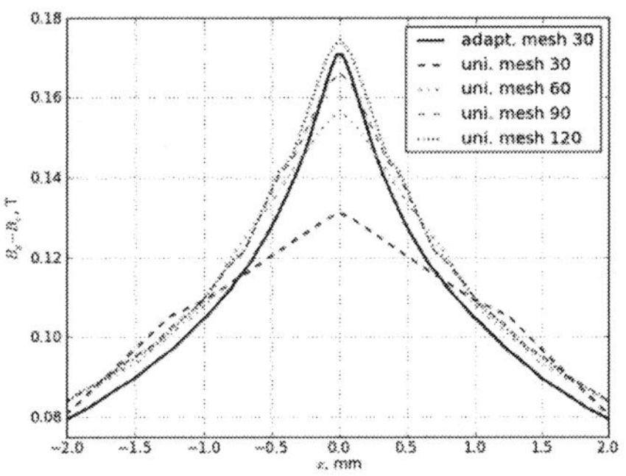

Figure 3. $B_y - B_{\text{coils}}$ distribution along the line for different meshes

5 Calculation of the magnetic field in the corner domain

We present numerical calculations of a magnet system using the obtained method. We consider the magnet system depicted in Figure 2. This is a conical ferromagnetic placed between two identical coils. A boundary-value problem corresponding to the formulation of the magnetostatic problem was solved in the domain under consideration. The efficiency of the differential mesh condensing method described in Section 4 was estimated by the following calculations: the solution of the boundary value problem was found on the sequence of uniform meshes $(30k) \times (30k) \times (30k)$, where $k = 1, 2, 3, 4$. Then the same problem was calculated on a $30 \times 30 \times 30$ mesh, but the mesh spacing was chosen using the mesh condensing method. The number of node points in the domain was the same as in the case when $k = 1$ for a uniform mesh, and only their distribution changed. Then all the calculated results were compared in order to assess how many nodes are needed by the uniform mesh to provide a result similar to the result obtained using the adaptive mesh. Figure 3 shows the distribution $B_y - B_{\text{coils}}$, which is the ferromagnetic field without contribution from coils, along the x-directed straight line, which starts at $x = -2$, $y = 0.1$, $z = 0$ and finishes at $x = 2$, for different meshes. It is evident from Figure 3 that the adaptive mesh with 30 nodes in linear dimension give a comparable result to a uniform mesh with 120 nodes. Thus, it follows that the proposed constructing mesh method in the corner domain is worth using and yields results comparable in accuracy with the results obtained on meshes with the number of nodes in each axis three to four times larger than in the initial mesh.

6 Conclusion

1. The upper estimate for the magnetic field growth $\vec{H}(s)$ in the corner domain $H(s) \leq C_0 \ln \frac{1}{r_s} + w(s)$, where C_0 is a constant, $w(s)$ is a bounded function, and r_s is the distance to the corner, is asymptotically obtained for the case of $\mu(H) \to 1$ when $H \to \infty$.
2. The condensing mesh method in the corner domain has been proposed, which allows one to reduce the number of mesh nodes at the numerical simulation of the magnetostetic problem.

References

[1] E.E. Perepelkin, R.V. Polyakova, A.D. Kovalenko, P.N. Sysoev, M.B. Sadovnikova, A.A. Tarelkin, and I.P. Yudin, RUDN Journal of MIPh. **25**, 3, 253–265 (2017)

[2] E.E. Perepelkin, R.V. Polyakova, A.D. Kovalenko, L.A. Nyanina, P.N. Sysoev, M.B. Sadovnikova, and I.P. Yudin, Computer Technologies in Physics **13**, 6, 782–786 (2016)

Modified Method of Adaptive Artificial Viscosity for Solution of Gas Dynamics Problems on Parallel Computer Systems

Igor Popov[1,2,⋆] and *Sergey* Sukov[1,⋆⋆]

[1] *Keldysh Institute of Applied Mathematics of RAS*
4 Miusskaya square, 125047, Moscow, Russia
[2] *National Research Nuclear University MEPhI*
31 Kashirskoe shosse, 115409, Moscow, Russia

Abstract. A modification of the adaptive artificial viscosity (AAV) method is considered. This modification is based on one stage time approximation and is adopted to calculation of gasdynamics problems on unstructured grids with an arbitrary type of grid elements. The proposed numerical method has simplified logic, better performance and parallel efficiency compared to the implementation of the original AAV method. Computer experiments evidence the robustness and convergence of the method to difference solution.

Introduction

The paper describes a one-step modification of the adaptive artificial viscosity (AAV) method [1] with generalization for the case of spatial discretization using unstructured grids with an arbitrary type of elements. The implementation of the one-step method has simplified logic, better performance and parallel efficiency compared to the implementation of the original AAV method. Based on the modified AAV method, a parallel algorithm and a program for simulating gas dynamics problems for multiprocessor systems with multi-core general-purpose processors are developed. The paper presents parallel software efficiency and examples of the possibility of using it for large-scale computational experiments.

1 Statement of the problem and approximation of equations

Let us consider a system of Euler equations [2] with time Lax – Vendroff corrections. The system in the difference-differential form is written as

$$\frac{\rho^{n+1}-\rho^{n}}{\tau} + \operatorname{div}(\rho\mathbf{v}) = \frac{\tau}{2}\left[\operatorname{div}\left(\sum_{\alpha}\mathbf{e}_{\alpha}\operatorname{div}(I_{\alpha}\mathbf{v}) + \nabla p\right) + \operatorname{div}(\mu\nabla\rho)\right], \tag{1}$$

$$\begin{cases} \dfrac{I_{\alpha}^{n+1}-I_{\alpha}^{n}}{\tau} + \operatorname{div}(I_{\alpha}\mathbf{v}) + \nabla_{\alpha}p = \operatorname{div}(\mu\nabla_{\alpha}p) + \dfrac{\tau}{2}\left[\operatorname{div}\mathbf{A}_{\alpha} + \nabla_{\alpha}B\right], \\ \mathbf{A}_{\alpha} = \displaystyle\sum_{\beta}\mathbf{e}\left(\operatorname{div}\left(v_{\beta}I_{\alpha}\mathbf{v}\right) + \frac{I_{\alpha}}{\rho}\nabla_{\beta}p + v_{\beta}\nabla_{\alpha}p\right), \qquad B = \operatorname{div}(p\mathbf{v}) + (\gamma-1)\,p\operatorname{div}\mathbf{v}, \end{cases} \tag{2}$$

⋆e-mail: piv2964@mail.ru
⋆⋆e-mail: ssoukov@gmail.com

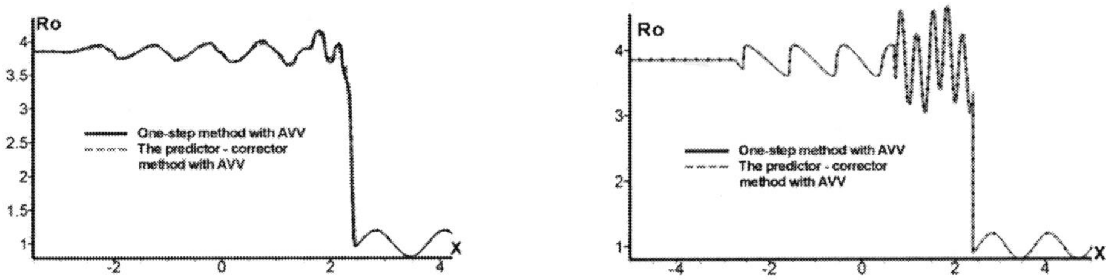

Figure 1. Profiles of density. Left: discretization continuing 401 points. Right: ibid., 20001 points.

$$\begin{cases} \dfrac{E^{n+1} - E^n}{\tau} + \operatorname{div}\Big((E + p)\,\mathbf{v}\Big) = \operatorname{div}(\mu \nabla E) + \dfrac{\tau}{2} \operatorname{div}\mathbf{C}, \\ \mathbf{C} = \sum\limits_{\alpha} \mathbf{e}_\alpha \left(\operatorname{div}(v_\alpha (E + p)\,\mathbf{v}) + \dfrac{(E + p)}{\rho} \nabla_\alpha p + v_\alpha \left((\gamma - 1)\, p \operatorname{div}\mathbf{v} + \operatorname{div}(p\mathbf{v}) \right) \right). \end{cases} \tag{3}$$

Here $\mathbf{e}_\alpha$, $\mathbf{e}_\beta$ are basis vectors in the corresponding coordinate directions $\alpha, \beta = 1, \ldots, D$, D is the space dimension, ρ is the density, $\mathbf{v}$ is the velocity vector, I is the momentum vector, $E = \rho\left(0.5|v|^2 + \varepsilon\right)$ is the energy density, $\rho\varepsilon$ is the internal energy density related to the pressure p by the equation of state $p = (\gamma - 1)\,\rho\varepsilon$, γ is the degree index of the Poisson adiabat.

The Lax–Vendroff corrections ensure the second order of accuracy in time when constructing the difference scheme. The difference approximation of the remaining differential operators can be found in the monograph [1]. In the same paper an algorithm for finding the artificial viscosity is given. For a one-step modified AAV method, only the maximum artificial viscosity is used, both on shock waves and on oscillations of the numerical solution, which has the form:

$$\mu_i^n = \mu_{\max} = \max \frac{\Delta n_i}{2D} \sqrt{v_{n,i}^2 + c_i^2} \left(1 - \frac{\tau^2 D^2}{\Delta n_i^2} \left(v_{n,i}^2 + c_i^2 \right) \right), \tag{4}$$

where c_i is the sound speed, Δn_i is the projection of the vector between the centers of the cells on the normal vector $\mathbf{n}_i$ to the face in question, $v_{n,i}$ is the projection of the velocity vector defined at the center of the cell onto the normal vector $\mathbf{n}_i$ edges.

The proposed numerical algorithm has three stages:

1. Determination of the regions of shock waves (SW), compression waves (CW), rarefaction waves (RW), and oscillations of a numerical solution (OSC) from the n-th time layer.
2. Setting the artificial viscosity maximum value inside SW and OSC regions and equating it to zero outside these two regions.
3. Computation of the fluxes together with all the unknown functions on the $(n + 1)$-th time layer.

2 Numerical experiments

To check the proposed modified method, all the 1D and 2D tests from [3], [4] were solved within proposed approach. In this paper, we discuss the numerical calculation of the Shu–Osher problem [5] with initial data on the interval [0, 1]: $\rho_L = 3.857143$, $u_L = 2.629369$, $p_L = 10.333333$, $\rho_R = 1 + 0.2\sin(50x)$, $u_R = 0$, $p_R = 1$. The initial point of discontinuity was located at the point $x_0 = 0.1$. The calculations was carried out up to the time $T = 0.18$.

Fig. 1 left shows numerical outputs for a number of 401 calculated points. In Fig. 1 right, the number of calculated points is 20001. With such a number of points, the lines obtained by the modified and original AAV method coincide, from which it can be concluded numerical outputs converge

Figure 2. Isolines of density. Left: overall view. Right: enlarged view showing discontinuity interactions.

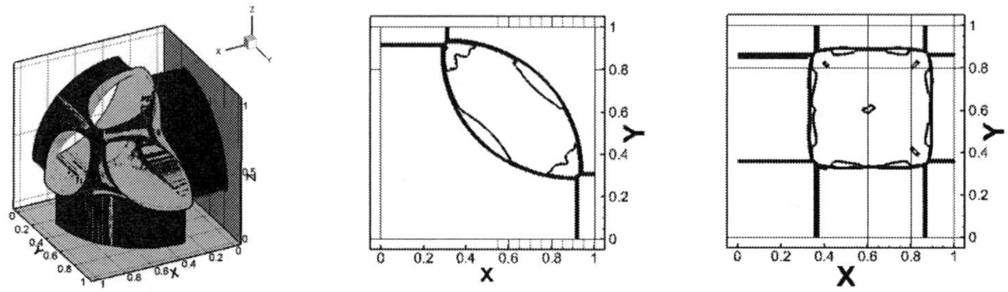

Figure 3. Isolines of density in 3D. Left: 3D view, Center: inside plane $z = 0$, Right: inside plane $z = 0.612$

to the same solution. The solid black line shows the output of the modified AAV method, and the solid gray line the output of the original AAV method. Calculations were carried out assuming the Courant number equal to 0.1.

Figs. 2 show the results of the calculation of the test problem No. 15, respectively, on the decays of the 4 discontinuities given by the initial data, which, after the disintegration of the discontinuity, form two contact discontinuities, a shock wave and a rarefaction wave. Fig. 2 right shows an enlarged image of the interaction of these four discontinuities. The solid black curve shows the results of the calculation using the modified scheme, and the gray dotted curve shows the outputs of the original predictor-corrector AAV method.

The Riemann problem [6] was taken as a three-dimensional test problem. A cube with a side equal to one was divided into eight equal cubes. Each of the cubes has its own velocity and pressure density parameters, see [6]. In this problem, the boundary conditions assumed vanishing normal derivatives of all quantities at cube surfaces. The task was solved up to $T = 0.25$. The comparison of the results produced by the modified and original AAV methods on uniform 3D grids 200^3. Fig. 3 left shows the isosurfaces at time $T = 0.25$ and density $\rho = 1.5$. Fig. 3 center shows isolines inside the plane $z = 0$, and Fig. 3 right – isolines inside the plane $z = 0.612$. The two results are visually indistinguishable.

3 Parallel program package

Based on the proposed modification of the AAV method, a parallel program for simulating gas dynamics problems for high-performance cluster architecture systems with multi-core general-purpose processors was developed. The synchronization and interaction of parallel processes occurs within the framework of a hybrid model of a concurrency control for the message transfer (MPI + OpenMP). The distribution of calculations between processes is performed by the decomposition of the calculation domain into subdomains (the method of geometric parallelism). The data exchange area includes grid elements of two levels of connectivity over a dual graph. At the same time, duplication of calculations concerns only elements of the first level. Preprocessor data processing is implemented using the library functionality of the hm4PreProcessorLib [7] routines, designed for distributed processing of meshes with elements like tetrahedron, triangular prism, quadrangular pyramid and hexahedron, containing a total of up to $5 \cdot 10^8$ elements. The tests of this package on hybrid large volume meshes confirmed high productivity and parallel efficiency of the parallel program.

4 Conclusion

The obtained numerical evidence points to the agreement of the outputs of the modified AAV method with those of the original AAV method. Under the refinement of the discretization mesh, the convergence to the same solutions have been found to occur. The proposed modified method of adaptive artificial viscosity on discontinuous solutions can compete with modern numerical methods.

Acknowledgments

This work was supported by the Russian Foundation for Basic Researches (projects No. 15-01-04620-a, 15-07-04213-a, 16-07-00519-a, 16-29-15095-ofi-m).

References

[1] I.V. Popov, I.V. Fryazinov, *Method of Adaptive Artificial Viscosity for Numerical Solution of Gas Dynamics Equations* (KRASAND, Moscow, 2015) 288 pp.
[2] L. Landau, E. Lifshitz, *Fluid Mechanics: Course of Theoretical Physics*, Vol. 6. (Pergamon Press, Oxford, 1987) 554 p.
[3] I.V. Popov, I.V. Fryazinov, Math. Models Comput. Simul., **1**(4), 493–502 (2009)
[4] I.V. Popov, I.V. Fryazinov, Math. Models Comput. Simul., **2**(6), 724–732 (2010)
[5] C. Shu, S. Osher, J. Comput. Phys., **83**(1), 32–78 (1989)
[6] M. Kucharik, R. Liska, S. Steinberg, B. Wendroff, Applied Numerical Mathematics, **56**, 589–607 (2006)
[7] S.A. Sukov, *Unstructured mesh preprocessing library hm4PreprocessorLib* (KIAM Preprint, No. 108, 2016)

EPJ Web of Conferences **173**, 03021 (2018) https://doi.org/10.1051/epjconf/201817303021
Mathematical Modeling and Computational Physics 2017

Computational Algorithm for Covariant Series Expansions in General Relativity

Ivan Potashov[1,⋆] and *Alexander* Tsirulev[1,⋆⋆]

[1] *Faculty of Mathematics, Tver State University, Sadovyi per. 35, Tver, Russia, 170002*

Abstract. We present a new algorithm for computing covariant power expansions of tensor fields in generalized Riemannian normal coordinates, introduced in some neighborhood of a parallelized k-dimensional submanifold ($k = 0, 1, \ldots < n$; the case $k = 0$ corresponds to a point), by transforming the expansions to the corresponding Taylor series. For an arbitrary real analytic tensor field, the coefficients of such series are expressed in terms of its covariant derivatives and covariant derivatives of the curvature and the torsion. The algorithm computes the corresponding Taylor polynomials of arbitrary orders for the field components and is applicable to connections that are, in general, nonmetric and not torsion-free. We show that this computational problem belongs to the complexity class *LEXP*.

1 Introduction

Covariant series expansions in Riemann normal coordinates [1, 2], in Fermi normal coordinates [3, 4], and in some other geodesic coordinate systems (e.g., see Ref. [5]) find important applications in general relativity [6–9]). Explicit formulas and recipes for coefficients of the series are given in Refs. [10–15], so that, from a formal point of view, the problem is solved completely. However, the complexity of calculations increases rapidly with increasing the order of a covariant expansion and there obviously arises the question whether we can organize these calculations on a computer. To avoid confusion, it is essential to note that the problem, in the context of calculating a covariant series, can be set in different ways depending on the explicit form of the expansion. We employ the mathematical framework of Ref. [15] where the problem is considered with a sufficient degree of generality: one needs only a connection to form the series; the connection is not assumed to be metric or torsion-free; the covariant series are defined in some normal neighborhood of an arbitrary submanifold. In our setting, having established a connection (or spacetime metric), one needs to compute the covariant derivatives of the curvature (and, in general, of the torsion), to collect like terms, and then to compute coefficients of coordinate monomials.

Sec. 2 contains some necessary mathematical preliminaries and a short description of the computational algorithm. In Sec. 3 we make the estimations of numbers of multiplications at various stages of the algorithm and show that the complexity class of the algorithm is *LEXP*.

⋆e-mail: potashov.im@tversu.ru
⋆⋆e-mail: tsirulev.an@tversu.ru

2 General expansions and computational algorithm

We deal with a real analytic manifold H, $\dim H = n+m$, $n > m \geqslant 0$, and a connected parallelizable submanifold $M \subset H$, $\dim M = m$; M is a point if $m = 0$. At each point $p \in M$ the tangent space $T_p(H)$ splits into the direct sum $T_p(M) \oplus N_p$, where the subspace N_p is normal (transverse, if the metric is not specified) to M. Let $\{e_\alpha\}$ and $\{e_a\}$ be bases for N_p and $T_p(M)$ respectively, so that their union is a basis for $T_p(H)$. We assume that the Latin indices i, j, k, l take the values $1, \ldots, n+m$, while the Greek indices α, β, γ run from 1 to n, and the Latin indices a, b run from $m+1$ to $n+m$; they are used to denote tensor components in $T_p(H)$, N_p, and $T_p(M)$ respectively. For *these* indices, we also adopt Einstein's summation convention, that is, a pair of identical indices, with one lower index and one upper index, denotes summation over their range. Let ∇ be a linear connection on H and $\nabla_{e_m} e_k = \Gamma^i_{km} e_i$, $\nabla_{e_m} X = X^k_{;m} e_k$; $R^k_{mij} = \left\langle e^k, \nabla_i \nabla_j e_m - \nabla_j \nabla_i e_m - \nabla_{[e_i, e_j]} e_m \right\rangle$ are the curvature components. In what follows, for the sake of shortness, we restrict our consideration to the case of torsion-free connections.

If γ_X is the geodesic in the direction of a vector $X = X^\alpha e_\alpha \in N_p$, such that $\gamma_X(0) = p$, $\gamma_X(1) = q \in H$, then $X^1, \ldots, X^n$ are normal coordinates of the point q (the *exp* map). Now let Q be a tensor field of type (r, s) (of type (0,2), in the particular case of a metric g) in a normal neighborhood of the submanifold M, then the covariant series expansion for Q has the form [15]

$$\left(Q^{j_1 \ldots j_s}_{i_1 \ldots i_r}\right)_q = \sum_{\sigma+|\mu|+|\nu| \geq 0} \frac{1}{\sigma!} X^{\gamma_1} \ldots X^{\gamma_\sigma} \left(Q^{l_1 \ldots l_s}_{k_1 \ldots k_r; \gamma_1 \ldots \gamma_\sigma}\right)_p \underset{(\mu_1)}{u}{}^{k_1}_{i_1} \ldots \underset{(\mu_r)}{u}{}^{k_r}_{i_r} \underset{(\nu_1)}{v}{}^{j_1}_{l_1} \ldots \underset{(\nu_s)}{v}{}^{j_s}_{l_s}, \tag{1}$$

where $\sigma, \mu_1, \ldots, \mu_r, \nu_1, \ldots, \nu_s \geq 0$, $|\mu| = \sum_{t=1}^{r} \mu_t$, $|\nu| = \sum_{t=1}^{s} \nu_t$, $\underset{(0)}{u}{}^k_i = \underset{(0)}{v}{}^k_i = \delta^k_i$, $\underset{(1)}{u}{}^k_\alpha = 0$, $\underset{(1)}{u}{}^k_a = X^\beta \left(\Gamma^k_{\beta a}\right)_p$,

$$\underset{(\mu)}{u}{}^k_i = \frac{1}{\mu(\mu+\varepsilon(i))} \sum_{\sigma=2}^{\mu} \frac{1}{(\sigma-2)!} X^{\alpha_1} \ldots X^{\alpha_\sigma} \left(R^k_{\alpha_1 \alpha_2 l; \alpha_3 \ldots \alpha_\sigma}\right)_p \underset{(\mu-\sigma)}{u}{}^l_i \; (\mu \geq 2), \quad \underset{(\mu)}{v}{}^k_i = -\sum_{\sigma=1}^{\mu} \underset{(\mu-\sigma)}{v}{}^k_l \underset{(\sigma)}{u}{}^l_i \; (\mu \geq 1),$$

$\varepsilon(\alpha) = 1$ and $\varepsilon(a) = -1$ in the ranges of α and a. We come now to the problem of algorithmic reducibility of the expansion (1). The number of coordinate monomials in (1) is equal to the number of terms of the corresponding Taylor series. The most difficult step is the computation of the coefficients of monomials in matrices u^i_j; if it has been done, the monomials in (1) can be isolated by well-known methods. Therefore, we restrict the discussion to these computations.

For each $\mu \geq 2$ we generate the lexicographically ordered set of the sequences of n whole numbers

$$S_\mu = \left\{(A_1, \ldots, A_n) \,|\, A_1 + \cdots + A_n = \mu\right\} = \left\{S_\mu[1], \ldots, S_\mu[N]\right\}, \quad N = \left|S_\mu\right| = \binom{n+\mu-1}{\mu-1},$$

and for each element $S_\mu[p] = (A_1, \ldots, A_n) \in S_\mu$, $1 \leq p \leq N$, we consider the set

$$S^p_\mu = \left\{(\alpha_1, \ldots, \alpha_\mu) \in \{1, 2, \ldots, n\}^\mu\right\} = \left\{S^p_\mu[1], \ldots, S^p_\mu[M]\right\}, \quad M = \left|S^p_\mu\right| = \frac{\mu!}{A_1! A_2! \ldots A_n!}, \tag{2}$$

the elements of which (sequences of length μ) are the permutations of the sequence which consists of A_1 numbers 1, A_2 numbers 2, ..., A_n numbers n. We also need to form the set C_μ of the sequences of variable length and the collection of the corresponding rational constants $h : C_\mu \to \mathbb{Q}$:

$$C_\mu = \left\{(\sigma_1, \ldots, \sigma_\tau) \,|\, \sigma_1 + \cdots + \sigma_\tau = \mu, \, \sigma_1, \ldots, \sigma_\tau \geqslant 2, \, 1 \leqslant \tau \leqslant [\mu/2]\right\}, \quad N_1 = \left|C_\mu\right| = \sum_{\tau=1}^{[\mu/2]} \binom{\mu-\tau-1}{\tau-1}, \tag{3}$$

$$h(\sigma_1, \ldots, \sigma_\tau) = \prod_{r=1}^{\tau} \frac{\xi(\sigma_r + \sigma_{r+1} + \ldots + \sigma_\tau)}{(\sigma_r - 2)!}, \quad \xi(\sigma) = \frac{1}{\sigma(\sigma+1)}. \tag{4}$$

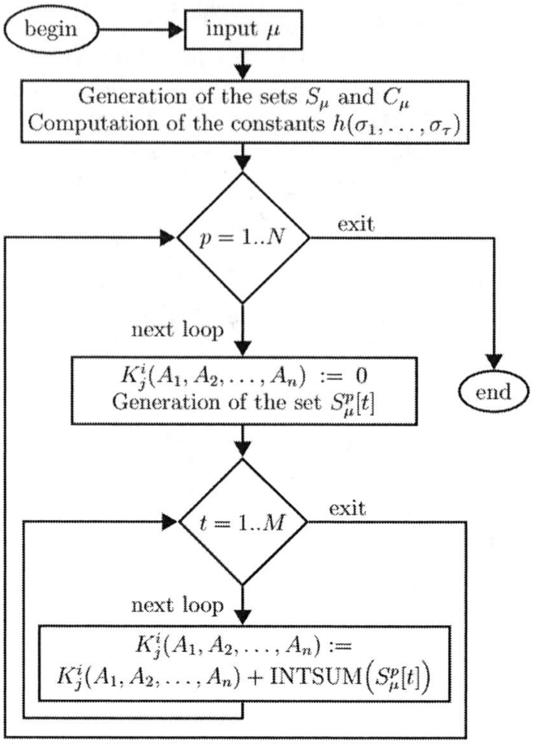

Figure 1. Block diagram of the algorithm

For the matrix u^j_i of order μ, we have

$$\underset{(\mu)}{u}{}^i_j = \sum_{S_\mu} K^i_j(A_1, \ldots, A_n) \left(X^1\right)^{A_1} \ldots \left(X^n\right)^{A_n}, \tag{5}$$

where all the $K^i_j(A_1, \ldots, A_n)$ should be expressed in terms of the curvature. The algorithm is shown schematically in Fig. 1. We compute K^i_j for the sequence $S_\mu[p] = (A_1, \ldots, A_n)$ at the pth step of the external loop, where p runs from 1 to N. In doing this we generate the set C_μ and compute, for each of its elements $(\sigma_1, \ldots, \sigma_\tau)$, the constants (4). The covariant derivatives for S^p_μ are computed by a special iterative procedure. Let $n = 3$, $\mu = 4$, and $S_4[7] = (1, 2, 1)$. Thus $C_\mu = \{(4), (2, 2)\}$ and, e.g., if $(\alpha_1, \alpha_2, \alpha_3, \alpha_4) = (2, 1, 2, 3)$, the procedure gives explicit expressions for $R^i_{21j;23}$, R^i_{21j}, and R^k_{23j}. In the internal loop, the algorithm calls the special procedure INTSUM. Finally, for the coefficients in (5) we obtain

$$K^i_j(A_1, \ldots, A_n) = \sum_{t=1}^{M} \mathrm{INTSUM}\left(S^p_\mu[t]\right), \tag{6}$$

$$\mathrm{INTSUM}\left(S^p_\mu[t]\right) = \sum_{C_\mu} h(\sigma_1, \ldots, \sigma_\tau) R^i_{\alpha_1\alpha_2 l_1;\alpha_3\ldots\alpha_{\delta_1}} R^{l_1}_{\alpha_{\delta_1+1}\alpha_{\delta_1+2} l_2;\alpha_{\delta_1+3}\ldots\alpha_{\delta_2}} \ldots R^{l_{\tau-1}}_{\alpha_{\delta_{\tau-1}+1}\alpha_{\delta_{\tau-1}+2} j;\alpha_{\delta_{\tau-1}+3}\ldots\alpha_{\delta_\tau}},$$

where $S^p_\mu[t] = (\alpha_1, \ldots, \alpha_\mu)$ and $\delta_r \equiv \sum_{s=1}^{r} \sigma_s$.

3 Computational complexity of the algorithm

There are two ways of summation in the formula (6) for $K^i_j(A_1, \dots, A_n)$: one is over the set S^p_μ and the other is over the set C_μ in the procedure INTSUM. In addition, each summand in the latter sum is a matrix product with $\tau - 1$ summation indices; it gives $(n+m)^{\tau-1}$ multiplications. The total number of multiplications in INTSUM is the sum of N_1 multiplications by the factors h and the number N_2 of matrix multiplications. From (2) and (3), using the generating functions for the sums (particularly the Binet's Fibonacci number formula), we find

$$N_1 = \sum_{\tau=0}^{[\mu/2]-1} \binom{\mu-\tau-2}{\tau} = F_{\mu-1} = \frac{1}{\sqrt{5}}\left[\left(\frac{1+\sqrt{5}}{2}\right)^{\mu-1} - \left(\frac{1-\sqrt{5}}{2}\right)^{\mu-1}\right] \leqslant \frac{1}{\sqrt{5}}\left(\frac{1+\sqrt{1+4(n+m)}}{2}\right)^{\mu-1},$$

$$N_2 + 1 = \sum_{\tau=1}^{[\mu/2]} \binom{\mu-\tau-1}{\tau-1}(n+m)^{\tau-1} = \frac{\left(1+\sqrt{1+4(n+m)}\right)^{\mu-1}}{2^{\mu-1}\sqrt{1+4n+4m}} - \frac{\left(1-\sqrt{1+4(n+m)}\right)^{\mu-1}}{2^{\mu-1}\sqrt{1+4n+4m}}.$$

From here we obtain the estimation

$$N_1 + N_2 = O\left(\left(\frac{1+\sqrt{1+4(n+m)}}{2}\right)^{\mu}\right), \quad \mu \to \infty, \quad n+m \text{ is fixed.}$$

For the number of multiplications on the right hand side in formula (6) and the total number of multiplications in computing all the coefficient $K^i_j(A_1, \dots, A_n)$ we have, respectively, the estimations

$$(N_1 + N_2) \cdot M = (N_1 + N_2) \cdot \frac{\mu!}{A_1! \dots A_n!}, \qquad (N_1 + N_2) \cdot \sum_{\substack{A_1 + \dots + A_n = \mu \\ A_1, \dots, A_n \geqslant 0}} \frac{\mu!}{A_1! \dots A_n!} = (N_1 + N_2) \cdot n^\mu.$$

Finally, we have $O\left(\frac{n+n\sqrt{1+4(n+m)}}{2}\right)^\mu$ for the estimation of computational complexity of the algorithm. The algorithm belongs to the complexity class $LEXP \equiv E \subset EXPTIME$.

References

[1] L.P. Eisenhart, *Riemannian Geometry* (Princeton University Press, Princeton, 1926)
[2] A.Z. Petrov, *Einstein Spaces* (Pergamon Press, Oxford, 1969)
[3] F.K. Manasse and G.W. Misner, J. Math. Phys. **4**, 735–745 (1963)
[4] E. Poisson, A. Pound and I. Vega, Living Rev. Relativ. **14**, 7 (2011)
[5] V. Iyer and R.M. Wald, Phys. Rev. D **50**, 846–864, (1994)
[6] D. Bini, A. Geralico and R.T. Jantzen, Gen. Relativ. Grav. **44**, 1837–1853 (2011)
[7] A.I. Harte, Class. Quant. Grav. **27**, 135002 (2010)
[8] M.N.R. Wohlfarth and C. Pfeifer, Phys. Rev. D **87**, 024031 (2013)
[9] K.-P. Marzlin, Gen. Relativ. Grav. **26**, 619–636 (1994)
[10] P. Mukhopadhyay, Reviews in Math. Phys. **26**, 1, 1350019 (2013)
[11] L. Brewin, Class. Quantum Grav. **26**, 175017 (2009)
[12] D. Klein and P. Collas, J. Math. Phys. **51**, 022501 (2010)
[13] B.Z. Iliev, *Handbook of Normal Frames and Coordinates* (Birkhäuser Verlag, Berlin, 2006)
[14] U. Müller, C. Schubert and A.E.M. van de Ven, Gen. Relativ. Grav. **1**, 1759–1765 (1999)
[15] A.N. Tsirulev, Theor. Math. Phys. **102**, 3, 245–250 (1995)

EPJ Web of Conferences **173**, 03022 (2018) https://doi.org/10.1051/epjconf/201817303022
Mathematical Modeling and Computational Physics 2017

A Variational Solution of the Schrödinger Equation in an Inhomogeneous Central Field

Ilya Tolstov[1,3,*], *Boris* Freinkman[2,**], and *Sergey* Polyakov[3,***]

[1] *Keldysh Institute of Applied Mathematics Miusskaya sq., 4 Moscow, 125047, Russia*

Abstract. The present work is devoted to the computer modeling of emission processes from the graphene surface which is a promising material for modern applications. We examine the effect of the ion field inhomogeneity on the variational solution of the Schrödinger equation for the ground state of a loosely bound electron of a hydrogen-like carbon atom, which simulates the graphene model. This shows a significant impact of the allowance for the field inhomogeneity to the ground state of the electron.

1 Introduction

Currently, there is a huge amount of publications about field emission properties of various carbon nanostructures. Almost in all papers, the authors note the discrepancy between the experimental data obtained on the geometric form and parameters of field emission [1]. One of the important problems of the use of graphene and nanotubes on its basis in the nanotechnology is the lack of reliable data on the quantum states of atoms of graphene lattice in the presence of external field sources. Thus, for promising and potentially possible appliances of nanotubes as a cantilever in tunneling microscopes or as an electron source [2] in probe electron-optical systems, mathematical modeling is needed to determine their emission characteristics. For this, it is necessary to develop a model describing the interaction of the electron with the surface. In the previous work [3], the graphene model in the form of the lattice of hydrogen-like atoms with a screened ion in the Brandt model was used to calculate the ground state of atoms of the graphene surface [4]. Also, for calculation the ground state of an electron we used a variational solution of the Schrodinger equation, based on minimizing the electron potential energy in the ion field, that assumes homogeneity of the ion field. However, the field of the shielded and the singly ionized carbon atom in the Brant model was not homogeneous.

Within the hydrogen-like atom model used to describe the carbon atom in the graphene lattice the problem is regarded as the electron interaction with the ion lattice having some effective charge q. The Brandt-Kitagawa model [4] is an appropriate a model describing the inhomogeheous interaction.

*e-mail: tolstov_ilya@inbox.ru
**e-mail: bgfreink@gmail.com
***e-mail: polyakov@imamod.ru

2 Allowance for the inhomogeneity of the field for the hydrogen-like atom model

In the previous work [3] the ground state of an electron in the singly ionized hydrogen-carbon atom was determined by finding an effective homogeneous field $\tilde{U} = -\frac{q}{r} + A$ from the condition of residual error minimum of the electron potential energy in the s-state in this field and in the ion field in the Brandt model [4]

$$U(r) = \frac{1}{r} + \frac{Z-1}{r}\exp\left(-\frac{r}{\lambda}\right), \qquad \lambda = g\frac{N_e/Z^{2/3}}{1 - N_e/7Z}\frac{a_0}{Z^{1/3}}, \tag{1}$$

where $g = 0.48, Z = 6, N = 5, \lambda = 0.2655$.

The effective charge of the ion q was determined from the numerical solution of the equation:

$$J(q) = \frac{q^2}{2n^2} + \int_0^\infty R_{ns}^2(r,q)V(r)r^2 dr \tag{2}$$

However, the ion field in the Brandt model is not homogeneous along the radius since

$$\frac{\partial}{\partial r}U_i(r) = -\frac{U_i(r)}{r} - \frac{Z-1}{\lambda_i} \neq \chi\frac{U_i(r)}{r}. \tag{3}$$

According to [5], the total energy of a finite motion in an inhomogeneous central field is

$$E = \frac{r}{2}\frac{\partial U}{\partial r} + U, \tag{4}$$

therefore, the effective charge q of an equivalent homogeneous field for the s-status should be determined from the condition of residual error minimum of an electron total energy in the s-state in this field and in the ion field in the Brandt model.

In this case, the effective charge q will be determined from the solution of the variational equation

$$J_2(q) = \left\langle R_s^*(x)\left[\frac{r}{2}\frac{\partial}{\partial r}U_i(r) + U_i(r)\right]R_s(x)\right\rangle - \frac{q^2}{2n^2}, \qquad x = \frac{2q}{n}r. \tag{5}$$

3 Results

A numerical results are proposed on figures 1–3. Figure 1 shows the distribution of the values calculated in this paper. Figure 2 shows the numerical solution of (2) and (5). Figure 3 shows the distribution of the effective homogeneous field in these two solutions.

As can be seen from the graphs, the results are significantly different. The calculated values of the effective charge are given in Table 1.

Table 1. Calculated values for charge and minimal energy

	minimal potential energy	minimal total energy
Q	0.71218754	0.4473579
A	0.0836779	0.02791427
E	2.4214	1.5208

The variation of the values computed for the lowest energy within the two formulations of the problem reaches 37%.

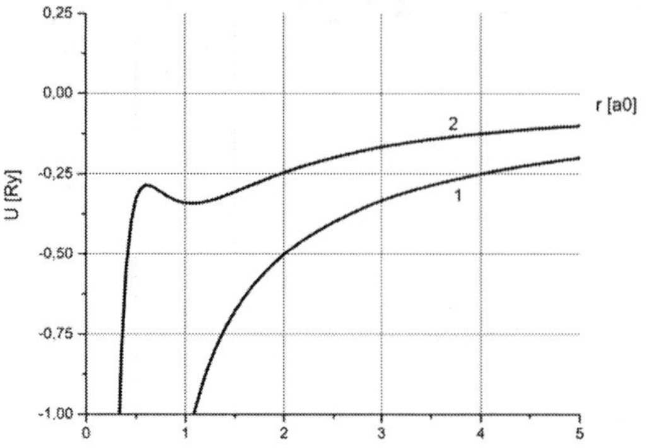

Figure 1. The distribution of the potential created by the effective charge of the ion (line 1) and the mean total energy (line 2). Here and below, the potential is measured in Rydberg.

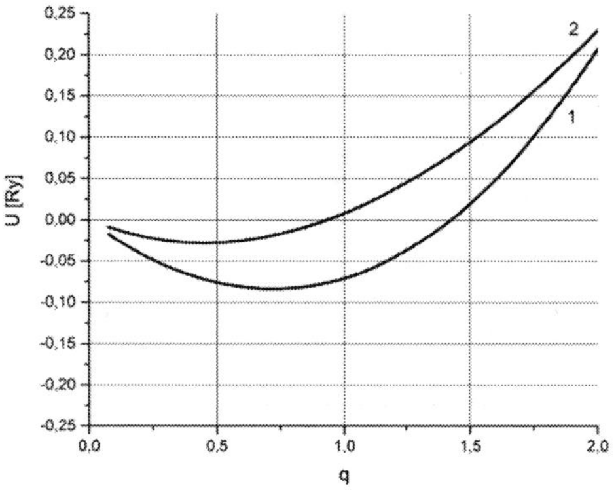

Figure 2. Solution of the variational problem for a homogeneous (line 1) and inhomogeneous (line 2) case

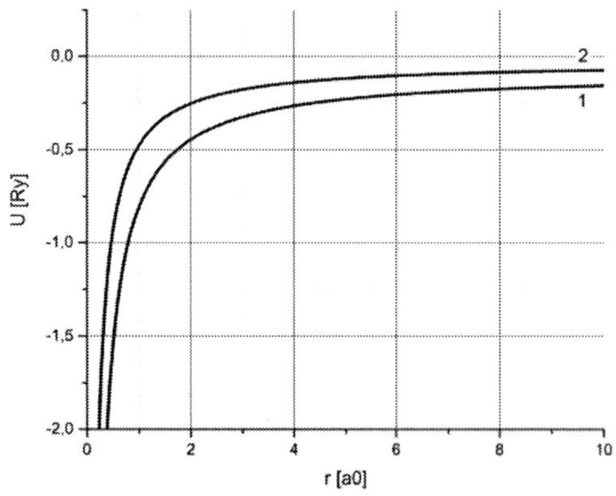

Figure 3. Distribution of a homogeneous field of a hydrogen-like atom without taking into account (line 1) and taking into account (line 2) the virial theorem

4 Conclusion

These results indicate that during calculation of the ground state of an electron of the hydrogen-like atom with a shielded ion in the Brandt model, we should take into account the inhomogeneity of the ion field in the Brandt model.

The model under development will allow to predict the emission and transport characteristics of the devices based on graphene, which is necessary for the further progress of the electron microscopy.

The main advantage of the model under development is its flexibility: the Brandt-Kitagawa potential allows to describe atomic charge and different degrees of ionization. Therefore, it is possible to calculate the surface potential for silicon.

Acknowledgments

This work was supported by the Russian Foundation for Basic Researches (project No. 17-01-00973-a).

References

[1] K. Asano, Journal of Electrostatics, **68**, 132–137 (2010)
[2] A.V. Eletskii, Physics Uspekhi, **53**, 863–892 (2010)
[3] B.G. Freinkman, Matematicheskoe modelirovanie **27**, 122–128 (2015) (in russ.)
[4] W. Brandt, M. Kitagawa, Phys. Rev. B **25**, 5631–5637 (1982)
[5] A.C. Phillips, *Introduction to Quantum Mechanics* (Wiley, New York, 2003)

EPJ Web of Conferences **173**, 03023 (2018) https://doi.org/10.1051/epjconf/201817303023
Mathematical Modeling and Computational Physics 2017

On a Class of Hermite Interpolation Polynomials for Nonlinear Second Order Partial Differential Operators

Leonid A. Yanovich[1,*] and *Marina V.* Ignatenko[2,**]

[1] *Institute of Mathematics, National Academy of Sciences of Belarus, Surganova Str. 11, 220072 Minsk, Belarus*

[2] *Faculty of Mechanics and Mathematics, Belarus State University, Nezavisimosti Ave. 4, 220030 Minsk, Belarus*

Abstract. This article is devoted to the problem of construction of Hermite interpolation formulas with knots of the second multiplicity for second order partial differential operators given in the space of continuously differentiable functions of two variables. The obtained formulas contain the Gateaux differentials of a given operator. The construction of operator interpolation formulas is based on interpolation polynomials for scalar functions with respect to an arbitrary Chebyshev system of functions. An explicit representation of the interpolation error has been obtained.

1 Introduction

The interpolation of the functions provides the basis for the construction and research of approximate and numerical methods for the solution of many classes of problems. In particular, it is widely used for the approximate representation and calculation of functions, the numerical integration and differentiation, the construction of approximate methods for solving various classes of linear and nonlinear equations, etc. It is natural and necessary an extension of the interpolation theory of the functions to other mathematical objects. The operator interpolation, as one of the sections of applied functional analysis and the general theory of approximate methods, significantly generalizes the problem of function interpolation and is the foundation for the construction of approximate methods and algorithms to solve applied problems. The main problems of the operator interpolation are formulated similarly to the classical problems of the function interpolation.

A number of other interpolation formulas for operators given in the spaces of functions and matrices are given in [1]–[4].

2 Interpolation operator formulas of the Hermite type

Let $X = X(T \times S)$ be a given space of smooth functions on $T \times S \subseteq R^2$. By $\delta F[x; h]$ we denote the Gateaux differential at the point $x = x(t, s)$ in the direction $h = h(t, s)$ $(x, h \in X)$ for an operator $F(x)$

*e-mail: yanovich@im.bas-net.by
**e-mail: ignatenkomv@bsu.by

defined on the set X. We consider the operator polynomials of the form

$$P_{2n+1}(x) = \sum_{k=0}^{2n+1} a_k(\xi, t, s)\varphi_k(x(t, s)), \tag{1}$$

where $a_k(\xi, t, s)$ are arbitrarily given functions of the variables ξ, t and s, and $\{\varphi_k(x)\}_{k=0}^{2n+1}$ is some arbitrary Chebyshev system of functions.

We consider the Hermite interpolation operator formulas with respect to knots of the second multiplicity. In [5] we have constructed the operator polynomial of the Hermite type containing the Gateaux differentials of a given operator of the form

$$H_{2n+1}(F; x) = F(x_0) + \sum_{k=1}^{n} \int_0^1 \delta F\left[x_0 + \tau(x_k - x_0); \frac{h_{n,k}(x)}{\sigma_n(x)}(x_k - x_0)\right] d\tau + \sum_{k=0}^{n} \delta F\left[x_k; \frac{q_{n,k}(x)}{\sigma_n(x)}\right], \tag{2}$$

where $h_{n,k}(x)$ and $q_{n,k}(x)$ are the Hermite fundamental polynomials of the $(2n + 1)$-degree of the form (1) in the case of second multiplicity knots $x_0, x_1, \ldots, x_n$ with respect to some Chebyshev system $\{\varphi_k(x)\}_{k=0}^{2n+1}$ of functions, for which $h_{n,k}(x_j) = q'_{n,k}(x_j) = \delta_{kj}$, $h'_{n,k}(x_j) = q_{n,k}(x_j) = 0$ $(k, j = 0, 1, \ldots, n)$, and $\sigma_n(x) = \sum_{k=0}^{n} h_{n,k}(x)$ is a constant or a variable value. For example, if $\left\{\phi_k(x) = x^k\right\}_{k=0}^{2n+1}$ is an algebraic system of functions, then the fundamental polynomials $h_{n,k}(x)$ and $q_{n,k}(x)$ are defined by the rules

$$h_{n,k}(x) = l_{n,k}^2(x)\left[1 - \frac{\omega_n''(x_k)}{\omega_n'(x_k)}(x - x_k)\right], q_{n,k}(x) = l_{n,k}^2(x)(x - x_k), k = 0, 1, \ldots, n,$$

where $l_{n,k}(x) = \prod_{j=0;\, j\neq k}^{n} \frac{x - x_j}{x_k - x_j}$ are algebraic Lagrange fundamental polynomials. In this case $\sigma_n(x) \equiv 1$, and from the interpolation nodes it is required that $x_k(t, s) - x_j(t, s) \neq 0$ $(k \neq j)$ for all $(t, s) \in T \times S$.

The polynomial (2) is the interpolation operator for a given operator $F(x)$ on the set X relative to the knots $x_0, x_1, \ldots, x_n$ of the second multiplicity. It satisfies the following conditions:

$$H_{2n+1}(F; x_k) = F(x_k), \delta H_{2n+1}[x_k; h] = \delta F[x_k; h] \, (k = 0, 1, \ldots, n; h \in X). \tag{3}$$

For the interpolation error $r_{2n+1}(x) = F(x) - H_{2n+1}(F; x)$, where $H_{2n+1}(F; x)$ is the interpolation polynomial (2), the following representation holds:

$$r_{2n+1}(x) = \sum_{k=1}^{n+1} \int_0^1 \delta F\left[x_0 + \tau(x_k - x_0); \left(\frac{h_{n+1,k}(x)}{\sigma_{n+1}(x)} - \frac{h_{n,k}(x)}{\sigma_n(x)}\right)(x_k - x_0)\right] d\tau + \sum_{k=0}^{n+1} \delta F\left[x_k; \frac{q_{n+1,k}(x)}{\sigma_{n+1}(x)} - \frac{q_{n,k}(x)}{\sigma_n(x)}\right], \tag{4}$$

where $x_{n+1} = x$, $h_{n,n+1}(x) = q_{n,n+1}(x) \equiv 0$.

3 Interpolation formulas of the Hermite type for second order partial differential operators

There exists a rich literature on the theory of partial differential equations and operators. See, for example, [6]–[9] and others. We consider the second order partial differential operators $F : C^2(T \times S) \to Y$ of the form

$$F(x) = f\left(t, s, x(t, s), x_t'(t, s), x_s'(t, s), x_{t^2}''(t, s), x_{t,s}''(t, s), x_{s^2}''(t, s)\right), \tag{5}$$

where $x_t'(t, s) = \frac{\partial x(t,s)}{\partial t}$, $x_{t,s}''(t) = \frac{\partial^2 x(t,s)}{\partial t \partial s}$, $x_{t^2}''(t, s) = \frac{\partial^2 x(t,s)}{\partial t^2}$, the derivative $x_{t,s}''(t, s) = x_{s,t}''(t, s)$, $C^2(T \times S)$ is the space of two times continuously differentiable functions $x(t, s)$ on $T \times S \subseteq R^2$, the function

$y = f(t, s, u_0, u_1, \dots, u_5)$ is defined on the rectangle $\Omega = T \times S \times T_0 \times T_1 \times \cdots \times T_5$, T_i are sets of the number line $(i = 0, 1, \dots, 5)$, and Y is a function space.

We construct the interpolation polynomial for the operator (5) using the equality (2). The differential operator (5) depends on the one functional variable $x(t, s)$, for it the Gateaux differential $\delta F[x; h]$ at the point $x = x(t, s)$ in the direction $h = h(t, s)$ $\left(x, h \in C^2(T \times S)\right)$ is calculated according to the rule

$$\delta F[x; h] = \frac{\partial f}{\partial x} h(t, s) + \frac{\partial f}{\partial x'_t} h'_t(t, s) + \frac{\partial f}{\partial x'_s} h'_s(t, s) + \frac{\partial f}{\partial x''_{t^2}} h''_{t^2}(t, s) + \frac{\partial f}{\partial x''_{t,s}} h''_{t,s}(t, s) + \frac{\partial f}{\partial x''_{s^2}} h''_{s^2}(t, s) =$$

$$= \sum_{i,j=0;\, i+j \le 2}^{2} \frac{\partial f\left(t, s, x(t, s), x'_t(t, s), x'_s(t, s), x''_{t^2}(t, s), x''_{t,s}(t, s), x''_{s^2}(t, s)\right)}{\partial x^{(i+j)}_{t^i, s^j}} h^{(i+j)}_{t^i, s^j}(t, s), \tag{6}$$

where $x^{(i+j)}_{t^i, s^j}(t, s) = \frac{\partial^{i+j} x(t,s)}{\partial t^i \partial s^j}$ $(i, j = 0, 1, 2;\ i + j \le 2)$. Bearing in mind the rule (6), the formula (2) for the operators (5) takes the form

$$H_{2n+1}(F; x) = F(x_0) + \sum_{k=0}^{n} \sum_{i,j=0;\, i+j \le 2}^{2} \frac{\partial F(x_k(t, s))}{\partial \left(\frac{\partial^{i+j} x_k}{\partial t^i \partial s^j}\right)} \frac{\partial^{i+j}}{\partial t^i \partial s^j} \frac{q_{n,k}(x(t, s))}{\sigma_n(x(t, s))} +$$

$$+ \sum_{k=1}^{n} \int_0^1 \sum_{i,j=0;\, i+j \le 2}^{2} \frac{\partial}{\partial \left(\frac{\partial^{i+j} v_k}{\partial t^i \partial s^j}\right)} f\left(t, s, v_k(t, s, \tau), \frac{\partial v_k(t, s, \tau)}{\partial t}, \frac{\partial v_k(t, s, \tau)}{\partial s}, \frac{\partial^2 v_k(t, s, \tau)}{\partial t^2},\right.$$

$$\left.\frac{\partial^2 v_k(t, s, \tau)}{\partial t \partial s}, \frac{\partial^2 v_k(t, s, \tau)}{\partial s^2}\right) \frac{\partial^{i+j}}{\partial t^i \partial s^j} \left\{ \frac{h_{n,k}(x(t, s))}{\sigma_n(x(t, s))} (x_k(t, s) - x_0(t, s)) \right\} d\tau, \tag{7}$$

where $v_k = v_k(t, s, \tau) = x_0(t, s) + \tau (x_k(t, s) - x_0(t, s))$, $k = 1, 2, \dots, n$.

The polynomial (7) satisfies the interpolation conditions (3), where $X = C^2(T \times S)$.

We may obtain the representation of the interpolation error of the operator $F(x)$ by the polynomial (7) and taking into account the equality (4). We have

$$r_{2n+1}(x) = \sum_{k=0}^{n+1} \sum_{i,j=0;\, i+j \le 2}^{2} \frac{\partial F(x_k(t, s))}{\partial \left(\frac{\partial^{i+j} x_k}{\partial t^i \partial s^j}\right)} \frac{\partial^{i+j}}{\partial t^i \partial s^j} \left(\frac{q_{n+1,k}(x(t, s))}{\sigma_{n+1}(x(t, s))} - \frac{q_{n,k}(x(t, s))}{\sigma_n(x(t, s))} \right) +$$

$$+ \sum_{k=1}^{n+1} \int_0^1 \sum_{i,j=0;\, i+j \le 2}^{2} \frac{\partial}{\partial \left(\frac{\partial^{i+j} v_k}{\partial t^i \partial s^j}\right)} f\left(t, s, v_k(t, s, \tau), \frac{\partial v_k(t, s, \tau)}{\partial t}, \frac{\partial v_k(t, s, \tau)}{\partial s}, \frac{\partial^2 v_k(t, s, \tau)}{\partial t^2},\right.$$

$$\left.\frac{\partial^2 v_k(t, s, \tau)}{\partial t \partial s}, \frac{\partial^2 v_k(t, s, \tau)}{\partial s^2}\right) \frac{\partial^{i+j}}{\partial t^i \partial s^j} \left\{ \left(\frac{h_{n+1,k}(x(t, s))}{\sigma_{n+1}(x(t, s))} - \frac{h_{n,k}(x(t, s))}{\sigma_n(x(t, s))} \right) (x_k(t, s) - x_0(t, s)) \right\} d\tau,$$

where $x_{n+1} = x$, $h_{n,n+1}(x) = q_{n,n+1}(x) \equiv 0$.

Example. Let us consider the second order partial differential operator

$$F(x) = \phi(t, s) + a(t, s)\, x(t, s) + b(t, s)\, x^p(t, s) + c(t, s)\, x'_t(t, s) + h(t, s)\, x''_{s^2}(t, s), \tag{8}$$

where p is a fixed non-negative integer number, and $\phi(t, s), a(t, s), b(t, s), c(t, s), h(t, s)$ are arbitrarily given functions of the variables t and s. We construct the Hermite interpolation polynomial $H_3(F; x)$ of the form (7) for the operator (8). As fundamental interpolation polynomials $h_{1,k}(x)$, $q_{1,k}(x)$ $(k = 0, 1)$ we choose the algebraic polynomials

$$h_{1,k}(x) = l^2_{1,k}(x)\,(1 + 2\, l_{1,1-k}(x)), \quad q_{1,k}(x) = l^2_{1,k}(x)\,(x - x_k),$$

where $l_{1,k}(x) = (x - x_{1-k})/(x_k - x_{1-k})$, $k = 0, 1$, and we take as interpolation nodes $x_k(t, s)$, $k = 0, 1$, the system of functions $x_0(t, s) \equiv 3$, $x_1(t, s) = \sin t$. In this case the function $v_1(t, s, \tau) = 3 + \tau(\sin t - 3)$, $\sigma_1(x) \equiv 1$.

Bearing in mind that $\int_0^1 pv_1^{p-1}(t,s,\tau)(\sin t-3)d\tau = \sin^p t - 3^p$ the Hermite interpolation formula (7) in the case $n=1$ for the operators (8) takes the form

$$H_3(F;x) = \phi(t,s) + 3a(t,s) + 3^p b(t,s) + \left(a(t,s) + p3^{p-1}b(t,s)\right) q_{1,0}(x(t,s)) +$$
$$+\left(a(t,s) + pb(t,s)\sin^{p-1}t\right) q_{1,1}(x(t,s)) + \sum_{k=0}^{1}\left\{c(t,s)\frac{\partial}{\partial t}q_{1,k}(x(t,s)) + h(t,s)\frac{\partial^2}{\partial s^2}q_{1,k}(x(t,s))\right\} +$$
$$+a(t,s)(\sin t-3)h_{1,1}(x(t,s)) + b(t,s)(\sin^p t - 3^p)h_{1,1}(x(t,s)) +$$
$$+c(t,s)\frac{\partial}{\partial t}\{h_{1,1}(x(t,s))(\sin t-3)\} + h(t,s)\frac{\partial^2}{\partial s^2}\{h_{1,1}(x(t,s))(\sin t-3)\}. \tag{9}$$

By direct calculations, we verify the validity of the interpolation conditions

$$H_3^{(j)}(F;x_i) = F^{(j)}(x_i)\ (i,j=0,1). \tag{10}$$

Really, in view of the equalities $h_{1,1}(3)=0$, $q_{1,0}(3)=0$ and $q_{1,1}(3)=0$ we obtain $H_3(F;3) = \phi(t,s) + 3a(t,s) + 3^p b(t,s) = F(3)$. As $q_{1,0}(\sin t) = q_{1,1}(\sin t) = 0$, and $h_{1,1}(\sin t) = 1$, then $H_3(F;\sin t) = \phi(t,s) + 3a(t,s) + 3^p b(t,s) + b(t,s)(\sin^p t - 3^p) + a(t,s)(\sin t - 3) + c(t,s)\cos t = \phi(t,s) + a(t,s)\sin t + b(t,s)\sin^p t + c(t,s)\cos t = F(\sin t)$.

Further, since $q'_{1,0}(3) = q'_{1,1}(\sin t) = 1$, $q'_{1,1}(3) = h'_{1,1}(3) = 0$, $q'_{1,0}(\sin t) = h'_{1,1}(\sin t) = 0$, then $H'_3(F;3) = a(t,s) + p3^{p-1}b(t,s) = F'(3)$, $H'_3(F;\sin t) = a(t,s) + pb(t,s)\sin^{p-1}t = F'(\sin t)$. Thus, the Hermite interpolation formula (9) satisfies the condition (10).

4 Conclusion

The obtained results may be used in theoretical research as a basis for constructing approximate methods of solving some nonlinear operator equations that occur in nonlinear dynamics and mathematical physics.

References

[1] L. A. Yanovich and M. V. Ignatenko, Mathematical Models and Computer Simulations **26**, 11, 90–96 (2014)

[2] L. A. Yanovich and M. V. Ignatenko, Proc. of NAS of Belarus, Series of Physical-Mathematical Sciences **2**, 7–23 (2017) (in Russian)

[3] V. L. Makarov, V. V. Khlobystov, and L. A. Yanovich, *Methods of Operator Interpolation* (Institute of Mathematics of NAS of Ukraine, Kiev, 2010) 517 pp.

[4] L. A. Yanovich and M. V. Ignatenko, *Basics of the Interpolation Theory of Functions of Matrix Variables* (in Russian, Belaruskaya Navuka, Minsk, 2016) 281 pp.

[5] M. V. Ignatenko and L. A. Yanovich, Proc. of NAS of Belarus, Series of Physical-Mathematical Sciences **1**, 5–11 (2004) (in Russian)

[6] L. Hormander, Acta Mathematica **94**, 161–248 (1955)

[7] R. Courant, *Partial Differential Equations* (Vol. II of *Methods of Mathematical Physics* by R. Courant and D. Hilbert, Interscience Publishers, New York, 1962) 811 pp.

[8] O. A. Ladyzhenskaya, *The Boundary Value Problems of Mathematical Physics* (Applied Mathematical Sciences **49**, Berlin–Heidelberg–New York: Springer Verlag, 1985) 322 pp.

[9] S. Mizohata, *The Theory of Partial Differential Equations* (Cambridge University Press, 1973) 504 pp.

Generating Function Approach to the Derivation of Higher-Order Iterative Methods for Solving Nonlinear Equations

Tugal Zhanlav[1,⋆], *Ochbadrakh* Chuluunbaatar[1,2,⋆⋆], and *Vandandoo* Ulziibayar[1,3,⋆⋆⋆]

[1] *Institute of Mathematics, National University of Mongolia, Mongolia*
[2] *Joint Institute for Nuclear Research, Dubna, 141980 Moscow region, Russia*
[3] *School of Applied Sciences, Mongolian University of Science and Technology, Mongolia*

Abstract. In this paper we propose a generating function method for constructing new two and three-point iterations with p ($p = 4, 8$) order of convergence. This approach allows us to derive a new family of optimal order iterative methods that include well known methods as special cases. Necessary and sufficient conditions for p-th ($p = 4, 8$) order convergence of the proposed iterations are given in terms of parameters τ_n and α_n. We also propose some generating functions for τ_n and α_n. We develop a unified representation of all optimal eighth-order methods. The order of convergence of the proposed methods is confirmed by numerical experiments.

1 Introduction

Solving nonlinear equations is important in many applied mathematics and theoretical physics problems. In recent years, a number of higher-order iterative methods have been developed and analyzed on this issue, see [1–11] and references therein. Motivated by the recent results in [11], in this paper we introduce a generating function method for the construction of new two and three-point iterations with p-th order of convergence. This paper is organized as follows. Section 2 is devoted to the construction of a generating function for the optimal fourth-order method. We then present some choices for the parameters τ_n and α_n. Some iterations are proposed among which some are already well known. In Section 3 we propose a family of optimal eighth-order methods, that include many well-known methods as particular cases. In our previous paper [11] we have considered two and three-point iterative methods of solving nonlinear equation $f(x) = 0$

$$y_n = x_n - \frac{f(x_n)}{f'(x_n)}, \qquad x_{n+1} = y_n - \bar{\tau}_n \frac{f(y_n)}{f'(x_n)}, \tag{1}$$

and

$$y_n = x_n - \frac{f(x_n)}{f'(x_n)}, \qquad z_n = y_n - \bar{\tau}_n \frac{f(y_n)}{f'(x_n)}, \qquad x_{n+1} = z_n - \alpha_n \frac{f(z_n)}{f'(x_n)}. \tag{2}$$

We have proved in [11] the following theorems:

⋆e-mail: tzhanlav@yahoo.com
⋆⋆e-mail: chuka@jinr.ru
⋆⋆⋆e-mail: ulzii@jinr.ru

Theorem 1. *Assume that $f(x)$ is smooth enough function with a simple root $x^* \in I$ and the initial approximation x_0 is close enough to x^*. Then the iterative method (1) has fourth-order of convergence if and only if the parameter τ_n is given by*

$$\tau_n = 1 + \theta_n + 2\theta_n^2 + O(\theta_n^3), \qquad \theta_n = \frac{f(y_n)}{f(x_n)}. \tag{3}$$

Theorem 2. *Assume that all assumptions of Theorem 1 are fulfilled. Then the three-point iterative methods (2) has an eighth-order of convergence if and only if the parameters $\bar{\tau}_n$ and α_n are given by*

$$\bar{\tau}_n = 1 + 2\theta_n + \beta\theta_n^2 + \gamma\theta_n^3 + \cdots, \qquad \left(\bar{\tau}_n = \frac{\tau_n - 1}{\theta_n}\right), \tag{4}$$

and

$$\alpha_n = 1 + 2\theta_n + (\beta + 1)\theta_n^2 + (2\beta + \gamma - 4)\theta_n^3 + (1 + 4\theta_n)\frac{f(z_n)}{f(y_n)} + O(\theta_n^4). \tag{5}$$

Our approach in [11] is constructive in the sense that it proposes a new way to obtain optimal order iterations (see [11] for details). An extended version of the present paper will be published elsewhere.

2 Construction of optimal fourth-order methods

The Theorems 1 and 2 not only give sufficient conditions for iterations of p-th ($p = 4, 8$) order of convergence, but they also allow us to construct new iterations with p order of convergence. Obtaining new optimal methods of order four is still important, because they combine higher-order of convergence and low computational cost. We consider the following choice of the parameter τ_n

$$\tau_n = H(\theta_n), \tag{6}$$

where $H(\theta)$ is a real function to be determined properly. Obviously τ_n will satisfy the condition (3) if

$$H(0) = 1, \qquad H'(0) = 1, \qquad H''(0) = 4. \tag{7}$$

We call the function $H(\theta)$ satisfying conditions (7) a generating function for the iteration (1). The construction of the generating function allows us to derive a new optimal order family of iterations. The following theorem is a consequence of Theorem 1.

Theorem 3. *Assume that all assumptions of Theorem 1 are fulfilled. Then the optimal fourth-order two-point iterations (1) are obtained by the generating function (6) satisfying the conditions (7).*

Many different variants of the generating function $H(x)$, satisfying condition (7) are possible. We cite here one simple form, namely

$$H(x) = \frac{1 + (1 - m\alpha)x + (2 - m\alpha + \frac{m(m-1)}{2}\alpha^2)x^2 + \omega x^3}{(1 - \alpha x)^m}, \qquad \alpha, m, \omega \in \mathbb{R}. \tag{8}$$

The optimal two-point iterations (1) with $\tau_n = H(\theta_n)$ given by (8) include many well-known iterations as special cases. If $\omega = 0$, $m = 1$ and $\alpha = 2 - b$, $b \in \mathbb{R}$, then (1) leads to King's method [5]. If $\alpha = 0$, $m = 1$ and $\omega = 1$ in (8), then (1) yields a modification of Potra-Ptak's method [4]. If $\alpha = m = 1$ and $\omega = -1$ in (8), then (1) leads to Maheshwari's method [7]. If $\alpha = 1$, $m = 2$ and $\omega = 0$ in (8), then (1) leads to Chun and Lee's method [2]. Recently, Behl et al [12] proposed a general class of fourth-order optimal methods that includes the well-known Ostrowski's and King's family as special cases. We note that this general class of optimal fourth-order iterations is also included in our methods with

$\tau_n = H(\theta_n)$ given by (8) as a special case. Namely, if $m = 3$, α replaced by $-\alpha$ and $\omega = \alpha^2 + \frac{5}{3}\alpha + \frac{4}{3}$ or $\omega = \left(1 - \frac{\beta}{6}\right)\alpha^3 + \alpha^2 - 2\alpha$, then the iterations (1) with $\tau_n = H(\theta_n)$ given by (8) reduce to (3.8) and (3.10) in [12], respectively. This shows that our class of optimal fourth-order methods is wider than that of [12]. So, we have obtained an optimal fourth-order convergence family of iterative methods with three degrees of freedom based on the generating function method.

3 Proper representation of the optimal order three-point iterative methods

Recently, based on optimal fourth-order methods some higher-order, in particular eighth order three-point methods have been proposed for solving nonlinear equations. It is easy to show that $\tau_n = H(\theta_n)$ given by (8) satisfies the condition (4) provided

$$\beta = \omega + 2m\alpha - \frac{m(m-1)}{2}\alpha^2 + \frac{m(m-1)(m-2)}{6}\alpha^3, \tag{9a}$$

$$\gamma = \omega m\alpha + m(m+1)\alpha^2 - \frac{(m-1)m(m+1)}{3}\alpha^3 + \frac{(m-2)(m-1)m(m+1)}{8}\alpha^4. \tag{9b}$$

Table 1. Optimal order three-point iterative methods recovered at special parameter values

	m	$\alpha_n - (1+4\theta_n)[f(z_n)/f(y_n)]$	Methods
1	0	$1 + 2\theta_n + (\beta+1)\theta_n^2 + (2\beta+\gamma-4)\theta_n^3$	$\beta = \gamma = 1$, [14] Maheshwari-based $\beta = 4,\ \gamma = 8$, Method 1 in [14] see [11] $\beta = 3,\ \gamma = 4$, Chun Lee [2]
2	1	$(2-\theta_n)/(6\theta_n^2 - 5\theta_n + 2)$	$p = -3,\ d = 0,\ q = 5/2$, Maheshwari-based optimal methods [16]
3	1	$[2\beta - 1 + 2\beta(\beta-2)\theta_n]/$ $[2\beta - 1 + 2(\beta^2 - 4\beta + 1)\theta_n + (1+4\beta)\theta_n^2]$	$d = 0,\ q = 2(\beta^2 - 4\beta + 1)/(1-2\beta)$, $p = (1+4\beta)/(1-2\beta)$, King-based optimal methods [17]
4	1	$1/(1 - 2\theta_n - \theta_n^2)$	$\beta = 4,\ \gamma = 8,\ q = 2,\ p = 1,\ d = \omega = 0$, method 4 in [14]

The following theorem is a consequence of Theorem 2.

Theorem 4. *Assume that all assumptions of Theorem 1 are fulfilled. Then the family of three-point iterative methods (2) has an eighth-order of convergence if and only if the parameters τ_n and α_n are given by (6), (8) and*

$$\alpha_n = \left(H(\theta_n) + \theta_n + (\beta-1)\theta_n^2 + (\beta+\gamma-4)\theta_n^3\right) + (1+4\theta_n)\frac{f(z_n)}{f(y_n)}. \tag{10}$$

Thus, we propose the families of three-point iterative methods (2) with generating function $\tau_n = H(\theta_n)$. They include many well-known eighth-order methods, as particular cases (see Table 1). The expression in brackets in (10) can be approximated by a simple rational function without loss of generality. Then α_n can be represented as

$$\alpha_n = \frac{1 - (2 - mq)\theta_n + c\theta_n^2 + \omega\theta_n^3}{\left(1 - \theta_n(d\theta_n^2 + p\theta_n + q)\right)^m} + (1+4\theta_n)\frac{f(z_n)}{f(y_n)}, \qquad q, p, d, m \in \mathbb{R}, \tag{11}$$

where

$$c = \beta + 1 - m\left(p + 2q + \frac{1}{2}(m-1)q^2\right),$$

$$\omega = (2\beta + \gamma - 4) - m\left(d + 2p + \left(\beta + 1 + (1-m)p\right)q - (m-1)q^2 + \frac{(m-1)(m-2)}{6}q^3\right). \tag{12}$$

We call the optimal order three-point iterative methods (2), with parameters τ_n and α_n given by (6), (8) and (11) respectively, proper representations. It is easy to show that all the well-known optimal order three-point iterative methods can be represented uniquely in the proper form (see [1–3, 6, 8–11, 13–20] and references therein). It should be mentioned that Wu and Lee in [10] first used a proper representation of (2). Thus, by means of (6), (8) and (11) we find a unified representation of all optimal order three-point iterations. It should be mentioned that the order of convergence of the proposed methods was confirmed by numerical experiments.

Conclusions

The construction of the generating function for τ_n and α_n allows us to derive new optimal order family of iterations. This family includes many known iterations as special cases. We develop a unified and proper representation of optimal eighth-order three-point methods. The sufficient and necessary conditions for iterations (2) to be p ($p = 4, 8$) order of convergence are also given in term of parameters τ_n and α_n.

Acknowledgments

The work was supported partially by the Foundation of Science and Technology of Mongolia under grant SST_007/2017. O. Ch. acknowledges support from the Hulubei-Meshcheryakov program JINR-Romania.

References

[1] W. Bi, H.Ren, and Q. Wu, J. Comput. Appl. Math. **225**, 105–112 (2009)
[2] C. Chun and M.Y. Lee, Appl. Math. Comput. **223**, 506–519 (2013)
[3] A. Cordero, J.R. Torregrosa, and M.P. Vassileva, J. Comput. Appl. Math. **235**, 3189–3194 (2011)
[4] A. Cordero, J.L. Hueso, E. Martinez, and J.R. Torregrosa, J. Comput. Appl. Math. **234**, 2969–2976 (2010)
[5] R.F. King, SIAM Journal on Numerical Analysis **10**, 876–879 (1973)
[6] T. Lotfi, S. Sharifi, M. Salimi, and S. Siegmund, Numer. Algor. **68**, 261–288 (2015)
[7] A.K. Maheshwari, Appl. Math. Comput. **211**, 383–391 (2009)
[8] S. Sarifi, M. Ferrara, M. Salimi, and S. Siegmant, Open Mathematics **14**, 443–451 (2016)
[9] J.R. Sharma and R. Sharma, Numer. Algor. **54**, 445–458 (2010)
[10] X. Wang and L. Liu, J. Comput. Appl. Math. **234**, 1611–1620 (2010)
[11] T. Zhanlav, V. Ulziibayar, and O. Chuluunbaatar, Comput. Math. Math. Phys. **57**, 1090–1100 (2017)
[12] R. Behl, A. Cordero, S.S. Motsa, and J.R. Torregrosa, Appl. Math. Comput. **271**, 89–101 (2015)
[13] T. Zhanlav and V. Ulziibayar, Amer. J. Comp. Appl. Math. **6(5)**, 177–181 (2016)
[14] R. Thukral and M. S. Petkovic, J. Comput. Appl. Math. **233**, 2278–2284 (2010)
[15] C. Chun and B. Neta, Appl. Math. Comput. **245**, 86–107 (2014)
[16] C. Chun and B. Neta, Appl. Math. Comput. **253**, 294–307 (2015)
[17] C. Chun and B. Neta, American Journal of Algorithms and Computing **2(1)**, 1–17 (2015)
[18] C. Chun and B. Neta, Numer. Algor. **74(4)**, 1169—1201 (2017)
[19] C. Chun and B. Neta, Appl. Math. Comput. **274**, 762—773 (2016)
[20] J.R. Sharma and H. Arora, Appl. Math. Comput. **273**, 924—933 (2016)

EPJ Web of Conferences **173**, 04001 (2018) https://doi.org/10.1051/epjconf/201817304001
Mathematical Modeling and Computational Physics 2017

Time Based Global Track Reconstruction in CBM

Timur Ablyazimov[1,2,⋆], *Volker* Friese[1,⋆⋆], and *Victor* Ivanov[2,3,⋆⋆⋆]

[1] *GSI Helmholtzzentrum für Schwerionenforschung GmbH, Darmstadt, Germany*
[2] *Laboratory of Information Technologies, Joint Institute for Nuclear Researches, Dubna, Russia*
[3] *National Research Nuclear University "MEPhI", Moscow, Russia*

Abstract. The paper describes the approaches, implemented by the authors, for the track reconstruction in data, subdivided to time slices, in the CBM experiment.

1 Introduction

CBM ("Compressed Baryonic Matter") [1] is an experiment being prepared to operate at the future Facility for Anti-Proton and Ion Research (FAIR) in Darmstadt, Germany. The CBM experiment will explore the high-density region of the QCD phase diagram by investigating nuclear collisions from 2 to 45 GeV beam energy per nucleon. Its main focus is the measurement of very rare probes (e.g. charmed hadrons), which requires interaction rates of up to 10 MHz. A specific feature of the experiment will be the free-running data acquisition fashion, without a hardware trigger. It differs from the traditional event based approach, and demands taking into account the time-coordinate measurements in track reconstruction, and using smart approaches to cope with an enormous combinatorics. For the sake of brevity, and also because the time measurements are taken into account in the described reconstruction algorithms, this approach is further referred as "time based". Physical studies in the CBM experiment are planned to be based on the the so called "global" track reconstruction – which means the reconstructed tracks comprise data, obtained from different detectors. The CBM software is developed under the CBMROOT framework, which in turn extensively uses the CERN ROOT infrastructure. Before the authors started their work, the only time based track reconstruction component in the CBMROOT software repository was the track finder for the Silicon Tracking System (STS) detector, called L1. It was also available a facility for the global tracking, called Littrack, in the CBMROOT software package, but it was not optimized for work with timely sliced data – it treated input data from time slices the same way as for events, which, presumably, gave it only limited reconstructing capabilities. The local reconstruction – clustering – was also implemented without taking into account time for the most of the CBM detecting systems. The main goal of authors' efforts was to link measurements from the Time of Flight (ToF) detector to the reconstruction chain. There were two possible directions of development: adapting the existing global tracking software (Littrack) to the time slice based reconstruction and creating a dedicated ToF hits linking software from scratch. When the work started, there was no working measurements clustering software for ToF, optimized to work with time

⋆e-mail: t.ablyazimov@gsi.de
⋆⋆e-mail: v.friese@gsi.de
⋆⋆⋆e-mail: ivanov@jinr.ru

slices. So the first step was to implement a clustering algorithm for ToF, taking time into account. As a final result both global track reconstruction approaches have also been implemented.

2 Measurements clustering in ToF

Before global (track) reconstruction can be done, the local entities, hits, have to be reconstructed in all the involved detectors. There were already time based cluster finders for the STS and the Muon Chamber (MuCh) detectors, but was no such a finder for ToF. So it was implemented by the authors. ToF detector is a "wall", consisting of detecting "bricks" – Multi-gap Resistive Plate Chambers (MR-PCs) – of several different types, having nevertheless common conceptual structure, which is depicted on the figure 1.

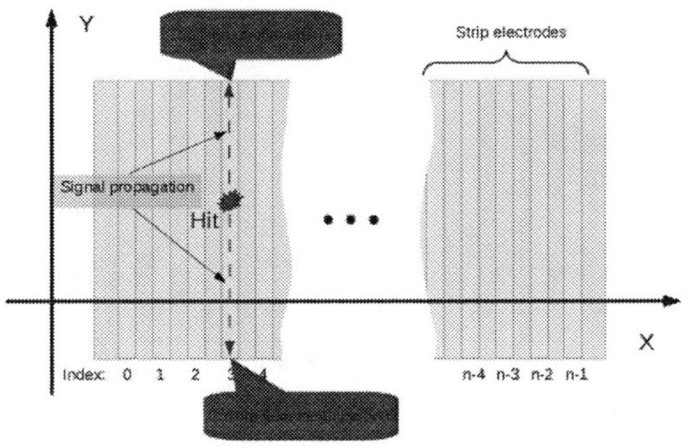

Figure 1. ToF measurements are obtained on the top and the bottom ends of the ToF MRPC conducting strips, and provide time information with high precision

ToF hits are represented as points in a 4-dimensional space: 3 spatial coordinates, and one time coordinate. The later coordinate is obtained in the most straightforward way – as an average of the corresponding top and bottom time measurements. One spatial coordinate ("horizontal" – X) is calculated in the MRPC local frame of reference from the measurement strip number and the known MRPC pitch. Another ("vertical" – Y) is calculated by the difference between measurements on the top and the bottom of an MRPC channel. The last space coordinate (Z) is determined by the position of an MRPC self, and X and Y coordinates of a point inside the MRPC. Measurements from different neighbouring MRPC channels are clustered together, because a single particle can fire multiple channels at once.

3 Dedicated ToF tracker

This tracker uses tracks, reconstructed in STS, extrapolating them into the area of the ToF wall. Then it tries to link them with the ToF hits, located in the vicinity. The time coordinate is taken into account in the vicinity definition. The track extrapolation uses the routines, implemented in the Littrack library. In the initial version of the dedicated tracker the time was handled separately, but, after upgrading Littrack (as described in the next section), the time was added as another component of the state vector, describing the dynamical model of tracking. To keep combinatorics on moderate level in situations, when time slices contain much of data, the ToF measurements data are subdivided to addressable space-time regions of moderate size – cuboids. The ToF station self is represented as a bigger cuboid in the 4-dimensional space-time. The time span is equal to the time slice duration. The spans in space corresponds to the dimensions of the 3-dimensional cuboid, circumscribing the ToF detector. The detecting station cuboid is subdivided to the smaller cuboids, used then for the acceleration of hit searches, as shown of the figure 2. The smaller cuboids, used for hit searches, are later referred as "bins".

Figure 2. Hierarchical binning in 3D space

As the influence of the magnetic field is negligible in the ToF wall area, the tracks are modeled as a straight lines. It makes the process of locating bins appropriate to a particular track easy: these bins are bins, which are intersected by the track, as shown on the left half of the figure 3. All the intersected bins are inspected for hits, and the closest with respect to all space and time coordinates hit is assigned to the track (if its distance from the track, measured in the units of errors for space and time coordinate, is not too big) – this is depicted on the right side of the same figure 3.

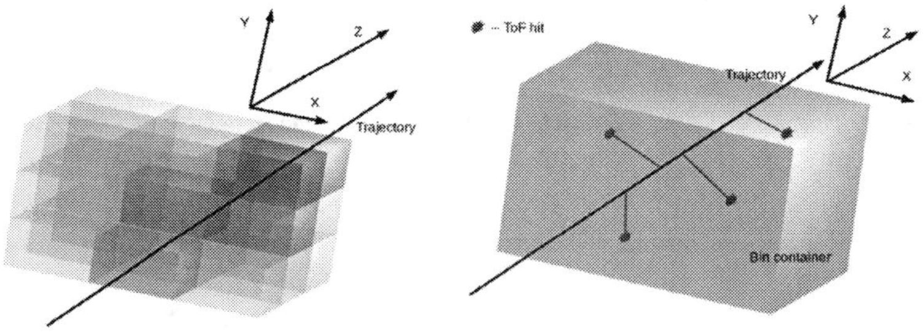

Figure 3. Selecting a subset of bins for a track and selecting the closest hit in this subset

4 Littrack upgrade

The Littrack library is available in the CBMROOT framework for many years and intended for reconstructing tracks from measurements, obtained concurrently from several different detecting systems (global tracking) [2]. Track reconstruction in the Littrack library is based on the track following approach and uses Kalman filter formalism for calculation of particles physical properties. It uses track parameters from the L1 track finder as seeds. One of the detectors, which data the Littrack library can use in its reconstruction chain, is ToF. Because of its important role, it was decided to try to upgrade the Littrack library for the time based reconstruction as well. It was done, by carefully upgrading the underlying Kalman's formalism: time coordinate was added to the state vector, the appropriate members were added to the covariance matrix, and the Runge-Kutta method implementation was upgraded with calculation for the time related terms. One of the other detectors, supported by the Littrack library, is MuCh. The currently existing clustering algorithm in MuCh is also able to work with time based data. So Littrack has been tested for the ability to use MuCh and ToF hits in the time based reconstruction.

5 Results

All the three described here software components were tested with respect to their efficiency. The first two of them were also tested for their performance. The testing was conducted on an Intel(R) Xeon(R) CPU E5-1607 v3 @ 3.10GHz machine. For the tests, where only ToF hits were linking to reconstructed L1 tracks, input data, produced with URQMD generator, simulating minimum bias Au+Au collisions at 10 AGeV, were used. The Littrack efficiencies were tested with input data, produced by the Box generator, simulating a pure stream of muons (20 muons per event), having energies in the range 4-10 GeV. Monte Carlo tracks were considered reconstructible according the criterion of having at minimum the following number of points in detecting systems: 4 in STS, 4 in MuCh and 1 in Tof – the efficiencies were calculated for this kind of tracks. The matching criterion fulfilled, a reconstructed track had at least 70% common points with a Monte Carlo. The results are presented in the table 1, where the figures for reconstruction in time slices are compared with the efficiencies of the appropriate tools in the event by event mode.

Table 1. Algorithms testing results

Algorithm	Efficiency (%)	Runtime (per event)
ToF clustering (time slices)	97	4ms
Dedicated ToF tracker (time slices)	93	14ms
Littrack (ToF) (time slices)	93	N/A
Littrack (MuCh) (time slices)	98	N/A
Littrack (MuCh+ToF) (time slices)	89	N/A
Littrack (ToF) (event by event)	93	N/A
Littrack (MuCh) (event by event)	100	N/A
Littrack (MuCh+ToF) (event by event)	86	N/A

It can be seen that the Littrack reconstruction efficiency was a little bit worse for MuCh (98% vs nearly 100%). Probably the main reason for this was the inconsistency between the MuCh detector simulation, where the electronics delay was taken into account, and the MuCh hit reconstruction, where this delay was neglected. The mentioned delay was about 17 ns in average. It was partially compensated by shifting the time measurement values during loading the MuCh input date into Littrack by 17 ns. But the correct solution must be later done in the MuCh hit finder.

References

[1] B. Friman, P. Senger et al., *The CBM Physics Book* (Lecture Notes in Physics, Vol. 814, 1st Edition, 2011) 960 pp.

[2] A. Lebedev, *The Software and Algorithms for the Track Reconstruction in the Transition Radiation and the Muon Chamber Detectors in the CBM Experiment* (Ph.D. thesis, Dubna, 2011) 97 pp.

EPJ Web of Conferences **173**, 04002 (2018) https://doi.org/10.1051/epjconf/201817304002
Mathematical Modeling and Computational Physics 2017

Time-based Reconstruction of Free-streaming Data in CBM

Valentina Akishina[1,3,⋆], *Ivan* Kisel[1,2,3], *Iouri* Vassiliev[3], and *Maksym* Zyzak[3]

[1] *Goethe-Universität Frankfurt, Frankfurt am Main, Germany*
[2] *Frankfurt Institute for Advanced Studies, Frankfurt am Main, Germany*
[3] *GSI Helmholtzzentrum für Schwerionenforschung GmbH, Darmstadt, Germany*

Abstract. Traditional latency-limited trigger architectures typical for conventional experiments are inapplicable for the CBM experiment. Instead, CBM will ship and collect time-stamped data into a readout buffer in a form of a time-slice of a certain length and deliver it to a large computer farm, where online event reconstruction and selection will be performed. Grouping measurements into physical collisions must be performed in software and requires reconstruction not only in space, but also in time, the so-called 4-dimensional track reconstruction and event building. The tracks, reconstructed with 4D Cellular Automaton track finder, are combined into event-corresponding clusters according to the estimated time in the target position and the errors, obtained with the Kalman Filter method. The reconstructed events are given as inputs to the KF Particle Finder package for short-lived particle reconstruction. The results of time-based reconstruction of simulated collisions in CBM are presented and discussed in details.

1 Introduction

The physics program of the CBM experiment [1] requires studying the multiplicity, distribution in phase space and collective effects of all interesting particles, including very rare short-lived particles. In order to collect enough statistics, the fixed-target experiment will be run at extreme interaction rates of up to 10 MHz. The complicated topology of the trigger decays and observables does not allow to build a hardware trigger based on raw detector data. Thus, the experiment will operate with free-streaming data and self-triggered front-end electronics. Within such a complicated operation scenario detector measurements corresponding to particles produced in different collisions cannot be separated based on their time information in a simple way.

The experiment will ship and collect time-stamped data into a readout buffer in the form of a time-slice of a certain length. The collisions will be resolved in the online mode by the First Level Event Selection (FLES) package [2] during the reconstruction of tracks and short-lived particles. The reconstruction algorithms have been developed to operate with the time-slices taking into account the time information. The time-based algorithms demonstrate performance similar to the conventional event-by-event mode.

⋆e-mail: V.Akishina@gsi.de

Table 1. Track reconstruction efficiency performance for one hundred minimum bias AuAu collisions at 25 *A*GeV in the cases of event-by-event reconstruction and time-slice-based reconstruction at 0.1 MHz, 1 MHz and 10 MHz interaction rates, presented for different track categories.

Track category, %	E-by-E	10^5Hz	10^6Hz	10^7Hz
All tracks	92.1	92.6	92.6	92.2
Primary high-p	97.9	98.2	98.2	97.9
Secondary high-p	92.0	92.7	92.7	92.0
Clone level	2.8	0.3	0.3	3.1
Ghost level	4.9	3.5	3.5	4.2

2 Reconstruction of particle trajectories

The track finder algorithm used to reconstruct charged particle trajectories in the tracking detectors of CBM is based on the Cellular Automaton method (CA) [3, 4]. Since the experiment is designed to work with free-streaming data, the algorithm takes as input hits from a time-slice, which include time and spacial measurements. During the first stage the CA track finder combines the hits, delivered by the reconstruction algorithms of the main tracking detector of CBM – the Silicon Tracking System (STS), into combinations of three hits on adjacent detector stations, so-called triplets. The Kalman Filter method [5] allows to estimate the momentum of such track segments before constructing the whole track.

In order to take into account detector inefficiency, a missing hit in one station is tolerated for a reconstructed track. The triplets are combined into track candidates. The track candidates which pass through a special selection procedure according to their χ^2 and length are taken for reconstructed tracks. To get a faster and more efficient algorithm, the track reconstruction is done in several iterations. In the first iteration only high-momentum primary tracks are reconstructed, in the second one – low-momentum primary tracks, and then – all other tracks. After each iteration the used hits are removed from further consideration, since it significantly reduces the combinatorics. This is particularly important for later iterations when searching for low-momenta and secondary tracks [6].

The resulting track reconstruction efficiency for different track sets and ratios of clones and ghost tracks for the event-by-event analysis as well as for the 4D CA track finder in case of reconstructing time-slices, produced out of one hundred AuAu minimum bias event at 25 *A*GeV, are presented in table 1. For evaluation purposes the particle is called reconstructable if it has crossed at least 4 stations of the STS detector. A reconstructed track is assigned to a simulated particle, if at least 70% of its hits have been caused by the same particle. If a certain particle is assigned to more than one track, such additionally reconstructed track candidates are regarded as clones. If a track was not assigned to any particle according to the 70% criterion it is called a ghost. The tests have been performed on an Intel Xeon E7-4860 CPU server.

The inclusion of the time and optimization of the event-based track finder algorithm into the time-based reconstruction have made it possible to achieve an efficiency performance comparable to that of the event-by-event analysis. Moreover, the track reconstruction efficiency has been improved after taking into account the STS time measurement even for the case of the highest interaction rate scenario of CBM. It can be explained by the presence of slow particles, which create random combinations of hits and cannot be rejected in the absence of time information.

The CA track finder is both vectorized (using SIMD instructions) and parallelized (between CPU cores with the help of OpenMP interface). The algorithm shows strong scalability on many-core systems. The speed-up factor of 10.6 has been achieved on an Intel Xeon E7-4860 CPU with 10 hyper-threaded physical cores [7].

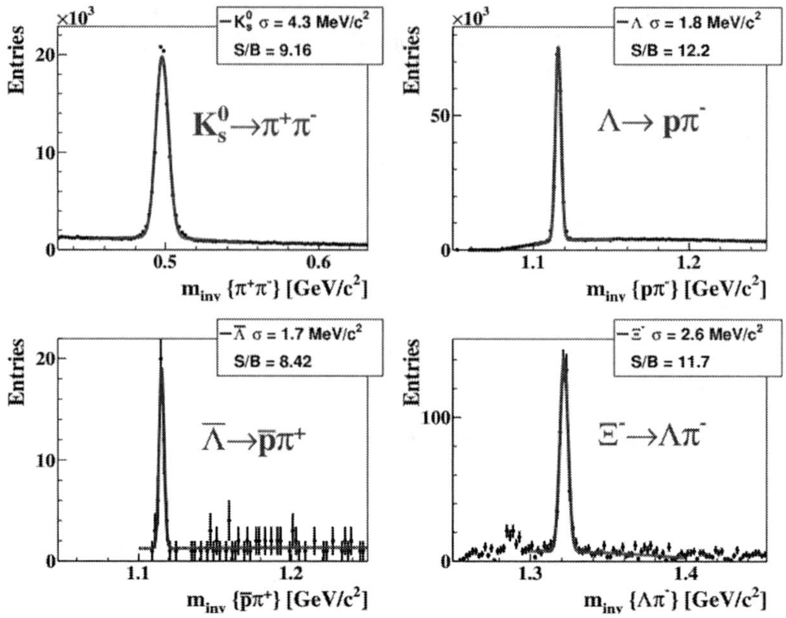

Figure 1. Mass spectra of strange particles reconstructed in 300k AuAu minimum bias UrQMD events at 10 *A*GeV energy and 10 MHz interaction rate using event builder and KF Particle Finder. For particle identification of daughter products the Monte Carlo information was used as the first step.

3 Event building

The highest interaction rate scenario of CBM requires a special event building procedure at the track level, which will allow to resolve information on different collisions from each other and deliver isolated events for the physics analysis. The input for the event building algorithm are tracks, reconstructed with the CA track finder, spacial and time parameters of which are estimated with the Kalman Filter procedure. The first implementation of the event builder [6] algorithm splits the reconstructed tracks into event-corresponding groups based on the information of fitted time parameter in the region of primary vertex and the estimated errors, obtained from the track fit [7].

The results of the event building have shown, that 83% of events were reconstructed without event merging and 17% of events were merged into double event clusters. The avoidance of event splitting was achieved. The presence of event merging shows that the time information in the STS alone is not enough to resolve all events from each other at the interaction rate of 10 MHz and the remaining events are to be resolved later via searching for several vertices with the help of event topology.

The currently implemented procedure is based on the information from the STS detector alone. The addition of information obtained by the Time-of-Flight (ToF) detector, the time resolution of which is by several orders of magnitude better, is expected to drastically improve the ability to resolve different events, containing ToF time information, from each other [6].

4 Reconstruction of short-lived particles in time-slices

Tracks grouped into events are provided as an input for the next step of the reconstruction chain — reconstruction of short-lived particles. The properties of short-lived particles contain a large part of the CBM physics. These particles cannot be registered directly by the detector system, but only reconstructed from their decay products. Short-lived particles were reconstructed by the KF Particle Finder package [2, 8] in each event, provided by the event builder algorithm. As the first approach, one primary vertex in each reconstructed event was assumed. Further, the multi-primary-vertex analysis will be implemented, that is required to improve event separation and signal to background ratio of the reconstructed particles for the extreme case of 10 MHz interaction rate.

Tests of 300k AuAu minimum bias UrQMD [9, 10] events at 10 *A*GeV energy were simulated assuming the interaction rate of 10 MHz. The simulated events were reconstructed using the event builder and the KF Particle Finder package. As the first step, the long-lived daughter particles were identified by the Monte Carlo information. The reconstructed mass spectra of the most abundant strange particles and anti-particles are shown in figure 1. Further addition of the time information from the fast ToF detector and implementation of the multi-primary-vertex analysis will allow to bring the reconstruction performance even closer to the results of event-by-event analysis.

Conclusions

The first version of the FLES reconstruction chain for the CBM experiment has been developed. It contains stages of track finding, track fitting, event building, short-lived particle finding, and event selection. The package allows to reconstruct free-streaming data collected by the detector and obtain reconstructed particles from individual collisions in the end. The resulting efficiencies and combinatorial background are similar to the event-based analysis, showing high values for the signal to background ratio for short-lived particle reconstruction even for the 10 MHz interaction rate.

Acknowledgements

This work was supported by HICforFAIR, FIAS and BMBF.

References

[1] B. Friman, C. Hohne, J. Knoll, S. Leupold, J. Randrup, R. Rapp, and P. Senger, *The CBM physics book: Compressed baryonic matter in laboratory experiments* (Lect. Notes Phys. 814, 2011) 980

[2] I. Kisel, I. Kulakov, and M. Zyzak, IEEE Transactions on Nuclear Science **60** (2013)

[3] I. Kisel, Nucl. Instr. and Meth. **A566**, 85–88 (2005)

[4] V. Akishina, *4D Event Reconstruction in the CBM Experiment* (Dissertation thesis, Goethe university, Frankfurt am Main, 2017) 181 pp.

[5] R. Frühwirth *et al.*, "Data Analysis Techniques for High-Energy Physics. Second Edition," Cambridge Univ. Press (2000) 421 p.

[6] V. Akishina and I. Kisel, "Online Event Reconstruction in the CBM Experiment at FAIR," this journal

[7] V. Akishina and I. Kisel, IEEE Trans. Nucl. Sci. **62**, 6, 3172 (2015)

[8] M. Zyzak, *Online Selection of Short-Lived Particles on Many-Core Computer Architectures in the CBM Experiment at FAIR* (Dissertation thesis, Goethe university, Frankfurt am Main, 2016) 159 pp.

[9] S. A. Bass, *et al.*, Prog. Part. Nucl. Phys. **41** (1998)

[10] M. Bleicher, *et al.*, J. Phys. G: Nucl. Part. Phys. **25** (1999)

Fitting by Orthonormal Polynomials of Silver Nanoparticles Spectroscopic Data

Nina Bogdanova[1,3,*] and *Mihaela* Koleva[2,**]

[1] *Institute of Nuclear Research and Nuclear Energy, BAS, Sofia 1784, bul.Tzarigradsko chaussee 72*
[2] *Institute of Electronics, BAS, Sofia 1784, bul.Tzarigradsko chaussee 72*
[3] *Joint Institute of Nuclear Research, 141980 Dubna, Moscow region, Russia*

Abstract. Our original Orthonormal Polynomial Expansion Method (OPEM) in one-dimensional version is applied for first time to describe the silver nanoparticles (NPs) spectroscopic data. The weights for approximation include experimental errors in variables. In this way we construct orthonormal polynomial expansion for approximating the curve on a non equidistant point grid. The corridors of given data and criteria define the optimal behavior of searched curve. The most important subinterval of spectra data is investigated, where the minimum (surface plasmon resonance absorption) is looking for. This study describes the Ag nanoparticles produced by laser approach in a ZnO medium forming a AgNPs/ZnO nanocomposite heterostructure.

1 Introduction

The metal nanostructures have attracted considerable attention due to their optical properties. It is related to the efficient excitation of collective electron oscillations, plasmons, which define the particle response to external electromagnetic field. For some metals, as silver, the plasmon resonance is realized in the near UV or visible spectral range. This makes the metals good candidates for resonance plasmon excitation [1]. These properties of metal nanoparticles are used in techniques for applications in optical, electronic, catalytic, sensing and biomedical devices [2–5]. The laser annealing leads to decomposition of the layer into nanoparticles by a dewetting mechanism [2]. The evolution of the dewetting process is a function of a thin film composition and dictates a size distribution and spacing of the nanoparticles. The properties of nanostructures of noble metals are strictly related to a material in which they are embedded. The NPs incorporation into dielectric or semiconductor matrices can lead to emergence of new features of composite materials showing properties different from those of individual components [3–5]. The application of methods for precise study of the resonance absorption band position of noble metal nanoparticles is of a particular interest.

2 Physical data

The silver nanoparticles are produced by pulsed laser deposition (PLD) on quartz substrates SiO_2 (001) in a vacuum chamber. The films are deposited by a laser of type Nd:YAG (λ = 355 nm,

*e-mail: nibogd@inrne.bas.bg
**e-mail: mihaela_ek@yahoo.com

Figure 1. SEM images of Ag/ZnO before (left) and after the laser nanostructuring

τ = 18 ns, ν=10 Hz) at fluence of F = 1.5 J/cm^2 at room temperature. The films are post-deposition annealed for surface nanostructuring by laser-induced decomposition of the film into nanoparticles with diameters of **few tens of nanometers**. The deposited films are laser annealed in air by the same laser system with a fluence of 200 mJ/cm^2. **The transmission T** of the Ag nanoparticles as a function of **wavelength λ** is analyzed using a UV-VIS spectrometer (HR 4000 Ocean optics) in the range of 220–800 nm. We have chosen one subinterval with M = 94 points and λ in [468.9,493.2] and the smaller one with λ in [468.9, 481.7]. The selected curve corresponds to the transmission spectrum of AgNPs after the annealing by 10 laser pulses. The lower number of pulses leads to incomplete decomposition of the layer into nanoparticles. The Ag nanoparticles with a mean size of 30 nm are described here. The PLD grown thin film is transformed into a discontinuous structure of small particles. Changes in the resonance absorption are associated with changes in size, shape and interparticle distances. as well as with the dielectric constant of the surrounded medium [3–5].

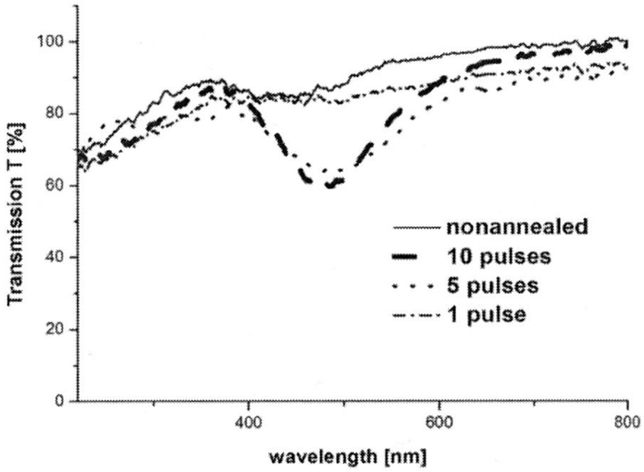

Figure 2. Transmission spectrum of Ag/ZnO nanocomposites before and after the laser annealing

The problem is to find a position of a plasmon resonance. For this aim we have to define a minimum of transmission curve. The mathematical task is to find the best fitting curve and its minimum.

3 Mathematical approach

One defines a new variance at the i-th given point $(\lambda_i, T_i, \sigma_{T_i}, \sigma_{\lambda_i})$, i = 1, 2, ..., M, following [6]: $S_i^2 = \sigma_{T_i}^2 + (\partial T_i/\partial\lambda_i)^2\,\sigma_{\lambda_i}^2$. Here we use the Bevington's (1969) proposal to combine both variable uncertainties and assign them to dependent one. **The generalized OPEM [7]:** Our principal relation for one-dimensional generation of orthonormal polynomials by Forsythe [8] $\{P_i^{(0)}, i = 1, 2, \ldots\}$ and their derivatives $\{P_i^{(m)}, m = 1, 2, \ldots\}$ in an **arbitrary** discrete set in M points is as follows:

$$P_{i+1}^{(m)}(\lambda) = 1/\nu_{i+1}\left[(T - \mu_{i+1})P_i^{(m)}(\lambda) - (1 - \delta_{i0})\nu_i P_{i-1}^{(m)}(\lambda) + mP_i^{(m-1)}(\lambda)\right]. \quad (1)$$

The generalization of Forsythe procedure in one-dimensional case is with involving arbitrary weights in every points, evaluating derivatives ($m > 0$) or integrals ($m < 0$) and normalizing polynomials. Here the normalization coefficient $1/\nu_i$ and the recurrence coefficients μ_i, ν_i are given as scalar products of the polynomials in the given data in M points in our paper [9]. We developed some features of our algorithm. One can generate $P_i^{(m)}(\lambda)$ recursively. The polynomials satisfy the following orthonormality relations $\sum_{i=1}^{M} w_i P_k^{(0)}(\lambda_i)P_l^{(0)}(\lambda_i) = \delta_{k,l}$ over the discrete point set $\{\lambda_i, i = 1, 2, \ldots\}$, where $w_i = 1/(\sigma_{T_i}^2)$ are the corresponding weights. An approximation function $T^{\rm appr}$ is constructed with orthonormal a_k and usual c_k coefficients. The coefficient matrix in the least square method becomes an identity one, and the coefficients a_k are computed by known values $\{P_k, \lambda_k, w_i\}$ due to orthonormality:

$$T^{\rm appr\,(m)}(\lambda) = \sum_{k=0}^{L} a_k P_k^{(m)}(\lambda) = \sum_{k=0}^{L} c_k \lambda^k, \quad \text{where} \quad a_k = \sum_{i=1}^{M} T_i w_i P_k^{(m)}(\lambda_i). \quad (2)$$

Let us write polynomials in the ordinary basis (see [10])

$$P_k = \sum_{j=0}^{k} c_j^{(k)} \lambda^j, k = 0, \ldots, L, \text{ then} \quad c_j = \sum_{i=j}^{L} a_i c_j^{(i)}, \quad j = 0, \ldots, L \quad (3)$$

The knowledge of a_i enables to calculate c_j from Eq. (3). The inherited errors in usual coefficients are given by the inherited errors in orthonormal ones:

$$\Delta c_j = \sqrt{\sum_{i=j}^{L} \left(c_j^{(i)}\right)^2 \Delta a_i}\ , j = 0, 1, 2 \ldots, L, \text{ where } \Delta a_i = \sqrt{\sum_{k=1}^{M} P_i^2(\lambda_k) w_k \left(T_k - T_k^{\rm appr}\right)^2}\ , i = 0, \ldots, L.$$

The OPEM advantages are: a) We use unchanged the coefficients of the lower-order polynomials to calculate the higher ones. In this way we shorten the computing time. b) We avoid the inversion of the coefficient matrix to obtain the solution [9–11]. c) We define two criteria for evaluating of an optimal polynomial degree.

First criterion (i): Here one neglects the errors in λ variable, the graph of the fitting curve lies inside the "old" error corridor $[T - \sigma, T + \sigma]$. (ii): After calculating the derivatives at any point λ_i using Eqs. (1), (2), (3) the fitting curve has to lie inside the total error corridor $[T - S, T + S]$.
Second criterion We extend the above algorithm to include S_i^2 in OPEM in two stages: (i): By minimizing the following $\chi^2 = \sum_{i=1}^{M} w_i[T^{\rm appr}(\lambda_i) - T(\lambda_i)]^2/(M - L - 1)$, with weights $w_i = 1/\sigma_{T_i}^2$; (ii): The next approximation is done with the weight function $w_i = 1/S_i^2$. The preference is given to the first criterion and when it is satisfied, the search for the minimal χ^2 stops. Based on the above features the OPEM selects the optimal solution for a given set $\{T, \lambda\}$.

4 Approximation results

The main results are given in Table 1. and Fig. 3. We use one subinterval with $M = 94$ points around a supposed minimum of T and other subinterval with $M = 50$ points in it. We present the approximation with orthonormal coefficients. There are two different type of weights. The table shows: number of

Table 1. OPEM approximations results for different given subintervals

M	L	$\sqrt{\chi^2}$	W	$\Delta T_{max}[\%]$	$\lambda(\Delta T_{max})$ [nm]	$T^{appr}_{min}[\%]$	$\lambda(T^{appr}_{min})$ [nm]
50	2	0.12	$10^3/T^2$	0.57	481.7	59.67	481.7
94	2	0.27	$10^3/T^2$	0.84	474.9	60.32	481.9
94	3	0.23	1.	0.73	493.2	59.85	485.6

points M, optimal number of degrees of polynomials L, $\sqrt{\chi^2}$, maximal deviation between the given and the fitted values $\Delta T_{max} = \max|T^{appr} - T|$, λ [nm] at ΔT_{max}, and the interesting results: T^{appr}_{min} and the corresponding $\lambda\left(T^{appr}_{min}\right)$. The best results by the given criteria are: in the second row with $T^{appr}_{min} = 60.32$, in 481.9 nm at $L = 2$ and in the third row with $T^{appr}_{min} = 59.85$, in 485.6 nm at $L = 3$.

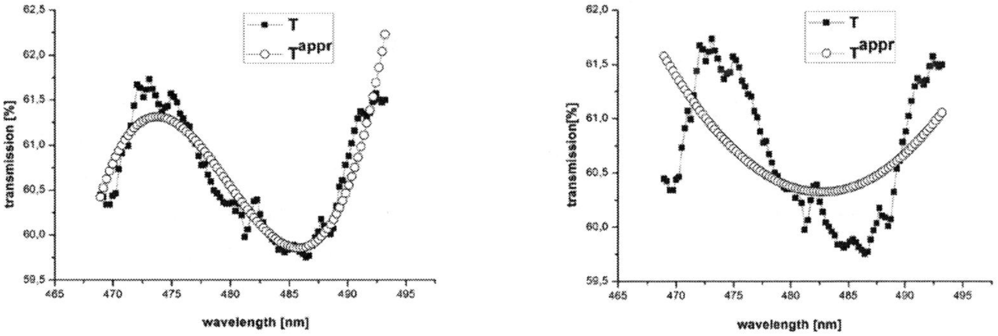

Figure 3. a) b)
Given and OPEM approximated values of T by the third a) and the second b) order of polynomials (Table 1).

5 Main Conclusions

The results in Table 1. and Fig. 3. present smooth approximations by the second and the third number of degrees of polynomials in non-equidistant grid and show the minimums by two approaches. The OPEM gives numerical estimations for interpretations of the optical properties of laser produced NPs.

References

[1] O. Zhang, Y.N. Tan, J. Xie, and J.Y. Lee, Plasmonics **4** (2009)
[2] S.J. Henley, J.D. Carey, and S.R.P. Silva, Phys. Rev. **B 72**, 195408 (2005)
[3] M. Koleva, A. Dikovska, N. Nedyalkov, P. Atanasov, and I. Bliznakova, Applied Surface Science **258**, 23, 9181–9185 (2012)
[4] M. Koleva, N.N. Nedyalkov, P.A. Atanasov, J.W. Gerlach, D. Hirsch, A. Prager, B. Rauschenbach, N. Fukata, and W. Jevasuwan, J. Alloy. Compd. **665**, 282–287 (2016)
[5] M. Koleva, N. Nedyalkov, N. Fukata, W. Jevasuwan, and S. Amoruso, Proc. of SPIE **10226**, 1022609 (2017)
[6] P.R. Bevington, *Data Reduction and Error Analysis for the Phys. Sci.* (McGrow-Hill, NY, 1969)
[7] N. Bogdanova, J. Comp. Appl. Math. **14**, 345 (1986)
[8] G. Forsythe, J. Soc. Ind. Appl. Math. **5**, 74 (1957)
[9] V. Gadjokov and N. Bogdanova, Commun. JINR, P11-12860 (1979)
[10] N. Bogdanova, AIP proc., ed. S.A. Cetin and I. Hikmet ,978-0-735400404-5/07, p. 377 (2007)
[11] N. Bogdanova, S. Todorov, and G.A. Ososkov, Communication JINR E10-2015-67 (2015)

EPJ Web of Conferences **173**, 04004 (2018) https://doi.org/10.1051/epjconf/201817304004
Mathematical Modeling and Computational Physics 2017

Towards $J/\psi \to e^+e^-$ Decays Triggering with TRD in CBM Experiment

Olga Derenovskaya[1,⋆], *Timur* Ablyazimov[1,2,⋆⋆], and *Victor* Ivanov[1,3,⋆⋆⋆]

[1] *Joint Institute for Nuclear Research (JINR), Dubna, Russia*
[2] *Gesellschaft fü Schwerionenforschung mbH (GSI), Darmstadt, Germany*
[3] *National Research Nuclear University "MEPhI", Moscow, Russia*

Abstract. The paper presents an efficient Cellular Automaton based algorithm for trajectory reconstruction in the Transition Radiation Detector of the CBM experiment. The comparison of the different electron identification methods is also given.

1 Introduction

The Compressed Baryonic Matter (CBM) collaboration [1] conducts dedicated heavy-ion experiments to investigate the properties of highly compressed baryonic matter as it is expected to be produced in nucleus–nucleus collisions at the Facility for Antiproton and Ion Research (FAIR) in Darmstadt, Germany.

The CBM experimental setup for studying dielectron decays is shown in Fig. 1. Behind the target between the poles of the superconducting dipole magnet there is a Silicon Tracking System (STS). The STS detectors are intended to reconstruct trajectories and momenta of charged particles, as well as to reconstruct primary and secondary vertices. The electron–positron identification system includes the Ring Imaging CHerenkov (RICH) and Transition Radiation Detectors (TRD). The TRD is also used to reconstruct the trajectories of charged particles registered by the detector. The detector for particle Time-Of-Flight (TOF) measurement is intended for hadron identification. The Electromagnetic CALorimeter (ECAL) serves to identify photons. The Projectile Spectator Detector (PSD) calorimeter serves to the determination of the reaction plane.

The study of the charmonium production is one of the key objectives of the CBM experiment. To register them via the dielectron decay channel, one needs a reliable electron–positron identification in the conditions of a dominant hadronic background, primarily from pions. The TRD is the most suitable for solving the above-mentioned task. TRD should yield reliable electron identification, a high pion suppression level, a reconstruction of trajectories of charged particles passing through the detector in the condition of intense fluxes (up to 10^7 collisions per second), and a high multiplicity of secondary particles (from 100 to 1000 particles per nucleus–nucleus collision).

⋆e-mail: odenisova@jinr.ru
⋆⋆e-mail: abl@jinr.ru
⋆⋆⋆e-mail: ivanov@jinr.ru

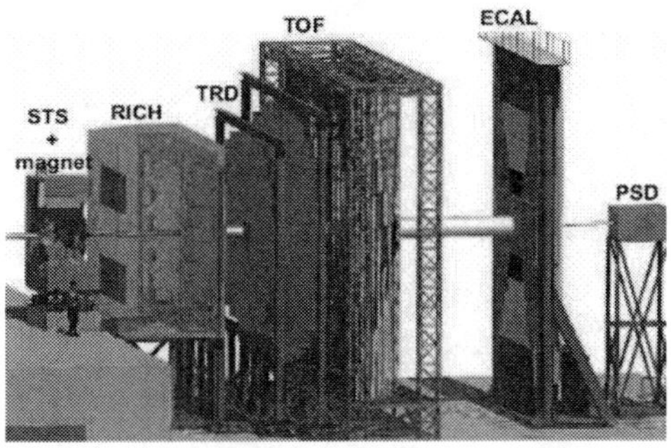

Figure 1. General view of the CBM setup

2 Track reconstruction algorithm

First step towards the $J/\psi \to e^+e^-$ decays triggering is trajectories reconstruction in TRD. The cellular automaton (CA) based algorithm was developed for TRD. The following assumptions have been made to simplify and accelerate the algorithm [2]:

1. Because of secondaries from the J/ψ decay have high momenta and the influence of the magnetic field on the area of the TRD stations is negligible, the particle trajectories can be approximated by segments of straight lines.

2. Only those tracks are considered which have hits in all TRD stations.

The track reconstruction algorithm includes two main stages: a) segment set formation, b) segments binding and track construction.

The segments are fragments of straight lines connecting the neighbouring points of the consecutive TRD stations (Fig. 2(a)). These lines do not differ much from the lines connecting one of the segment ends and the target center. Limits for the segment inclinations have been determined by Monte Carlo (MC) simulation. After the segments are built, consecutive segments are bound if they have a common

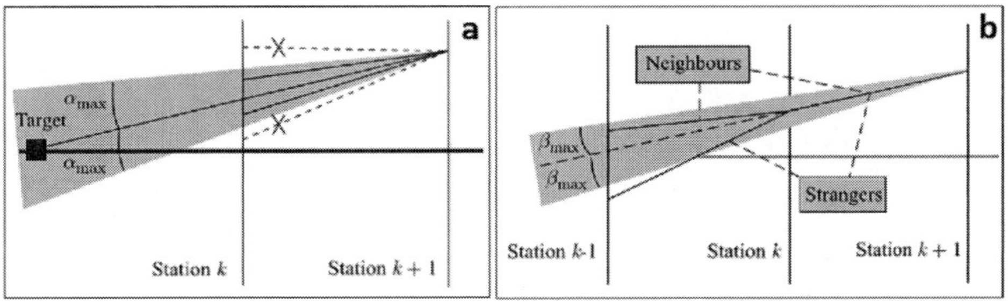

Figure 2. (a) Segment set formation and (b) segments binding procedure

point and the angle between them does not exceed the limit found by simulation (Fig. 2(b)). If two segments are bound, they are referred as "neighbours", if not – as "strangers".

A track candidate is formed from the segment sequence from right to left (upstream the beam direction) by joining neighbouring segments. The first segment is taken on the last TRD station and the last must end on the first station. If during the track candidate building several alternatives appear, the one with minimum χ^2 is chosen.

3 Electron identification

Each track is associated with a set of measurements of the particle energy losses (see Fig. 3(a)).

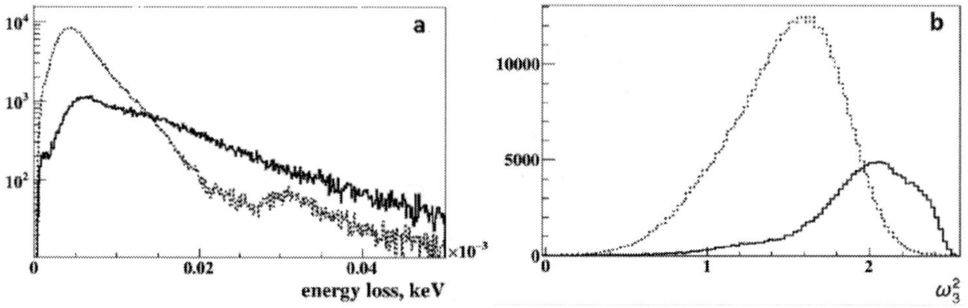

Figure 3. Distributions of the energy losses (a) and ω_3^2 values (b) for $\pi^{+/-}$ (dash line) and $e^{+/-}$ (solid line)

With the help of various mathematical methods, it is possible to determine to which distribution (electrons or pions) these losses are related.

3.1 Goodness-of-fit ω_n^k-criterion

To apply the goodness-of-fit ω_n^k-criterion [3], one needs to calculate the formula:

$$\omega_n^k = -\frac{n^{\frac{k}{2}}}{k+1}\sum_{j=1}^{n}\left\{\left[\frac{j-1}{n} - \phi(\lambda_j)\right]^{k+1} - \left[\frac{j}{n} - \phi(\lambda_j)\right]^{k+1}\right\}, \tag{1}$$

where k is the criterion degree, $\phi(\lambda)$ is the Landau distribution function (which describes the pion energy losses) in terms of the station dependent values of the variable λ_j:

$$\lambda_j = \frac{\Delta E_j - \Delta E^j_{\mathrm{mp}}}{\xi_j} - 0.225, \qquad j = 1, 2, \ldots, n, \tag{2}$$

ΔE_j is the energy loss in the j-th TRD layer, ΔE^j_{mp} is the value of most probable energy loss, $\xi_j = \frac{1}{4.02}$ FWHM of distribution of the energy losses for pions, and n is the number of TRD layers.

Fig. 3(b) shows the distributions of ω_3^2 values for pions (dash line) and electrons (solid line).

3.2 Artificial neuron network (ANN)

In [3], the possibility of electron identification using an artificial neural network has been investigated. A three-layered perceptron from the ROOT package is currently used in the CBM experiment. Fig. 4(a) shows the distributions of the ANN output signals.

3.3 Likelihood function ratio method

While applying the likelihood test[4] to the problem considered, the value

$$L = \frac{P_e}{P_e + P_\pi}, \qquad P_e = \prod_{i=1}^{n} p_e(\Delta E_i), \qquad P_\pi = \prod_{i=1}^{n} p_\pi(\Delta E_i), \tag{3}$$

is calculated for each event (see Fig. 4(b)), where $p_\pi(\Delta E_i)$ is the value of the density function p_π in the case when the pion loses the energy ΔE_i in the i-th station, and $p_e(\Delta E_i)$ is a similar value for the electron.

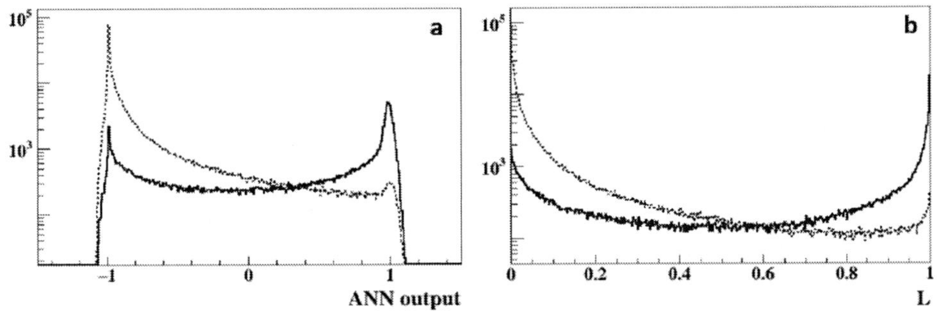

Figure 4. Distributions of the ANN output (a) and the variable L (b) for $\pi^{+/-}$ (dash line) and $e^{+/-}$ (solid line)

4 Conclusion

An algorithm for the trajectory reconstruction in the TRD based on the CA has been developed. The efficiency of the signal track reconstruction calculated by the formula

$$\mathrm{Eff} = \frac{N_{\mathrm{rec}}}{N_{\mathrm{ref}}} \times 100\% \tag{4}$$

is about 92%. Here N_{ref} is the number of reference tracks corresponding to the electron or the positron from the J/ψ decay and which have hits in all layers of the TRD. N_{rec} is the number of reference tracks that have been matched to the reconstructed track. The track is matched if at least three hits of the track coincide with three points of the reference track.

Table 1 shows the results of comparison of the given methods: background suppression factor corresponding to 11% of electron/positron losses. The best pion suppression level is achieved using the

Table 1. Background suppression factor corresponding to 11% of electron/positron losses

method	ω_n^k	ANN	LFR
suppression factor	5.4	9	9

LFR method and ANN with transformation (2) of energy losses in the TRD stations. The bottleneck of these methods is the requirement to know the density functions of energy losses for both pions and electrons. The distribution of pion ionization losses in a material is well studied, whereas the energy losses of electrons/positrons in the TRD staions are complex by nature. For ω_n^k application one should only know the parameters of the pion losses distribution.

References

[1] B. Friman, P. Senger et al., *The CBM Physics Book* (Lecture Notes in Physics **814**, 1st Edition, 2011) 960 pp.

[2] T.O. Ablyazimov and V.V. Ivanov, Particles and Nuclei, Letters **14**, No 3, 504–511 (2017)

[3] O.Yu. Derenovskaya, *Methods and Algorithms for $J/\Psi \to e^+e^-$ Decays Recognition in the CBM Experiment* (Ph.D. thesis, Dubna, 2015) 108 pp.

[4] F. E. James, *Statistical Methods in Experimental Physics: 2nd edition* (World Scientific Publishing Company, 2006)

Simulations in the Analysis of Experimental Data Measured by BM@N Drift Chambers

Ján Fedorišin[1,2,⋆]

[1] *Veksler and Baldin Laboratory of High Energy Physics, Joint Institute for Nuclear Research, 141980 Dubna, Moscow region, Russia*
[2] *Institute of Experimental Physics SAS, Watsonova 47, 040 01 Košice, Slovakia*

Abstract. The drift chambers (DCH's) are an important part of the tracking system of the BM@N experiment designed to study the production of baryonic matter at the Nuclotron energies.
The method of particle hit and track reconstruction in the drift chambers has been already proposed and tested on the BM@N deuteron beam data. In this study the DCH's are first locally and globally aligned, and subsequently the consistency of the track reconstruction chain is tested by two methods. The first one is based on the backward extrapolation of the DCH reconstructed deuteron beam to a position where its deflection in the BM@N magnetic field begins. The second method reconstructs the deuteron beam momentum through its deflection angle. Both methods confirm correctness of the track reconstruction algorithm.

1 Introduction

The BM@N (Baryonic Matter at Nuclotron) [1, 2] experiment is designed to study the strange matter production in collisions of heavy nuclei. BM@N is a fixed target experiment using heavy-ion beams extracted from JINR Nuclotron with the beam energies ranging from 1 to 6 AGeV. The proposed target and projectile nuclei include p, d, C, Ar, Cu, Xe, Au, i.e., they cover a wide range of nuclear masses.

The drift chambers [3] constitute an important part of the BM@N tracking system. There are two identical chambers, DCH1 and DCH2, placed outside the magnetic field of the SP41 dipole magnet [4]. Each DCH consists of four segments measuring respectively y, x, u, v track coordinates where u and v stand for the x coordinate rotated by angles $\pm 45°$ around the z axis. These coordinates are used to produce charged particle tracks passing through all sixteen planes of both the drift chambers. The track reconstruction is a sophisticated process involving the detector calibration, the hits calculation as well as the track finding and fitting algorithm (all the data processing chain is thoroughly described in [5] or [6]). Furthermore, the experimental data is corrected and refined by the autocalibration procedure and the detector alignment within which, first, the DCH1 is globally aligned to the initial beam direction and position, and then the DCH2 is locally aligned with respect to DCH1. The correction procedures are tuned up on pure beam data where it is less complicated due to low multiplicities of secondary particles.

⋆e-mail: fedorisin@jinr.ru

2 The method consistency checks

Because of the complexity of the DCH track reconstruction algorithm, it is necessary to verify its accuracy and reliability. There are two basic consistency checks applied in the presented analysis, both of them try to recover the kinematic characteristics of the initial deuteron beam. Furthermore, the first method is able to reconstruct even the beam geometric parameters at an arbitrary moment of beam deflection in the magnetic field.

2.1 Check #1 – beam backward extrapolation

Using Geant4 [7], the reconstructed experimental tracks are extrapolated backwards to the point where their deflection in the magnetic field begins (see Fig. 1). If everything is correct, then the following conditions hold for the coordinates: x, $y \to 0$ and the momentum components: p_x, $p_y \to 0$, $p_z \to p_{\text{beam}}$.

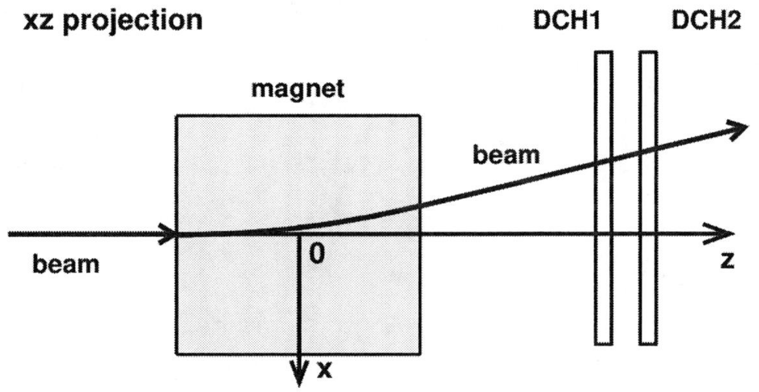

Figure 1. Layout of BM@N magnet and drift chambers

The instantiation of the results produced by the beam backward extrapolation for a typical deuteron beam run with $p_{\text{beam}} = 8.68$ GeV/c is shown in Figs. 2 and 3. All the kinematic and the geometric characteristics of the extrapolated beam agree within the errors with the expected values.

2.2 Check #2 – momentum reconstruction of the deuteron beam

The deflection of the beam by a magnetic field in the xz plane is described by the angle φ_{xz} which depends on the magnetic field integral and the beam momentum p:

$$\varphi_{xz} = q\frac{\int Bdl}{p} \quad [\text{rad}] \qquad \Longrightarrow \qquad p = q\frac{\int Bdl}{\varphi_{xz}},$$

where B stands for the B_y component of the magnetic field and q stands for the beam particle charge [8]. These relations are valid for small deflection angles when $\tan\varphi \simeq \varphi$. Fig. 4 displays the dependences of the angle φ_{xz} and the beam momentum p on the magnetic field integral. The reconstructed momentum 8.6 ± 0.1 GeV/c is consistent with the nominal value of the deuteron momentum of 8.68 GeV/c.

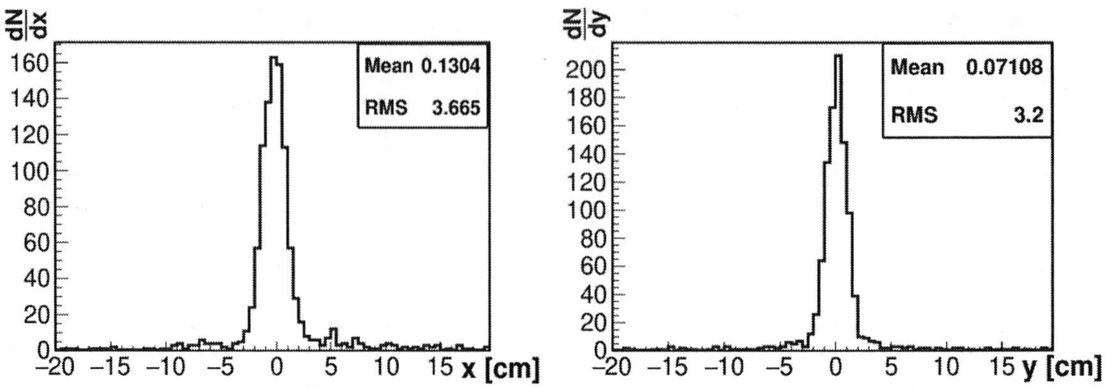

Figure 2. Restored x and y coordinates of the reconstructed beam at the entrance into the magnetic field

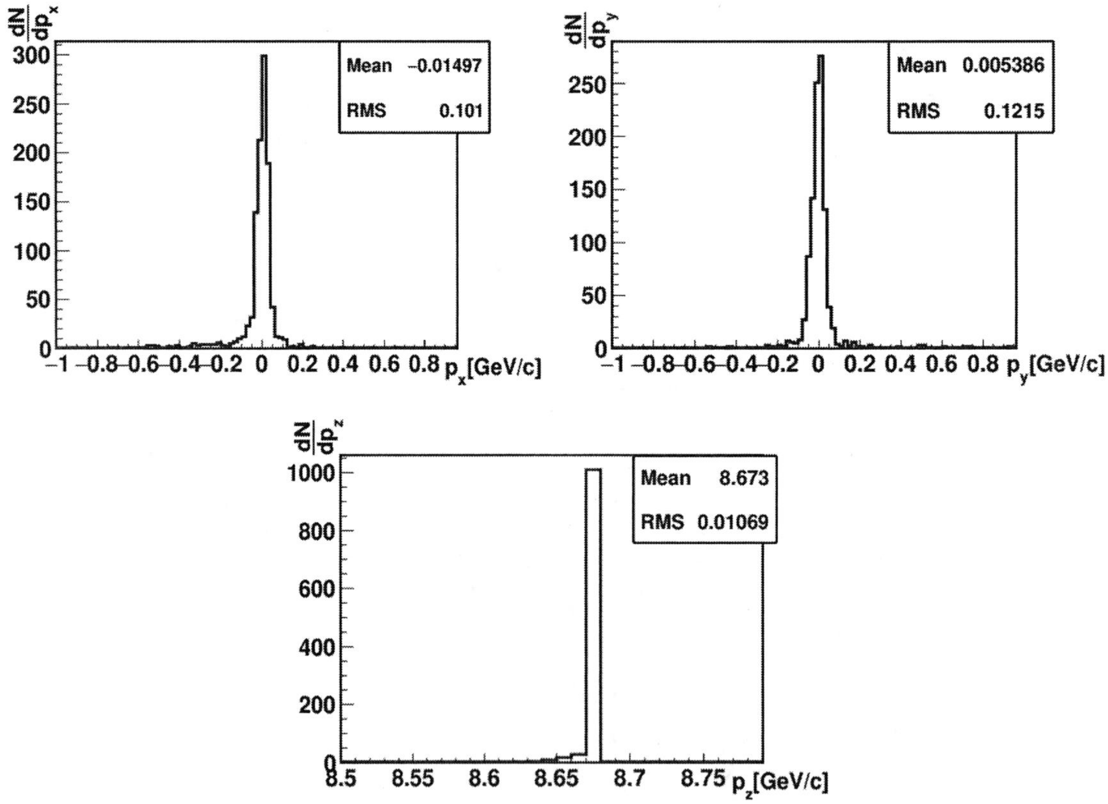

Figure 3. Restored components of the beam momentum p_x, p_y, and p_z, at the initial deflection point

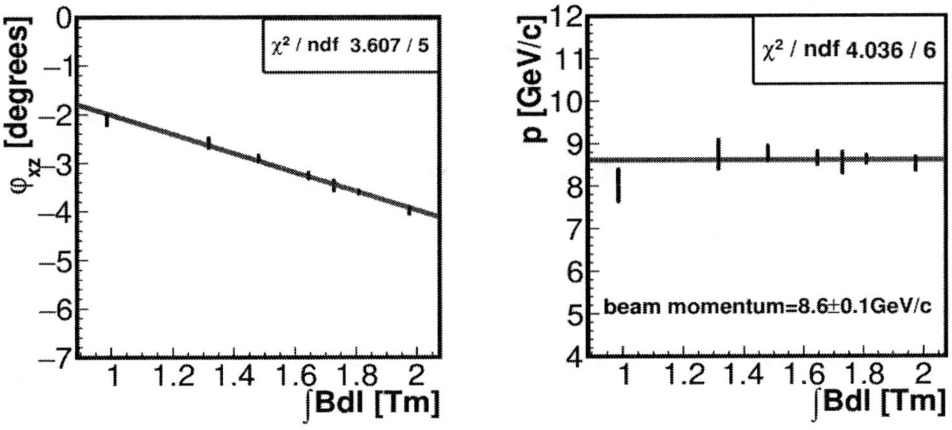

Figure 4. Deflection angle of deuteron beam in xz plane (on the left) and deuteron momentum (on the right) as functions of magnetic field integral

3 Summary

1. A track reconstruction algorithm involving two independent consistency checks was proposed, programmed and tested to verify its accuracy and efficiency.
2. Geant4 based track extrapolations were used in the testing procedures.
3. The two developed checks were based on the reconstruction of the initial deuteron beam kinematic and geometric characteristics – the momentum components and the x, y coordinates.
4. The reconstructed values are consistent to a high degree with the corresponding initial ones.

References

[1] T.O. Ablyazimov et al., BM@N Collaboration, *BM@N – Baryonic Matter at Nuclotron* (Conceptual Design Report), http://nica.jinr.ru/files/BM@N/BMN_CDR.pdf (2012)
[2] M. Kapishin, Eur. Phys. J. A **52**: 213 (2016)
[3] D. Bédérède et al., Nucl. Instrum. Meth. A 367, 88–91 (1995)
[4] P.G Akishin et al., physics.ins-det arXiv:1407.7096v1 (2014)
[5] J. Fedorišin, EPJ Web of Conferences **108**, 02021 (2016)
[6] J. Fedorišin, EPJ Web of Conferences **138**, 11005 (2017)
[7] J. Allison et al., Nucl. Instrum. Meth. A **835**, 186–225 (2016)
[8] Tables of Physical & Chemical Constants (16th edition 1995). Kaye & Laby Online. Version 1.0 (2005) http://www.kayelaby.npl.co.uk

Mock Data Challenge for the MPD/NICA Experiment on the HybriLIT Cluster

Konstantin Gertsenberger[1,*] and *Oleg* Rogachevsky[1,2]

[1] *Laboratory of High Energy Physics, Joint Institute for Nuclear Research, Russia, Dubna*
[2] *Dubna State University, Russia, Dubna*

Abstract. Simulation of data processing before receiving first experimental data is an important issue in high-energy physics experiments. This article presents the current Event Data Model and the Mock Data Challenge for the MPD experiment at the NICA accelerator complex which uses ongoing simulation studies to exercise in a stress-testing the distributed computing infrastructure and experiment software in the full production environment from simulated data through the physical analysis.

1 Introduction

According to the program of the Joint Institute for Nuclear Research on the creation of the ion accelerator complex for the range of collision energy $\sqrt{S_{NN}}$ from 4 to 11 GeV, the Nuclotron-based ion collider facility (NICA) is being constructed. Two interaction points are foreseen at NICA for two detectors. The MultiPurpose Detector (MPD) [1] located at one of the two interaction points is optimized for a comprehensive study of the properties of hot and dense nuclear matter in heavy-ion collisions and search for the critical point of the phase transition to the quark–gluon plasma. The MPD experiment would be capable of unique measurements of heavy-ion collisions including high-precision tracking and particle identification as well as of very accurate event characterization.

The software for simulation, reconstruction and analysis of particle physics data is an essential part of each high-energy physics experiment. To support the MPD experiment, the MpdRoot [2] framework is under development. It is based on the ROOT environment [3] and the FairRoot framework [4] of the GSI Institute and serves for simulation of the MPD detector, the reconstruction of simulated and experimental data and the following physical analysis of particle collisions registered by the MPD detector. The MPD software covers all stages, such as the simulation process of the particle interactions with media and detector materials, digitization – translating the interactions with detectors into clusters of signals, reconstruction of events and physical data analysis.

The Mock Data Challenge (MDC) is performed to simulate the offline processing of the data produced by the MPD detector before obtaining the first experimental data. Its main purpose is to exercise in a stress testing the distributed computing infrastructure and experiment software in the full production environment from simulated data through a physics analysis. The following sections of the article give an overview of how the MPD event data model is constructed and event data are processed in a distributed manner, and show the first mock data challenge for the MPD experiment.

*e-mail: gertsen@jinr.ru

2 Event Data Model in the MPD experiment

Fig. 1 provides an overview of the experimental data processing chain in the MpdRoot software. Raw data in a special *MPD* format using the TLV (type-length-value) architecture from the DAQ system are digitized and converted into ROOT format by an event builder. A reconstruction macro restores the information about the particles registered by the detectors, their tracks and other parameters. The target size of a reconstructed file is about 1 MB per central Au – Au event. Then physics analysis of the reconstructed data is performed to investigate the physical properties of the nuclear matter produced in the collisions. For the experimental data it is planned to use the customary two formats in the high-energy experiments: event summary data (ESD) containing the detailed output of the reconstruction, and analysis object data (AOD) with sufficient information for physics analyses. The

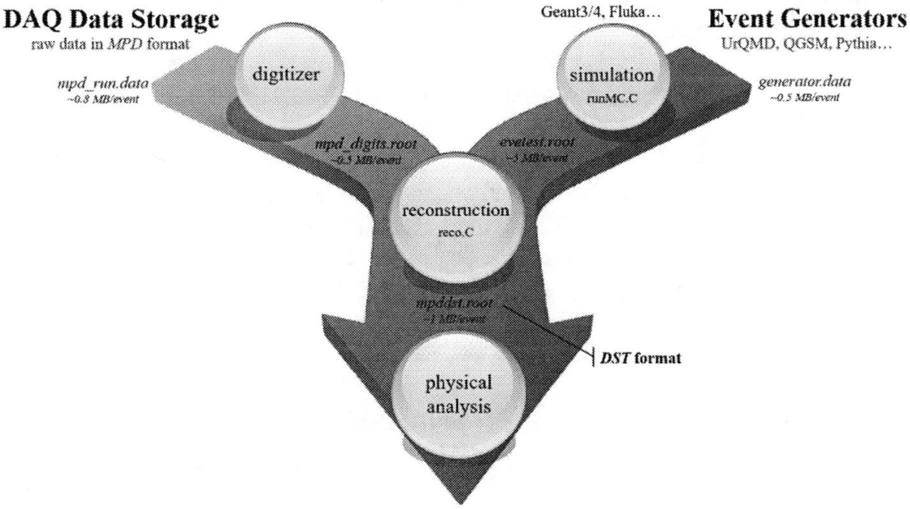

Figure 1. MpdRoot offline data processing chain

processing chain with simulation data containing the full information about the particles obtained by the event generators is used at this stage. After simulating the passage of the particles through the detectors using particle transport packages, the data are converted to detector responses. The MpdRoot simulation macro uses Geant 3 (or 4) as a particle transport package. The event data are stored in ROOT files in a hierarchical tree structure. The size of an event resulting from heavy-ion collisions is directly proportional to the multiplicity, which is a function of the interaction centrality. Furthermore, the amount of data taken in experimental runs varies with the trigger conditions. The approximate event sizes estimated in MpdRoot and shown in Fig. 1 correspond to central heavy ion collisions.

For the offline processing of the MPD data, a Unified Database [5] was developed on PostgreSQL database management system. In addition to other data, it also stores information on simulated files obtained by different event generators. We use a wide list of event generators with proper physics effects, for example, UrQMD and QGSM to provide realistic particle distributions or Hybrid UrQMD and Three Fluid Hydrodynamics to predict new phenomena in nuclear collisions. At present, the data storage contains about 34,000 simulated files of 1.2 TB. These initial simulation data were used for the Mock Data Challenge of the MPD experiment.

3 Distributed MPD event processing

It is assumed that the MPD experiment with an interaction rate up to 7 kHz will produce about twenty billion events per year. An event of a central Au + Au collision at the NICA energies contains up to 1,000 charged particles. The high interaction rate, the high particle multiplicity and the long sequential processing time of millions of events in MpdRoot are the main reasons to develop a distributed computing system for the MPD experiment.

There are two main approaches to the distributed MPD event processing now: the Parallel ROOT Facility (PROOF) tool of the ROOT environment to parallelize the processing of event trees in ROOT macros on parallel architectures and the MPD-Scheduler developed for user task distribution on cluster nodes.

The PROOF uses data independent parallelism following from the lack of event correlations to process different MPD events in parallel. The PROOF support was added to the MpdRoot software and the reconstruction code was rewritten according to the PROOF rules and classes.

The MPD scheduling system was developed to distribute user jobs among computing clusters. MPD-Scheduler was implemented in C++ language with the ROOT classes support. It simplifies the parallel execution of user macros without asking for the knowledge of different batch systems. At present, it supports SLURM, Sun Grid Engine and Torque scheduling systems to run user jobs on cluster nodes in parallel and can use the Unified Database of the experiment. Jobs for multi-threading execution on one multi-core machine or distributed execution on clusters are described and passed to the MPD-Scheduler as XML files. To execute a user job on the cluster, the MPD-Scheduler parses the job description and runs scripts by the given batch system. It defines free workers of the cluster and performs data processing in parallel. To perform the Mock Data Challenge for the MPD experiment on the HybriLIT cluster, the MPD-Scheduler was used.

4 MPD Mock Data Challenge on the HybriLIT cluster

The Mock Data Challenge aims at providing large-scale simulation production, exercising the full spectrum of the experiment software and hardware from simulation through to physics analysis. It is good for stress-testing of the distributed computing infrastructure of the experiment, identifies potential issues before first data, helps to estimate offline computing requirements, checks data quality and prepares quick turnaround from data taking to publication.

The Mock Data Challenge was performed on the HybriLIT cluster [6]. The HybriLIT heterogeneous cluster is a computation component of the Multifunctional Information and Computing Center of the Laboratory of Information Technologies of JINR, which contains a multi-core component and computation accelerators: NVIDIA graphic processors and Intel Xeon Phi co-processors. It is designed for parallel computations asked by a wide range of tasks arising in the scientific and applied research conducted by JINR. The HybriLIT cluster stores the MPD data on the distributed file system NFS and uses SLURM to distribute jobs among the cluster nodes. All external packages for MpdRoot and MpdRoot itself were installed and configured on the cluster for parallel event data processing. The distributed HybriLIT NFS storage contains files with 3 million simulated events obtained by the LAQGSM generator for the MPD experiment.

An XML description of the job used for the MPD mock data processing was created for the MPD-Scheduler, which distributes the processing of the files among cluster nodes by a SLURM batch system. The job includes dependent tasks of simulation, reconstruction and physics analysis executed at the same time on the HybriLIT cluster. Femtoscopy was chosen as a physics analysis task for reconstructed data. The list of input files consists of three hundred twenty simulated files for the MPD collision energies from 5 to 11 GeV. Fig. 2 presents the time and speedup of the MPD mock data processing in terms of the number of processor cores used.

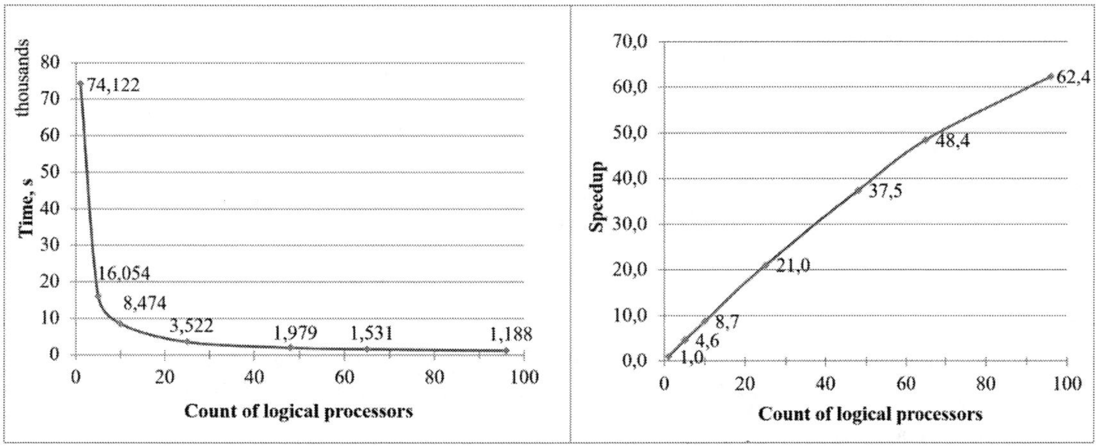

Figure 2. The dependence of the time (left graph) and the speedup (right graph) of the Mock Data Challenge on the number of processor cores

As a result, the MPD Mock Data Challenge was successfully performed on the HybriLIT cluster. All of the steps were successfully completed, and software and hardware errors did not occur during the execution. Now the HybriLIT storage includes four versions of the reconstructed data for the generated events, depending on the simulation settings.

5 Conclusions

The MpdRoot environment with required packages (Fairsoft, ROOT/PROOF, MpdRoot, the MPD-Scheduler and others) was deployed on the HybriLIT cluster for MPD data processing. Two methods were implemented to process event data of the MPD experiment on distributed systems in parallel: using the PROOF system and the MPD-Scheduler. The MPD-Scheduler tool was developed to automate running MpdRoot macros concurrently and it was used in the MPD Mock Data Challenge (*simulation-reconstruction-analysis* chain) to test the software and computing infrastructure of the HybriLIT cluster. Practical values of time and speedup for MPD event processing were obtained.

Acknowledgement

The authors would like to thank the HybriLIT team for the opportunity to use the computational resources of the HybriLIT cluster and support.

References

[1] MPD Collaboration, *The MultiPurpose Detector – MPD. Conceptual Design Report* (JINR, Dubna, Russia, 2012) 259 pp.

[2] K. Gertsenberger, S. Merts, O. Rogachevsky, and A. Zinchenko, European Physical Journal A **52**, 8, p. 5 (2016)

[3] CERN, *ROOT – An Object-Oriented Data Analysis Framework* (CERN, 2013) 628 pp.

[4] M. Al-Turany, D. Bertini, T. Karabowicz, et al., Journal of Physics : Conference Series **396**, 022001 (2012)

[5] K. Gertsenberger, Physics of Particles and Nuclei Letters **13**, 5, 634–639 (2016)

[6] HybriLIT, *Heterogeneous computing cluster HybriLIT of LIT JINR* (JINR, Dubna, Russia, 2016)

GEM Detector Performance Assessment in the BM@N Experiment

Mikhail Kapishin[1,⋆], *Vladimir* Karjavin[1], *Elena* Kulish[1], *Vasilisa* Lenivenko[1,⋆⋆], *Alexander* Makankin[1], *Anna* Maksymchuk[1], *Vladimir* Palichik[1], and *Sergey* Vasiliev[1]

[1] *Joint Institute for Nuclear Research, Joliot Curie 6, 141980 Dubna, Moscow region, Russia*

Abstract. The Gas Electron Multiplier (GEM) chambers are developed for modern purposes in the elementary particle physics. In the BM@N experiment, a GEM system is used for the reconstruction of the trajectories of the charged particles. The investigation of GEM performance (efficiency and spatial resolution) is presented.

1 Introduction

1.1 Motivation

The relativistic heavy ion collisions provide a unique opportunity to study the nuclear matter under extreme density and temperature. If the energy density in the formed fireball is sufficiently large the quark-gluon substructure of nucleons becomes visible. The existing theoretical models suggest different possible scenarios to describe the features of the strongly interacting matter. Experimental data at high detector resolution are needed in order to disentangle among the different theoretical predictions [1, 2] and to get the proper description of these phenomena.

1.2 BM@N experiment

The fixed target experiment BM@N (Baryonic Matter at Nuclotron) is the first step of the realization of the NICA project [3]. The main goals of the experiment are [4] the investigation of the nuclear equation-of-state, the study of the in-medium properties of the hadrons, the production of the multi-strange hyperons at the threshold and the search for hypernuclei.

The experiment combines high precision track measurements with time-of-flight information for particle identification. The magnetic field of the analyzing magnet can be varied up to 1 T to get the optimal detector acceptance and momentum resolution for different reactions and beam energies. The charged particle momentum and multiplicity are measured with the set of the Gas Electron Multipliers (GEM) [5] located inside the analyzing magnet and by the Drift Chambers (DCH) and the Cathode Strip Chambers (CSC) situated outside the magnetic field.

⋆e-mail: kapishin@jinr.ru
⋆⋆e-mail: vasilisa@jinr.ru

1.3 Structure of the Gaseous Electron Multiplier

Detectors based on the GEM technology have been identified as appropriate for the BM@N tracking system for their capability of stable operation in conditions of high beam intensities up to 10^6 ions per second and in a strong magnetic field up to 1 T. The BM@N GEM chambers consist of three GEM multipliers, with the following gaps between the electrodes: the drift gap of 3 mm, the first transfer gap of 2.5 mm, the second transfer gap of 2 mm and the induction gap of 1.5 mm.

Presently six GEM chambers are installed into the magnet: four "middle" chambers with the size of 66×41 cm^2 and two "large" chambers with the size of 163×45 cm^2. Each chamber has two-coordinate readout: along horizontal X-axis and along X' at angles ± 15 degrees to the X-axis. The pitch of readout strips is 800 microns.

2 Track-segment building algorithm (magnetic field off) and testing with Monte-Carlo data

Two layers (X-plane and X'-plane) compose chamber. Each chamber layer consists of anode strips. In each fired strip charge from electron avalanches is collected. A group of sequence strips is united into one cluster. The center of gravity of the cluster provides the coordinate of the reconstructed hit.

$$x_c = \frac{\sum_{i=1}^{n} x_i A_i}{\sum_{i=1}^{n} x_i},$$

where n is the cluster width, x_i and A_i are the coordinate and amplitude of the signal of the strip with signal respectively, x_c is the center of gravity. A charged particle track-segment is built collecting the X and X' hits which are located along the straight line. In algorithm GEM hit efficiency per layer is calculated for events where track-segments were reconstructed and defined as the ratio:

$$\frac{\text{Number of the reconstructed hits in all layers}}{\text{Number of layers}}.$$

For testing the efficiency of calculation algorithm simple generator was used with established efficiency per layer 90%. Obtained averaged efficiencies are in a good agreement with the established number. Generator with background = 5 hits per layer and dependence of efficiency vs background are shown in the figure 1.

3 Nuclotron data

3.1 GEM hit efficiency per layer

The GEM hit efficiency per layer with Nuclotron data (C beam without target, 4.5 GeV/n) found for different gas mixtures is shown on the figure 2. The efficiency is high (92–95%), the difference between the two investigated gas mixtures is 2–3%.

For the two gas mixtures the cluster (a group of strips with signal) width is different. From Ar + CO_2 cluster width is 2.3 strips in average. Most of clusters have only one strip. Cluster width for two gas mixtures is presented in figure 3.

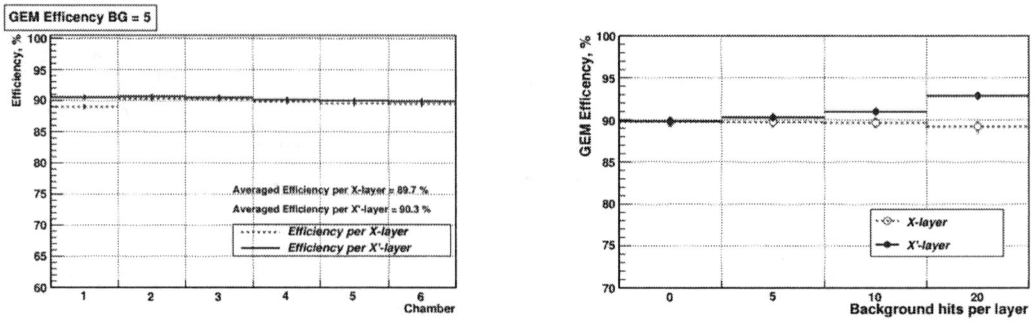

Figure 1. Resolved efficiency per chamber (left), dependence of efficiency vs background hits per layer (right) for simulated data. Simulated efficiency is 90% per layer.

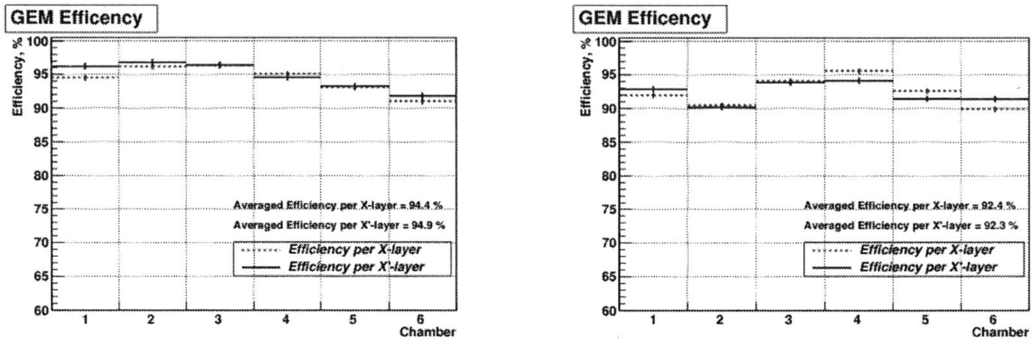

Figure 2. Efficiency per layer for two gas mixtures: Ar + CO_2 (left) and Ar + Isobutane (right).

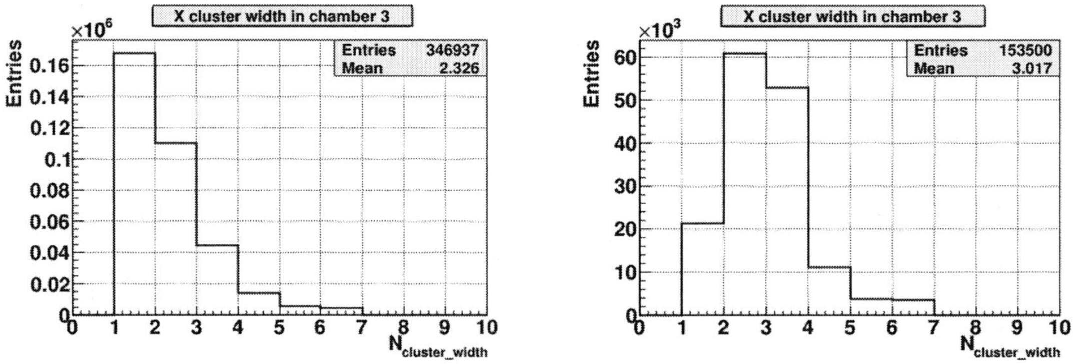

Figure 3. Cluster width in chamber 3 for two gas mixtures: Ar + CO_2 (left) and Ar + Isobutane (right).

3.2 GEM spatial resolution

The spatial resolution was calculated from the residuals between the measurements and the fitted predictions in each layers. Results obtained for the Ar + CO_2 gas mixture are presented in the figure 4. Obtained resolution about 200 μm is similar to the one-strip clusters resolution, that should be $d/\sqrt{12} =$ 231 μm with strip pitch equal 800 μm. This leads to the drop in presicion of the X and X' coordinates reconstruction. Figure 5 illustrates the spatial resolution for the Ar + Isobutane gas mixture. The obtained resolution of 90–120 μm is much better than that for Ar + CO_2 gas mixture, because the cluster width for the Ar + Isobutane gas mixture is in average 3 strips (mainly two- and three-strips clusters).

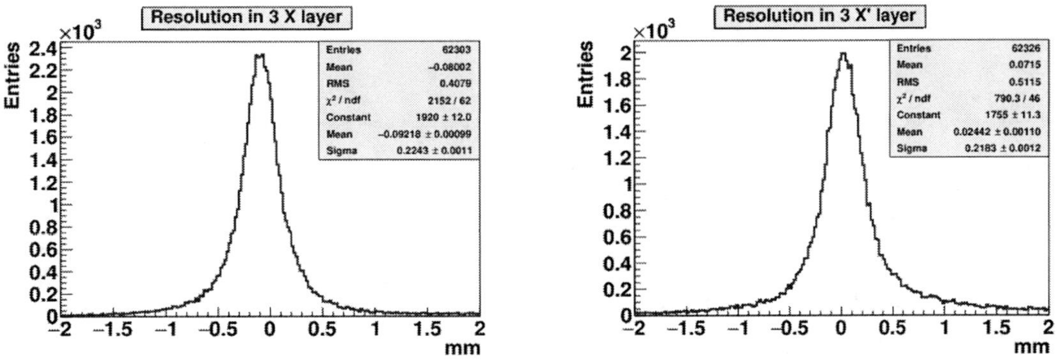

Figure 4. GEM chamber 3 *X* and *X'* resolution for Ar + CO_2 gas mixture. RMS are 0.3306 mm and 0.3113 mm accordingly.

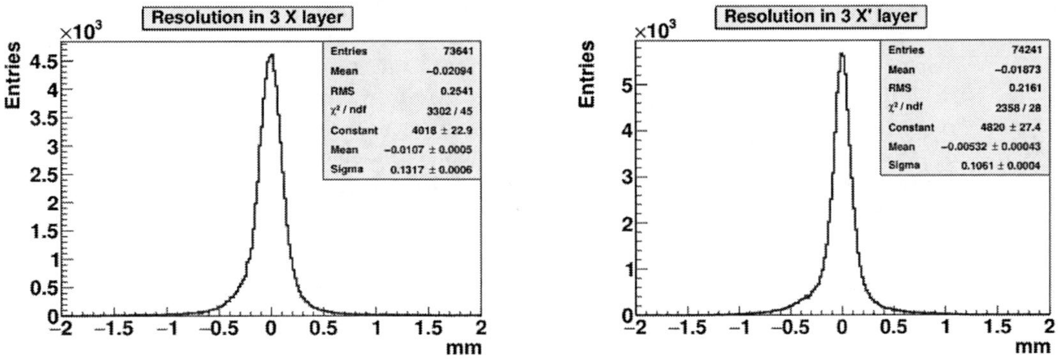

Figure 5. GEM chamber 3 *X* and *X'* resolution for Ar + Isobutane gas mixture. RMS are 0.2541 mm and 0.2161 mm accordingly.

4 Conclusions

The GEM chambers performance in the BM@N experiment was assessed for the Nuclotron data (C beam without target, 4.5 GeV/n): satisfactory hit efficiency per layer 92–95%, a spatial resolution of around 200 μm for Ar + CO_2 gas mixture and a good spatial resolution of 90–120 μm for Ar + Isobutane gas mixture. Due to better spatial resolution at almost the same efficiency Ar + Isobutan gas mixture will be chosen to fill GEM chambers.

References

[1] J. Adams et al, Nucl. Phys. **A 757**, 102–183 (2005)
[2] K. Adcox et al., Nucl. Phys. **A 757**, 184–283 (2005)
[3] A.N. Sissakian et al. (NICA Collaboration), J. Phys. **G36** 064069 (2009)
[4] M. Kapishin, Eur. Phys. J. **A 52**, 213–219 (2016)
[5] F. Sauli, Nucl. Instrum. Meth. **A 805**, 2–24 (2016)

Nuclotron Beam Momentum Estimation Using Multiwire Proportional Chambers and Drift Chambers in the BM@N Experiment

Mikhail Kapishin[1], *Vasilisa* Lenivenko[1], *Vladimir* Palichik[1], and *Nikolay* Voytishin[1,⋆]

[1] *Joint Institute for Nuclear Research, Joliot Curie 6, 141980 Dubna, Moscow region, Russia*

Abstract.
The BM@N experiment is considered as the first phase of NICA Mega science project. The energy of the beam will vary from 1 to 6 GeV/u. The beams delivered by Nuclotron will be of different types from protons to Au. The ability to reconstruct the beam momentum with high precision is one way for showing that the tracking detectors are tuned in the right way and the reconstruction procedure performs well.
A quick overview of the experimental setup is given in the work along with the description of some of the main tracking detectors. The beam momentum reconstruction procedure is described and results are presented for different values of the beam energy.

1 Introduction

The Baryonic Matter at Nuclotron (BM@N) is the first step of the realization of the Nuclotron-based Ion Collider fAcility (NICA) mega-science project [1]. The purpose of the BM@N experiment is to study the interactions of relativistic heavy ion beams delivered by the Nuclotron with fixed targets [2]. The Nuclotron facility is able to provide various types of beams with kinetic energies from 1 to 6 GeV per nucleon. The experimental setup is a complex structure. Its main parts are the analyzing magnet, a set of tracking detectors and a set of identification detectors. All of them are meant to become a precise tool for the study of strange hyperon and hyper-nuclei production yields and ratios.

The first cycles of physical data gathering passed. The accuracy assessment of the main detector systems performance is one of the main tasks at the moment. This paper presents the results of the beam momentum reconstruction procedure using two tracking detector systems: Multi-Wire Proportional Chambers (MWPC) and Drift Chambers (DCH). The accuracy of this reconstruction provides a measure to the performance capabilities of these two detector systems.

2 MWPC and DCH tracking systems

The MWPC are situated in front of the analyzing magnet and are mostly used for the alignment of the entire experimental setup and for the incoming beam trajectory positioning. They represent standard proportional chambers [3]. They consist of three double coordinate planes with wires positioned at

⋆e-mail: voitishinn@gmail.com

angles of 0° and ±60°. The wire pitch is d = 2.5 mm and the estimated spatial resolution is $d/\sqrt{12}$ = 720 μm. The reconstruction algorithm for these chambers [4] was developed and implemented into the official experiment software package. The analysis of recent experimental data showed that the chambers reconstructed tracks with high efficiency. The beam alignment procedure was performed, providing the remaining part of the detector systems the beam position used for global alignment.

The DCH detectors were granted from the finalized NA48 experiment [5]. Two DCH are used in the experimental setup. They form the outer tracker. Each drift chamber consists of four double coordinate planes with wires positioned at angles of 0°, 90° and ±45°. The wire pitch is 10 mm. The spatial resolution achieved by using the experimental data varies from 180 to 300 μm for different planes. The reconstruction efficiency for all planes is above 90 %. Good spatial resolution and high reconstruction efficiency shows that the work aimed at restoring the detectors to their initial state was done with a high degree of professionalism.

3 Beam momentum estimation procedure

The beam trajectory passing through the magnetic field is shown schematically in figure 1. Its momentum can be estimated out of the experimental data [6] by using the formula:

$$\frac{P_{\text{beam}}}{Z_{\text{beam}}} = \frac{0.3 \cdot \int B\,dl}{\sin\left(\alpha_{\text{out}} - \alpha_{\text{in}}\right)}, \tag{1}$$

where $\int B\,dl$ is the magnetic field integral, α_{in} is the particle trajectory angle before the magnet given by MWPC, α_{out} is the particle trajectory angle after the magnet given by DCH and 0.3 is the coefficient for transferring Tesla · meter to GeV/c.

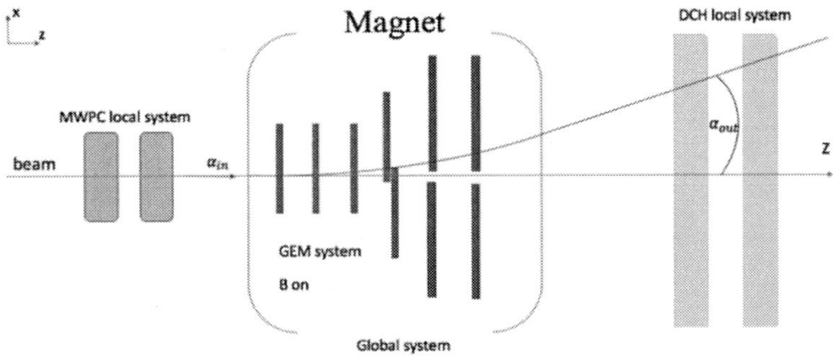

Figure 1. BM@N tracking system

The nominal beam momentum value can be established from the formula:

$$\frac{P_{\text{beam}}}{Z_{\text{beam}}} = \frac{A}{Z}\sqrt{\left[E + M_p\right]^2 - M_p^2}, \tag{2}$$

where A is the mass number of the ion, Z is the number of protons, E is the beam kinetic energy per nucleon and M_p is the proton mass.

4 Results and analysis

The most recent experimental data were gathered with carbon beams with different energies. We considered two different values of energies of 3.5 and 4.5 GeV/nucleon in this paper. The beam momentum was estimated for diverse values of the magnetic field integrals in order to get a complete illustration of the method performance.

The results for the 3.5 GeV/nucleon carbon beam are presented on figure 2.

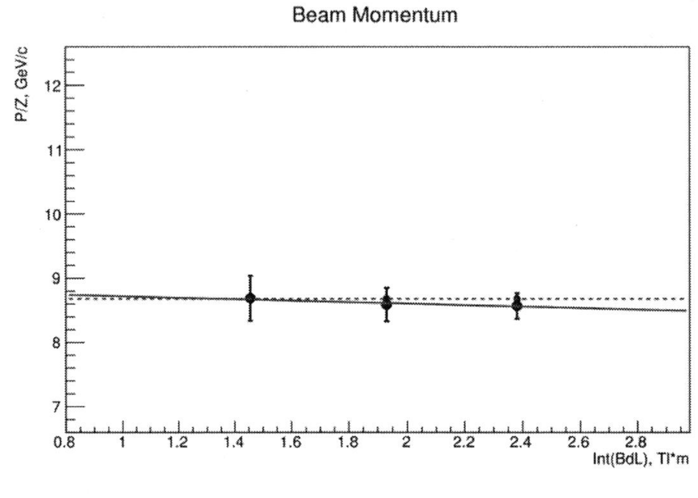

Figure 2. C beam energy 3.5 GeV/nucleon momentum estimation. The dashed line is the nominal beam momentum value. The points are the beam momentum values estimated from experimental data.

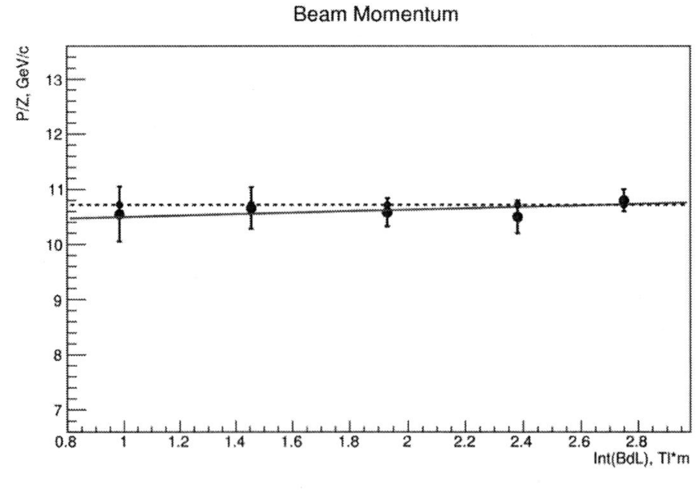

Figure 3. C beam energy 4.5 GeV/nucleon momentum estimation. The dashed line is the nominal beam momentum value. The points are the beam momentum values estimated from experimental data.

The dashed line represents the nominal beam momentum value calculated by the formula 2 and it is at 8.7 GeV/c. The three points fitted with a straight solid line are the values estimated from the experimental data. They satisfy the nominal values within errors. The resolution gets improved with the increase of the magnetic field integrals values from 4% to 2.2%. The obtained errors have contributions from different factors: the multiple scattering in the detectors' material and the precision with which the magnetic field, the incoming and the outcoming angles were measured. Minor influence has the multiple scattering in the material of the three tracking systems: MWPC, GEM and DCH. The magnetic field was measured by the accelerator team, but we do not have any information on its precision. The beam trajectory was positioned by the MWPC system at 0° with an error less than 1%. The outcoming angle values given by the DCH system increase with the increase of the magnetic field integral while their errors decrease from 4% for the lowest value of the integral down to 2% for the highest values of the field integral. The same procedure was performed for carbon beam with the energy of 4.5 GeV/nucleon and the results are shown in figure 3. All the estimated values and their errors satisfy the expected results and the desired resolution of 2% was achieved for the working values of the magnetic field integral.

5 Conclusion

The BM@N experiment is operating and gathering experimental data as the first step of NICA project. All of its detector systems were commissioned and showed very good performance through last physical runs, in particular, the two detector systems described in detail in this paper demonstrated high efficiency and spatial resolution in track reconstruction. All these along with the global alignment procedure performed give us the possibility to estimate the beam momentum value with a precision of 2% for the working values of the magnetic field integral. The high precision and accuracy of the momentum estimation certifies that this system of detectors is advantageous for the track reconstruction in the sophisticated runs that involve targets and multiple scattering which are of the highest interest for the BM@N experiment.

References

[1] NICA Collaboration (Sissakian, A.N. et al.), J. Phys. **G36** 064069 (2009)

[2] M. Kapishin, Eur. Phys. J. **A 52**, 213–219 (2016)

[3] F. Sauli, Nucl. Instrum. Meth. **A 805**, 2–24 (2016)

[4] E. Lenivenko and V. Palichik, JINR preprint P1-2017-26 [in Rus.], 3–7 (2017)

[5] G.D. Barr et all, CERN/SPSC/90-22/P253

[6] R. Frühwirth, M. Regler, K. Bock, H. Grote, and D. Notz, *Data Analysis Techniques for High-Energy Physics* (Cambridge Monographs on Particle Physics, Nuclear Physics and Cosmology, No. 11, 2nd ed., 2000)

Strange Particle Reconstruction by the Missing Mass Method

Pavel Kisel[1,3,4,⋆], *Ivan* Kisel[1,2,3], *Peter* Senger[3], *Iouri* Vassiliev[3], and *Maksym* Zyzak[3]

[1]*Goethe University Frankfurt, Frankfurt am Main, Germany*
[2]*FIAS Frankfurt Institute for Advanced Studies, Frankfurt am Main, Germany*
[3]*GSI Helmholtz Center for Heavy Ion Research, Darmstadt, Germany*
[4]*JINR Joint Institute for Nuclear Research, Dubna, Russia*

Abstract. The main goal of modern heavy-ion experiments is a comprehensive study of the QCD phase diagram, in a region of Quark-Gluon Plasma (QGP) and possible phase transition to QGP phase.

Strange particles produced in the collision are sensitive probes of the created media. Reconstruction of Σ particles together with other strange particles completes the picture of strangeness production. Σ^+ and Σ^- have all decay modes with at least one neutral daughter, which can not be registered by the CBM detector.

For their identification the missing mass method is proposed: a) tracks of the mother (Σ^-) and the charged daughter (π^-) particles are reconstructed in the tracking system; b) the neutral daughter particle (n) is reconstructed from these tracks; c) a mass constraint is set on the reconstructed neutral daughter; d) the mother particle is constructed of the charged and reconstructed neutral daughter particles and the mass spectrum is obtained, by which the particle can be identified.

The method can be applied for other strange particles too. In total 18 particle decays with neutral daughter are now included into physics analysis.

1 Introduction

The Compressed Baryonic Matter (CBM) experiment is future fixed-target heavy ion experiment, which will focus on a comprehensive study of the phase diagram of strongly interacting matter and a study of the equation of state of the matter at extremely high baryonic densities [1]. Strangeness production is a sensitive probe of the media created in the collision. Being abundant particles, Σ^+ and Σ^- carry out large fraction of produced strange quarks. Reconstruction of Σ together with other strange particles completes the picture of strangeness production and allows to compare yields of Σ and Σ^*. Reconstruction of Σ-particles will open a possibility to investigate H-dybarion objects, if such exist, by the decay channel $\Sigma^- p$, which is expected to be the dominant one.

All strange mesons and hyperons, except Σ-particles, can be either registered directly by the Micro Vertex Detector (MVD) and the Silicon Tracking System (STS) of the CBM experiment and identified by the particle identification (PID) detectors, including Ring Image Cherenkov, Transition Radiation, Muon Chambers, and Time of Flight (ToF) detectors, or have decay modes with all charged particles and, as a result, can be reconstructed through their daughter particles. Σ^+ and Σ^- have all decay

⋆e-mail: P.Kisel@gsi.de

modes with at least one neutral daughter, which can not be registered by the CBM detector at all (decay modes with n or ν) or can be registered with an efficiency of the 10^{-7} level (decay modes with π^0).

The tracking detectors of the CBM experiment will be positioned close enough to the collision point to detect Σ-hyperons: $c\tau_{\Sigma^+} = 2.4$ cm and $c\tau_{\Sigma^-} = 4.4$ cm, the first station of MVD will be placed at about 5 cm and the first station of the STS detector — at 30 cm from the target position downstream the beam. Thus, the CBM experiment provides perfect capabilities for registration of the Σ-hyperons, however, these particles can not reach the ToF detector to be identified. The missing mass method is proposed to be applied for their identification, that expands the physics potential of the CBM experiment.

2 Implementation of the missing mass method

The missing mass method is based on the conservation laws of the energy and momentum.

$$M^2_{\text{miss}} = (P_{\text{mother}} - P_{\text{ch. daughter}})^2, \quad (1)$$

$$P = (E, p_x, p_y, p_z). \quad (2)$$

the P stay for 4-momentum vectors. Using these formulas the 4-vector of the missing particle can be reconstructed, thus, its mass can be determined.

A typical instance is the decay $\Sigma^- \to n\pi^-$. The lifetime $c\tau_{\Sigma^-} = 4.4$ cm, but with the relativistic boost the average length of flight can reach up to 15–20 cm. This is enough for Σ^- to cross 3–4 detector layers producing measurements there. Thus, the Σ^- hyperon can leave the track in the detector system, that can be reconstructed using the Cellular Automaton [2] and the Kalman Filter [3], the main algorithms of the First Level Event Selection (FLES) software package [4]. Since the decay take place in this case insight the detector system, a pion produced in the decay can can also cross several layers and can be detected. Moreover, it can reach the ToF detector and be identified.

The invariant mass of the neutron can be found from two reconstructed Σ^- and π^- particles:

$$M^2_n = (P_\Sigma - P_\pi)^2. \quad (3)$$

The track of Σ^- hyperon can not be identified directly. Therefore the table mass of the Σ^- hyperon is assumed for all particles, which are considered as candidates to be Σ^-. Under such an assumption a 4-momentum vector is constructed. Currently, only primary particles are considered as Σ^- candidates, that are produced directly in the collision or as a result of the resonance decays. Two cases can occur (see Fig. 1):

- if it was a real Σ^-, then the reconstructed particle candidate will result a mass, which will form a peak in the region of the neutron mass 0.939 GeV/c^2;
- otherwise, if the hypothesis is wrong, it will form a continuous combinatorial background and physics background created by other particle decays, for example, $\Xi^- \to \Lambda\pi^-$.

Then the mass constraint is set to the neutron candidate using the table mass of the neutron. In such a way, the whole 4-momentum vector of the candidate is modified, that improves the momentum resolution for true particles. In order to get a fully constructed Σ^- candidate, the obtained neutron candidate is added with the π^- daughter particle candidate. As a result, a fully fitted Σ^- candidate is constructed. In such a way, the reconstructed Σ^- particles contain information not only from the track they leaved in the detector, but from the daughter π^- track as well through the reconstructed decay vertex, that allows to improve the resolution of Σ^- momentum. Like in case of the neutron candidates,

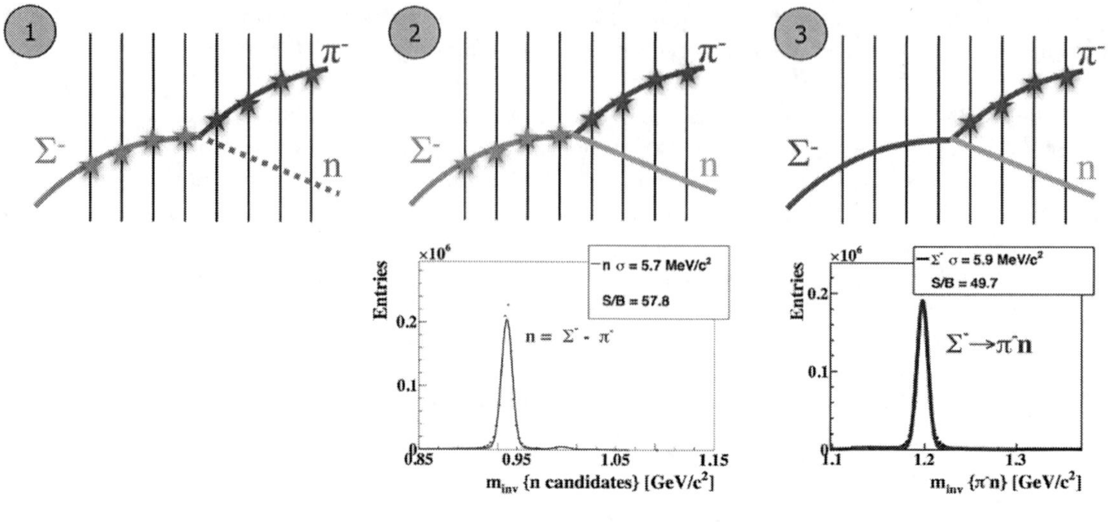

Figure 1. Three stages of the missing mass method implementation in the CBM experiment with illustration of the mass spectra of the reconstructed neutron and Σ^- candidates obtained in 5 millions simulated AuAu central UrQMD events at 10 *A*GeV.

true particles form a peak in the region of 1.197 GeV/c^2, that corresponds to Σ^-, while incorrect candidates form a continuous background. And Σ^- particles are identified by the reconstructed mass.

Thus, the algorithm for reconstruction of Σ-hyperons by the missing mass method consists of three stages (Fig. 1):

1. find tracks of Σ^- and its charged daughter in the tracking system;
2. reconstruct a neutral daughter from the mother and the charged daughter;
3. reconstruct Σ^- mass spectrum from the charged and obtained neutral daughters.

The method has been added to the KF Particle Finder package [5] for short-lived particles reconstruction, that is based on the Kalman filter mathematics.

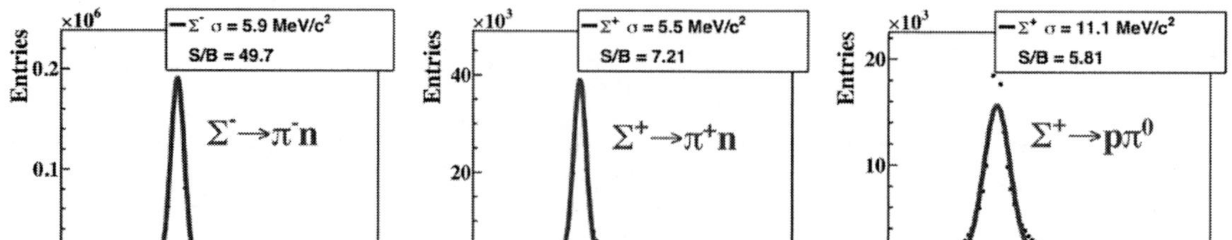

Figure 2. Mass distributions of reconstructed $\Sigma^- \to n\pi^-$, $\Sigma^+ \to n\pi^+$ and $\Sigma^+ \to p\pi^0$ in 5 millions simulated AuAu central UrQMD events at 10 *A*GeV.

The topology of decays allows to require the decay point to be positioned after the last hit of a mother track and before the first hit of a daughter track. A cut on z position of the reconstructed decay point reduces a significant part of the combinatorial background.

The comprehensive study of the Σ decays has shown that a large fraction of the residual background comes from misinterpretation of the tracks, which were reconstructed in two parts (clones). Since the reconstructed momentum of both parts is the same, the resulting values of rapidity and momentum for a neutral daughter candidate are close to 0 and are well separated from the signal region. Thus, such fake candidates can be strongly suppressed by cutting on the low momentum of the neutral candidate.

The method has been developed and tested with 5 millions simulated AuAu central UrQMD [6, 7] events at 10 *A*GeV. Corresponding mass distributions of Σ decays with ToF PID are shown in Fig. 2. Clean topology and accurate mathematics based on the Kalman filter method allow to achieve high signal to background ratios and efficiencies [8].

3 Summary

The missing mass method has been realised as an extension of the KF Particle Finder package, which is widely used for searching and reconstructing short-lived particles in several heavy-ion experiments. For the time being 18 decays with neutral daughter have been added to the package.

Acknowledgement

Work supported by HICforFAIR, FIAS and BMBF.

References

[1] B. Friman *et al.*, *The CBM physics book: Compressed baryonic matter in laboratory experiments*, Lect. Notes Phys. **814** (2011)
[2] V. Akishina, Dissertation thesis, Goethe university, Frankfurt am Main (2017)
[3] S. Gorbunov, U. Kebschull, I. Kisel, V. Lindenstruth, and W.F.J. Müller, Computer Physics Communications, **178**, No. 5, 374–383 (2008)
[4] I. Kisel, I. Kulakov, and M. Zyzak, IEEE Transactions on Nuclear Science **60**, No. 5, 3703 (2013)
[5] M. Zyzak, Dissertation thesis, Goethe university, Frankfurt am Main (2016)
[6] S. A. Bass, *et al.*, Prog. Part. Nucl. Phys. **41**, 225–370 (1998)
[7] M. Bleicher, *et al.*, J. Phys. G: Nucl. Part. Phys. **25**, 1859–1896 (1999)
[8] V. Akishina, I. Kisel, P. Kisel, P. Senger, I. Vassiliev, and M. Zyzak, XXVI international conference on ultrarelativistic heavy-ion collisions "Quark Matter" (Chicago, USA, February 5–11, 2017)

EPJ Web of Conferences **173**, 04010 (2018) https://doi.org/10.1051/epjconf/201817304010
Mathematical Modeling and Computational Physics 2017

Heavy Ion Fragmentation Reactions at Energies of 35–140 MeV in a Combined Transport and Statistical Approach

Tatiana Mikhailova[1,⋆], *Batchuluun* Erdemchimeg[1,2], *Anatoliy* Artukh[1], *Yuriy* Sereda[1], *Massimo* Di Toro[3], and *Hermann H.* Wolter[4]

[1]*Joint Institute for Nuclear Research, Dubna, 141980 Russia*
[2]*Mongolian National University, NRC, P.O.B 46A/305, Ulaanbaatar, Mongolia*
[3]*Lab. Naz. del Sud, INFN, I-95123 Catania, Italy*
[4]*Faculty of Physics, University of Munich, Garching, 85748 Germany*

Abstract. Fragment formation in heavy ion collisions at low to intermediate energy is described by a combined application of transport theory of the Boltzmann type and of a statistical program for the decay of the fragments at the late stage. The transport equations are solved by simulations using the test particle method as a finite element representation of the phase space distribution. The description of experimental data is reasonable overall, but the fragment velocity distributions point to the presence of other mechanisms and the role of fluctuations.

1 Introduction

The phase diagram of matter interacting by the strong force as a function of density and temperature has a rich variety of phases and phase transitions. Heavy ion collision at various energies can be used to explore this. Here we investigate the liquid-gas phase transition in nuclear matter, which is closely connected to the process of fragmentation in heavy ion reactions, where in the final expansion phase a decomposition into intermediate mass fragments and a nucleon gas may take place. This phase transition is of theoretical interest, but fragment formation also has practical applications in the production of, possibly exotic, isotopes, and generally for accelerator-driven technologies (ADS). However, heavy ion collisions are non-equilibrium processes and have to be interpreted by non-equlibrium approaches. Hydrodynamical or statistical methods can be used in parts of the process, but a universal description requires transport theory, which is a microscopic description using effective interactions. The solution of transport theories is a challenging computational problem in many part of physics. We discuss fragment production in a two-stage approach consisting of the formation of primary hot fragments, and the de-excitation to cold fragments, yielding isotope distributions and their energy or velocity spectra. Thus we attempt to understand the mechanism of the fragment production in a microscopic and systematic way in reactions of various systems and incident energies.

⋆e-mail: tmikh@jinr.ru

2 Theoretical and Computational Considerations

The BNV, also often called Boltzmann-Uehling-Uhlenbeck (BUU), transport approach describes the time evolution of the one-body phase space distribution function $f(\vec{r}, \vec{p}, t)$ under the influence of a self-consistent mean field $U([f])$ and a Boltzmann two-body collision term, which includes the effect of Pauli blocking [1].

$$\frac{\partial f_1}{\partial t} + \frac{\vec{p_1}}{m}\nabla_{\vec{r}} f_1 - \nabla_{\vec{r}} U \nabla_{\vec{p_1}} f_1 = \tag{1}$$
$$(\frac{2\pi}{m})^3 \int d\vec{p}_2 d\vec{p}_3 d\vec{p}_4 |\vec{v}_1 - \vec{v}_2| \sigma_{NN}(\Omega_{12}) \delta(\vec{p}_1 + \vec{p}_2 - \vec{p}_3 - \vec{p}_3)(f_3 f_4 \bar{f}_1 \bar{f}_2 - f_1 f_2 \bar{f}_3 \bar{f}_4).$$

Here $f_i = f(\vec{r}, \vec{p}_i, t)$, $\bar{f}_i = (1 - f_i)$, $v_i = p_i/m$ are velocities, $\sigma_{NN}(\Omega)$ is the in-medium NN cross-section, and m is the nucleon mass. The solution of this non-linear integro-differential equation is achieved by simulations using the test particles (TP) method, where the distribution function is represented in terms of finite elements as $f(\vec{r}, \vec{p}; t) = 1/N_{TP} \cdot \sum_{n=1}^{AN_{\mathrm{TP}}} g(\vec{r} - \vec{r}_n(t))\tilde{g}(\vec{p} - \vec{p}_n(t))$, where N_{TP} is the number of TP per nucleon (of the order of 50–100; 50 in this work), $\vec{r}_n$ and $\vec{p}_n$ are the time-dependent coordinates and momenta of the test particles, and g and $\tilde{g}$ are the shape functions in coordinate and momentum space (e.g. δ-functions or Gaussians), respectively. Upon inserting this ansatz into the l.h.s. of eq. (1) Newtonian equations of motion for the test particles follow $m\,\frac{\mathrm{d}^2\vec{r}_n}{\mathrm{d}t^2} = -\vec{\nabla}_{r_n} U$, which are solved by the leapfrog method (here with a time step of 0.5 fm/c). The collision term is simulated stochastically, by performing test particle collisions with a probability depending on the cross section and checking the Pauli principle for the final state. The potential $U([f])$ and the cross-section are either derived from an energy density functional, or are parametrized in order to test them relative to the data (here a Skyrme functional and parametrized cross sections [2]).

Fragments are identified in terms of a coalescence criterion in coordinate and momentum space. The time evolution of the reaction is followed until the freeze-out time when the different fragments are sufficiently isolated so that nuclear forces between them become negligible. Then the intrinsic energy of each fragment, and thus its excitation energy, is calculated consistently using the same energy functional that determines the potential U in the transport equation. These 'primary' fragments are hot fragments, i.e. they are excited by several MeV per nucleon depending on the impact parameter. They are propagated to infinity analytically by the Coulomb interaction between projectile- and target-like fragments. Then the de-excitation is calculated using a statistical decay code, in our case the Statistical Multifragmentation Model (SMM) [3], where we consider evaporation, Fermi-breakup, and multifragmentation. Thus we obtain the isotope distributions and velocity spectra of the final cold fragments, which are compared to experimental data.

3 Results

Here we intend to show to which extent we are able to describe isotope distributions and their velocity spectra using the microscopic transport theory sketched above. We consider reactions, which cover a range of incident energies from 35 to 140 AMeV per nucleon and a variety of projectiles always on the same target, namely ^{181}Ta. In particular, these are the projectiles ^{18}O at 35 AMeV [4,2]; ^{40}Ar at 57 AMeV [5, 6]; ^{40,48}Ca at 140 AMeV [7, 8], and ^{64}Ni at 140 AMeV [7]. Here the first reference refers to the data, the second to our previous analyses. The calculation for the last reaction is presented here.

In Fig. 1 we show the isotope distributions for the element with charge number $Z_{\mathrm{proj}} - 2$ (in panel (f) also for the light element Mg from the ^{64}Ni reaction) for the hot fragments (BNV, circles) and the cold fragments (BNV+SMM, squares) in comparison to the corresponding data (triangles). For

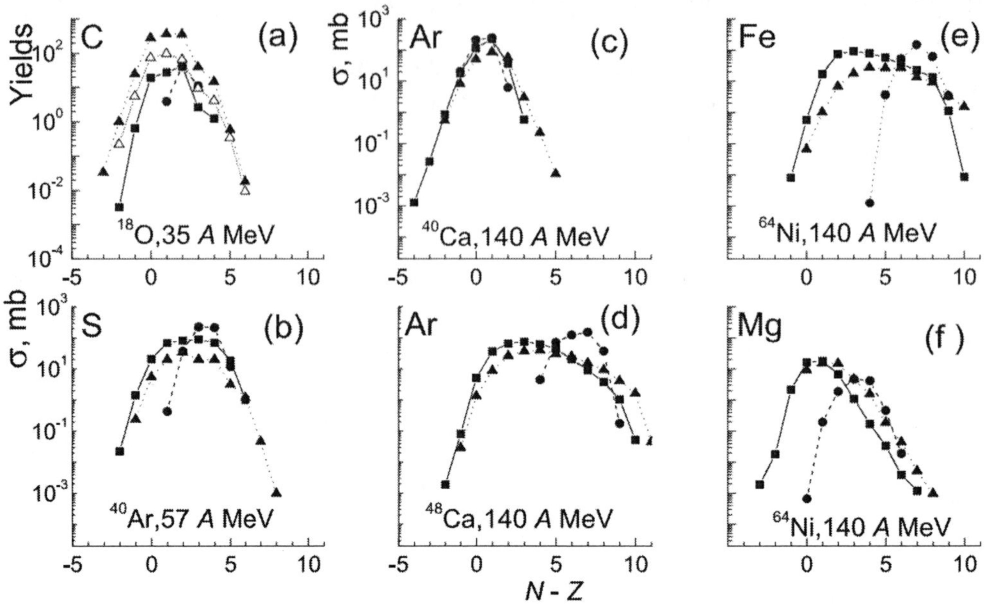

Figure 1. Isotope distributions for different reactions on the target ^{181}Ta. The projectile and the incident energy are identified in each panel at the bottom. For each reaction we give the isotope distribution for one particular element identified in the upper left corner of each panel. The lines are: BNV (solid circle, dashed line), BNV+SMM (solid square, solid line), experiment (solid triangle, dotted line). In panel (a) also the experimental dissipative component is given (see text, open triangle, dotted line)

^{18}O we show an extra dissipative curve explained below. The hot fragments have a rather narrow distribution, which is centered around the $N - Z$ of the projectile. In the de-excitation of the fragments predominantly neutrons are evaporated, and the distribution of the cold fragments is shifted towards the line of stability. The agreement with experiment is reasonable considering that this is a microscopic calculation without fitted parameters.

In Fig. 2 we give, for the same reactions and the same elements as in Fig. 1, the velocity spectra as a function of the reduced velocity $v_{\text{red}} = v/v_{\text{proj}}$, summed over the isotopes of the given element. The experimental distributions peak at v_{red}=1, while the calculations generally peak below 1, as particularly evident for the ^{18}O reaction in panel (a). We assume that the experiment contains a component with almost no loss of velocity, as could be expected in a direct abrasion-like process or in multi-nucleon transfer, which a semi-classical transport theory cannot describe. To separate this contribution from the experiment we subtract a Gaussian which is fitted to the distribution for $v_{\text{red}}>1$. The calculations are seen to be in better agreement with this dissipative curve. The direct component decreases with higher energy as seen in the other panels, where we have not subtracted it, but where it is still somewhat apparent. The width of the BNV distributions, i.e. the velocity fluctuations, is considerably narrower than the data. By including the de-excitation it widens and is now in better agreement especially on the the low velocity side, though still usually too narrow. This is particularly evident for the C spectrum for the ^{18}O reaction in panel (a), which has a large low velocity tail. Thus the fluctuations in the transport calculations are not large enough. It is known that the Boltzmann theory does not contain realistic physical fluctuations, which should be described by an explicit fluctuation term in the Boltzmann-Langevin theory.

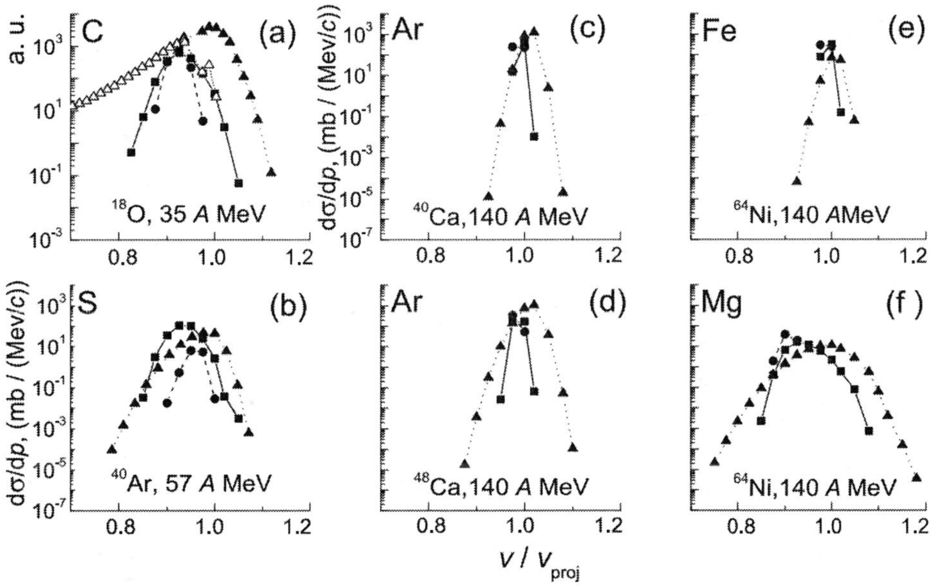

Figure 2. Velocity spectra of the same reactions for the same elements as in Fig. 1 as a function of the reduced velocity. The signatures are also the same as in Fig. 1.

4 Conclusion

We present a systematic study of heavy ion fragmentation reactions using microscopic transport theory, combined with a statistical code for the de-excitation of the primary hot fragments. Generally this can describe the isotope distributions fairly well. The velocity spectra show finer details of the reaction mechanism, but are too narrow, probably due to too small fluctuations. However, the approach appears to be useful to estimate fragment production in heavy ion collisions.

Acknowledgements

The authors thank A. Botvina for providing us the computer code SMM and for usefull advice. This work was supported by the Heisenberg-Landau Program. TM is supported by the Russian Foundation for Basic Research under Grant N 17-01-00661-a, and HHW by the Universe Cluster of Excellence of the DFG Germany.

References

[1] G. F. Bertsch, and S. Das Gupta, Phys. Rep. **160**, 189 (1988)
[2] T. I. Mikhailova et al., Phys. Part. Nucl. Lett. **12**, 409 (2015)
[3] J. P. Bondorf, A. Botvina, et al. Phys. Rep. **257**, 133 (1995); A. Ergun et al. Phys. Rev. C **92**, 014610 (2015)
[4] A. G. Artukh et al., Nucl.Phys. A **701**, 96c (2002)
[5] X. H. Zhang et al., Phys. Rev. C **85**, 024621 (2012)
[6] T. I. Mikhailova et al., Bull. of RAS: Physics, **78**, 1131 (2014)
[7] M. Mocko et al., Phys. Rev. C **78**, 024612 (2008)
[8] T. I. Mikhailova et al., Phys. Atomic Nuclei, **79**, 604 (2016)

Light Ion Beams for Energy Production in ADS

Mihaela Paraipan[1,2,★], *Anton A.* Baldin[1,3], *Elina G.* Baldina[1,3], and *Serguey I.* Tyutyunikov[1]

[1] *Joint Institute for Nuclear Research, Dubna, Russia*
[2] *Institute of Space Science, Bucharest-Magurele, Romania*
[3] *Institute for Advanced Studies "OMEGA", Dubna, Russia*

Abstract. A comparative study of the energy efficiency of proton beams with an energy from 0.5 GeV to 4 GeV and light ion beams (^{7}Li, ^{9}Be, ^{11}B, and ^{12}C) with energies from 0.25 AGeV to 1 AGeV in natural and enriched quasi-infinite U target is presented. The numerical results on the particle transport and interaction are obtained using the code Geant4. The following target optimization issues are addressed: the beam window dimensions, and the possibility to use a core from low Z materials. The best solution for ADS from the point of view of the energy gain and miniaturization is obtained for ^{7}Li or ^{9}Be beam with an energy of 0.3–0.4 AGeV and a target with Be core.

1 Introduction

The use of proton beams for energy production in accelerator-driven systems (ADS) was extensively investigated and the general opinion is that optimal for ADS is a proton beam with energy $\sim$ 1 GeV [1]. The possibility to use heavy ion beams was less analyzed and the conclusions of authors are contradictory [2]–[3]. Our results presented in [4] show that one can get a higher efficiency by accelerating heavier ions. In the present work we analyse in more detail the use of light ion beams at energies below 1 AGeV.

2 The energetic efficiency of the beam

The beam efficiency is described by the energy gain factor G calculated as the ratio of the produced electrical power P_{prod} to the power spent to accelerate the beam P_{spent}:

$$G = P_{\text{prod}}/P_{\text{spent}} \,. \tag{1}$$

The produced electrical power is:

$$P_{\text{prod}} = \eta_{\text{el}} \cdot E_{\text{dep}} \cdot I_{\text{beam}} \,, \tag{2}$$

where η_{el} represents the conversion coefficient from thermal to electrical power, E_{dep} is the energy deposited in the target, and I_{beam} is the beam intensity. The power spent for the beam acceleration P_{spent} is analysed as the sum of two terms. The first term P_{beam} represents the power transmitted to the

★e-mail: mihaela_paraipan@yahoo.com

beam. The second term P_{acc} contains the energy necessary to ensure the functioning of the accelerator (magnets, cooling system, etc.)

$$P_{\text{spent}} = P_{\text{beam}} + P_{\text{acc}} = A \cdot E \cdot I_{\text{beam}} + P_{\text{acc}} \,. \tag{3}$$

The term P_{acc} depends on the type of the accelerator used. In a synchrotron this term scales as:

$$P_{\text{acc}} = \frac{A \cdot Z_0 \cdot p}{A_0 \cdot Z \cdot p_0} P_{\text{acc}_0} \,. \tag{4}$$

In a linac P_{acc} scales as:

$$P_{\text{acc}} = \frac{A \cdot Z_0 \cdot E}{A_0 \cdot Z \cdot E_0} P_{\text{acc}_0} \tag{5}$$

and in a cyclotron:

$$P_{\text{acc}} = \left(\frac{A \cdot Z_0 \cdot p}{A_0 \cdot Z \cdot p_0} \right)^2 P_{\text{acc}_0} \,. \tag{6}$$

In formulas (3)–(6) A, Z, p and E are the mass number, the atomic number, the momentum per nucleon and the energy per nucleon of the analysed particle. The symbols with the subscript 0 represent the corresponding values for the reference particle. These formulas allow to calculate the power spent for any beam if we know the accelerator efficiency for a reference beam.

3 The energy produced in the target

The energy released in the target was calculated with the code Geant4. The hadronic inelastic interaction of protons, neutrons and pions was modelled with the Bertini cascade model. For neutrons with energies below 20 MeV high precision neutron models based on a detailed implementation of the data from ENDF (Evaluated Nuclear Data File) were used. The inelastic interaction of ions was modelled with the binary cascade model.

Cylindrical targets with dimensions large enough in order to ensure quasi-infinite conditions (the saturation of the energy deposited in both radial and longitudinal directions) were used. In the case of ^{nat}U target the radius is 60 cm and the length 160 cm. The ^{nat}U target has a low critical coefficient (0.48) and needs very high beam intensities for efficient power production. An enriched target with critical coefficient 0.96 ensures a safe exploitation and needs beam intensities 50 times lower. In such target (5.7 % ^{235}U) the quasi-infinite conditions are reached at higher dimensions. In this case a cylinder with radius 100 cm and length 200 cm was used. The influence of the the beam window length on the energy released was investigated. The presence of beam window increases the energy deposited by including the effect from backscattered particles. The simulation shows that the optimum length of the beam window is ~ 40 cm.

For ion beams at low energies the possibility to use low Z materials in the central part of the target was considered. Two phenomena with opposite effects concur in this case. On one side in a low Z material the energy loss through ionization decreases and the probability of inelastic interaction at higher energy per nucleon increases, generating higher multiplicity and a harder spectrum of the secondaries. On the other side a low mass number of the target diminishes the cross section. The overall result depends on which of these effects dominates. The simulation predicts that a core from very low Z materials (Li, Be, C) increases the energy released for light ions at low energies with a factor of 1.3–2. The maximum effect is obtained with a core of Be.

4 Results and discussion

The first set of simulations was performed in a ^{nat}U target irradiated with protons and ions with masses until ^{40}Ca and energies from 0.3 AGeV to 10 AGeV. The dependence of the energy released on the projectile mass and energy per nucleon is presented in figure 1a. The relative beam efficiency for different types of accelerators is also given (figure 1b-d). As reference beam we use protons with energy 1 GeV for which a value of 0.1 for the accelerator efficiency was assumed. When accelerated in a synchrotron or linac ion beams starting with ^{7}Li demonstrate higher energetic efficiency than protons.The maximum efficiency for ions with higher masses is reached at higher energies. That does not represent a problem when the beam is accelerated in a synchrotron, but such energies are difficult to reach in a linear accelerator, needing accelerator lengths of a few km. Unfortunately, the beam intensities achieved in a synchrotron are by orders of magnitude lower than those necessary for ADS. Linear accelerators look more promising for obtaining the required beam intensity.

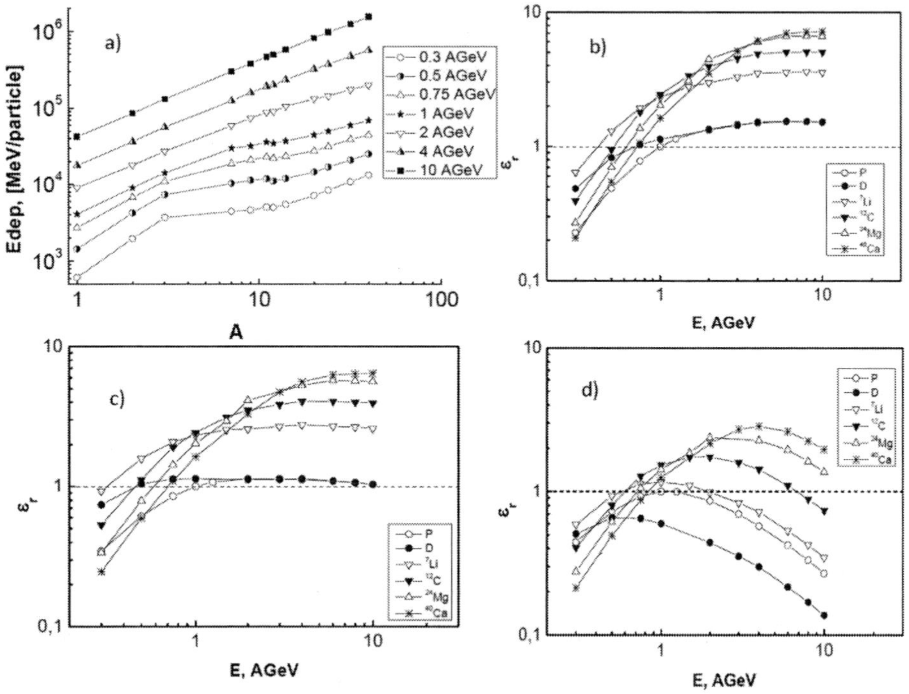

Figure 1. The energy released as function on the projectile type and energy (a), the relative efficiency for beams accelerated in a synchrotron (b), linear accelerator (c), and cyclotron (d)

For this reason in the second set of simulations we realized in enriched U target irradiated with light ions from ^{7}Li to ^{12}C at low energies (0.3 AGeV–1 AGeV), accelerated in a linac. We use the data from the European Spallation Source (ESS) project for the reference particle (protons with final energy 2.5 GeV, beam intensity I_0 of $1.25 \cdot 10^{16}$ p/s, and a value of 0.18 for the accelerator efficiency). The total energy deposited per projectile, the energy gain and the net electrical power obtained with different beams at the same intensity I_0 are presented in table 1 for a target entirely from enriched U and with core of Be.

In target with Be core beams of ^{7}Li or ^{9}Be are equivalent from the point of view of the energy produced with a beam of 1.5 GeV protons (at maximum proton beam efficiency) but with higher energy gain and needing accelerator with lower dimensions.

Table 1. The energy released E_{dep}, the gain factor G and the net electrical power P_{el} produced in enriched U target irradiated with protons and light ions

		Fuel			Core Be		
P	E_{beam} AGeV	E_{dep} MeV	G	P_{el} MW	E_{dep} MeV	G	P_{el} MW
proton	1	1.09e5	7.3	75.6	1.25e5	8.64	88.1
	1.5	1.85e5	8.23	130.2	2.18e5	9.68	156.3
	2	2.49e5	8.32	175.6	2.94e5	9.8	211.2
^{7}Li	0.3	8.49e4	6.06	56.7	1.79e5	12.8	132.1
	0.4	1.54e5	8.19	107.3	3.21e5	17.2	242.1
	0.5	2.39e5	10.3	172.9	4.65e5	19.9	353.1
	0.75	4.99e5	14.3	371.4	7.7e5	22	588.1
	1	7.19e5	15.4	537.7	1.07e6	22.8	814.7
^{9}Be	0.4	1.5e5	7.39	120	3.34e5	16.5	267
	0.5	2.5e5	9.86	200	5.3e5	20.9	424
^{11}B	0.5	2.26e5	8.17	158	5.21e5	18.9	417
	0.75	5.22e5	12.7	385.1	1.08e6	26.2	863
^{12}C	0.3	5.5e4	3.3	30.8	1.22e5	7.4	84.7
	0.4	1.08e5	4.9	68.6	2.62e5	11.9	191.9
	0.5	1.93e5	7	32.5	4.68e5	17	352.1
	0.75	4.8e5	11.6	351.3	1.03e6	25	793.4
	1	8.07e5	14.7	601.7	1.5e6	27.3	1155.2

5 Conclusions

The optimal energy of proton beam depends on the accelerator type and is ~ 3 GeV in synchrotron, 1.5 GeV in linac, and 1 GeV in cyclotron. The optimal energy for ion beams depends on the type of the ion (1.5–2 AGeV for ^{7}Li, 2 AGeV for ^{12}C, 4 AGeV for ^{40}Ca) and the efficiency is significantly higher (more than 2 times) than for protons.

Light ions ^{7}Li and ^{9}Be with energy 0.3–0.35 AGeV realize the same energy release as a beam of proton 1.5 GeV. This allows one to obtain the same electrical power with lower energy consumption and an accelerator with ~ 2 times lower dimensions. The acceleration of ^{11}B, and ^{12}C at 0.7–0.75 AGeV needs an accelerator with the same dimensions as for proton beam 1.5 GeV but produces a net electrical power about 5 times higher.

The best solution from the point of view of the energy gain and miniaturization is the ^{7}Li beam with an energy of 0.3 AGeV and a target with core of Be.

References

[1] C. Rubbia et al., preprint CERN/AT/95-44 (ET) (1995)

[2] D. G. Koshkarev, N. M. Sobolevski, and A. V. Barhudarian, Atomic Energy **105** (3), 137–139 (2008)

[3] S. R. Hashemi-Nezhad, W. Westmeier, M. Zamani-Valasiadou, B. Thomauske, and R. Brandt, Ann. Nucl. Energy **38**, 1144–1155 (2011)

[4] A. A. Baldin, A. I. Berlev, M. Paraipan, and S. I. Tyutyunikov, Phys. Part. Nucl. Let. **14** (1), 113–119 (2017)

Stability Analysis of the IBR-2M Pulsed Reactor in Automatic Regulated Regime at Various Levels of Average Power

Yuriy N. Pepelyshev[1], *Alexander K.* Popov[1], and *Sumkhuu* Davaasuren[1,2,*]

[1] *Frank Laboratory of Neutron Physics, Joint Institute for Nuclear Research, Dubna, Russia*
[2] *Institute of Physics and Technology, Mongolian Academy of Sciences, Ulaanbaatar, Mongolia*

Abstract. The IBR-2M pulse reactor is characterized by a high level of reactivity fluctuations and consequently by a high level of fluctuations in the regulated parameter as well as in the amplitude of the power pulses. Fluctuations of the regulated parameter relative to the average level in the standard stabilization regime usually fall within the range of 20% with certain peaks up to 40%, which is close to the limits of the emergency protection operation (50%). Under these conditions the automatic regulator may receive contradictory requirements. In this paper, the results of stability analysis of the IBR-2M reactor in the automatic regulating regime at various levels of average power are presented. It is shown that stability margins of the amplitude and the phase are enough, especially for the IBR-2M reactor operation with automatic regulator.

1 Introduction

In contrast to stationary reactors characterized by a constant neutron flux, the IBR-2M reactor generates periodic neutron pulses. The IBR-2M reactor generates short neutron pulses (200 μs at half width) with period 0.2 s and amplitude of 1830 MW. In the IBR-2M reactivity pulses are produced by the reactivity modulators rotating near the core. When the two reflectors pass the core simultaneously, a reactivity pulse develops and for a short time (450 μs) the reactor stops being in a super-critical state with prompt neutrons. As the reflectors move away from the reactor core, the reactor becomes deeply sub-critical. The controlled parameter of the reactor is the energy deviation of the power pulse E_p from its basic value E_p^0 in relative units ($\Delta e_p = (E_p - E_p^0)/E_p^0$) [1]. The basic value is the average amplitude corresponding to the stationary regime, i.e., the regime in which the reactor average power is generally constant [2]–[4].

The IBR-2M reactor is equipped with a power automatic regulator (AR). The principal aim of the AR is to automatically bring the reactor to the average power preset level and to stabilize the reactor power at the preset level. The structure and parameters of the AR have essential influence on the power transition processes. At the stage of IBR-2M design, an algorithm of the AR reactivity forming was derived. The algorithm provided a statistically efficient power level, i.e., the attainment of a minimum of the mean-square energy deviation for the future power pulse based on information obtained from the previous pulses [5]. It was assumed when deriving the AR algorithm that random reactivity excitations with a normal distribution law affects the reactor. It was as well assumed that the

*e-mail: dsumkhuu@jinr.ru

AR estimates the reactivity values corresponding to previous power pulses with a certain statistical measurement error which is also characterized by a normal distribution law. In addition, a concept of the information "aging" rate in the AR was introduced. It was shown as a result that the AR must represent an integrating element to the input of which energy deviation of the current power pulse from its base value is applied. The conventional AR of the IBR-2M, besides the regulation rod with the step motor and its control unit, includes also the smoothing unit. The output signal of the smoothing unit arrives to the control unit of the step motor and is shaped by smoothing power pulse energy deviation relatively to the base value. The smoothing unit is a time-lag element and the step motor with the regulation rod and control unit form the integrating element [6].

In this paper, results of a study of the influence of the AR parameters values (Δ = 0.05, 0.1, 0.15, 0.2 is the parameter of the step motor velocity AR and $q = 1, 2, \ldots, 32$ is the smoothing coefficient) on the stability of the IBR-2M reactor in automatic regime of operation are reported.

2 Stability analysis of the IBR-2M reactor operation with the automatic regulator

The model of the IBR-2M reactor is designed to simulate the transient processes, calculate the frequency characteristics of the reactor for the system as a whole and the constituent components, and assess the stability of the reactor. If a reactor operating at power undergoes a small perturbation in reactivity, however, it is possible to linearize the equations for the point reactor [7]. The IBR-2M reactor with automatic regulator is represented as a linearized closed single-loop system with negative feedback (Fig. 1). The reactivity deviation Δr_n is taken as the input signal of the closed system, and the relative deviation Δe_{pn} of the energy of the power pulses from the basic value is taken as the output signal. The main dynamic properties of a pulsed reactor can be determined if its amplitude-phase-frequency characteristics (APFCs), or Nyquist plots, are known. The APFC is a plot of the pulse frequency transfer function of a reactor $W(j\omega) = U(\omega) + jV(\omega)$ on the complex plane.

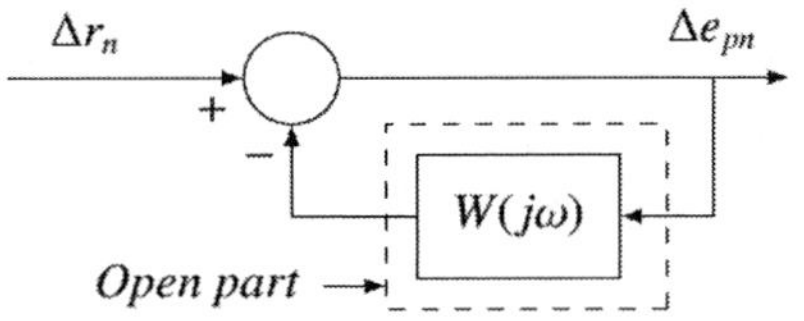

Figure 1. A linearized model of IBR-2M in the self-regulation regime, represented as a single-loop closed system

The pulse transfer function $W(j\omega)$ is given by

$$W(j\omega) = \frac{W_A(j\omega)}{1 + \frac{1-k_b}{1-k_b\sum_{i=1}^{6} W_{si}(j\omega)}\left[E^0 \sum_{j=1}^{3} W_{Tj}(j\omega) - \sum_{i=1}^{6} W_{si}(j\omega)\right]},$$

where k_b and E^0 are a constant parameters [1]. $W_A(j\omega)$, $W_{si}(j\omega)$ and $W_{Tj}(j\omega)$ denote the transfer functions of the automatic regulator, the i-th delayed neutrons group and the j-th power feedback component, respectively.

Here, $W_A(j\omega)$ denotes the Δ and q dependent transfer function of the AR which is given by

$$W_A(j\omega) = \frac{v_\Delta m_A T_p}{\Delta} \frac{\exp(-j\omega)}{[q-(q-1)\exp(-j\omega)][1-\exp(-j\omega)]},$$

where $v_\Delta m_A = 0.08 \cdot 1.54 \cdot 10^{-4}\ \mathrm{s}^{-1}$ is a constant factor, $T_p = 0.2$ s is the period of the power pulse, Δ is the parameter of the automatic regulator speed (0.05, 0.1, 0.15 or 0.2) and q is the smoothing coefficient of the automatic regulator $(1, 2, \ldots, 32)$.

Further, $W_{si}(j\omega)$ denotes the transfer function of the i-th delayed neutrons group

$$W_{si}(j\omega) = \frac{\mu_i\lambda_i}{\sum_{i=1}^{6}\mu_i\lambda_i\frac{1}{\exp(\lambda_i T_p)}}\frac{\exp(-j\omega)}{\exp(\lambda_i T_p) - \exp(-j\omega)},$$

where, μ_i and λ_i denote the delayed neutron fraction and the decay constant of the source of i-th delayed neutrons group, respectively [8].

Finally, $W_{Tj}(j\omega)$ denotes the transfer function of the j-th component of the power feedback:

$$W_{Tj}(j\omega) = \frac{k_{Tj}}{T_{Tj}}\frac{\exp(-j\omega)}{\exp(T_p/T_{Tj}) - \exp(-j\omega)}, \tag{1}$$

where k_{Tj} is the transfer coefficient and T_{Tj} is the time constant of the j-th components of the power feedback. The parameters of power feedback, which are introduced into the model of the reactor dynamics, were determined by analyzing experimental transients of power using a special search algorithm [9].

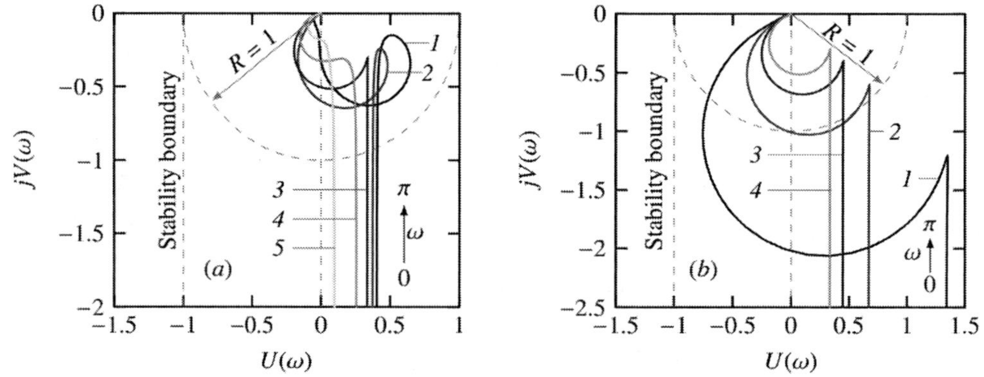

Figure 2. The amplitude-phase frequency characteristic of the open part of the IBR-2M reactor in the automatic regulated regime at 2 MW of average power. *a*) $\Delta = 0.2$ and $q = 1$ (*1*), 4(*2*), 8(*3*), 16(*4*), 32(*5*), *b*) $q = 8$ and $\Delta = 0.05$(*1*), 0.1(*2*), 0.15(*3*), 0.2(*4*).

The oscillation frequency in the pulse transfer function $W(j\omega)$ varies from 0 to $0.5f_p$, where $f_p = 1/T_p$ is the power pulse frequency, and the dimensionless frequency $\omega = 2\pi f_p T_p$ varies from 0 to π. The length of the vector from the origin of coordinates to the point of frequency ω ($|W(j\omega)|$) is the amplitude–frequency characteristic (gain), and the angle of rotation of the vector from the real positive semi axis $\varphi(\omega) = \arctan[V(\omega)/U(\omega)]$ is the phase–frequency characteristic of the reactor. For example, for the pulsed reactor APFC (this characteristic will be considered in more detail below, Fig. 2) the length of the vector drawn from the origin to any point on the curve corresponding to a frequency ω is equal to the ratio between the amplitude of the sine power pulse energy oscillation in relative units of $\Delta e_p = (E_p - E_p^0)/E_p^0$ (output value) and the amplitude of the sine reactivity oscillation (input value) expressed in terms of parts of β_p. Here E_p and E_p^0 are the energy of the current power pulse and its basic value. The angle of rotation of the vector with respect to the real positive semi axis is a phase shift between the power and reactivity oscillations.

The assessment of the reactor stability using the Nyquist stability criterion is done as follows. The reactor is represented as a closed one-loop system. While the open system is stable or neutral, the closed system is stable provided that the APFC of the open system does not include the point with the coordinates $(-1, 0)$. The amplitude (or gain) stability margin (a) shows how many times this coefficient must be increased to bring the system from the stable state to the stability boundary $(-1, 0)$. The phase margin (φ) is defined as the angle of rotation of the unit-length vector at which its end turns out to be at the point with the coordinates $(-1, 0)$ and the APFC of the open system goes through this point, i.e., the closed system reaches the stability boundary. In the Fig. 2 the phase margin at the different values of the AR parameters is defined as the dashed unit circle. The phase margin reflects possible influence of the additional neglected inertia of the system.

3 Conclusions

An analysis was reported of the amplitude and phase frequency characteristics of the open part of the closed loop system which corresponds to the IBR-2M reactor operation in the automatic regulated regime. Estimated stability margins of the amplitude (a) and phase (φ) of the reactor IBR-2M at different level of average power have been found. In Fig. 3 the dependence of the stability margins on the parameters of AR (Δ – parameter of the step motor velocity AR and q – smoothing coefficient) at the average power 2 MW are shown. It is shown that the stability margins of the amplitude and the phase are well within the safety regime, especially for the automatic regulated regime, characterized by $\Delta = 0.05$ and $4 < q < 20$ at 2 MW of average power.

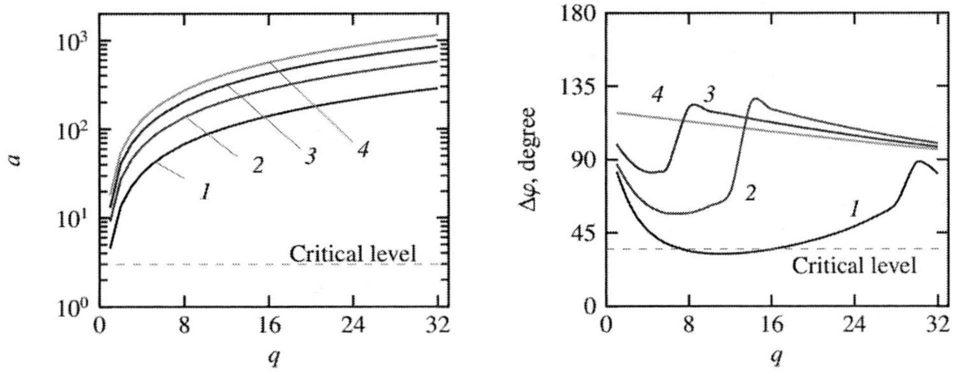

Figure 3. Dependence of the stability margins of the amplitude (*left*) and the phase (*right*) on the parameters values (q and Δ) of the automatic regulator of the IBR-2M reactor at the average power of 2 MW. The labels *1*, *2*, *3*, *4* point to curves obtained for $\Delta = 0.05$, $\Delta = 0.1$, $\Delta = 0.15$, and $\Delta = 0.2$, respectively.

References

[1] Yu.N. Pepelyshev, A.K. Popov, and D. Sumkhuu, Annals of Nuclear Energy **85**, 488–493 (2015)
[2] Yu.N. Pepelyshev, A.K. Popov, D. Sumkhuu, and D. Sangaa, Physics of Particles and Nuclei Letters **12** (3), 435–438 (2015)
[3] E.A. Bondarchenko, Yu.N. Pepelyshev, and A.K. Popov, Fiz. Elem. Chastits At. Yadra **35** (4), 928–983 (2004)
[4] A.K. Popov, *Fundamentals of Nuclear Reactor Control* (in Russian, Mosk. Gos. Univ., Moscow, 2012)
[5] A.K. Popov, Atomic Energy **31** (3), p. 269 (1971)
[6] E.A. Bondarchenko, Yu.N. Pepelyshev, and A.K. Popov, Annals of Nuclear Energy **28** (1), 63–78 (2001)
[7] G.J. Bell and S. Glasstone, *Nuclear reactor theory* (Van Nostrand Reinhold Company, 1970)
[8] G.R. Keepin, *Physics of Nuclear Kinetics* (Addison-Wesley, 1965)
[9] Yu.N. Pepelyshev, A.K. Popov, and D. Sumkhuu, Atomic Energy **122** (2), 63–67 (2017)

Study of the Possibility of Reducing the Slow Fluctuations of the Reactivity and Thermal Power of the IBR-2M Reactor

Yuri Pepelyshev[1,⋆] and *Tsolmon* Tsogtsaikhan[1,2,⋆⋆]

[1] *Joint Institute for Nuclear Research, Dubna, Russia*
[2] *Institute of Physics and Technology, Ulaanbaatar, Mongolia*

Abstract. This paper presents an artificial neural network method for long-term prediction of the thermal dynamic parameters of the IBR-2M reactor. Attention is focused mainly on the prediction of the temperature and sodium flow at the entry into the core as well as the thermal power. It is shown that the prediction makes it possible to reduce by a factor of 3 the influence of slow fluctuations of reactivity on the power and thereby reduce the operational requirements for the automatic power stabilization system.

1 Introduction

The random variations of the thermodynamic parameters of the primary cooling circuit of the sodium cooling system of the core in the IBR-2M reactor in normal operation cause large fluctuations in reactivity, which lead to changes in pulse energy and power, respectively. The automatic stabilization system compensates for these changes and maintains the power at a prescribed level. The concept of power stabilization is based on the knowledge of the energy of the running pulse and of some number of preceding pulses. The value of the previous information decreases in time. It is clear that the knowledge of the energy of pulses subsequent to the current power pulse or subsequent changes in reactivity can improve the stabilization and facilitate at high-load an operating regime securing the stabilization of the system itself. The aim of the present work is to investigate the possibility of predicting fluctuations of the thermodynamic parameters of the cooling system of the IBR-2M reactor core and the associated reactivity fluctuations. The possibility of reducing the reactivity fluctuations according to the prediction is analyzed.

2 Experimental Data and Method

The IBR-2M reactor is located at the Joint Institute for Nuclear Research (Dubna, Russia) and operates with the design power of 2 MW. The IBR-2M core capacity is 69 fuel elements, which are sleeve-like PuO_2 pellets. The coolant is liquid sodium, pumped through the emergency protection blocks by two induction pumps. There are two blades rotating with different speeds past one of the core faces. They are the main moving reflector and the auxiliary moving reflector of the reactivity modulator. The reactivity level is adjusted by the control and protection system elements, which are

⋆e-mail: pepel@nf.jinr.ru
⋆⋆e-mail: tsolmonuc@gmail.com

movable tungsten blocks in an array of fixed steel reflectors. The reactor core is surrounded by water moderators enabling the creation of thermal neutrons for users.

IBR-2M operates in cycles: two weeks at 2 MW followed by shutdown for one week (power close to zero). The cycles are repeated periodically. The initial time series comprises an array of successive values of the sodium flow G_Σ and the sodium temperature T_{in} at the inlet into the core, the thermal power W_T, and the ratio W_T/G_Σ proportional to the heating. The sodium temperature at the core inlet and outlet T_{out}, the total sodium flow $G_\Sigma = G_A + G_B$ in the circuits A (G_A) and B (G_B) in the primary cooling circuit were measured during several reactor cycles in 2014. The data survey period was 0.1 sec for all parameters. The length of the time series in the measurements reached $\sim 6\times10^6$ values. The thermal power was estimated as $W_T = G_\Sigma(\Delta T - 0.252)/3.15$, where $\Delta T = (T_{out} - T_{in})$ is the heating of the sodium in the reactor core. This sample of the parameters values are the basis for prediction. A neural net model for predicting the course of the time series was chosen as the prediction system [1–3].

3 Results

The analysis showed that a slow change (at the level of daily fluctuations) in the main parameters of the primary cooling circuit of the cooling system of the IBR-2M core can be predicted with satisfactory accuracy for reactor operation up to 5–7 days (Fig. 1).

The changes in reactivity $\rho_T(t)$ and sodium flow $\rho_{G_\Sigma}(t)$ accompanying a deviation of the temperature from the average values G_0 and T_0, respectively, are characterized in terms of the ratios $\Delta K/\Delta G_\Sigma$ and $\Delta K/\Delta T$ which denote the flow and temperature coefficients of reactivity, respectively [4–6]. The predicted reactivity $\Delta\rho_\Sigma(t)$ is a previously known process, which, in principle, can be excluded from the overall change in reactivity (Fig. 2). The method for eliminating it may be different, for example, the introduction of a reactivity $\Delta\rho_\Sigma(t)$ of the opposite sign using an addition control rod. In this case, this is not fundamental because only the possibility of reactivity compensation and not the method of its implementation is under study.

4 Conclusion

We presented predictions, within five percent accuracy, obtained by means of nonlinear auto regression neural net software concerning changes in the thermodynamic parameters of the IBR-2M primary cooling circuit (liquid-sodium). The input of the method uses values of measured corresponding quantities at previous primary cycles.

On the basis of the predicted parameters we have determined the needed compensating reactivity. Up to now this was hardwired in the automatic power stabilization system. Although we predict only the slow-component of the compensation, the residual (that the stabilization system is left to determine on its own) is approximately three times smaller.

The prediction eliminates stabilization system wear out, thereby increasing the reliability and safety of the reactor. Our future work will be directed to the study of the reliability of the software under (unexpected) shocks.

References

[1] G. Ososkov, Yu. Pepelyshev, and Ts. Tsogtsaikhan, European Physical Journal **108**, 02036 (2016)

[2] Yu. Pepelyshev, Ts. Tsogtsaikhan, and G. Ososkov, Physics of Particles and Nuclei Letters **13**, 704–707 (2016)

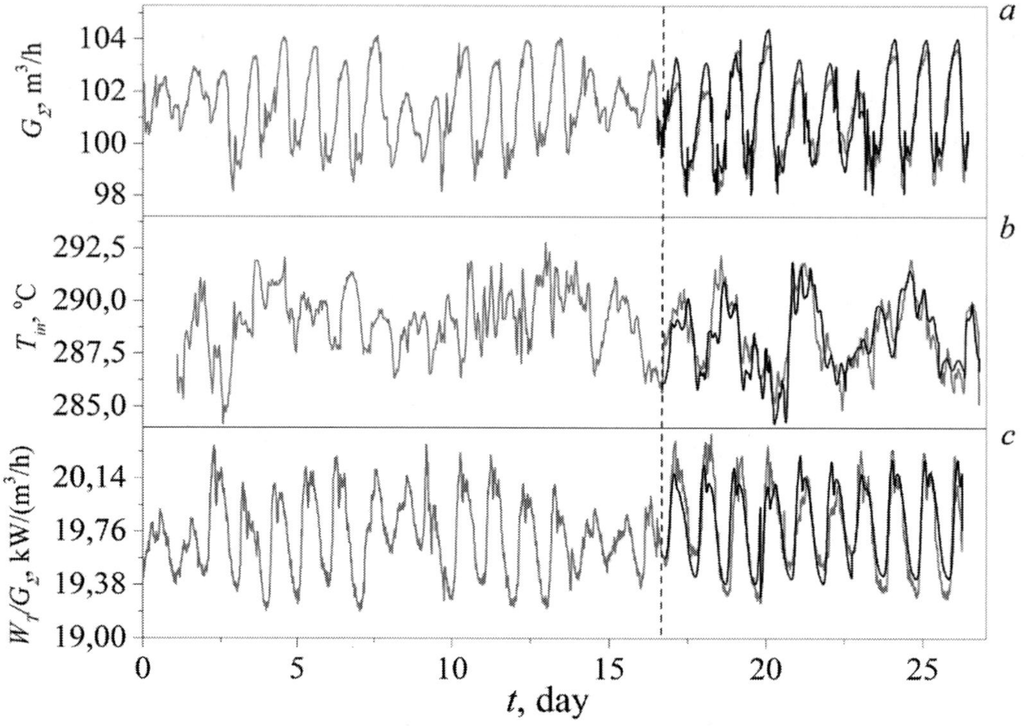

Figure 1. Change in the sodium flow G_Σ (a), sodium temperature at the inlet into the core T_{in} (b), W_T/G_Σ proportional to the heating (c) versus the operating time of the reactor (t). The experimental data (grey lines) extend over the whole variation range of the operating time. Predictions (black lines) start at the broken separation line and are in qualitative agreement with the corresponding measured data set at a given time

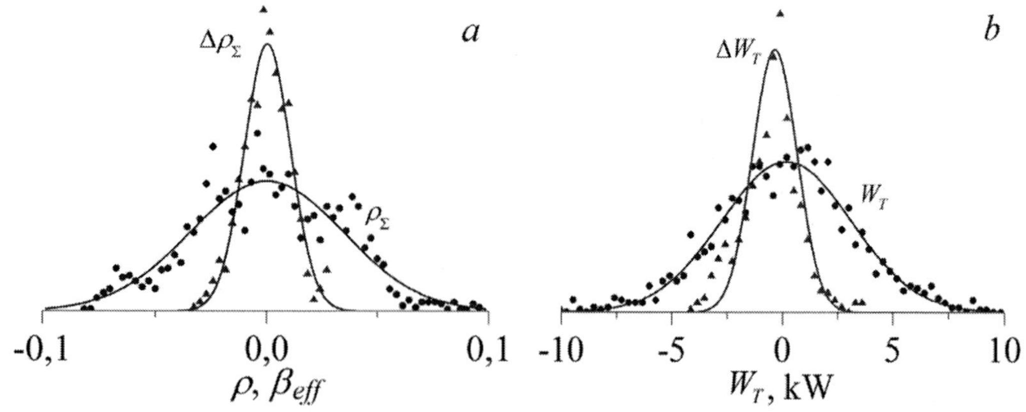

Figure 2. Distribution of the reactivity fluctuations before (ρ_Σ) and after ($\Delta\rho_\Sigma$) the prediction (a) and thermal power before (W_T) and after (ΔW_T) the prediction (b)

[3] Yu. Pepelyshev and Ts. Tsogtsaikhan, Atomic Energy **121**, 173–178 (2017)
[4] Yu. Pepelyshev and Ts. Tsogtsaikhan, Preprint JINR **R13-2014-61**, 1–16 (2014)
[5] V. Anan'ev, A. Vinogradov, A. Dolgikh, et al., Preprint JINR **R13-2012-41**, 1–21 (2012)
[6] V. Anan'ev, A. Vinogradov, A. Dolgikh, et al., Preprint JINR **R13-2012-42**, 1–24 (2012)

EPJ Web of Conferences **173**, 04014 (2018) https://doi.org/10.1051/epjconf/201817304014
Mathematical Modeling and Computational Physics 2017

Radioactivity Registered With a Small Number of Events

Victor Zlokazov[1,⋆] and *Vladimir* Utyonkov[2,⋆⋆]

[1]*Joint Institute for Nuclear Research Laboratory of Information Technologies*
[2]*Joint Institute for Nuclear Research Laboratory of Nuclear Reactions*

Abstract. The synthesis of superheavy elements asks for the analysis of low statistics experimental data presumably obeying an unknown exponential distribution and to take the decision whether they originate from one source or have admixtures. Here we analyze predictions following from non-parametrical methods, employing only such fundamental sample properties as the sample mean, the median and the mode.

The experiments on the synthesis of the superheavy elements yield chains of the sequential decays of the elements E_i, following a scheme: $E_i \rightarrow E_{i+1} \rightarrow E_{i+2} \rightarrow E_{i+3}$ etc. Each examined chain consists of registered decay times t_i of the corresponding element although some chains can be incomplete. These experiments can be performed by different groups using somewhat different methods of producing the nuclei under study, registration, and analysis of the obtained data. The goal of the present analysis is to estimate the statistical behavior of the detected elements and to test the results of different groups for compatibility.

First, we have to answer the following question: should we analyze each individual component in the chains or analyze the chain as a whole? To make it more comprehensive, let us recall the basic properties of the radioactive decays [1].

- Any k decays belong to the same type, are independent, occur in non-intersecting time points, and their probability within any time interval $[t_0, t_0 + \Delta t]$ does not depend on the choice of t_0;
- The decays are labeled as 'rare events', i.e., the probability of 2 and more events for any small Δt is infinitesimal compared with the probability of 0 or 1 events;
- Markov property of decays – the probability of observing any number of decays after any moment t does not depend on the origin of the element (it means on its prehistory).

Thus, our problem is: being given the experimental events t_{iEc}, where i is the number of event, E is the name of the isotope, c is the number of team performing the experiment,

- estimate the parameters of each group of data t_{iEc}, $i = 1, \ldots, n_{Ec}$ for each couple Ec
- test these estimates $\hat{t}_{Ec}$ for each E and c for compatibility.

However, do all the methods follow the above rules? For instance, if the third condition (Markov property) is not satisfied, we should not analyze separate events, but consider the chain as a whole (for each E and c), which may complicate the analysis. Therefore, we must first test the definitely uncorrelated data t_{iEc} of E_j and the corresponding t_{iEc} of E_{j+1}.

⋆e-mail: zlokazov@jinr.ru
⋆⋆e-mail: utyonkov@jinr.ru

Given two simulated samples of uncorrelated data $A_1 = (t_{11}, t_{12}, \ldots, t_{1n})$ and $A_2 = (t_{21}, t_{22}, \ldots, t_{2n})$ with the same exponential distribution function, their sample correlation coefficient r is

$$r = \sum_{i=1}^{n}(t_{1i} - m_1)(t_{2i} - m_2)/(\sigma_1 \cdot \sigma_2)/n, \tag{1}$$

where $m_1, m_2, \sigma_1, \sigma_2$ are expectations and sigmas (standard deviations) of the both distributions.

In an ideal case (infinite n, absolutely exact $m_1, m_2, \sigma_1, \sigma_2$) we should have $r = 0$. However, in our case of finite data, the calculations of r will result in a nonvanishing value. The Monte-Carlo tests showed that the main factor defining the spread of the of sample values of r is the size n. Table 1 collects the depending on n minimum, maximum, mean and unspecified values of r, respectively.

n	Min	Max	Mean	MAbs
10	−0.1335	0.3646	0.0865	0.1132
20	−0.3781	0.2673	−0.0200	0.1024
30	−0.1011	0.1312	0.0054	0.0367
50	−0.1043	0.2531	0.0091	0.0489
80	−0.0661	0.1659	0.0184	0.0375
100	−0.0623	0.1895	0.0269	0.0394

The table demonstrates occurrence of significant deviations from zero of the correlation coefficients for finite samples (especially at values $n = 10$ and 20).

A rigorous mathematical study of the distribution of the sample correlation coefficients can be found in [2], but it covers only asymptotics of normal event distributions and is described by a very cumbersome formula, including infinite series, that can be hardly used in practice.

However, based on results of Table 1, we can specify an interval (e.g. $[-0.4, 0.4]$) and say that if the obtained sample correlation coefficient does not exceed this range, the tested sample does not contradict the hypothesis that the data is uncorrelated.

To verify if the following radioactive decays are correlated or not, we can use the data with rather good statistics, e.g. ^{288}Mc $\longrightarrow$ ^{284}Nh $\longrightarrow$ ^{280}Rg; 63 events for each decay step were selected from [3, 4]. The performed analysis gave the estimate $r = -0.18$. So the real practice shows examples that do not reject the Markov property of the successive decays.

As for the data discussed in [5] (11 events in each sample) for the transition 'element 115 $\longrightarrow$ element 113', we obtained $r = -0.11$ and for the samples 'element 113 $\longrightarrow$ element 111', $r = +0.30$. Therefore, we can accept the hypothesis that the successive decays are uncorrelated and we can use for our analysis separate events in each data group rather than the fixed chains of events.

Furthermore, we can estimate the parameters T of all the exponential probability distributions of decay times t_i:

$$P(t_i < t) = P(t, T) = 1 - \exp(-t/T), \tag{2}$$

where the expectation of t_i is T and its variance is T^2.

The sum of the times of m decays $S = \sum_{i=1}^{m} t_i$ has the real (m, T)-gamma distribution [2]. Its density function is

$$g(t, m, T) = \begin{cases} \frac{t^{m-1}}{T^m(m-1)!}\exp(-t/T) & \text{for } t \geq 0; \\ 0 & \text{otherwise,} \end{cases} \tag{3}$$

where m is a positive integer, and a T is positive real. Let's consider a quantity $S_m = S/m$. The density of the S_m distribution is $g_m(t) = m \cdot g(mt, m, T)$, and its mean and the variance are equal to T and to T^2/m, respectively. The maximum of $g_m(t)$ (let it be t_x) is reached at the root of the equation

$$m\left(\frac{(m-1)(mt)^{m-2}}{T^m(m-1)!} - \frac{(mt)^{m-1}}{T^{m+1}(m-1)!}\right) \cdot \exp(-mt/T) = 0,$$

from which we obtain $t_x = (m-1)T/m$.

If m is small, this maximum can be rather far from the mean T. Therefore, the gamma-distribution is not favorable to the low statistics – it is the case where 'the most probable is not the most expected'. If m tends to the infinity, $g_m(mt, m, T)$ tends slowly to the normal density. The statistic S_m is sufficient for (2) [5, 6] and can serve as an estimate of the parameter T. At the same time it is the maximum likelihood estimate of T and it is efficient, i. e., having the variance asymptotic $1/n$.

For the two sets of decays $A_1 = (t_{11}, t_{12}, \ldots, t_{1m})$, $A_2 = (t_{21}, t_{22}, \ldots, t_{2m})$, two hypotheses are tested

1. $H_1 : A_1$ and A_2 have the same distribution;
2. $H_2 : A_1$ and A_2 have different distributions.

We can use a conventional approach to solve this problem by building the confidence intervals for the estimated means T (Student's test) and for the variances σ^2 (χ^2 test):

$$\hat{T} - t_{\alpha/2,m-1} * \hat{s}/\sqrt{m} < T < \hat{T} + t_{\alpha/2,m-1} * \hat{s}/\sqrt{m},$$

$$(m-1)\hat{s}^2/\chi^2_{\alpha/2,m-1} < \sigma^2 < ((m-1)(\hat{s}^2/\chi^2_{1-\alpha/2,m-1},$$

where $\hat{T} = \sum_{i=1}^{m} t_i/m$, $\hat{s} = \sqrt{\sum_{i=1}^{m}(\hat{T} - t_i)^2/(m-1)}$, $t_{\alpha,m-1}$ and $\chi^2_{\alpha,m-1}$ are the values of the Student's and χ^2 standard random quantity, that corresponds to the α probability and the number of degrees of freedom $m-1$.

It is reasonable enough to accept H_1 if the confidence intervals I_1 and I_2 overlap and the probability of the interval overlapping is sufficiently large. Otherwise, we can calculate the so called Type I and Type II errors for rejecting a true hypothesis and accepting the wrong one, and, after analyzing them, make the corresponding decision.

We shall now consider the concept of the optimal confidence interval $[a, b]$ (OCI) for testing the hypotheses [7]. Such an OCI should have the maximum accuracy of a testing criterion which means: the minimal difference $b - a$ and, at the same time, the probability of the events to belong to the interval $[a, b]$ should be maximum. Since these conditions contradict each other, an OCI should be determined from a compromise between two requirements [7]:

- for a given length $b - a$ to find an interval with the best ratio 'pro/contra';
- for a given ratio 'pro/contra' to find an interval of the shortest length $b - a$.

Very often one takes the interval $[T - \sigma, T + \sigma]$, the most popular for the normal distribution. For the gamma distribution it will be $[T - T/\sqrt{n}, T + T/\sqrt{n}]$; however, the convergence of the gamma distribution to the normal one is slow so that this interval will be asymmetric even at large n.

Apart from these criteria, we can build an OCI from the requirement that the physical meaning of the interval $[a, b]$ and its bounds a and b should be the most obvious and natural.

One of the ways to build such an OCI is to make use of the so called order statistics, which for the distribution (3) can be derived as easily integrable analytical functions [7]. Following [2] we define the following order statistics. Denote the minimum value in the sample S as u_1; and denote the maximum value in the sample S as u_m; their expectations $\hat{u}_1 = \hat{E}u_1$ and $\hat{u}_m = \hat{E}u_m$ are the order statistics of the 1st and the last rang, the meanings of which are:

1. $\hat{u}_1$ is the average minimum value in the sample;
2. $\hat{u}_m$ is the average maximum value in the sample. They can be used as the most 'natural' bounds of the average decay time in the sample.

An additional means to test the two distributions for congruence is to take the expectation E_d and the variance (sigma) σ_d of the normalized differences of both sets t_{1i} and t_{2i} and build the confidence interval $[-\sigma, \sigma]$ and see whether the E_d falls into it. Unfortunately, the formula for the variance of the difference of the two gamma-distributed random quantities has no simple closed form, so that it is difficult to estimate the trustworthiness of this test.

Now, we describe a method to solve a very important problem: is the analyzed data really a sample from a single exponential distribution or from a mixture?

This method is based on checking the sample relations between such fundamental characteristics of any event distribution as the mode, the median and the mean. In the case of exponentials these relations are uniquely defined and are different for a single component and a mixture of components.

The relation between the mean E and the median M for a single exponential is $M = E \cdot \ln 2$. For the ratio K of 'median / mean' in finite samples (e.g. $n = 15$) the statistical test gives $K = 0.7 \pm 0.17$ for a confidence interval of 67 %. However, if the data consist of the decays of two (or more) sources, K can differ from $\ln 2$ by several orders. Therefore, K is both a simple and a rather reliable indicator of whether the data originate from a single source or from a mixture of components.

Discussion and Conclusions

The literature covering the methods for the problem solution is very large (e.g. one can note the excellent paper [8]), but usually they are oriented to the analysis of large data volume size about which there is a certain amount of a priori information available. Such are all the methods based on the use of the regression techniques or likelihood relations. To advance in the data analysis with low statistics the accent was made on the non-parametric methods, employing only such fundamental sample properties as the sample mean, the median and the mode.

The report confirms that a radioactive process really possesses at least the weak form of the Markov property - the uncorrelatedness of the decays that is the necessary condition for a correct use of the mathematical techniques. It provides an analysis of the various methods of constructing of optimal and easily interpretable confidence intervals, using the order statistics and suggests a simple non-parametric criterion for testing the exponential data for admixtures by comparing the sample median with the sample mean. As example, this criterion applied to the data [9] reveals that some of them are most probably not pure.

Acknowledgments

We acknowledge the support by the JINR Directorate grant and the RFBR Grants Nos. 16-52-55002 and 14-12-00896. The authors thank the two anonymous referees for valuable criticism.

References

[1] B. V. Gnedenko, *Theory of Probability* (English translation, Mir Publishers, 1978)
[2] J. E. Freund, *Mathematical Statistics* (Prentice-Hall, Inc. Englewood Cliffs, N. J., 1962)
[3] Yu. Ts. Oganessian et al., Phys. Rev. C 87, 014302 (2013); D. Rudolph et al., Spectroscopy of Element 115 Decay Chains, Phys. Rev. Lett. **111**, 112502 (2013)
[4] J. M. Gates et al., Phys. Rev. C **92**, 021301(R) (2015)
[5] U. Forsberg et al., Nucl. Phys. A **953**, 117 (2016)
[6] S. S. Wilks, *Mathematical Statistics* (Princeton, N.J., Princeton University Press, 1947)
[7] V. B. Zlokazov, Physics of Particles and Nuclei Letters **12**, 2, 262–268 (2015)
[8] M. Stehlik and H. Wagner, IFAS Research Paper Series 2009-39 (January 2009)
[9] V. B. Zlokazov and V.K. Utyonkov, J. Phys. G: Nucl. Part. Phys. 44 (2017) 075107 (13pp)

EPJ Web of Conferences **173**, 05001 (2018) https://doi.org/10.1051/epjconf/201817305001
Mathematical Modeling and Computational Physics 2017

Symbolic and Numerical Modeling of Nonlinear Dynamics of Particles in Accelerators

Sergey Andrianov[1,*] and *Ekaterina* Sboeva[1,**]

[1] *Saint Petersburg State University, Faculty of Applied Mathematics and Control Processes*

Abstract. This paper is devoted to one of the methods of symbolic computation, which based on perturbation theory and constructed in the framework of matrix formalism. This method can be used for solving optimization problems and preliminary modeling in accelerator physics. There are presented the main theoretical positions and demonstration the result of this method on example of one of classical problems.

1 Introduction

The main instruments of high-energy physics are accelerators, the effective work of which depends both on the mathematical models and the corresponding software. Nowadays there are several widely used packages for this kind of tasks, for example, MAD [1], COSY Infinity [2] and MaryLie [3], which allow getting consistent results with the data. The first two (MAD, COSY Infinity) use phase variables to describe the beam, and both are based on numerical methods. The main approach used in them is the trajectory analysis. But numerical methods do not have the functionality that is required in the modern world, for example, a feature of numerical methods is the need to repeat all calculations under system parameter changing, it is time-consuming. The MaryLie package, unfortunately, does not support the Lego-ideology, and therefore requires new calculations for the entire system when replacing any of its elements.

Symbolic calculations in turn are cheap in time; they can support the Lego-ideology, and also taking into account symplecticity. In this paper, we propose to use a matrix approach [4] to modeling of nonlinear dynamical systems. It allows the use of symbolic calculations, such that cumbersome calculations are performed just once to yield final formulas which can be stored and used later. Also, the matrix approach allows us consider trajectories beam as a single object. Particular this allows us to move from coordinates and impulses to quantities that can be measured experimentally.

2 Theoretical basis

In this section we describe shortly a schematic description of the theory of constructing a solution of a system of ordinary differential equations in explicit form using perturbation theory. Consider the equation of motion in the form $d\mathbf{X}/ds = \mathbf{F}(\mathbf{X}, s)$, $\mathbf{F}(0, s) = 0$, where $\mathbf{X}$ is a vector of phase

*e-mail: sandrianov@yandex.ru
**e-mail: e.o.sboeva@gmail.com

moments and the arbitrary analytic function $\mathbf{F}(\mathbf{X}, s)$ is defined within a neighborhood of $\mathbf{X} = 0$, $\mathbf{X} \in R^n$ and is measurable at $s \in R^n$. The following definition for the vector of phase moment is necessary: $\mathbf{X}^{[k]} = \underbrace{\mathbf{X} \otimes ... \otimes \mathbf{X}}_{k \text{ times}}$ is a vector of phase moments of the k-th order, representing the k-th Kronecker degree of the phase vector $\mathbf{X}$. Further $\mathbb{P}^{1k}(s)$ is a matrix of dimension $(n \times d[n, k])$, where $d[n, k] = \binom{n + k - 1}{k}$, the elements of which are the derivatives of the components of the vector-valued function $\mathbf{F}(0, s) = 0$ of the k-th order. Note that $\mathbb{P}^{1k}(s)$ are matrices which include the control parameters of the system. Thus, taking into account the assumption that $\mathbf{F}(0, s) = 0$, we can write:

$$\frac{d\mathbf{X}}{ds} = \sum_{k=1}^{\infty} \mathbb{P}^{1k}(s)\mathbf{X}^{[k]}. \tag{1}$$

The solution of such a nonlinear system in the form of 2-dimensional matrices, which can be calculated according to the algorithm presented in [4], is sought in the form:

$$\mathbf{X} = \sum_{k=1}^{\infty} \mathbb{R}^{1k}(s)\mathbf{X}_0^{[k]}. \tag{2}$$

Let us note that the matrices $\mathbb{R}^{1k}(s)$ contain coefficients for the corresponding orders of the elements of the vector $\mathbf{X}_0$, where $\mathbf{X}_0 = \mathbf{X}(s_0)$. A finite cut-off of this infinite series can be defined which follows from the properties of the object under consideration (accelerator).

However, multi-rotation in cyclic machines and the necessity to make a huge number of integration steps leads to the need to preserve the integrals of motion, for example, the law of conservation of energy, and symplecticity. The requirements for modern cyclic systems lead to the necessity of using nonlinear control fields up to some k-th order, which in turn leads to a violation of the symplectic property. This is a critical moment for this kind of tasks. In terms of the matrix formalism, the symplecticity condition can be ensured using the Jacobi matrix $\mathbb{M}(\mathbf{X}, s|s_0)$ of the transformation $M(\mathbf{X}, s|s_0; H)$ of the dynamical system as follows:

$$\begin{gathered} \mathbb{M}^*(\mathbf{X}, s|s_0)\mathbb{J}(\mathbf{X})\mathbb{M}(\mathbf{X}, s|s_0) = \mathbb{J}(\mathbf{X}), \\ \mathbb{M}(\mathbf{X}, s|s_0; M) = \mathbb{M}(\mathbf{X}, s|s_0) = \frac{\partial M(\mathbf{X}, s|s_0; H) \circ \mathbf{X}}{\partial \mathbf{X}^*}, \end{gathered} \tag{3}$$

where H is the Hamiltonian of the system.

Traditionally, symplectic properties can be restored using various path-based methods, in particular, the methods of canonical transformations [5]. This approach is used for each individual trajectory of particles (10^9–10^{12}) and needs a lot of time. In this paper it is proposed to make corrections not in the trajectory, but in the matrices $\mathbb{R}^{1k}$. The proposed approach is based on the correction of the elements of the transformation matrices $\mathbb{R}^{1k}(s)$ to the desired order of nonlinearity [6]. As a result, the amount of computations performed is much less than on-trajectory analysis. Note that the simplification procedure must be performed sequentially for all orders (from 1 to the k-th), because the elements of each current transformation matrix have a relationship with the elements of the previous one.

3 Demonstration example

The general approach described above is demonstrated using the example of the dynamics of a pendulum with a cubic nonlinearity in comparison with the solution of this equation by the fourth-order Runge-Kutta method. The analytic solution presented in [7] was used to check the numerical results. The system of differential equations corresponding to the dynamics of such a 2-dimensional pendulum is Hamiltonian and has the form:

$$\begin{cases} q'(s) = p(s), \\ p'(s) = -q(s) + \alpha \dfrac{q^3(s)}{6}, \end{cases} \qquad H(s) = \frac{q^2(s) + p^2(s)}{2} - \alpha \frac{q^4(s)}{24} = const, \tag{4}$$

where α is a variable parameter of the system, s is length along the reference trajectory, H is the Hamiltonian of the system.

The solution of the system (4), describing dynamics of the 2-dimensional pendulum, i.e. $\mathbf{X} = (q, p)^T$ was obtained under initial conditions $q(0) = 0.8$, $p(0) = 0$. Taking into account symplectic corrections, final solutions were obtained, the phase portraits of which are consistent with the ones presented [7]:

$$\begin{cases} q(s) = 0.8\cos(s) + \alpha \dfrac{0.512}{24} \cos(s)(-1 + \cos^4(s)), \\ p(s) = -0.8\sin(s) + \alpha \dfrac{0.512}{24} \sin(s)(-1 + \cos^4(s)). \end{cases} \tag{5}$$

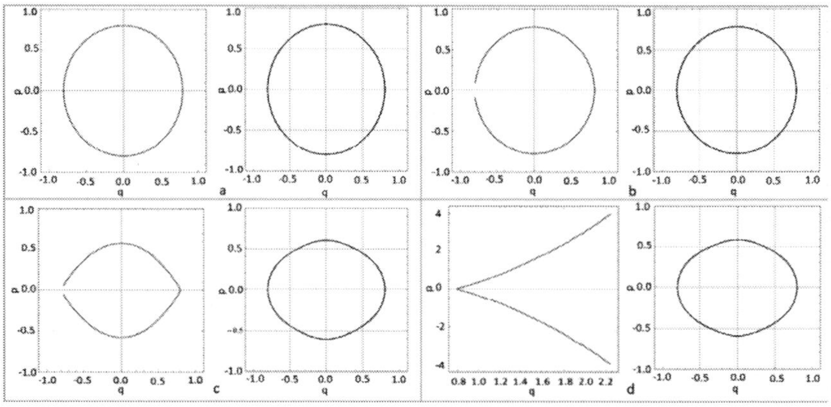

Figure 1. Plots represent phase portraits on the interval $s \in [-\pi, \pi]$ for a) $\alpha = 0$, b) $\alpha = 1$, c) $\alpha = 9$, d) $\alpha = 9.8$. Left: generated in the framework of the matrix formalism. Right: solutions obtained by fourth-order Runge-Kutta.

Fig. 1 provides plots of the phase portraits under various α. The parameter α is responsible for the compression along the p axis: the larger it is, the stronger the compression. At $\alpha = 9.8$ the Runge-Kutta method of the 4th order fails to generate the right solution, whereas within framework of the matrix formalism the true solution is found. Let's pay attention to fact that phase portraits of the numerical solution are closed for different values of s. The solutions obtained within framework of matrix formalism always close at $s = 2\pi$, due to the correction procedure performed.

Fig. 2 shows the time dependence for numerical calculations of the integrals used to obtain the analytical solution with different accuracies. It can be seen up to an accuracy of $\pm 10^{-3}$ or less, the analytical method allows almost instantaneous derivation of the solution (0.14 seconds). To reach an accuracy of $\pm 10^{-4}$, the numerical method already requires significant time costs – 13 seconds.

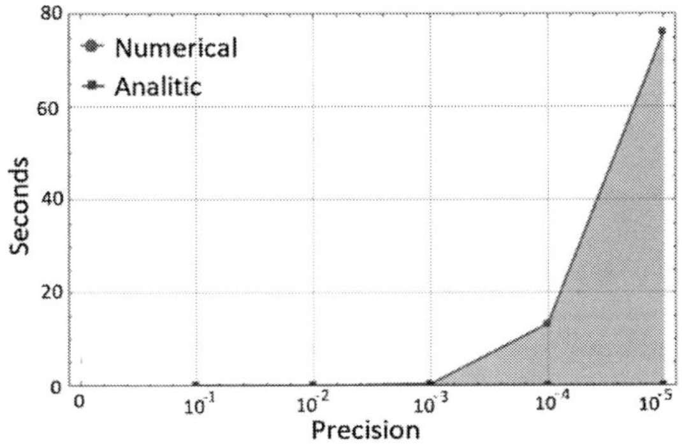

Figure 2. Comparing the time spent on obtaining a solution without symplectic corrections analytically and numerically.

4 Conclusion

This work is for the most part demonstrative of the main ideas of our approach. The features of the proposed approach can be summarized as follows: the correction of the mappings does not depend on the initial state of the system. The computational cost is decreased due to the possibility to perform only once and reuse a similar correction which is needed at each order of the step by step algorithm. In addition, these operations with matrices are performed independently of the initial dynamic system and corresponding results can be stored in a special database.

In the case described in this paper, the laboriousness of symbolic computations is not too high, however, in the case of dynamics of particle beam it is. In addition to all of the above, we would like to emphasize once again the intrinsic property of this method of being independent of the concrete representation of dynamic (Hamiltonian) system.

References

[1] H. Grote and F.C. Iselin, *The MAD program (Methodical Accelerator Design), User's Reference Manual* (CERN/SL/90-13, 1994)
[2] M. Berz., Nuclear Instruments and Methods **A363**, 100–104 (1995)
[3] A. J. Dragt, L. M. Healy, F. Neri, and R. Ryne, IEEE Trans. Nucl. Sci. **NS-32:5**, 2311–2313 (1985)
[4] S.N. Andrianov, *Dynamic Modeling of Particle Beam Control Systems*, (Saint Petersburg State University, 2002), 98–103
[5] B. Erdélyi and M. Berz, Physical Review Letters **87**, 11, 114302 (2001)
[6] S.N. Andrianov, AIP Conf. Proc. **N391**, 355–360 (1997)
[7] G. Rangarajan, Pramana-Journal of Physics **48**, 1, 129–142 (1997)
[8] S. Blanes and F. Casas, *A Concise Introduction to Geometric Numerical Integration* (CRC Press Book, 2016)
[9] D.M. Himmelblau, *Applied Nonlinear Programming* (The University of Texas, Austin, Texas, 1972)

User Software for Numerical Study of Josephson Junction with Magnetic Momenta

Pavlina Atanasova[1,⋆], *Stefani* Panayotova[1], *Yury* Shukrinov[2,3], *Ilhom* Rahmonov[2,4], and *Elena* Zemlyanaya[2,3]

[1] *University of Plovdiv Paisii Hilendarski, 24 Tzar Asen, 4000 Plovdiv, Bulgaria*
[2] *Joint Institute of Nuclear Research, Dubna, Moscow Region, 141980, Russia*
[3] *Dubna State University, Dubna, 141980, Russia*
[4] *Umarov Physical Technical Institute, TAS, Dushanbe, 734063, Tajikistan*

Abstract. A user software for numerical study of a Josephson junction model with magnetic momenta is presented. Computer implementation has been done by means of *Wolfram Mathematica* using the extensive capabilities of this system to create interactive dynamic objects. Two methods for numerical solution of the respective system of ordinary differential equations are implemented: the four-step Runge-Kutta algorithm and the Runge-Kutta-Fehlberg method with predetermined accuracy. Results of numerical simulation are presented to confirm the correctness of the calculations done with the developed software.

1 Introduction

Josephson junctions in combination with magnetic moments occupy an important place in "hot" fields of science like spintronics, nanotechnology, superconductivity, condensed matter physics, low temperature physics [1]. They represent interest in a number of practical applications and are subjects in many research areas [2]. Usually such problems cannot be solved analytically and the only way of investigation is by numerical studies. The numerical analysis of such multiparameter physical problems in a difficult undertaking. Thus, an important task is to create a suitable user software which implements the corresponding numerical methods as well as the visualization of the results. The chosen programming language for the creation of such software is *Wolfram Mathematica* [3] which is known as a finished product for the ultimate technical application and is convenient for research and experimentation.

In this paper we describe a user software for Josephson Junctions Simulations (USJJS) in the numerical study of the Josephson junctions with magnetic momenta. Using this software one can compute accurate values of some important physical parameters of the system under electric current pulse. We show that our method can demonstrate a magnetization reversal effect [4]. This effect is very important for superconducting spintronics and could be useful in the frame of applications in the quantum computing field [1].

⋆e-mail: atanasova@uni-plovdiv.bg

2 Mathematical model

The geometry of the system of superconductor/ferromagnetic/superconductor Josephoson junction (SFS JJ) under consideration is presented in Fig. 1. The ferromagnetic easy-axis is directed along the z-axis, which is also the direction n of the gradient of the spin-orbit interaction. The magnetization component m_y is coupled with the Josephson current I_s, which is along the x-axis. The magnetic moment **M** is specified by its three spatial components along the axes. The dynamics of the magnetic

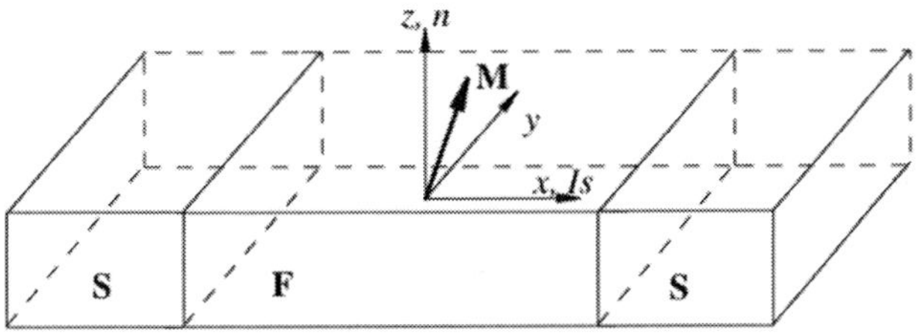

Figure 1. Geometry of the considered SFS JJ system. Here S is superconductor and F – ferromagnetic.

moment and phase difference in the SFS JJ is described by the following problem [4] of four ordinary first-order differential equations with initial conditions in dimensionless form:

$$\frac{dm_x}{dt} = \frac{\omega_F}{1+\alpha^2}\Big\{-m_y m_z + Grm_z\sin(\varphi - rm_y) - \alpha[m_x m_z^2 - Grm_x\; m_y\sin(\varphi - rm_y)]\Big\},$$

$$\frac{dm_y}{dt} = \frac{\omega_F}{1+\alpha^2}\Big\{m_x m_z - \alpha[m_y m_z^2 - Gr(m_z^2 + m_x^2)\sin(\varphi - rm_y)]\Big\},$$

$$\frac{dm_z}{dt} = \frac{\omega_F}{1+\alpha^2}\Big\{-Grm_x\sin(\varphi - rm_y) - \alpha[Grm_y m_z\sin(\varphi - rm_y) - m_z(m_x^2 + m_y^2)]\Big\},$$

$$\frac{d\varphi}{dt} = \frac{1}{w}\Big(I_{\text{pulse}}(t) - \sin(\varphi - rm_y)\Big), \qquad I_{\text{pulse}}(t) = \begin{cases} A_s, & t \in [t_0 - 1/2\Delta t, t_0 + 1/2\Delta t]; \\ 0, & \text{otherwise}, \end{cases}$$

$$m_x(0) = 0, \qquad m_y(0) = 0, \qquad m_z(0) = 1, \qquad \varphi(0) = 0, \qquad t \geq 0,$$

where ω_F is the ferromagnetic resonance frequency, α is the damping parameter, G is a parameter of the phase difference coupling, r is the spin-orbit coupling parameter, $w = V_F/(I_c R) = \omega_F/\omega_R$, $V_F = \hbar\omega_F/(2e)$, I_c is the critical current, R is the resistance of JJ, $\omega_R = 2eI_c R/\hbar$ is the characteristic frequency. The parameters of the amplitude A_s, the middle time of influence t_0 and the time interval of influence Δt characterise the electric current pulse. The time dependent unknown functions are the magnetic moment components $m_x(t), m_y(t), m_z(t)$ and the phase difference $\varphi(t)$.

The superconducting current $I_s(t)$ is calculated via the function $\varphi(t)$ by the following formula $I_s(t) = I_c\sin\big(\varphi(t) - rm_y(t)\big)$, $t \geq 0$.

3 Numerical approach and user software USJJS

The above system of ordinary differential equations with initial conditions was solved by the four-step Runge-Kutta method and independently by the Runge-Kutta-Fehlberg method, which ensures a predetermined accuracy of the calculations. Several modules have been implemented for the visualisation of the results within of *Wolfram Mathematica* using the extensive capabilities of this system to

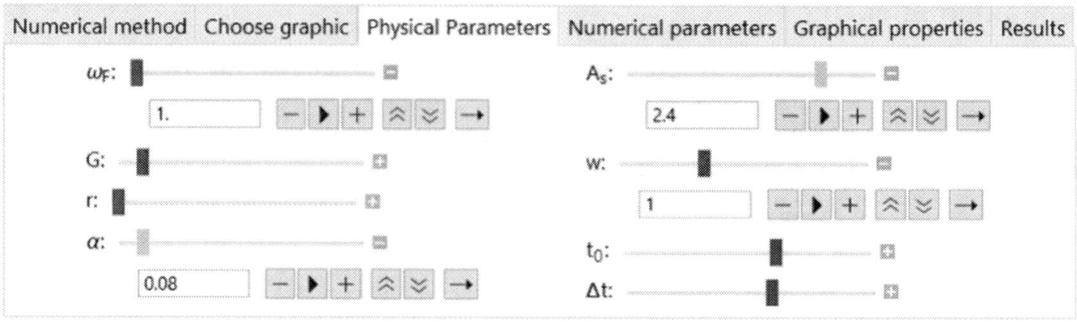

Figure 2. Menu of the USJJS where "Physical parameters" Tab is selected

create interactive dynamic objects. The results can be displayed in different ways depending on what is needed to the user.

The user software USJJS for the SFS JJ system study provides both graphical and tabular formatting options. Fig. 2 depicts how the USJJS menu looks. The menu is organized with the help of the following Tab Views: "Numerical method", "Choose Graphic", "Physical parameters", "Numerical parameters", "Graphical properties" and "Results". In "Numerical Method", the user can select one of two numerical simulation methods (the classical four-step Runge-Kutta algorithm or the Runge-Kutta-Fehlberg method). In "Choose Graphic", the user can choose one of nine types of graphics, each of them invokes a respective module that implements it. For the first eight types, each graph is displayed individually, and the ninth option "All Graphic" shows all graphs together. In the "Physical parameters" submenu (Fig. 2), specific parameter values can be entered and a change of function can be traced. In "Numerical parameters", the system's capabilities allow the manipulation of numerical parameters in an analogous way. Here, the "Animation Time Mode" Check Box is available to trace the dynamics of the functions. In the "Graphical properties" Tab, the user can "play" with the settings of the graphics. The last Tab "Results" propose a button that shows the user a table with all the numerical results of the calculations.

4 Numerical results and conclusions

To verify the correctness of the USJJS, we recovered all the numerical results from [4]. Excellent qualitative and quantitative agreement between our simulations and the simulations in [4] has been obtained. The instant in Fig. 3 reproduces the results in Fig. 2(b) of [4]. The parameters in this simulation are: $\omega_F = 1$, $G = 10$, $r = 0.1$, $\alpha = 0$, $A_s = 1.5$, $w = 1$, $t_0 = 25$, $\Delta t = 6$.

To analyze the influence of the pulse electric current amplitude A_s, we make a simulation with the above values of parameters in the USJJS animation mode over the parameter A_s. One can observe how the function $m_z(t)$ is moving while the parameter A_s is gradually growing. In the interval $A_s \in (0; 0.3]$, the current pulse changes the direction of the magnetic moment from up ($m_z = 1$) to down ($m_z = -1$) only for a small time interval. After that the magnetic moment becomes again up to $m_z = 1$ and stays that way. In the interval $A_s \in (0.3; 0.7]$, a gradual transition from upper direction of the magnetic moment to the down position is observed. Between values of $A_s = 0.7$ and $A_s = 0.9$, m_z stays down to -1. At the value of $A_s = 1.2$, the magnetic moment again keeps stable its upper position $m_z = 1$.

The magnetization reversal effect is demonstrated in Fig. 4 for the parameter values $\omega_F = 1$, $G = 50\pi$, $r = 0.1$, $\alpha = 0.1$, $A_s = 2.1$, $w = 1$, $t_0 = 25$, $\Delta t = 6$. On this figure the dynamics of the magnetic moment components, phase difference, phase diagrams $m_z - m_x$, $m_z - m_y$, $m_y - m_x$ and superconducting current are displayed. After the influence of signal in $t \in [22, 28]$ the transition of

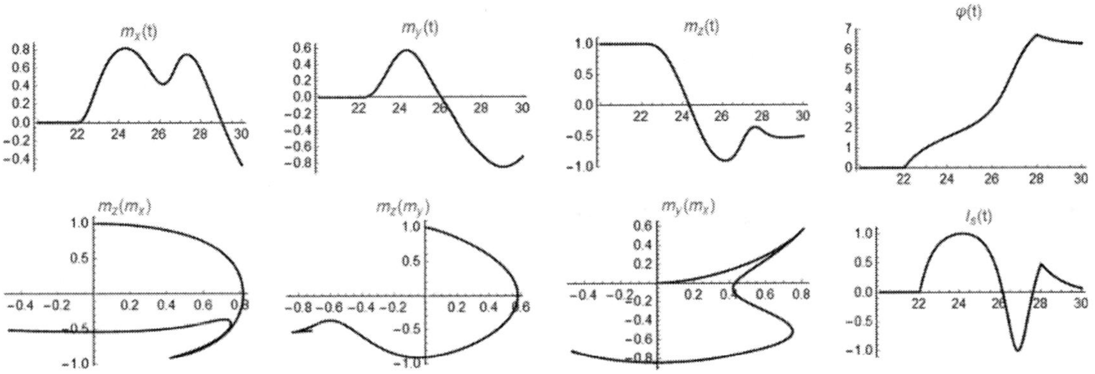

Figure 3. Results of the simulations with the same parameter values as in Fig. 2(b) of [4]

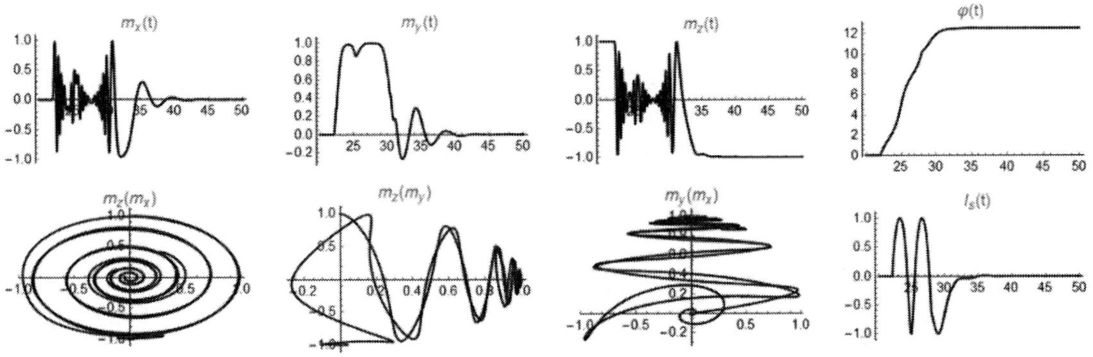

Figure 4. The magnetization reversal effect simulation in case of parameters $\omega_F = 1$, $G = 50\pi$, $r = 0.1$, $\alpha = 0.1$, $A_s = 2.1$, $w = 1$, $t_0 = 25$

the m_z component from the upper direction $m_z = 1$ to the down one $m_z = -1$ is clearly seen while the components m_x and m_y come to zero.

An analogous investigation can be done with the help of our USJJS over the influence of the other physical parameters on the SFSJJ system behaviour. Thus, by means of the S, one can find accurate values of some important physical parameters of the electric current pulse and Josephson junction, in which the magnetization reversal effect is observed. This could be useful in the investigation of quantum computing models considering the two directions of magnetization as bits of information. Also, this effect could be evidenced in physical experiments.

Acknowledgement.

The work is supported by project FP17-FMI-008, Bulgaria, by the JINR–Bulgaria cooperation program, and by the RFBR (projects 15-29-01217, 16-52-45011, 17-01-00661).

References

[1] Jacob Linder and W.A.Jason Robinson, Nature Physics **11**, 307 (2015)

[2] X.Waintal and P.W.Brouwer, Phys. Rev. B **65**, 054407 (2002)

[3] http://reference.wolfram.com

[4] Yu.M.Shukrinov, I.R.Rahmonov, K.Sengupta, and A.Buzdin, Appl.Phys.Lett. **110**, 182407 (2017)

Determination of the Vesicular Systems Parameters: Parallel Implementation and Analysis of the PTNS Vesicle Structure

Maxim Bashashin[1,2,⋆], *Elena* Zemlyanaya[1,2], *Evgeniya* Zhabitskaya[1,2], *Mikhail* Kiselev[1,2], and *Tatyana* Sapozhnikova[1]

[1] *Joint Institute for Nuclear Research, Joliot-Curie str. 6, 141980 Dubna, Russia;*
[2] *Dubna State University, University str. 19, 141980 Dubna, Russia*

Abstract. Parameters of the phospholipid vesicular systems are determined by means of minimization of the discrepancy between experimental data on the small angle scattering intensity and the numerical results within the frame of the separated form factors method. Effectiveness of parallel implementation is tested at the cluster HybriLIT. Results of numerical analysis of small angle neutron scattering data on polydispersed population of unilamellar vesicles of phospholipid transport nanosystem are presented.

1 Introduction

Phospholipid vesicles (liposomes, nanospheres) have a wide range of applications in modern biochemistry and pharmacology. One of the important trends is the development of the drug delivery systems on the basis of phospholipds [1, 2]. The small angle scattering of neutrons (SANS) and of X-rays (SAXS) are known as powerful tools for the nanosystems structure investigation. SANS and SAXS measurements are well suited for the analysis of the structure of the polydispersed populations of unilamellar vesicles (ULVs) of phospholipids in water solutions of disaccharides [3–6]. For this analysis, we utilize the separated form factors method (SFF) [6–8]. Parameters of ULVs are adjusted by means of the weighted least square method

$$\chi^2 = \frac{1}{N-k} \sum_{j=1}^{N} \left(\frac{I^{\mathrm{theor}}(q_j) - I^{\mathrm{exper}}(q_j)}{\delta^{\mathrm{exper}}(q_j)} \right)^2, \tag{1}$$

where $I^{\mathrm{theor}}(q_j)$ and $I^{\mathrm{exper}}(q_j)$ are, respectively, theoretical and experimental small angle intensity values at the points q_j of the scattering vector, N is the number of experimental points, k is the number of parameters to be fitted, δ are the experimental errors.

In this contribution, we discuss the effectiveness of parallel implementation of the SFF-parameters fitting procedure. Also, results of the SFF analysis of SANS data on polydispersed population of ULVs of phospholipid transport nanosystem (PTNS) are presented.

2 Separated form factors method

The SFF method for analysis of SANS data is determined by the following formulas (details are given in [3, 4, 6–8]). The expression of the intensity $I(q)$ has the following form:

⋆e-mail: bashashinmv@jinr.ru

$$I(q) = I_m(q) + \frac{1}{2}\Delta^2 \frac{d^2 I_m(q)}{dq^2} + I_B, \quad I_m(q) = \left[\int_{R_{\min}}^{R_{\max}} I_{\text{SFF}}(q, R) G\left(R, \bar{R}\right) dR \right] \times \left[\int_{R_{\min}}^{R_{\max}} G\left(R, \bar{R}\right) dR \right]^{-1}, \quad (2)$$

where q is the scattering vector, I_B characterizes the incoherent background, Δ^2 is the second momentum of spectrometer resolution function. I_m is the ULV system SANS intensity with average radius $\bar{R}$. Polydispersity σ is accounted for by the Schulz distribution $G(R, \bar{R})$ with coefficient m:

$$G\left(R, \bar{R}\right) = \frac{R^m}{m!} \left(\frac{m+1}{\bar{R}}\right)^{m+1} \exp\left[-\frac{(m+1)R}{\bar{R}}\right], \quad \sigma = \frac{1}{\sqrt{m+1}}. \quad (3)$$

I_{SFF} is the intensity of monodispersed system of ULVs:

$$I_{\text{SFF}}(q) = nI_o F_s(q) F_b(q) S(q), \quad F_s(q) = \left(4\pi R^2 \frac{\sin(qR)}{qR}\right)^2, \quad F_b(q) = \left(\int_{-d/2}^{+d/2} \rho_c(x) \cos(qx)\, dx \right)^2, \quad (4)$$

where I_o is the intensity of the incident beam, n is a number of vesicles in cm^3, F_s is the form factor of the spherical surface with radius R, F_b is the form factor of the symmetric lipid bilayer of thickness d, $\rho_c(x)$ is the contrast between the scattering length density of the lipid bilayer and the density of the solvent, $S(q) \sim 1$ is a structural factor. Here, we use the hydrophilic-hydrophobic model (HH) of the density distribution across bilayer suggested in [8] and shown in Fig. 1(a).

3 Parallel implementation

In our calculations, we employ both global and local minimization for the ULVs SFF-parameters adjustment. The *global* minimization is based on the asynchronous differential evolution algorithm (ADE). The ADE method incorporates the classical differential evolutions algorithm into a parallel steady-state strategy. ADE minimiser with adaptive correlation matrix automatically adapts to the landscape of the optimized objective function, see [11] and references therein for details. ADE approach was successfully employed for the SFF analysis of SAXS data [7, 12]. High effectiveness of parallel implementation of SFF-ADE calculations is demonstrated in [12].

On the other hand, in many cases, the *local* minimization is also effective. The iterative minimization procedure implemented in the code DFUMIL[1] and FUMILIM [9], provides quick convergence to the minimum of the χ^2 expression (1), see [10] and references therein for details. In the parallel MPI-version of this method (PFUMILI[2]), the parallelism is based on a distribution of the set of experimental points between parallel MPI-processes, with the collection of intermediate results to the master process at each iteration. Because of the intensive interaction between the MPI-processes, significant acceleration of calculations on multi-processor systems (as we observe in the case of ADE calculations) is not available. Nevertheless, in case of relatively small number of parallel processes, one expects a noticeable decrease of computer time.

In order to test the effectivenes of the PFUMILI-based parallel implementation of the SFF-parameters fitting procedure, we have made methodical calculations for two sets of SANS experimental data on a polydispersed ULVs population of dimirostoilphosphatidilholin (DMPC) in D_2O.

[1] https://wwwinfo.jinr.ru/programs/jinrlib/d510
[2] https://wwwinfo.jinr.ru/programs/jinrlib/pfumili

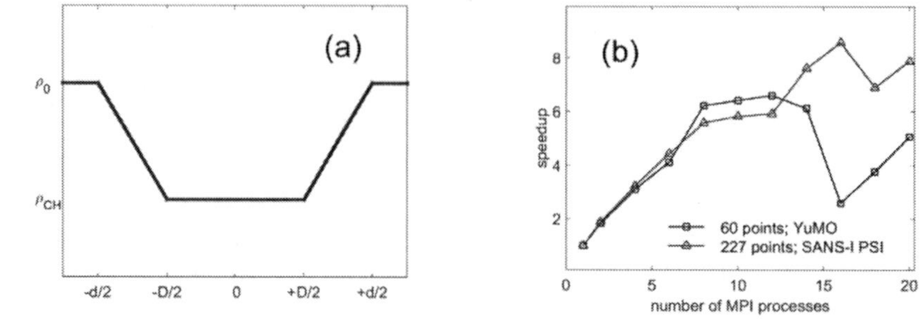

Figure 1. (a) HH-model of the the scattering length density across bilayer: d is the bilayer thickness, D is the thickness of hydrophobic region, $\rho_{CH} = -0.36 \cdot 10^{10}$ is the hydrocarbon chains scattering length density, ρ_0 is the solvent density. (b) Speedup of PFUMILI-SFF calculations versus the number of MPI-processes in case of analysis of the 60-point YuMO spectrum and the 227-points SANS-I spectrum.

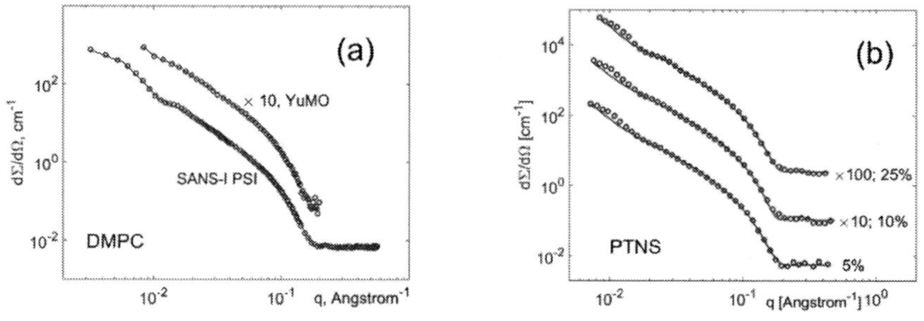

Figure 2. (a) Theoretical and experimental spectra of the polydispersed DMPC ULV system in $D_2 0$. SANS data have been obtained at spectrometers YuMO (Dubna) and SANS-I PSI (Villingen). (b) Theoretical and experimental spectra of polydispersed population of PTNS ULVs in the heavy water solvent of maltose with different concentrations.

The first spectrum with 60 experimental points was measured at the YuMO spectrometer (Dubna). The second one (227 experimental points) was obtained at the SANS-I PSI facility (Willingen). The speedup of the calculations versus the number of MPI-processes is shown in Fig. 1(b). The maximal speedup is about 6.6 times for the 60-point YuMO spectrum (12 MPI-processes) and about 8.6 time for the 227-points SANS-I spectrum (16 MPI-processes). Obviously the parallel implementation efficiency should be increased as the number of experimental points is growing.

The values of basic adjusted parameters (see Table 1) are in agreement with the results in [8]. Fig. 2(a) demonstrates the respective experimental and theoretical curves.

Table 1. Basic parameters of polydispersed populations of DMPC and PTNS ULVs. Adjusted parameters: the average radius $\bar{R}$, the polydispersity coefficient m, the bilayer thickness d, the hydrophobic region thickness D.

Experiment	Maltose	ρ_0, cm^{-2}	$\bar{R}$, Å	m	d, Å	D, Å	χ^2
YuMO; DMPC	0%	$6.37 \cdot 10^{10}$	277.0±5.0	9.8±0.7	49.0±2.0	20.0±3.0	1.1
SANS-I; DMPC	0%	$6.37 \cdot 10^{10}$	275.6±0.5	12.6±0.3	47.8±0.2	20.5±0.4	1.7
YuMO; PTNS	5%	$6.16 \cdot 10^{10}$	190.4±4	4.7±0.5	50.6±0.2	2.0±0.2	7.0
YuMO; PTNS	10%	$5.95 \cdot 10^{10}$	200.3±2.1	4.8±0.5	51.3±1.9	2.0±0.9	8.0
YuMO; PTNS	25%	$5.33 \cdot 10^{10}$	192.1±1.4	7.6±0.7	47.7±4.7	11.9±2.0	8.2

4 Analysis of PTNS ULVs structure

The phospholipid transport nanosystem (PTNS) with extremely small ULVs size has been obtained on the basis of the soybean phosphatidylcholine in the Orekhovich Research Institute of Biomedical Chemistry (Moscow, Russia). Incorporation of drugs into PTNS ULVs is found to increase sufficiently the therapeutic effectiveness. The structure of PTNS ULVs was analysed in [12, 13] on the basis of SAXS data collected at the Kurchatov Synchrotron Radiation Source of the NRC "Kurchatov Institute" (Moscow, Russia). Here, we present first estimations of basic parameters of PTNS ULVs in the heavy water solvent of maltose on the basis of the HH-SFF analysis of SANS data collected at YuMO spectrometer. Fig. 2(b) demonstrates a reasonable agreement of the SFF calculations with experimental data. Values of basic parameters of PTNS system are given in Table 1. The results of both local and global minimization are in very good agreement.

From Table 1 it follows that the average radius $\bar{R}$ of PTNS ULVs is 1.5 times smaller then the radius of "classical" DMPC ULVs. Note also the weak dependence of $\bar{R}$ on maltose concentration. The PTNS and DMPC bilayer thicknesses d are close to each other. At 5% and 10% concentration we obtained extremely small values of the hydrophobic region D in bilayer. This requires further analysis and explanation.

5 Summary

We have found that the PFUMILI-based parallel implementation of the SFF-parameters fitting procedure provides the 6-9 times speedup in comparison with the sequential calculations. First estimations of the PTNS ULVs structure within the frame of the HH-SFF analysis of SANS data collected at YuMO spectrometer have been done. Parallel versions of both ADE-based and PFUMILI-based fitting procedures are expected to accelerate the further study of PTNS ULVs structure on the basis of the SFF analysis of SANS and SAXS data.

Acknowledgement

The work has been performed under the grant of Russian Scientific Foundation (project No 14-12-00516).

References

[1] S. Das and A. Chaudhury, AAPS PharmSci. Tech. **12**, 62–76 (2011)
[2] S. Martins, B. Sarmento, D.C. Ferreira, E.B. Souto, Int. J. Nanomedicine **2**, 595–607 (2007)
[3] M.A. Kiselev *et al*, Nucl. Inst. & Method **A 470**, 409–416 (2001)
[4] M. Kiselev *et al*, J. Alloys and Compounds **328**, 71–76 (2001)
[5] M.A. Kiselev *et al*, Chemistry and Physics of Lipids **123**, 31–44 (2003)
[6] M.A. Kiselev *et al*, Colloids and Surfaces **A 256** 1–7 (2005)
[7] M.A. Kiselev, E.V. Zemlyanaya, E.I. Zhabitskaya, V.L. Aksenov, Crystallography Reports **60**, 143–147 (2015)
[8] M.A. Kiselev, E.V. Zemlyanaya, V.K. Aswal, R. Neubert, European Biophys. J **35**, 477–493 (2006)
[9] I.M. Sitnik, Comput. Phys. Comm. **209**, 199–2004 (2016)
[10] V.S. Kurbatov and I.N. Silin, Nucl. Instr. & Meth. **A 345**, 346–350 (1994)
[11] E.I. Zhabitskaya and M.V. Zhabitsky, Proc. 15th Ann. Conf. on Genetic and Evolutionary Comput., (USA, N.Y., 2013) 455–462
[12] E. Zhabitskaya, E. Zemlyanaya, M. Kiselev, A. Gruzinov, EPJ Web of Conf. **108**, 02047 (2016)
[13] M.A. Kiselev *et al*, J. Pharmaceutical and Biomedical Analysis **114**, 288–291 (2015)

EPJ Web of Conferences **173**, 05004 (2018) https://doi.org/10.1051/epjconf/201817305004
Mathematical Modeling and Computational Physics 2017

Radiation Damage to Nervous System: Designing Optimal Models for Realistic Neuron Morphology in Hippocampus

Munkhbaatar Batmunkh[1,2,*], *Alexander* Bugay[1], *Lkhagvaa* Bayarchimeg[1], and *Oidov* Lkhagva[2]

[1] *Laboratory of Radiation Biology, Joint Institute for Nuclear Research, 141980, Dubna*
[2] *Division of Natural Sciences, National University of Mongolia, 210646, Ulaanbaatar*

Abstract. The present study is focused on the development of optimal models of neuron morphology for Monte Carlo microdosimetry simulations of initial radiation-induced events of heavy charged particles in the specific types of cells of the hippocampus, which is the most radiation-sensitive structure of the central nervous system. The neuron geometry and particles track structures were simulated by the Geant4/Geant4-DNA Monte Carlo toolkits. The calculations were made for beams of protons and heavy ions with different energies and doses corresponding to real fluxes of galactic cosmic rays. A simple compartmental model and a complex model with realistic morphology extracted from experimental data were constructed and compared. We estimated the distribution of the energy deposition events and the production of reactive chemical species within the developed models of CA3/CA1 pyramidal neurons and DG granule cells of the rat hippocampus under exposure to different particles with the same dose. Similar distributions of the energy deposition events and concentration of some oxidative radical species were obtained in both the simplified and realistic neuron models.

1 Introduction

Recent experimental studies at particle accelerators have shown the occurrence of neurochemical alterations, oxidative and radical damage, modifications to dendritic morphology, electrophysiological alterations in different parts of the brain of the laboratory animals after exposure to energetic protons and heavy ion beams at doses matching the real fluxes in the energy spectrum of galactic cosmic rays and solar particle events [1–3]. The analysis of these measurements suggests that one of the most radiation sensitive-brain structure is the hippocampus. The hippocampus plays an important role in learning, memory consolidation as well as the integration of the information received simultaneously from diverse sources. The precise mechanisms of the radiation-induced brain injury remains mostly unclear. The development of models of the radiation-induced structural and functional variations in different brain structures (especially the hippocampus) is important for the understanding of the long-term changes to the neuronal electrophysiological states as a result of prolonged deep space missions [3]. The present work focuses at designing an optimal model of the hippocampal neurons irradiation with the aim at understanding the electrophysiological behaviour of the hippocampal neural network.

*e-mail: batmunkh@jinr.ru

2 Simplified neuron models in the Monte Carlo track-structure simulations

Neuronal cells exist in different shapes and sizes, and can be classified by their morphology and function. In the different regions of hippocampus, there are two main types of neuronal cells: 1.2 million granule cells in dentate gyrus region (DG) and 615000 pyramidal cells in cornu ammonis region (CA3/CA1). A soma of a granule cell rests in the cell layer of the DG region, while its dendrites in the overlying molecular layer of the DG region where they receive synaptic connections mainly from the entorhinal cortex (EC) via the perforant pathway. A soma of a pyramidal cell rests in the cell layer of the CA region, while its dendrites extend across the strata from oriens layer (basal portion) to radiatum layer and molecular layer (apical portion) where they are making synapses with DG and EC.

In our previous modeling, we calculated the microscopic energy depositions and production of water radiolysis species in the vicinity of neuron structures using the Geant4/Geant4-DNA Monte Carlo radiation transport code [4–7]. In the present work, simplified models of a DG granule and a CA pyramidal cell were constructed from models of realistic neuron morphologies and adapted into standardized SWC file format. The total volumes of the simplified models were calculated with 11279.7 μm^3 for a DG granule cell and 31689.6 μm^3 for CA3/CA1 pyramidal cell. The total volumes of the realistic models of granule and pyramidal cell were found as 12455.5 ± 4395.2 μm^3 and 34112.5 ± 9693.7 μm^3, which are estimated from all the hippocampal neurons in the Claiborne archive at the NeuroMorpho.Org repository. The realistic and simplified models of granule cell are represented as 451 ± 228 and 9 compartments of soma and dendrites. Pyramidal cell soma and dendrites are represented as 2259 ± 563 compartments for realistic model and 64 compartments for simplified model.

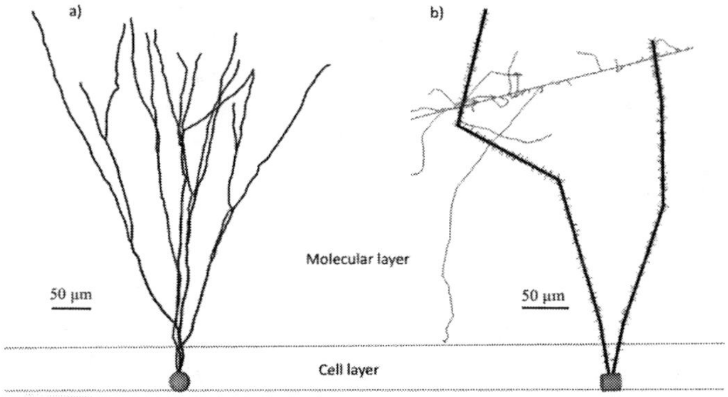

Figure 1. Realistic (a) and simplified (b) model of a granule cell in different layers of DG region of rat hippocampus. An example of a simulated track structure (red online, thin long lines in print) of 290 MeV/u carbon ion crossing simplified model of a granule cell.

In order to build dendritic spines along the dendritic branches of the developed neuron models, we used the measured spine density [8, 9], which is defined as the total number of spines per unit length (10 μm) of the total dendritic length. The spine density was found to be 8.2 ± 0.8 for the molecular layer of a DG granule cell, 8.0 ± 0.6 for the oriens layer and 11.3 ± 0.5 for the radiatum and molecular layers of the CA3/CA1 pyramidal cell. Each spine is described by means of a small cylindrical volume that is randomly pointing perpendicular to the dendrite surface between 0–2π, and located on dendrites as a Gaussian distribution with given spine density. The spine sizes were generated from a Poisson distribution in a range between 0.3–0.76 μm for diameter and 0.9–4.2 μm for the total length, which were derived from literature [10, 11]. In the simplified models, 810 spines for a DG granule cell and 3582 spines for a CA3/CA1 pyramidal cell were constructed using Poisson and Gaussian distributions with the above-mentioned spine densities.

https://doi.org/10.1051/epjconf/201817305004

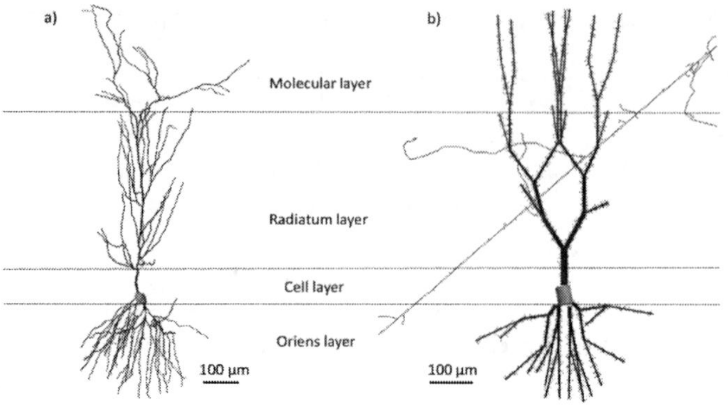

Figure 2. Realistic (a) and simplified (b) model of a pyramidal cell in different layers of CA3/CA1 region of rat hippocampus. An example of a simulated track structure (red online, thin long lines in print) of 290 MeV/u carbon ion crossing simplified model of a pyramidal cell.

3 Comparison of simple and realistic neuron models for calculation of energy deposition and production of oxidative radical species

The results of the three-dimensional construction of the developed neuron models with irradiation of particle track structure are illustrated in Figs 1 and 2. Similar distributions of the energy deposition events were obtained in the dendrites and soma of the two neuron models under an exposure to 290 MeV/u carbon ion with a dose of 0.1 Gy (Figure 3). It should be noted that the error bars were estimated from the energy depositions in the individual neuron models in the Claiborne archive at the NeuroMorpho.Org repository.

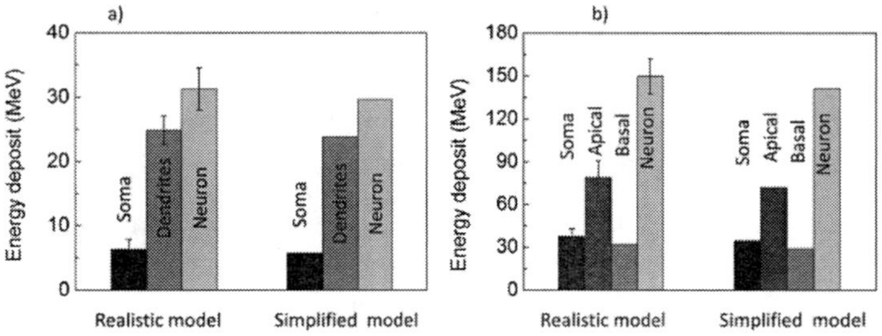

Figure 3. Total energy deposition in realistic and simplified model of a DG granule cell (a) and a CA3/CA1 pyramidal cell (b) under exposure to 290 MeV/u carbon ions with a dose of 0.1 Gy.

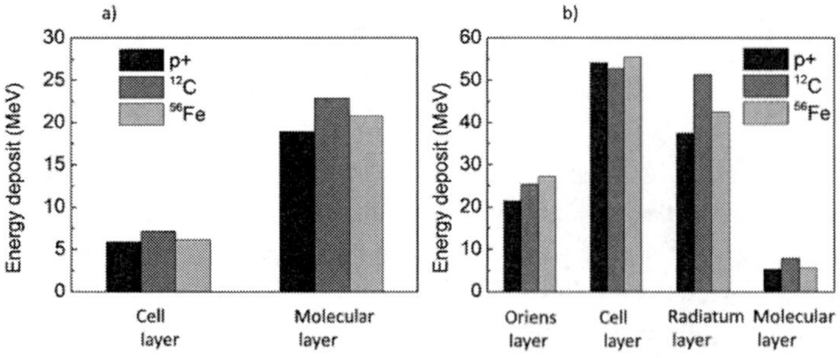

Figure 4. Distribution of energy deposition in different layers of the simplified model of a DG granule cell (a) and CA3/CA1 pyramidal cell (b) under exposure to 170 MeV protons, 290 MeV/u carbon and 600 MeV/u iron ions with the same dose of 0.1 Gy.

Under exposure to protons, carbon and iron ions with the same dose, the microscopic energy depositions in different layers of the granule and pyramidal cells were estimated and compared (Figure 4). We also calculated the molar concentration of molecules, which is the total number of oxidative radical species in a given volume of developed neuron models after exposure to different particles with the same dose (Figure 5).

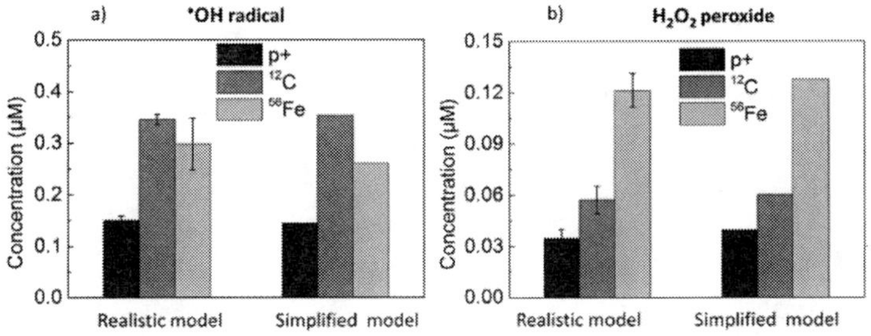

Figure 5. Total concentration of •OH and H_2O_2 in models of a DG granule cell (a) and CA3/CA1 pyramidal cell (b) at 1 nanosecond after exposure to 170 MeV protons, 290 MeV/u carbon and 600 MeV/u iron ions with the same dose of 0.1 Gy.

4 Conclusions

This study presents the results of developments of simplified neuron models with properties equivalent to realistic neuron morphology for microdosimetric calculations, in order to investigate radiation-induced effects in the electrophysiology of hippocampal neurons. The geometric structure of a simple compartmental model and a complex model with realistic morphology extracted from experimental data were constructed in the Geant4/Geant4-DNA Monte Carlo transport code. This simulation was applied to build the uniform random distribution of spines along the dendritic branches of the developed neuron models, which were adapted into standardized SWC file format. Then, we compared the distribution of microscopic energy depositions and production of oxidative radical species in the simple and realistic neuron models under exposure to protons, carbon and iron ions with the same dose. The microdosimetric distributions were obtained to be similar in both the simplified and the realistic neuron models subject to irradiation. We should emphasize that the implementation of the realistic morphology meets difficulty to develop biological neural networks and the calculation of electrical activity requires much more computation time-consuming than in the case of the developed simplified models. New prospects are thus open for further development and understanding of the electrophysiological behaviour of irradiated hippocampal neurons on the basis of simplified models.

Acknowledgements

The study was conducted as part of the joint research (Protocol No. 4695-9-17/17) between the Laboratory of Radiation Biology of JINR and the National University of Mongolia. Authors acknowledge the financial support from the JINR Theme 04-9-1077-2015/2017.

References

[1] D. Greene-Schloesser et al., Front Oncol. **2**, 73 (2012)
[2] A. G. Belyaeva et al., Neurochemical Journal **11**, 2, 168–175 (2017)
[3] I.V. Sokolova et al., Radiat. Res. **183**, 208–218 (2015)
[4] M. Batmunkh et al., J. Radiat. Res. Appl. Sci. **8**, 498–507 (2015)
[5] M. Batmunkh, O.V. Belov, L. Bayarchimeg, and O. Lkhagva, Mong. J. Phys. **2**, 317–323 (2016)
[6] O. V. Belov, M. Batmunkh, S. Incerti, and O. Lkhagva, Physica Medica **32**, 1510–1520 (2016)
[7] L. Bayarchimeg, M. Batmunkh, O. Belov, and O. Lkhagva, EPJ Web of Conferences (this issue)
[8] E. Gould et al., J. Neurosci. **10** (4), 1286 (1990)
[9] E. Gould, M.D. Allan, and B.S. McEwen, Brain Res. **525** (2), 327–329 (1990)
[10] M. Papa, et al., J. Neurosci. **15**, 1–11 (1995)
[11] A. Belly et al., J. Cell Sci. **123**, 2943–2954 (2010)

EPJ Web of Conferences **173**, 05005 (2018) https://doi.org/10.1051/epjconf/201817305005
Mathematical Modeling and Computational Physics 2017

Simulation of Radiation Damage to Neural Cells with the Geant4-DNA Toolkit

Lkhagvaa Bayarchimeg[1,*], *Munkhbaatar* Batmunkh[1,2,**], *Oleg* Belov[1,***], and *Oidov* Lkhagva[2,****]

[1] *Laboratory of Radiation Biology, Joint Institute for Nuclear Research, 141980, Dubna*
[2] *Division of Natural Sciences, National University of Mongolia, 210646, Ulaanbaatar*

Abstract. To help in understanding the physical and biological mechanisms underlying effects of cosmic and therapeutic types of radiation on the central nervous system (CNS), we have developed an original neuron application based on the Geant4 Monte Carlo simulation toolkit, in particular on its biophysical extension Geant4-DNA. The applied simulation technique provides a tool for the simulation of physical, physico-chemical and chemical processes (e.g. production of water radiolysis species in the vicinity of neurons) in realistic geometrical model of neural cells exposed to ionizing radiation. The present study evaluates the microscopic energy depositions and water radiolysis species yields within a detailed structure of a selected neuron taking into account its soma, dendrites, axon and spines following irradiation with carbon and iron ions.

1 Introduction

The investigation of radiation damage to the central nervous system (CNS) has been an on-going challenge for the last decades primarily due to the issues of brain radiotherapy and radiation protection for astronauts during space travel. Although recent findings revealed a number of molecular mechanisms associated with radiation-induced impairments in behaviour and cognition, some uncertainties exist in the initial neuronal cell injury leading to the further development of CNS malfunction. As usual, these initial stages of neuronal injury are hardly accessible to experimental measurements. Many events cannot be investigated experimentally at all. In this regard, development of computational methods for assessing these early stages of radiation damage to CNS is of great interest.

The irradiation with heavy charged particles is known to be a cause of numerous neurological disturbances in the CNS, leading to various behaviour and cognitive impairments [1]. The effects therefore cannot be completely understood without examining the earliest stages of the radiation injury to individual neurons. The initial radiation-induced events depend on the kinetic energy of the incident particles, and some of this energy can be transferred to the atoms and molecules during the incident particle pass through the target volume. The ionized or excited molecules produce new charged ions, free radicals and highly reactive chemical species that can break the chemical bonds and result thus

*e-mail: bayarchimeg@jinr.ru
**e-mail: batmunkh@jinr.ru
***e-mail: dem@jinr.ru
****e-mail: dorjtoinol@gmail.com

in cells and tissues damage. Discrete Monte Carlo techniques are efficiently used to address this issue and to perform detailed modeling of the stochastic energy depositions within the trajectory of the incident particle and all secondary low energy particles. The neural cells have complex morphologies and narrow-tree-like structures with large area that makes them unique targets for particle irradiation.

The one of the most suitable tools for the simulation of the radiation transport inside and outside neural cells is the general-purpose Geant4 Monte Carlo toolkit, which is used in a wide range of applications, including particle physics, accelerator design, space engineering, medical physics, education and industrial domains [2, 3]. The Geant4 extension to the very low energies encompasses the Geant4-DNA models serving to the calculation of initial radiation damage at the molecular level and showing good agreement with the experimental cross sections for nanodosimetry [4, 5].

We have extended a Geant4-based original application called NEURON to the investigation of the initial mechanisms of radiation-induced events in CNS, involving detailed modeling of the physico-chemical processes in the vicinity of the neurons [6–8]. The purpose of the present work is to describe the capabilities of this radiobiology application for the simulation of single neuron irradiation with energetic heavy ions.

2 The NEURON user application

NEURON is a new radiobiology application of the Geant4 Monte Carlo simulation toolkit which reads any standardized neuromorphometric format (SWC) file representing the neuron morphology and estimates the microscopic energy depositions and production of water radiolysis species in the vicinity of the neural cells. The SWC file format is an output file defining the most realistic geometrical description of the neuron structure generated by digitally tracing tools based on 3D confocal microscopy images. Many laboratories are freely sharing their tracing reconstructions within SWC formats via online archives such as the NeuroMorpho.Org repository. A typical neuron morphology is composed of a cell body (soma), a single axon and dendritic trees with thousands of spines. These structures are simulated through combination of spherical and cylindrical volumes representing the smallest neuronal elements known as compartments (two interconnected morphological points). The source code of NEURON is written in C++, where classes are defined which implement joining morphological points, geometrical transformations, rotations, calculation of the volume and the overall dimensions of the neuron, a bounding volume etc. Before computation, the user loads a standardized SWC file of a neuron and generates a bounding volume and a homogeneous spherical medium with dimensions automatically extrapolated using the SWC file describing 3D coordinates of a neuron morphology. The homogeneous medium contains both volumes of neural cell and bounding volume, that are all filled with liquid water. Geant4 condensed electromagnetic models are used outside the neuron structure and Geant4-DNA discrete physicochemical models are activated inside the neuron structure, which is declared as a *G4Region* (Figure 1a). Nevertheless, the user can change the physical and chemical processes in the simulated media via simple UI commands, such as *G4EmLivermorePhysics* (condensed) or *G4EmDNAPhysics* (discrete) for all the medium, or a combination of condensed and discrete processes outside and inside the neuron structure. The primary particle is starting at a random position on the sphere surface (homogeneous medium) and uniformly directed towards the neuron structure (Figure 1b).

Neuronal morphological data of a single granule cell (GranuleCell-Nr2.CNG.swc file) from NeuroMorpho.Org has been selected as an example for our application which is found in the dentate gyrus region of mouse hippocampus (Figure 1). The overall dimensions (width × height × depth) of the selected neuron are calculated ($252 \times 317 \times 64 - \mu m^3$) from coordinates of morphological points. The total volume and length of the selected neuron are 5048.43 μm^3 and 2191.54 μm, respectively.

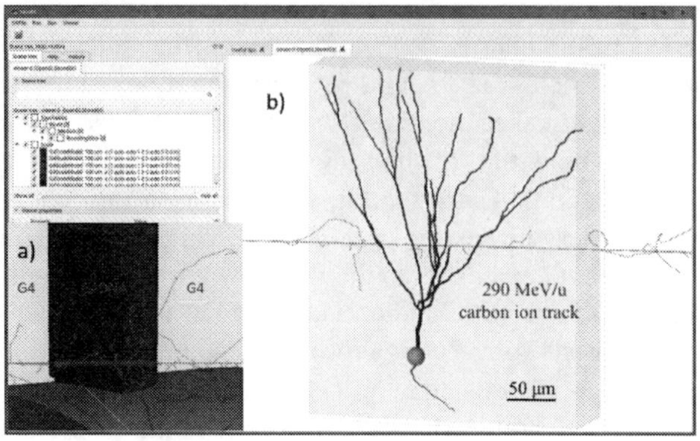

Figure 1. Screenshot of the NEURON application in the Geant4 (version 10.2-patch02) environment with Qt GUI. (a) – demonstration of Geant4 and Geant4-DNA models of particle track passing through a compartment of neuron. (b) – irradiation of a single granule cell with track structure of a carbon ion beam.

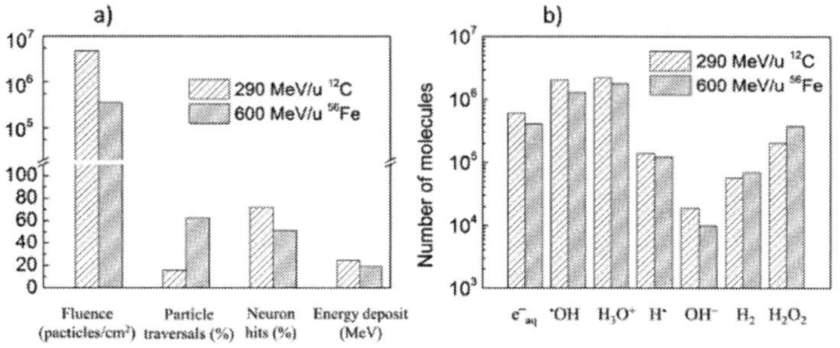

Figure 2. Summary of simulation results. (a) – particle fluence at the given dose, particle neuron traversal probability, hit percentage of neuron and total energy deposition. (b) – total number of water radiolysis species in neuron at 1 ns after exposure to 0.1 Gy carbon and iron ions.

3 Calculation of microscopic energy depositions and production of water radiolysis species in the neuron model

In the present study, a neuron is assumed to be exposed to 0.1 Gy carbon and iron ions with kinetic energies of 290 MeV/u and 600 MeV/u, respectively. The fluence dose of 0.1 Gy is equivalent to 12809 carbon ions and 938 iron ions in the simulation medium. The simulation results are reported in Figure 2 where simulations of the direct and indirect effects on a single neuron under irradiation with carbon and iron ions with same fluence dose are reported.

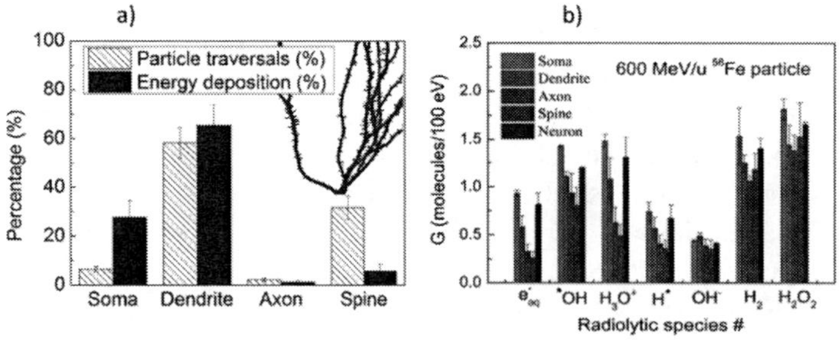

Figure 3. Analysis of energy depositions (a) and production of water radiolysis species (b) in single neuron's components, including soma, dendrite, axon and spines after irradiation with 600 MeV/u iron ions.

In the case of 0.1 Gy iron ions, particle traversals are longer than those of the carbon ions, however, the total number of neuron hits and the energy deposition are larger for the carbon ions as compared to the iron ions. The most widespread products of the water radiolysis are the solvated electrons e^-_{aq}, hydroxyl radical $^\bullet OH$, hydronium ion H_3O^+, hydrogen radical $H^\bullet$, hydroxide ion OH^-, dihydrogen H_2, and hydrogen peroxide H_2O_2. Under exposure to iron ions, the total number of H_2 and H_2O_2 species are larger than those of the carbon ions at a same given dose (Figure 2b). In the current study we have obtained morphological points of dendritic spines from original experimental data of selected neurons [9] that included coordinates and sizes of only spine heads, do not include neck information and spines morphology. The diameter and length of the spine neck were generated from ratio of spines head/neck diameter and total length derived from literature [10]. Thus, 2446 spines were simulated and each spine defined as two solids of sphere and cylinder, neck diameter depending on various size of spines head as ratio of 1.1. Figure 3 presents detailed results of the analysis of energy depositions and production of chemical species in the neutron structure, taking into account such cell components as soma, dendrites, axon and spines. In this case, most particles are traversing the structure of dendrites and spines. In addition, the energy deposition occurred mainly in dendrites rather than in other structures at the given dose of iron ions. As a result (Figure 3b), the hydrogen peroxide is the most observed product in soma and dendrites after irradiation.

4 Conclusions

The paper presented simulation results of the new NEURON application of the Geant4 and Geant4-DNA toolkits. The applied simulation tool allows for the modeling of neuron cell irradiation, including physical, physico-chemical and chemical processes (e.g. production of water radiolysis species in the vicinity of the neurons). It uses realistic geometrical models of the neurons generated from a standardized SWC file representing the neuron morphology and a multi-scale discrete/non-discrete combination of Geant4-DNA and Geant4 electromagnetic processes. We have obtained accurate estimates for the probability of particle traversals, distribution of microscopic energy depositions, hit percentage of neuron structures and number of water radiolysis species in chosen neuron and such cell components as soma, dendrites, axon and spines after irradiation of energetic carbon and iron ions. In the case of carbon ions (290 MeV/u), the total energy deposition and number of most water radiolysis in a neuron were observed larger than those of the iron ions (600 MeV/u) at a same dose. The NEURON application helps to understand the initial physico-chemical events of ionizing radiation action in the CNS in the context of space radiobiology problems.

Acknowledgements

The study was conducted as part of the joint research protocol between the Laboratory of Radiation Biology of the Joint Institute for Nuclear Research (JINR) and the National University of Mongolia. Authors acknowledge the financial support from the JINR Theme 04-9-1077-2015/2017.

References

[1] A. G. Belyaeva et al., Neurochemical Journal **11**, 2, 168–175 (2017)
[2] S. Agostinelli et al., Nucl Instrum Methods Phys Res Section A **506**, 250–303 (2003)
[3] J. Allison et al., Nuclear Instruments and Methods A **835**, 186–225, (2016)
[4] S. Incerti et al., Med. Phys. **37**, 4692–4708 (2010)
[5] M. A. Bernal et al., Phys. Med. **31**, 861–874 (2015)
[6] M. Batmunkh et al., J. Radiat. Res. Appl. Sci. **8**, 498–507 (2015)
[7] O. V. Belov, M. Batmunkh, S. Incerti, and O. Lkhagva, Physica Medica **32**, 1510–1520 (2016)
[8] M. Batmunkh, O.V. Belov, L. Bayarchimeg, and O. Lkhagva, Mong. J. Phys. **2**, 317–323 (2016)
[9] M. Vuksic et al., HIPPOCAMPUS **18**, 36, 4–375 (2008)
[10] A. Rodriguez et al., PLOS one **3(4)**, e1997 (2008)

EPJ Web of Conferences **173**, 05006 (2018) https://doi.org/10.1051/epjconf/201817305006
Mathematical Modeling and Computational Physics 2017

The Automation of Stochastization Algorithm with Use of SymPy Computer Algebra Library

Anastasya Demidova[1,⋆], *Migran* Gevorkyan[1,⋆⋆], *Dmitry* Kulyabov[1,2,⋆⋆⋆], *Anna* Korolkova[1,⋆⋆⋆⋆], and *Leonid* Sevastianov[1,3,†]

[1] *Department of Applied Probability and Informatics, Peoples' Friendship University of Russia (RUDN University), 6 Miklukho-Maklaya str., Moscow, 117198, Russia*
[2] *Laboratory of Information Technologies, Joint Institute for Nuclear Research, 6, Joliot-Curie St., Dubna, Moscow region, 141980, Russia*
[3] *Bogoliubov Laboratory of Theoretical Physics, Joint Institute for Nuclear Research, 6, Joliot-Curie St., Dubna, Moscow region, 141980, Russia*

Abstract. SymPy computer algebra library is used for automatic generation of ordinary and stochastic systems of differential equations from the schemes of kinetic interaction. Schemes of this type are used not only in chemical kinetics but also in biological, ecological and technical models. This paper describes the automatic generation algorithm with an emphasis on application details.

1 Introduction

To describe models, where the intensity of the interaction between components depends on the concentration levels of the components, one can use the schemes of kinetic interactions [1–4]:

$$\sum_{i=1}^{n} N_i^s X_i \underset{k_s^-}{\overset{k_s^+}{\longleftrightarrow}} \sum_{i=1}^{n} M_i^s X_i, \qquad s = 1, \dots, S.$$

Here X_i represents the quantity of i-th component, the matrices $\mathbf{M} = [M_i^s]$ and $\mathbf{N} = [N_i^s]$ are *system state matrices* which determine the number of interacting components of the system at each stage of the reaction. k_s^+ and k_s^- are *coefficients of interaction* in direct and reverse directions, n is system dimension, S is the number of interactions.

Kinetic schemes are often used to model chemical reactions, biological and ecological systems. From these diagrams one can derive a system of ordinary differential equations (ODEs) which gives the evolution of variables $X_i(t)$ over time line.

Our research team has generalized the method of stochastisation of deterministic models [5]. This method allows to obtain the stochastic model, based on implicit stochastic properties of the system.

⋆e-mail: demidova_av@rudn.university
⋆⋆e-mail: gevorkyan_mn@rudn.university
⋆⋆⋆e-mail: kulyabov_ds@rudn.university
⋆⋆⋆⋆e-mail: korolkova_av@rudn.university
†e-mail: sevastianov_la@rudn.university

All necessary information for the stochastisation process can be extracted from the kinetic interaction schemes. For multi-dimensional systems the stochastisation method requires a large amount of routine work. However, it is possible to program all the steps of the method and obtain the drift vector and the diffusion matrix. We carry out the implementation of the algorithm by using the Python language (version 3) and SymPy library for computer algebra.

2 An algorithm of obtaining ODE and SDE systems

The initial data are the system state matrices **M**, **N** and the coefficients of the interaction between system components k_s^+ and k_s^-. At the first algorithm stage one has to find a matrix $\mathbf{R} = \mathbf{M}^T - \mathbf{N}^T$, the columns of which are denoted by vectors $\mathbf{r}^j$. Next, one has to calculate the auxiliary values s_j^+ and s_j^-:

$$s_j^+(\mathbf{x}) = k_j^+ \prod_{i=1}^{n} \frac{x^i!}{(x^i - N_i^j)!} = k_j^+ \times \frac{x^1!}{(x^1 - N_1^j)!} \times \frac{x^2!}{(x^2 - N_2^j)!} \times \cdots \times \frac{x^n!}{(x^n - N_n^j)!},$$

$$s_j^-(\mathbf{x}) = k_j^- \prod_{i=1}^{n} \frac{x^i!}{(x^i - M_i^j)!} = k_j^+ \times \frac{x^1!}{(x^1 - M_1^j)!} \times \frac{x^2!}{(x^2 - M_2^j)!} \times \cdots \times \frac{x^n!}{(x^n - M_n^j)!}.$$

Using these values one can obtain *the drift vector* $\mathbf{f}(\mathbf{x})$ and *the diffusion matrix* $\mathbf{G}(\mathbf{x})$. The drift vector is calculated by the formula:

$$\mathbf{f}(\mathbf{x}) = \mathbf{r}^1(s_1^+(\mathbf{x}) - s_1^-(\mathbf{x})) + \mathbf{r}^2(s_2^+(\mathbf{x}) - s_2^-(\mathbf{x})) + \cdots + \mathbf{r}^s(s_s^+(\mathbf{x}) - s_s^-(\mathbf{x})),$$

and it can be used to write out the system of ODEs

$$\frac{\mathrm{d}\mathbf{x}}{\mathrm{d}t} = \mathbf{f}(\mathbf{x}).$$

To get the system of stochastic differential equations (SDEs) in Itô form, it is necessary to calculate also the diffusion matrix $\mathbf{G}(\mathbf{x})$:

$$\mathbf{G}(\mathbf{x}) = \mathbf{r}^1(\mathbf{r}^1)^T(s_1^+(\mathbf{x}) - s_1^-(\mathbf{x})) + \mathbf{r}^2(\mathbf{r}^2)^T(s_2^+(\mathbf{x}) - s_2^-(\mathbf{x})) + \cdots + \mathbf{r}^s(\mathbf{r}^s)^T(s_s^+(\mathbf{x}) - s_s^-(\mathbf{x})).$$

After that it is possible to obtain the system of Itô SDE:

$$\mathrm{d}\mathbf{x} = \mathbf{f}(\mathbf{x})\,\mathrm{d}t + \sqrt{\mathbf{G}(\mathbf{x})}\mathbf{W}(t),$$

where **W** is the n-th dimensional Wiener process.

2.1 Brief description of the software implementation

For the automation of the above method we used Python [6] with SymPy [7] module. The program takes symbolic vectors X and K and numeric matrices N and M on input, and gives the drift vector and the diffusion matrix on output. For numerical matrices we use NumPy arrays [7]. As interface for our module we provide two functions:

```
f = drift_vector(X, K, N, M),
G = diffusion_matrix(X, K, N, M).
```

These functions return the symbolic drift vector and diffusion matrix. It is possible to use built-in SymPy methods to convert **f** and **G** to different formats, such as LaTeX formula notation or expression in any supported programming language. SymPy provides support for a large number of programming languages ranging from popular ones such as C/C++, Fortran, Java etc., to new growing languages such as Julia language.

3 Examples and program verification

Let us consider two examples. The first example is the Verhulst model describing the growth of population in conditions of limited resources (logistic curve), and the second example is the model of brusselator [8]. Ilya Prigogine has proposed this model as a simple example of Belousov–Zhabotinsky reaction type. These models are well known and we use them to test our program and to show algorithm in action.

3.1 Verhulst model

The model of logistic growth is described by the following scheme:

$$X \xrightarrow{k_1} 2X, \qquad X + X \xrightarrow{k_2} X.$$

In this case, the matrices **N** and **M** turn into vectors: $N_1^1 = 1$, $N_1^2 = 2$, $M_1^1 = 2$, $M_1^2 = 1$, and vectors **r** turn into scalars $r^1 = 1$, $r^2 = -1$. Because the reaction is not reversible, it is enough to compute only s^+:

$$s_1^+ = k_1 \cdot \frac{x!}{(x-1)!} = k_1 x,$$
$$s_2^+ = k_2 \cdot \frac{x!}{(x-2)!} = k_2 x(x-1) = k_2 x^2 - k_2 x.$$

$$f(x) = 1 \cdot k_1 x - 1(k_2 x^2 - k_2 x) = k_1 x + k_2 x - k_2 x^2,$$
$$g(x) = 1 \cdot k_1 x + (k_2 x^2 - k_2 x) = k_1 x - k_2 x + k_2 x^2.$$

Usually the term $k_2 x$ is discarded due to the fact that the x^2 value is large and the x value is small:

$$\frac{\mathrm{d}x}{\mathrm{d}t} = k_1 x - k_2 x^2,$$

and

$$\mathrm{d}x(t) = (k_1 x - k_2 x^2)\,\mathrm{d}t + \sqrt{k_1 x + k_2 x^2}\,\mathrm{d}W.$$

3.2 The Brusselator model

The Brusselator model is completely determined by the following scheme:

$$A \xrightarrow{k_1} X, \qquad 2X + Y \xrightarrow{k_2} 2X,$$
$$B + X \xrightarrow{k_3} Y + D, \qquad X \xrightarrow{k_4} E.$$

Besides the usual reagents X and Y, the inexhaustible reagents A, B, D and E are also involved into reaction. Their number is considered to be time invariant. They cannot be taken into account while constructing the matrices M and N. However, even if we include them in M and N we will obtain correct results. Let use to demonstrate that fact.

$$\mathbf{N} = \begin{bmatrix} 0 & 0 & 1 & 0 & 0 & 0 \\ 2 & 1 & 0 & 0 & 0 & 0 \\ 1 & 0 & 0 & 1 & 0 & 0 \\ 1 & 0 & 0 & 0 & 0 & 0 \end{bmatrix}, \quad \mathbf{M} = \begin{bmatrix} 1 & 0 & 0 & 0 & 0 & 0 \\ 3 & 0 & 0 & 0 & 0 & 0 \\ 0 & 1 & 0 & 0 & 1 & 0 \\ 0 & 0 & 0 & 0 & 0 & 1 \end{bmatrix}, \quad \mathbf{x} = \begin{bmatrix} x \\ y \\ a \\ b \\ d \\ e \end{bmatrix}, \quad \mathbf{k} = \begin{bmatrix} k_1 \\ k_2 \\ k_3 \\ k_4 \end{bmatrix}.$$

Based on the above matrices, our Python functions return the following drift vector and diffusion matrix:

$$\begin{pmatrix} ak_1 - bk_3x + k_2xy(x-1) - k_4x \\ bk_3x - k_2xy(x-1) \\ -ak_1 \\ -bk_3x \\ bk_3x \\ k_4x \end{pmatrix},$$

$$\begin{bmatrix} ak_1 + bk_3x + k_2xy(x-1) + k_4x & -bk_3x - k_2xy(x-1) & -ak_1 & bk_3x & -bk_3x & -k_4x \\ -bk_3x - k_2xy(x-1) & bk_3x + k_2xy(x-1) & 0 & -bk_3x & bk_3x & 0 \\ -ak_1 & 0 & ak_1 & 0 & 0 & 0 \\ bk_3x & -bk_3x & 0 & bk_3x & -bk_3x & 0 \\ -bk_3x & bk_3x & 0 & -bk_3x & bk_3x & 0 \\ -k_4x & 0 & 0 & 0 & 0 & k_4x \end{bmatrix}.$$

Now it is possible to write out the ODE system:

$$\begin{cases} \dot{x} = ak_1 - bk_3x + k_2xy(x-1) - k_4x, \\ \dot{y} = bk_3x - k_2xy(x-1). \end{cases}$$

4 Conclusion

In this paper, we described an algorithm of obtaining ODE and SDE systems from the interaction schemes. Taking into account the inhibitory interactions, will result in algorithm improvement.

Acknowledgement

This work is partially supported by RFBR grants No's 15-07-08795 and 16-07-00556. The publication was also financially supported by the "RUDN University Program 5-100".

References

[1] M. Hnatič, E.G. Eferina, A.V. Korolkova, D.S. Kulyabov, L.A. Sevastyanov, EPJ Web of Conferences **108**, 58 (2015), [arXiv]1603.02205

[2] E.G. Eferina, M. Hnatich, A.V. Korolkova, D.S. Kulyabov, L.A. Sevastianov, T.R. Velieva, in *Distributed Computer and Communication Networks. DCCN 2016. Communications in Computer and Information Science*, edited by V.M. Vishnevskiy, K.E. Samouylov, D.V. Kozyrev (Springer, Cham, 2016), Vol. 678, pp. 483–497

[3] A.V. Demidova, M.N. Gevorkyan, T-Comm **8**, 43 (2014)

[4] M. Hnatich, J. Honkonen, Physical Review E **61**, 3904 (2000)

[5] A. Demidova, M.N. Gevorkyan, A.D. Egorov, D.S. Kulyabov, A.V. Korolkova, L.A. Sevastyanov, Bulletin of Peoples' Friendship University of Russia. Series Mathematics. Information sciences. Physics. pp. 71–85 (2014)

[6] G. Rossum, Tech. rep., Amsterdam, The Netherlands, The Netherlands (1995)

[7] E. Jones, T. Oliphant, P. Peterson et al., *SciPy: Open source scientific tools for Python* (2001)

[8] C.W. Gardiner, *Handbook of Stochastic Methods: for Physics, Chemistry and the Natural Sciences* (Springer Series in Synergetics, 1985)

Web Platform for Sharing Modeling Software in the Field of Nonlinear Optics

Julia Dubenskaya[1,*], *Alexander* Kryukov[1], and *Andrey* Demichev[1]

[1] *Skobeltsyn Institute of Nuclear Physics, Lomonosov Moscow State University, Leninskie gory, GSP-1, Moscow 119991, Russian Federation*

Abstract.
We describe the prototype of a Web platform intended for sharing software programs for computer modeling in the rapidly developing field of the nonlinear optics phenomena. The suggested platform is built on the top of the HUBZero open-source middleware. In addition to the basic HUBZero installation we added to our platform the capability to run Docker containers via an external application server and to send calculation programs to those containers for execution. The presented web platform provides a wide range of features and might be of benefit to nonlinear optics researchers.

1 Introduction

Nonlinear optics is a rapidly evolving area of modern physical research and engineering with many important applications. Successful development of optical devices requires complicated modeling of physical processes that occur in components of devices. Many developers of application software have the intention to share their products among other researchers.

At the present time dedicated web resources are developed and widely used in the research community all over the world. There are a number of web resources that provide software on nonlinear optics [1–3]. Most of them are intended for commercial distribution of desktop applications. These applications need to be downloaded to the local computer, and then could be launched offline.

The architecture of an online web platform for nonlinear optics was proposed in [4, 5]. A distinguishing feature of the platform in question is that it provides web access to the already installed and configured software. A web platform of that kind will be called "a closed web platform". It is easily manageable, but sometimes its fixed functionality is too limited. By contrast, a platform where users can install and share their own programs will be called "an open web platform". In our studies we came across that there is a need for such a platform for nonlinear optics.

We developed a web platform for sharing users' software in the field of nonlinear optics using HUBzero middleware [6] as a basis. The HUBzero middleware provides services for application software installation and includes a set of tools for simplifying the interaction between software developers, regular users and resource administrators. In particular, the HUBzero middleware comprises tools for transforming the application software into software as a service (SaaS).

*e-mail: jdubenskaya@gmail.com

The basic HUBzero installation provides a working web server and a local computational resource. By further customizing it is possible to set up access to GRID resources [7]. We added to our HUBzero installation the capability to execute applications in a specially prepared Docker container [8] which is launched on the external application server. From a technical perspective our platform is a single frontend for remote hardware resources where special middleware services are installed and run.

Thus, the presented open web platform is a user-friendly product with extended functionality that combines the power of the HUBzero middleware with the flexibility of the external Docker containers.

In the section 2, we will take a closer look at the architecture of the web platform. In the section 3, the operation of the web platform will be shown on a concrete example. In the Conclusion, we will summarize the accomplished work, and discuss practical applications and future improvements of the presented platform.

2 The web platform architecture

The HUBzero is a complex product that integrates many software and is based on various technologies. The basic HUBzero installation provides access to a powerful all-in-one server with rich functionality. In general, this functionality includes software lifecycle management, software launching, searching, access rights management, logging, software codes repository management, and so on. By further customizing it is possible to set up access to GRID computing resources, but the customizing process takes a lot of time, is difficult to perform and the type of external resources is restricted to GRID only.

We propose a lightweight solution for remote tasks execution. The basic idea is that in addition to the HUBzero server we deploy an external application server capable of receiving calculation tasks, running these tasks in a dedicated Docker container, and returning the result in a predefined form. A detailed description of the architecture and protocol of operation of such a server is presented in [9, 10]. This external application server is logically detached from the HUBzero platform, all the interactions are conducted via standard HTTP requests (REST API).

In order to run the calculation program to be shared on external resources a user should prepare two programs:

1. A frontend program to be installed and started via HUBzero. Upon starting, it receives input parameters from the user, sends a computational request to the application server, waits for the results to be received and displays them in a graphical form. Exactly this program can be shared among other users using standard HUBzero tools.
2. A calculation program to be run on external resources. The calculation program will be launched in the Docker container by the application server after receiving a request from the frontend program. Upon starting, it performs the requested calculations and returns results.

To ensure the transfer of a remote program to a Docker container a user needs to pre-allocate it in the image repository of the application server, while the frontend program call must contain the corresponding program identifier. Thus, running tasks on external resources is implemented using standard HUBzero tools, and the overall architecture of the system becomes quite flexible. As a result, the users can seamlessly mix launching basic HUBzero tools with running external Docker containers.

3 An example of using the web platform for nonlinear optics

As stated above, we used the HUBZero middleware to deploy a prototype of a web platform for the library of programs in nonlinear optics. As an example we pick up a program which calculates the angle of phase matching [11] for a large number of crystal types stored in a database. The "angle of phase matching" is a geometrical feature which specifies conditions for nonlinear optical effects in crystal to become volume effects rather than point effects. Volume effect means that many atoms contribute to the effect.

This program calculates the angle of phase matching for the second-harmonic generation as a function of temperature, crystal parameters, and frequency of the incident radiation. The example program initially is a command-line script written in python, while its graphical interface is created by the special HUBzero module called the Rappture [13].

This program has been installed by its authors on our prototype of a web platform [12], after that the authors granted access to this program to a certain group of users. Further developments are considered to meet supplementary user needs. A frontend program concerning the use of Docker containers can be installed and shared in a similar way.

On the initial screen of the program there is a list of input parameters and a special button that should be pressed to perform the calculation. After the button is pressed the output results can be seen on the screen. In our example results are shown on a two-dimensional plot (see figure 1). The horizontal axis represents the frequency of the incident radiation and the vertical axis represents the calculated angle of phase matching. Holding a mouse pointer over the particular graph point a user can see the precise calculated value. Also there is a very convenient way to download all the results to the user's local computer using the "Download" button located on the right side of the screen. After that a user can choose another set of parameters, run calculation one more time and download results again.

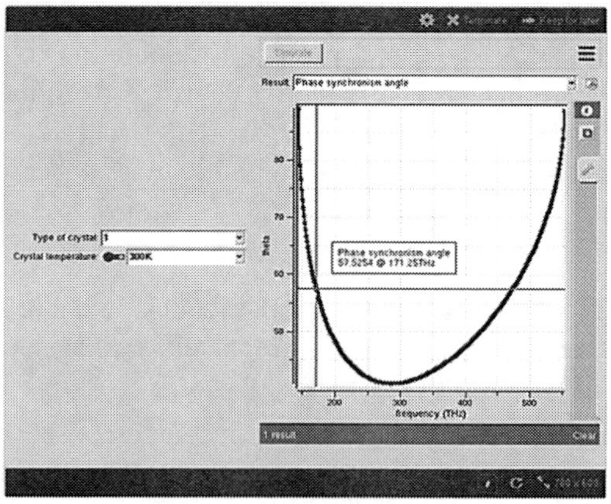

Figure 1. Screenshot illustrating the dependence of the angle of phase matching on the laser frequency

The above example shows that our HUBZero-based web platform is an open one, as users themselves can install new programs and share them among each other. This is an important difference from the already existing systems in the field of nonlinear optics. With the help of such a platform, users can share also the data, for example, the parameters of the crystals, which are used in the calculations.

It ought to be remarked that using Docker containers for external computation somewhat complicates the task of the author who wants to share the program. As for a regular user, nothing will change, and this is one of the most important features of the proposed solution.

4 Conclusion

Online web platforms is a fast-growing and promising area thanks largely to their usability and seamlessness. Unlike the solutions that offer a pre-installed set of programs, an open web platform allows its users to install their own programs and share them among others.

The HUBzero middleware is a very convenient tool for creating web platforms for scientific research, which was shown in this article by the example of a program for the angle of phase matching calculation. We also added to the basic HUBzero installation the capability to execute a calculation program on a specially created Docker container which is launched on the remote application server. This lightweight solution enhances the functionality of the platform.

As for the directions for our future research, we plan to improve our existing installation by increasing the computational capacity and providing some additional applications.

Summarizing, the proposed web platform provides shared access to the programs in nonlinear optics developed by the different users, which, in turn, stimulates wider use and faster development of that software.

Acknowledgement

The authors are grateful to the Russian Foundation for Basic Research (grant No 15-07-09309), with the financial support of which this work was carried out.

References

[1] NLOsource, http://www.nlosource.com/index.html
[2] RP Photonics, https://www.rp-photonics.com/
[3] AS-Photonics, http://www.as-photonics.com/index.php
[4] Polyakov S., et al., Procedia Computer Science **66** 602–608 (2015)
[5] Prikhod'ko N., et al., Journal of Physics: Conference Series 681 (2016) 012052
[6] McLennan M. and Kennell R., Comput. Sci. Eng. **12**, 48–52 (2010)
[7] Thompson C. S., et al., Proc. of the 2015 XSEDE Conf.: Scientific Advancements Enabled by Enhanced Cyberinfrastructure, 36:1–36:7 (2015)
[8] The docker software, http://www.docker.com
[9] Polyakov S., et al., Selected Papers of the 7th Int. Conf. Distributed Computing and Grid-technologies in Sci. and Edu. **1787**, 557–560 (2017)
[10] Polyakov S., et al., In Proc. CSP2017, Journal of Physics: Conference Series (2018), to be published
[11] Ed. Trager F., *Springer Handbook of Laser and Optics* (Spr.-Verlag, 2012) 1694 pp.
[12] Publishing of a tool, https://help.hubzero.org/resources/173
[13] The Rappture, http://nanohub.org/infrastructure/rappture/

EPJ Web of Conferences **173**, 05008 (2018) https://doi.org/10.1051/epjconf/201817305008
Mathematical Modeling and Computational Physics 2017

Symbolic-Numerical Modeling of the Influence of Damping Moments on Satellite Dynamics

Sergey A. Gutnik[1,2,*] and *Vasily A.* Sarychev[3,**]

[1] *Moscow State Institute of International Relations 76, Prospekt Vernadskogo, Moscow, 119454, Russia*
[2] *Moscow Institute of Physics and Technology, 9 Institutskiy lane, Dolgoprudny, 141700, Russia*
[3] *Keldysh Institute of Applied Mathematics (RAS) 4, Miusskaya Square, Moscow, 125047, Russia*

Abstract. The dynamics of a satellite on a circular orbit under the influence of gravitational and active damping torques, which are proportional to the projections of the angular velocity of the satellite, is investigated. Computer algebra Gröbner basis methods for the determination of all equilibrium orientations of the satellite in the orbital coordinate system with given damping torque and given principal central moments of inertia were used. The conditions of the equilibria existence depending on three damping parameters were obtained from the analysis of the real roots of the algebraic equations spanned by the constructed Gröbner basis. Conditions of asymptotic stability of the satellite equilibria and the transition decay processes of the spatial oscillations of the satellite at different damping parameters have also been obtained.

1 Introduction

In this paper, a symbolic-numerical study of the satellite dynamics under the influence of gravitational and active damping torques is presented. The gravity oriented satellite with different moments of inertia in the central Newtonian force field in a circular orbit has 24 equilibrium orientations and four of them are stable [1]. An important property of the gravity orientation systems is that these systems can operate for a long time without needing fuel spending. In the present work we analyze the behavior of a satellite acted upon by the gravity gradient and active damping torques. The action of damping torques can provide the asymptotic stability of the equilibria of the gravity oriented satellites. Therefore, it is important to investigate the joint action of the gravitational and active damping torques and to analyze the necessary and sufficient conditions for asymptotic stability of the satellite equilibria on a circular orbit.

In the present work, the problem of determination of the classes of equilibrium orientations and the conditions for asymptotic stability of defined equilibria for the general values of damping torques is considered. The investigation of equilibria was performed by application of the Computer Algebra Gröbner basis methods. The types of transition decay processes of spatial oscillations of the satellite at different damping parameters have been investigated numerically.

The computer algebra methods for the investigation of the asymptotic stability of the equilibrium orientations of a satellite were successfully used earlier for another model of damping torques [2].

*e-mail: s.gutnik@inno.mgimo.ru
**e-mail: vas31@rambler.ru

2 Equations of Motion

Consider the motion of the satellite-rigid body subjected to gravitational and active damping torques on a circular orbit. We assume that the active damping torques are proportional to the projections of the angular velocity of the satellite. To write the equations of motion we introduce two right-handed Cartesian coordinate systems with origin at the satellite center of mass O. The orbital coordinate system $OXYZ$; the axis OZ is directed along the radius-vector from the Earth center of mass and the satellite center of mass; the axis OX is along the direction of the satellite orbital motion. The satellite body coordinate system $Oxyz$; Ox, Oy, and Oz are the principal central axes of inertia of the satellite. The orientation of the satellite body coordinate system $Oxyz$ with respect to the orbital coordinate system is determined by means of the aircraft angles of pitch (α), yaw (β), and roll (γ). The direction cosines of the transformation matrix between the orbital coordinate system $OXYZ$ and $Oxyz$ are given in [3].

Let the satellite be acted upon by the moments of active damping, their integral vector projections on the axis Ox, Oy, and Oz are equal respectively to: $M_x = \bar{k}_1 p_1$, $M_y = \bar{k}_2 q_1$, and $M_z = \bar{k}_3 r_1$. Here $\bar{k}_1$, $\bar{k}_2$, and $\bar{k}_3$ are the damping coefficients, p_1, q_1, and r_1 are the projections of the satellite angular velocity onto the axes Ox, Oy, and Oz; ω_0 is the angular velocity of the orbital motion of the satellite center of mass. Then, the equations of satellite attitude motion can be written in the dimensionless Euler form:

$$
\begin{aligned}
&\theta_A \dot{p} + (\theta_C - 1)qr - 3(\theta_C - 1)a_{32}a_{33} + \tilde{k}_1 p = 0,\\
&\dot{q} + (\theta_A - \theta_C)rp - 3(\theta_A - \theta_C)a_{31}a_{33} + \tilde{k}_2 q = 0,\\
&\theta_C \dot{r} + (1 - \theta_A)pq - 3(1 - \theta_A)a_{31}a_{32} + \tilde{k}_3 r = 0,
\end{aligned} \tag{1}
$$

$$
p = (\dot{\alpha} + 1)a_{21} + \dot{\gamma}, \quad q = (\dot{\alpha} + 1)a_{22} + \dot{\beta}\sin\gamma, \quad r = (\dot{\alpha} + 1)a_{23} + \dot{\beta}\cos\gamma.
$$

Here, $\theta_A = A/B$, $\theta_C = C/B$ denote dimensionless inertial parameters, A, B, and C are the principal central moments of inertia of the satellite, $p = p_1/\omega_0$, $q = q_1/\omega_0, r = r_1/\omega_0$, $\tilde{k}_1 = \bar{k}_1/B\omega_0$, $\tilde{k}_2 = \bar{k}_2/B\omega_0$, $\tilde{k}_3 = \bar{k}_3/B\omega_0$, $\tau = \omega_0 t$. The dot denotes differentiation with respect to τ.

3 Equilibrium Orientations of the Satellite

Putting in (1) $\alpha = \alpha_0 = \text{const}, \beta = \beta_0 = \text{const}, \gamma = \gamma_0 = \text{const}$, we obtain the stationary equations

$$
a_{22}a_{23} - 3a_{32}a_{33} + k_1 a_{21} = 0, a_{21}a_{23} - 3a_{31}a_{33} + k_2 a_{22} = 0, a_{21}a_{22} - 3a_{31}a_{32} + k_3 a_{23} = 0, \tag{2}
$$

which specifies the equilibrium orientations of the satellite in the orbital coordinate system at $A \neq B \neq C$, such that the orthogonality conditions for the direction cosines are fulfilled

$$
a_{i1}a_{j1} + a_{i2}a_{j2} + a_{i3}a_{j3} - \delta_{ij} = 0, \tag{3}
$$

where δ_{ij} is the Kronecker delta and $(i, j = 1, 2, 3)$. Here $k_1 = \tilde{k}_1/(C - B)$, $k_2 = \tilde{k}_2/(A - C)$, $k_3 = \tilde{k}_3/(B - A)$. We will consider the case when the damping coefficients k_1, k_2 and k_3 are positive.

Equations (2) and (3) form a closed system of equations with respect to the six direction cosines identifying the satellite equilibrium orientations. For this system of equations we formulate the following problem: given k_1, k_2, and k_3, determine all the nine directional cosines, i.e., all satellite equilibrium orientations in the orbital coordinate system. After a_{21}, a_{22}, a_{23}, a_{31}, a_{32}, and a_{33} are found, the direction cosines a_{11}, a_{12} and a_{13} can be determined from the conditions of orthogonality.

To find solutions of the algebraic system (2)–(3) we used the algorithm for constructing the Gröbner bases [4], implemented as the package `Groebner[Basis]` in the computer algebra system Maple

15 [5]. We constructed the Gröbner basis of the system of six second-order polynomials (2)–(3) with six variables a_{ij} ($i = 2, 3; j = 1, 2, 3$), with respect to the lexicographic ordering of variables by using the option `plex`. The list of polynomials F includes the left-hand sides of algebraic equations. (2), (3): `G:=map(factor, Groebner[Basis](F, plex(a31, ... a22)))`. Below we write down the polynomial in the Gröbner basis that depends only on one variable $x = a_{22}$. This polynomial has the simple form

$$a_{22}(a_{22}^2 - 1) = 0. \tag{4}$$

The polynomials from the Gröbner basis that depend on the variables a_{21}, a_{23} have similar forms $(a_{21}(a_{21}^2 - 1) = 0, a_{23}(a_{23}^2 - 1) = 0)$. The polynomial from the Gröbner basis that depends on the variable a_{33} is: $a_{33}(a_{33}^4 - 9a_{33}^2 + k_2^2)((a_{33}^4 - 9a_{33}^2 + k_1^2) = 0$.

To determine all equilibria, it is required to consider separately the following three cases: $a_{22} = 1$, $a_{22} = -1$, and $a_{22} = 0$. In the first and second cases, i.e., $a_{22} = \pm 1$ ($a_{21} = a_{23} = 0$) the system (2), (3) results in the following equilibrium solutions ($k_2 < 9/4$):

$$a_{22} = \pm 1, a_{21} = a_{22} = a_{32} = 0; a_{31}^2 = (3 \mp \sqrt{9 - 4k_2^2})/6, a_{33}^2 = (3 \pm \sqrt{9 - 4k_2^2})/6. \tag{5}$$

In the third case, $a_{22} = 0$, it is easy to find from the system (2)–(3) two sets of simple equilibrium solutions.

4 Necessary and Sufficient Conditions of Asymptotic Stability of the Equilibrium Orientations of a Satellite

In order to study the necessary and sufficient conditions of asymptotic stability of the determined above equilibrium orientations of the system (2)–(3) we use the linearized system of equations (1) in the vicinity of the equilibrium solution (5)

$$\alpha_0 = \pm \arcsin\left(\left[\left(3 - \sqrt{9 - 4k^2}\right)/6\right]^{1/2}\right), \quad \beta_0 = \gamma_0 = 0. \tag{6}$$

The quantities α, β, and γ are written in the form $\alpha = \alpha_0 + \bar{\alpha}, \beta = \beta_0 + \bar{\beta}, \gamma = \gamma_0 + \bar{\gamma}$, where $\bar{\alpha}, \bar{\beta}$ and $\bar{\gamma}$ are small deviations from the equilibrium orientation of the satellite $\alpha = \alpha_0, \beta = \beta_0, \gamma = \gamma_0$, satisfying the system of equations (2)–(3).

Now let us consider the small oscillations of the satellite in the vicinity of the specific equilibrium orientation (6) in the case $k_1 = k_2 = k_3 = k$. The linearized equations in this case take the form

$$\begin{gathered}
\ddot{\bar{\alpha}} + k\dot{\bar{\alpha}} + 3(\theta_A - \theta_C)(\cos^2\alpha_0 - \sin^2\alpha_0)\bar{\alpha} = 0, \\
\theta_C\ddot{\bar{\beta}} + k\dot{\bar{\beta}} - (\theta_A + \theta_C - 1)\dot{\bar{\gamma}} + (1 - \theta_A)(1 + 3\sin^2\alpha_0)\bar{\beta} + [3(1 - \theta_A)\sin\alpha_0\cos\alpha_0 - k]\bar{\gamma} = 0, \\
\theta_A\ddot{\bar{\gamma}} + k\dot{\bar{\gamma}} + (\theta_A + \theta_C - 1)\dot{\bar{\beta}} + (1 - \theta_C)(3\cos^2\alpha_0 + 1)\bar{\gamma} + [3(1 - \theta_C)\sin\alpha_0\cos\alpha_0 + k]\bar{\beta} = 0.
\end{gathered} \tag{7}$$

The characteristic equation of the system (7)

$$[\lambda^2 + k\lambda + 3(\theta_A - \theta_C)\sqrt{9 - 4k^2}](A_0\lambda^4 + A_1\lambda^3 + A_2\lambda^2 + A_3\lambda + A_4) = 0. \tag{8}$$

decomposes into quadratic and 4th degree equations. Here, the following notations are used:

$$\begin{aligned}
A_0 &= \theta_A\theta_C, \quad A_1 = k(\theta_C + \theta_A), \quad A_4 = (1 - \theta_A)(1 - \theta_C)(3 + k^2) + (\theta_A + \theta_C - 1)k^2. \\
A_2 &= k^2 + (\theta_A + \theta_C - 1)^2 + [\theta_A(1 - \theta_A)(5 - \sqrt{9 - 4k^2}) + \theta_C(1 - \theta_C)(5 + \sqrt{9 - 4k^2})]/2, \\
A_3 &= k[(\theta_A + \theta_C - 1)(\theta_A - \theta_C + 2) + [(1 - \theta_C)(5 + \sqrt{9 - 4k^2}) + (1 - \theta_A)(5 - \sqrt{9 - 4k^2})]/2].
\end{aligned}$$

The necessary and sufficient conditions for the asymptotic stability (Routh-Hurwitz criterion) of the equilibrium solution (6) take the following form:

$$k > 0,\ \theta_A - \theta_C > 0,\ \Delta_1 = A_1 > 0,\ \Delta_2 = A_1A_2 - A_0A_3 > 0,$$
$$\Delta_3 = A_1A_2A_3 - A_0A_3^2 - A_1^2A_4 > 0, \Delta_4 = \Delta_3A_4 > 0, (A_4 > 0).$$

The detailed analysis of the regions where necessary and sufficient conditions of stability hold was studied in the plane of the two dimensionless inertia parameters (θ_A, θ_C) at different values of the damping coefficient k. It is evident that along with the conditions of stability the triangle inequalities for a rigid body should also be satisfied, $\theta_A + \theta_C > 1$, $\theta_C + 1 > \theta_A$, $\theta_A + 1 > \theta_C$. One may disregard the third triangle inequality, since it holds automatically at $\theta_A > \theta_C$. Thus, the region is limited by the straight lines $\theta_C = 1 - \theta_A$, $\quad \theta_C = \theta_A$, $\quad \theta_C = \theta_A - 1$.

An example of such region and also all the lines at which the inequalities convert into equalities for $k = 0.5$ is shown in Figure 1 left. The region where the necessary and sufficient conditions of stability are satisfied is bounded by the above defined straight lines and by the hyperbola $\Delta_4 = 0$ and are marked out by black color.

The numerical integration of system (1) has been done for different values of the parameter k. The different types of transition decay processes of the spatial oscillations of the satellite at different damping parameters has been investigated numerically. Figure 1 right shows an example of transition decay processes of spatial oscillations for $k = 0.5$ at inertia parameters $\theta_A = 0.9, \theta_C = 0.2$, where the conditions of asymptotic stability hold. The system in this case reaches the equilibrium position (6) at all three angles, at the τ value equal to 25, which correspond to the 4 satellite turnover in the orbit.

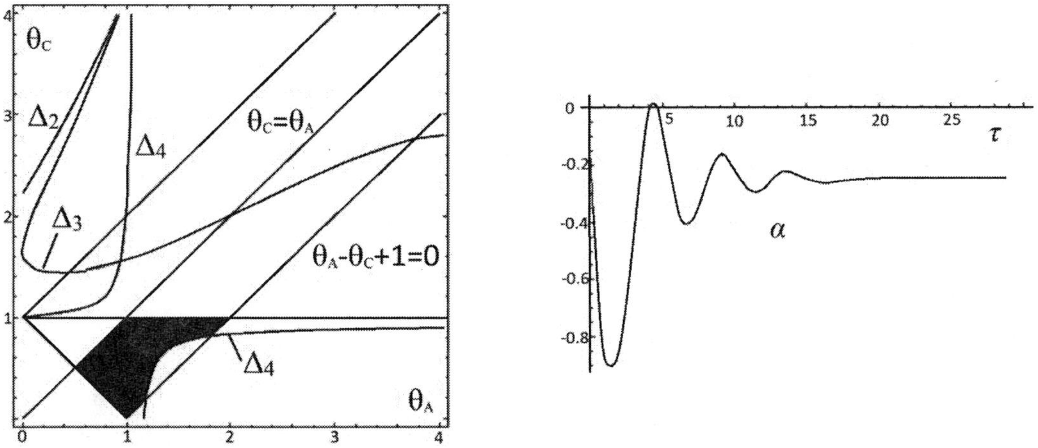

Figure 1. Left: The region of fulfillment of the asymptotic stability conditions for $k = 0.5$.
Right: The transitional process of damping oscillations for $k = 0.5, \alpha_0 = -0.2$

References

[1] Beletsky V.V., *Motion of Satellite in Gravitational Field* (MGU Press, Moscow, 1975) 416
[2] Gutnik, S.A., Sarychev, V.A., LNCS **10490**, 167–182 (2017)
[3] Sarychev V.A., *Problems of orientation of satellites. Itogi Nauki i Tekhniki. Ser. "Space Research"* **11**, (VINITI , Moscow, 1978) 222
[4] Buchberger B., SIGSAM Bulletin **3**, 19–29 (1976)
[5] *Maple online help* (http://maplesoft.com/support/help/)

EPJ Web of Conferences **173**, 05009 (2018) https://doi.org/10.1051/epjconf/201817305009
Mathematical Modeling and Computational Physics 2017

Application of Artificial Neural Networks and Singular-Spectral Analysis in Forecasting the Daily Traffic in the Moscow Metro

Victor Ivanov[1,2,⋆] and *Evgenii* Osetrov[1,3,⋆⋆]

[1]*Joint Institute for Nuclear Research (JINR), Dubna, Moscow region, Russia*
[2]*National Research Nuclear University "MEPhI", Moscow, Russia*
[3]*Federal Treasury Institution "Rostransmodernizatsia", Moscow, Russia*

Abstract. In this paper, we investigate the possibility of applying various approaches to solving the problem of medium-term forecasting of daily passenger traffic volumes in the Moscow metro (MM): 1) on the basis of artificial neural networks (ANN); 2) using the singular-spectral analysis implemented in the package "Caterpillar"-SSA; 3) sharing the ANN and the "Caterpillar"-SSA approach. We demonstrate that the developed methods and algorithms allow us to conduct medium-term forecasting of passenger traffic in the MM with reasonable accuracy.

1 Introduction

The Moscow metro is the main transport system in Moscow, carrying up to 10 million passengers a day. Given the dynamics of the MM development and the construction of modern transport infrastructure in Moscow, the role of the MM in passenger transportation will grow. To ensure effective functioning of the MM, constant monitoring of passenger traffic and its near- and medium-term forecasting are necessary.

2 Preparation of initial data for forecasting

The initial time series, reflecting the dynamics of the daily volume of passenger transportation by the Moscow metro over the past 5 years, contains 1,793 observations. Our analysis has shown [1] that the intensity of passenger traffic varies greatly between working days and weekend-holidays. Moreover, the corresponding distributions of passenger traffic practically do not overlap [1]. This observation made it possible to break the initial series into two series: 1) working days, 2) weekend-holidays. Our research is connected with analysis of the traffic in the Moscow metro during working days.

It is obvious that the forecasting of a time series is possible only when there is a link between subsequent values of the series and the previous ones. Therefore, at the preliminary stage, a high-frequency component was excluded from the initial series by means of wavelet filtration (see Fig. 1). As an estimate of the correlation interval (forecasting horizon), the value of the first intersection of the autocorrelation function of the time axis in the confidence interval corresponding to the white

⋆e-mail: ivanov@jinr.ru
⋆⋆e-mail: esosetrov@gmail.com

noise was used [2]. We have shown that realizations of the noisy component are independent of each other, symmetric with respect to zero, and that their distribution agrees well with the Gaussian law. In this way we isolated the noisy component from the source data, and the filtered series was used for forecasting.

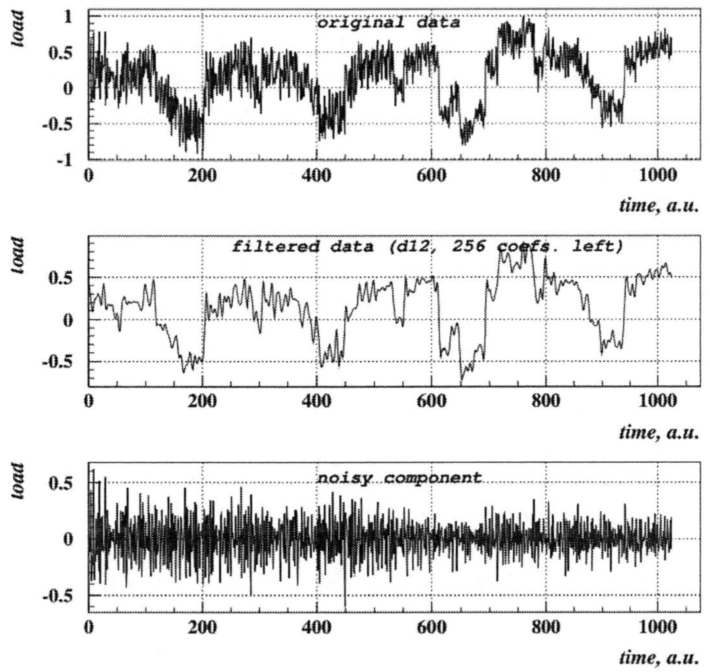

Figure 1. From top to bottom: 1) the original time series (transformed to the interval [−1, 1]) containing data on 1024 working days; 2) this series after application of wavelet-filtering; 3) excluded noisy component

3 Forecasting passenger's traffic using ANN

As an ANN model we used a multilayered perceptron (MLP), at the input of which 6 variables were applied as factors affecting the passenger's traffic in MM: Var1 – year of observation, Var2 – month, Var3 – day of week, Var4 – type of day, Var5 – daily consumption of electricity in the Moscow region. Var6 assumes different values: 1) daily traffic in the MM at the ANN training stage; 2) forecast for the previous day at the forecast stage based on the ANN; 3) forecast for the current day based on the SSA (sharing ANN and SSA approaches). These variables are due to the seasonality of traffic in the metro, they can be used to obtain predicted values using alternative methods of obtaining information, and these variables do not exhibit the phenomenon of multicollinearity [3]. The MLP contained 6 neurons at the input, two hidden layers of 16 and 8 neurons, respectively, and one output neuron, which gives the predicted value. We conducted a series of experiments using the TMVA 4.2.0 package in the ROOT environment [4] and the Fletcher-Reeves method of learning was adopted as optimal. The training procedure included 1500 epochs, upon its completion the state of the network was fixed and it was used for forecasting. At the forecasting stage, the value of the passenger's traffic predicted by the network at the current step was applied to the network input (as variable Var6).

The results of ANN predictions for 30 and 50 days are shown in Figs. 4 and 5. We have found that the relative error ($R_{\text{error}} = \frac{\text{Real}-\text{Forecast}}{\text{Real}}$) is symmetric with respect to zero and does not exceed 3–5%. This result means that using ANN, we can quite accurately predict the traffic of passengers in the metro.

4 Forecasting passenger's traffic using the 'Caterpillar'-SSA approach

For forecasting, we used the latest version of the CaterpillarSSA program (version 3.40, Professional M Edition, for details see [5]). As input information for the CaterpillarSSA program, the filtered time series was used. Before converting to multivariate form, the analyzed time series was standardized by means of CaterpillarSSA [5]. The length of the caterpillar (the number of components for decomposition) was set equal to 492. This is approximately half of the analyzed series, allowing to distinguish all its characteristic features. In the reconstruction of a one-dimensional series, 13 main components were used. Their total contribution was 99.91%. At the prediction stage, the confidence interval was set equal to 0.25.

Figure 2 shows the graphs of the first 13 components of the decomposition of the analyzed series when predicted for 30 days of observation. It can be seen from the graphs that the components under

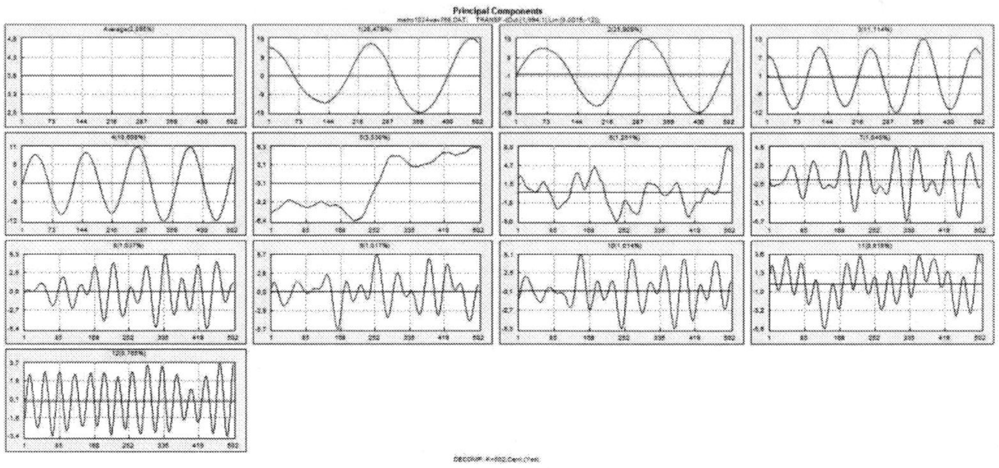

Figure 2. Graphs of first 13 components when predicting 30 days of observation

consideration are responsible for the trend and periodic (seasonal) components of the analyzed series.

Figure 3 presents: 1) above – the original series with its approximation, reconstructed on the basis of the first 13 components, 2) at the bottom – a series rebuilt on the basis of the discarded components.

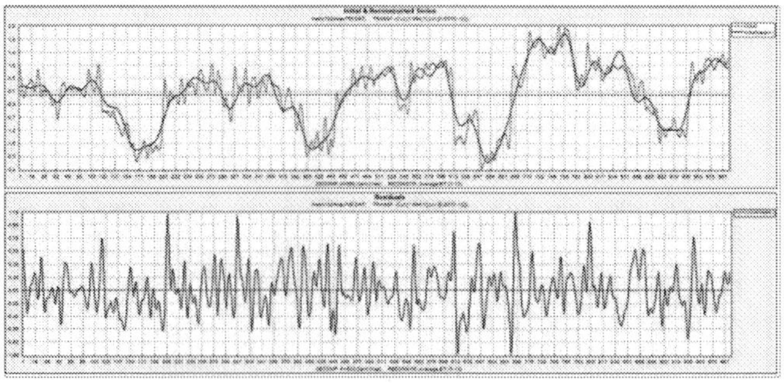

Figure 3. Above – the original series with its approximation, at the bottom – the series corresponding to the discarded components

We found that the discarded data are in good agreement with the normal distribution, and it can be assumed that the responding process is close to Gaussian noise in its behavior. The results of forecasting by the SSA method for 30 and 50 days are shown in Figs. 4 and 5.

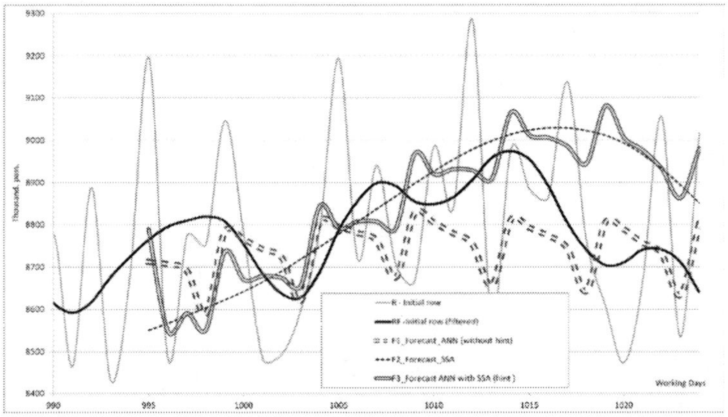

Figure 4. Comparative predictive results for 30 days of observation using approaches developed in this paper

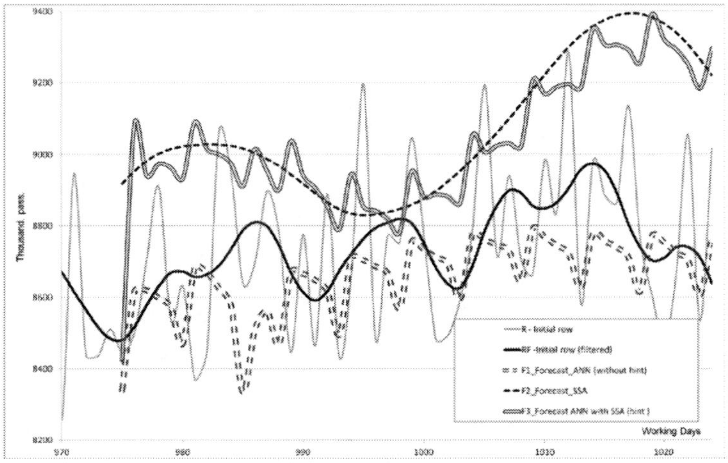

Figure 5. Comparative predictive results for 50 days of observation using approaches developed in this paper

5 Discussion of results and conclusion

The best version of the forecast was achieved with the help of an ANN of the feed-forward type when a number of factors influencing the traffic of passengers in the metro were submitted to the ANN entrance. The prediction by the SSA method turned out to be shifted relative to the forecasted data in the region of large values of passenger traffic. The same was the reason for the displacement of the forecast in the region of large volumes of passenger traffic, when ANN and SSA were used together.

The technique for forecasting traffic in the MM, developed in this paper, will enhance the efficiency and speed of making management decisions in transport.

References

[1] V.V. Ivanov, E.S. Osetrov, Vestnik MIFI **5**, No 1, 65–74 (2016) [in Russian]
[2] V.V. Ivanov, E.S. Osetrov, Vestnik MIFI **5**, No 2, 162–169 (2016) [in Russian]
[3] http://www.theignatpost.ru/magazine/index.php?mlid=2905
[4] R. Brun and F. Rademakers, Nucl. Inst. Meth. in Phys. Res. A **389**, 81 (1997)
[5] http://www.gistatgroup.com/cat/

EPJ Web of Conferences **173**, 05010 (2018) https://doi.org/10.1051/epjconf/201817305010
Mathematical Modeling and Computational Physics 2017

Application of SLURM, BOINC, and GlusterFS as Software System for Sustainable Modeling and Data Analytics

Vladislav V. Kashansky[1,*] and *Igor L.* Kaftannikov[1,**]

[1] *School of Electrical Engineering and Computer Science, South Ural State University, Russia*

Abstract.
Modern numerical modeling experiments and data analytics problems in various fields of science and technology reveal a wide variety of serious requirements for distributed computing systems. Many scientific computing projects sometimes exceed the available resource pool limits, requiring extra scalability and sustainability. In this paper we share the experience and findings of our own on combining the power of SLURM, BOINC and GlusterFS as software system for scientific computing. Especially, we suggest a complete architecture and highlight important aspects of systems integration.

1 Introduction

Distributed environments have become very influential in the contemporary data processing and analytics. During several decades, we had been observing architecture changes emerging to support computation trends of various forms. Among them, there were hardware-defined platforms, SSI-based systems [1] and clustered multi-node solutions, orchestrated by different batch schedulers. Contemporary computing architectures [2] involve clustering approach, managing computing processes among the nodes of a large distributed computing platform. Many papers have been written about the general nature of the distributed computing. In this paper, we are going to discuss and focus on more "local" and application-specific tasks.

The problem of infrastructure deployment arose during 2015–2016, when we decided to process medium-sized data sets and perform corresponding computations for several projects led by Dr. I.L. Kaftannikov. At early stages, we were using volunteer computing resources managed by the BOINC system. The more data we wanted to process, the more inner resources we had to involve. The problem is that with the growth of the project, we had to redefine the local infrastructure, making it more reliable and stable. We also had some computational jobs, which required additional control and it was no longer possible to use it in the context of BOINC, as we required lower latencies and higher degree of security. At the same time, we wanted to save the existing approach, allowing volunteers to integrate their computational resource base.

*e-mail: vladislav.kash@gmail.com
**e-mail: kil7491@gmail.com

2 Early Stages Architecture. BOINC System

As we stated above, we decided to use the BOINC system as initial solution. It is a free software platform [3] developed by the Berkeley University, implementing a volunteer public-resource model of distributed computations. BOINC allowed us to involve different types of computing powers, connecting many individual computers to a common computational network. Owners of physical or virtual computing resources were able to join us publicly via the BOINC client program and to help by contributing some resources.

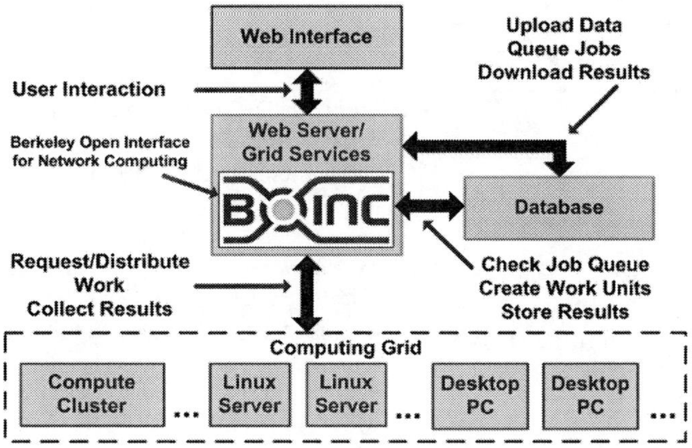

Figure 1. Architecture of the BOINC project. The platform is centralized and includes server and user-side parts. User-side part consists of a client and a graphical program – BOINC Manager. The manager brings the GUI for controlling of one or many BOINC clients. Proper details of BOINC project can be found in [3], improvements in [4]. The most recent changes and documentation can be obtained from the Github repository and the official website.

Many popular BOINC projects [3] demonstrate significant computational performance, but with some strict limitations and only over a certain class of jobs. This of course allows scientific groups to carry out resource-intensive calculations without the use of expensive hardware. But with this approach we faced two major architectural problems:

1. The architecture of data storage is undefined as the BOINC system operates at a higher layer of data abstraction.

2. The scheduling model within BOINC system is inconsistent with the local jobs distribution, which asks for a fast acting system with low transition rates.

Based on Refs. [5–7] we decided to use the SLURM as a local batch scheduler and GlusterFS as a distributed file system for storing experimental data sets.

3 Redefined Architecture. GlusterFS and SLURM

GlusterFS is a distributed, parallel, linearly scalable, failure-tolerant file system. We used it in the context of TCP/IP networking to export data "bricks" from different servers, forming a single parallel distributed data space. GlusterFS operates in user space, based on FUSE technology and works on the top of the existing file systems, so we didn't have to face the kernel support problem. Unlike other distributed file systems, such as Lustre and Ceph [5], GlusterFS does not require a separate server for storing metadata[1], providing a great degree of flexibility and scalability. By bringing GlusterFS to our BOINC stack we changed the storage mechanisms, allowing storing our data in a more efficient way.

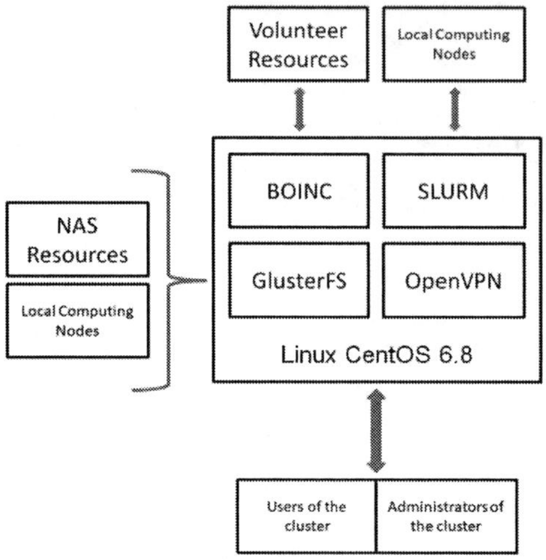

Figure 2. Diagram of the new architecture. The whole solution runs on CentOS 6.8. BOINC and SLURM form computational core. Custom meta-scheduler controls the job distribution process. It classifies jobs and sends them into the correct queue according to the specific scientific workflow. GlusterFS provides extensible single parallel distributed data space. OpenVPN forms a secure overlay network for communication of the inner services. We used it to connect several storage nodes that weren't accessible directly via IPv4 address. Administrators and users interact with the system via SSH commands and GUI.

The other valuable part of our new architecture is SLURM. The developers describe it as scheduling system for large and small Linux clusters. It requires no kernel modifications for its operation and is relatively self-contained [7]. However, the integration with SLURM required jobs classification which was done within the following scheme:

1. Jobs with low data exchange intensity, possibility of execution pausing and reassignment[2] are running on the volunteer resources.

2. Jobs with intensive data exchange or strict deadlines are running on the local cluster and scheduled via SLURM.

After performing the integration depicted in figure 2, SLURM filled an important gap in our architecture, providing the ability for starting, executing, and monitoring jobs on the set of locally allocated nodes. Now, when a user registers the specific job in system's workflow, meta-scheduler can run it locally, or assign it to any possible volunteer's computing resource, depending on the set of preferences. The distribution scheme [8–10] as well as whole infrastructure tuning, monitoring [11] and maintenance can turn to really complex solution. This is beyond the scope of this work, but we suggest papers and books mentioned in our references for interested readers.

[1] GlusterFS uses Elastic Hashing Algorithm (EHA).
[2] Volunteers are usually able to pause or even cancel assigned jobs, making execution time unpredictable.

4 Conclusion

The use of Distributed File Systems and Batch Schedulers is a common practice for clusters, data-centers, and supercomputers. In this paper, we tried to highlight a radical infrastructure redefinition, joining high-latency loosely-coupled volunteer public resources with local high-performance cluster. The use of SLURM, BOINC and GlusterFS as a platform for our computations allowed us:

1. To get a horizontally scalable computing system with minimal costs for organization and subsequent maintenance. We used the available local and volunteer computing resources and spent around 1200 man-hours on planning and development.

2. To solve the resource idleness problem. We connected 20 Intel Core I7-4770 2x4096DDR3 SATA 1TB via BOINC and 10 Intel Xeon-X5680 1x4096DDR3 SAS 1TB nodes via SLURM.

3. To socialize research of the department, allowing outside participants offer their own resources. Volunteers donated more than 200 heterogeneous nodes, including several server farms.

4. To process wider spectrum of jobs due to combining local high-performance cluster with BOINC system.

5. To automate functions related to the jobs distribution, results aggregation, and reporting data creation via a meta-scheduler and a set of custom scripts.

References

[1] L. Renaud, P. Gallard, G. Vallée, Ch. Morin, and B. Boissinot, *IEEE International Symposium on Cluster Computing and the Grid (CCGrid 2005)* **2** (Cardiff, Wales, UK, 2005) pp. 1016–1023

[2] S. H. Fuller and L. I. Millett, *The Future of Computing Performance: Game Over or Next Level?* (National Academy Press, Washington DC USA, 2011)

[3] D. P. Anderson, *Fifth IEEE/ACM International Workshop on Grid Computing* (Pittsburgh PA USA, 2004) pp. 4–10

[4] V. V. Kashansky and I. L. Kaftannikov, *Selected Papers of the 7th International Conference Distributed Computing and Grid-technologies in Science and Education* (Dubna, Russia, 2016) pp. 284–288

[5] B. Depardon, G. Le Mahec, and C. Séguin, *Analysis of Six Distributed File Systems* (Research Report, The open archive HAL, 2013)

[6] A. Davies and A. Orsaria, Linux Journal **2013** (235), 72–82 (2013)

[7] A. B. Yoo, A. J. Morris, and M. Grondona, *Job Scheduling Strategies for Parallel Processing* (Springer, Berlin, Heidelberg, 2003) pp. 44–60

[8] V. Berten, J. Goossens, and E. Jeannot, IEEE Transactions on Parallel and Distributed Systems **17** (2), 275–278 (2006)

[9] N. A. Balashov, A. V. Baranov, I. S. Kadochnikov, V. V. Korenkov, N. A. Kutovskiy, A. V. Nechaevskiy, and I. S. Pelevanyuk, *Selected Papers of the 7th International Conference Distributed Computing and Grid-technologies in Science and Education* (Dubna, Russia, 2016) pp. 114–118

[10] M. Combarro, A. Tchernykh, D. Kliazovich, A. Drozdov, and G. Radchenko, *2016 International Conference on Engineering and Telecommunication (EnT)* (Moscow, Russia, 2016) pp. 29–33

[11] I. S. Kadochnikov and I. S. Pelevanyuk, *Selected Papers of the 7th International Conference Distributed Computing and Grid-technologies in Science and Education* (Dubna, Russia, 2016) pp. 275–278

EPJ Web of Conferences **173**, 05011 (2018) https://doi.org/10.1051/epjconf/201817305011
Mathematical Modeling and Computational Physics 2017

Parallel Implementation of Numerical Solution of Few-Body Problem Using Feynman's Continual Integrals

Mikhail Naumenko[1,*] and *Viacheslav* Samarin[1,2,**]

[1] *Joint Institute for Nuclear Research, Dubna, Russia*
[2] *Dubna State University, Dubna, Russia*

Abstract. Modern parallel computing algorithm has been applied to the solution of the few-body problem. The approach is based on Feynman's continual integrals method implemented in C++ programming language using NVIDIA CUDA technology. A wide range of 3-body and 4-body bound systems has been considered including nuclei described as consisting of protons and neutrons (e.g., ^{3,4}He) and nuclei described as consisting of clusters and nucleons (e.g., ^{6}He). The correctness of the results was checked by the comparison with the exactly solvable 4-body oscillatory system and experimental data.

1 Introduction

The wave functions of the ground states of ^{3,4,6}He nuclei were calculated in Refs. [1, 2] by Feynman's continual integrals method in Euclidean time [3, 4]. The present work is devoted to studying hydrogen, helium, lithium, and beryllium nuclides using the same approach. The few-body nuclei ^{2,3}H, ^{3,4}He were considered as consisting of protons and neutrons, whereas the nuclei ^{6}He, ^{6}Li, ^{9}Be were considered as α-cluster nuclei. The algorithm allowing us to perform calculations directly on GPU was developed and implemented in C++ programming language. The energy and the square modulus of the wave function of the ground states for several few-body nuclei have been calculated using NVIDIA CUDA technology. The results show that the use of GPU is very effective for these calculations.

2 Theory and computing

The energy E_0 and the square modulus of the wave function $|\Psi_0|^2$ of the ground state of a system of few particles with coordinates q may be calculated by Feynman's continual integrals method using the propagator $K_E(q,\tau;q,0)$ in Euclidian time τ [4]

$$K_E(q,\tau;q,0) = \sum_n |\Psi_n(q)|^2 \exp\left(-\frac{E_n\tau}{\hbar}\right) + \int_{E_{\text{cont}}}^{\infty} |\Psi_E(q)|^2 \exp\left(-\frac{E\tau}{\hbar}\right) g(E)\, dE. \quad (1)$$

*e-mail: anaumenko@jinr.ru
**e-mail: samarin@jinr.ru

Here $g(E)$ is the density of states with the continuous spectrum $E \geq E_{\text{cont}}$. For the system with a discrete spectrum and finite motion of particles the square modulus of the wave function of the ground state may be found in the limit $\tau \to \infty$ together with the energy E_0

$$K_E(q, \tau; q, 0) \to |\Psi_0(q)|^2 \exp\left(-\frac{E_0 \tau}{\hbar}\right), \qquad \tau \to \infty. \tag{2}$$

The theoretical approach is described in detail in Ref. [2]. The calculation of $K_E(q, \tau; q, 0)$ for the fixed τ was performed by parallel calculation of exponentials F

$$F = \exp\left[-\frac{\Delta\tau}{\hbar} \sum_{k=1}^{N} V(q_k)\right] \tag{3}$$

for every random trajectory $q_k = f(q, k\Delta\tau)$, where $\Delta\tau$ is time step, $N = \tau/\Delta\tau$. The same effective pairwise nucleon-nucleon, nucleon-cluster and cluster-cluster interaction potentials $V(r)$ were used for all the studied nuclei.

The Monte Carlo algorithm for numerical calculations was developed and implemented in C++ programming language using NVIDIA CUDA technology. Calculations were performed on the NVIDIA Tesla K40 accelerator installed within the heterogeneous cluster [5] of the Laboratory of Information Technologies, Joint Institute for Nuclear Research, Dubna. The code was compiled with NVIDIA CUDA version 7.5 for architecture version 3.5.

The code implementing Feynman's continual integrals method was initially written for CPU. The calculation time of the ground state energy for ^{3}He with statistics 10^6 using Intel Core i5 3470 (double precision, 1 thread) and NVIDIA Tesla K40 (single precision) was ~18377 sec and ~47 sec, respectively, which corresponds to the impressive performance gain of ~400 times [2]. The typical calculation time of the square modulus of the wave function for the ground state of ^{3}He with statistics 10^6 on the mesh of 60·60·12 points using NVIDIA Tesla K40 was ~10 hours, which yields an estimation of the calculation time using Intel Core i5 3470 (double precision, 1 thread) ~170 days. This performance gain allows us to increase the statistics and the accuracy of calculations, reduce the mesh step, and may even enable calculations impossible before.

To check the correctness of the calculation of the propagator the comparison with the exactly solvable 4-body oscillatory system has been performed. For four particles with masses $m_1 = m_2 = m_3 = m_4 = m$ interacting with each other by oscillator potentials

$$V_{ij}(r_{ij}) = \frac{m\omega^2}{2} r_{ij}^2, \qquad V = -U_0 + \sum_{i<j} V_{ij}(r_{ij}) = -U_0 + \frac{m\omega^2}{2}\left(r_{12}^2 + r_{13}^2 + r_{14}^2 + r_{23}^2 + r_{24}^2 + r_{34}^2\right), \tag{4}$$

the kinetic energy T and the potential energy V may be written using the normalized Jacoby coordinates x, y, z

$$T = \frac{1}{2}\left(\dot{x}^2 + \dot{y}^2 + \dot{z}^2\right), \qquad V = -U_0 + \frac{1}{2}(2\omega)^2\left(x^2 + y^2 + z^2\right), \tag{5}$$

and the energy E_0 of the ground state is given by

$$E_0 = -U_0 + (2\hbar\omega)\left(\frac{3}{2} + \frac{3}{2} + \frac{3}{2}\right) = -U_0 + 9\hbar\omega. \tag{6}$$

Assuming $\omega = 1, \hbar = 1$, we obtain $E_0 = -U_0 + 9$. For example, in the cases of $U_0 = 0, E_0 = 9$ and $U_0 = 15, E_0 = -15+9 = -6$, whereas Feynman's continual integrals method with statistics $N = 7\cdot10^7$ yields $E_0 = 9.05 \pm 0.1$ and $E_0 = -5.98 \pm 0.02$, respectively, which is close to the exact result.

3 Results

In the calculation of the propagator for the nuclei ^{2,3}H, ^{3,4}He nucleon-nucleon two-body effective strong interaction potentials $V_{i-j}(r)$ $(i, j = n, p)$ with repulsive cores, similar to the M3Y potential [6], have been used (Figure 1*a*)

$$V_{i-j}(r) = \sum_{k=1}^{3} u_k \exp\left(-r^2/b_k^2\right). \tag{7}$$

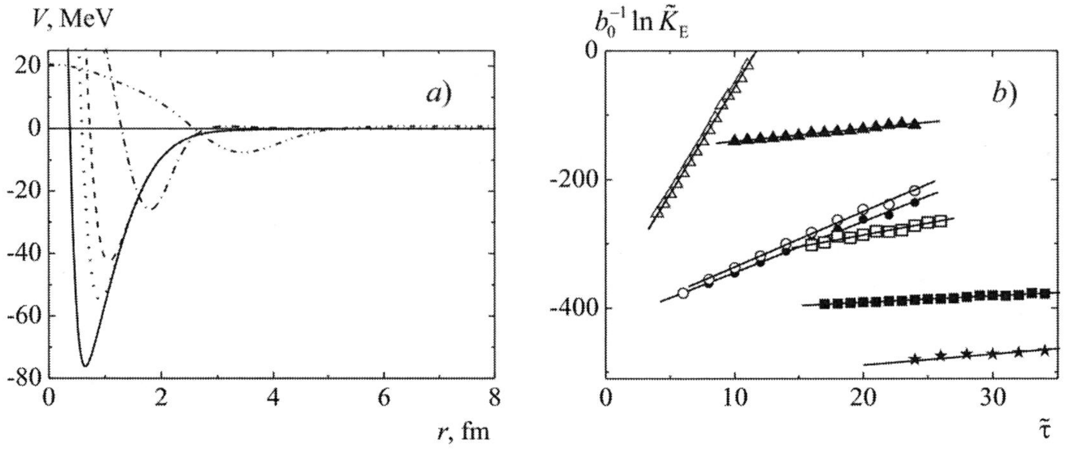

Figure 1. *a*) Two-body interaction potentials $V_{i-j}(r)$ for the following pairs of particles: neutron-proton with parallel (solid line) and antiparallel spins (dotted line), neutron-neutron (dashed line), α-cluster-neutron (dash-dotted line) and α-cluster-α-cluster (dash-dot-dotted line). *b*) The dependence of the logarithm of the dimensionless propagator $\tilde{K}_E$ on the dimensionless Euclidean time $\tilde{\tau}$ for ^{2}H (filled triangles), ^{3}H (circles), ^{3}He (dots), ^{4}He (empty triangles), ^{6}He (filled squares), ^{6}Li (empty squares), ^{9}Be (stars); lines are the results of linear fitting of the data; b_0 is a coefficient [2].

The calculated logarithm of the propagator for nuclei ^{2,3}H and ^{3,4}He is shown in Figure 1*b*. The theoretical binding energies $E_B = -E_0$ obtained using formula (2) are listed in Table 1 together with the experimental values taken from the NRV knowledge base [7]. It is clear that the theoretical values are close enough to the experimental ones. The observed difference between the calculated binding energies of ^{3}H and ^{3}He is also in agreement with the experimental values.

Momentum distributions after breakup in nuclear reactions show that ^{6}He nucleus consists of an α-cluster core and a two-neutron cluster (e.g., [8]). Similarly, the ^{6}Li nucleus consists of an α-cluster core and a deuteron cluster. The ^{9}Be nucleus consists of two α-clusters and a neutron. There are several different approaches to the approximate analysis and solution of the three-body problem (e.g., [9]). Feynman's continual integrals method provides a simpler possibility of calculating the energy and the probability density for the ground state of the few-body system, because it does not require expansion of the wave function in a system of functions. The calculated dependence of logarithm of the propagator on the Euclidean time for nuclei ^{6}He $(\alpha + n + n)$, ^{6}Li $(\alpha + n + p)$ and ^{9}Be $(\alpha + n + \alpha)$ is shown in Figure 1*b*. The α-cluster-nucleon and α-cluster-α-cluster strong interaction potentials $V_{i-j}(r)$ $(i, j = n, p, \alpha)$ were used in the form of the combination of Woods-Saxon potentials (Figure 1*a*)

$$V_{\alpha-j}(r) = \sum_{i=1}^{s} U_i \left[1 + \exp\left[(r - R_i)/a_i\right]\right]^{-1}, \tag{8}$$

Table 1. Comparison of theoretical and experimental binding energies for the ground states of the studied nuclei

Atomic nucleus	Experimental value [7], MeV	Theoretical value, MeV
^{2}H	2.225	2.22 ± 0.15
^{3}H	8.482	8.21 ± 0.3
^{3}He	7.718	7.37 ± 0.3
^{4}He	28.296	30.60 ± 1.0
^{6}He	0.97542	0.96 ± 0.05
^{6}Li	3.637	3.87 ± 0.2
^{9}Be	1.573	1.7 ± 0.1

where $s = 2, 3$. The values of parameters are given in Ref. [10]. The obtained theoretical energies of separation into cluster(s) and nucleon(s) $E_S = -E_0$ are listed in Table 1 together with the experimental values taken from the NRV knowledge base [7]. It can be seen that the theoretical values are close enough to the experimental ones.

The results show that even simple central two-body potentials provide good agreement with experimental data and thus may be used for calculations of the ground states. Moreover, we would like to emphasize the fact that good agreement with experimental data was obtained using the same unified set of potentials with the same parameters for all the studied nuclei, which makes the approach universal and self-consistent.

4 Conclusion

In this work an attempt is made to use modern parallel computing solutions to speed up the calculations of ground states of few-body nuclei by Feynman's continual integrals method. The developed parallel algorithm provided significant increase of the speed of calculations. The method was applied to the nuclei consisting of nucleons and cluster nuclei. The results of calculations demonstrate that the obtained theoretical values are close enough to the experimental ones for the studied nuclei. The obtained probability densities may be used for the correct definition of the initial conditions in the time-dependent calculations of reactions with the considered nuclei. The results may also serve as a useful addition to the results obtained by the expansion in hyperspherical functions.

References

[1] V. V. Samarin and M. A. Naumenko, Bull. Russ. Acad. Sci.: Phys. **80**, 283 (2016)
[2] M. A. Naumenko and V. V. Samarin, Supercomp. Front. Innov. **3**, 80 (2016)
[3] R. P. Feynman and A. R. Hibbs, *Quantum Mechanics and Path Integrals* (McGraw-Hill, New York, 1965)
[4] E. V. Shuryak and O. V. Zhirov, Nucl. Phys. B **242**, 393 (1984)
[5] Heterogeneous cluster of LIT, JINR, http://hybrilit.jinr.ru/
[6] M. A. G. Alvarez, L. C. Chamon, D. Pereira, et al., Nucl. Phys. A **635**, 187 (1999)
[7] V. I. Zagrebaev, A. S. Denikin, A. V. Karpov, A. P. Alekseev, M. A. Naumenko, V. A. Rachkov, V. V. Samarin, and V. V. Saiko, *NRV web knowledge base on low-energy nuclear physics*, http://nrv.jinr.ru/
[8] Yu. E. Penionzhkevich, Phys. At. Nucl. **74**, 1615 (2011)
[9] M. V. Zhukov, B. V. Danilin, D. V. Fedorov, et al., Phys. Rep. **231**, 151 (1993)
[10] V. V. Samarin and M. A. Naumenko, Phys. At. Nucl. **80**, 877 (2017)

A Scientific Workflow System for Satellite Data Processing with Real-Time Monitoring

Minh Duc Nguyen[1]*

[1] *Skobeltsyn Institute of Nuclear Physics, Lomonosov Moscow State University, Moscow, Russia*

Abstract. This paper provides a case study on satellite data processing, storage, and distribution in the space weather domain by introducing the Satellite Data Downloading System (SDDS). The approach proposed in this paper was evaluated through real-world scenarios and addresses the challenges related to the specific field. Although SDDS is used for satellite data processing, it can potentially be adapted to a wide range of data processing scenarios in other fields of physics.

1 Introduction

The ultimate goal of the research in space weather is to create complex operational magnetosphere-ionosphere models which allow predicting the radiation risk for space satellites in different orbits, and also estimating the occurrence of risks of technological disasters due to magnetic storms and charged particle precipitation. The models use space-related data sets acquired by multiple satellites of various purposes and capabilities as well as ground stations to produce forecasts. Such models require all data sets to be collected at one place and also to be converted into a unified data format used by them as the input. Such task is challenging and time-consuming because data formats, storage methods, datasets and values of registered factors differ between satellites. The size of data sets can be huge so that the disk capacity of a personal computer might be insufficient for data storage. There is a high demand for creating an automatic system for collecting data from various satellites, processing them, converting them to a unified canonical format and making them available in a standardized machine-readable format.

Currently, there are several existing solutions to the problem such as SWX [1], xv_SAT [2, 3], the Herschel Data Processing System [4], and NJOY [5]. The Satellite Data Downloading System (SDDS) presented in this paper has several advantages in comparison with the systems listed above. Firstly, SDDS does not act as a specific data processing system that targets a particular data source and has just one workflow. SDDS is a framework that eases the task of creating such systems that can have different workflows in data processing. Secondly, SDDS provides a web interface for data scientists, a RESTful API and a Python library for developers so they can build their systems upon SDDS. Thirdly, from the first day, SDDS has been developed as a modular system that can scale easily. Components of SDDS can run either on the same machine or different machines connected via a local network or the Internet. Different versions of a component can run interchangeably. Each component can be replaced with a new one just by editing the proper configuration file as long as the new component

*e-mail: nguyendmitri@gmail.com

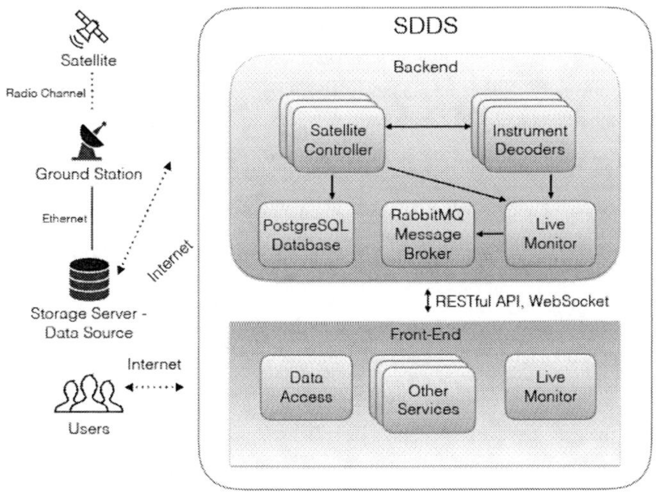

Figure 1. The architectural overview of SDDS

follows the convention. And finally, SDDS has a built-in real-time monitoring subsystem that helps operators to identify the scope of an error quickly and prevent it in the future. To our knowledge, SDDS is the first framework dedicated to satellite data processing that provides API for developers.

2 Overview

SDDS has been designed to implement the whole workflow of a satellite data processing cycle. A cycle consists of six steps: 1) Connecting to data sources. A data source is a server where data of a satellite are collected and can be downloaded; 2) Downloading actual data (so-called raw data). Raw data of a satellite can be a big one file in which data of all instruments installed on the satellite are packed or a set of files each of which contains data of an instrument; 3) Extracting binary (level-0) data of each instrument from the raw data; 4) Processing level-0 data and producing scientific (level-1) data that are ready to be loaded into a database; 5) Loading level-1 data to a database; 6) Compressing and moving raw, level-0, and level-1 data to storage.

SDDS consists of two parts: a backend and a frontend. A detailed view of the architecture is shown in figure 1. Backend components have been developed using the Django Framework written in Python. Frontend components have been developed using the React library created by Facebook, Javascript, HTML, and CSS.

The backend part consists of the following components:

- A set of satellite controllers, each of which is responsible for a satellite and implement the whole data processing cycle. Each satellite controller uses decoders provided by satellite developers to decode level-0 data and produce level-1 data;
- The Live Monitor subsystem which is responsible for monitoring all activities of all components in real time. The Live Monitor subsystem records each event during the data processing cycle and sends it to the RabbitMQ message broker;
- The DBIface interface which is responsible for database interaction. The DBIface interface supports the following engines: PostgreSQL, MySQL, and Oracle;

- The RabbitMQ message broker which is responsible for broadcasting all recorded events (short text messages) to all subscribed clients. A subscribed client is an instance of the Live Monitor's web interface that has established a WebSocket connection to the RabbitMQ message broker. The messages are sent using the STORM protocol;
- A web server, which could be either Apache or Nginx, to serve RESTful requests.

On the frontend, there are two base components available currently:

- The Live Monitor's web interface shows recorded events (or current conditions) during the data processing cycle;
- The Data Access services allow users to acquire data from the database. Data can be retrieved either by using the web interface or the RESTful API.

Each component of SDDS is a micro service with a RESTful API. Components can work together locally on the same computer or can be distributed across different machines connected via a local network or the Internet. Almost every component of SDDS has been developed as a library. Hence it gives users of SDDS the possibility to customize certain components to meet their requirements. It is also possible to change the behavior of a component by editing its configuration file which is written in the JSON format.

3 Workflow automation

The main component of SDDS is the satellite controller which is purely a Python class called SDDownloader. Each satellite controller in SDDS is dedicated for processing data of a certain satellite. Each step of the data processing cycle is implemented as a function of the SDDownloader class. Each function, in turn, is a composition of smaller functions (routines). Using the object oriented programming principles, one can create a customized class which is inherited from the SDDownloader class and redefine certain functions to meet their requirements.

Usually, data of a satellite are duplicated in several data sources. In this case, it is enough to download and process data from the first data source. But in many cases, data are distributed across data sources, and they need to be merged after processing. The SDDownloader class can handle all described scenarios. If a data source is temporarily or permanently unavailable, the satellite controller will retry and continue downloading data when the data source becomes available.

Different data sources have different security constraints and thus use different types of connection. At the moment the following connection types are supported by SDDS out-of-box: HTTP(S), FTP(S), PPTP, IPsec, HTTP(S)/FTP(S) over PPTP or IPsec. To support a new connection type one just needs to implement a function for this purpose and pass it as a parameter to the function of the SDDownloader class used for establishing connections to data sources. The SDDownloader class also supports adding additional routes to the routing table after a connection has been established. This kind of operation is required mostly when the data source is located in another subnet that differs from the subnet leased by the VPN gateway.

Currently, the SDDownloader class supports decoders written in Python, PHP, and Javascript programming languages. C and Fortran programs compiled statically are also supported. To call a certain decoder, one must describe how the decoder should be executed in the configuration file. The SDDownloader class also supports post-processing. The mechanism of executing a post-processing decoder is the same as the decoder execution. To control how a decoder runs, the SDDownloader class records the whole output produced by the decoder and its return code. All the recorded events are then passed to the RabbitMQ message broker so that the satellite operators could watch the execution process in real time.

The SDDownloader class uses the meta information from the instrument configuration files to access and load data into the database using the DBIface interface. But one can implement a customized way to load data by passing a program as a parameter to the SDDownloader class. The execution mechanism is the same as the decoder execution.

The SDDownloader class requires the storage to be mounted to the file system. By default, the SDDownloader class compresses data before moving them to the storage. All raw, level-0, and level-1 data are stored. Currently, this behavior can be changed only by editing the global configuration file of the satellite.

4 Conclusion

A data processing, storage, and distribution system has been created using the SDDS framework and deployed at the Space Monitoring Data Center of Skobeltsyn Institue of Nuclear Physics, Lomonosov Moscow State University. The system is available at https://downloader.sinp.msu.ru. During eighteen months of operation, the system has been functioning correctly and showing its effectiveness and stability. Using the API provided by SDDS to add support for new satellites to the system is easy and requires a minimum of coding. In particular, adding support for the DSCOVR satellite and three satellites of the GOES-series has taken 14 hours and 12 hours accordingly. SDDS has been approved at SINP MSU as the base framework to be used in satellite service development in the future.

Acknowledgment

I would like to thank the Space Monitoring Data Center team for collaboration in the development and support of RDB, metadata, and data access services. I would like to thank Dr. Vladimir Kalegaev for helpful discussions on satellite data processing and clear problem statements. Also, I would like to thank Dr. Alexander Kryukov for his support in all aspects. This project is supported by RSF grant #16-17-00098.

References

[1] V.O. Barinova, O.G. Barinov, S.Y. Bobrovnikov, S.A. Dolenko, V.V. Kalegaev, D. Mottl, E.A. Muraveva, L.R. Mukhametdinova, and I.N. Myagkova, *Proceedings of the 8th Solar System Plasma Conference* (Space Research Institute of RAS, Moscow, Russia, 2013) p. 132

[2] V.A. Egorov, V.O. Ilyin, E.A. Loupian, A.A. Mazurov, A.A. Proshin, and E.V. Flitman, Current problems in remote sensing of the earth from space **1**, 1, 431–436 (2004)

[3] D.A. Kobets, A.M. Matveev, A.A. Mazurov, and A.A. Proshin, Current problems in remote sensing of the earth from space **12**, 1, 145–155 (2015)

[4] S. Ott, J. Bakker, J. Brumfitt, N. de Candussio, L. Dwedari, A. M.Heras, S. Leeks, A.P. Marston, J.-J. Mathieu, J. Pizarro, and H. Siddiqui, Astronomical Data Analysis Software and Systems XV ASP Conference Series **351**, 516–519 (2006)

[5] R. E. MacFarlane, D. W. Muir, R. M Boicourt, A. C. Kahler, and J. L. Conlin, *The NJOY Nuclear Data Processing System* (https://github.com/njoy/NJOY2016-manual/raw/master/njoy16.pdf, Version 2016)

EPJ Web of Conferences **173**, 05013 (2018) https://doi.org/10.1051/epjconf/201817305013
Mathematical Modeling and Computational Physics 2017

SaaS Platform for Time Series Data Handling

Ekaterina Oplachko[1,⋆], *Stanislav* Rykunov[1,⋆⋆], and *Mikhail* Ustinin[1,⋆⋆⋆]

[1] *Institute of Mathematical Problems of Biology RAS - the Branch of Keldysh Institute of Applied Mathematics, Pushchino, Moscow Region, Russia*

Abstract. The paper is devoted to the description of MathBrain, a cloud-based resource, which works as a "Software as a Service" model. It is designed to maximize the efficiency of the current technology and to provide a tool for time series data handling. The resource provides access to the following analysis methods: direct and inverse Fourier transforms, Principal component analysis and Independent component analysis decompositions, quantitative analysis, magnetoencephalography inverse problem solution in a single dipole model based on multichannel spectral data.

1 Introduction

Cloud platforms become more and more popular for different fields. This fact is determined by sophisticated requirements toward computing machines which scientists use to analyze their data. Usually, experimental data demand big capacity of hardware to perform all the desired computations. The cloud technologies can help in this task due to their ability to provide scalable capacity, operating system independency, and simple access through the Internet [1]. The scientist can use them as much as he needs and when he needs. This concept makes the analysis easier and more accessible [2].

Our cloud resource MathBrain provides users with tools for time series analysis. The most of methods are dedicated to magnetic- and electro encephalography (MEG, EEG) analysis which contains big amounts of data to compute. The methods of brain analysis are noninvasive, and the process looks like a registration of electro-magnetic activity. During the procedure, the magnetic encephalograph registers a magnetic filed for several minutes, in hundreds of channels. Thus, as a result of these experiments, the specialists get big amounts of data with complex structure. We have chosen the Software as a Service (SaaS) model to give users a complete set of tools to analyze their data.

2 Architecture and technical realization

The purpose of this work is the description of a tool which can handle big amounts of encephalography data. This tool should meet the following technical requirements: it should work without local installation; no additional licensing; only opensource libraries/technologies; compatibility with .mat files; it should allow to perform big calculations simultaneously for several users; operating system

⋆e-mail: oplachkoe@gmail.com
⋆⋆e-mail: rykunov@psn.ru
⋆⋆⋆e-mail: ustinin@impb.ru

independency. To meet these requirements, we have chosen the SaaS model of cloud resource. The main benefit here is the resources scalability, i.e. tasks can be balanced between nodes and the users would not be affected by performance issues. At the same time, the cloud model does not need installation process expertise and licensing, and, also, we receive operating system independent resource.

The architecture of this resource contains the layer of abstraction which helps to share hardware between containers (tasks) and can be used for balancing the load. As a virtualization platform, we are using the Docker [3] technology which allows to create containers and to balance them between nodes depending on CPU and RAM load. The architecture of MathBrain resource contains three nodes. One of them is a Manager host, the others are nodes (Workers) for user's tasks execution. The Manager host contains repository of images, these images can be implemented to container and Portainer – user web interface. MathBrain, as a web resource, contains a web server and a database. These functions are also being placed in different containers on the Manager node. The cloud flexibility and scalability can be affected if all physical resources are asked simultaneously. In such a situation, a supplementary public cloud capacity can be used to expand the local hardware resources. There are three main vendors of public cloud who foster work with containers: Google Cloud Platform [4], Amazon AWS and Microsoft Azure [5]. All of them have their own services which help to build hybrid solutions (mix of the on-premises and public resources usage). Without connecting to public cloud resources Docker Swarm manager will put tasks in a queue till local hardware capacity is available. The engine of the resource which handle encephalography data is written in Python language. All the corresponding scripts were packed in an API container and can be used according to request.

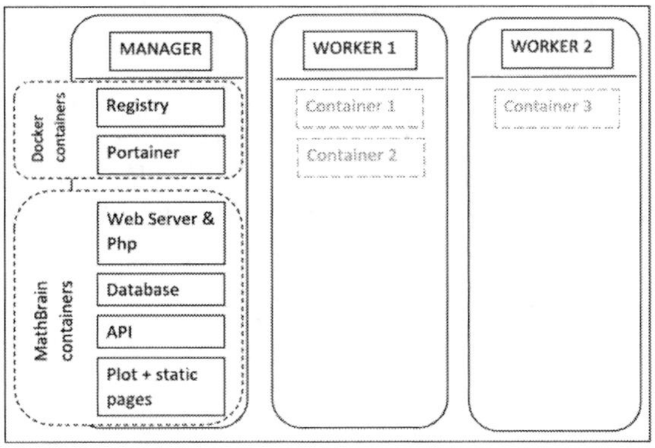

Figure 1. MathBrain architecture

3 Functionality

The MathBrain resource provides several analysis methods to the users: direct and inverse Fourier transforms, Principal Component Analysis (PCA) [6] and Independent Component Analysis (ICA) [7] decompositions, quantitative analysis, inverse MEG problem solution in a single dipole model based on multichannel spectrum data [8]. Direct and inverse Fourier transforms are the basic methods for the analysis of MEG time-series [9]. Direct Fourier transform can be calculated on a selected frequency band as well as on a whole frequency range of input data. The obtained spectrum can be used in further analysis, e.g. for inverse problem solution. Inverse Fourier transform is used for restoration of time-series from spectrum. It can be calculated on selected time-frame or on the whole length of the original time-series. Also at this stage one can select frequency band for restoration. Restored time-series can be used in further PCA and ICA analysis. The PCA and ICA are multivariate signal processing techniques widely used in neuroimaging to separate (as much as possible) independent subcomponents from linearly mixed signals. In MathBrain service they are applied to multi-channel MEG time-series in a "temporal" variant, i.e. independence is intended in the domain of temporal observations whereas the number of channels defines the original dimensionality of the multivariate data set. The quantitative analysis is a direct Fourier transform of MEG time-series, calculated in moving time window. The length and shift of this time window are defined by user. As the result of this analysis, one-dimensional power spectrogram (sum of powers in all channels in corresponding frequency bin) is produced, allowing one to evaluate spectral changes in signal during the time of measurement. The inverse problem solution in MEG addresses the finding of the magnetic field sources from the known values of magnetic induction at some sensors on the head surface. To solve this problem, the following function depending on the magnetic field sources is minimized

$$f = \sum_{i=1}^{n} \omega_i (B_i - B_i^0)^2 \rightarrow \min .$$

B_i^0 are the values of the magnetic induction measured by the sensors, B_i are the relevant values from forward field modeling, ω_i are the sensors' weights, and n is the number of sensors. In terms of spectral-based approach B_i^0 are the values of restored MEG channels at selected frequency at given moment of time. Provided with the initial guess, the dipole location is determined by standard mathematical methods designed for searching the local minimum of the function of several variables. Since in this case the information on the derivatives of the function being minimized is difficult to obtain, the zero-order methods were selected. Namely, the Nelder-Mead simplex method is used for minimization. As a forward field model we use equivalent current dipole model in conducting sphere [10]. The results of this procedure are dipole coordinates and direction. These results are shown superimposed onto subject's MRI.

The example of the MathBrain resource work is illustrated in Figure 2 and 3. In this example, the inverse problem is solved at the selected frequency. User chooses a spectrum and MRI. First, the scientist gets a one-channel chart where he can choose an interesting frequency and then use it for calculations. When the necessary frequency is identified, the user receives a multichannel chart with details. As a result, user sees the source of the signal at the MRI.

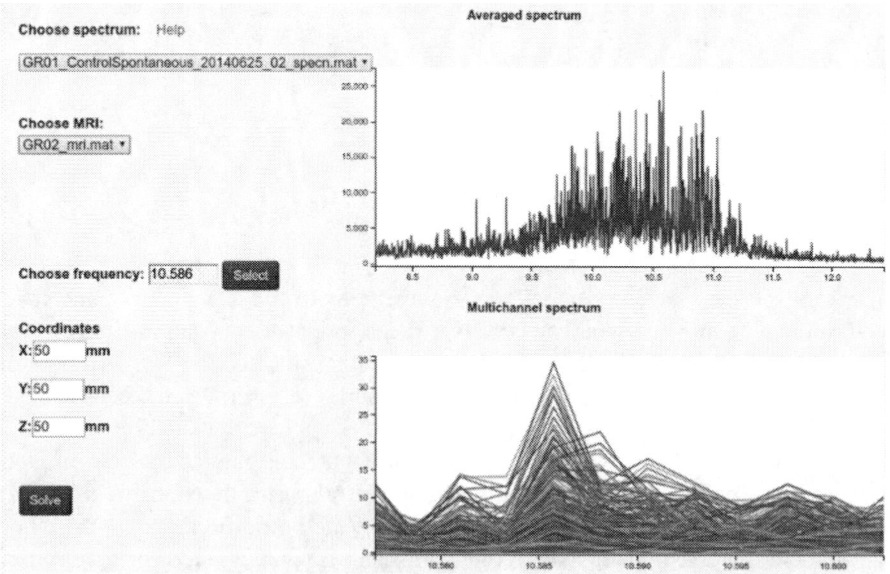

Figure 2. "Inverse MEG problem" method input data screen

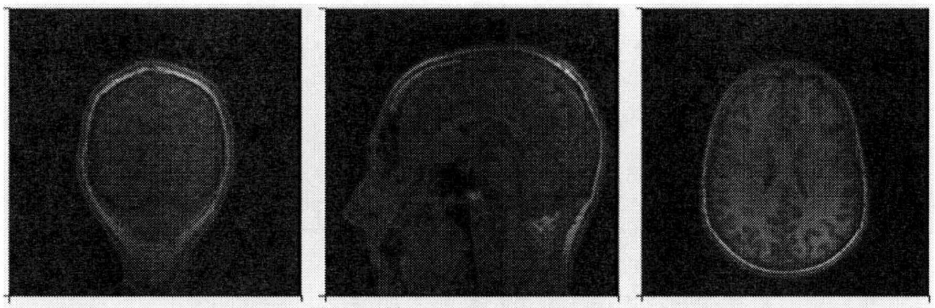

Figure 3. Example of the "Inverse MEG problem" method work

4 Conclusion

A cloud based solution for time series data handling was created. It meets all technical pre-requisites. The resource provides an access to widely-used analysis techniques. Plans for the further development: to add magnetic field map visualization to ICA, PCA and inverse problem solution; to extend inverse problem solution to two-dipole model; to develop more complex analysis workflows, e.g. inverse problem solution, based on ICA decomposition of spectral data; to create the semi-automatic MEG data conversion program.

Acknowledgement

This work was partially supported by the Russian Foundation for Basic Research (grants 16-0700937, 16-07-01000, 17-07-00677, 17-07-00686, 17-29-02178), by the Program I.33P for Fundamental Research of the Russian Academy of Sciences, and by the CRDF Global (USA) (grants CRDF RB1-2027 and RUB-7095-MO-13).

References

[1] A. Mu-Hsing Kuo, J Med Internet Res **13 (3)**, e67 (2011)

[2] E.S. Oplachko, D.M. Ustinin, M.N. Ustinin, Mathematical Biology and Bioinformatics **8 (2)**, 449–466 (2013)

[3] T. Adufu, C. Jieun, and K. Yoonhee, Network Operations and Management Symposium (APNOMS), 507–510 (2015)

[4] Z. Liu and S. Cho, 8th International Workshop on Scheduling and Resource Management for Parallel and Distributed Systems (SRMPDS'12), 397–403 (2012)

[5] V. A. Gandhi and C.K. Kumbharana, International Journal of Advanced Networking Applications (IJANA), 117–123 (2010)

[6] I.T. Jolliffe, *Principal Component Analysis, Second Edition* (Springer-Verlag, New York, 2002) pp. 299–337

[7] A. Hyvärinen, J. Karhunen, and E. Oja, *Independent Component Analysis* (Wiley, New York, 2001) pp. 147–164

[8] S.D. Rykunov, E.S. Oplachko, M.N. Ustinin, and R.R. Llinás, Mathematical Biology and Bioinformatics **12**, 176–185 (2017)

[9] B.J. Jansen, J.R. Bourne, J.W. Ward, IEEE Trans. Biomed **28**, 836–838 (1981)

[10] J. Sarvas, Physics in Medicine and Biology **32**, 11–22 (1987)

EPJ Web of Conferences **173**, 05014 (2018) https://doi.org/10.1051/epjconf/201817305014
Mathematical Modeling and Computational Physics 2017

Concept of a Cloud Service for Data Preparation and Computational Control on Custom HPC Systems in Application to Molecular Dynamics

Dmitry Puzyrkov[1,★], *Sergey* Polyakov[1,2,★★], *Viktoriia* Podryga[1,3,★★★], and *Sergey* Markizov[1,★★★★]

[1] *Keldysh Institute of Applied Mathematics of RAS, 4, Miusskaya sq., 125047, Moscow, Russia*
[2] *National Research Nuclear University MEPhI (Moscow Engineering Physics Institute), 31, Kashirskoe shosse, 115409, Moscow, Russia*
[3] *National Research Center "Kurchatov Institute", 1, Akademika Kurchatova pl., 123182, Moscow, Russia*

Abstract. At the present stage of computer technology development it is possible to study the properties and processes in complex systems at molecular and even atomic levels, for example, by means of molecular dynamics methods. The most interesting are problems related with the study of complex processes under real physical conditions. Solving such problems requires the use of high performance computing systems of various types, for example, GRID systems and HPC clusters. Considering the time consuming computational tasks, the need arises of software for automatic and unified monitoring of such computations. A complex computational task can be performed over different HPC systems. It requires output data synchronization between the storage chosen by a scientist and the HPC system used for computations. The design of the computational domain is also quite a problem. It requires complex software tools and algorithms for proper atomistic data generation on HPC systems. The paper describes the prototype of a cloud service, intended for design of atomistic systems of large volume for further detailed molecular dynamic calculations and computational management for this calculations, and presents the part of its concept aimed at initial data generation on the HPC systems.

1 Introduction

The study of the evolution of complex processes without ad hoc simplifications is a scientific problem of the highest interest. A powerful mathematical method for the description of the evolution of systems consisting of large numbers of particles (molecules, atoms) obeying the laws of the classical mechanics is the method of molecular dynamics (MD) [1]. The number of particles in the investigated systems can vary from dozens of individual particles to billions of particles. These particles interact "each with each" and this generates a high computational complexity in the large-scale MD systems. Such problems can be solved on high-performance computing (HPC) systems like supercomputers

★e-mail: dpuzyrkov@gmail.com
★★e-mail: polyakov@imamod.ru
★★★e-mail: pvictoria@list.ru
★★★★e-mail: arh11msn@yandex.ru

and GRID systems. In this paper, we discuss the concept and implementation of a dedicated web service intended to help scientists in HPC MD simulations especially as it concerns the data preparation and control of the computing process.

2 Formulation of the problem

The geometry of the system of particles under study can be very complicated. Not every MD simulator provides a means for describing such a geometry, and in most cases the task of consistent particle distribution rests on the shoulders of third-party applications.

The question arises of the method of designing such a particle system. In particular, to create meta-description files of geometry and physical conditions, a special editor is required, which will allow creating the necessary geometric model and immersing the necessary particle conglomerates into it.

In addition to the design problem, there is also the problem of managing the available computational resources and automated task launch in the user-accessible environment. This is required in order to maximize the efficiency and ease of use of the HPC systems and provide to the user a single access point for a set of such computers. This task has already been partially solved with the KIAM Job_Control [2] job management system, but a close integration of the design and accounting environment with this system is required.

In order to successfully implement a service which allows the user to solve the tasks of creating convenient computational domains for MD modeling as well as managing such calculations, the following components are necessary: a graphical geometry designer; applications for processing meta-description files; a computational tasks management system.

3 Existing solutions

As a system for HPC access and monitoring we can mention the "FlyElephant" platform [3]. FlyElephant provides access to different HPC clusters and helps users to send their computing tasks to them. For now, the platform support has access to 3 clusters and it can work with SLURM. This platform is not aimed on MD calculations in general, so any special tasks, such as data generation for simulators should be done by other software.

In the KIAM MolSDAG web service the HPC systems will be accessible by user credentials enabling the use of any cluster resources the researcher has access to. Apart from SLURM, the KIAM Job_control system supports the SUPPZ queue control.

4 The concept of cloud service architecture

In this paper, the concept of constructing and implementing the KIAM MolSDAG cloud service (Molecular Structure Designer and Generator) is presented, which in the long term will solve the problems posed above.

The main idea of this software package is to give to the final user a simple and universal system which will provide him with an interface for creating complex geometry, creating initial data sets based on this geometry and performing MD simulations.

The main components of the service are the editor of the geometry of the computational domain, the database of microstructures and their properties, a set of various parallel applications for generation atomistic data, as well as access modules to various file storages.

4.1 CAD-application

The graphical geometry designer is a CAD-like web application using WebGL (Web Graphics Library) for drawing a three-dimensional design area. In its function, first of all, the construction of a geometric model of a projected microsystem. This service has direct connection with the database of molecular dynamic calculations (DBMC), which allows the use of metadata of the already existing materials in the database.

4.2 MD data generators

In view of the possible large scale of the microsystem being created, at the stage of designing the geometry of the computational domain, there are no real procedures for generating and filling the region with particles which are processed. After the user has finished editing the structure, the output of the editor is a text-friendly, human-readable file in the YAML format. This file contains a meta-description of the projected particle system, including all properties and parameters necessary for its generation and simulation. Using this file, the KIAM MolSDAG scenario management system initiates the process of creating a script for generating initial data. The output script of this system is shown in fig. 1. It is a python script that describes the interactions between multiple generation programs. On the calculation run it will be executed by Job_control task runner module, providing actual generators execution on a target computational resource.

```
#job_control management system import
import wmcs
nitrogen_gen = wmcs.get_application("mxwgen")
nikel_gen = wmcs.get_application("fccgen")
unite_gen = wmcs.get_application("unite")
def applications():
    return [nitrogen_gen, nickel_gen, unite_gen]
def run():
    nitrogen_gen.run()
    nickel_gen.run()
    nitrogen_gen.wait()
    nickel_gen.wait()
    unite_gen.run()
    unite_gen.wait()
```

Figure 1. Scenario file for KIAM Job_control management subsystem

As a result of the execution of this task on the target system, the computational domain shown in fig. 2 will be created. The simulation task will use this data as an initial dataset.

4.3 Database of molecular dynamic calculations

The main purpose of such a database is to reduce the time required to generate and calculate typical structures. For example, for the task of simulating a binary metal-gas microsystem of nickel and nitrogen till the thermodynamic equilibrium is achieved, the materials, already reduced to thermodynamic equilibrium, can be used as initial data. This approach helps significantly reduce the time required for

Figure 2. Visualized output of the generator scenario

simulation. By saying "database" in this context, we mean a set of programs for accounting various geometries, binary files and meta-descriptions of materials. The main components of this service are the database itself, a management system that provides API for interaction with the database, a set of modules for accessing distributed data [4], and a server-side client application. Replenishment of DBMC is performed by user calculations done with the help of the developed service.

5 Conclusion

The paper is devoted to the development of the KIAM MolSDAG cloud service and presents part of its concept, aimed for initial data generation on the HPC systems. The service is being developed for graphical design of large-scale microsystems and performing calculations. The purpose of creating such microsystems is a detailed molecular dynamic studying the processes of their evolution and the properties appearing in this operation. After graphical design of large-scale microsystems, the proposed system generates a Python language scenario and executes a generator programs based on this script. The generation process assumes both the creation (generation) of microsystems on the basis of new data, and their arrangement based on previously generated and studied microstructures. The developed service will contain in the future a large number of microstructure generators and programs for filling areas of complex shapes with particles of various types, as well as programs for combining pre-calculated data from various sources. Individual elements of the service have already been implemented and confirmed the effectiveness of the overall service concept.

Acknowledgments

The research was supported by RFBR (projects no. 15-07-06082-a, 15-29-07090-ofi_m) and by the Academic Excellence Project of the NRNU MEPhI under contract with the Ministry of Education and Science of the Russian Federation No. 02.A03.21.0005.

References

[1] D. Frenkel and B. Smit, *Understanding Molecular Simulation. From Algorithm to Applications* (Academic Press, New York, 2002)
[2] D. Puzyrkov, V. Podryga, S. Polyakov, and M. Iakobovskii, CEUR Workshop Proceedings **1787**, 416–422 (2017)
[3] *FlyElephant – All-in-one automated workflow platform*, https://flyelephant.net/
[4] D.V. Puzyrkov, V.O. Podryga, and S.V. Polyakov, Vestnik UGATU **20**, 175–186 (2016)

Primary Data Treatment Software for Position-Sensitive Detector of Small-Angle Neutron Scattering Spectrometer in the Isotropic Pattern Scattering Case

Alexei Soloviev[1,⋆], *Sergei* Kutuzov[2,⋆⋆], *Olexander* Ivankov[3,⋆⋆⋆], and *Alexander* Kuklin[3,⋆⋆⋆⋆]

[1] *Laboratory of Information Technologies, Joint Institute for Nuclear Research, Dubna, Russia*
[2] *Dzhelepov Laboratory of Nuclear Problems, Joint Institute for Nuclear Research, Dubna, Russia*
[3] *Frank Laboratory of Neutron Physics, Joint Institute for Nuclear Research, Dubna, Russia*

Abstract. A new data converter has been created for the new position-sensitive detector (PSD) of small-angle neutron scattering (SANS) spectrometer YuMO. In the isotropic pattern scattering case, it provides the possibility for processing PSD data with the SAS data processing program that has already been in use.

1 Introduction

The modern trends in use of the small-angle neutron scattering (SANS) method consist in varying the sample environment, e.g., temperature, pressure, magnetic or electric fields [1–5]. The objects under investigation can be oriented in the field or by the flow of a liquid. In this case the need for the position-sensitive detectors (PSD) arises.

A PSD of a new type has been created for the YuMO spectrometer. Currently, this spectrometer has a two-detector system with coaxial ring detectors with a central hole [6, 7] and lacking the azimutal angle sensitivity. The SAS package [8–10] is used for its data treatment. The PSD is planned to be mounted either in the place of one of the YuMO detectors, or in addition to them (turning it into a three-detector system). The PSD data processing is of the utmost importance.

In the present paper we discuss a technique of data processing which significantly improves the statistics of the PSD dependent data and makes possible the use of the allready existing on-line data processing package SAS.

2 Current state of primary data treatment for YuMO spectrometer

The key element of the modernized time-of-flight SANS YuMO spectrometer is its two-detector system. It provides a unique dynamic range of the neutron momentum transfer ($Q_{\max}/Q_{\min}$ is about 90) and asks for short data acquisition times. The main view of the setup of the two-detector modernized YuMO spectrometer as well as an instantiation of a SANS experiment graphic user interface (GUI)

⋆e-mail: solovjev@jinr.ru
⋆⋆e-mail: kutuzov@jinr.ru
⋆⋆⋆e-mail: aleksandr.ivankov@gmail.com
⋆⋆⋆⋆e-mail: alexander.iv.kuklin@gmail.com

are presented in figure 1. For a detailed description of the operating mode of the spectrometer in the two-detector variant, see [6, 7].

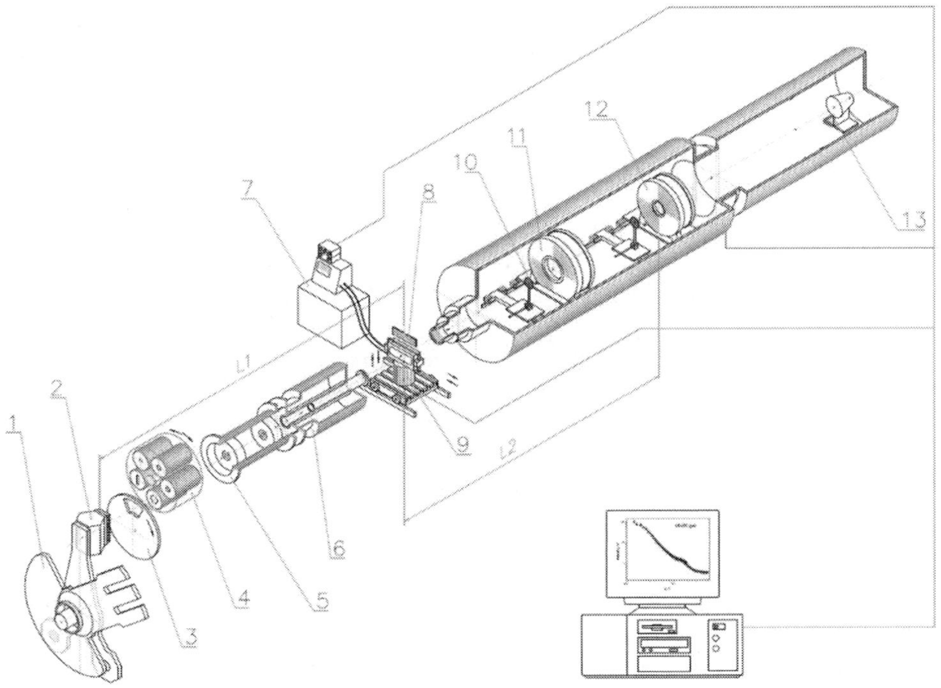

Figure 1. The setup for small-angle scattering of neutrons YuMO (4-th channel IBR-2): 1 – reflectors, 2 – active zone of reactor with moderator, 3 – chopper, 4 – changeable collimator, 5 – vacuum tube with additional background collimators, 6 – justified collimator, 7 – thermostat, 8 – holder with samples on the table with thermos boxes, 9 – table of samples, 10 – the vanadium standard, 11, 12 – detectors ("OLD" and "NEW" accordingly), 13 – direct beam detector.

2.1 SAS package

The SAS package [8–10] is used to process the spectra measured on the YuMO spectrometer. In general, processing with SAS includes two main stages:

1. normalization of the time spectra over the scattering from the vanadium standard;
2. conversion of the time spectra into the neutron momentum transfer space.

SAS does not operate with the "OLD" and "NEW" detectors separately, and does not sew spectra from them. Instead, it operates with a set of their rings and produces the final spectrum in a natural way [9].

The main SAS window is shown in figure 2. One can see that the statistics at the end of the time spectra is poor. It negatively impacts on the normalization procedure and has to been improved. Within SAS, the combination of the neighbouring cells of the time histograms is used to this purpose.

3 Primary data treatment for PSD

The PSD of a new type has been created and tested for the YuMO spectrometer [11]. Typical patterns of scattering are shown in figure 3. For most of the experiments one needs the isotropic pattern of scattering only.

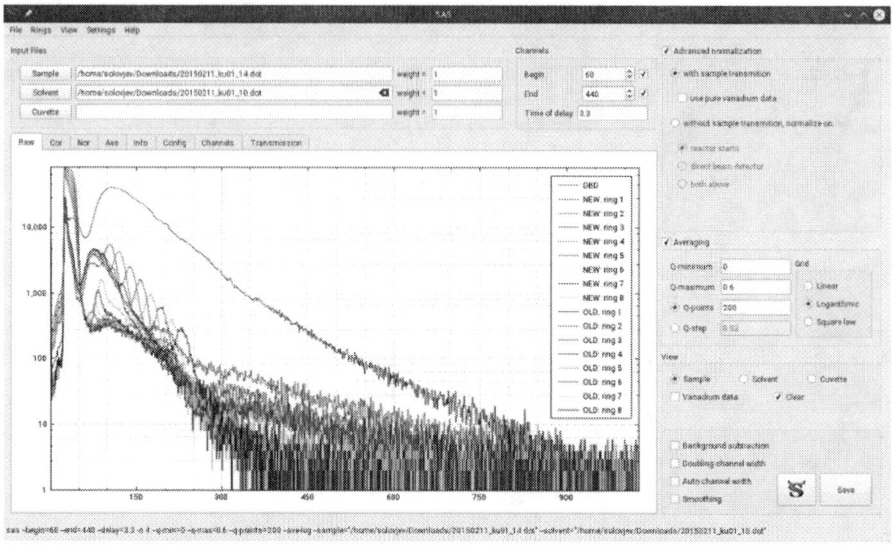

Figure 2. Main SAS window. Time spectra on detector rings.

Due to the small size of the PSD cells the statistics collected on them is often poor. That fact negatively impacts on the data treatment. On the other hand, in most of the experiments one needs an isotropic pattern of the scattering only, when objects in the sample are not oriented. In this case, it is possible to combine the PSD cells into concentric rings characterized by a single data value. This has two desireable features: statistics improvement and straightforward applicability of the on-line SAS package implementation for the data processing.

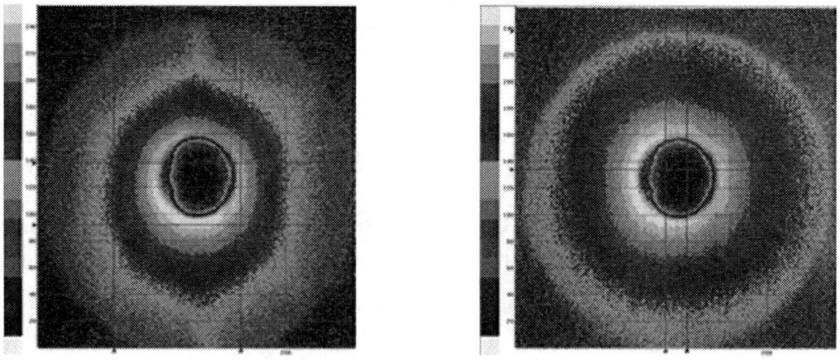

Figure 3. Anisotropic (left) and isotropic (right) patterns of scattering.

3.1 PSD2SAS data converter

The data converter for PSD has been created [12]. It is implemented as ROOT-based [13] Qt Application. The intersection area of a cell with the resulting ring is taken into account. The entire area of the detector is in use. OpenMP parallelism is used both for reading data and for ring construction.

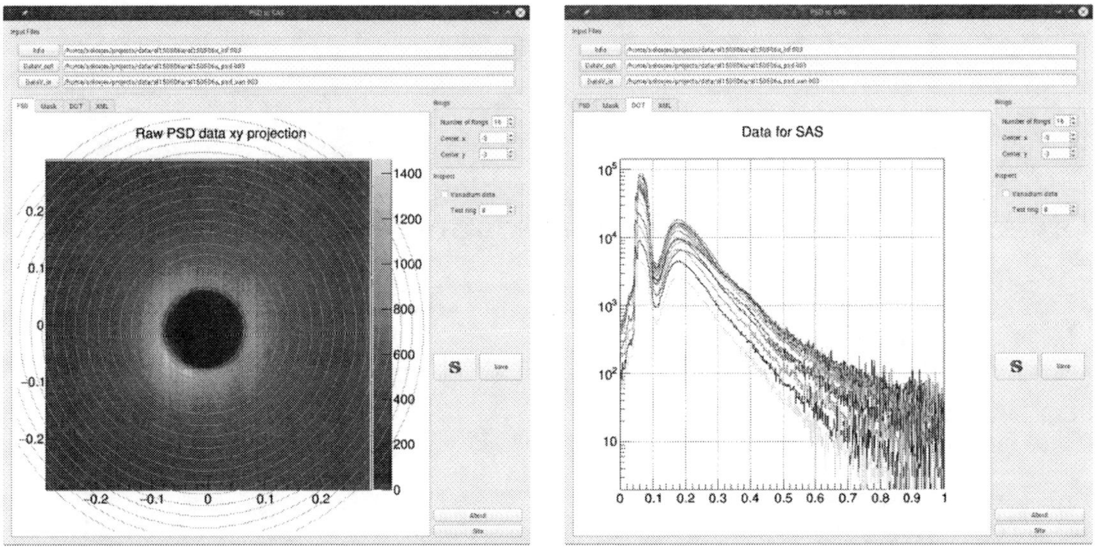

Figure 4. PSD rings pattern (left) and time spectra (right) for these rings, ready for processing in SAS

4 Conclusion

- The data converter for position-sensitive detector has been created.
- The Q-resolution provided by the detector is practically fully available.
- The Q-range provided by the detector is not truncated due to the use of detector edges.
- The converter is fast, the time is comparable to the processing time in SAS.

References

[1] D.V. Soloviov et al., Journal of Physics: Conference Series **351**, 012010 (2012)
[2] D.V. Solov'ev et al., Journal of Surface Investigation. X-ray, Synchrotron and Neutron Techniques **5**, 7 (2011)
[3] H. Gibhardt, C. Haramagatti, A. Islamov, O. Ivankov, A. Ivanovich Kuklin, G. Eckold, **228**, 769 (2014)
[4] M. Balasoiu, A. Kuklin, **61** (2016)
[5] H. Gibhardt, C. Haramagatti, A. Islamov, O. Ivankov, A. I. Kuklin, G. Eckold (2015)
[6] A.I. Kuklin, A.K. Islamov, V.I. Gordeliy, Neutron News **16**, 16 (2005)
[7] A.I. Kuklin et al., Journal of Physics: Conference Series **291**, 012013 (2011)
[8] A.G.Soloviev, T.M.Solovieva, A.V.Stadnik, O.I.Ivankov, A.I.Kuklin, Communication of JINR **P10-2003-86** (2003)
[9] A.G. Soloviev, T.M. Solovjeva, O.I. Ivankov, D.V. Soloviov, A.V. Rogachev, A.I. Kuklin, Journal of Physics: Conference Series **848**, 012020 (2017)
[10] `https://wwwinfo.jinr.ru/programs/jinrlib/sas/indexe.html`
[11] A.I.Kuklin et al., Frank Laboratory of Neutron Physics, Annual Report (2007)
[12] `https://wwwinfo.jinr.ru/programs/jinrlib/psd2sas/indexe.html`
[13] `https://root.cern.ch/`

EPJ Web of Conferences **173**, 05016 (2018) https://doi.org/10.1051/epjconf/201817305016
Mathematical Modeling and Computational Physics 2017

Object Classificators Using the AdaBoost Algorithm and Neural Networks

Alexey V. Stadnik[1,⋆], *Pavel S.* Sazhin[1,⋆⋆], and *Slavomir* Hnatic[2,3,⋆⋆⋆]

[1] *OOO «Videointellect», 33 Pravda St, office 11, 141981 Dubna, Moscow Region, Russian Federation*

[2] *The Laboratory of Information Technologies, JINR, 6 Joliot-Curie, 141980 Dubna, Moscow Region, Russian Federation*

[3] *Institute of Experimental Physics, Slovak Academy of Sciences Watsonova 47, 04001 Košice, Slovak Republic*

Abstract. The construction of image object detectors is still a relevant task, due to dynamic developments in the field of computer vision. In this work, we combined neural network technologies with existing data processing algorithms to obtain effective object classifiers. We demonstrate our approach on the example of face detection.

1 Introduction

One of the standard methods used in building of fast and effective classifiers in computer vision is the Viola-Jones cascade (also called Haar cascade). The method was developed in 2001 by Paul Viola and Michael Jones [1] and it is still considered fundamental in real-time objects detection [2]. This method proved itself to be particularly effective in detection of faces in images, which in turn became a reference problem for construction and testing of object detectors. The method is implemented in the open computer vision library OpenCV [3]. We used the images from the CMU Face Database [4]. AdaBoost algorithm is used widely in computer vision to obtain effective classifiers [5]. It can be used to combine set of weak classifiers to obtain strong decision rules [6]. When using it in computer vision it is important to consider computational complexity. It is convenient to use threshold decision functions or Haar features as classifiers, because of their small computational cost. But, as AdaBoost is a resource-expensive algorithm, it is not always giving the effective combination of classifiers.

This work presents the approach to the construction of classifiers with comparable effectiveness [7].

⋆e-mail: alexey.stadnik@intellect.video
⋆⋆e-mail: pavel.sazhin@intellect.video
⋆⋆⋆e-mail: drhnatic@drhnatic.com

2 Demonstration of the approach

To demonstrate our approach we considered the following classification model task. On the left of the Fig. 1 one can see two initial sets with schematically shown threshold decision rules to which AdaBoost converged. The algorithm resulted in 10 decision rules, of which 6 are independent.

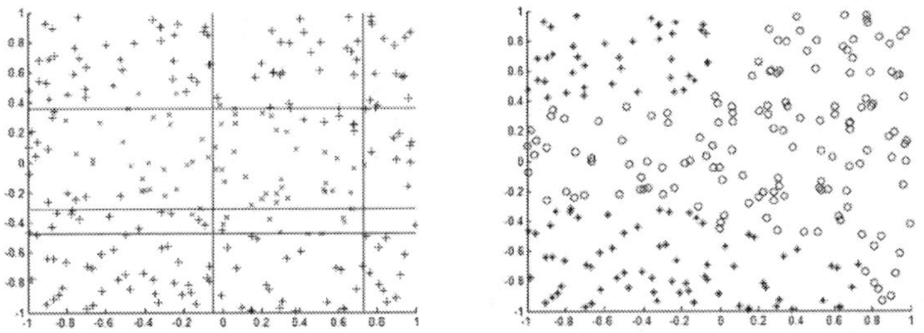

Figure 1. Model task and AdaBoost classification

The plot on the right of the Fig. 1 shows the result of the classification with corresponding weights. One can see that the decision rules are selected quite effectively, but the weight coefficients forming the final strong classifier are not optimal.

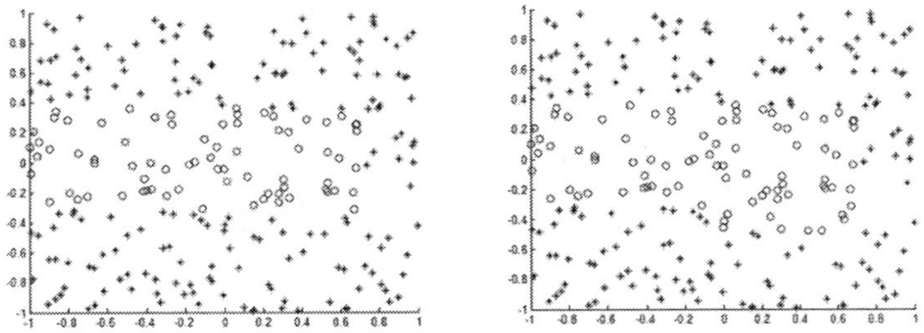

Figure 2. Classification using single-layer (left) and double-layer (right) neural networks

In Fig. 2 the result of the classification of the model data are shown for the cases when the threshold decision rules determined by AdaBoost are combined with use of single layer (left) and double layer (right) neural network. This means that the alternative approach to obtain effective classifiers is: stage 1 – AdaBoost, stage 2 – single or double layer artificial neural network. Neural network is trained on results of AdaBoost classifiers, thus they can be combined in a more effective way.

3 The working principle of the Viola-Jones cascade

This approach for image object detection uses combination of four key concepts:

1. Simple rectangular functions, also called rectangular Haar features.
2. Integral image for the fast detection of features.
3. Machine learning method AdaBoost.
4. Cascading classifier for the effective combination of Haar features.

The features used by Viola and Jones are based on Haar features. Machine learning algorithm AdaBoost is used for the selection of specific Haar features and to determine the threshold levels. AdaBoost selects the set of weak classifiers in order to combine them and assigns a weight to each classifier. This weighted combination is a strong classifier. In resulting cascade the series of AdaBoost classifiers are combined as a sequence of filters, which is particularly effective for the classification of image regions. Each filter itself is an AdaBoost classifier containing low number of weak classifiers.

The threshold at each level is set to be low enough, so that all (or nearly all) images in the training sample are accepted. The filters at each level are trained to classify training images that passed all of the previous stages. If any of those filters is not accepting some image region, then this region is classified as "not face". When the filter accepts the image region, then this region is passed to the next filter for processing. The region of the image accepted by all filters is classified as "face".

4 Detectors based on neural networks

Classifiers used in cascade are quite simple. Functionally, they are linear maps with decision threshold, or single layer linear perceptron. This means, that the number of such classfiers is quite high. In OpenCV cascades there are more than 3000 of them. We used following principles to build a detector:

1. Constant Haar basis giving constant vector of features in a form of input image descriptor.
2. Using multilayer perceptron as a "strong" classifier.

The basis from 112 Haar features satisfying such requirements and used in this work is presented as a single image in Fig. 3.

Figure 3. The basis from 112 Haar features

Figure 4. Left image – result of the descriptor with basis from the Fig. 3 and ANN (112-16-2); center image – result of the detection using first 110 rectangular features from Viola-Jones cascade trained on the sample taken from [4]; right image – analogical result when using two-layer neural network

The chosen basis for the construction of the descriptor along with artificial neural network with configuration 112-16-2 (112 inputs, 16 neurons of the hidden layer, 2 neurons of the output layer) gave a face detector with ratio $false\ positive\ rate/detection\ rate \sim 1.4 . 10^{-5}$. In given context, it is important to have low rate of false positives, or maximal value of ratio of detection rate and false positive rate as the criterion of detector optimization. Fig. 4 shows the detector outputs.

5 Detectors based on Haar eigenvalues

The idea to reduce the dimension of the image descriptor, which in turn gives smaller dimension of the neural network, resulted in use of Haar feature base having lower number of features, which describes better characteristic properties of objects. Such base was obtained using Karhunen-Loeve transform and formed from 48 eigenvectors. To obtain Haar features, these eigenvectors were quantized, where the pixels not deviating by more than half of standard deviation were given value 0, and the pixels deviating more than half of standard deviation were given value of 1 with appropriate sign. Obtained eigen Haar-features formed the base to obtain the descriptor of the image. It appears to be reasonable to use the approximation of this base containing only essential areas of each feature. This reduces the number of rectangular areas necessary for calculations. The results are presented in fig.5.

Figure 5. Eigen-Haar basis (top), results, comparison with the Viola-Jones cascade (bottom): "false positive rate / detection rate" ~ 3.7×10^{-6}, ANN (48-16-1)

6 Conclusion

We analyzed the method combining the AdaBoost algorithm and neural networks. The examples we considered show that this combination is fully suitable for application in real cases. Such method can be used to build a high quality object detector reducing the time needed to train the neural network.

References

[1] P.Viola and M.J.Jones, *IEEE Conference on Computer Vision and Pattern Recognition* (IEEE, USA, 2001) pp. 511–518

[2] P.Viola and M.J.Jones, International Journal of Computer Vision **57 (2)**, 137–154 (2004)

[3] http://opencv.org

[4] http://vasc.ri.cmu.edu/idb/html/face/

[5] Y. Freund, R. Schapire, Journal of Computer and System Sciences **55**, 119–139 (1997)

[6] Y. Freund, R. Schapire, Journal of Japanese Society for Artificial Intelligence **14(5)**, 771–780 (1999)

[7] Henry A. Rowley, Shumeet Baluja, Takeo Kanade, IEEE Transactions on Pattern Analysis and Machine Intelligence **20**, 23–28 (1998)

EPJ Web of Conferences **173**, 05017 (2018) https://doi.org/10.1051/epjconf/201817305017
Mathematical Modeling and Computational Physics 2017

Roadmap for Computer-Aided Modeling of Theranostics and Related Nanosystems

Jozef Ulicny[1] and *Tibor* Kozar[2,3,4,⋆]

[1] *Department of Biophysics, Institute of Physics, Safarik University, Kosice, Slovakia*
[2] *Center for Interdisciplinary Biosciences, TIP-UPJS, Safarik University, Kosice, Slovakia*
[3] *Institute of Experimental Physics, Slovak Academy of Sciences, Kosice, Slovakia*
[4] *Laboratory of Information Technologies JINR, 141980 Dubna, Moscow Region, Russia*

Abstract. Detailed understanding of the interactions of novel metal-containing nanoparticles with biological membranes, macromolecules and other molecular targets of the living cell is crucial for the elucidation of the biological actions of such functionalized nanosystems. We present here the construction and modeling of thiolate-protected gold clusters and the prediction of their static and dynamic properties.

1 Introduction

Nanotechnologies have significant influence on the recent research and development in many application areas. The biomedicine belongs to the most influenced fields. Thanks to the introduction of a new kind of nanoparticles (NPs) entitled theranostics (NP systems designed for combined diagnostics and therapeutic use), a new horizon was opened in imaging and treatment. Theranostic systems can be constructed from metallic core NPs functionalized with monolayer-covered bio-active molecules. Despite the recent "state of the art" level of molecular modeling methodologies and the number of articles published annually on noble-metal nanostructures, this particular field is still in the discovery phase [1]. In addition, the detailed understanding of the interactions of functionalized NPs with biological membranes, macromolecules and other entities of the living cells has crucial importance for the elucidation of the mechanisms governing the biological actions of such nanosystems. Insights into molecular details as to how these nanomaterials interact with subcellular nano-machinery of the cells can facilitate the design and engineering of a new generation of NPs [2]. We were interested in building and modeling thiolate-protected gold clusters and gold clusters functionalized with other biomolecules. We have computed their static and dynamic properties. Due to the system size and the inherent difficulties of the metallic NP modeling, such a task is a formidable challenge. As a part of our modeling roadmap, we aim to build models for NP – chromatin fragments, resembling interactions of such particles with higher-order DNA structures of living cells. Reliable computer modeling of intra- and intermolecular interactions in NP-chromatin models is a very complex task due to the size of such systems. Accordingly, we proceeded in several modeling steps which are described below.

⋆e-mail: tibor@saske.sk;tibor.kozar@upjs.sk

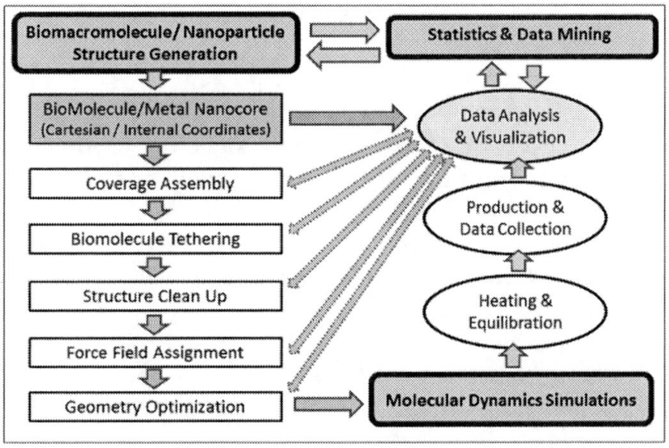

Figure 1. Schematic roadmap starting from structure generation. Due to the number of atoms and the complexity of the systems, the permanent data analysis and visualization are of key importance to the derivation of solutions

2 Computational methodology

Several different computational protocols were needed to build, compute, and analyse the nanosystems under study. Figure 1 represents the roadmap – interconnections between different tasks starting from structure generation (upper left box). Understanding the structural complexity of the studied systems (starting from gold-containing NPs to model theranostics up to building DNA-related biomacromolecular systems) required very different approaches (all-atom (AA) versus coarse-grained (CG) modeling). Accordingly, different programs (running on Windows, Linux, MacOS) and web services were used in order to fill the boxes shown in Figure 1. Table 1 summarizes the most important programs used in the presented study. Taking into account our former experience with GPGU computing [3], we preferred the GPGU versions of the listed programs where available.

3 Results and discussion

We were interested in functionalization of differently sized gold-core NPs (from 25 to more than 100 Au atoms). Based on the experimental availability of Au-related structural data the data of Murray et al [4] were used for the construction of Au-25 clusters and the data of Kornberg et al [5] for the Au-102 clusters. The results of geometry optimization (carried out by the Forcite program of Materials Studio, see Table 1) of two differently functionalized $Au_{25}S_{18}$ clusters are presented in Figure 2.

MD simulations (Forcite program of Materials Studio) of these nanoclusters confirmed the conformational stability of the systems. In addition to molecules shown in Fig. 2, we built and calculated several other NPs in two ways: by enlarging the number of core Au atoms and, changing the functionalizing coverage as well. Even more, we have included doxorubicin into the NP [6] as a known way of facilitating drug delivery.

In the next step, we were interested in the development of advanced models for biomacromolecular systems where we predict possible bindings of the generated NPs. For detailed atomistic modeling of functionalized gold NP-nucleosome interactions, we downloaded the structure of nucleosome core particle (NCP) resolved at 2.5 Å resolution [7]. In accordance with our roadmap-scheme (Fig. 1), we ended up with minimized structure of NCP to which the gold-core NPs were docked. Figure 3 represents the lowest-energy structure obtained in this way.

Table 1. Summary of computational tools used in presented study

PROGRAM or WEB SERVICE:	**Reference or provider:**
Discovery Studio (a) **Materials Studio** (b)	(a) BIOVIA, Discovery Studio Modeling Environment (2015) (b) Accelrys/BIOVIA Materials Studio v7.0, 2013 Dassault Systèmes, San Diego, USA
HEX	D.W. Ritchie, *Evaluation of protein docking predictions using Hex 3.1 in CAPRI rounds 1 and 2,* Proteins **52** (1) 98–106 (2003)
ICM	R.C. Stolz and T.C. Bishop, *ICM Web: the interactive chromatin modeling web server,* Nucleic Acids Res. **38** (Web Server issue) W254–61 (2010)
MAESTRO	*Maestro*, version 10.1.012, in Schrödinger Release 2015-12015, Schrödinger, LLC: New York, NY., USA
NAMD (c) **VMD** (d)	(c) L. Kale et al., *NAMD2: Greater scalability for parallel molecular dynamics,* Journal of Computational Physics **151** (1) 283–312 (1999) (d) W. Humphrey, A. Dalke, and K. Schulten, *VMD: visual molecular dynamics,* Journal of Molecular Graphics **14** (1) 27–28, 33–38 (1996)
PACKMOL	L. Martinez et al., *PACKMOL: a package for building initial configurations for molecular dynamics simulations,* Journal of Computational Chemistry **30** (13) 2157–2164 (2009)
VDNA	T.C. Bishop, *VDNA: the virtual DNA plug-in for VMD,* Bioinformatics **25** (23) 3187–3188 (2009)

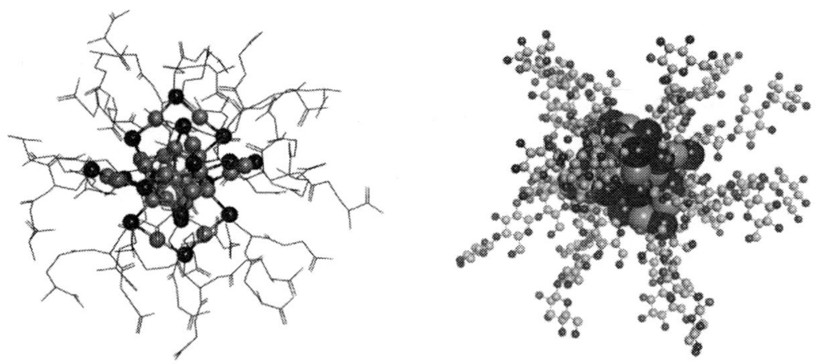

(A) (B)

Figure 2. Glutathion and chitotriose functionalized NPs: (A) $Au_{25}Glut_{18}$ with the Au-S core shown in ball&stick and glutathions in line representation; (B) $Au_{25}Chito_{18}$ with the Au-S shown as CPK models and S-linked chitotriose molecules shown in ball&stick representation. The S atoms are colored black.

The computer-aided modeling of chromosome fragments is a complex task due the size and related number of atoms of these systems. There are several computational resources helping the task, and we ended up with plenty of interesting coarse-grained structures. One of these structures is shown in Figure 3. Part of the molecule was updated by atomistic representation of histone in order to have better option to realistically dock the functionalized NP and get deeper insight into atomistic details of the intermolecular interactions. Due to the complexity/size of the system this modeling and the corresponding computations are not finalized yet.

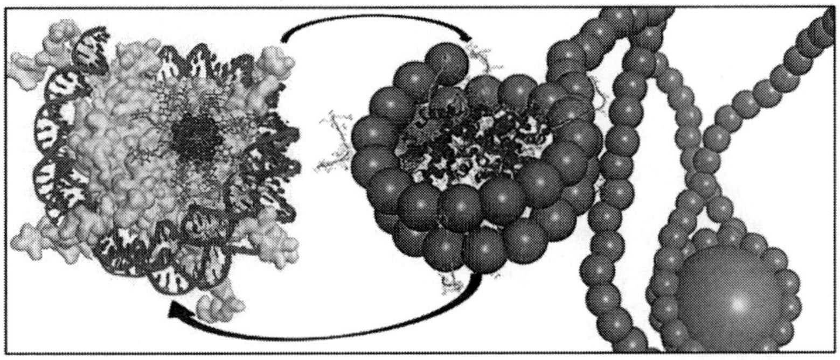

Figure 3. NCP in interaction with chitotriose functionalized gold NP (left; the NP is shown in black) and coarse-grained chromatin model with embedded all-atom histone (right). The arrows indicate the corresponding structural parts of NCP/chromatin. In coarse-grained model the medium size circles correspond to nucleotides. The largest circle corresponds to the second histone. Real molecular surface is shown for histone proteins (left).

4 Conclusion

Based on hardware and software availability, together with the gained experience in the generation of complex supramolecular assemblies, we were able to construct functionalized gold-NPs and calculate their static and dynamic properties. We proceeded to the modeling of their interactions with more complex macromolecular entities present in living cells like nucleosome particles. Such approaches will be of key importance for computer-aided modeling of biomarkers, as well as the interpretation of data for upcoming use in free-electron X-ray experiments.

Acknowledgements

This work was funded by the EU 7FP grant CELIM 316310. Supports from EU projects 2622012033, 26210120002, 26110230097, and 26210120012 are gratefully acknowledged. We also would like to acknowledge Dr. Eva Valusova and Prof. Marian Antalik (IEP SAS) for experimental support and discussions on the functionalized NPs.

References

[1] S. Malola and H. Häkkinen, Europhysics News **46**, 23 (2015)
[2] N. Yanamala, V. E. Kagan, and A. A. Shvedova, Adv. Drug. Deliv. Rev. **65**, 2070 (2013)
[3] T. Kozar, in *Lecture Notes in Computer Science: Mathematical Modeling and Computational Sciences* **7125**, (Springer Verlag, Heidelberg, Dordrecht, London, New York 2012) pp. 276
[4] M. W. Heaven, A. Dass, P. S. White, K. M. Holt, and R. W. Murray, J. Am. Chem. Soc. **130**, 3754 (2008)
[5] P. D. Jadzinsky, G. Calero, C. J. Ackerson, D. A. Bushnell, and R. D. Kornberg, Science **318**, 430 (2007)
[6] F. Wang, Y. C. Wang, S. Dou, M. H. Xiong, T. M. Sun, and J. Wang, ACS Nano **5**, 3679 (2011)
[7] J. M. Harp, B. L. Hanson, D. E. Timm, and G. J. Bunick, Acta Crystallographica Section D, Biological Crystallography **56**, 1513 (2000)

EPJ Web of Conferences **173**, 05018 (2018) https://doi.org/10.1051/epjconf/201817305018
Mathematical Modeling and Computational Physics 2017

Computing Gröbner and Involutive Bases for Linear Systems of Difference Equations

Denis Yanovich[1,⋆]

[1] *Laboratory of Information Technologies, Joint Institute for Nuclear Research, 141980, Dubna, Russia*

Abstract. The computation of involutive bases and Gröbner bases for linear systems of difference equations is solved and its importance for physical and mathematical problems is discussed. The algorithm and issues concerning its implementation in C are presented and calculation times are compared with the competing programs. The paper ends with consideration on the parallel version of this implementation and its scalability.

1 Introduction

Recently there was considerable interest in applying difference Gröbner bases to the reduction of multiloop Feynman integrals to minimal sets of basis integrals [1]. The related systems of linear difference polynomial equations are difficult to transform into the Gröbner basis form with available software. In this note we present a competitive software implementing the involutive approach to compute the basis form.

1.1 Elements of Difference Algebra [2]

Let $\{y^1, \ldots, y^m\}$ be the set of indeterminates – m functions (dependent variables) of n variables $x_1, \ldots, x_n$ (independent variables), and $\{\theta_1, \ldots, \theta_n\}$ be the set of mutually commuting difference operators:

$$(\theta_i \circ y^j)(x_1, \ldots, x_n) = y^j(x_1, \ldots, x_i + 1, \ldots, x_n).$$

A difference ring R with differences $\theta_1, \ldots, \theta_n$ is a commutative ring:

$$\theta_i\theta_j = \theta_j\theta_i, \qquad \theta_i \circ (f + g) = \theta_i \circ f + \theta_i \circ g, \qquad \theta_i \circ (fg) = (\theta_i \circ f)(\theta_i \circ g).$$

Let $\mathcal{R}$ be the difference ring of polynomials over field $\mathcal{K}$ in variables

$$\{\theta^\mu \circ y^k \mid \mu \in \mathbb{Z}^n_{\geq 0},\ k = 1, \ldots, m\}.$$

Let $\mathcal{R}_\mathcal{L}$ be the set of linear polynomials from $\mathcal{R}$ and let

$$\Theta = \{\theta^\mu \mid \mu \in \mathbb{Z}^n_{\geq 0}\}, \qquad \deg_i(\theta^\mu \circ y^k) = \mu_i, \qquad \deg(\theta^\mu \circ y^k) = \sum_{i=1}^n \mu_i.$$

⋆e-mail: yan@jinr.ru

A difference ideal is an ideal $\mathcal{J} \subseteq \mathcal{R}$ closed under the action of any operator from Θ. For $F \in \mathcal{R}$, the smallest difference ideal containing F will be denoted by Id(F). If for an ideal $\mathcal{J}$ there is $F \in \mathcal{R}_\mathcal{L}$ such that $\mathcal{J}$ =Id(F), then $\mathcal{J}$ is a linear difference ideal.

A total ordering $\succ$ on the set of $\theta^\mu \circ y^j$ is a ranking if for all i, j, k, μ, ν the following holds:

$$\theta_i\theta^\mu \circ y^j \succ \theta^\mu \circ y^j,$$
$$\theta^\mu \circ y^j \succ \theta^\nu \circ y^k \Longleftrightarrow \theta_i\theta^\mu \circ y^j \succ \theta_i\theta^\nu \circ y^k.$$

If $|\mu| > |\nu|$ implies $\theta^\mu \circ y^j \succ \theta^\nu \circ y^k$ for all j, k, then the ranking is orderly. If $j > k$ implies $\theta^\mu \circ y^j \succ \theta^\nu \circ y^k$ for all μ, ν, then the ranking is elimination.

Given a ranking $\succ$, a linear polynomial $f \in \mathcal{R}_\mathcal{L} \setminus \{0\}$ has the leading term $a\vartheta \circ y^j$, $\vartheta \in \Theta, a \in \mathcal{K}$, where $\vartheta \circ y^j$ is maximal w.r.t. $\succ$ among all $\theta^\mu \circ y^k$ which appear with nonzero coefficient in f. $\mathrm{lc}(f) = a \in \mathcal{K} \setminus \{0\}$ is the leading coefficient and $\mathrm{lm}(f) = \vartheta \circ y^j$ is the leading monomial.

A ranking acts in $\mathcal{R}_\mathcal{L}$ as a monomial order. If $F \subseteq \mathcal{R}_\mathcal{L} \setminus \{0\}$, $\mathrm{lm}(F)$ will denote the set of the leading monomials and $\mathrm{lm}_j(F)$ will denote its subset for the indeterminate y^j. Thus,

$$\mathrm{lm}(F) = \bigcup_{j=1}^{m} \mathrm{lm}_j(F).$$

1.2 Gröbner and Involutive Bases of a Difference Ideal [2–4]

Given a nonzero linear difference ideal $\mathcal{J} = \mathrm{Id}(G)$ and a ranking $\succ$, the ideal generating the set $G = \{g_1, \ldots, g_s\} \subset \mathcal{R}_\mathcal{L}$ is a *Gröbner basis* of $\mathcal{J}$ with respect to $\succ$ if for all $f \in \mathcal{J} \cap \mathcal{R}_\mathcal{L} \setminus \{0\}$:

$$\exists g \in G,\ \theta \in \Theta : \mathrm{lm}(f) = \theta \circ \mathrm{lm}(g).$$

It follows that $f \in \mathcal{J} \setminus \{0\}$ *is reducible modulo G*:

$$f \underset{g}{\rightarrow} f' := f - \mathrm{lc}(f)\theta \circ (g/\,\mathrm{lc}(g)), \quad f' \in \mathcal{J}.$$

If $f' \neq 0$, then it is again reducible modulo G, and, by repeating the reduction, in finitely many steps we obtain

$$f \underset{G}{\rightarrow} 0.$$

Similarly, a nonzero polynomial $h \in \mathcal{R}_\mathcal{L}$, the terms of which are reducible (if any) modulo a set $F \subset \mathcal{R}_\mathcal{L}$, can be reduced to an irreducible polynomial $\bar{h}$, which is said to be *in normal form modulo F* (denotation: $\bar{h} = NF(h, F)$).

Replacing the conventional division with the involutive one [3] are obtained definitions of the *involutive basis* and *involutive normal form* which have several major advantages as compared to the Gröbner basis computation: automatic avoidance of some unnecessary reductions, smoother arithmetics growth during calculation and natural parallelism.

The author used the Janet division as involutive, but there exist theoretically better divisions, for example Janet-like [2].

2 Algorithm

In the Algorithm 1 below the computations at the lines 3–8, which consume about 90% of programm runtime are mutually independent, so they may be started simultaneously. This is a source of natural parallelism of the involutive algorithm [5–7].

Algorithm 1 Sequental Computation of Involutive Basis

Input: F, a set of polynomials from $\mathcal{R}_{\mathcal{L}}$, L, an involutive division, $\succ$, a monomial ordering
Output: T, an involutive basis of F with respect to $\succ$

1: $T := \emptyset;\ \ Q := F$
2: **while** $Q \neq \emptyset$ **do**
3:   **for all** $\{\ f \in Q \mid \mathrm{lm}(f) = min_{\succ}(\mathrm{lm}(Q))\ \}$ **do**
4:     $Q := Q \setminus \{f\};\quad f' := NF_L(f, T)$
5:     **if** $f' \neq 0$ **then**
6:       $Q := Q \cup \{f'\}$
7:     **fi**
8:   **od**
9:   $T := T \cup \{\ p \mid \mathrm{lm}(p) = min_{\succ}(\mathrm{lm}(Q))\ \}$
10:   $Q := Q \setminus \{p\};\ \ Q := Q \cup \{\ \theta^{\mu} \circ p,\ \mu \in NM_L(p, T)\ \}$
11:   **if** $\mathrm{lm}(p)$ was reduced on step 4 **then**
12:     **for all** $\{\ r \in T \mid \mathrm{lm}(r) = \theta^{\nu} \circ \mathrm{lm}(p)\ \}$ **do**
13:       $Q := Q \cup \{r\};\quad T := T \setminus \{r\}$
14:     **od**
15:   **fi**
16: **od**

3 Implementation Issues

This algorithm was implemented as a program written in C language. The initial system of linear difference equations rewrites the input data into the form of a polynomial system with parametric coefficients (independent variables) and power products of the forward shift operators $\theta^{\mu}, |\mu| \geq 0$ applied to the dependent variable (sec. 1.1). For computations of the greatest common divisor for reduced polynomials with parametric coefficients our implementation uses the well-known multi-modular Brown algorithm [8], arbitrary precision arithmetics uses *libgmp*.

4 Benchmarking

Due to their high practical value for the computation of Feynman integrals [1] we shall take some systems of difference equations as benchmarks and examples. As a competitive software, the Maple package LDA [2] was chosen. The comparative timings of sequental programs are shown in Table 1. The most difficult computed example, *Smirnov3*, was used as a benchmark for the parallel version of our program (Table 2).

Smirnov1 [1]: Integral relations for the one-loop massless propagator

$$
\begin{aligned}
&(d - 2a_1 - a_2)f(a_1, a_2) - a_2 f(a_1 - 1, a_2) + a_2 q^2 f(a_1, a_2 + 1),\\
&(a_2 - a_1)f(a_1 + 1, a_2 + 1) - a_1 q^2 f(a_1 + 1, a_2) + a_1 f(a_1 + 1, a_2 - 1)\\
&\quad - a_2 f(a_1 - 1, a_2 + 1) + a_2 q^2 f(a_1, a_2 + 1).
\end{aligned}
$$

Smirnov2 [1]: Integral relations for the one-loop propagator with mass

$$
\begin{aligned}
&(d - 2a_1 - a_2)f(a_1, a_2) - 2m^2 a_1 f(a_1 + 1, a_2) - m^2 a_2 f(a_1, a_2 + 1)\\
&+ q^2 a_2 f(a_1, a_2 + 1) - a_2 f(a_1 - 1, a_2 + 1),\\
&(a_2 - a_1)f(a_1, a_2) - m^2 a_1 f(a_1 + 1, a_2) - q^2 a1 f(a_1 + 1, a_2)\\
&- m^2 a_2 f(a_1, a_2 + 1) + q^2 a_2 f(a_1, a_2 + 1) - a_2 f(a_1 - 1, a_2 + 1) + a_1 f(a_1 + 1, a_2 - 1).
\end{aligned}
$$

Smirnov3 [1] consists of six integral relations for two-loop massless propagators and utilizes five independent variables and two parameters.

Table 1. Sequental programs, time in seconds@Intel Core i5-3470

Benchmark	JB (present)	LDA
Smirnov1	16559.10	∞
Smirnov2	0.02	0.22
Smirnov3	3629.42	∞

Table 2. Parallel program, *Smirnov3* example, time in seconds@Hybrilit cluster [9], Intel Xeon E5-2695 v2

threads	compute time	speedup	CPU-utilization
1	2585.3		
3	1460.65	1.77	0.59
8	1555.57	1.66	0.21
16	507.05	5.10	0.32

5 Conclusion

Using a similar algorithmic background, our implementation demonstrates significantly better timings in comparison with the Maple implementation. It has thus the potential for computing relatively large problems. The parallel version shows rather good CPU-utilization, but it is not the final result. Hope to achieve better results in the future.

Acknowledgements

The work is supported in part by the Russian Foundation for Basic Research (grant 16-01-00080). The author thanks Vladimir Gerdt for the problem statement and useful remarks.

References

[1] A.V. Smirnov and V.A. Smirnov, Journal of High Energy Physics **0601**, 001 (2006)
[2] V.P. Gerdt, and D. Robertz, Computer Science Journal of Moldova **20**, no. 2 (59), 203–226 (2012)
[3] V.P. Gerdt and Yu.A. Blinkov, Mathematics and Computers in Simulation **45**, 519–542 (1998)
[4] V.P. Gerdt, Physics of Particles and Nuclei Letters **5**, no. 3, 248–254 (2008)
[5] D.A. Yanovich, Programming and Computer Software **28**, no. 2, 66–69 (2002)
[6] V.P. Gerdt and D.A. Yanovich, Programming and Computer Software **31**, no. 2, 73–80 (2005)
[7] D.A. Yanovich, Vestnik Rossijskogo universiteta druzhby narodov. Seriya Matematika, informatika, fizika **3.2**, 19–24 (2010)
[8] W.S. Brown, Journal of the ACM **18**, 476–504 (1971)
[9] `http://hybrilit.jinr.ru`

Modeling Thermal Effects in Metals Irradiated by Copper Nanoclusters

Balt Batgerel[3], *Stefka* Dimova[2], *Temenujka* Kupenova[4], *Igor* Puzynin[1], *Taisia* Puzynina[1], *Ivan* Hristov[2], *Radoslava* Hristova[2], *Zafar* Tukhliev[1], and *Zarif* Sharipov[1,★]

[1] *Joint Institute for Nuclear Research, 141980, Dubna, Russia*
[2] *Sofia University "St. Kliment Ohridski", Sofia, Bulgaria*
[3] *Mongolian State University of Science and Technology, Ulaanbaatar, Mongolia*
[4] *Institute for Nuclear Research and Nuclear Energy, Bulgarian Academy of Sciences, Sofia, Bulgaria*

Abstract. One of the promising areas in condensed matter physics is the study of the interaction of nanoclusters with various metallic targets. In this paper we investigate the thermal processes in metals irradiated with nanoclusters by the molecular dynamics method. In the framework of this method wave effects of the heat transfer are obtained, a feature which is absent in the thermal spike model. We repost results of computer simulation of the structural changes in irradiated copper targets of different sizes depending on the energy (in the range of 10–50 eV/atom) of copper nanoclusters.

1 Introduction

Over the past few decades, a growing number of theoretical and experimental works has been done to study the interaction of heavy ions, pulsed ion beams and nanoclusters with condensed media. These studies are motivated by the possibility of applications to the creation of radiation-resistant materials, thin coatings, new materials and technologies. Among these studies one of the most important and promising is the simulation of the interaction of nanoclusters with metallic targets [1]. When nanoclusters interact with a metal target, a large amount of energy is released in a small volume when calculated per unit volume, which leads to structural changes. The changes can be a consequence of thermal processes or of elastic interactions (repulsion of target atoms). The experimental studies give the final result of structural changes in the targets. The application of methods of mathematical modeling makes it possible to see the development of the whole process of the target irradiation in time. Note that in the real life experiments the target irradiation with nanoclusters consists of two distinct stages: the formation and the acceleration of the nanoclusters. This is technically a very difficult task. Therefore, the mathematical modeling remains an important tool and feasible alternative for such studies. In particular, the problem of modeling the accelerated nanoclusters considered in [2]. The purpose of this work is to study by the molecular dynamics (MD) method [1, 3] the thermal processes in metals irradiated with nanoclusters. The modeling of the irradiation with nanoclusters of various targets is quite similar to modeling the irradiation by other sources, e.g., heavy ions of high energies and/or pulsed beams. When heavy ions of high energies are used, a large amount of energy

★e-mail: zarif@jinr.ru

per unit volume is released in the local target region. Due to the inelastic interactions at high energies of the incident heavy ion, the direct application of the MD method for the problem of high-energy ion irradiation results in unreliable predictions. Nevertheless, the irradiation of targets with pulsed ion beams can approximately be simulated as irradiation with nanoclusters sequences and in this case the problem can be also tackled within the framework of the MD method.

2 Formulation of the problem

The methods of classical molecular dynamics are based on the model representation of a multiparticle system in which all particles (atoms or molecules) are material points. The behavior of an individual particle is described by the classical Newton's equations of motion [1, 3], which can be written as:

$$m_i \frac{d^2\mathbf{r}_i}{dt^2} = \mathbf{f}_i. \tag{1}$$

Here i is the particle number ($1 \le i \le N$); N is the total number of particles; m_i is the particle mass; $\mathbf{f}_i$ is the resultant of all forces acting on the particle, having the following representation:

$$\mathbf{f}_i = -\frac{dU(\mathbf{r}_1, \ldots, \mathbf{r}_N)}{d\mathbf{r}_i} + \mathbf{f}_i^{\mathrm{ex}},$$

where U is the interaction potential between the particles; $\mathbf{f}_i^{\mathrm{ex}}$ is the force due to external fields (not present in the reported simulations).

In this work we study the thermal processes in a copper target irradiated with copper nanoclusters $Cu_{(141)}$ of energies 10–50 eV / atom.

3 Method of numerical solution and results

Currently, there is a number of existing and widely used software packages for the application of the methods of molecular dynamics. In our investigations and computer experiments we have used LAMMPS [4].

The simulation of the process of interaction of copper nanoclusters with the target was carried out in the following way. Within the framework of the molecular-dynamic approach, we solve the problem of irradiating a copper target with nanoclusters of energy 10–50 eV/atom. Further, using the obtained solution, we calculate the temperature profiles of the target as a function of time. To do this, the target material is divided into cells and inside each cell the temperature at each instant time t_k is computed. It is to be observed that the number of particles in the cell must be larger than 100. Thus, we get the temperature distribution in the form $T_{\mathrm{md}}(x_i, y_i, z_i, t_k)$. Within the framework of molecular dynamics, the computational domain was a parallelepiped with sides $22 \times 22 \times 16$ nm, the number of particles in the target was 439200 and in the nanoclusters it was 141. As the interatomic potential, the universal EAM (embedded atom, [4]) potential for copper, integrated into LAMMPS package, was used. The calculation area for the temperature is divided into $55 \times 55 \times 30$ cells (in the x-, y-, z-directions respectively).

Figure 1 shows the crater dynamics formation as a function of time in a target irradiated with copper nanoclusters of energy 50 eV/atom. It can be seen from the figure that within the time frame from 5 ps to 20 ps the shape of the crater does not change significantly.

Figure 2 presents the dynamics of the changes (in time) of the target temperature field at a depth of 5 nm. The figure illustrates the development of a thermal process. Namely, at time 0.7 ps at a depth of 5 nm in a circle with a diameter of 7 nm, the temperature increases (the highest temperature rise is

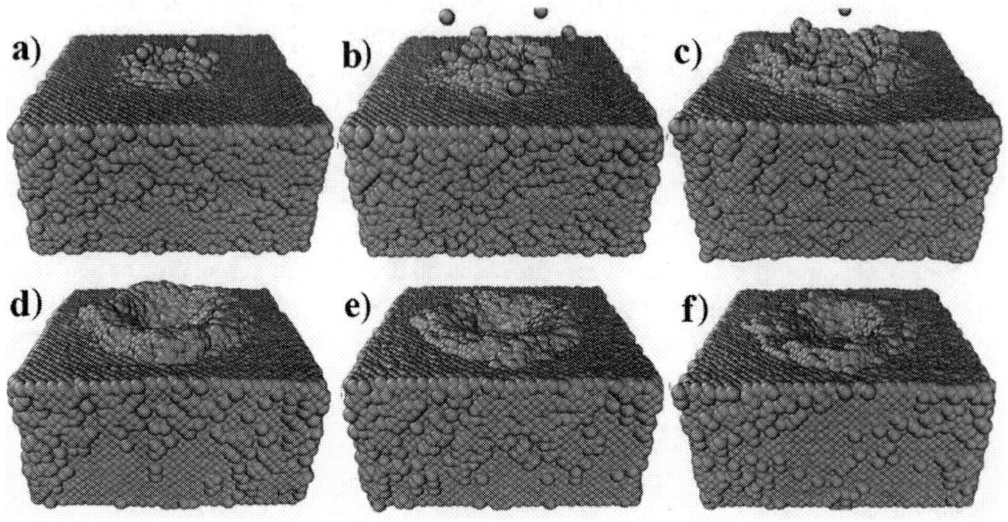

Figure 1. Dynamics of crater formation as a function of time: a) 0.5 ps; b) 1 ps; c) 2 ps; d) 5 ps; e) 10 ps; f) 20 ps

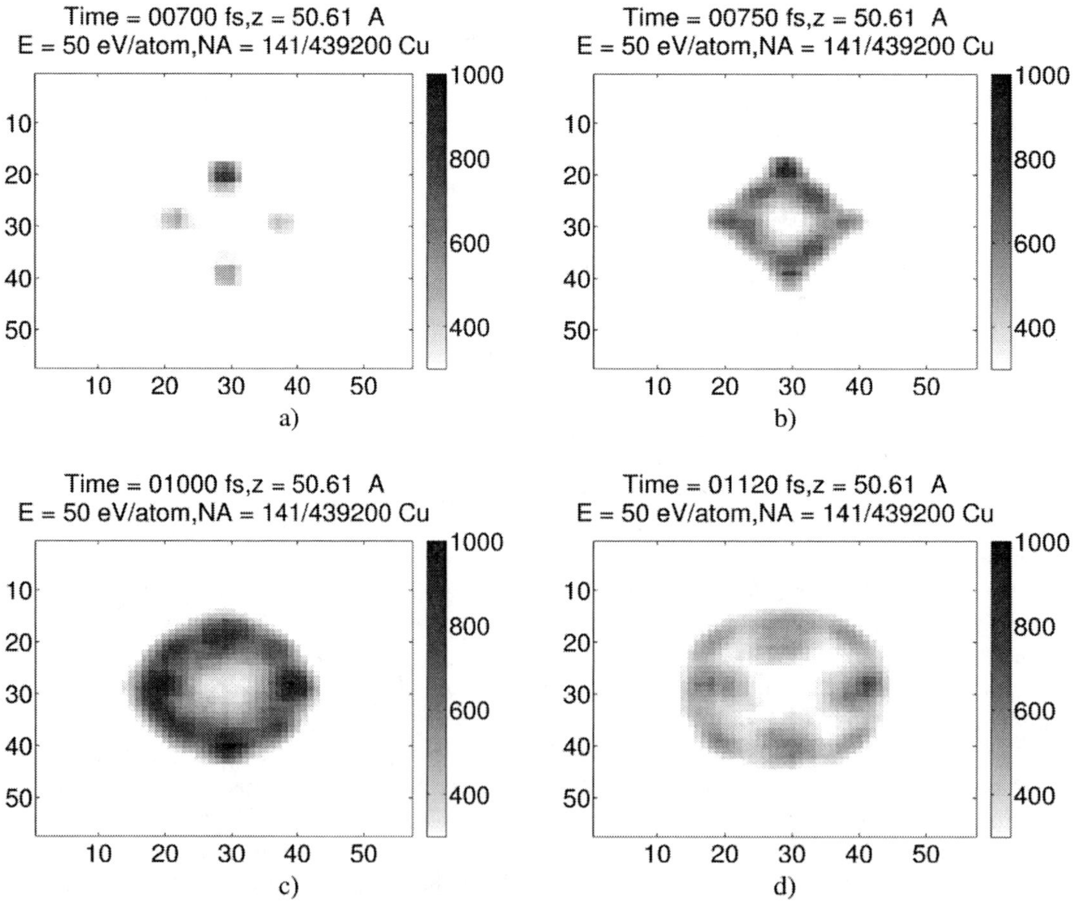

Figure 2. Dynamics of changes in the temperature field of the target at a depth of 5 nm as a function of time a) 700 fs; b) 750 fs; c) 1000 fs; d) 1120 fs

along the OX and OY axes). Similar results are observed at different depths and times. They point to the occurrence of a wave-like thermal conductivity directed along the principal crystallographic axes of the lattice. Fractal analysis methods, [5], were used for processing the results.

4 Conclusion

We simulated the interaction of copper nanoclusters of energy 10–50 eV / atom with a copper target in the framework of the molecular dynamics method and using the software package LAMMPS. Based on the computer simulations, we can draw two main conclusions:

1. The molecular dynamics approach predicts the occurrence of heat transfer wave effects, phenomena which are absent when heat equations of parabolic types are solved.
2. The resulting heat transfer wave effects are of anisotropic: the wave transport of the heat depends on the principal crystallographic axes of the crystal lattice.

Acknowledgements

The work was financially supported by the RFBR grant No. 17-01-00661-a and partially supported by a grant of the Plenipotentiary Representative of the Republic of Bulgaria at the JINR. The work of I. Hristov, R. Hristova, S. Dimova is partially supported by the NSF of BMSE under Grant DFNI-I02/9 and Grant DFNI-I02/8.

References

[1] B.Batgerel, A.Yu.Didyk, I.V.Puzynin et al., Journal of Surface Investigation **9**, 1026–1030 (2015)
[2] B.Batgerel, E.G.Nikonov, I.V.Puzynin, Vestnik RUDN, Journal of Mathematics, Information Sciences and Physics **1**, 47–51 (2014)
[3] H.T. Holmurodov, I.V.Puzynin et al., PEPAN **34**, 2, 472 (2003)
[4] S. Plimpton, Journal of Computational Physics **117**, 1–19 (1995)
[5] B.B. Mandelbrot, *The Fractals Geometry of Nature* (N. Y., Freeman, 1982) 468 pp.

EPJ Web of Conferences **173**, 06002 (2018) https://doi.org/10.1051/epjconf/201817306002
Mathematical Modeling and Computational Physics 2017

Numerical Investigation of the Time Discretization Impact on the Accuracy of a Point Target Localization by UWB Radar

Ján Buša[1,⋆], *Dušan* Kocur[1,⋆⋆], and *Mária* Švecová[1,⋆⋆⋆]

[1] *Faculty of Electrical Engineering and Informatics, Technical University of Košice, Košice, Slovakia*

Abstract. UWB radar technologies enable localization of moving persons (targets) situated behind nonmetallic obstacles. Under exact knowledge of the propagation times, from the transmitting to the receiving antennas, of the radar emitted electromagnetic wave (time of arrival, TOA), a highly accurate target localization can be achieved. Since TOA estimates only are available, their use for target localization may result in a sizeable target localization error. In this paper we study the influence of TOA quantization on the point target localization accuracy using numerical simulation methods.

1 Introduction

In the last years, great efforts have been made to develop methods for the localization of human beings based on monitoring their respiratory motion [1, 5]. For that purpose, ultra-wideband (UWB) sensors (radars) operating in the frequency band DC-5 GHz can be used with advantage. Electromagnetic waves occupying such frequency band can penetrate most nonmetallic materials with a small enough attenuation, and hence persons located behind such obstacles can be detected. On the other hand, due to the employment of the ultra-wide bandwidth, UWB sensors can provide fine range resolution (in order of centimetres), and hence a high accuracy of the target localization as well [2].

Deep analyses of the problem of the person localization by UWB sensor has shown that there are many factors affecting the accuracy of person localization. The accuracy of the so-called time-of-arrival (TOA) estimation (time necessary for electromagnetic wave emitted by the radar to travel from transmitting to receiving antennas) and the UWB radar antenna layout belongs among them. In this paper, the impact of TOA quantization (TOA estimation problem) and the geometry of the radar antennas (defining the radar antenna layout) on the accuracy of the localization of a point target behind a wall will be studied using numerical simulation methods.

2 TOA based localization of a static point target behind a wall

Let us consider a measurement equipment consisting of one transmitting antenna T placed at the origin $(0, 0, 0)$ of the Cartesian coordinate system, and 4 receiving antennas R_i, $i = 1, 2, 3, 4$, placed

⋆e-mail: jan.busa@tuke.sk
⋆⋆e-mail: dusan.kocur@tuke.sk
⋆⋆⋆e-mail: maria.svecova@tuke.sk

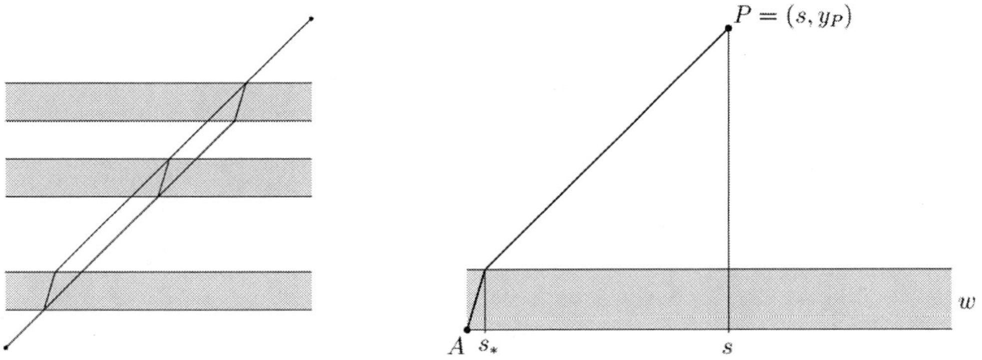

Figure 1. The wave propagation time does not depend on the wall position (left). The beam trajectory plane containing a point target P and an antenna A (right).

at the distance d from the transmitting antenna at the points

$$R_1 = (d, 0, 0), \qquad R_2 = (-d, 0, 0), \qquad R_3 = (0, 0, d), \qquad R_4 = (0, 0, -d). \tag{1}$$

A concrete wall of width w is assumed to be parallel to the plane $y = 0$ (O_{xz}). The speed of light in the air, resp. within the wall we denote c_a, resp. c_w. Further let us suppose that a point target is at a point $P = (x_P, y_P, z_P)$.

An electromagnetic wave propagates from the transmitting antenna T through the wall and air reaching the point target P, and after the reflection it propagates through the wall and air to the receivers R_i, $i = 1, \ldots, n$ (in this paper we consider case $n = 4$, see (1)). The time of the electromagnetic wave propagation from a transmitting to receiving antenna is denoted as **the time of arrival** (TOA). For these times, the notation such as TOA_i or $\mathrm{TOA}(P)_i$, $i = 1, \ldots, n$, will be used throughout this paper.

An algorithm for the localization of a point target behind a wall based on the information about TOA to the receivers has been presented in [3]. Under the assumption of 4 or more available **exact TOA values** it is possible to determine not only **the target position**, but also **the wall width** (or the material relative permittivity).

In [4] the results of through-floor localization of a person using an ultra-wide band (UWB) radar are presented. A motionless person has been lying on the base floor at the position $P = (-0.77; 2.84; 0.35)$ [m] (5 antennas were placed on the first floor, and coordinate axis y was directed down). Note that the respiratory motion is usually the only visible form of a movement for a static motionless person. For the TOA_i estimation, a reduced WP-STAPELOC method was employed [5]. The estimated position was $P_{\mathrm{est}} = (-0.91; 2.71; 0.61)$ [m], getting the localization precision 32.265 cm.

Using some sampling frequency of the measurement one gets only the TOA values $\overline{\mathrm{TOA}}_i$ rounded (at least) to the closest upper sample time values. From a point of view of the rounded TOA values a **points set of equivalency** $\mathcal{E}(P) = \{\tilde{P} \in \mathbb{R}^3 \,|\, \overline{\mathrm{TOA}}(\tilde{P})_i = \overline{\mathrm{TOA}}(P)_i\}$ (equivalency domain) can be defined. The impact of the time quantization (discretization) on the target localization accuracy – the shape and size of an equivalency domain $\mathcal{E}(P)$ – is presented below.

3 Time of arrival calculation

The TOA calculation in our simulations was based on the **Fermat principle of least time**, which leads to the **Snell–Descartes law** of refraction describing the relationship between the angles of incidence and refraction, when referring to the waves passing through a boundary between two different isotropic

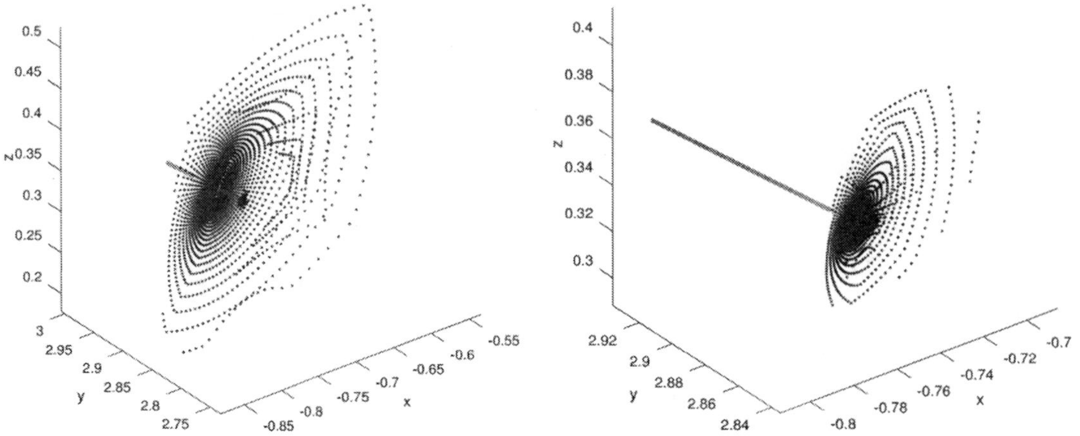

Figure 2. Equivalency domain $\mathcal{E}(P)$ boundary points for a point $P = (-0.77; 2.84; 0.35)$ [m] for 3 GHz (left) and 12 GHz (right) sampling frequency (lengths of the arrows in radius vector P direction are 10 cm)

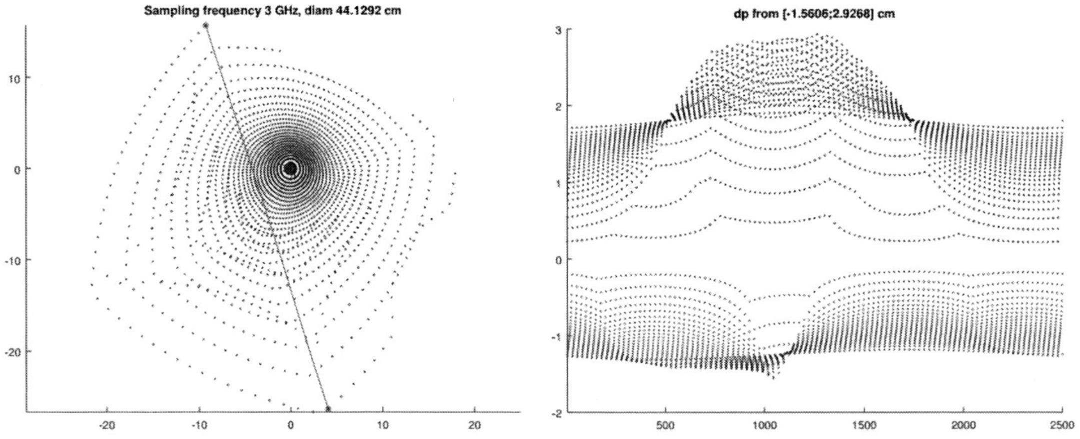

Figure 3. Projections of $\mathcal{E}(P)$ boundary points onto the plane orthogonal to the radius vector P (left), distribution of the projections onto the line parallel to the radius vector P (right) – sampling frequency 3 GHz

media, such as water, glass, or air. From the figure 1 left it is evident, that the wave propagation time does not depend on the wall position, so one side of the wall has been placed at the plane $y = 0$. Applying the Fermat principle, minimizing the wave propagation time (see [3]), we get the following algebraic equation

$$[c_a^2 - c_w^2] \cdot s_*^4 - 2s[c_a^2 - c_w^2] \cdot s_*^3 + [c_a^2 y_P(y_P - 2w) + (c_a^2 - c_w^2)(s^2 + w^2)] \cdot s_*^2 + 2c_w^2 \cdot s \cdot w^2 \cdot s_* - c_w^2 \cdot w^2 \cdot s^2 = 0$$

for the determination of a value s_*, where $s = \sqrt{x_P^2 + z_P^2}$ (Figure 1 right).

4 Numerical results

For selected points P discretized values $\overline{\mathrm{TOA}}_i$, $i = 1, \ldots, 4$ have been calculated. Relative permittivity 7.7 for reinforced concrete, receivers distance $d = 0.8$ m, and the wall width $w = 41$ cm have been used. Afterwards for 5102 selected directions on a unit sphere using bisection method boundary points of the equivalency domain $\mathcal{E}(P)$ have been determined (Figure 2).

The shape of the equivalency domain $\mathcal{E}(P)$ is flattened – its diameter for 3 GHz sampling frequency size was 44.13 cm (the first row of Table 1), it was close to the diameter of the $\mathcal{E}(P)$ projection onto the plane orthogonal to the radius vector P (Figure 3 left), and the deviations in the radius vector P direction were from $d_{P_{\min}} = -0.938$ upto $d_{P_{\max}} = 3.827$ cm getting the equivalency domain width in the P direction 4.765 cm (Table 1 and Figure 3 right).

Table 1. Precision dependence on the sampling frequency for a point $P = (-0.77; 2.84; 0.35)$ [m]

frequency [GHz]	$d_{\min}$ [cm]	$d_{\max}$ [cm]	$d_{P_{\min}}$ [cm]	$d_{P_{\max}}$ [cm]	diam [cm]
3	0.016	26.833	−0.938	3.827	44.129
6	0.016	17.542	−0.938	1.277	22.747
12	0.016	7.106	−0.055	0.945	9.570

Table 2. Precision dependence on the sampling frequency for a point $P = (1; 10; -2)$ [m]

frequency [GHz]	$d_{\min}$ [cm]	$d_{\max}$ [cm]	$d_{P_{\min}}$ [cm]	$d_{P_{\max}}$ [cm]	diam [cm]
3	0.414	93.829	−3.777	1.253	142.245
6	0.052	41.214	−0.623	0.663	52.287
12	0.052	28.612	−0.568	0.609	36.641
3*	0.710	60.564	−2.881	1.777	78.774
6*	0.100	27.562	−1.509	0.789	35.999
12*	0.100	11.794	−0.770	0.306	18.473

* – receivers distance from the transmitting antenna 1.6 m

Minimal, resp. maximal distances of boundary points from the point P are denoted by $d_{\min}$, resp. $d_{\max}$ (Tables 1 and 2). The table 2 contains values for a point $P = (1; 10; -2)$ [m].

5 Conclusions

Our investigation has shown that for a higher sampling frequency 12 GHz the localization accuracy represented by the equivalency domain diameters is acceptable for practical using. For larger distances the accuracy is smaller. Larger receiving antennas distances lead to a higher localization precision. It would be interesting to consider equipment with 2 receivers, e.g., R_3 and R_4 shifted in the x direction.

Acknowledgments

This work has been supported by the Scientific Grant Agency (VEGA) under the contract No. 1/0772/17 and by the Slovak Research and Development Agency under the contract No. APVV-15-0692.

References

[1] E. Zaikov and J. Sachs, *Ultra Wideband* (InTech, Croatia, 2010) 323–346

[2] J. Sachs, *Handbook of ultra-wideband short-range sensing: theory, sensors, applications* (Wiley-VCH Verlag GmbH & Co. KGaA., 2012) 843 pp.

[3] J. Buša, *Numerical Methods and Applications* (Springer, LNCS **8962**, 2015) 254–261

[4] M. Švecová, D. Kocur, J. Demčák, J. Buša, and M. Gamcová, Electronics Letters (submitted 2017).

[5] D. Novák, M. Švecová, and D. Kocur, *Microwave systems and applications* (InTech, 2017) 399–422

EPJ Web of Conferences **173**, 06003 (2018) https://doi.org/10.1051/epjconf/201817306003
Mathematical Modeling and Computational Physics 2017

Multilane Traffic Flow Modeling Using Cellular Automata Theory

Antonina Chechina[1,*], *Natalia* Churbanova[1,**], and *Marina* Trapeznikova[1,***]

[1] *Keldysh Institute of Applied Mathematics RAS, 4 Miusskaya Square, Moscow, Russia*

Abstract. The paper deals with the mathematical modeling of traffic flows on urban road networks using microscopic approach. The model is based on the cellular automata theory and presents a generalization of the Nagel-Schreckenberg model to a multilane case. The created program package allows to simulate traffic on various types of road fragments (T or X type intersection, strait road elements, etc.) and on road networks that consist of these elements. Besides that, it allows to predict the consequences of various decisions regarding road infrastructure changes, such as: number of lanes increasing/decreasing, putting new traffic lights into operation, building new roads, entrances/exits, road junctions.

1 Introduction

There are two main approaches to the traffic flow modeling: macroscopic, in which the vehicle flow is presented (in the dynamic case) as a compressible fluid flow, and microscopic, where each car is considered separately. The macroscopic approach is characterized by its simpicity – the computational complexity and the calculation time do not depend on the amount of cars in the network. It uses average characteristics of the traffic flow, such as the flow density and velocity. But there are serious restrictions: this approach can't be applied if the density is too low; also, the modeling can be carried out only on long (in comparison with car sizes) road fragments, which is not suitable for traffic simulation on crossroads and small road networks of city districts. The microscopic approach is free of these restrictions and can be applied to a flow with any density and on any road fragment. It also can provide much more detailed information about the traffic dynamics. It can accommodate a great many nuances of the real flows, such as the difference in driver's behavior and strategies, as well as the reactions on other drivers' actions, various car properties and so on. Modern computational facilities, such as high-performance supercomputer systems, allow real-time calculations on large-scale road networks, that is why microscopic approach becomes more and more popular nowadays.

2 Two-dimensional microscopic model based on cellular automata theory and the Nagel-Schreckenberg model

The model created by the present authors is based on the cellular automata theory: the road is divided into equal cells, each cell can either contain a vehicle, or be empty. A cell is 7.5 m long and one

*e-mail: chechina.antonina@yandex.ru
**e-mail: nataimamod@mail.ru
***e-mail: mtrapez@yandex.ru

lane wide; the distance in the model is measured in the number of cells, speed of a vehicle – in the number of cells that a vehicle can cover during one time step, time – in time steps, one second long each. As all the cells are equal in size and can fit in only a light car, the model cannot represent heterogeneous traffic flows, that include heavy vehicles, such as lorries and trucks. As opposed to the classic Nagel-Schreckenberg model [1], the developed model is two-dimensional and allows to depict multilane traffic.

Each car has a set of parameters: unique ID, maximum speed, current speed and final destination. A road fragment with cars is shown in Fig. 1. The car color in the picture corresponds to different destination points (road exits, turns, etc.)

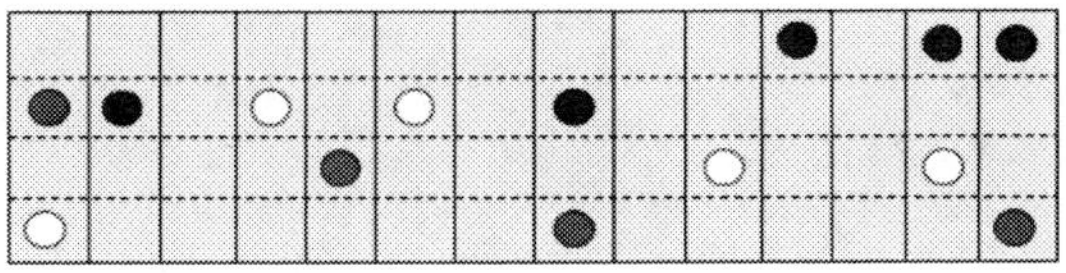

Figure 1. Multilane road fragment with cars

Each time step cell state update is carried out according to the following rules:

- a vehicle changes a lane if it is necessary (to reach the desired destination or to drive around an obstacle), it is advantageous for a driver (leads to speed increase and/or density decrease) and it is possible (i.e., if the lane change is allowed and the target cell is empty);
- a vehicle moves along the road according to the classic rules for the one-lane traffic.

For more details, see [2]. The works featuring the comparison between the created model and other approaches can be found in the list of references in [2].

3 Modeling different driving strategies

Each driver has his own driving style: ones are more determined, change lanes more easily, others are more cautious, they wait for a larger gap to move to the neighboring lane. Their behavior influences the overall situation on the road. To reflect that, the following driver types were included in the model:

- "Cautious" drivers change lanes only if the gap between the target cell and the first occupied cell upstream is larger than the maximum speed;
- "Aggressive" drivers change lanes if the gap between the target cell and the first occupied cell upstream is larger than the actual speed of the vehicle which is situated in the first occupied cell;
- "Cooperative" drivers can be either cautious or aggressive. The percentage of cooperative drivers in the system can vary. These drivers:
  - slow down ($v = v - 1$, $v \geq 1$), if they see a traffic jam before an obstacle or a road bottleneck on the neighboring lane;
  - if there is a jam before the obstacle on the neighboring lane, and there are drivers that want to change their lane, cooperative drivers stop and let them pass;
  - wait for several time steps if the car from the other lane can't move immediately, because the target cell is occupied by another car.

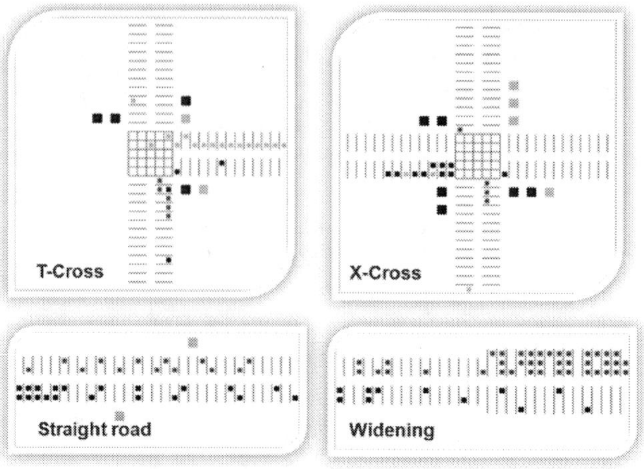

Figure 2. Simulation of traffic on different road elements

4 Modeling traffic flows on road elements and small road networks

The traffic flow modeling on different kinds of road elements can be executed using the created microscopic model. The numerical realization is represented as a module-structured program package implemented in C++ and a the visualization block (based on the Glut library), which allows to see the step-by-step evolution of the system along with the computations. The result of the work of the package is shown in Fig. 2, featuring a T-crossroad and an X-crossroad with traffic lights, a straight road fragment with a traffic light for pedestrians, and a road fragment with a widening. The larger squares represent traffic lights (darker ones indicate red light, lighter ones – green light), the smaller squares point to cars with different final destinations. Special algorithms are developed for crossroad overcoming, widening/narrowing of the road, forming an entrance queue [2].

Being able to model traffic flows on different types of road elements, we can combine them, building various road networks. The computations for each element can be carried out in parallel with data exchange on the borders: cars leave one fragment and enter another one. The example of small road network that consists of two intersections: a T-cross and an X-cross, is shown in Fig. 3.

Different kinds of average characteristics (such as, for instance, the capacity of the crossroad) can be also obtained using the program package.

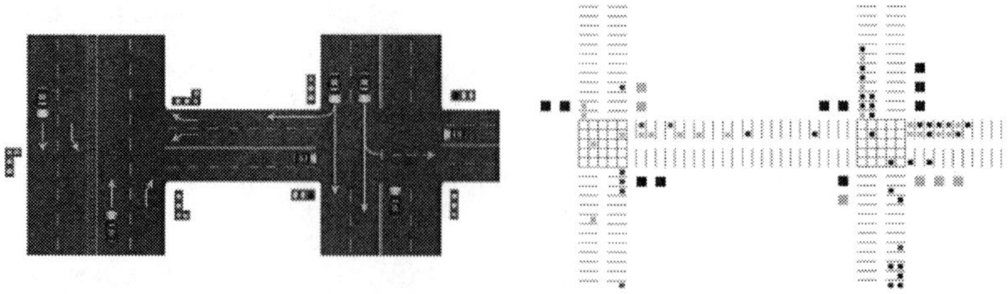

Figure 3. Road network: a scheme (on the left) and a simulation result (on the right)

5 Modeling a road accident

Road accidents can not happen inside the model due to the safety rules: a car always slows down approaching an obstacle or another car in the lane, and it doesn't change its lane if the gap in the target lane is not enough to do it safely. But it is possible to model consequences of an accident (see Fig. 4), that might be necessary for real-time online modeling of city traffic in order to develop the measures to avoid traffic jams and evaluate the efficiency of these measures. It is also possible to evaluate how much the capacity of a road will decrease after the accident. In Fig. 4 the influence of driving strategies on the road capacity in case of the accident is shown. Black squares represent cars that stopped due to the accident, and grey squares – moving cars that have to drive around them. On the left picture all drivers in the system are "cautious" drivers, and on the right picture all drivers are "aggressive". As is seen from the pictures, there is a congestion before the obstacle in both cases, but the density of cars in case of "cautious" drivers is larger, this behavior decreases the road capacity more significantly and leads to a traffic jam faster than in the case of "aggressive" driving.

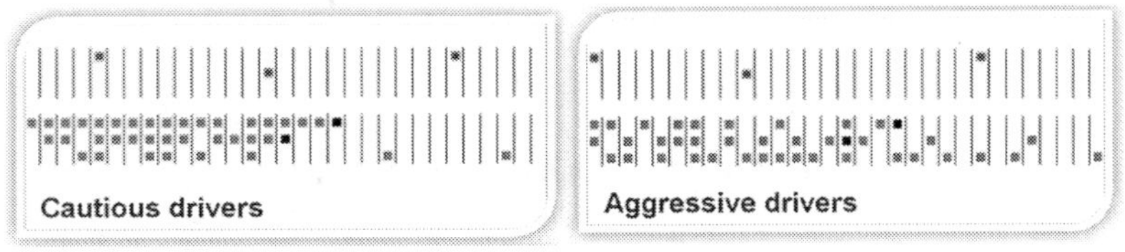

Figure 4. Road accident: influence of driving strategies ("cautious" or "aggressive") on jam formation

6 Conclusions

The model and algorithms developed showed results comparable with the simulation by the standard program software. The created program package can be used for traffic modeling on city road networks, taking into consideration different driving strategies. Due to the inner parallelism and the simplicity of the numerical algorithm the calculations can be carried out using high performance supercomputers on large-scale road networks in real time.

Acknowledgment

The work is supported by RFBR projects 16-31-00087, 16-01-00347, 18-01-00405.

References

[1] K. Nagel and M. Schreckenberg, J. Phys. I France **2**, 2221 (1992)

[2] A. Chechina, N. Churbanova, and M. Trapeznikova, *Traffic and Granular Flow'13* (M. Chraibi et al. Eds., Springer, 2015) 361–368

EPJ Web of Conferences **173**, 06004 (2018) https://doi.org/10.1051/epjconf/201817306004
Mathematical Modeling and Computational Physics 2017

On the Thermodynamic Equilibrium Distribution of a Charge in a Homogeneous Chain with a Defect

Nadezhda Fialko[1,*], *Maxim* Pyatkov[1], and *Victor* Lakhno[1]

[1] *Institute of Mathematical Problems of Biology RAS – the Branch of Keldysh Institute of Applied Mathematics of Russian Academy of Sciences*

Abstract. Based on the semi-classical Holstein Hamiltonian we consider charge transfer along a DNA chain of sites at different thermostat temperatures. Recently, using the computer simulation, it has been shown that the charge distribution in homogeneous chains in thermodynamic equilibrium depends not only on the temperature, but also on the length of the chain. We have studied numerically the case of polyadenine fragments with a defect site in the middle of the chain. The results demonstrate qualitatively similar behavior of thermodynamic equilibrium quantities in the case of the homogeneous chain and of the chain with a defect. Insertion of a trap-site enhances the stability of polaron states.

Introduction

The problem of charge or energy transfer in quasi-one-dimensional biomolecules is very interesting for such fields as biophysics and nanobioelectronics. In the 1970s, A.S. Davydov suggested that vibrational intramolecular excitations can be transferred along α-helix of protein molecules by solitons without loss of energy [1]. The polaron mechanism of charge transfer attracts considerable attention of researchers due to its stability. Presently it is believed that the charge carriers in biopolymers, such as DNA, are polarons or solitons (see e.g. [2–4] and refs. therein). This view has been formed mainly due to the fact that DNA is a quasi-one-dimensional system in which excess charges – electrons or holes – evolve into polaron states in the case of their strong interaction with oscillatory degrees of freedom of a molecule. Today, there is a huge amount of modeling results, which demonstrate solitory waves moving along an unperturbed chain. On the temperature dependent polaron propagation, numerous computational experiments and theoretical investigations [5–8] allow one to conclude that a polaron disappears at a temperature lower then its binding energy.

In the papers [9, 10] the values of the total energy, electronic part of the energy and delocalized parameter in the thermodynamic equilibrium state were calculated for 1D homogeneous chains, and some common features were found. It was shown that it is not the temperature, but the thermal energy of the lattice subsystem that is the critical parameter that determines the polaron disruption. In this work we consider a homogeneous polyadenine chain with one defect site at the center, which plays the role of a trap for the charge. We simulate a model defect site and consider a simple but important case of a DNA fragment with an A... AGA... A sequence, where G plays the role of the 8-oxoguanine which is one of the most common DNA lesions resulting from hole migration along the chain [11].

*e-mail: fialka@impb.ru

The resultant damage to DNA bases may be a significant source of mutations that lead to cancer and other human pathologies.

1 Charge transfer model

The model is based on the Holstein Hamiltonian for a discrete chain of sites [12]. In the semiclassical approximation, choosing the wave function Ψ in the form $\Psi = \sum_{n=1}^{N} b_n|n\rangle$, where b_n is the amplitude of the probability of the charge (electron or hole) occurrence at the n-th site ($n = 1, \dots, N$, N is the chain length), we write down the averaged Hamiltonian:

$$\langle\Psi|\hat{H}|\Psi\rangle = \sum_{m,n} \nu_{nm} b_m b_n^* + \frac{1}{2}\sum_n M\dot{\tilde{u}}_n^2 + \frac{1}{2}\sum_n K\tilde{u}_n^2 + \sum_n \alpha' \tilde{u}_n b_n b_n^*. \tag{1}$$

Here ν_{mn} ($m \neq n$) are matrix elements of the transition between the m-th and the n-th sites (depending on overlapping integrals), ν_{nn} is the electron energy at the n-th site. We use the nearest neighbor approximation, i.e. $\nu_{mn} = 0$ if $m \neq n \pm 1$; the intrasite fluctuations $\tilde{u}_n$ are assumed to be small and can be considered to be harmonical; suppose that the probability of charge's occurrence at the sites depends linearly on the sites displacements $\tilde{u}_n$, α' is the constant of coupling between quantum and classical subsystems, M is the effective mass of the site, K is the elastic constant.

Motion equations for Hamiltonian (1) have the form

$$i\hbar\, \dot{b}_n = \nu_{n,n-1} b_{n-1} + \nu_{n,n+1} b_{n+1} + \nu_{n,n} b_n + \alpha' \tilde{u}_n b_n, \tag{2}$$

$$M\,\ddot{\tilde{u}}_n = -K\tilde{u}_n - \alpha'|b_n|^2 - \gamma\,\dot{\tilde{u}}_n + \xi_n(t). \tag{3}$$

To model a thermostat, subsystem (3) involves the term with friction (γ is a friction coefficient) and the random force $\xi_n(t)$ such that $\langle\xi_n(t)\rangle = 0$, $\langle\xi_n(t)\xi_m(t+s)\rangle = 2k_B T\gamma\delta_{mn}\delta(s)$ (T is the temperature [K]). This way of imitating the environmental temperature with the use of Langevin equations (3) is well known [7, 13].

We calculated a set of samples using 2o2s1g-method [14]. Each sample is a trajectory of system (2)–(3) with its own initial data and random values of $\{xi_n(t)\}$. Having obtained a number of samples we find averaged "over ensemble" functions of time, and for big time interval of integration we can estimate thermodynamic parameters. We are interested in what the delocalization parameter

$$\langle R\rangle = \left\langle \frac{1}{\sum_n |b_n|^4} \right\rangle \tag{4}$$

will be when the system reaches thermodynamical equilibrium. Obviously, if a charge is localized at one k-th site, i.e. charge probability $|b_k|^2 \sim 1$, then $R \sim 1$. For homogeneous polyA chains, the parameter values of which correspond to a small-radius polaron, in thermodynamic equilibrium at low T $\langle R\rangle \sim 1$ and for high T $\langle R\rangle \sim N/2$.

For the homogeneous chains the electron energy values are the same for all sites, and we can take $\nu_{nn} = 0$. When the chain has one defect site in the center, this site has negative electron energy $\nu_{nn} < 0$ and serves as a trap for the charge. We have computed chains A...A X A...A with a model defect site X in the center for chain lengths $N = 19$ and $N = 40$, and a more realistic case: the Guanine is incorporated in the center of a polyA fragment. The model parameters corresponding to nucleotide pairs are chosen as follows [15, 16]. The effective mass of a site is $M = 10^{-21}$ g; the rigidity $K = 0.062\,\text{eV/Å}^2$ and coupling constant $\alpha' = 0.13\,\text{eV/Å}$. The matrix elements of the transition between sites are: $\nu_{AA} = 0.030\,\text{eV}$, we choose the same $\nu_{AX} = \nu_{AA}$; also we choose the electron energy on A $\nu_A = 0$; model traps are $\nu_X = -0.066\,\text{eV}$ and $\nu_X = -0.132\,\text{eV}$. For A...A G A...A fragments $\nu_{AG} = 0.049\,\text{eV}$, $\nu_{GA} = 0.089\,\text{eV}$, and the electron energy $\nu_G = -0.45\,\text{eV}$.

2 Results

A qualitative picture of the $\langle R(T)\rangle$ values for such chains in thermodynamic equilibrium is similar to the results for a homogeneous chain without a defect [10]. When the thermal energy of the classical chain $E_{\text{class}} = Nk_BT$ is less than the critical value E_{crit}, the charge is in the polaron state, $\langle R(T)\rangle \sim 1$. When $E_{\text{class}} > E_{\text{crit}}$, the charge is delocalized, $\langle R(T)\rangle \sim N/2$.

If we rescale abscissa, i.e., choose the thermal energy of the classical chain instead of the temperature T as variable x, $x = E^*NT$ (E^* is a dimensionless coefficient related to k_B), the values of $\langle R(x)\rangle$ for different values of N are found to be close to the same curve. Figure 1 a) shows the $\langle R(x)\rangle$ curves for homogeneous polyA and for chains with the defect site. Low parts of graphs of Figure 1 a) correspond to polaron state, $\langle R\rangle \sim 1$. The greater is the depth of the defect ν_X, the greater is the value of E_{crit}; i.e. for fixed N the stability of the polaron state increases with increasing electron energy on the defect site $|\nu_X|$.

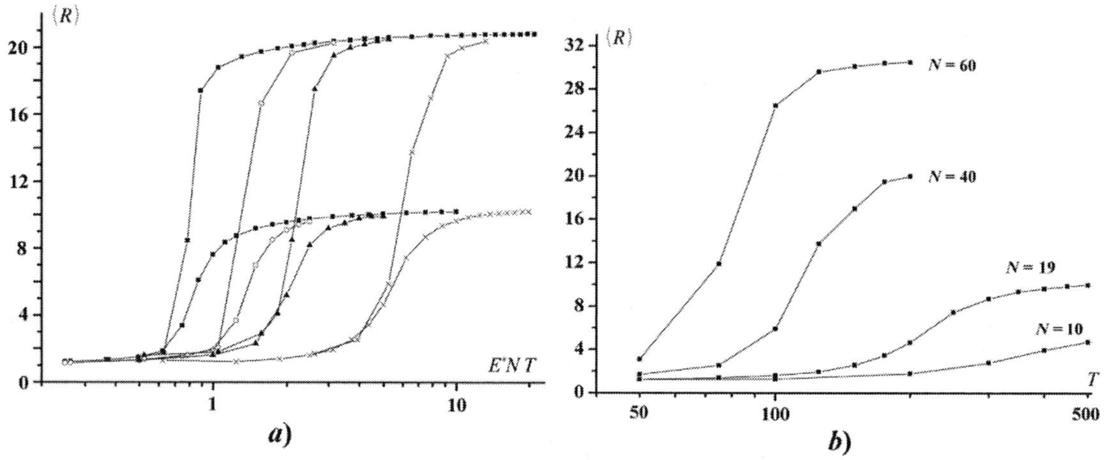

Figure 1. Left: Dependencies of delocalization parameter on the energy of the classical chain. The graphs show two chain lengths $N = 19$ and $N = 40$ sites. Squares denote $\langle R\rangle$ values for homogeneous chains, circles show $\langle R\rangle$ for polyA with defect site $\nu_X = -0.066$ eV, triangles – for polyA with defect site $\nu_X = -0.132$ eV, and crossed lines correspond to A...A G A...A fragments, $\nu_G = -0.45$ eV. Right: $\langle R(T)\rangle$ for the A...A G A...A fragments of different lengths N.

Figure 1 b) shows the temperature dependencies $\langle R(T)\rangle$ for the A...A G A...A fragments of different lengths. It is shown that, e.g. for the same temperature $T = 200$: the polaron state exists in the "short" chain of 10 sites; the charge is in an intermediate state in the chain of 19 sites (the charge is localized about half the time in a sample); and the charge is delocalized in the "long" chains $N = 40$ and $N = 60$. For homogeneous polyA fragments, in TDE the polaron state exists at low temperatures, less than 5Θ (Θ is Debye temperature, below which the semiclassical approximation is inapplicable). For the oligonucleotide AAAAGAAAAA ($N = 10$), the critical temperature T_{crit} of polaron disruption is $T_{\text{crit}} \sim 400$ K, and for A...A G A...A fragment of 19 sites $T_{\text{crit}} \sim 300$ K lies in the region of biological temperatures.

Conclusions

The qualitative picture for a homogeneous chain with a defect site is the same as for the chain without defect. The temperature stability of polarons in a chain with a defect site increases in comparison with the case of a homogeneous chain. The simulation results demonstrate that as the length of the chain increases, the type of charge distribution changes. Apparently, in short chains, the charge is in the polaron state, and its transfer occurs slowly by a hopping mechanism [3, 4, 17]. In long chains at the same temperature, the charge is in delocalized state, and the transfer process is more rapid.

The polaron localized on guanine can create 8-Hydroxyguanine, an abundant form of oxidative DNA damage, causes G→T and C→A substitutions [11]. The results obtained demonstrate that we can reduce the damaged guanine by changing the length of the chain or the sequence which leads to a decrease in the critical temperature of its delocalization. This can be interesting for making protectors from a DNA damage.

Acknowledgments

We are grateful to the HybriLIT group of JINR for computational resources.

The work is partially supported by Russian Foundation for Basic Research, grants no. 16-07-00305, 17-07-00801, and Russian Science Foundation, grant 16-11-10163.

References

[1] A.S. Davydov, Journal of Theoretical Biology **66** (2), 379–387 (1977)
[2] E.M. Conwell, PNAS **102** (25), 8795–8799 (2005)
[3] G.B. Schuster (ed.), *Long-Range Charge Transfer in DNA II*, Topics in Current Chemistry **237** (Springer, 2004), 245 p.
[4] E.B. Starikov, S. Tanaka, and J.P. Lewis (eds.), *Modern Methods for Theoretical Physical Chemistry of Biopolymers* (Elsevier Scientific, Amsterdam, 2006) 461 pp.
[5] S. Komineas, G. Kalosakas, and A.R. Bishop, Phys. Rev. E **65** (6), 061905 (2002)
[6] L. Cruzeiro-Hansson and S. Takeno, Phy. Rev. E **56** (1), 894 (1997)
[7] P.S. Lomdahl and W.C. Kerr, Phys. Rev. Lett. **55** (11), 1235–1238 (1985)
[8] V.D. Lakhno, International Journal of Quantum Chemistry **108** (11), 1970–1981 (2008)
[9] V.D. Lakhno and N.S. Fialko, JETP **120**, 125–131 (2015)
[10] N.S. Fialko, E.V. Sobolev, and V.D. Lakhno, JETP **124** (4), 635–642 (2017)
[11] K.C. Cheng, D.S. Cahill, H. Kasai, S. Nishimura, and L.A. Loeb, J. Biol. Chem. **267** (1), 166–172 (1992)
[12] T. Holstein, Annals of Physics **8**, 325–342 (1959)
[13] E. Helfand, J. Chem. Phys **69**, 1010–1018 (1978)
[14] H.S. Greenside and E. Helfand, Bell System Tech. J. **60**, 1927–1940 (1981)
[15] A. A. Voityuk, J. Jortner, M. Bixon, and N. Rosch, J. Chem. Phys. **114** (13), 5614–5620 (2001)
[16] F.D. Lewis and Ya.Wu, J. Photochem. Photobiol. C **2** (1), 1–16 (2001)
[17] M. Bixon and J. Jortner, J. Phys. Chem. B. **104** (16), 3906 (2000)

EPJ Web of Conferences **173**, 06005 (2018) https://doi.org/10.1051/epjconf/201817306005
Mathematical Modeling and Computational Physics 2017

Simulation of AntiMatter–Matter Interactions in Geant4

Aida Galoyan[1,⋆], *Vladimir* Uzhinsky[2], and *Alberto* Ribon[3]

[1] *VBLHEP, JINR, Dubna, Russia*
[2] *LIT, JINR, Dubna, Russia*
[3] *CERN, Geneva, Switzerland*

Abstract. Cross sections of antiproton and antinucleus interactions with nuclei are calculated using stochastic averaging method. A new implementation of the Quark-Gluon-String Model (QGSM) is proposed for simulation of multi-particle production in antinucleus-nucleus collisions. A combination of the cross sections and the new implementation of QGSM allows experimental data on antiproton and antinucleus interactions with nuclei to be described. The combination is included in the well-known Geant4 simulation toolkit.

1 Introduction

One of the most exciting puzzles in cosmology is connected with the question of the existence of antimatter in the Universe. Some cosmic ray experiments (PAMELA, BESS, AMS, CAPRISE etc.) aim to search for antinuclei. Antinuclei have also been observed in nucleus–nucleus and proton-proton collisions at accelerators by the RHIC and LHC Collaborations. An experimental study of the antinucleus production and reactions requires the knowledge of antinucleus interaction cross sections with the matter. The cross sections are needed to estimate various experimental corrections, especially those due to particle losses which reduce the detected rate. We calculate the cross sections using the Glauber approach [1–3].

Many particles (pions, K-mesons, baryons and so on) are produced in the antiproton-proton and antinucleus–nucleus interactions. To simulate their production, we use the main ideas of the Quark-Gluon-String model [4, 5]. The main ingredient of the model is the estimation of various process cross sections of the quark–gluon string productions. The simulation of the string fragmentation is based on the LUND fragmentation model. All of these allow the known experimental data to be described successfully.

The combination of these models is included in the Geant4 simulation toolkit [6]. Geant4 is a software toolkit for the simulation of the passage of particles through matter. It is used by a large number of experiments and projects in a variety of application domains, including high energy physics, astrophysics and space science, medical physics and radiation protection.

⋆e-mail: galoyan@lxpub01.jinr.ru

2 Cross section of antinucleus–nucleus interactions

In the Glauber approach, the amplitude of the elastic scattering of an antinucleus containing B antibaryons on a target nucleus with the mass number A is given as:

$$F_{BA}(\mathbf{q}) = \frac{i}{2\pi} \int d^2b\, e^{i\mathbf{qb}} \left\{1 - \prod_{i=1}^{B}\prod_{j=1}^{A}\left[1 - \gamma(\mathbf{b} + \tau_\mathbf{i} - \mathbf{s_j})\right]\right\} |\Psi_B|^2|\Psi_A|^2 \left(\prod_{i=1}^{B} d^3\, t_i\right)\left(\prod_{j=1}^{A} d^3\, r_j\right) \quad (1)$$

$$= i \int_0^\infty bP_{BA}(b)\, J_0(qb)\, db,$$

where γ is the amplitude of the elastic antinucleon-nucleon scattering in the impact parameter representation averaged over the spin and isospin degrees of freedom, $\gamma(\mathbf{b}) = \frac{1}{2\pi i}\int d^2q\, e^{i\mathbf{qb}} F_{\bar{N}N}(\mathbf{q})$. Ψ_A (Ψ_B) is the wave function of the nucleus (antinucleus) in the ground state. Taking the origins of the coordinate systems at the centers of the nuclei, the nucleon coordinates ($\{\mathbf{r_A}\}$, $\{\mathbf{t_B}\}$) are decomposed into longitudinal ($\{z_i\}$) and transverse ($\{\mathbf{s_j}\}$, $\{\tau_\mathbf{i}\}$) components. The z-axis is directed along the projectile momentum. $\mathbf{b}$ is the impact parameter vector orthogonal to the momentum. $P_{BA}(b)$ is the profile function and J_0 is the Bessel function of zeroth order.

Quite often γ is parameterized as $\gamma(\mathbf{b}) = \sigma^{\rm tot}_{\bar{N}N}(1 - i\rho)/(4\pi\,\beta) \cdot e^{-\mathbf{b}^2/2\beta}$, where $\sigma^{\rm tot}_{\bar{N}N}$ is the total cross section of the antinucleon-nucleon interactions, ρ is the ratio of the real to imaginary parts of the $\bar{N}N$ elastic scattering amplitude at zero momentum transfer, and β is the slope parameter of the $\bar{N}N$ differential elastic scattering cross section.

Using these expressions, one can calculate various quantities characterizing the antinucleus-nucleus interactions: total, elastic and inelastic cross sections and so on (see details in [7]). In Fig. 1, the experimental absorption cross sections of the antiproton and antideuteron interactions with various nuclei are presented and compared with our calculations. As one can see, there is a good agreement between the calculations and the experimental data.

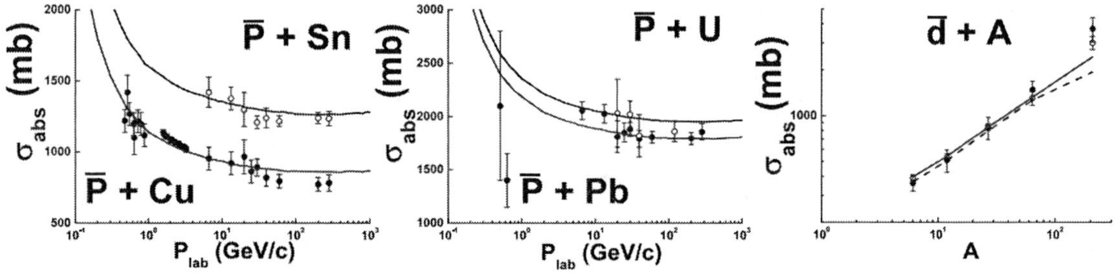

Figure 1. Absorption cross sections of the antiproton and antideuteron interactions with nuclei. Points are experimental data. The lines are our calculations. The experimental data for $\bar{d} + A$ interactions [8] are given for two momenta – 13.3 and 25 GeV/c (open and close circles, correspondingly). The calculation results at the energies are presented by solid and dashed lines.

The complexity of the Glauber expressions prevents their direct calculations. A way to the solution is offered by the stochastic averaging method [9] in which a quantity is found as an average over various samples of the nucleon coordinates. The nucleon coordinates are sampled according to the measure $|\Psi_A|^2|\Psi_B|^2$. The method consists of multiple sampling of the nucleon coordinates according to the function $|\Psi_A|^2|\Psi_B|^2$, calculation of needed expression for each sampling, and an averaging of the calculation results over the samples. Results obtained in such manner were parameterized, and the parameterizations were implemented in Geant4.

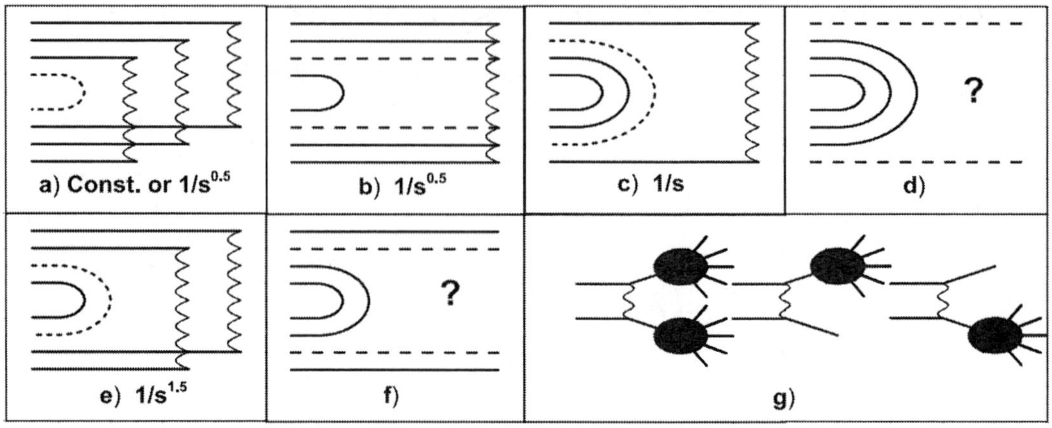

Figure 2. Quark flow diagrams of $\bar{p}p$ interactions

3 Particle production in antinucleus-nucleus interactions

Usually, the list of diagrams shown in Fig. 2 is considered in the high energy phenomenology for antiproton–proton interactions. The diagram *a* corresponds to string junction annihilation with 3 quark–antiquark string creation. The string junction is a gluonic object which couples together the quarks in a baryon. The diagram *b* represents the quark–antiquark annihilation leading to a diquark–antidiquark string creation. The diagram *c* shows a process of diquark–antidiquark annihilation and one quark–antiquark string formation. The diagram *e* corresponds to quark–antiquark and string junction annihilations with the creation of 2 quark–antiquark strings. The diagrams *d* and *f* are responsible for exotic meson production. The last diagrams at the bottom of the figure are important at high energies. They are connected with the pomeron exchange in the t-channel. The energy dependencies of the cross sections of the processes are predicted in some cases by the reggeon phenomenology. They are also shown in the figure. An elaborated scheme for the calculation of the cross sections in the reggeon phenomenology was proposed in [10, 11]. As shown there, the approach is valid at $P_{\rm lab}$ > 3 GeV/c. Since the experimental studies require much less energies, we undertook a new attempt to estimate the cross sections. It is presented in [12].

$$\sigma_a = \frac{16}{\sqrt{s-4m^2}}\left[(s-4m^2)^{-0.175} + 3.125\cdot(1-1.88/\sqrt{s})\right] \quad \text{(mb)}, \tag{2}$$

$$\sigma_b = 3.13 + 140\cdot(M_{\rm th} - \sqrt{s})^{2.5} \quad \text{(mb)}, \qquad \sqrt{s} \le M_{\rm th} = 2.172 \quad \text{(GeV)}, \tag{3}$$

$$\sigma_b = 6.8/\sqrt{s} \quad \text{(mb)}, \qquad \sqrt{s} > M_{\rm th} = 2.172 \quad \text{(GeV)},$$

$$\sigma_c = 23.3/s \quad \text{(mb)}, \quad \sigma_e = \sigma_d = \sigma_f = 0, \quad \sigma_g = 35\cdot(1. - 2.1/\sqrt{s}). \tag{4}$$

The parameterizations (2)–(4) of the process cross sections and the LUND string fragmentation algorithm implemented in the FTF generator of Geant4 allow one to reproduce a wide set of experimental data on the $\bar{p}p$ and $\bar{p}A$ interactions [12]. Of course, the generalization of the approach given above for the antiproton and antinucleus interactions with the nuclei is not a simple procedure. Some details of the procedure were considered in [12, 13]. The most important issue of the moment is that we can reproduce correctly the experimental data on the antideuteron–deuteron interactions (Fig. 3).

The figure 3 plots the π^+, π^- and proton distributions in terms of transverse momentum squared (P_T^2) respectively the rapidity (y). Fig. 3 points to a very good agreement between the experimental data and our calculations.

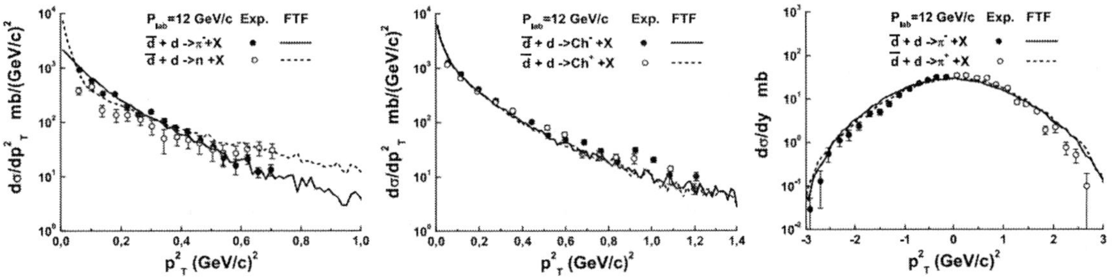

Figure 3. Distributions of particles on kinematical variables in the antideuteron-deuteron interactions at 12.2 GeV/c. Points are experimental data [14]. Lines are our calculations.

Conclusion

A new implementation of the Quark-Gluon-String model is proposed. Estimations of annihilation cross sections in the Glauber model and the LUND string fragmentation algorithm allow consistent description of a wide set of experimental data on the antiproton and antinucleus interactions with the nuclei. The proposed approach is included in the Geant4 simulation package. It can be used in the FTF_BERT physics list of Geant4.

Acknowledgement

The authors are grateful to the heterogeneous computing team of LIT JINR (HybriLIT) for support of our calculations and to Professor Cheuk-Yin Wong for helpful discussions.

References

[1] R.J. Glauber, *Lectures in Theoretical Physics* (Ed. W.E.Brittin et al., v. 1, Interscience Publishers, N.Y., 1959)

[2] R.J. Glauber, *Proc. of the 2nd Int. Conf. on High Energy Physics and Nuclear structure* (Ed. G.A.Alexander, Rehovoth, North-Holland, Amsterdam, 1967)

[3] V. Franco, Phys. Rev. **175**, 1376 (1968)

[4] A. Capella, U. Sukhatme, C-I. Tan, and J. Tran Thanh Van, Phys. Rept. **236**, 225 (1994)

[5] A.B. Kaidalov and K.A. Ter-Martirosian, Phys. Lett. **B117**, 247 (1982)

[6] Geant4 Collaboration (J. Allison et al.), Nucl. Instrum. Meth. **A835**, 186 (2016)

[7] V. Uzhinsky, J. Apostolakis, and A. Galoyan et al., Phys. Lett. **B705**, 235 (2011)

[8] S.P. Denisov et al., Nucl. Phys. **B31** 253 (1971)

[9] S. Shmakov, V. Uzhinsky, and A.Zadorozhny, Comp. Phys. Comm. **54**, 125 (1989)

[10] V.V. Uzhinsky and A.S. Galoyan, arXive:0212369 [hep-ph] (2012)

[11] A. Galoian and V. Uzhinsky, AIP Conf. Proc. **796**, 79 (2005)

[12] A. Galoyan, A. Ribon, and V. Uzhinsky, PoS BaldinISHEPPXXII, 049 (2015)

[13] Geant4 Hadronics Working Group (A. Galoyan et al.) Hyperfine Interact. **215**, 69 (2013)

[14] B.V. Batyunya et. al, JINR Preprint P1-87-849 (1987)

EPJ Web of Conferences **173**, 06006 (2018) https://doi.org/10.1051/epjconf/201817306006
Mathematical Modeling and Computational Physics 2017

Hidden Attractors in a Model of a Bubble Contrast Agent Oscillating Near an Elastic Wall

Ivan Garashchuk[1,⋆], *Dmitry* Sinelshchikov[1,⋆⋆], and *Nikolay* Kudryashov[1,⋆⋆⋆]

[1] *NRNU MEPhI, 115409, Russia, Moscow, Kashirskoe shosse, 31*

Abstract. A model describing the dynamics of a spherical gas bubble in a compressible viscous liquid is studied. The bubble is oscillating close to an elastic wall of finite thickness under the influence of an external pressure field which simulates a contrast agent oscillating close to a blood vessel wall. Here we investigate numerically the coexistence of chaotic and periodic attractors in this model. One of the tools applied for seeking coexisting attractors is the perpetual points method. This method can be helpful for localizing coexisting attractors, occurring in various physically realistic ranges of variation of the control parameters. We provide some examples of coexisting attractors to demonstrate the importance of the multistability problem for the applications.

1 Introduction

Ultrasound contrast agents are gas-filled bubbles of the order of 1–10 μm in diameter. Often they are encapsulated in a shell which can be a lipid, protein or polymer. Contrast agents are used in diagnostic ultrasound imaging because they can improve echogenecity. Bubble response to external pressure field, which is generated by the ultrasound, creates scattering sound waves. Some types of bubble response generate signals with recognizable signature. There are various works about the bifurcation structure of the contrast agent models, see e.g. [1, 2] and references therein. However, the modified Rayleigh–Plesset equation for the description of a bubble oscillation near an elastic wall (see [3, 4]) has not been studied numerically in detail. Moreover, the coexisting attractors governed by this equation have not been investigated at all. The choice of the initial data becomes a non-trivial problem when attractors can coexist. We apply the perpetual points method to overcome this problem [5, 6]. It does not require calculations of the attractors basins and, thus, is much less CPU consuming. The system of equations determining the perpetual points consists of rational equations with respect to $R, \dot{R}$ and can be efficiently solved numerically.

2 Main equation and results

We investigate the following equation [3, 4]:

$$\ddot{R}\left[R\left(1-\frac{1}{2}\frac{(\rho_1-\beta)R}{(\rho_1+\beta)d}-\frac{1}{2}\frac{(\beta-\rho_3)R}{(\beta+\rho_3)(d+h)}+\frac{1}{2}\frac{(\rho_1-\beta)(\beta-\rho_3)R}{(\rho_1+\beta)(\beta+\rho_3)h}-\frac{\dot{R}}{c}\right)+\frac{4\eta}{\rho_1 c}\right]+$$

⋆e-mail: ivan.mail4work@yandex.ru
⋆⋆e-mail: disine@gmail.com
⋆⋆⋆e-mail: nakudr@gmail.com

$$+\frac{3}{2}\dot{R}^2\left[1-\frac{2}{3}\frac{(\rho_1-\beta)R}{(\rho_1+\beta)d}-\frac{2}{3}\frac{(\beta-\rho_3)R}{(\beta+\rho_3)(d+h)}+\frac{2}{3}\frac{\rho_1-\beta}{\rho_1+\beta}\frac{(\beta-\rho_3)R}{(\beta+\rho_3)h}-\frac{\dot{R}}{3c}\right]= \quad (1)$$

$$\frac{1}{\rho_1}\left[\left(1+\frac{(1-3\gamma)\dot{R}}{c}\right)\left(P_0+\frac{2\sigma}{R_0}\right)\left(\frac{R_0}{R}\right)^{3\gamma}-\frac{2\sigma+4\eta\dot{R}}{R}-\left(1+\frac{\dot{R}}{c}\right)(P_0+P_{ac}\sin(\omega t))-\frac{R\omega}{c}P_{ac}\cos(\omega t)\right],$$

where R is the bubble radius, $R_0 = 10\ \mu$m is the equilibrium bubble radius, $P_{stat} = 100$ kPa is the static pressure, $P_v = 2.33$ kPa is the vapor pressure, the notation $P_0 = P_{stat} - P_v$ is used to simplify the equation view, P_{ac} is the magnitude of the pressure of the external field, $\sigma = 0.0725$ N/m is the surface tension, $\rho_1 = 1000$ kg/m^3 is the density of the liquid inside the blood vessel, $\rho_2 = 1060$ kg/m^3 is the blood vessel wall density, $\rho_3 = 1000$ kg/m^3 is the density of the fluid surrounding the blood vessel, $\eta = 0.001$ Ns/m^3 is the viscosity of the liquid, $c = 1500$ m/s is the sound speed, $\gamma = 4/3$ is the polytropic exponent (the process is assumed to be adiabatic), ν is the Poisson's ratio, $\nu = 0.5$ for an elastic wall and $\nu = 0$ for a rigid wall, $\beta = \rho_2\nu/(1-\nu)$ is a characteristic of the wall, $h = 1$ mm is the thickness of the wall, d denotes the distance between the wall and the center of the bubble and is close to R_0.

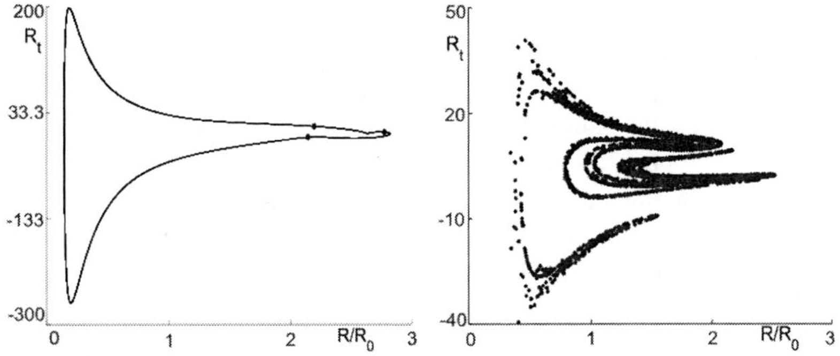

Figure 1. Phase portrait of the 3-periodic attractor (left) and Poincare section of the chaotic attractor at $P = 300$ kPa, $\omega/2\pi = 600$ kHz (right). R_t denotes the time derivative of the bubble radius.

Let us numerically investigate the dynamics of the gas bubbles governed by equation (1). An attractor is called hidden if its basin of attraction does not intersect with small neighborhoods of equilibria. The dynamical system corresponding to (1) does not have any fixed points. Thus, all the attractors of system (1) are hidden according to this definition. However, here we focus on the scenarios of multistability, which are the most interesting. The perpetual points are defined by the system of equations $\ddot{R} = 0$, $\dddot{R} = 0$. These points can be used as the initial conditions to achieve different attractors of the system [5, 6]. Often, all attractors of a dynamical system can be obtained this way. However, sometimes we cannot use some of these points as the initial data, because the problem will be ill–posed. For example, one of the points can have $\|\dot{R}\| > c$ or $R \gg 1$. Thus, the method does not always completely solve the problem of localizing all of the attractors.

The dynamics described by equation (1) can be quite complex. It can be shown that there can be a single periodic attractor, single chaotic attractor, coexisting periodic attractors and a chaotic attractor coexisting with a periodic attractor. It seems that the most interesting scenario is the one when periodic and chaotic attractors coexist. Examples of this behavior are given in Figs. 1 and 2. Points in the Poincare section are defined by the crossing of the phase trajectory with planes $t = Tk$, where $k \in \mathbb{N}$ and $T = 2\pi/\omega$. The original phase space is 3-dimensional, but here we demonstrate 2-dimensional projection on the $(R/R_0, \dot{R})$ plane. We also place points of the Poincare section onto the phase portraits to demonstrate the period of the solution with respect to the period of the driving force (bold dots on the phase portraits).

 https://doi.org/10.1051/epjconf/201817306006

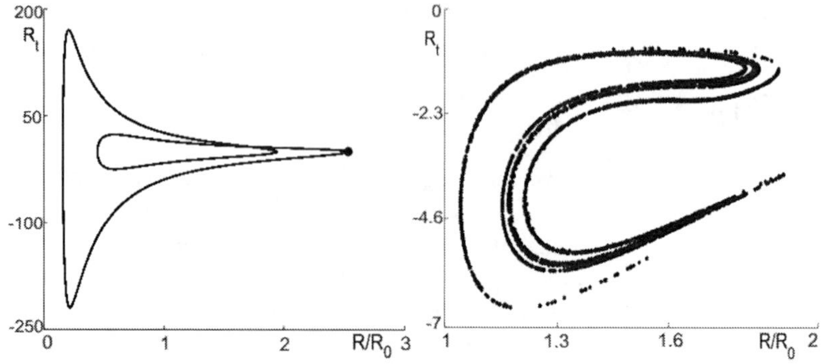

Figure 2. Phase portrait of the 1-periodic attractor (left) and Poincare section of the chaotic attractor at P = 90 kPa, $\omega/2\pi$ = 115 kHz (right).

Another interesting scenario is happening in the following parameters range: P = 90 kPa, $\omega/2\pi \in [110, 125]$ kHz. Over the entire range there exists a 1-periodic attractor, similar to that shown in Fig. 2. At $\omega/2\pi \geq 120$ kHz it coexists with a 2-periodic attractor (see Fig. 3, plate a). If the driving frequency is decreasing below 125 kHz, the 1-periodic attractor remains almost the same in the whole range, but the 2-periodic attractor goes through the period-doubling cascade and becomes a chaotic one, and then at approximately 110 kHz it becomes 2-periodic again (see Fig. 3, plate b). The corresponding bifurcation diagram is shown in Fig. 3, plate c. For any fixed control parameters in the set being discussed, there exists only one perpetual point that can be used as initial data. The diagram in Fig. 3, plate c, was obtained by applying this point.

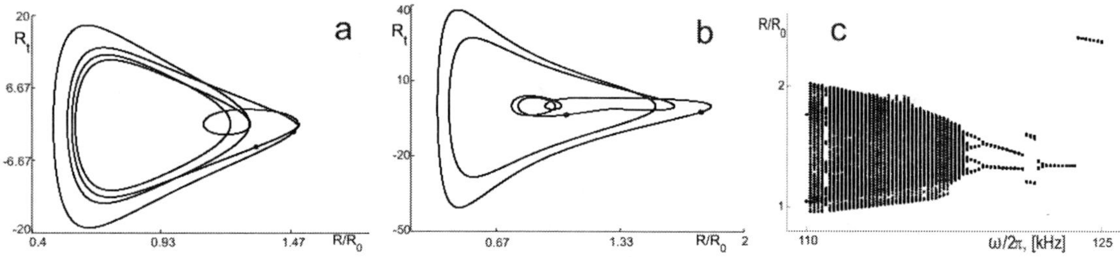

Figure 3. Plate a) 2-periodic attractor P = 90 kPa, $\omega/2\pi$ = 120 kHz; plate b) 2-periodic attractor P = 90 kPa, $\omega/2\pi$ = 110 kHz; plate c) period doubling cascade for the 2-periodic attractor

Now we consider other bifurcation diagrams for the same parameters region (see Fig. 4), but for another initial conditions. For these diagrams we use the state of the system at the last time step of the solution for the current frequency as the initial condition for the next step with respect to the frequency. For the diagram in Fig. 4, plate a, we use the increasing frequency sequence, and for plate b – the decreasing frequency sequence. For the first step with respect to the frequency, we use the perpetual point mentioned above as the initial data.

It can be seen that there are a lot of abrupt shifts from chaotic to periodic behavior and vice versa. This happens because the basins of the attractors deform under the change of the parameters of equation (1). If such a change is large enough, a point which was in a small neighborhood of one of the coexisting attractors can appear in the basin of another attractor at the next value of the parameter. The step size in the frequency axis, for which the diagrams are calculated, are about ~1% of the whole range. From the physical point of view, these frequency changes can be treated

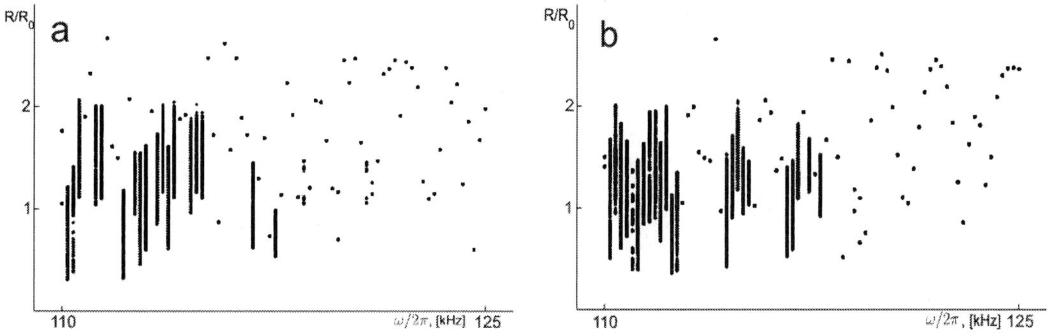

Figure 4. Bifurcation diagrams for: plate a) increasing driving frequency; plate b) decreasing driving frequency

as finite disturbances in the system. Such finite perturbations always exist in real physical systems along with finite measurement errors in the control parameters. This fact highlights a problem for the applications: one can see that a disturbance of the order of $\sim 1\%$ can cause unexpected shifts from a chaotic motion to a periodic one and vice versa. It is worth noting that signficant reduction of the step size will cause the system to stay on the same attractor. If we make the frequency steps $\sim$ 3–4 times smaller, we can achieve diagrams, completely according to the 1-periodic attractor, or to the another one (like in Fig. 3 (c)), if we start from the necessary attractor at the first step.

3 Conclusion

In this work we have studied an equation describing oscillations of a spherical bubble contrast agent near an elastic or rigid wall. We have numerically studied the considered model and have demonstrated that is exhibits complex dynamics. We have discussed some regions of parameters where chaotic and periodic attractors coexist. We have demonstrated that this problem can be important for applications, because the multistability can lead to unpredictable behavior in the bubble oscillations. These effects are undesirable for applications. Consequently, the corresponding areas of the control parameters should be avoided in practice.

Acknowledgement

This work is supported by Russian Science Foundation project 17-71-10241.

References

[1] U. Parlitz, V. Engisch, C. Scheffczyk, and W. Lauterborn, The Journal of the Acoustical Society of America **88**, 1061–1077 (1990)

[2] J.M. Carroll, M.L. Calvisi, and L.L. Luderbaugh, The Journal of the Acoustical Society of America **133**, 2641–2649 (2013)

[3] A.A. Doinikov, L. Aired, and A. Bouakaz, Physics in Medicine and Biology **56**, 6951 (2011)

[4] A.A. Doinikov and A. Bouakaz, Journal of Applied Mechanics and Technical Physics **54**, 867–876 (2013)

[5] D. Dudkowski, A. Prasad, and T. Kapitaniak, arxiv.org **1702.03419v1** (2017)

[6] D. Dudkowski, S. Jafari, T. Kapitaniak, N.V. Kuznetsov, G.A. Leonov, and A. Prasad, Physics Reports **637**, 1–50 (2016)

EPJ Web of Conferences **173**, 06007 (2018) https://doi.org/10.1051/epjconf/201817306007
Mathematical Modeling and Computational Physics 2017

Nonlinear Wave Simulation on the Xeon Phi Knights Landing Processor

Ivan Hristov[1,2,⋆], *Goran* Goranov[3,⋆⋆], and *Radoslava* Hristova[1,2,⋆⋆⋆]

[1] *JINR, LIT, Dubna, Russia*
[2] *Sofia University, Bulgaria*
[3] *Technical University of Gabrovo, Bulgaria*

Abstract. We consider an interesting from computational point of view standing wave simulation by solving coupled 2D perturbed Sine-Gordon equations. We make an OpenMP realization which explores both thread and SIMD levels of parallelism. We test the OpenMP program on two different energy equivalent Intel architectures: 2× Xeon E5-2695 v2 processors, (code-named "Ivy Bridge-EP") in the Hybrilit cluster, and Xeon Phi 7250 processor (code-named "Knights Landing" (KNL). The results show 2 times better performance on KNL processor.

1 Introduction

The second generation *Intel®XeonPhi™* processors code-named Knights Landing (KNL) are expected to deliver better performance than general purpose CPUs like *Intel®Xeon®* processors for applications with both high degree of parallelism and well behaved communications with memory [1]. Compute-bound applications run better on KNL due to its larger (512-bit) vector registers. Bandwidth-bound applications also run better on KNL due to its high bandwidth memory (HBM).

In this work we consider an interesting from computational point of view example of numerical solution of coupled 2D perturbed Sine-Gordon equations. We really need serious computational resources because in some cases the computational domain size may be very large – 10^6–10^8 mesh points and very long time integration is also needed – 10^8–10^9 time steps. Usually applications with stencil operations (like those in the presented work) are bandwidth-bound. The calculation of the transcendental *sine* function however makes our application to be closer to the compute-bound case and hence it benefits both from using HBM and from vectorization.

The considered systems of coupled 2D perturbed Sine-Gordon equations are of practical interest because it is well known that they model the dynamics of the so called Intrinsic Josephson Junctions (IJJ) [2].

The goals of the work are:

- To make an OpenMP realization of a finite difference scheme for solving systems of 2D perturbed Sine-Gordon equations. We want this realization to explore both thread and SIMD levels of parallelism.

⋆e-mail: christov_ivan@abv.bg
⋆⋆e-mail: ph.d.g.goranov@gmail.com
⋆⋆⋆e-mail: radoslava@fmi.uni-sofia.bg

- To test the OpenMP program on two different energy equivalent Intel architectures: 2× Xeon E5-2695 v2 processors with 24 cores, 48 threads, (code-named "Ivy Bridge-EP") in the Hybrilit cluster, and Xeon Phi 7250 processor with 68 cores, 272 threads, (code-named "Knights Landing" (KNL)).

2 Mathematical model and numerical scheme

We consider the following systems of 2D perturbed Sine-Gordon equations :

$$S(\varphi_{tt} + \alpha\varphi_t + \sin\varphi - \gamma) = \Delta\varphi, \qquad (x, y) \in \Omega \subset R^2. \tag{1}$$

Here Δ is the 2D Laplace operator. Ω is a given domain in R^2. S is the $N_{\text{eq}} \times N_{\text{eq}}$ cyclic tridiagonal matrix:

$$S = \begin{pmatrix} 1 & s & 0 & . & 0 & s \\ s & 1 & s & 0 & . & 0 \\ . & . & . & . & . & . \\ . & . & . & . & . & . \\ 0 & . & 0 & s & 1 & s \\ s & . & 0 & 0 & s & 1 \end{pmatrix},$$

where $-0.5 < s \leq 0$ and N_{eq} is the number of equations. The unknown is the column vector $\varphi(x, y, t) = (\varphi_1, \ldots, \varphi_{N_{\text{eq}}})^T$. Neumann boundary conditions are considered:

$$\left.\frac{\partial\varphi}{\partial\vec{n}}\right|_{\partial\Omega} = 0. \tag{2}$$

In (2) $\vec{n}$ denotes the exterior normal to the boundary $\partial\Omega$. To close the problem (1)–(2) appropriate initial conditions are posed. The model (1)–(2) describes very well the dynamics of N_{eq} periodically stacked IJJs [2]. The parameter s represents the inductive coupling between adjacent Josephson junctions, α is the dissipation parameter, γ is the external current. All the units are normalized as in [2].

We follow the approach of construction of the numerical scheme from [3]. We solve the problem numerically in rectangular domains by using second order central finite differences with respect to all of the derivatives. As a result at every mesh point of the domain we have to solve a system with the tridiagonal matrix S. Because of the specific tridiagonal structure of S we need only $(9N_{\text{eq}} - 12)$ floating point operations for solving one system. So the algorithm complexity is $O(N_{\text{eq}} \cdot N_x \cdot N_y \cdot N_{\text{time\_steps}})$, where N_{eq} is the number of equations, N_x is the number of mesh points in x-direction, N_y is the number of mesh points in y-direction and $N_{\text{time\_steps}}$ is the number of steps in time.

3 Parallelization strategy and performance scalability results

OpenMP Fortran realization of the above numerical scheme is made. We store the unknown solution of two consecutive time levels in two multidimensional arrays $U1(N_{\text{eq}}, N_x, N_y)$, $U2(N_{\text{eq}}, N_x, N_y)$. To ensure a good data locality, the main loop over indexes of $U1$ and $U2$ follows the column major order of multidimensional arrays in Fortran language.

To utilise the computational capabilities of the considered processors we explore two levels of parallelism: SIMD parallelism (vectorization) at the inner level and thread parallelism at the outer level. The smallest piece of work which we consider consists of solving 8 successive linear systems with the cyclic tridiagonal matrix S. Such pieces of work are distributed between OpenMP threads. Both the calculation of the right-hand sides for each linear system and solving 8 linear systems at once are vectorized. Our parallelization strategy consists of a parallelization of the main nested DO loop by using the OMP DO directive [4]:

```
!$OMP DO [ Clauses ]
    DO I=1,Ny
       DO J=1,Nx,8
          DO K=J,J+7
          .................
          Vectorized loops for calculation of the right-hand sides
          of 8 successive linear systems
          .................
          ENDDO
          ...................
          Vectorized loops for solving 8 systems at once
          ...................
       ENDDO
    ENDDO
!$OMP END DO
```

The use of -O2 optimization flag for compiling ensures an automatic vectorization of the innermost loops. Therefore the indexes of the outermost loop are distributed between the OpenMP threads and all the innermost loops (all with length 8) are vectorized. Let us mention that 8 double precision words correspond to the length of one vector register in KNL processors – 512 bit and two vector registers (256 bit each) in Ivy Bridge-EP processors. As a result we achieve about 2 times better performance from vectorization on "Ivy Bridge-EP" and about 4 times better performance from vectorization on KNL processors. The achievement of 50 % effectiveness from vectorization can be explained by the fact that our application is somewhere between bandwith-bound and compute-bound cases. On the one hand the application is of stencil type which type is by rule bandwidth-bound. On the other hand a calculation of the transcendental *sine* function is needed at every mesh point of the computational domain, which makes our application closer to the compute-bound case.

In the following figure the computational domain size is: $N_{eq} \times N_x \times N_y = 8 \times 4096 \times 4096$. As seen from this figure we achieve good performance scalability on both architectures and 2 times better performance on KNL processor.

4 Numerical example of a nonlinear standing wave

To check that the realized OpenMP program really works, we repeated the numerical results (in 2D case) from the classical works [5, 6]. As explained in these papers the powerful THz radiation from IJJs reported in [7] corresponds to a new type of standing wave solutions with excited so called cavity modes. For certain parameters α and γ the phase $\varphi(x, y, t)$ in a particular equation (junction) is a sum of three terms: a linear with respect to the time term, vt, a static π kink term $pi\_kink(x, y)$ and an oscillating term:

$$\varphi(x, y, t) = vt + pi\_kink(x, y) + oscillating\_term(x, y, t)$$

The oscillating term is approximately a solution of the linear wave equation:

$$\varphi_{tt} = \frac{1}{1 + 2s} \Delta\varphi, \qquad (x, y) \in \Omega \subset R^2$$

with Neumann boundary conditions.

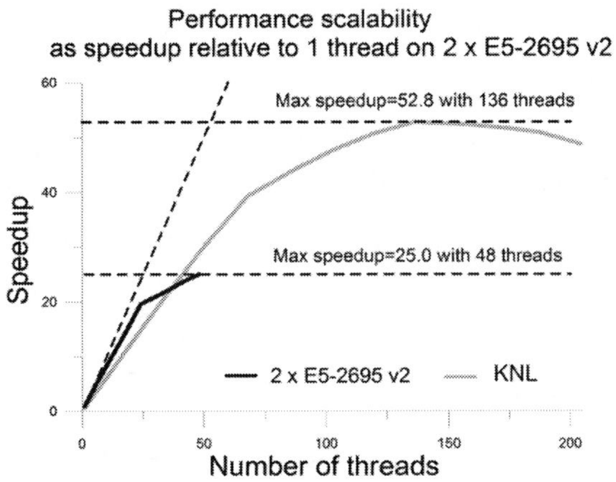

Figure 1. Performance scalability as speedup relative to one vectorized thread on 2 × "Ivy Bridge-EP" processors

As opposite to the oscillating term, which is the same for each junction (equation), the static term (in our case static π kinks) have alternative character, i.e. opposite static π kinks alternate in odd and even junctions (equations).

5 Conclusions

An OpenMP program for solving systems of 2D perturbed Sine-Gordon equations is realized and tested on two different Intel architectures: one computational node in the Hybrilit cluster consisting of two Ivy Bridge-EP processors and a KNL processor provided by RSC Group, Moscow. The results shows 2 times better performance on the KNL processor.

6 Acknowledgements

We greatly thank to the opportunity to use the computational resources of the HybriLIT cluster and to use KNL processors provided by RSC Group, Moscow. This work is supported by the National Science Fund of Bulgaria under Grant DFNI-I02/8 and by the National Science Fund of BMSE under grant I02/9/2014.

References

[1] J. Jeffers, J. Reinders, and A. Sodani, *Intel Xeon Phi Processor High Performance Programming: Knights Landing Edition* (Morgan Kaufmann, 2016)
[2] S. Sakai, P. Bodin, and N.F. Pedersen, Journal of Applied Physics **73** (5), 2411–2418 (1993)
[3] G.S. Kazacha and S.I. Serdyukova, Zhurnal Vychislitelnoi Matematiki i Matematicheskoi Fiziki **33** (3), 417–427 (1993)
[4] B. Chapman, G. Jost, and R. Van Der Pas, *Using OpenMP: portable shared memory parallel programming* (MIT press, 2008)
[5] S. Lin and X. Hu, Physical Review Letters **100** (24), p. 247006 (2008)
[6] A.E. Koshelev, Physical Review B **78** (17), p. 174509 (2008)
[7] L. Ozyuzer, A.E. Koshelev, et al., Science **318** (5854), 1291–1293 (2007)

EPJ Web of Conferences **173**, 06008 (2018) https://doi.org/10.1051/epjconf/201817306008
Mathematical Modeling and Computational Physics 2017

Short-Range Interaction Impact on Two-Dimensional Dipolar Scattering

Eugene A. Koval[1,2,⋆] and *Oksana A.* Koval[3,⋆⋆]

[1] *Bogoliubov Laboratory of Theoretical Physics, Joint Institute for Nuclear Research, Dubna, Moscow Region 141980, Russia*
[2] *Department of fundamental problems of micro-world physics, Dubna University, Dubna, Moscow Region, Russia*
[3] *Obukhov Institute for Atmospheric Physics, Moscow, Russia*

Abstract. We report numerical investigation of the short range interaction influence on the two-dimensional quantum scattering of two dipoles. The model simulates two ultracold polar molecules collisions in two spatial dimensions. The used algorithm allows us to quantitatively analyse the scattering of two polarized dipoles with account for strongly anisotropic nature of dipolar interaction. The strong dependence of the scattering total cross section on the short range interaction radius was discovered for threshold collision energies. We also discuss differences of calculated scattering cross section dependencies for different polarisation axis tilt angles.

1 Introduction

In recent years, two-dimensional (2D) systems with anisotropic interactions have been of increasing interest. One can mention a gas of dipolar atoms or ultracold polar molecules in quasi-2D geometry of magneto-optical traps [1] and dipolar Rydberg atoms [2]. Beside the gas of polar molecules long range dipolar interaction also arises in a gas of 2D dipolar excitons in 2D quantum well of semiconductor heterostructures. Dipolar exciton gas quantum dynamics [3–5] and Bose-Enstein Condensate [6] have been also intensively investigated in recent years. Investigation of the dipolar diatomic molecules are also actual due to the prospects of their possible applications as cubits for the quantum computing [7].

Dipole-dipole interactions are long-range and strongly anisotropic, that leads to the possibility of the creation of exotic long-range quantum systems with dipolar gases [8, 9]. Particularly, anisotropic superfluidity [10], 2D dipolar fermions [11] and dipolar few-body complexes [12] have been investigated. For a recent progress in the physics of few-body systems with long-range interactions we refer the reader to the comprehensive review [13].

2 Two-dimensional dipole-dipole scattering

We model the polar molecules collisions in a layer of a pancake-shaped optical trap with dipole-dipole scattering in two spatial dimensions. The quantum scattering in the plane is described by the

⋆e-mail: e-cov@yandex.ru
⋆⋆e-mail: kov.oksana20@gmail.com

2D Schrödinger equation in polar coordinates (ρ, ϕ)

$$\left[-\frac{\hbar^2}{2\mu}\left(\frac{1}{\rho}\frac{\partial}{\partial\rho}\left(\rho\frac{\partial}{\partial\rho}\right)+\frac{1}{\rho^2}\frac{\partial^2}{\partial\phi^2}\right)+U(\rho,\phi)\right]\Psi(\rho,\phi)=E\Psi(\rho,\phi) \tag{1}$$

with the scattering boundary conditions in the asymptotic region $\rho \to \infty$:

$$\Psi(\rho,\phi) \to e^{iq\rho} + f\left(q,\phi,\phi_q\right)\frac{e^{iq\rho}}{\sqrt{-i\rho}}. \tag{2}$$

The interaction potential $U(\rho,\phi)$ reads

$$U(\rho,\phi) = V_{SR}(\rho) + V_{dd}(\rho,\phi), \tag{3}$$

$$V_{SR}(\rho) = \begin{cases} \infty, & \rho \leqslant \rho_{SR}, \\ 0, & \rho > \rho_{SR}, \end{cases} \tag{4}$$

where the isotropic short-range potential $V_{SR}(\rho)$ is modeled by an infinitely high potential barrier at the origin, that was used in a number of papers of other the authors (see [14]). $V_{dd}(\rho,\phi)$ denotes the *long-range and strongly anisotropic* interaction potential of two arbitrarily oriented dipoles:

$$V_{dd}(\rho,\phi;\alpha,\beta,\gamma) = \frac{d_1 d_2}{\rho^3}[\sin(\alpha)\sin(\gamma)\cos(\beta)+\cos(\alpha)\cos(\gamma)-3\sin(\alpha)\sin(\gamma)\cos(\phi)\cos(\phi-\beta)], \tag{5}$$

where $\boldsymbol{d}_i, (i = 1, 2)$ are dipole moments. The angles α, γ denote to the tilt of each dipole $\boldsymbol{d}_1, \boldsymbol{d}_2$ with respect to the Z axis (that is perpendicular to the plane of the motion) and β is the angle between the dipoles polarisation planes $Z\boldsymbol{d}_1, Z\boldsymbol{d}_2$ (see Ref. [15]).

If scattering amplitude is calculated the differential scattering cross section and the total cross section are defined as

$$\sigma(q)=\frac{1}{2\pi}\int_0^{2\pi}\int_0^{2\pi}\frac{d\sigma}{d\Omega}d\phi_q\,d\phi \equiv \frac{1}{2\pi}\int_0^{2\pi}\int_0^{2\pi}\left|f(q,\phi,\phi_q)\right|^2 d\phi_q\,d\phi, \tag{6}$$

where $d\Omega = d\phi\,d\phi_q$ and the angle ϕ_q defines the direction $\boldsymbol{q}/q$ of the incident wave.

To integrate the problem (1)–(2) we use the algorithm applied in Refs. [15, 16] for the anisotropic quantum scattering in two dimensions and in Refs. [17, 18] for the investigation of bound state properties of 2D Hydrogen atom and 2D exciton in a tilted magnetic field. Following the ideas of the method suggested in Ref. [19] we choose the eigenfunctions $\xi_m(\phi) = e^{im(\phi-\pi)}/\sqrt{2\pi} = (-1)^m e^{im\phi}/\sqrt{2\pi}$ of the operator $h^{(0)} = \partial^2/\partial\phi^2$ as a Fourier basis for the angular-grid representation of the searched wave-function $\Psi(\rho,\phi)$. We introduce the uniform grid $\phi_j = 2\pi j/(2M+1)$ (where $j = 0, 1, \ldots, 2M$) over the ϕ- and ϕ_q-variables and search the wave function as the expansion

$$\Psi(\rho,\phi) = \frac{1}{\sqrt{\rho}}\sum_{j=0}^{2M}\sum_{m=-M}^{M}\xi_m(\phi)\xi^{-1}_{mj}\psi_j(\rho) = \frac{1}{(2M+1)\sqrt{\rho}}\sum_{j=0}^{2M}\sum_{m=-M}^{M}e^{im(\phi-\phi_j)}\psi_j(\rho), \tag{7}$$

where $\xi^{-1}_{mj} = 2\pi\xi^*_{jm}/(2M+1) = \sqrt{2\pi}e^{-im(\phi_j-\pi)}/(2M+1)$ is the inverse matrix to the $(2M+1)\times(2M+1)$ square matrix $\xi_{jm} = \xi_m(\phi_j)$ defined on the angular grid.

This expansion reduces the 2D Schrödinger equation (1) to a system of $2M+1$ coupled ordinary differential equations of the second order:

$$\frac{d^2\psi_j(\rho)}{d\rho^2} + \frac{2\mu}{\hbar^2}\left(E - U(\rho,\phi_j) + \frac{\hbar^2}{8\mu\rho^2}\right)\psi_j(\rho) + \frac{1}{\rho^2}\sum_{j'} h^{(0)}_{jj'}\psi_{j'}(\rho) = 0 \qquad (8)$$

with the "left-side" boundary condition $\psi_j(\rho \to 0) \to const \times \sqrt{\rho}$ $(j = 0, 1, \ldots, 2M)$ and with the scattering boundary condition in the asymptotic region $\rho \to \infty$ of the form

$$\frac{1}{(2M+1)\sqrt{\rho}}\sum_{j=0}^{2M}\sum_{m=-M}^{M} e^{im(\phi-\phi_j)}\psi_j(\rho) = \exp\{iq\rho\cos(\phi)\} + f(q,\phi,\phi_q)\frac{e^{iq\rho}}{\sqrt{-i\rho}}. \qquad (9)$$

The key advantages of the algorithm are summarized below. Any local interaction in this representation is diagonal. In Ref. [15] much faster convergence over the angular grid nodes number than partial wave representation was proved. After the finite-difference approximation of the boundary-value problem (8–9) the obtained block matrix can be stored in a compressed form, which allows optimal resource using. The resulting algebraic system is efficiently solved by a fast implicit matrix algorithm based on the idea of the block sweep algorithm.

3 Short-range interaction influence on 2D dipolar scattering

The present study evidenced the strong dependence of the total cross section (6) on the short range cut-off radius ρ_{SR} for threshold energies of dipole-dipole collisions. When the polarization axis is aligned along the z-axis the dipole-dipole interaction is isotropic and repulsive, no resonances occurs, because repulsive dipolar term prevents particles to getting too close to each other.

For a tilted ($\alpha = \gamma = 0.25\pi$) polarization axis as well as for extremal case ($\alpha = \gamma = 0.5\pi$) of polarization axis lying in the plane of particles motion, we discovered strong dependence of the total cross section on the cut-off radius ρ_{SR} (4), which is illustrated in Fig. 1. The calculations were carried out at the energy $E = 10^{-5}$ a.u. and a dipolar length $D = \mu d_1 d_2/\hbar^2 = 1$. Note, that we study the case of distinguishable particles collisions.

The sharp increases in total cross section appear due to the virtual bound states arising as the interplay of the short-range (4) and long-range (5) potentials. It should be noted the existence of small cusps for the following values of $\rho_{SR} = 0.035$ a.u., 0.0895 a.u., 0.42 a.u. For small values of ρ_{SR} a resonance width decrease with simultaneous increase of their frequency are observed.

Acknowledgements

This work was supported by the Russian Foundation for Basic Research, Grant No. 16-32-00865.

References

[1] S. Ospelkaus, K.K. Ni, D. Wang, M. De Miranda, B. Neyenhuis, G. Quéméner, P. Julienne, J. Bohn, D. Jin, J. Ye, Science **327**, 853 (2010)

[2] K. Afrousheh, P. Bohlouli-Zanjani, D. Vagale, A. Mugford, M. Fedorov, J.D.D. Martin, Phys. Rev. Lett. **93**, 233001 (2004)

[3] R. Rapaport, G. Chen, S.H. Simon, Physical Review B **73**, 033319 (2006)

[4] V.B. Timofeev, A.V. Gorbunov, A.V. Larionov, Journal of Physics: Condensed Matter **19**, 295209 (2007)

[5] V. Timofeev, A. Gorbunov, Journal of applied physics **101**, 081708 (2007)

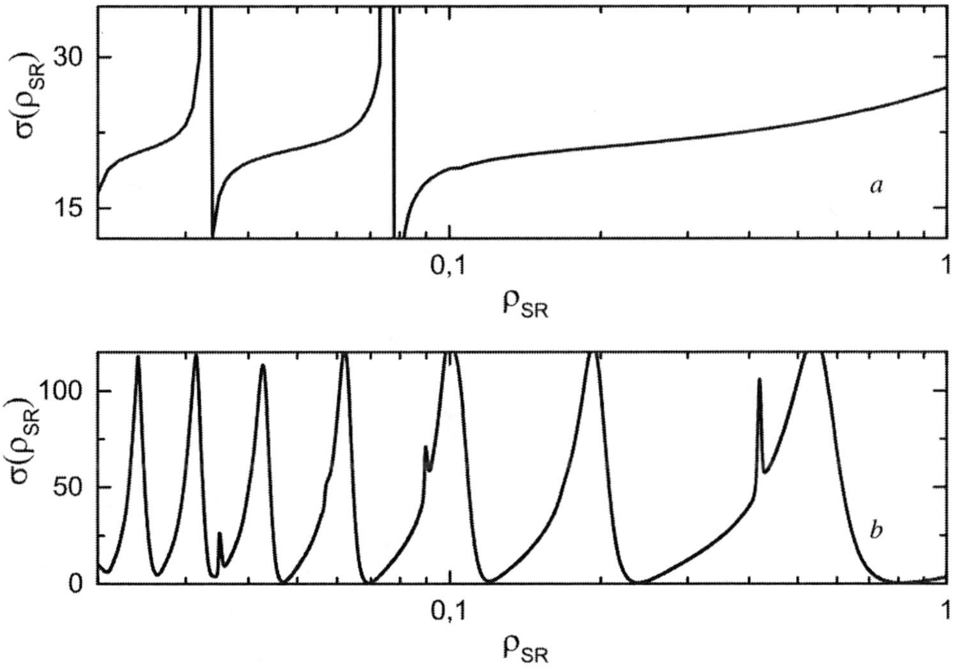

Figure 1. The full cross section $\sigma(\rho_{SR})$ (6) dependence on the short range interaction radius ρ_{SR} (4) for the case of polarized dipoles $\beta = 0$ for $\alpha = \gamma = 0.25\pi$ (a) and $\alpha = \gamma = 0.5\pi$ (b) at the energy $E = 10^{-5}$ a.u. and a dipolar length $D = \mu d_1 d_2/\hbar^2 = 1$. The quantities are given in units $\hbar = \mu = 1$.

[6] V. Timofeev, A. Gorbunov, Physica status solidi (c) **5**, 2379 (2008)
[7] D. DeMille, Physical Review Letters **88**, 067901 (2002)
[8] M.A. Baranov, Phys. Rep. **464**, 71 (2008)
[9] I. Bloch, J. Dalibard, W. Zwerger, Reviews of modern physics **80**, 885 (2008)
[10] C. Ticknor, Physical Review A **84**, 032702 (2011)
[11] G.M. Bruun, E. Taylor, Physical review letters **101**, 245301 (2008)
[12] J.C. Cremon, G. Bruun, S. Reimann, Physical review letters **105**, 255301 (2010)
[13] Y. Wang, P. Julienne, C.H. Greene, *Few-body physics of ultracold atoms and molecules with long-range interactions* (World Scientific, 2015), chap. Chapter 2, pp. 77–134
[14] K. Kanjilal, D. Blume, Physical Review A **73**, 060701 (2006)
[15] E.A. Koval, O.A. Koval, V.S. Melezhik, Physical Review A **89**, 052710 (2014)
[16] E. Koval, O. Koval, V. Melezhik, Physics of Particles and Nuclei Letters **12**, 448 (2015), [Pis'ma v Zh. Fiz. Elem. Chastits At. Yadra **12**, 702 (2015)]
[17] E.A. Koval, O.A. Koval, Journal of Experimental and Theoretical Physics **125**, 35 (2017), [Pis'ma Zh. Eksp. Teor. Fiz. **152**, 45 (2017)]
[18] E.A. Koval, O.A. Koval, Physica E **93**, 160 (2017)
[19] V. Melezhik, Journal of Computational Physics **92**, 67 (1991)

Molecular Dynamic Simulation of Water Vapor and Determination of Diffusion Characteristics in the Pore

Eduard G. Nikonov[1,⋆], *Miron* Pavluš[2,⋆⋆], and *Mária* Popovičová[2,⋆⋆⋆]

[1] *Joint Institute for Nuclear Research, 141980 Dubna, Moscow Region, Russia*
[2] *University of Prešov, str. Konštantínova 16, 080 01 Prešov, Slovakia*

Abstract. One of the varieties of pores, often found in natural or artificial building materials, are the so-called blind pores of dead-end or saccate type. Three-dimensional model of such kind of pore has been developed in this work. This model has been used for simulation of water vapor interaction with individual pore by molecular dynamics in combination with the diffusion equation method. Special investigations have been done to find dependencies between thermostats implementations and conservation of thermodynamic and statistical values of water vapor – pore system. The two types of evolution of water – pore system have been investigated: drying and wetting of the pore. Full research of diffusion coefficient, diffusion velocity and other diffusion parameters has been made.

1 Molecular dynamics model

In classical molecular dynamics, the behavior of an individual particle is described by the Newton equations of motion [1]. For a simulation of particle interaction we use the Lennard-Jones potential [2] with $\sigma = 3.17$Å and $\varepsilon = 6.74 \cdot 10^{-3}$ eV. It is the most used to describe the evolution of water in liquid and saturated vapor form. Equations of motion were integrated by Velocity Verlet method [3]. Berendsen thermostat [4] is used for temperature equilibrating and control. The coefficient of the velocity recalculation $\lambda(t)$ at every time step t depends on the so called 'rise time' of the thermostat τ_B which belongs to the interval $[0.5, 2]$ ps. The Berendsen algorithm is simple to implement and it is very efficient for reaching the desired temperature from far-from-equilibrium configurations.

2 Computer simulation of microscopic model

We made simulation for a pore of dimensions $l_x = 500$ nm, $l_y = 50$ nm, $l_z = 50$ nm with integration time step $\Delta t = 0.016$ ps and evolution time 65.3 ns. Otherwise, we have considered the following input data for the drying process:

- 1000 H_2O molecules in the pore volume $500 \times 50 \times 50$ nm^3 form a saturated water vapor at temperature $T_0 = 25°$ C and pressure $p_0 = 3.17$ kPa;

⋆e-mail: e.nikonov@jinr.ru
⋆⋆e-mail: miron.pavlus@unipo.sk
⋆⋆⋆e-mail: maria.popovicova@unipo.sk

- 1800 molecules in the outer area that form 20 % of the saturated water vapor

and input data for the wetting process:

- 200 H_2O molecules in the pore volume $500 \times 50 \times 50$ nm^3 are 20 % of saturated water vapor;
- 9000 molecules in the outer area form a saturated water vapor at temperature $T_0 = 25°$ C and pressure $p_0 = 3.17$ kPa.

The diffusion coefficients for drying process (left) and for wetting process (right) are shown in Fig. 1. The left figure depicts diffusion coefficients for pore (upper curve), for outer area (down curve), for mean value of previous (middle curve) and for constant value $D = 1779.1$ [nm^2/ps] (middle dashed line). The right figure shows diffusion coefficients for pore (upper curve), for outer area (down curve), for mean value of previous (middle curve) and for constant value $D = 536.33$ [nm^2/ps] (middle dashed line).

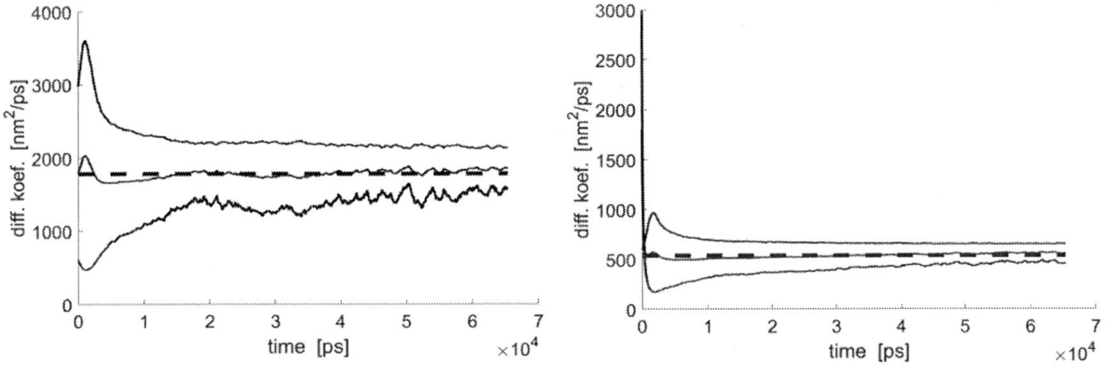

Figure 1. Diffusion coefficients for drying (left) and for wetting (right) processes (upper curves in pore, down curves in outer space, middle curves – the mean of pore and outer space, and dashed lines – constant values)

3 Macroscopic diffusion model

Let us denote the water vapor concentration as $w_v(x, y, z, t)$ [ng/(nm)^3] where x, y, z are space independent variables and t is time independent variable. Then, we consider the following macroscopic diffusion model

$$\frac{\partial w_v}{\partial t} = D\left(\frac{\partial^2 w_v}{\partial x^2} + \frac{\partial^2 w_v}{\partial y^2} + \frac{\partial^2 w_v}{\partial z^2}\right), \tag{1}$$

$$0 < x < l_x, \quad 0 < y < l_y, \quad 0 < z < l_z, \quad t > 0,$$

$$w_v(x, y, z, 0) = w_{v,0}, \qquad 0 \le x \le l_x, \quad 0 \le y \le l_y, \quad 0 \le z \le l_z, \tag{2}$$

$$\frac{\partial w_v}{\partial n}(t)\bigg|_{(x,y,z)\in\Gamma_2\cup\Gamma_3\cup\Gamma_4\cup\Gamma_5\cup\Gamma_6} = 0, \qquad t > 0, \tag{3}$$

$$-D\frac{\partial w_v}{\partial x}(t)\bigg|_{(l_x,y,z)\in\Gamma_1} = \beta[w_v(l_x, y, z, t) - w_{v,out}(t)], \tag{4}$$

$$0 \le y \le l_y, \quad 0 \le z \le l_z, \quad t > 0,$$

where D is the diffusion coefficient [(nm)^2/ps]; l_x, l_y, l_z are 3D pore dimensions [nm]; Γ_1, Γ_2, Γ_3, Γ_4, Γ_5, Γ_6 are boundaries of 3D pore (Γ_1 is free boundary while the rest boundaries are isolated; $w_{v,0}$ is the initial concentration of water vapor, $w_{v,0} = 2.304 \cdot 10^{-17}$ for the drying process, and $w_{v,0} = 0.461 \cdot 10^{-17}$ for the wetting process [ng/(nm)^3]; $w_{v,out}(t)$ is the water vapor concentration in outer area [ng/(nm)^3]; β is the coefficient of water vapor transfer from pore space to outer space, $\beta = 50000$ [nm/ps].

We suppose that the outer area water vapor concentration is expressed as $w_{v,out}(t) = \varphi_0 \cdot w_{sv}(T_0)$, where φ_0 is the relative humidity of outer space ($0 \le \varphi_0 \le 1$) and $w_{sv}(T_0)$ is saturated water vapor concentration at outer temperature T_0.

The linear problem (1)–(4) can be solved exactly using the variables separation method [5] and the result of the solution is

$$w_v(x,y,z,t) = w_{sv}(T_0) \cdot \varphi_0 + \left[w_{v,0} - w_{sv}(T_0) \cdot \varphi_0\right] \cdot \tag{5}$$

$$\cdot \sum_{m=1}^{\infty} \sum_{n=0}^{\infty} \sum_{p=0}^{\infty} e^{\lambda_{mnp} D t} c_{mnp} \cos(\alpha_{xm} x) \cos(\alpha_{yn} y) \cos(\alpha_{zp} z),$$

$$0 \le x \le l_x, \qquad 0 \le y \le l_y, \qquad 0 \le z \le l_z, \qquad t > 0.$$

Here, c_{mnp} are coefficients of unity expansion

$$c_{mnp} = \begin{cases} \dfrac{4 \sin(\alpha_{xm} l_x)}{2 l_x \alpha_{xm} + \sin(2 \alpha_{xm} l_x)} & \text{if } n = 0;\ p = 0;\ m = 1, 2, 3, \ldots \\ 0 & \text{if } m, n, p = 1, 2, 3, \ldots \end{cases}$$

and λ_{mnp} are eigenvalues, where

$$\lambda_{mnp} = -\alpha_{xm}^2 - \alpha_{yn}^2 - \alpha_{zp}^2,$$

$$\alpha_{yn} = \frac{n\pi}{l_y}, \qquad n = 0, 1, 2, \ldots \qquad \alpha_{zp} = \frac{p\pi}{l_z}, \qquad p = 0, 1, 2, \ldots$$

and α_{xm} are solutions of the equation

$$\alpha_{xm} \cdot \tan(\alpha_{xm} l_x) = \beta / D, \qquad m = 1, 2, 3, \ldots$$

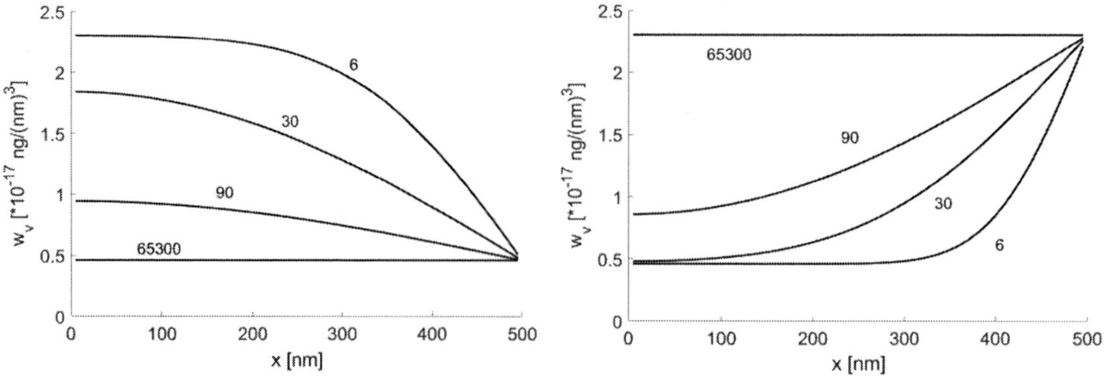

Figure 2. The dynamics of water vapor concentration at different time moments $t = 6$, 30, 90, 65300 ps (cross section at $y = l_y/2$ and $z = l_z/2$), drying process (left) from top to bottom, and wetting process (right) from bottom to top

Results of macro model calculations are presented in Fig. 2.

4 Comparison of microscopic and macroscopic models

Finally, we compare the space mean value

$$\frac{1}{l_x l_y l_z} \int_0^{l_x} \int_0^{l_y} \int_0^{l_z} w_v(x, y, z, t)\, dx\, dy\, dz$$

of water vapor concentration (5) for macro model with the density obtained by micro model. The results are in the Fig. 3.

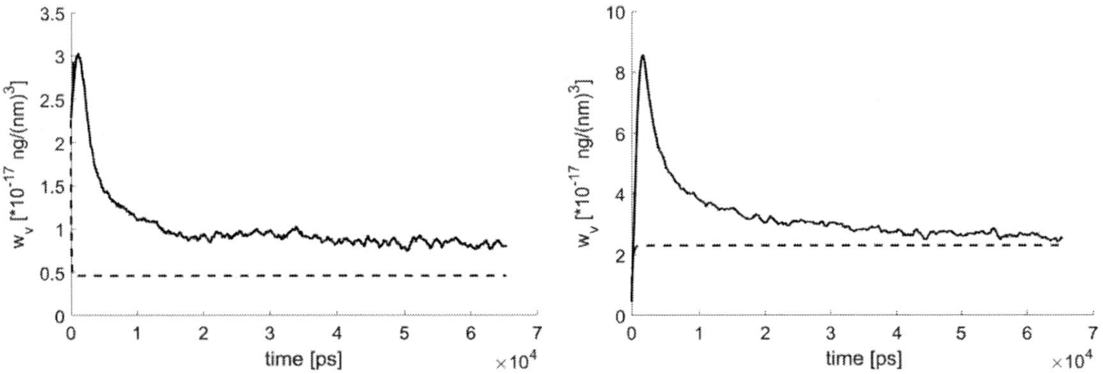

Figure 3. The dynamics of water vapor concentration (WVC) obtained by diffusion equation (dashed curves) versus WVC obtained by molecular dynamics (solid curves). The drying process (left) and the wetting process (right).

Our investigations allow to affirm that an approach based on combination of diffusion coefficients determination by means of molecular dynamics and further application of these coefficients in macro model computations is useful for accuracy increasing of water-pore interaction description.

References

[1] H. Gould, J. Tobochnik, and W. Christian, *An Introduction to Computer Simulation Methods* (Addison Wesley, San Francisco, Chapter 8, Third edition, 2007) pp. 267–268

[2] J. E. Lennard-Jones, Proc. Roy. Soc. **A 106**, 463–477 (1924)

[3] L. Verlet, Phys. Rev. **159**, 98–103 (1967)

[4] H. J. C. Berendsen, J. P. M. Postma, W. F. van Gunsteren, A. DiNola, and J. R. Haak, J. Chem. Phys. **81**, 3684–3690 (1984)

[5] A. V. Bitsadze and D.F. Kalinichenko, *A Collection of Problems on the Equations of Mathematical Physics* (in Russian, Alians Publisher, Moscow, 2007), pp. 106–310

EPJ Web of Conferences **173**, 06010 (2018) https://doi.org/10.1051/epjconf/201817306010
Mathematical Modeling and Computational Physics 2017

Parallel Evolutionary Optimization Algorithms for Peptide-Protein Docking

Sergey Poluyan[1,*] and *Nikolay* Ershov[2,**]

[1]*Institute of System Analysis and Control, Dubna State University, Universitetskaya str. 19, Dubna 141980, Moscow Region, Russian Federation*
[2]*Faculty of Computational Mathematics and Cybernetics, Lomonosov Moscow State University, Leninskie Gory 1, Bldg. 52, Moscow 119991, Russian Federation*

Abstract. In this study we examine the possibility of using evolutionary optimization algorithms in protein-peptide docking. We present the main assumptions that reduce the docking problem to a continuous global optimization problem and provide a way of using evolutionary optimization algorithms. The Rosetta all-atom force field was used for structural representation and energy scoring. We describe the parallelization scheme and MPI/OpenMP realization of the considered algorithms. We demonstrate the efficiency and the performance for some algorithms which were applied to a set of benchmark tests.

1 Introduction

Protein-peptide docking (PPeD) aims to predict peptide binding sites on protein surfaces and the associated binding affinities. Traditional experimental methods for binding site determination include crystallography, nuclear magnetic resonance and site-directed mutagenesis and other techniques [1]. Despite their accuracy, efficiency and the huge amount of details they can provide, they are expensive and very labour and skill demanding. Moreover, protein-peptide complexes are more difficult to crystallize than individual proteins. While docking is relatively cheap, it is a prediction method only. Thus, using computational methods have become a popular research endeavor in recent years. Most of them involve stochastic optimization at different stages [2]. The main advantage of using stochastic methods is the possibility of including various knowledge-based information. Furthermore, it is more computationally attractive rather than molecular dynamics simulations.

The current approaches to PPeD are based on Anfinsen's hypothesis [3] that native-like complex conformations represent unique, low-energy, thermodynamically stable conformations. Therefore, the PPeD problem can be considered as a global optimization problem where the objective is to find the complex conformation with the lowest energy. The main motivation of this study is to compare different evolutionary algorithms and identify the most effective strategies within a certain force-field. Solving PPeD problems typically involve the use of combined methods which require a number of various steps and special techniques. However, such approaches are beyond the scope of the current study.

*e-mail: svpoluyan@gmail.com
**e-mail: ershovnm@gmail.com

2 Protein-peptide docking

The interaction between peptide and protein can be described by an objective function calculated with respect to three components representing degrees of freedom: (1) the translation of the peptide, involving the three axis values (x, y, z) in cartesian coordinate space; (2) the peptide orientation, modeled as a four variables quaternion; and (3) the flexibilities, represented by the free rotation of torsion (dihedral angles) of the peptide and side-chain of the protein.

As illustrated in Figure 1, the problem solution is encoded by a real-valued vector of $n + 7$ variables. The first three values correspond to the peptide translation. The next four values correspond to the peptide orientation with range of $[-1, 1]^3$ including the angle slope w with range of $[0, \pi]$. The remaining n values are peptide backbone and side-chain dihedral angles (ϕ, ψ, ω, and χ_{1-4}) and protein side-chain dihedral angles within search area. The backbone and side-chain torsion variables are measured in radians and encoded in the range of $[-\pi, \pi]$, except for peptide backbone angle ω with range of $[\pi - \delta, \pi + \delta], \delta = 0.2$ which is locked in the trans-state. The remaining degrees of freedom (valence angles and bond lengths) are idealized with respect to force-field values and do not change during the conformational search.

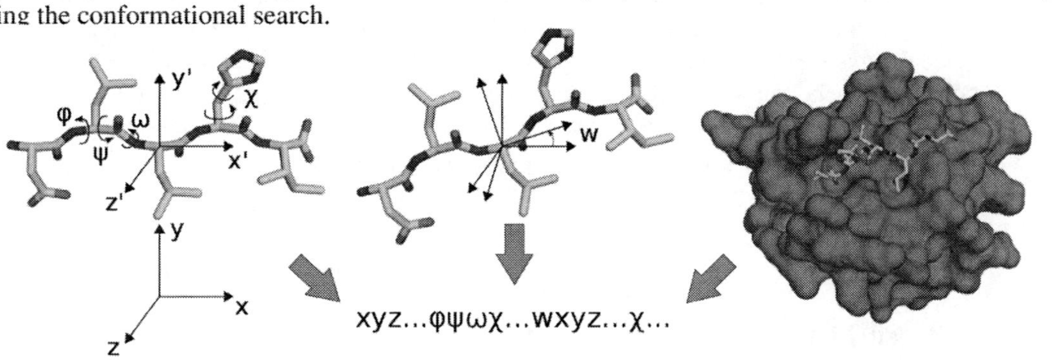

Figure 1. Solution encoding: peptide degrees of freedom, translation, rotation; protein side-chain dihedral angles

The optimization problem is formulated as minimization of the binding energy (BE). This energy (shown in eq. (1)) is defined as the energy of the bound state minus the energy of the unbound state.

$$E_{\mathrm{BE}} = E_{\mathrm{complex}} - (E_{\mathrm{protein}} + E_{\mathrm{peptide}}). \tag{1}$$

The Rosetta 3.8 [4] framework was used for full-atom complex structure representation and scoring (energy evaluation). The Rosetta all-atom force-field is a dedicated structure prediction and docking force-field. It has borrowed much from classical molecular-mechanics force-fields: Lennard-Jones 6-12 potential, E_{vdw}, Lazaridis-Karplus implicit solvation model, E_{sol}, Coulombic electrostatic potential, E_{elec}, etc. The main feature of this force-field is that it uses several knowledge-based terms. For instance, probability of backbone ϕ, ψ angles, E_{rama}. For that reason the energy score has not a direct conversion to physical energy units like kcal/mol. The *talaris2014* energy function is computed from a linear combination of 16 energy terms E_i which are calculated as a function of geometric degrees of freedom, Θ, chemical identities, aa, and scaled by a weight on each term, w, as shown in eq. (2).

$$E_{\mathrm{total}} = \textstyle\sum_i w_i E_i(\Theta_i, aa_i) = \underbrace{w_v E_{\mathrm{vdw}} + \cdots + w_e E_{\mathrm{elec}} + w_s E_{\mathrm{sol}}}_{\text{physics-based terms (kcal/mol)}} + \underbrace{w_r E_{\mathrm{rama}} + \cdots + w_d E_{\mathrm{dunbrack}}}_{\text{knowledge-based terms}}. \tag{2}$$

There are a lot of statistics about degrees of freedom: neighbor-dependent Ramachandran plots [4], backbone-independent [5] and backbone-dependent [6] libraries for side-chain dihedral angles. These libraries have been used to exclude impossible conformations by creating 1–4 dimensional cumulative distribution functions that had been derived from given probability density functions.

3 Algorithms and Implementation

In general, evolutionary algorithms have a similar structure with operators like mutation, recombination, and selection. However, in this study we focus on three algorithms with radically different structure: Competitive Swarm Optimizer (CSO) [7], Particle Swarm Optimization (PSO) [8], and Adaptive Differential Evolution with Optional External Archive (JADE) [9]. These techniques have been selected as they are widely used to solve various real-coded optimization problems, both model and real-world [2].

One of the many advantages of the evolutionary algorithms is that they are easy to parallelize. It is possible to parallelize specific operations, or to parallelize the evolutionary process itself. We considered algorithms where function evaluation for each individual is out from main operators and presented as an independent operation which consists in function evaluation for a whole population. Then the simplest parallelization scheme can be used. First, the population must be divided into groups proportionally to number of cluster nodes. Second, all groups must be sent to cluster nodes using MPI technology according to scheme: one process – one node. Third, objective function must be evaluated for individuals in groups using OpenMP technology within each MPI process.

The calculations were done using the HybriLIT (LIT JINR) cluster with two Intel Xeon twelve-core processor at each node. The effect of parallelization for JADE is presented in table 1. All the presented algorithms have similar to JADE result due to the equal level of complexity.

Table 1. Performance of the applied scheme of parallel calculations

Threads / Nodes	4 / 1	8 / 1	12 / 1	16 / 1	24 / 1	24 / 2	48 / 2
Speed-up	3.31	6.28	9.03	11.84	17.1	15.85	31.1
Efficiency	0.82	0.78	0.75	0.74	0.71	0.66	0.64

4 Results and Discussion

The performance of the selected algorithms has been assessed on the set of structures from [1] which are presented in table 2. There were 10 independent runs for each algorithm with the number of energy evaluations equal to 10^7 per run. Typical execution time for one run using one thread vary from 15 hours to 22 hours. It depends on number of degrees of freedom. In the case of local docking FlexPepDock (FPD) [10] protocol from the Rosetta framework was used with comparable run time. It performs a high-resolution PPeD using a Monte Carlo-Minimization-based approach to refine all the peptide's degrees of freedom (rigid body orientation, backbone and side chain flexibility) as well as the protein receptor side chains conformations. For the case of global docking we compare results with CABS-dock [11] and pepATTRACT [12]. The preparation of structures was performed using the Rosetta Relax protocol.

Table 2. Protein-peptide complexes with their accession codes from the RCSB Protein Data Bank

Protein		Peptide	Problem		
PDB id:Chain	Length	Sequence	Docking type	Search space	Dimension
2CYH:A	164	AP	Local	sphere, R = 5Å	25
1JWG:B	140	DLLHI	Local	two spheres $R_1 = R_2$ = 5Å	54
2HO2:A	33	P_9L	Global	sphere with radius of 60Å	93

The FPD protocol requires the initial starting position of the peptide. We considered two starting states relative to the native state: a random 3d rotation with small translation (FPD2) and rotation along one axis in the binding spot (FPD1). It should be noted that in JADE the probability of a mutation and crossover are adaptive parameters. However, at any iteration the mutation probability was high while the crossover probability was low for all the tasks. However, the experiments [13] showed that the crossover operator is a crucial step for the global search. This emphasizes a poor adaptation scheme for the crossover operator.

The obtained docking results are shown in Figure 2. As can be seen, the JADE outperforms other algorithms in two cases and achieves a satisfactory sub-angstrom precision. In case of the 1JWG complex JADE outperforms FPD2, which correspond to blind docking with a random start position. The set of FPD values is achieved with similar to JADE run time. The error is specified in Angstroms.

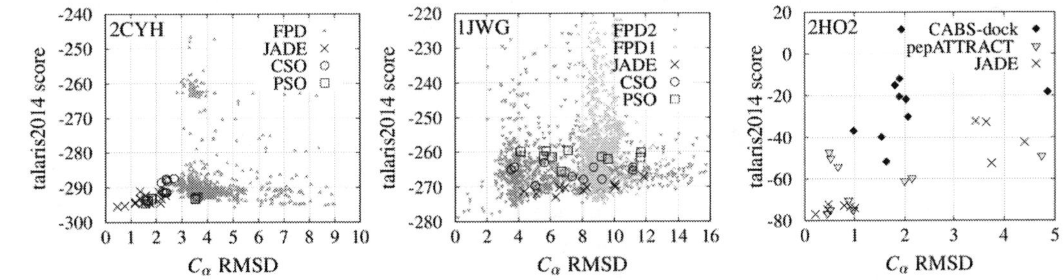

Figure 2. Energy score against alpha carbon Root-Mean-Square Deviation from the native conformation

5 Conclusions

The results of this study show that JADE provides the best overall performance. However, it shows poor results even for search space of about 50 parameters. Unfortunately, it is hard to achieve much better performance due to the heuristic nature of the algorithm. It is important to note that in relevant PPeD tasks it is necessary to consider peptides with lengths of 10–15 amino acid. With prior knowledge of the binding area and peptide structure the number of parameters will grow up to 250–300. This makes impossible to use such algorithms with all-atom resolution. However, using other evolutionary approaches like estimation of distribution algorithms [14], which is presently the current cutting edge, can show better performance. This will be the subject of future studies.

References

[1] R. Rentzsch and B.Y. Renard, Briefings in Bioinformatics **16**, 1045–1056 (2015)
[2] E. Lopez-Camacho, M.J. Garcia Godoy et al., Applied Soft Computing **28**, 379–393 (2015)
[3] C. Anfinsen, Science **181**, 330–331 (1973)
[4] R.F. Alford et al., Journal of Chemical Theory and Computation **13**, 3031–3048 (2017)
[5] B.J. Hintze et al., Proteins: Structure, Function, and Bioinformatics **84**, 1177–1189 (2016)
[6] M. Shapovalov and R.L. Dunbrack, Structure **19**, 844–858 (2011)
[7] R. Cheng and Y. Jin, IEEE Transactions on Cybernetics **45**, 191–204 (2015)
[8] M. Clerc and J. Kennedy, IEEE Transactions on Evolutionary Computation **6**, 58–73 (2002)
[9] J. Zhang and A. Sanderson, IEEE Transactions on Evolutionary Computation **13**, 945–958 (2009)
[10] B. Raveh, N. London, L. Zimmerman, and O. Schueler-Furman, PLoS ONE **6**, (2011)
[11] M. Kurcinski, M. Jamroz et al., Nucleic Acids Research **43**, 419–424 (2015)
[12] S.J. de Vries, J. Rey et al., Nucleic Acids Research **45**, 361–364 (2017)
[13] S.V. Poluyan, N.M. Reinhard, and N.M. Ershov, Vestnik Rossijskogo universiteta druzhby narodov, seriia: Matematika. Informatika. Fizika **2**, 415–418 (2014)
[14] B. Moradabadi and H. Beigy, Genetic Programming and Evolvable Machines **15**, 169–193 (2014)

https://doi.org/10.1051/epjconf/201817306011

Simulation of Collective Excitations in Long Josephson Junction Stacks

Ilhom Rahmonov[1,2,★], *Yury* Shukrinov[1,3,★★], *Pavlina* Atanasova[4], *Elena* Zemlyanaya[3,5], *Oksana* Streltsova[5,6], *Maxim* Zuev[5], *Andrej* Plecenik[7], and *Akinobu* Irie[8]

[1] *BLTP, Joint Institute for Nuclear Research, Dubna 141980, Russia*
[2] *Umarov Physical Technical Institute, Tajik Academy of Science, Dushanbe 734063, Tajikistan*
[3] *Dubna State University, Dubna 141980, Russia*
[4] *University of Plovdiv Paisii Hilendarski, Plovdiv 4000, Bulgaria*
[5] *LIT, Joint Institute for Nuclear Research, Dubna 141980, Russia*
[6] *Peoples' Friendship University of Russia (RUDN University), Moscow, Russia*
[7] *Department of Experimental Physics, Comenius University, Bratislava, Slovakia*
[8] *Utsunomiya University, Utsunomiya 321-8585, Japan*

Abstract. The phase dynamics of a stack of long Josephson junctions has been studied. Both inductive and capacitive couplings between Josephson junctions have been taken into account in the calculations. The IV-curve, the dependence on the bias current of the radiation power and dynamics of each JJs of the stack have been investigated. The coexistence of the charge traveling wave and fluxon states has been observed. This state can be considered as a new collective excitation in the system of coupled Josephson junctions. We demonstrate that the observed collective excitation leads to the decrease of radiation power from the system.

1 Introduction

The layered high-Tc superconducting materials such as $Bi_2Sr_2CaCu_2O_{8+\delta}$ (BSCCO) can be considered as stacks of coupled Josephson junctions (JJs) [1]. The interest to the investigation of this system is caused by its rich nonlinear properties and different applications, particularly this system is one of the promising object of superconducting electronics [2]. The JJs stack demonstrates a series of interesting properties such as parametric resonance [3–5], chaotic features [6], and the occurrence of the fluxons [7–9] and collective excitations [10, 11]. Coherent terahertz electromagnetic radiation from this system provides wide possibilities for various applications [12, 13]. The effect of collective excitations on radiation power from the stack is not studied in detail yet. In this paper, we investigate the IV-curve of stacked JJs and radiation power from the stack taking into account the inductive and capacitive couplings [14, 15] and the diffusion current [5, 16, 17].

The system of equations which describes the phase dynamics of the coupled long JJs stack in the normalized quantities can be written as follows [14, 15, 18].

★e-mail: rahmonov@theor.jinr.ru
★★e-mail: shukrinv@theor.jinr.ru

$$\begin{cases} \dfrac{\partial \varphi_l}{\partial t} = D_c V_l + s_c V_{l+1} + s_c V_{l-1}, \\ \dfrac{\partial V_l}{\partial t} = \displaystyle\sum_{k=1}^{N} \pounds_{lk}^{-1} \dfrac{\partial^2 \varphi_k}{\partial x^2} - \sin\varphi_l + \beta \dfrac{\partial \varphi_l}{\partial t} + I, \end{cases} \tag{1}$$

where φ_l and V_l are the phase difference and voltage of lth JJ, respectively. I is the bias current normalized to the critical current j_c, β is a dissipation parameter, $D_c = 1 + (2\lambda_e/d_I)\coth(d_s/\lambda_e)$ is the effective electrical thickness of JJ normalized to the insulating layer thickness d_I, $s_c = -\lambda_e/[d_I \sinh(d_s/\lambda_e)]$ is the capacitive coupling parameter, d_s is the thickness of the superconducting layer, and λ_e is Debye screening length. The voltage is normalized to $V_0 = \hbar\omega_p/(2e)$, where $\omega_p = \sqrt{8\pi d_I e j_c/(\hbar\varepsilon)}$ is the plasma frequency of JJ, and ε is the dielectric constant of the insulating layer. The time t and coordinate x are normalized to the plasma frequency ω_p^{-1} and the Josephson penetration depth λ_J, respectively. The matrix of inductive coupling $\hat{\pounds}$ has the form

$$\hat{\pounds} = \begin{pmatrix} 1 & S & 0 & \cdots & & & 0 & S \\ S & 1 & S & 0 & & & & 0 \\ & \ddots & \ddots & \ddots & \ddots & & & \\ \cdots & 0 & S & 1 & S & 0 & \cdots & \\ & & & \ddots & \ddots & \ddots & \ddots & \\ 0 & & & 0 & S & 1 & S & \\ S & 0 & & & 0 & S & 1 & \end{pmatrix},$$

where $S = s_\pounds/D_\pounds$ is the inductive coupling parameter, $s_\pounds = -\lambda_L/\sinh(d_s/\lambda_L)$, $D_\pounds = d_I + 2\lambda_L\coth(d_s/\lambda_L)$ is the effective magnetic thickness of JJ, and λ_L is the London penetration depth. Valid values of the inductive coupling parameter S are in the range $S \in (-0.5, 0)$. The initial conditions for the system of equations (1) are $\varphi_l(x,0) = 0$ and $V_l(x,0) = 0$. The boundary conditions in the x dirrection are given by the external magnetic field $\partial\varphi_l/\partial x|_{x=0,L} = B_{\text{ext}}$. The magnetic field is normalized to $B_0 = \hbar c/2eD_\pounds$. In the z direction we use periodic boundary conditions: at $l = N$, $\varphi_{l+1} = \varphi_1$, $V_{l+1} = V_1$; at $l = 1$, $\varphi_{l-1} = \varphi_N$, $V_{l-1} = V_N$.

Simulations are based on the numerical solution of the system (1) by the fourth order Runge–Kutta method, the finite-difference approximation of the spatial derivatives, and the MPI technique for parallel implementation. The details of simulation of the IV-curve and other characteristics are discussed in our previous papers [5, 18, 19]. The power of radiation from the stack is simulated using the expression $P = V_{\text{AC}}^2/R_Z$ [18, 20], where $V_{\text{AC}} = d_I E_{\text{AC}}$, E_{AC} is the AC part of the electric field, $R_Z = (d_I/W)Z$, W is the width of JJ, Z is the impedance of the radiation. Details of simulation of the radiation power are discussed in Ref. [18, 20]. The magnetic field in the JJs is calculated using the expression $B_l = B_0\sum_{k=1}^{N}\pounds_{lk}^{-1}\partial\varphi_k/\partial x$. The charge in superconducting layers is calculated using the expression $Q_l(x,t) = Q_0[V_l(x,t) - V_{l-1}(x,t)]$ [5], where $Q_0 = \varepsilon V_0/4\pi d_s d_I$.

2 Results and discussion

In Fig. 1 the IV-curve and power of radiation as a function of bias current for the stack of ten JJs are presented. Calculation is provided for a stack with dissipation parameter $\beta = 0.2$, capacitive coupling $s_c = -0.05$ and effective electric thickness $D_c = 1.1$. The typical BSCCO parameter S is close to -0.5 [21]. However, in order to clarify the influence of the inductive coupling we choose the weak coupling parameter $S = -0.05$. The IV-curve demonstrates seven zero field steps (ZFS) [7–9], which are formed due to the appearance of fluxons (kinks and antikinks). The vertical dashed lines mark the boundaries of the regions of IV-curve where fluxons appear. For each region the corresponding number of fluxons is indicated. Between the regions of the IV-curve with one and two fluxons near some ZFSs in comparison with the case of single JJ there appears an additional branch due to the different number of fluxons in each JJs of the stack. This feature was discussed in [18]. In the case of single JJ a significant radiation power is observed in the regions of IV-curve corresponding to the ZFS [18]. In the case of the stacked JJs the radiation power is close to zero near some ZFSs in comparison with the case of a single JJ (in our case ZFS with 5 and 7 fluxons).

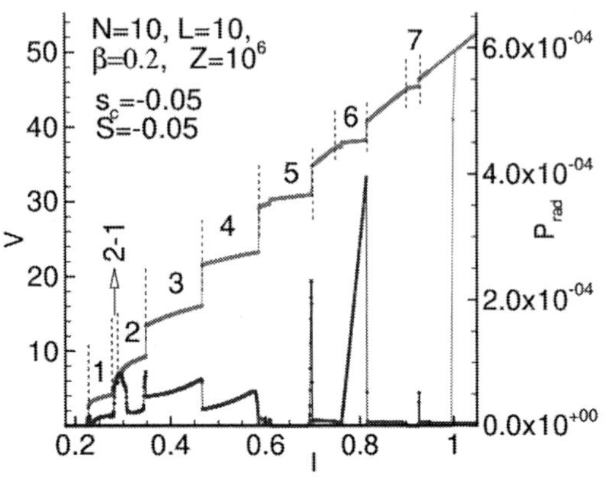

Figure 1. IV-curve together with bias current dependence of the radiation power calculated for a stack of $N = 10$ JJs with model parameters $D_c = 1.1$, $s_c = -0.05$, $S = -0.05$ and $\beta = 0.2$. Vertical dashed lines show the boundaries of ZFS. For each ZFS the number of fluxons is indicated.

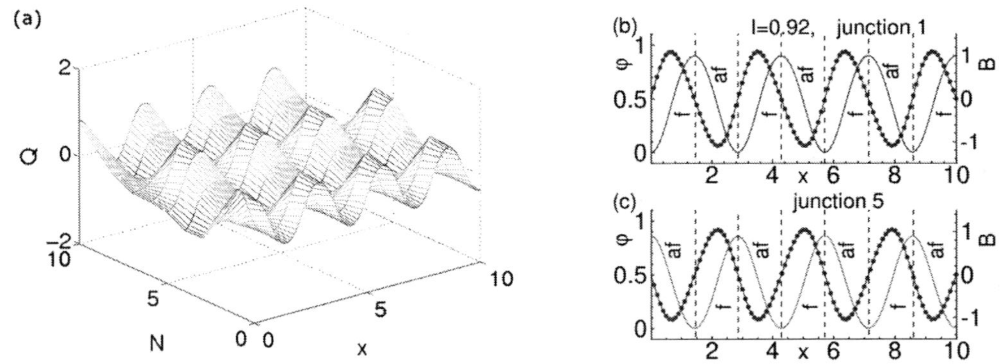

Figure 2. (a) Charge distribution in superconducting layers along the coordinate and stack of JJs at $I = 0.92$ at a fixed time moment; (b) Distribution of phase difference (solid line) and magnetic field (doted line) along the coordinate at $I = 0.92$ at a fixed time moment for first JJ. The boundaries of fluxons and antifluxons are shown with the vertical dashed line; (c) The same as in the case (b) for fifth JJ.

In order to explain such behavior of stacked JJs we have investigated the dynamics of each JJs of the stack and calculated the charge dynamics in superconducting layers in the above mentioned region of the IV-curve. Figure 2(a) demonstrates the distribution of charge in superconducting layers along the stack and coordinate, for fixed time moment at $I = 0.92$. It can be seen that along the stack a wave–like behavior of the charge is realized. Along the coordinate we observe the fluxons. The analysis of the dynamics of each JJs of the stack demonstrates that along the stack a charge traveling wave (CTW) appears, i.e. the CTW and fluxons coexist. Such dynamical state can be considered as a new collective excitation in the stacked JJs.

The question arises why the observed collective excitation leads to the decrease of the radiation power. In order to explain this phenomena we have analyzed the dynamics of the phase difference and magnetic field for each JJs of the stack. In Fig. 2(b) the spatial distribution of the phase difference (solid line) and magnetic field (doted line) at the fixed time moment in the first JJ at the current value

I = 0.92 are shown, which corresponds to the seventh ZFS. This dependence demonstrates seven fluxons (four fluxons and three antifluxons). The same characteristics for the middle of stack, i.e., in the fifth JJ are presented in Fig. 2(c). In comparison with the first JJ (Fig. 2(b)), here we can see four antifluxons and three fluxons, i.e., the opposite situation. This lets us conclude that the first and fifth JJs are of opposite phases. This circumstance results in the zero value of the average radiation power.

3 Conclusions

In this paper we have investigated the structure of the IV-curve of JJs and the radiation power emitted by a stack of JJs. We demonstrate that in the stack of JJs a charge traveling wave can appear in the ZFS, i.e. fluxons coexist with the charge traveling wave. This points to the appearance of a new collective excitation in the system of coupled JJs. We note that such collective excitation in the stack of JJs has a significant influence on the magnitude of the radiation power from the stack of JJs. Particularly, it leads to a zero radiation power from the stack. The obtained results are of interest to the analysis of the experimental IV-curves.

Acknowledgement

The study was funded by the RFBR, according to the research project 15–29–01217, Heisenberg–Landau program, JINR–Slovakia collaboration, JINR–Bulgaria cooperation program, project FP17-FMI-008, and by the Ministry of Education and Science of the Russian Federation (the Agreement number 02.a03.21.0008).

References

[1] R. Kleiner, F. Steinmeyer, G. Kunkel, and P. Müller, Phys. Rev. Lett. **68**, 2394 (1992)
[2] A.A. Yurgens, Supercond. Sci. Technol. **13**, R85 (2000)
[3] Yu. M. Shukrinov and F. Mahfouzi, Phys. Rev. Lett. **98**, 157001 (2007)
[4] Yu. M. Shukrinov, F. Mahfouzi, and M. Suzuki, Phys. Rev. B **78**, 134521 (2008)
[5] I. R. Rahmonov, Y. M. Shukrinov, and A. Irie, JETP Letters **99**, 632 (2014)
[6] Yu. M. Shukrinov, A. E. Botha, S. Yu. Medvedeva, M. R. Kolahchi, and A. Irie, Chaos **24**, 033115 (2014)
[7] T. A. Fulton and R. C. Dynes, Solid St. Commun. **12** 57 (1972)
[8] N. F. Pedersen and D. Welner, Phys. Rev. B **29**, 2551 (1984)
[9] D. W. McLaughlin and A. C. Scott, Phys. Rev. A **18**, 1652 (1978)
[10] R. Kleiner, T. Gaber, and G. Hechtfischer, Phys. Rev. B **62**, 4086 (2000)
[11] Y. Matsuda, M.B. Gaifullin, K. Kumagai, K. Kadowaki, and T. Mochiku, Phys. Rev. Lett. **75**, 4512 (1995)
[12] L. Ozyuzer, A.E. Koshelev, C. Kurter et al., Science **318**, 1291 (2007)
[13] U. Welp, K. Kadowaki and R. Kleiner, Nature Photonics **7**, 702 (2013)
[14] S. Sakai, P. Bodin, and N.F. Pedersen, J. Appl. Phys. **73**, 2411 (1993)
[15] M. Machida and S. Sakai, Phys. Rev. B **70**, 144520 (2004)
[16] Y. M. Shukrinov and F. Mahfouzi, Physica C **434**, 6 (2006)
[17] Y. M. Shukrinov and I. R. Rahmonov, JETP **115**, 289 (2012)
[18] I. R. Rahmonov, Yu. M. Shukrinov, P. Kh. Atanasova, E. V. Zemlyanaya, and M. V. Bashashin, JETP **124** 131 (2017)
[19] Yu. Shukrinov, I. Rahmonov, and M. Hamdipour, LNCS **7125**, 234 (2012)
[20] V. M. Krasnov, Phys. Rev. B **82**, 134524 (2010)
[21] V. M. Krasnov, Phys. Rev. B, **83**, 174517, (2011)

An Explicit Algorithm for the Simulation of Fluid Flow through Porous Media

Marina Trapeznikova[1,⋆], *Natalia* Churbanova[1,⋆⋆], and *Anastasiya* Lyupa[1,⋆⋆⋆]

[1] *Keldysh Institute of Applied Mathematics RAS, 4 Miusskaya Square, 125047 Moscow, Russia*

Abstract. The work deals with the development of an original mathematical model of porous medium flow constructed by analogy with the quasigasdynamic system of equations and allowing implementation via explicit numerical methods. The model is generalized to the case of multiphase multicomponent fluid and takes into account possible heat sources. The proposed approach is verified by a number of test predictions.

1 Introduction

The development of mathematical methods and software for the simulation of complex fluid flow in the subsurface is one of urgent tasks of industrial mathematics. Among applications there are hydrocarbon recovery problems as well as ecological problems concerning the soil and groundwater contamination. Models and algorithms for their solution still need improvement to predict thermo-hydro-mechanical-chemical processes in strata with sufficient accuracy at reasonable computing time.

The work presents an original model of fluid flow through porous media constructed by analogy with the quasigasdynamic system of equations and allowing implementation via explicit numerical methods [1]. Large-scale oil recovery problems (e.g., combustion fronts, phase transitions, complicated functions of the relative phase permeability) require calculations with very small space steps what strictly constrains the time step. Then explicit schemes can surpass implicit schemes involved in standard solution methods like IMPES or SS [2]. Besides explicit algorithms are preferable for HPC.

2 Mathematical model and numerical implementation

For many problems of continuum mechanics the principle of minimal sizes holds: we assume the existence of minimal space and time scales which act as lower limits for description details. In porous media the minimal reference length l is a distance at which rock microstructure is negligible (l is of the order of hundred rock grain sizes), the minimal reference time τ is the time needed to reach inner equilibrium in the volume of linear size l. Starting from the basic flow equations for slightly compressible fluid [2], using the principle of minimal sizes and the differential approximation technique the next multiphase flow model has been derived ($\alpha = w$ (water), n (NAPL), g (gas); r denotes a rock):

$$\varphi \frac{\partial (\rho_\alpha S_\alpha)}{\partial t} + \tau \frac{\partial^2 (\rho_\alpha S_\alpha)}{\partial t^2} + \mathrm{div}\,(\rho_\alpha \mathbf{u}_\alpha) = q_\alpha + \mathrm{div} \frac{l c_\alpha}{2} \mathrm{grad}\,(\rho_\alpha S_\alpha), \quad (1)$$

⋆e-mail: mtrapez@yandex.ru
⋆⋆e-mail: nataimamod@mail.ru
⋆⋆⋆e-mail: nastenka.aesc@gmail.com

$$\mathbf{u}_\alpha = -K\frac{k_\alpha}{\mu_\alpha}\left(\text{grad}\, P_\alpha - \rho_\alpha \mathbf{g}\right), \tag{2}$$

$$\frac{\partial}{\partial t}\left[\varphi\sum_\alpha \rho_\alpha S_\alpha E_\alpha + (1-\varphi)\rho_r E_r\right] + \text{div}\left(\sum_\alpha \rho_\alpha H_\alpha \mathbf{u}_\alpha\right) = \text{div}\lambda_{\text{eff}}\text{grad}T + \sum_\alpha \text{div}\frac{lc_\alpha}{2}\rho_\alpha \text{grad}T, \tag{3}$$

$$\rho_g = \rho_{0g}\frac{P_g}{P_{0g}}\frac{T_0}{T}, \qquad \rho_\alpha = \rho_{0\alpha}\left[1 + \beta_\alpha\left(P_\alpha - P_{0\alpha}\right) - \eta_\alpha\left(T - T_0\right)\right], \qquad \alpha = w, n, \tag{4}$$

$$\sum_\alpha S_\alpha = 1. \tag{5}$$

Here S_α is the saturation, P_α is the pressure, ρ_α is the density, $\mathbf{u}_\alpha$ is the Darcy velocity, T is the temperature, E_α is the internal energy, q_α is the source of fluid, φ is the porosity, K is the absolute permeability, k_α is the relative phase permeability, μ_α is the dynamic viscosity, $\mathbf{g}$ is the gravity vector, c_α is the sound speed in α-phase, β_α is the coefficient of isothermal compressibility, η_α is the coefficient of thermal expansion, $P_{0\alpha}$, $\rho_{0\alpha}$ and T_0 are reference values, $\rho_r = const$ – the rock density.

The phase continuity equation (1) is modified: it gets a regularizing term and a second order time derivative with small parameters. In [1, 3] some estimations of these parameters are given. Particularly the value $\tau \sim h/c$ is satisfactory (h is the space grid step). The equation type is changed from parabolic to hyperbolic, consequently a three-level explicit scheme with a rather mild stability condition can be used, convective terms are approximated by central differences. As the temperature of all phases and of the rock is identical the system involves (3) as a single equation of the total energy conservation, also modified by analogy with the quasigasdynamic system. This equation includes the effective coefficient of heat conductivity $\lambda_{\text{eff}} = \varphi\sum_\alpha S_\alpha\lambda_\alpha + (1-\varphi)\lambda_r$, and the enthalpy which is calculated via the heat capacity at constant pressure C_{P_α}:

$$H_\alpha = H_{0\alpha} + \int_{T_0}^{T} C_{P_\alpha}(T)\,dT. \tag{6}$$

The connection between the internal energy and the enthalpy is expressed by the next relations:

$$E_\alpha = H_\alpha - \frac{P_\alpha}{\rho_\alpha}, \qquad E_r = H_r. \tag{7}$$

Capillary pressures are described by Parker's functions, the relative phase permeability is presented by Stone's Model I [2].

For numerical implementation of the above system an algorithm of the explicit type is proposed. P_w, S_w, S_n and T are chosen as primary variables in the three-phase case. Main stages of the algorithm are explicit approximations of (1) and (3) as well as the solution of a system of nonlinear algebraic equations by Newton's method locally at each computational point [3].

In order to account for the complicated multicomponent structure of the fluid it is necessary to pass from conservation laws for phases to conservation laws for components. The relative quantity of j-th component in phase α can be expressed as the mass concentration: $C_{j\alpha} = m_{j\alpha}/m_\alpha$, $\alpha = 1,\ldots,n_\alpha$, $j = 1,\ldots,n_c$. Then the next equation represents the mass conservation law of each component:

$$\varphi\frac{\partial}{\partial t}\left(\sum_\alpha \rho_\alpha S_\alpha C_{j\alpha}\right) + \tau\frac{\partial^2}{\partial t^2}\left(\sum_\alpha \rho_\alpha S_\alpha C_{j\alpha}\right) + \sum_\alpha \text{div}\left(\rho_\alpha C_{j\alpha}\mathbf{u}_\alpha\right) = Q_j + \sum_\alpha \text{div}\frac{lc_\alpha}{2}\text{grad}\left(\rho_\alpha S_\alpha C_{j\alpha}\right). \tag{8}$$

Equations (2) and (3) are valid for the multicomponent case but the dependences $\rho_\alpha = \rho_\alpha\left(P_\alpha, T, C_{j\alpha}\right)$ and $\mu_\alpha = \mu_\alpha\left(P_\alpha, T, C_{j\alpha}\right)$ should be noted.

The next closure relations are defined:

$$H_\alpha = \sum_j H_{j\alpha}, \quad \sum_j C_{j\alpha} = 1,$$

$$\frac{C_{j\alpha}}{C_{j\beta}} = K_{j\alpha\beta}\left(P, T, C_{j\gamma}\right), \qquad \alpha, \beta, \gamma = 1, \ldots, n_\alpha, \qquad j = 1, \ldots, n_c, \tag{9}$$

where $K_{j\alpha\beta}$ are constants of the phase equilibrium.

3 Test predictions

For verification of the proposed model and algorithm a drainage test problem [4] has been solved. Two-phase (water/air) isothermal infiltration due to the gravity is simulated. The geometry is illustrated in the figure 1(a). Initially the given thin column is filled by a fully water-saturated porous medium. The top boundary is open to the atmosphere with a no-flow condition on water, water drains down from the column, a no-flow condition on air exists at the bottom.

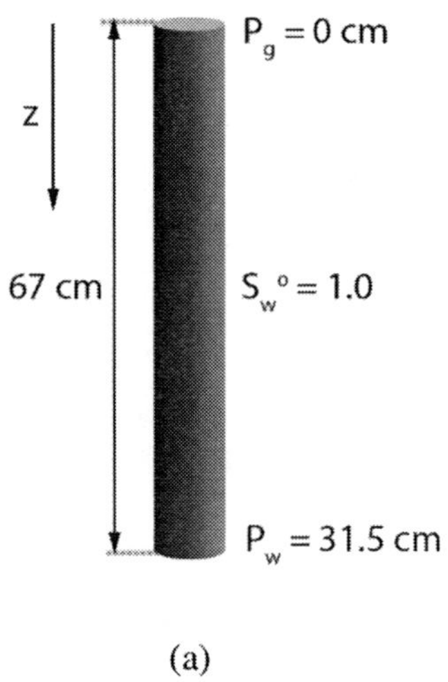

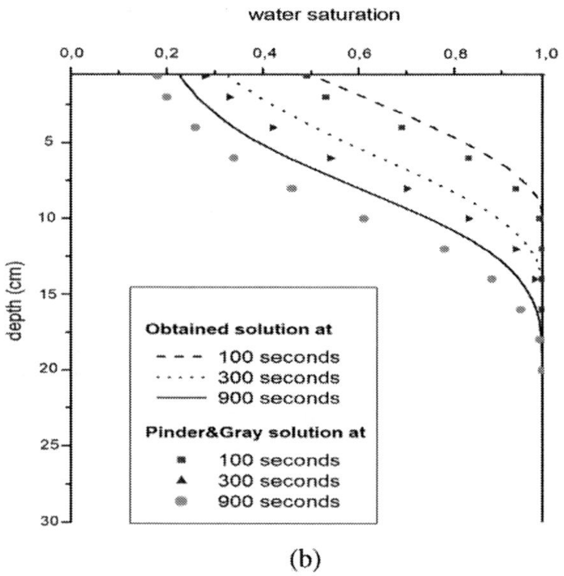

Figure 1. The drainage problem statement (a) and comparison of results (b)

In computations $l = 10^{-7}$ cm, $\tau = 10^{-4}$ s. The obtained distribution of water saturation with the depth at different time moments is compared with results by G.F. Pinder and W.G. Gray [4] (figure 1(b)). A good agreement is observed that testifies to adequacy of the developed approach.

To check the full mathematical model of nonisothermal three-phase flow, the 3D problem of infiltration with a hot source has been investigated (figure 2). Initially in the cube reservoir the residual water saturation is set, oil and gas are distributed periodically, the water pressure equals the atmospheric pressure, the temperature is 285 K. A source of water ($S_w = 0.6$, $S_n = 0.05$) with the temperature of 320 K is set on the top. The top boundary is open to the atmosphere, the bottom is impermeable, side boundaries are permeable. Results at the moment of 2000 seconds are shown in figure 2. Water gradually forces out oil and gas and tends to the bottom, the temperature front differs from saturation fronts. Physically correct dynamics of the process is observed.

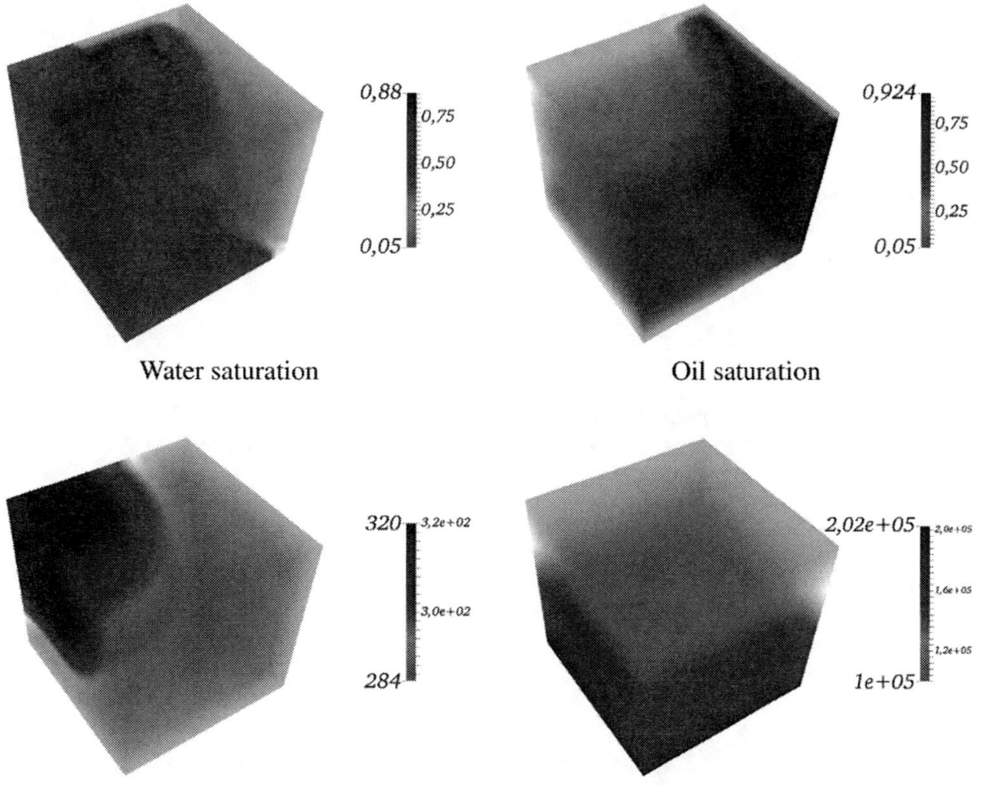

Figure 2. The problem of three-phase infiltration with a hot source: obtained results at 2000 seconds

4 Conclusions

The created approach can be used for the simulation of perspective thermal methods of oil recovery (such as heat carrier pumping into the stratum) aimed at increasing the oil production rate of difficult-to-recover hydrocarbon reserves.

Acknowledgements

The work is supported by the Russian Foundation for Basic Research (grant # 16-29-15095-ofi). Referee help for improving the presentation is acknowledged.

References

[1] B. N. Chetverushkin, et al., Mathematical Models and Computer Simulations **2** (6), 669–677 (2010).
[2] K. Aziz and A. Settari, *Petroleum Reservoir Simulation* (Applied Science Publ. Lmt., London, 1979) 476 p.
[3] B. N. Chetverushkin, et al., Rus. J. Numer. Anal. Math. Modelling **31** (4), 187–196 (2016).
[4] G. F. Pinder and W. G. Gray, *Essentials of Multiphase Flow and Transport in Porous Media* (John Wiley & Sons, Inc., Hoboken, NJ, 2008) pp. 199–205.

EPJ Web of Conferences **173**, 06013 (2018) https://doi.org/10.1051/epjconf/201817306013
Mathematical Modeling and Computational Physics 2017

Numerical Simulation of the Formation of Hydrated Electron States

Alina Volokhova[1,⋆], *Elena* Zemlyanaya[1], *Viktor* Lakhno[2], *Ilkizar* Amirkhanov[1], *Maxim* Bashashin[1], *Igor* Puzynin[1], and *Taisiya* Puzynina[1]

[1]*Joint Institute for Nuclear Research, 141980 Dubna, Russia;*
[2]*Institute of Mathematical Problems of Biology, Russian Academy of Science, 142290 Pushchino, Russia*

Abstract. Numerical simulation of the photoexcited electron states formation in water under the action of the ultraviolet range laser irradiation is carried out on the basis of the polaron model with a time-dependent calculation of the hydrated electron absorption band width. This framework is shown to reproduce well the experimental data on the light absorption by the hydrated electron in polarized water. Effectiveness of parallel implementation is tested on the HybriLIT cluster.

1 Introduction

The theoretical and experimental study of the solvated electron is motivated by its wide applicability to the theoretical explanations of complex processes in chemistry, biology and quantum physics [1–3]. For the numerical study of the hydrated electron, we start with the polaron model formulated in [4]. The numerical method and parallel computer code for solving the respective system of nonlinear partial differential equations, has been developed in [5, 6]. In [7], the results of numeral simulation of the hydrated electron formation process were shown to be in a reasonable agreement with experimental data from [8] on the absorption of the light in the polarized water. In this contribution, the model is extended to include time-dependent calculation of the hydrated electron absorption band width. Numerical simulations within the frame of the advanced model provide a noticeable improvement of agreement between experimental data and theoretical results in full range of time. Also, the results of methodical calculations demonstrating good efficiency of the parallel implementation on the basis of the partition algorithm [9] are presented.

2 Mathematical model and numerical approach

In the frame of the polaron model, the hydrated electron state is described by the following equations [4, 7, 10]:

$$i\hbar\dot{\Psi}(\vec{r},t) = -\frac{\hbar^2}{2m}\Delta\Psi(\vec{r},t) - e\Phi(\vec{r},t)\Psi(\vec{r},t), \tag{1}$$

$$\Delta\Phi(\vec{r},t) = \theta(\vec{r},t), \tag{2}$$

⋆e-mail: volokhova@jinr.ru

$$\frac{1}{4\pi\omega^2 c}\ddot{\theta}(\vec{r},t) + \gamma\dot{\theta}(\vec{r},t) + \frac{1}{4\pi c}\theta(\vec{r},t) + e|\Psi(\vec{r},t)|^2 = 0. \tag{3}$$

where m is the effective electron mass, $c = \varepsilon_\infty^{-1} - \varepsilon_0^{-1}$, ε_∞ and ε_0 are the high-frequency and static constants, e is the elementary charge, ω is the frequency of optical polarization medium oscillations, Φ is the potential appearing during the polarization process, θ describes the density of polarized charge induced by the electron. The relaxation (friction) γ is determined by the coefficient related to the dielectric relaxation time τ by the expression $\gamma^{-1} = \omega^2\tau c$. Expanding the functions $\Psi(\vec{r},t)$ $\Phi(\vec{r},t)$, and $\theta(\vec{r},t)$ in spherical harmonics, restricting our consideration to the spherically symmetric case, and coming to dimensionless variables, we obtain the following system of spatially one-dimensional partial differential equations with respect to the radial components $\psi(r,t)$, $\varphi(r,t)$ and $\Theta(r,t)$ of the functions Ψ, Φ and θ:

$$\begin{cases} \left[i2\bar{m}\frac{\partial}{\partial t} + \frac{\partial^2}{\partial x^2} + 2\bar{m}\frac{r_{00}}{\tilde{\epsilon}}\frac{\varphi(x,t)}{x}\right]\psi(x,t) = 0, \\ \frac{\partial^2}{\partial x^2}\varphi(x,t) = \Theta(x,t), \\ \left[\frac{\partial^2}{\partial t^2} + \bar{\gamma}\frac{\partial}{\partial t} + \bar{\omega}^2\right]\Theta(x,t) = -\bar{\omega}^2\frac{|\psi(x,t)|^2}{x} \end{cases} . \tag{4}$$

The boundary conditions have the form:

$$\varphi(0,t) = \varphi'(\infty,t) = 0, \;\; \psi(0,t) = \psi(\infty,t) = 0, \;\; \Theta(0,t) = \Theta(\infty,t) = 0. \tag{5}$$

Here, $\bar{m} = 2.692$, $\bar{\gamma} = 2.145$, and $\bar{\omega} = 1$ are the dimensionless parameters of the effective mass of the hydrated electron, the relaxation (friction), and the frequency; $\tilde{\epsilon} = 1.81$ is a constant coefficient; $r_{00} = 164.64$ is a scaling factor. The relations between the dimension and dimensionless variables are given in [7, 10].

For numerical solution, the system (4) is replaced with that of difference equations on a uniform discrete mesh in the coordinate x and time t. We employ the algorithm described in [5, 10] making it possible to sequentially calculate, at each time step, the solutions $\Theta(x,t)$, $\varphi(x,t)$, $\psi(x,t)$ at the x-nodes of the discrete mesh under given initial condition $\psi(x,0)$. The MPI/C++ parallel implementation of this scheme on the basis of partition algorithm [9] has been developed in [10, 11].

3 Numerical results

To reproduce the experimental values of the intensity of light absorption by a hydrated electron, we chose the initial condition of the wave function ψ in the Gaussian form with parameter σ which is varied to reproduce, in the computer simulation, the time dynamics of the light absorption by water.

The light absorption $I(t)$ is calculated as follows [4]:

$$I(t) = \frac{4\Omega^2\gamma_s^2}{(W(t)^2 - \Omega^2)^2 + 4\Omega^2\gamma_s^2}, \tag{6}$$

where γ_s is the width of the absorption band of the hydrated electron and Ω denotes of the light frequency of a scanning laser at which the light is absorbed by the hydrated electron.

In [7], we fixed γ_s to the constant value $\gamma_s = 0.38$ eV. The model provided a reasonable agreement with the experimental data except for the region of zero and small time, where a noticeable discrepancy between numerical and experimental data was observed [7, 10]. Here, $\gamma_s(t)$ at each step of time is calculated as follows:

$$\gamma_s(t) = \sqrt{\hbar\omega k_B T}\,\sqrt{2\ln 2}\,\sqrt{a(t)}, \qquad a(t) = \int_0^{r_{max}} dr \left[\int_0^r dr'\,|\psi(r',0)|^2 - \int_0^r dr'\,|\psi(r',t)|^2\right]^2, \tag{7}$$

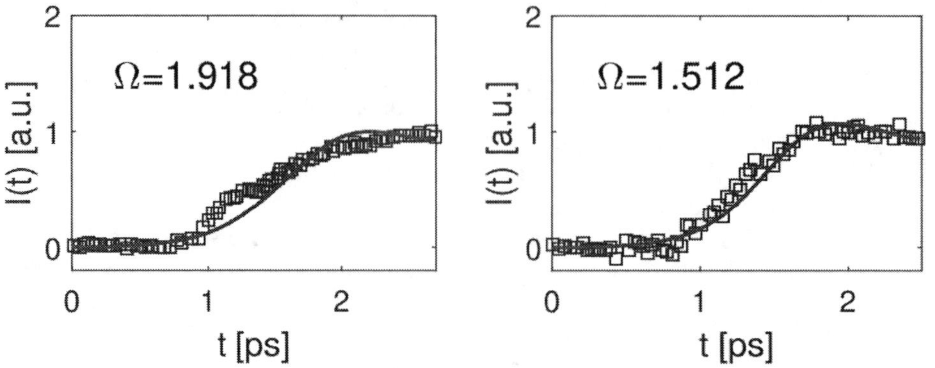

Figure 1. Numerical simulation of the light absorption by a hydrated electron for experimental data [8] derived at light frequencies of the scanning laser Ω = 1.918 eV (left) and Ω = 1.512 eV (right).

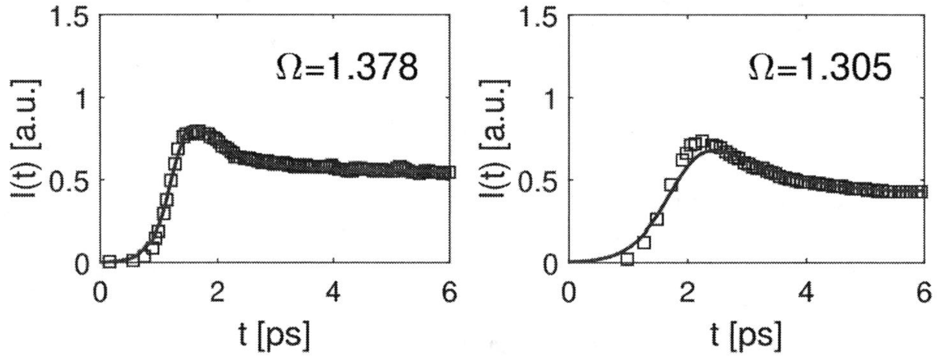

Figure 2. Same as Fig. 1 for experimental data [12] at Ω = 1.378 eV (left) and Ω = 1.305 eV (right).

where T is the absolute temperature, k_B is the Boltzmann constant. Results of the simulations are given in Fig. 1 and Fig. 2 in comparison with the experimental data from [8] and [12], respectively. Corresponding values of Gaussian parameter σ are given in the Table 1. It is seen that the modified approach with time-dependent calculation of γ_s ensures good agreement with the experimental data over all the range of the spectrum.

Table 1. Values Ω and fitted values of the Gaussian parameter σ providing good agreement of numerical results with experimental data from [8, 12].

Reference	[8]	[8]	[12]	[12]
Ω	1.918	1.512	1.378	1.305
σ	5.750	4.500	3.000	6.500

Methodical calculations with different number of computing nodes have been made on the heterogenous cluster HybriLIT (Laboratory of IT, Joint Institute for Nuclear Research, Dubna, http://hybrilit.jinr.ru/en/) with $2 \cdot 10^6$ steps in time. The execution time versus the number P of parallel computing nodes is shown in Fig. 3 (left panel) to demonstrate the effectiveness of the parallel C++/MPI code. It is seen that the execution time is decreasing hyperbolically while $P \leq 12$. In the left panel of Fig. 3, the maximal acceleration is given versus the number of x-nodes.

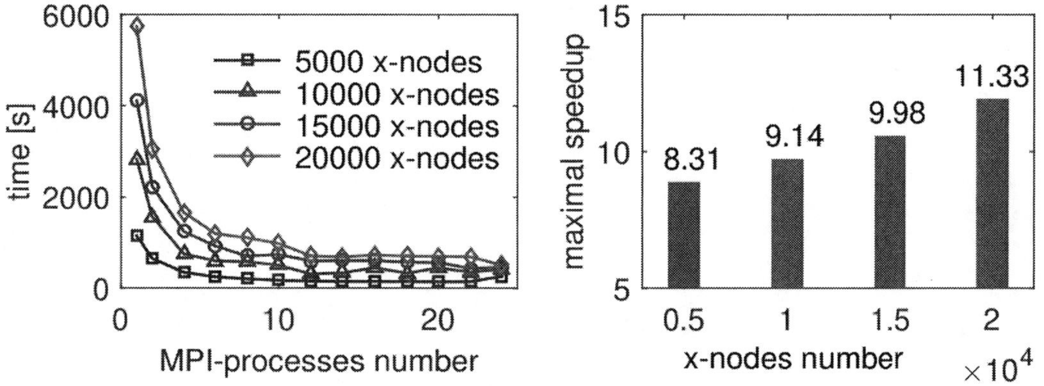

Figure 3. Left panel: dependence of the execution time on number of parallel MPI-processes for the cases of 5000, 10000, 15000, and 20000 x-nodes of the discrete grid. Right panel: Maximal acceleration in dependence on number of x-nodes. Calculations on the HybriLIT cluster (Dubna).

4 Summary

We have found that the polaron model with time-dependent calculation of γ_s provides realistic description of the light absorbtion dynamics by the photoexcited hydrated electron at light frequencies Ω of a scanning laser in the range 1.3–1.9 eV. The model can be used for further simulations and predictions of the polaron states dynamics in the aqueous medium and in other condensed media. An MPI-based parallel algorithm is shown to provide the 8–11 times acceleration in comparison with the serial calculations.

Acknowledgements

The work was supported by the RFBR (grants 17-01-00661a, 16-07-00305a).

References

[1] A.K. Pikaev, *The Solvated Electron in Radiation Chemistry* (World University, Jerusalim, 1970)
[2] K.R. Siefermann, Y. Liu, and E. Lugovoy, *et al.*, Nature chemistry **2**, 274–279 (2010)
[3] Y. Tang, H. Shen, K. Sekiguchi, *et al.*, Phys. Chem. Chem. Phys. **12**, 3653–3655 (2010)
[4] V.D. Lakhno, Chem. Phys. Lett. **437**, 198–202 (2007)
[5] I.V. Amirkhanov, E.V. Zemlyanaya, V.D. Lakhno, *et al.*, J. Surface Investigation **5**, 1, 60–64 (2011)
[6] Volokhova A.V. Zemlyanaya E.V. Lakhno V.D., *et al.* Komp'yut. Issled. Modelir. (Russian) **6**, 2, 253–261 (2014)
[7] V.D. Lakhno, A.V. Volokhova, E.V. Zemlyanaya, *et al.*, J. Surface Investigation **9**, 1, 75–80 (2015)
[8] F.H. Long, H. Lu, and K.B. Eisenthal, Phys. Rev. Lett. **64**, 1469–1472 (1990)
[9] H.H. Wang, ACM Trans. Math. Software **7**, 170–183 (1981)
[10] E.V. Zemlyanaya, A.V. Volokhova, V.D. Lakhno, *et al.*, AIP Conf. Proc. **1684** 100006 (2015)
[11] A.V. Volokhova, E.V. Zemlyanaya, and V.S. Rikhvitskij, Computational Methods and Programming [Russian] **16**, 281–289 (2015)
[12] R. Lian, R.A. Crowell, and I.A. Shkrob, J. Phys. Chem. **A 109**, 1510 (2005)

AUTHOR INDEX

AUTHOR INDEX

AUTHOR INDEX